Soft Rock Mechanics and Engineering

Milton Kanji • Manchao He
Luís Ribeiro e Sousa
Editors

Soft Rock Mechanics and Engineering

Springer

Editors
Milton Kanji
Polytechnical School
University of São Paulo
São Paulo, São Paulo, Brazil

Luís Ribeiro e Sousa
Department of Civil Engineering
University of Porto
Porto, Portugal

Manchao He
State Key Laboratory for GeoMechanics
and Deep Underground Engineering
China University of Mining & Technology
(Beijing), Tongji University
Beijing, China

ISBN 978-3-030-29476-2 ISBN 978-3-030-29477-9 (eBook)
https://doi.org/10.1007/978-3-030-29477-9

This Springer imprint is published by the registered company Springer Nature Switzerland AG
The registered company address is: Gewerbestrasse 11, 6330 Cham, Switzerland

Presentation

The science of rock mechanics was developed as a result of the activities of rock engineering, basically derived from tunnels, slopes, dam foundations, and mining workings in hard rock containing discontinuities. It developed specific theories and methods to include such features, being known as a science for "discontinuous" media, while soil mechanics, the parent science, basically considers "continuum" mechanics.

For the study of rock mechanics, specific equipment and apparatus were developed for laboratory and in situ investigation, requiring means to cut, perforate, sample, and test hard materials, to determine shear strength, deformability, physical properties, in situ stresses, and so on.

However, in many occasions, geotechnical and rock engineers have to deal with working sites dominated by rocks with impaired properties, such as rocks of low strength or weathered materials or rock masses with weakening aspects, as for instance, excessive jointing, faulting, solution cavities, weak inclusions, and other features which can be classified as *soft rocks* or *weak rock masses*. Its geotechnical behavior is typical of a material situated between a soil and a rock.

Such materials only a few decades ago started to be investigated and characterized more appropriately and deserve a special attention, since their properties are not well-known yet, leading to uncertainties in the design and construction of engineering workings dominated by these materials.

Under this framework, the publication of a book dealing with the different aspects of soft rocks and rock masses and their investigation, properties, and applications to the several types of rock engineering works is highly welcomed.

In this way, I am very glad to present this book that illustrates the characteristics and the behavior of soft rocks in many situations. The edition of this book is a result of the work developed by the ISRM Technical Commission on Soft Rocks chaired by Prof. Milton Kanji and by Prof. He Manchao as Vice Chairman.

The International Society for Rock Mechanics is deeply grateful to the editors, to the commission members, and to the invited specialists who collaborated in the production of this important book, which will certainly contribute to a more systematic knowledge of the behavior of soft rocks.

Eda Freitas de Quadros

President of the International Society for Rock Mechanics and Rock Engineering

Honorary Member (Posthumous)

Prof. Juan Jose Bosio Ciancio is here honored in acknowledgment to his pioneer work in soft rocks and cemented sands, started in Paraguay. He motivated other colleagues in South America to join him in the study of soft rocks of the Paraná Sedimentary Basin involving mostly Argentina, Brazil, Paraguay, and Uruguay. His enthusiasm led this group to establish firstly a Regional Commission on Soft Rocks which later turned out to be International Commissions at IAEG and ISSMGE when people from other countries joined the commission. His encouragement was the drive that ended up with the proposal for the settlement of a Technical Commission on Soft Rocks at ISRM, which was enthusiastically accepted by Prof. Xia-Ting Feng, former ISRM President, and supported by Dr. Eda Quadros, ISRM President.

Prof. Juan Jose Bosio was an active Geotechnical Engineer, Full Professor at the Faculty of Sciences and Technology in Asunción, Permanent Consultant for the Itaipu Binacional Project, Vice-President for South America of ISSMGE, Head of the Rock Mechanics Division of the Paraguayan Geotechnical Society, and Correspondent Member of the Argentinian Academy of Sciences, among many other important commitments and positions.

For these reasons, he was nominated Honorary Member of the ISRM Technical Commission on Soft Rocks.

Juan José Bosio Ciancio

Contents

Chapter 1
Introduction

Milton A. Kanji and Manchao He

This book, dealing with soft rocks and soft rock engineering, is a result of the dedication and effort made by several members of the ISRM Technical Commission on Soft Rocks, who contributed individually or in association with several coauthors.

Several chapters cover important subjects as laboratory and in situ investigation of soft rocks, problems and characteristics related to specific lithologies as for sandstones, mudstones, evaporites, and sedimentary rocks including coal; weathering and interaction of soft rocks with water, correlation of properties, and applications as for dams and rock fill embankments, civil and mining slopes, and underground workings. Some of the chapters deal with the mechanics and control of their deformation of underground openings. As much as possible, they cover the most important topics for rock mechanics, rock engineering, geomechanics, and engineering geology.

Along the many meetings of the commission and internal correspondence between the members, discussions were made on several topics, but the main one was about the understanding about "Soft Rocks." It was agreed to basically follow the understanding and definition of ISRM that it corresponds to rocks which uniaxial compressive strength of the intact material is limited to about 25 MPa. However, many members also considered that rock masses of rocks with intact strength higher than 25 MPa but affected by excessive jointing, or under high stresses or even undergoing fast weathering may behave in the same manner as rock masses of intact rocks with lower strengths than 25 MPa. Some typical examples have been presented by Prof. He et al. (1993, 2002) with respect to deep coal mining and other workings in China subjected to high stresses that undergo very large deformations.

M. A. Kanji (✉)
Polytechnical School, University of São Paulo, São Paulo, Brazil

M. He
State Key Laboratory for Geomechanics and Deep Underground Engineering, China University of Mining and Technology, Beijing, China

M. Kanji et al. (eds.), *Soft Rock Mechanics and Engineering*,
https://doi.org/10.1007/978-3-030-29477-9_1

In this way, the understanding of the commission is that the term "soft rocks" shall cover both intact rocks with strengths lower than about 25MPa and rock masses that present similar behavior, even if their intact strength is higher than the above-mentioned limit. This concept was for the first time put forth by Deere and Vardé (1986, General Report) at the 5th International Congress of IAEG. The importance of understanding the behavior of soft or weak rock was early stressed by two of the prominent researchers in geomechanics, namely, Derre (1975) and Rocha (1975).

Looking back to the development of rock mechanics, it could be divided in several stages:

1. Initial stage from the end of nineteenth century to the beginning of 20th century, represented by the hydrostatic theory (A. Heim; Rankine and Ginnick);
2. Empirical theory stage from the early 20th century to the 1930s, noted by the theory of natural balance arc (Protodyakonov; Terzaghi);
3. Classical theory stage in 1930s to 1960s, many monographs on rock mechanics were published in this period (e.g. "La Mécanique des Roches, by J. Talobre in 1957, "Principles of Engineering Geology and Geotechnics" by D.P. Krynine and W.R. Judd in 1957, and "Rock Mechanics" by H.R. Reynolds in 1961, among others).
4. Modern development stage from 1960s to 2000s, starting with the edition of "Rock Mechanics in Engineering Practice" (Ed. K.G. Stagg and O.C. Zienkiewicz, 1968), to the introduction of many modern scientific concepts (e.g., nonlinear theory, numerical methods, information and artificial intelligence technology) as applied to rock mechanics.
5. And recent stage from 2000s to present. In China, it started with the publication of the book of "Soft Rock Engineering Mechanics (He et al. 2002) and also with the establishment of the State Key Laboratory for Geomechanics and Deep Underground Engineering in Beijing in 2008. An assemblage of systems and equipment for soft rock tests has been developed.

We hope to believe that from now on the last stage may incorporate soft rock mechanics and soft rock engineering, since soft rocks dominates large areas of many countries and that the best places of gook rock have been already taken. The investigations on soft rocks are important mainly due to the fact that in designing important workings involving soft rocks, their properties are difficult to establish and most of the times their parameters are adopted on the conservative side, against the economy of the project. Because of soft rocks' complexity in structure, material property, and geological environment, knowledge and understanding about soft rocks cannot be included in a single book.

The authors of the introductory chapter would like to express their great expectations for continuously new research findings made by scholars working with soft rock problems on the basis of this book in the following suggested areas:

1. Better procedures for sample preparation and testing soft rocks (sometimes too hard for soil equipment or too soft for rock mechanics equipment); possible collaboration with the commission on Suggestion Methods for some specific testing methods;

2. Improvement of site investigation, mainly with attempts to improve core recovery for soft rocks of low strength. Some university could promote test fields to develop in situ investigation by new drilling techniques, in relation to geophysical methods, investigation pits, and lab testing of undisturbed samples.
3. Simulation of soft rocks through artificial materials (sand and Portland cement, clay with lime, clay with cementation material, etc.), to investigate their properties under controlled conditions.
4. Establishment of a data bank with parameters and properties of different kinds of soft rocks, to be available to the rock mechanics community.
5. Establishment of a data bank with problems at engineering workings in soft rocks and respective solutions that resulted successful.
6. Establishment of bibliographical references of publications related to soft rocks, with the possibility of making outstanding publications about soft rocks freely available.

The Technical Commission on Soft Rocks was formed in 2011 by the ISRM Board under the presidency of Prof. Xia-Ting Feng, who indicated Prof. Milton Kanji as Chairman and in 2015 Prof. He Manchao was elected vice-chairman, and the commission continued under the presidency of Dr. Eda Quadros. It was then natural that several authors of this book are from Brazil and from PR China, although many other authors belong to different other countries.

The technical collaboration between PR China and Brazil started some time ago. In 1978 Prof. M. Kanji organized an International Symposium on Rock Mechanics Related to Dam (Kanji and Abrahão 1978) being the first time a group of rock mechanics Chinese scholars participated in an international conference. Presently this collaboration continues in relation to CFRD dams.

Brazilian National Group is represented by the Brazilian Committee on Rock Mechanics which is within the Brazilian Association of Soil Mechanics and Geotechnical Engineering and is active since 1978. Together with the Argentinian and Paraguayan National Groups are organizing the 2019 ISRM Conference on Rock Mechanics.

In China, the first book on rock mechanics, "Rock Mechanics and Ground Support (in Chinese)" was published in 1956 (Yu 1956). The national group for ISRM was established in 1981, the Chinese Society for Rock Mechanics and Engineering (CSRME) was founded in 1985, and the Chinese Sub-society for Soft Rock Engineering & Deep disaster Control was set up in 1996, opening up a new era for the development of rock mechanics research within the country. Up to now, China has become the largest infrastructure construction site in the world. Correspondingly, more and more geological disasters associated with soft rocks have been encountered in the rock engineering projects such as rock slopes, tunnels, and deep mining. Soft rocks have complex structure, nonlinear constitutive relationship, and large-deformation behavior, constituting a major challenge to the community of rock mechanics, as a result of the effective solutions beyond the current knowledge of rock mechanics. Voices calling for new concept, theory, and technology in combating the soft rock problems appear as strong as ever.

1.1 A Few Works About the Editors of This Book

Prof. Dr. Milton Assis Kanji is a former vice president of ISRM and was inducted as Fellow of the Society. He is a geologist from the University of São Paulo, Brazil, with an M.Sc. in Engineering Geology and Rock Mechanics from the University of Illinois. He got his doctoral degree in Geology and later his postdoctoral degree on Earth Works from the University of São Paulo. He was an Associate Professor at the Polytechnic School of the University of São Paulo. In parallel, he worked for about 58 years in geotechnical and design companies, having participated in the design of several hydroelectric projects, as Itaipu, Agua Vermelha, Jaguara, and others, worked in slope investigation and stabilization, and in design and construction supervision of tunnels for gas and ore ducts. He has been working in forensic engineering for several adjusters and insurance companies for dams, tunnels, and rock slopes, and as expert witness in arbitrations and law suits related to accidents and failures of engineering works (as the Pinheiros Metro Station collapse, the causes of the failure of Fundão Dam by SAMARCO and presently working on the failure of Dam I in Brumadinho (Brazil), Pando Tunnel (Panamá), Sta Teresa Hydro Project (Peru), and others. He has intense international activity, mainly in Latin America. He has been general reporter or lecturer in many conferences and has more than 120 papers or chapters published in some book chapters.

As a senior researcher in geomechanics, Prof. M.C. He, Academician of the Chinese Academy of Sciences and the Argentinian National Academy of Engineering, has devoted his whole life in coping with the soft rock problems. Presently he is an ISRM vice president at large. As early as in 1993, in his first monograph, "Introduction to Soft Rock Roadway Engineering (in Chinese)," the concept of the "engineering soft rock" was put forward for defining the rocks that are "hard rocks" in the conventional sense but may exhibit a soft-rock-like behavior under high-stress environment found in deep mining and rock project. In the book "Soft Rock Engineering Mechanics (in Chinese)" Prof. M. C. He and co-authors (2002), presented detailed concepts on the engineering soft rock concept, classification system, large-deformation mechanism, and support strategy for soft rock engineering. That book has deserved more than ten thousand citations according to CNKI search results, promoting significantly the development of the subject for soft rock mechanics in China. Of particular importance is his work and development of constant-resistance-large-deformation bolt having a negative Poisson's ratio (NPR) structure as its working resistance generating device; as a result, it is referred oftentimes to CRLD bolt or NPR bolt for its extraordinary performance. Prof. He is also the Director of the State Key Laboratory for Geomechanics and Deep Underground Engineering in Beijing

Prof. Luís Ribeiro e Sousa has more than 45 years of professional experience. He is a retired Full Professor at the Porto University in Portugal where he worked on risk analysis and numerical analysis for geotechnical and rock mechanics workings. He was responsible for organizing the ISRM Rock Mechanics Congress in 2007 in Lisbon. He was president of the firm SKEC Engineering Consulting. His work encompasses rock mechanics applied to dam foundations, subways, and other types of underground structures, mining and petroleum applications, and field and laboratory testing. Prof. Sousa has broad international experience, having worked in about 16 different countries. His activities in PR China led him to be recognized as a scholar by the Chinese Foreign Experts Bureau in 2010, 2011, 2013, and 2014. He was a consultant to the State Key Laboratory for Geomechanics and Deep Underground Engineering in Beijing and Professor in the China University of Mining and Technology in Beijing, as well as Professor at the Tongii University in Shanghai. He is author or coauthor of about a hundred technical papers and 20 technical books.

Acknowledgements Funding for this work was provided by the Soft Rock Engineering Mechanics Subject Development Strategy Research Project of the Chinese Academy of Sciences (2017-JS-A-1).

References

Deere DU (1975) Applied rock mechanics of weak materials, general report, 5th Pan American conference on soil mechanics and foundation engineering, B. Aires, V. 4, p 479–492

Deere DU, Vardé O (1986) Engineering geological problems related to foundation and excavation in weak rocks, general report, 5th International congress, IAEG, B. Aires, V. 2.

He MC, Zou ZS, Zou YF (1993) Introduction to soft rock roadway engineering. China University of Mining and Technology Press, Xuzhou. (in Chinese)

He MC, Jing HH, Sun XM (2002) Soft rock engineering mechanics. Science Press, Beijing. (in Chinese)

Kanji MA, Abrahão RA (1978) International symposium on rock mechanics related to dam foundations, ISRM, ABMS, R. Janeiro, 2 Volumes

Rocha M (1975) Some problems related to rock mechanics of low strength materials (in Portuguese), panel report, 5th Pan American conference on soil mechanics and foundation engineering, B. Aires, V. 5, p 490–514

Yu XF (1956) Rock mechanics and ground support. Beijing Iron and Steel Institute (in Chinese), Beijing

Chapter 2
Engineering View of Soft Rocks

He Manchao and Sun Xiaoming

2.1 Concepts Related to Soft Rock

2.1.1 Introduction

With the increase of mining depth, disasters caused by large deformation of soft rock have been increased. It seriously threats to the safety and efficiency of deep resources utilisation. Soft rocks can be considered as complex rock medium with significant plastic deformation in specific environment. Deep soft rock engineering and disaster controlling have become a focus in rock mechanics research (He et al. 2010; He 2014). Figure 2.1 shows different ground failure cases in soft rock excavation.

Currently, there are several kinds of definition for soft rock (He 1992; He et al. 2002), namely descriptive definition, quantitative definition, and engineering definition.

2.1.1.1 Descriptive Definition

Descriptive definition of soft rock focuses on the lithology, structure and physico-chemical properties of the rock. Fractured, swelling or weathered rocks are often considered as soft rock (Zhu 1985). Compared with hard rock, the soft rocks have a low strength due to short diagenetic period, loose structure and poor cementation (Lu 1986). Soft rock refers to high fracture, expansible, strong weathering or in high stress state (Zheng 1988). AGI (1979), Sciotti (1990), Clerici (1992) also provided

H. Manchao (✉) · S. Xiaoming
State Key Laboratory of Geomechanics and Deep Underground Engineering (SKL-GDUE), China University of Mining and Technology, Beijing, China
e-mail: hemanchao@263.net

M. Kanji et al. (eds.), *Soft Rock Mechanics and Engineering*,
https://doi.org/10.1007/978-3-030-29477-9_2

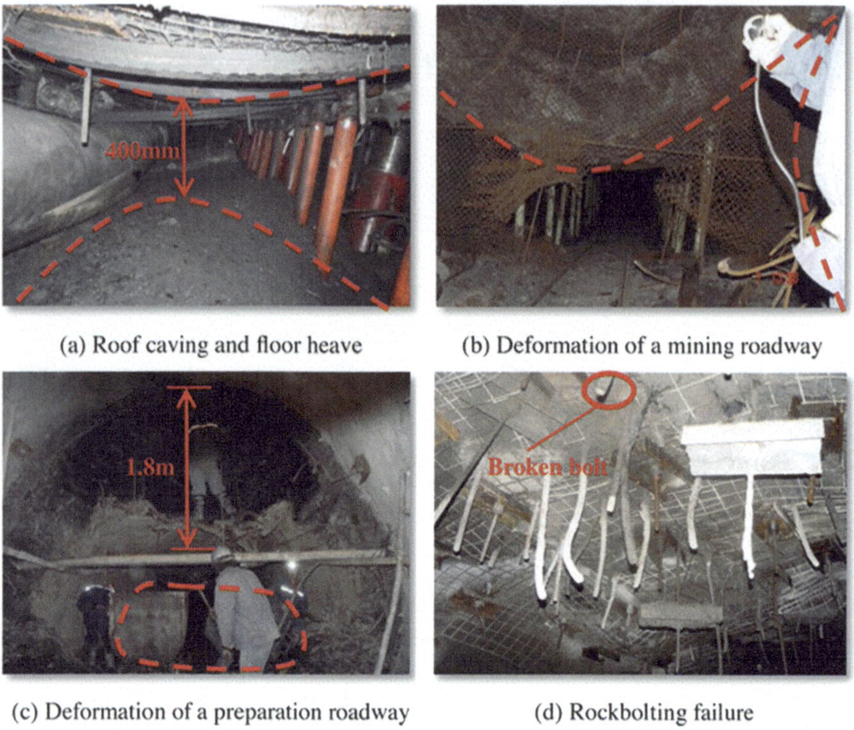

Fig. 2.1 Large deformation and ground failure of soft rock excavations (He et al. 2010). (**a**) Roof caving and floor heave, (**b**) Deformation of a mining roadway, (**c**) Deformation of a preparation roadway, (**d**) Rockbolting failure

classifications of soft rocks (weak rock) from the perspective of lithology and rock structure.

In China, the definition of soft rock was discussed in a coal mine ground control seminar, Kunming, in 1984 (He 1992), in which, the definition of soft rock is referred to rocks with low strength, large porosity and poor cementation. Soft rocks are often affected by systematic joints, weathering, or swelling clay minerals.

2.1.1.2 Quantitative Definition

Quantitative definition of soft rock is mainly related to the rock strength. Bieniawski (1984), Barla et al. (1990), and Russo (1994) defined soft rock by uniaxial compressive strength (UCS) less than 12.5 MPa, 15 MPa, and 17 MPa, respectively. Some scholars also define soft rock as $\mathrm{UCS}/(\gamma H) < 2$ where γ is rock specific weight and H is bury depth (Li and Wang 1987).

A definition was given by the International Society for Rock Mechanics (ISRM) in 1981. The ISRM describes rock with UCS from 0.25 to 25 MPa as "extremely weak" to "weak" (ISRM 1981).

2.1.1.3 Engineering Definition

Engineering definition of soft rock is mainly used in China. It is defined from view points of large deformation and ground failure in excavation. Dong et al. (1987) considered the surrounding rock as soft rock when the depth of disturbed rock mass around an excavation was more than 1.5 m. Lin (1999) suggested that soft rock was weak, loose or fractured rocks that caused large deformation.

2.1.2 Definition of Engineering Soft Rock

A proper definition of soft rock should cover a variety of lithological characteristics and reflect the basic deformational behaviour of soft rocks. Based on this point, a definition was proposed under considerations of geological soft rock and engineering soft rock (He 1992, 2014; He et al. 1993, 2002).

Geological soft rock refers to rocks characterised with low strength, large porosity, poor cementation, surface crushing or susceptible to weathering. It often contains swelling clayey minerals and/or loose, weak interlayer. The definition proposed by ISRM is within the scope of geological soft rock.

Engineering soft rock refers to rocks that undergo large deformation in engineering practice. The definition of engineering soft rock emphasises the relationship between material strength and engineering disturbance, it can be expressed as:

$$\begin{aligned} \sigma &\geq [\sigma] \\ U &\geq [U] \end{aligned} \tag{2.1}$$

where σ is engineering stress (MPa), $[\sigma]$ is strength of rock mass (MPa), U is rock deformation (mm), and $[U]$ is deformation allowance in practice (mm).

The key point of this concept is the engineering force, which causes large plastic deformation in engineering practice. Engineering force is the resultant force acting on the rock mass, including gravity force, swelling force (when exposed to water), and other engineering activity induced forces. Large deformation means deformation beyond designed allowance. It is noted that large deformation may be elasto-plastic deformation, visco-elasto-plastic deformation or continuous and discontinuous combined deformation.

The definition of soft rock reveals the relationship between engineering force and rock strength. If the rock strength is higher than the engineering force, the rock can be regarded as hard rock; if it is less than engineering force, it shows the mechanical characteristics of soft rock. Even for the same rock, it deforms as hard rock under low engineering force, but is regarded as soft rock under high engineering force (He et al. 2002; He 2014; Zhang et al. 2010).

The relationship between the geological soft rock and engineering soft rock is that, when the load is smaller than the strength of a geological soft rock (e.g. mud, shale), there is no significant plastic deformation in the geological soft rock, so it is geological soft rock but not engineering soft rock. When a geological hard rock (e.g. shale or cemented sandstone) is buried under critical depth associated with high ground stress, it would undertake large deformation as soft rock. In such case, it can be regarded as engineering soft rock.

2.2 Basic Mechanical Properties of Engineering Soft Rock

There are two basic indices associated with the deformational behaviour of engineering soft rocks: the critical softening load and the critical softening depth (He et al. 1993, 2002; He 2014).

2.2.1 Critical Softening Load

Creep test of soft rock shows that, when the applied load is small, the creep deformation does not change with time, and the rock is in stable state. When the applied load is greater than a certain value, the creep deformation of the rock begins to accelerate, that is, its plastic deformation increases noticeably and the rock is unstable. The threshold of the load is called critical softening load of the soft rock, which is the smallest load that leads unstable deformation of the rock.

2.2.2 Critical Softening Depth

Corresponding to the critical softening load of a soft rock, critical softening depth is also an objective parameter in a certain mining area for a certain kind of soft rock. Engineering experiences show that, if a roadway is developed less than a certain depth, the deformation of surrounding rock is small. However, if the roadway is developed beyond this depth, large deformation of surrounding rock occurs causing ground support difficulty. This threshold depth is called critical softening depth of the rock in this mining area.

The ground stress at critical softening depth is approximately equal to the critical softening load of the surrounding rock.

2.2.3 *Relationship Between Critical Softening Load and Critical Softening Depth*

According to the critical state definition of an engineering soft rock, critical softening load and critical softening depth can be derived from each other (He et al. 2002).

Ignoring tectonic and other stresses, the relationship between overburden load and rock strength is:

$$\sigma_c / \gamma H = 2$$

$$\sigma_{cs} = 2\gamma H_{cs} = 2\frac{\sum_{i=1}^{N} \gamma_i h_i}{H} H_{cs}$$

Transfer unit t/m^3 to KN/m^3, it is obtained (He et al. 1993):

$$\sigma_{cs} = \frac{\sum_{i=1}^{N} \gamma_i h_i}{50H} H_{cs} \tag{2.2}$$

$$H_{cs} = \frac{50H}{\sum_{i=1}^{N} \gamma_i h_i} \cdot \sigma_{cs} \tag{2.3}$$

where H_{cs} is the critical softening depth (m), σ_{cs} is the critical softening load (MPa), γ_i is the specific weight of i-th stratum in overlying strata (KN/m^3), H is the total thickness of overlying strata (m), h_i *is* the thickness of i-th stratum in overlying strata (m), and N is the total number of overlaying stratums.

In case of presence of tectonic or other stresses, the formulas are:

$$\sigma_{cs} = \frac{1}{50H} \cdot \left(\sum_{i=1}^{N} \gamma_i h_i \right) \cdot H_{cs} + \sum_{j=1}^{M} \Delta\sigma_{cs}^{j} \tag{2.4}$$

$$H_{cs} = \frac{50H}{\sum_{i=1}^{N} \gamma_i h_i} \cdot \left(\sigma_{cs} + \sum_{j=1}^{M} \Delta\sigma_{cs}^{j} \right) \tag{2.5}$$

where $\Delta\sigma_{cs}^{j}$ is the supplementary stress (MPa), j = 1 is stress caused by tectonic effect, j = 2 is stress by rock swelling, and j = 3 is dynamic stress.

2.2.4 Determinations of Critical Soften Load and Critical Soften Depth

To determine the critical soften load and critical soften depth of an engineering soft rock, following methods can be used.

2.2.4.1 Creep Test

The long-term strength of a soft rock can be determined by creep test, the critical creep load is roughly equal to its critical softening load.

2.2.4.2 Empirical Formula

The formula is:

$$\sigma_{cs} = KR_c \tag{2.6}$$

where R_c is the UCS of the rock, K is an empirical coefficient: for expandable soft rocks $K = 0.3–0.5$; for high strength soft rocks $K = 0.5–0.7$; and for jointed soft rocks $K = 0.4–0.8$.

2.2.4.3 Field Observation

The critical softening depth for a formation is the depth at which the surrounding rock begin to produce large plastic deformation.

2.3 Classification of Engineering Soft Rocks

2.3.1 Four Types of Engineering Soft Rocks

Based on the geological characteristics and deformational behaviours, engineering soft rocks can be classified into four categories, namely swelling soft rock (also called low strength soft rock), high strength soft rock, jointed soft rock and combined soft rock (He 1992, 2014; He et al. 2002), as shown in Table 2.1.

Table 2.1 Classification of engineering soft rock

Category	Condition	Dominant plastic deformation
Swelling soft rock (low strength)	Shale content >25%	Slipping along clay mineral of silicate; significant expansion under high water contents.
High strength soft rock	UCS ≥ 25 MPa	Slipping along flaky clay minerals.
Jointed soft rock	UCS < 25 MPa	Dilational slipping along the jointed surface.
Combined soft rock	Shale content ≤25%	Combination of above characteristics.

2.3.1.1 Swelling Soft Rock

Swelling soft rocks (abbr. S-type) refer to low strength rock (UCS < 25 MPa) containing large volume of expansible clay minerals, developing large deformation under low stress condition. Usually, the weak, loose, swelling, highly weathered rock mass associate with rock UCS less than 25 MPa all belong to low strength soft rock.

The plastic deformation of swelling soft rocks is closely related to slippage along clay mineral of silicate or volumetric expansion after absorbing water. In engineering practice, the site geological condition is often determined by the low strength rock that is usually composed of argillaceous rocks, which often contain large volume of clay minerals.

2.3.1.2 High Strength Soft Rock

High strength soft rock (abbr. H-type) refers to the medium to high strength rock mass (UCS > 25 MPa) that has large deformation under high stress condition. The stress level causing large plastic deformation to the rock is called high stress, which is normally great than 25 MPa based on the concept of soft rock defined by ISRM (UCS = 0.5–25 MPa). These rocks often have less shale contents and contain considerable sandy component, such as argillaceous silt and shaly sandstone. Their engineering characteristics are same as hard rock at shallow to medium bury depth. However, if bury depth increases to a critical value, it shows deformational characteristics of soft rock. Its plastic deformation procedure is often dilational shearing of the matrix material of the rock, causing micro-defects or cracks within the rock, and then slipping and expansion.

2.3.1.3 Jointed Soft Rock

Jointed soft rock (abbr. J-type) refers to rock mass with none or less shale contents but developed multiple joints. It shows hard rock mechanical properties in rock block. However, large plastic deformation may be generated under mining induced stresses, which is the characteristics of soft rocks. Its plastic deformation is shearing along the jointed plane and corresponding dilation in normal direction.

2.3.1.4 Combined Soft Rock

Combined soft rock is a combination of S-type, H-type and J-type soft rock. It has three kinds of rock, that is, high strength and swelling soft rock (abbr. HS-type), high-strength and jointed soft rock (abbr. HJ-type), high-strength, jointed and strong swelling soft rock (abbr. HJS-type).

2.3.2 *Softening Index*

The softening level of soft rock is a key for ground support. According to research results (He et al. 2002), a softening index has been developed to determine the softening level for engineering soft rocks.

2.3.2.1 Definition of Softening Index

The softening index (f_s) is defined as the ratio of the critical softening load (σ_{cs}) to the maximum stress (σ_{max}) in the surrounding rock mass,

$$f_s = \frac{\sigma_{cs}}{\sigma_{\max}} \tag{2.7}$$

2.3.2.2 Softening Index and Ground Support

The softening index (f_s) is related to the deformational behaviour of surrounding rock and corresponding underground excavation reinforcement (Table 2.2). For example, for Quasi-soft rock, f_s is from 0.8 to 1, indicating local plasticity deformation, consequently, anchorage and spraying on roof and two corners are advised for roadway support.

2.3.3 *Softening Procedure and Softening States*

According to softening procedure under mining induced stresses, the softening state of soft rock can be categorised into four types, namely, initial softening, strength softening, stress increasing, and combined strength reduction and stress increasing.

Table 2.2 Softening index classification

f_s	Soften level	Deformations	Advices on roadway support
≥1	Non	Elasticity	Local anchor and shotcrete on roof
1 to 0.8	Quasi	Local plasticity	Anchor and shotcrete on roof and two corners
0.8 to 0.5	Normal	Full cross-section plasticity	Full – section bolt, net and spraying
0.5 to 0.3	Super	Expansion-based plastic	Full – section bolt net and shotcrete + key point anchor
<0.3	Extremely	Expansion, and high stress caused plastic	Full – section bolt net and shotcrete + high strength key point anchor

2.3.3.1 Initial Softening

At the initial stage of excavation ($T = 0$), the critical soften load (σ_{cs}) of surrounding rock is just less than the maximum ground stress (σ_{max}), the roadway deforms continuously, but both of them are still constant. It indicates that the surrounding rock is entering into soft rock state, and is called initial softening. It can be expressed as

$$\begin{cases} T = 0 \\ \sigma_{max} = C_1 \\ \sigma_{cs} = C_2 \end{cases} \Rightarrow f_s = \frac{\sigma_{cs}}{\sigma_{max}} < 1 \Rightarrow \begin{cases} U = f_1(t) \\ \dfrac{\partial f_1}{\partial t} > 0 \end{cases} \tag{2.8}$$

where U is the total deformation, including elastic deformation, time-independent plastic deformation and time-dependent plastic deformation.

2.3.3.2 Strength Decreasing Softening

After a certain period of excavation ($T = t$), the maximum ground stress keeps constant ($\sigma_{max} = C$), but the critical softening load of surrounding rock decreases continuously due to weathering, crack propogation or fracture development ($\sigma_{cs} = f_2(t)$, $\partial f_2(t)/\partial t < 0$). The roadway deforms continuously, and is called strength softening. The softening state equation is:

$$\begin{cases} T = t \\ \sigma_{max} = C \\ \sigma_{cs} = f_2(t) \\ \dfrac{\partial f_2(t)}{\partial t} < 0 \end{cases} \Rightarrow f_s = \frac{\sigma_{cs}}{\sigma_{max}} < 1 \Rightarrow \begin{cases} U = f_1(t) \\ \dfrac{\partial f_1}{\partial t} > 0 \end{cases} \tag{2.9}$$

2.3.3.3 Stress Increasing

With the excavation, the maximum ground stress may increase ($\sigma_{max} = f_3(t)$, $\partial f_3(t)/\partial t > 0$), but the critical softening load of the surrounding rock remains unchanged ($\sigma_{cs} = C$). At some time ($T = t$), the surrounding rock is in soft rock state ($f_s < 1$) and the roadway deforms continuously ($U = f_1(t)$, $\partial f_1(t)/\partial t > 0$). This is called stress increasing, and the state equation is:

$$\begin{cases} T = t \\ \sigma_{max} = f_3(t) \\ \sigma_{cs} = C \\ \dfrac{\partial f_3(t)}{\partial t} > 0 \end{cases} \Rightarrow f_s = \frac{\sigma_{cs}}{\sigma_{max}} < 1 \Rightarrow \begin{cases} U = f_1(t) \\ \dfrac{\partial f_1(t)}{\partial t} > 0 \end{cases} \tag{2.10}$$

2.3.3.4 Strength Softening and Stress Increasing Combination

With continuous excavation, at times of $T = t$, the maximum ground stress increase ($\sigma_{max} = f_3(t)$, $\partial f_3(t)/\partial t > 0$) while the critical softening load of the surrounding rock decrease ($\sigma_{cs} = f_2(t)$, $\partial f_2(t)/\partial t < 0$). The rock mass in soft rock state ($f_s < 1$) and the roadway deforms continuously. This is called combination of strength softening and stress increasing, and the state equation is:

$$\begin{cases} T = t \\ \sigma_{max} = f_3(t) \\ \sigma_{cs} = f_2(t) \\ \dfrac{\partial f_3(t)}{\partial t} > 0 \\ \dfrac{\partial f_2(t)}{\partial t} < 0 \end{cases} \Rightarrow f_s = \frac{\sigma_{cs}}{\sigma_{max}} < 1 \Rightarrow \begin{cases} U = f_1(t) \\ \dfrac{\partial f_1(t)}{\partial t} > 0 \end{cases} \tag{2.11}$$

References

AGI (1979) Some Italian experiences on the mechanical characterization of structural complex formation. Associazione Geotecnica Italiana, Proc. of the ISRM Congress, Montreux, vol. 1, pp 827–846

Barla G, Forlati F, Zaninetti A (1990) Laboratory tests on rocks: issues and examples. MIR 90. Torino 4-1:4–47. (in Italian)

Bieniawski ZT (1984) Rock Mechanics design in mining and tunneling. Balkema, Boston

Clerici A (1992) Engineering geological classification of weak rocks: classification, sampling and testing. ISRM symposium. Eur Secur 92:179–184
Dong F, Lu S, Gao S (1987) Discussion on roadway surrounding rock condition and supporting theory. Ground Pressure 2:12–15. (in Chinese)
He M (1992) Concept, classification and supporting measures of soft rock. Coal Sci Technol Fengmei 2:7–10. (in Chinese)
He M, Jing H, Sun X (2002) Soft rock engineering mechanics. Science Press, Beijing. (in Chinese)
He M, Qian Q et al (2010) The basis of deep rock mechanics. Science Press, Beijing, 2010. (in Chinese)
He M (2014) Progress and challenges of soft rock engineering in depth. J China Coal Soc 39(8):1409–1417. (in Chinese)
He M, Zou Z, Zou Y (1993) Introduction to soft rock tunnel engineering. China University of Mining and Technology Press, Beijing. (in Chinese)
ISRM (1981) Rock characterization testing and monitoring. International Society for Rock Mechanics. Pergamon Press, New York, p 211
Li W, Wang J (1987) Study on deformation and failure mechanism and reasonable support of soft rock roadway. J China Coal Soc 3:55–62. (in Chinese)
Lin Y (1999) Discussion on several theoretical problems of engineering mechanics of soft rock. Chin J Rock Mech Eng 18(6):690–693. (in Chinese)
Lu J (1986) The combining support method of permanent opening in soft rock. Chinese J Geotechn Eng 8(5):50–57. (in Chinese)
Russo G (1994) Some considerations on the applicability of major geomechanical classifications to weak and complex rocks in tunnelling. GEAM 31:63–70
Sciotti M (1990) The rock formation of Catatteri. Lithology, and structural geology. MIR 90. Torino 1-1:1–18. (in Italian)
Zhang N, Wang B, Zheng X, Zhu X (2010) Analysis of grouting reinforcement results in secondary support of soft rock roadway in kilometer deep mine. Coal Sci Technol 38(5):34e8. (in Chinese)
Zheng Y (1988) Elastic-plastic theory of rock mechanics. Coal Industry Press, Beijing. (in Chinese)
Zhu X (1985) View of soft rock roadway support. Coal Sci Technol 1985(8):21–23. (in Chinese)

Chapter 3
The Geology of Soft Rocks

Georg R. Sadowski

3.1 Introduction

3.1.1 Definitions

The definition of soft rocks has been largely discussed in the literature (Terzaghi and Peck, 1967; Rocha, 1975; Dobereiner, 1984; Johnston, 1993; Gonzalez de Vallejo et al., 2002; Kanji, 2014; and so on). The lower limit of uniaxial compressive strength for soft or very soft rocks has been advocated by several authors as 2 MPa and the upper, around 20 MPa.

Kanji (2014) mentions, "…there is a practical coincidence that the upper limit of the strength of what is considered soft is about 25 MPa as unconfined compressive strength (UCS)".

A similar value (4000 psi = 27.6) was also considered by Deere and Miller (1966) in their engineering classification of intact rock concerning the Very Low Strength (E-type) rocks (By the way, this value may be considered in the lower strength range of regular concrete).

Deere (1968) mentioned that the geological classifications have been often considered inadequate for rock mechanical purposes although to be maintained in terms of lithologic terminology, understanding that for some rock types, the range of mechanical properties may be discouragingly large and bearing also in mind the different types of structural anisotropies (layering, cleavages, schistosity, and other types of foliations). The classification of Deere and Miller uses laboratory determined values and is based on the ratio of the tangent $E_{50\%}$ elasticity modulus to the uniaxial strength.

Martin et al. (2003) analyzing the maximum feasible Elasticity Module in several types of rocks characterized by the Deere and Miller intact rock classification

G. R. Sadowski (✉)
Institute of Geosciences of the University of São Paulo, São Paulo, SP, Brazil

M. Kanji et al. (eds.), *Soft Rock Mechanics and Engineering*,
https://doi.org/10.1007/978-3-030-29477-9_3

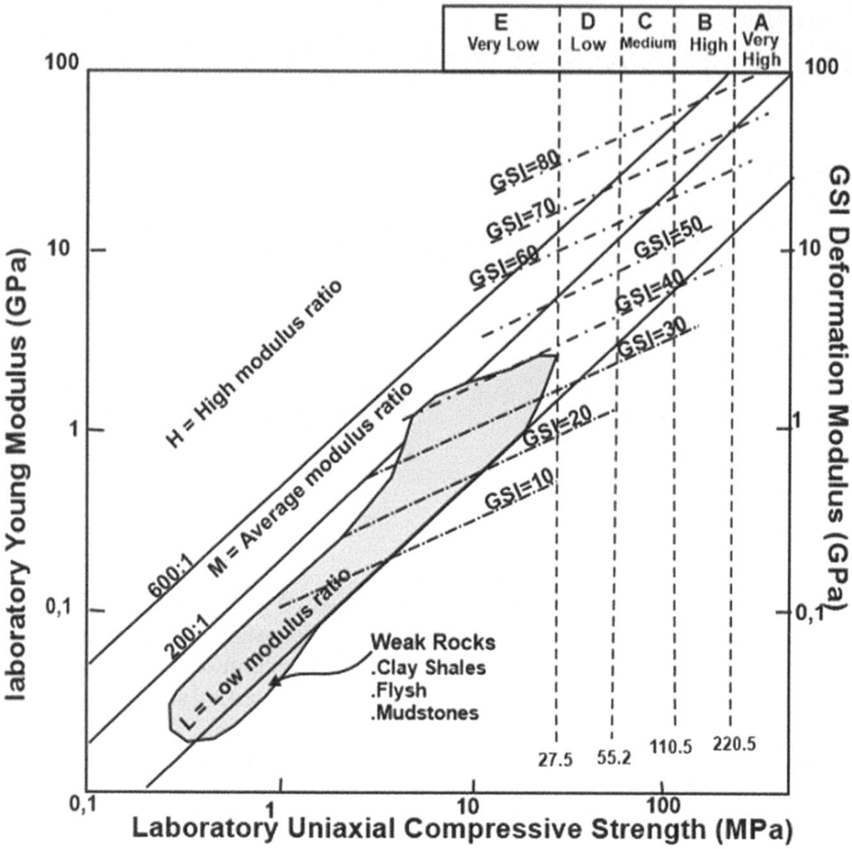

Fig 3.1 Relationships between Deere's classification for UCS vs Young's modulus and the Geological Strength Index (Martin et al. 2003)

related it to the GSI (Geological Strength Index) rock mass classes. The graph of Fig. 3.1 shows the relation used for Weak Rocks.

Bosio and Kanji (1998) using the Deere and Miller criterion presented a series of tests on sedimentary rocks of the Paraná Sedimentary Basin which show a low to medium modulus ratio (below 500:1) and a UCS below ca. 25 MPa allowing to classify them as EM and EL.

ISO14689-1 (2003) proposed a mechanical classification of rocks based on the uniaxial compressive strength (UCS) where weak to extremely weak rocks are considered in the range of <1 to 25 MPa.

Another decisive property for the definition of a soft rock is its slaking or behavior in water, as proposed by Nickmann et al. (2006), leads to the definition of five categories of durability suggesting that their properties result from a combination of several parameters such as compressive strength, texture or grain size distribution, mineralogical content such as clay minerals susceptible to water, and pore volume

Table 3.1 Most usual types of soft rocks

Basic types	Subclasses
Sedimentary rocks	Clastic: mudstones, shales, siltstones, sandstones, conglomerates and breccias, and marl; Evaporites: salt, carnalite, etc.; Soluble: limestone, dolomite and gypsum; and Coal
Igneous rocks	Volcanic conglomerates, breccias and lahar, Basaltic breccia; pyroclastic deposits, volcanic ash, tuff, and ignimbrite; and Weathering products of crystalline rocks
Metamorphic rocks	Slate, phyllite, schists, quartzite little cemented, metavolcanic deposits

(degree of the water absorption). This so-called "structural strength" would more accurately establish the border on the upper and lower bonds between the weak and soft rocks and soil respectively.

According to Kanji (2014) the most usual soft rocks would be expressed in Table 3.1.

3.1.2 Altered Rocks

Bearing in mind that many weathered hard rocks may present physical properties of soft rocks, their classification obviously tries to retrieve the properties of the sound matrixes however taking into account the "weathering products of crystalline rocks" which represents a very large range of products of weathering and their associated geomechanical characteristics having to be dealt under the term "softened rocks" instead of "soft rocks". The mixture of hard blocks of less weathered rocks inside the saprolitic soil medium refers us to materials something similar to what is called BIM rocks or Blocks in Matrix Rock (Medley 1994), however in smaller proportions than those of the tectonic mélanges with tenths to hundreds of meters large blocks (olistoliths) or transported whole layers (olistostromes).Those may be common in weathered granitic terrains in Brazil or in talus masses as found during the construction at the Caracas Metro in Venezuela and several other mountain slopes in the world.

Criteria to determine the degrees of weathering as the proposed by ISRM do not allow to precise the strength properties of all different types of the lithology subjected to alteration. One practical criterion to establish the degree of weathering is the proposed, for example, by Vaz (1996) which may work well for igneous and highly metamorphosed rocks but might not be adaptable to some of the soft sedimentary or low degree metamorphic rocks. In that criterion hard weathered rock can only be excavated with the use of explosives while so called soft weathered rocks are those too hard to be penetrated by the usual SPT tool but with properties of a very hard soil that may be excavated by usual scrapers not needing to be excavated by explosives.

The term "softened rock" should involve rocks softened by any type of alteration, be it through weathering, hydrothermal, or tectonic crushing processes. However, the original rock has to be mentioned. Remembering, for example, that weathered or hydrothermally affected granites differ substantially from some weathered phyllites which expand notoriously when stress-released in the presence of water. Mineralogical and structural features of the original soft rock still may dominate the mechanical properties of their weathering products being important to mention its sound name after the term "weathered" or "altered". Also, in the case of tectonic crushing, it might be important to mention which rock was reduced to "gouge", tectonic "breccia" or cataclasite be it a quartzite or an ultrabasite.

A relatively detailed description of the geological context of rocks related to construction engineering may be found in Goodman (1993) including several soft rocks not discriminated specifically under this designation.

I suggest that the best way to indirectly grasp their geomechanical properties and allow an idea of the correlated geotechnical risks is to understand their original genetic environment.

In terms of strength, the actual parameter to define hard and soft rocks or the degree of aggregation of the component grains or minerals is determined by practically the same diagenetic parameters which generate lithification in a soil, the main being compaction (packing) and degree of cementation, although others may be locally more important such as pressure-solution, replacement, recrystallization, hydration, and development of concretions. Mineralogical composition may be crucial not only to determine tectonic settings of the sediment but also its mechanical properties. Mudstones undergo compaction with serious and permanent loss of porosity and permeability. Carbonates compact to varying degrees depending on the proportion of plastic material. Most appear to compact by solution processes rather than by mechanical compaction.

For practical reasons it comes out that soft rocks comprise mainly sedimentary and low degree metamorphic rocks although some igneous lavas have to be considered.

3.2 The Main Types of Weak or Soft Rocks

3.2.1 Magmatic Rocks

Usually sound magmatic rocks belong to the hard to medium type. Highly porous lava flows and pumice may show low strength, light weight and high water absorption. Porous vesicular basalt, for example, which constitutes frequently the upper portion of some lava flows, may show low strength accordingly to the frequency of vesicles. Some with 10–25% volume of voids may show diminished strengths ranging between 15 and 55 MPa (Nace et al. 1972).

Basalts impregnated with smectite clays (nontronite) or similar expansive minerals that can concentrate in fissures and in the vesicles (amygdaloidal basalt) tend to slake when exposed to water. Some secondary impregnations by gypsum, anhydrite or by zeolites (laumontite) may provoke total disaggregation of apparently sound igneous rocks just after excavation or later, as confirmed at the La Higuera headrace tunnel in Chile.

Other soft volcanic rocks represented by different types of ignimbrites, tuffs and volcanic breccias may be considered as variations of sedimentary rocks and will be considered in the next item.

3.2.2 Sedimentary Rocks

Sedimentary rocks may be of epiclastic, chemical and of organic origin. When subjected to diagenetic processes sedimentary soils are transformed into rocks and those transformed into metamorphic rocks through solid state recrystallization. The scheme of Fig. 3.2 below helps to define the main types of sedimentary rocks.

The deep evolution of large masses of sediments into rock through diagenesis has been studied in deep buried clays as those deposited at the US Gulf Coast and at the sedimentary Welsh Basin in Great Britain (Larsen and Chilingar 1967). Such studies show that the upper border line between diagenetic and metamorphic transformation lays somewhere around the conditions which allow illites to be recrystallized into muscovite or the generation of slaty cleavage in mudstones(Merriman and Peacor 1999) (temperatures up to c. 200 °C and pressures of about 300 MPa).

Soft or even hard rocks may also be generated from soils by near surface processes such as cementation, dissection, direct precipitation of salts, or just formed through biogenic activity as in coral-reefs. Layers of caliche and sabkha may show properties of soft rocks although deposited in subaerial conditions. Iron oxide or carbonate impregnations of granular soils (duricrusts or ferricrusts) are not uncommon in tropical climates.

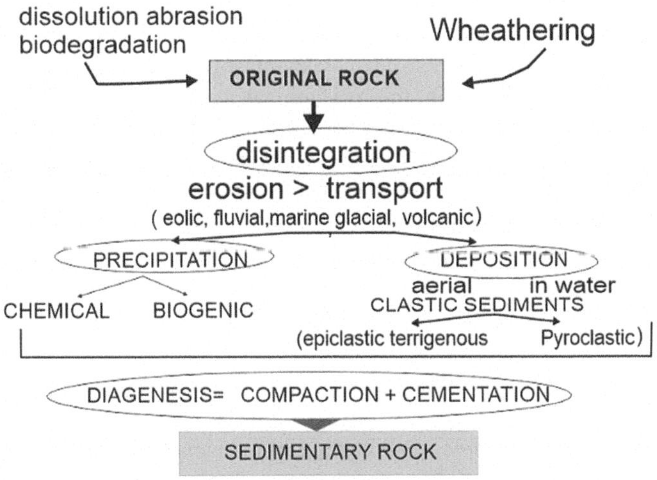

Fig. 3.2 Scheme of the formation of sedimentary rocks

3.2.3 Sedimentary Rocks of Igneous Ascent

Considering unweathered or unaltered matrixes, with the exception of vesicular volcanic lavas and some porous light lavas, soft igneous rocks are usually associated to sedimentary volcanoclastics or pyroclastics such as tuffs, lapilli layers, ignimbrites, cinerites or ash-types. Authigenic agglomerate lavas or volcanic breccias can be classified as hard rocks unless they include exogenic mixtures with different types of country rocks some subjected to former weathering. Other complex combinations are found in consolidated lahars and layered volcano-sedimentary assemblages.

Tunnels or underground caves cut in consolidated volcanic ashes as certain tuffs may present good construction performance if not subjected to late geochemical alterations [e.g. Karaj tunnel in Northern Iran, Yucca Mountain Radioactive Repository Complex in the USA, etc.]. Softening of volcanoclastic rocks due to disaggregation related to the presence of water-expansive minerals (smectites, anhydrates, etc.) is known and has caused problems in construction, as also the presence of certain meta-stable zeolites such as laumontite, mentioned above (La Higuera water tunnel in Chile). Attention must be paid to some glassy tuffs which may be transformed through post depositional alteration, producing bentonites (mixtures of smectite and devitrified glass). Tuffs from the Saint Helen volcano, Oregon, showed unit weights in the order of 1.38 ton/m^3 and triaxial saturated peak strength in the range of 40 MPa.

Practically contemporaneous transformations related to hardening by iron oxides, silica or other types of cementiceous minerals are common in volcanic environments and have to be taken into account since sudden hardened interlayers may be found in-between softer lithologies. However the opposite may also happen due to expansion of hydrothermal salts impregnating sediments (muds).

Another important peculiarity of some volcanic rocks is related to their external shape conditioned by paleotopographic features. Layers of tuffaceous rock may show a pseudofolding related to the original deposition of ashes on undulating hills as seen in Ecuador.

Modern hard lava flows and softer volcanoclastic rocks are frequently intercalated and may be observed in the volcanic arcs and shield volcanic terranes (Fig. 3.3).

The volcanoclastic or pyroclastic rocks related to gaseous and explosive volcanism may show hard rock fragments in-between ashes since country rock and solidified lava fragments may be thrown at enormous distances into different directions and later consolidated into a tuffaceous matrix (Fig. 3.4).

3.3 Terrigenous Epiclastics

The most common and simple classification of terrigenous epiclastics is based on the grainsize distributions and forms of particles (Table 3.2)

Fig. 3.3 Interstratified volcanoclastic layers and lavas on a road cut in Ecuador

Fig. 3.4 Basaltic lava fragment in a gray tuffaceous matrix-Chile

Table 3.2 Grainsize classification of sedimentary rocks

Clast diameter (mm)	Clast name	Rounded to subangular	If angular	Rock type: rounded to subangular	Rock type: angular
>256	Boulder	Block (≫256)	Bimrock (>)		>Olistoith?
64–256	Cobble		Rubble		
4–64	Pebble	Gravel	or	Conglomerate	Breccia
2–4	Granule		Rudite		
1/16–2	Sand	Sand	Sand	Sandstone	Gritstone
1/256–1/16	Silt	Mud	Silt	Siltstone	Siltstone
<1/256	Clay	Mud	Mud	Mudrock	Claystone

Grainsize distribution although important does not cover all the necessary parameters which condition the geomechanical properties of the rock and packing, porosity, imbrication, cementation and the relation matrix vs. grains have to be taken into account. Therefore a more detailed classification is needed concerning the general framework or texture of any rock. One of the parameters is the volumetric proportion between major clasts and their supporting matrix is very important regarding not only their relative percentages but also the type of the intergranular filling. The presence of a clast-supported or of a matrix-supported texture can determine the propagation of stress induced fissures and consequently the interparticle peak resistance of the sedimentary rock (Fig. 3.5).

The classification of the sedimentary rock will give a hint to the above reasoning.

In the case of sands with a percentage of fines (diameter < 0.06 mm) below 15% the dominant mineralogical composition of the grains allows to classify sandstones accordingly: when quartz grains are dominant the plain name sandstone might be sufficient, when dominant grains are feldspar we have an arkose and if there is a dominant percentage of rock clasts we will have lithic sandstones. If percentages of fines ranges from 15 to 75% the sedimentary rock will be called a wacke or greywacke and, above 75%, mudstone (encompasses siltstones and argillites). In the case a mudstone shows fine layering or fissility parallel to the bedding planes we will have a shale.

In sandstones and rudites several textural aspects developed during diagenesis increase the strength of the rock as shown in Fig. 3.6.

Another important aspect is mineralogical composition. If stress is not large enough to deform the grains of an uniformly selected quartz bearing sands, lithification will only advance significantly if there are other minerals flexible enough to deformed being around harder grains and so allow to diminish the porosities such as micas; or an addition of cementitious material (iron hydroxides, colloidal silica, carbonates, etc.). Sandstones of ancient eolic origin, as the Cretaceous Karroo or

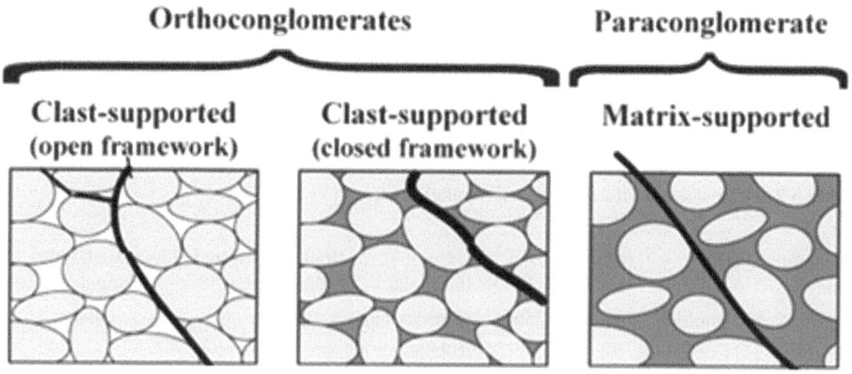

Fig. 3.5 The type of framework establishes the type of lines of weakness or potential fracture propagation in the rock when subjected to failure. On the right one can see that in wakes the strength of the rock may be controlled by the strength of the matrix

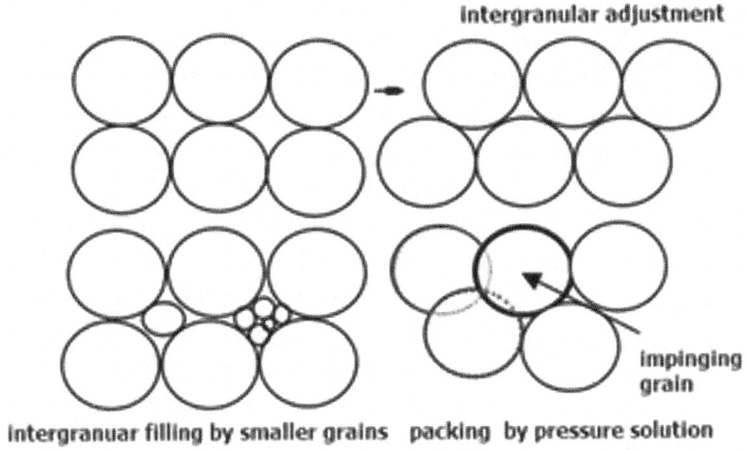

Fig. 3.6 Assuming uniform grain size distribution with semispherical grains (e.g. eolic sandstones) in a cubic packing of spheres results into a porosity of 47.65% while with the rhombohedral adjustment it turns into 25.95%. Low sorted sediments allow finer grains to fill the pores between larger grains and may significantly alter the packing. Under large loads pressure solution due to impinging grains may dissolve grains at the contacts and allow dissolved ions to be redeposited in the remaining voids

Botucatu sandstones originated in the Mesozoic deserts of Gondwana, will be only hardened by intergranular rearrangement (cubic to rhombohedral) or with the addition of some cement or by resizing and interlocking the grains by pressure solution, fragmentation, or recrystallization (Fig. 3.6).

Fine grained rocks or pelites such as siltstones, mudstones, lutites, claystones (argillites), shales, and slates are considered classical soft rocks. Shales and slates are usually fissile along the bedding surfaces and the last usually belong to the borderline of very low degree metamorphism. They are usually composed by more than 50% of clay-minerals or hydrated aluminum phyllosilicates which sometimes show grainsizes of silts although maintaining the typical properties of the clays.

They are frequently formed in deep burial basins through the gradual consolidation with reduction of their porosity at great depths (Fig.3.7).

Mudstones may present different minerals which tend to expand in the presence of water or through oxidation. Such are montmorillonitic clay minerals, vermiculite, or some very water friendly sulfate salts (e.g. anhydrite); sulphides as pyrrhotite, marcasite, and pyrite will increase the susceptibility to slake. Some known examples of this problem were identified in building on mudstones around Bogota in Colombia or on alu-shales in Oslo. If consolidated under heavy overburden they tend to develop slickensides and sole marks which substantially reduce their strength causing problems at high slope excavations or underground tunneling due to their susceptibility to squeeze. Strong anisotropy related to the bedding parallel orientation of platy minerals or squeezed aggregates may result into fissility of shales o even slates. With advancing depth and temperatures clay minerals may grow into

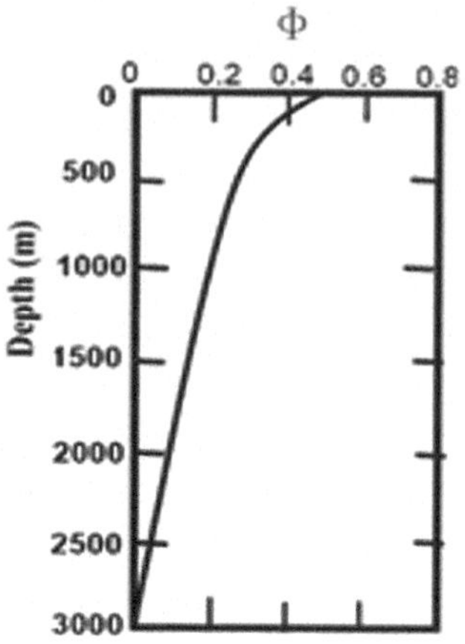

Fig. 3.7 Porosity of clay sediments based on data from Recent, Tertiary and Early Jurassic sediments (Von Engelhardt 1977)

Table 3.3 Classification of fine sedimentary rocks

Clay-size content (%)			0–32	33–65	66–100
Indurated	Beds	>10 mm thick	Bedded siltstone	Mudstone	Claystone
	Laminae	<10 mm thick	Laminated Siltstone	Mudshale	Claystone
Mata-morphosed	Degree meta-morphism	Low	Quartz Argillite	Argillite	
		High	Quartz slate	Slate	
			Phillite and/or micaschist		

large grain illites or moscovites or even chlorites when the rock is than considered as metamorphic. It has been verified that in this stage, bedding parallel cleavage is substituted by a slaty cleavage which crenellates the initially formed bedding parallel cleavage.

A classification of fine terrigenous rocks accordingly to grainsizes and fissility is shown in Table 3.3 modified from Potter et al. (1980). In this table we show in gray the transition zone to metamorphic rocks.

Shales also use to be frequently impregnated with biogenic substances as fossils, concretions, carbon and graphitic layers and even oil. For construction engineering their fissility and combinations with organic materials and sulphides may be a major problem for foundation or slope stability. Classical examples relating problems in these and other sedimentary rocks are described in a TVA (Tennessee Valley Authority) report (1949).

Major geotechnical problems are found specially where these rocks are folded or sheared due to their high anisotropy and susceptibility to slacking into soft expansive media. As some laminations show slickensides or polished surfaces from ancient sole marks and interlayer-slip their shear strength is related more to the residual than to the peak values.

Complex and specific terrigenous associations are found in different sedimentary environments. Glacial formations may be represented by matrix supported paraconglomerates or diamictites deposited by melting waters or breccia like tillites formed by glacier-scrapped and smeared fragments of country rock or by varvite alternating thin layers of silt to sandstone and dark clayey layers sometimes with organic remains. Dropstones fallen from melted floating icebergs may be immersed in fine lake-bottom sediments. Olystoliths or large blocks or lithoclasts ranging from few to tens of meters in size frequently related to submarine avalanches may be found inside siliciclastic turbidites. A whole layer composed of such clast is an olistrostrome.

Some turbidites related to cyclic deposition of mudflows towards the sea bottom show a repetitive sequence of gradational deposits, each starting from granular fining up graded sandstones or rudites passing into siltstones and mud. Such repetitive sequences may be of kilometric thickness usually designated as flysch (see, for example, Marinos and Hoek 2001).

3.3.1 Chemical and Biogenic Sedimentary Rocks

Chemical sedimentary rock forms when mineral constituents in solution become supersaturated and inorganically precipitate. Common chemical sedimentary rocks include oolitic limestone and rocks composed of evaporitic minerals, such as halite (rock salt), sylvite, carnalite, barite, anhydrite and gypsum (Fig.3.7). Carbonates such as limestones and dolostones are more widely known regarding construction engineering in contrast to other soluble salt rocks or evaporites which are targets of economic exploration and of studies related to petroleum exploration.

Carbonates such as limestones and dolomites are usually related to middle hard rocks (CMH and BM). Certain travertines may be related to soft rocks since they show many irregularities and voids. Some impure limestones intercalated with softer rocks or carbonaceous layers may show properties of soft rocks.

Evaporites show a large range of strength parameters. Some experiments in halite, glauberite, thenardite and rock salt (halite) (Liang et al. 2012), allow to classify them as EM (soft rocks) while data obtained from gypsite and anhydrite show that the former may be classified as EH (still soft rock with a high MR) and DH whilst the latter would be mainly in the CH field.

Sylvite and Carnalite which are important K-ores being intensively mined and should be considered in the range of soft rocks.

The majority of the evaporites are soluble in the presence of water. Sulfates are less soluble but may be highly unstable such as Anhydrite and Celestite that expand about 60% when allowed to hydrate into Gypsite. They usually show elastic-visco-plastic and tendency to secondary creep. Secondary creep takes place under 4–6 MPa in gypsite and 25–40 in anhydrite (Giambastiani 2005). Some deposits are show very complex structural geology typical to tectonic halokinesis.

Fig. 3.8 Salt cordillera in Atacama, Chile. White evaporite layers may be seen in-between mudstones

Evaporite salts have accumulated on the floors of continental brine lakes or in closed or semi-closed marine environments and as capillary efflorescences exposed arid land surfaces. Smaller deposits may be formed in volcanic environment, deserts (sabkha) or volcanic lakes. Some of large continental origin deposits are in Bolivia and Chile Cordillera del Sal (Fig. 3.8), Salar de Atacama and Salar de Uyuni. However the largest are of marine ascent such as the famous Zechstein basin in Central Europe or the buried type of deposits in the Mediterranean Sea bottom. Their geomechanical properties have been largely documented in the literature related to salt mining (Jeremic, 1994).

While the largest limestone and dolomite deposits are related to carbonates deposited through direct precipitation from sea water their origin may be chemically due to carbonate saturation or biogenic. Large biogenic carbonate formations were deposited by precipitation of skeletal remains of microorganisms such as foraminifera and construction of bioherms (coral reefs for example). There are also those initially generated from clastic deposits of shells. Many of the carbonates belong to the medium to high strength rock classes.

One of the best studied biogenic limestones is chalk which lays on the borderline between soil and soft or very soft rock. It is originated in deep sea water conditions from the gradual accumulation of the skeletal fossiliferous remains or minute aragonite/calcite shells (coccoliths) of Mesozoic age planktonic species. In-between the chalk there are fine bands of chert (flint) parallel to the bedding or as embedded nodules. Flint is a silica partially derived from the dissolution of sponge spicules or other siliceous organisms during compaction. Layers and seams of marls (carbonates plus clays) and some smectite bearing clays are also found some thick enough to be considered as Formations or Members of the whole sequence (e.g. Gault clay). Typical E_{50}/UCS) values in Britain are 100–125. Danish chalks near Copenhagen showed strengths UCS = 11.3 (lower bound) and UCS = 24.2 MPa (average). The averaged unconfined strength evaluation took also into account the role of very thin layers with high degree of induration (Kellezi et al. 2016). Studies performed during the Channel Tunnel Construction improved largely the knowledge about their mechanical properties (Harris et al. 1996).

3.3.2 Soft Metamorphic Rocks

There are some selected matrixes of sound metamorphic rock to be classified as soft rocks. Usually they are represented by low metamorphic degree slates or schists formed under different tectonic environments, usually subduction or collision related (Figs. 3.9 and 3.10 based on Tarbuk and Lutgens 2005). There are, however, some clayey soils/rocks subjected to deep burial inside large sedimentary basins where indications of low grade metamorphism was observed due to the increase of crystallinity and size of micaceous minerals under local geothermal gradients and lithostatic pressures. They may be already classified as soft metamorphic rock independently of the presence of an incipient tectonic crenulation cleavage affecting the bedding.

However, in usual metamorphic environments related to subduction or collision, tectonic cleavages are superimposed on the original layering or bedding in the form of slaty, strain-slip and crenulation cleavages and, further, with the increase of PT conditions, schistosity. These planar features introduce significant anisotropies and weakness surfaces in the metamorphic rock, mainly in pelites.

Micaceous bands when flattened into schistosity may develop expressive weakening surfaces in otherwise hard rocks such as banded gneisses. Also tectonic crushing expressed through cataclasite and gouges in shear zones or in tectonic mélanges of ultrabasic rocks may have turned into soft rocks. The presence of foliations intro-

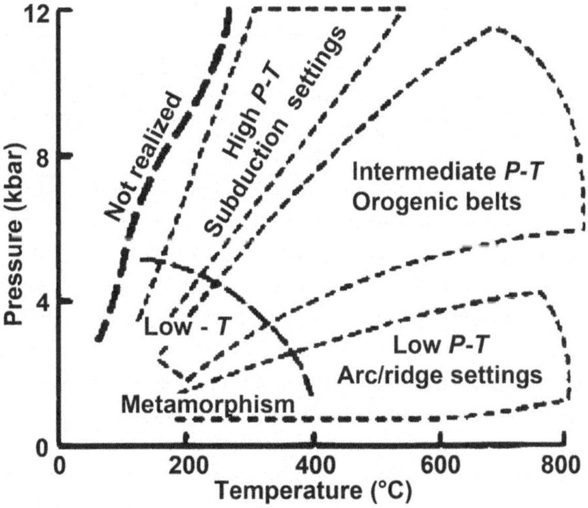

Fig. 3.9 Pressure—temperature diagram and tectonic environment (Tarbuk and Lutgens 2005)

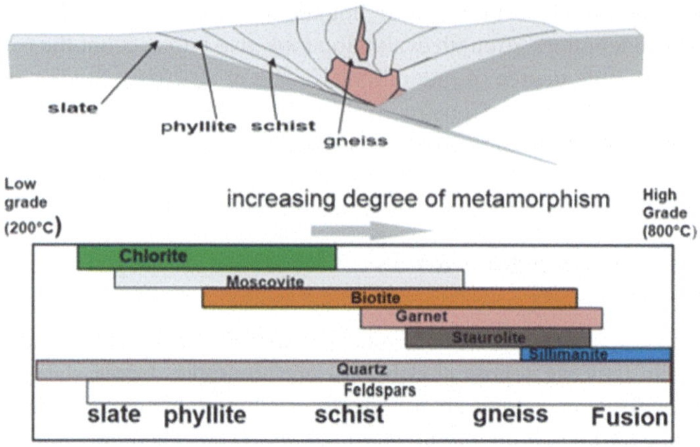

Fig. 3.10 Comprehensive sketch of the evolution of foliations and metamorphic degree along an orogenic belt (approximately based on Tarbuk and Lutgens 2005)

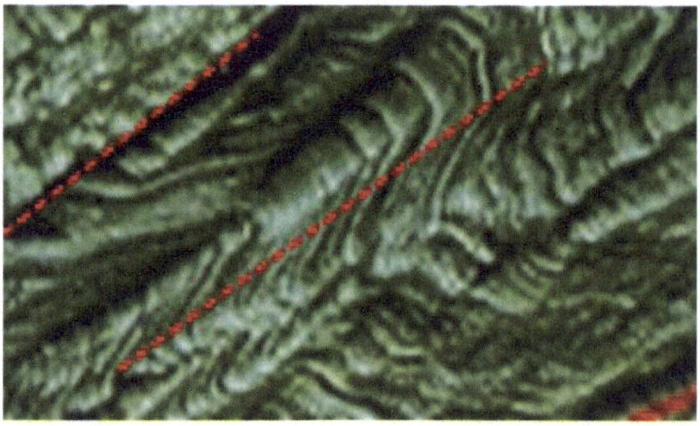

Fig. 3.11 Crenulation and strain-slip cleavage planes as pre sheared surfaces

duces the concept of pre-established weakness planes affecting the mineralogical matrix and strength may differ substantially accordingly to the direction parallel or perpendicular to the foliation.

Phyllites, slates and sericitic schists when rich in micaceous minerals may show expansion perpendicular to the foliation when excavated due to stress release generating an open foliated and relatively loose structure susceptible to infiltration of water and substantial decrease in shear strength parallel to the enhanced cleavage.

Some strain slip and crenulation cleavages generated through minuscule slip (Fig. 3.11) along their surfaces may show pre-shear structures such as small slickensides and therefore showing residual shear strength characteristics much below the expected peak (Mello et al. 2004).

3.4 Structural Geological Aspects of Soft Rocks

The variability of the strength parameters of the intact rock mass of soft rocks plays an important role in engineering mainly because it frequently ranges below that of the hard normal construction materials like concrete what turns their interaction with civil works more problematic for rock engineering than that of the medium to hard rocks.

In the harder rocks the engineer faces main problems usually related to the discontinuity patterns while in the soft rocks both the rock mass and its discontinuity patterns have to be taken jointly into careful account.

A chapter dealing with the structural geology of soft rocks is presently far beyond our purposes, but we suggest as a first step to try to understand the environmental conditions of their formation, to at least an idea of their possible spatial distribution and stratigraphy.

Anisotropies such as fine stratifications of certain lithologies like shales, glacial varves or flysch, and those related to tectonically imposed laminations comprising cleavages and schistosities, condition weakness directions in the rock matrix itself. However systems of discontinuities such as tectonic joints or syngenetic diaclases which occur systematically have to be necessarily surveyed to better characterize the geomechanics of the whole rock mass (Cosgrove and Hudson 2016; Sadowski 2014).

3.5 Final Considerations and Acknowledgements

The geology of soft rocks is basically related to low P T conditions and involves the comprehension of environmental processes occurring on the Earth's surface and in the first few kilometers of the crust.

Those are the rocks which represent the largest engineering challenges since their matrix has strength parameters frequently equal or below that of the main concrete built structures. When intermingled with hard to medium rocks they frequently constitute the weak bond of the chain.

I thank the coordinators of this edition and specially Prof. Milton Kanji for his encouragement and suggestions.

References

Bosio JJ, Kanji MA (1998) Soft rocks of the Rio de la Plata basin. In: AGI (ed) International symposium on indurated soils and soft rocks. A.A. Balkema, Rotterdam, pp 65–71

Cosgrove J, Hudson J (2016) Structural geology and rock engineering. Imperial College Press, London, p 554

Deere DU (1968) Chapter 1: geological considerations. In: Stagg KG, Zienkiewicz OC (eds) Rock mechanics in engineering practice. Wiley, New York, pp 1–20
Deere DU, Miller RP (1966) Engineering classification and index properties for intact rock. Tech report no AFWL-TR-65-116. Air Force Weapons Lab, New Mexico
De Mello LGFS, Sadowski GR, Nieble CM (2004) Shear resistance of biotite-schist: peak and post-peak behavior. In: Advances in geotechnical engineering: the Skempton conference, 1. ICE Press, London, pp 535–544
Dobereiner L (1984) Engineering geology of weak sandstones [PhD thesis]. University of London, p 471
Giambastiani M (2005) Comportamento dependente do tempo de rochas sulfáticas de Anidrita e Gipso. (in Portuguese). PhD dissertation. EESC University of São Paulo, p 465
Gonzales de Vallejo LI, Ferrer M, Ortuño L, Oteo C (2002) Ingeneria geologica. Pearson Education, Madrid, p 744
Goodman RE (1993) Engineering geology-rock in engineering construction. Wiley, New York, p 412
Harris CS, Hart MB, Varley PM, Warren CD (1996) Engineering geology of the channel tunnel. ICE, London
ISO 14689-1 (2003) (E). Geotechnical investigation and testing: identification and classification of rock. Part 1: identification and description. International Organization for Standardization, Geneva, pp 1–16
Jeremic ML (1994) Rock mechanics in salt mining. A.A.Balkema, Rotterdam, p 533
Johnston IW (1993) Soft rock engineering. In: Judson JA, Brown ET, Fairhurst C, Hoek E (eds) Comprehensive rock engineering: principles, practice & projects, vol 1, pp 367–393
Kanji MA (2014) Critical issues in soft rocks. J Rock Mechan Geotechnical Eng 6:186–195
Kellezi L, Hansen PB, Koreta O (2016) Interpretation of Danish chalk parameters applicable to foundation design. In: Proceedings 17th Nordic geotechnical meeting: challenges in Nordic Geotechnics IGS. NGM 2016 Reykjavik, Reykjavik, pp 597–604
Larsen G, Chilingar GV (1967) Diagenesis in sediments. In: Larsen G, Chilingar GV (eds) Developments in sedimentology 8. Elsevier, Amsterdam
Liang W, Zhang C, Gao H, Yang X, Xu S, Zhao Y (2012) Experiments in mechanical properties of salt rocks under cyclic loading. J Rock Mech Geotech Eng 4(1):54–61
Marinos P, Hoek E (2001) Estimating the geotechnical properties of heterogeneous rock masses such as flysch. Bull Eng Geol Environ 60:85–92
Martin CD, Kaiser PK, Christiansson R (2003) Stress, instability and design of underground excavations. Int J Rock Mech Mining Sci 40(7-8):1027–1047
Medley EW (1994) The engineering characterization of melanges and similar block-in-matrix rocks (Bimrocks) thesis (Ph.D. Civil Engineering), University of California, Berkeley, p 339
Merriman D, Peacor PR (1999) Chapter 2: very low grade metapelites. In: Frey M, Robinson D (eds) Low grade metamorphism. Wiley, New York, p 312
Nace RL, Deutsch M, Voegeli PT (1972) Physical environment of the national reactor testing station, Idaho—a summary. USGS Prof paper 725-A
Nickmann M, Spaun G, Thuro K (2006) Engineering geological classification of weak rocks IAEG 2006. Paper number 492
Potter PE, Maynard J, Pryor WA (1980) Sedimentology of shales. Springer Verlag, New York, p 303
Rocha M (1975) Some problems related to the rock mechanics of low strength natural materials. In Proceedings of the 5th Pan-American conference on soil mechanics and foundation engineering, Buenos Aires, pp 489–514. (in Portuguese)
Sadowski GR (2014) Basic elements of geology for civil geotechnical engineering (in Portuguese). Chapter 4. In De Mello, VFB Geotecnia do subsolo e de materiais terrosos-pedregosos construídos: primórdios, questionamentos, atualizações; Oficina de Textos-ABMS São Paulo. ISBN 978-85-7975-127-1:129-176
Tarbuk EJ, Lutgens FK (2005) Earth an introduction to physical geology. Pearson Education, New Jersey

Tennessee Valley Authority (1949) Geology and foundation treatment. TVA technical report 22. US Gov. Printing Office Apud Goodman, R.E. 1993 engineering geology: rock in engineering construction. Wiley, p 412
Terzaghi K, Peck R (1967) Soil mechanics in engineering practice. Wiley, New York
Vaz LF (1996) Classificação genética dos solos e dos horizontes de alteração de rocha em regiões tropicais Solos e Rochas, São Paulo, n. 19 v, vol 2, pp 117–136
Von Engelhardt W (1977) The origin of sediments and sedimentary rocks. Wiley, New York

Chapter 4
Mudrocks as Soft Rocks: Properties and Characteristics

Filipe Telmo Jeremias, Juan Montero Olarte, António B. Pinho, Isabel M. R. Duarte, Haris Saroglou, and Mario Camilo Torres Suárez

4.1 Introduction

Soft rocks can be grouped in a broad sense, as geological materials with poor mechanical characteristics, namely, high deformability and low strength that span the range between soils and hard rocks.

Mudrocks are part of the broader group of soft rocks, which correspond to a great diversity of geological materials, including not only those of sedimentary origin, like mudrocks, but also other types of rock materials, originally with high strength, which have suffered a deterioration of their geomechanical characteristics, after being subjected to weathering or tectonic processes.

Potter et al. (1980) recorded that mudrocks constitute more than 60% of all sedimentary rocks and as result of their wide distribution, mudrocks are often encountered in engineering projects, either as construction materials, (e.g. embankments) or in their natural undisturbed state (e.g. foundations, cut slopes, and underground excavations).

The engineering properties of mudrocks including durability are related to their composition, geological history, and degree of weathering. The influence of these aspects on the breakdown behaviour of mudrocks has been extensively

F. T. Jeremias (✉)
Laboratório Nacional de Engenharia Civil, Lisbon, Portugal
e-mail: ftelmo@lnec.pt

J. M. Olarte · M. C. Torres Suárez
Universidad Nacional de Colombia-Bogotá, Bogotá, Colombia

A. B. Pinho · I. M. R. Duarte
University of Évora, Évora, Portugal

H. Saroglou
National Technical University of Athens, Athens, Greece

M. Kanji et al. (eds.), *Soft Rock Mechanics and Engineering*,
https://doi.org/10.1007/978-3-030-29477-9_4

studied in the last few decades (Gamble 1971; Dick and Shakoor 1992; Taylor 1988; Bell et al. 1997; Venter 1998; Czerewko and Cripps 2012; Cripps and Czerewko 2017). The results of these studies demonstrated that the main processes involved in the breakdown of mudrocks relate to stress relief as a result of the removal of overburden, physical disintegration and mineralogical and chemical effects on contact with water and air, including the oxidation of pyrite breakdown and dissolution. These studies also showed that the progressive breakdown of the mudrocks from rock-like materials to soils is influenced by inter-granular diagenetic bonds, as well as by the presence of sedimentary structures and tectonic discontinuities, rock fabric, especially anisotropy and in these clay-rich materials, the types of clay minerals they contain.

Existing experience with mudrocks indicates that in many cases, they are prone to rapid degradation due to their high clay content (especially if expandable clay minerals are presented) and poor induration state. Changes to environmental conditions resulting from construction activities can result in significant changes in engineering properties of these materials during project design-lives or even periods as short of the construction phase of projects. The degradation of engineering properties of mudrocks has been associated with syn-and-post-constructional failures requiring costly additional unanticipated works during construction and additional expenditure on re-construction and/or maintenance works during the service phase. Depending on the situation, problems can arise even where mudrocks constitute only a minor part of the ground affected by the engineering works.

This chapter is organized to give a perspective of the mudrocks in the context of soft rocks in general, dealing with the main geological and engineering aspects and features that characterize these materials. In Sect. 4.1 the main controls on the formation of mudrocks are presented. This is followed in Sect. 4.2 by an explanation of the terminology used to classify fine-grained sedimentary rocks, together with a description of the main geological and geotechnical characteristics of mudrocks, including the relevant laboratory techniques. Section 4.3 reviews geological and engineering classifications of mudrocks and in Sect. 4.4 a set of case studies of mudrocks encountered in commonly occurring civil engineering works are presented. Finally, Sect. 4.5 is a summary of the main topics covered within this chapter.

4.2 Mudrock Cycle

To understand the behaviour of mudrocks it is necessary to consider where and how clay and silt (mud) occur in nature, as well as the conditions that accompany the formation of these rocks in sedimentary basins; especially the influence of burial on the diagenesis process by which mud becomes changed to mudrock.

4.2.1 Origin and Production of Mud and Silt in Nature

While most clay minerals (e.g. kaolinite, illite, smectite) are formed as a result of the actions of chemical weathering process on igneous and metamorphic crystalline rocks (granites, basalts, volcanoclasts, schists, phyllites, etc.) in humid regions; most silt is produced by the mechanical disintegration by crushing of hard rocks; attrition and gradation in the course of river transport; loess particles from wind erosion (deflation); rock flour from glacial erosion (abrasion) and contribution of fines by volcanoes.

Potter et al. (2005) analyze how *exposure time* of rocks to chemical weathering (*residence time*), *relief* and *climate* (*rainfall*) are involved when comparing *active-convergent tectonic margins that are characterized by high relief, deep valleys, and very unstable steep slopes*, in which earthquakes cause failure and volcanic activity results in the production of volcanic dust (e.g. the Andean mountain range) with *cratons and passive margins that usually possess, low relief and relatively gentle and stable slopes*. In active-convergent margins uplift (sustained by isostatic rebalancing) is rapid and is tectonically controlled, consequently, where the climate is wet the rate of denudation is high, such that large volumes of detritus are generated over long time spans. Although rocks subjected to weathering have short residence times and detritus are mineralogically immature, in regions such as the Andes the very high rainfall is such that the velocity of weathering outpaces the denudation rate, and torrents and rivers bring enormous volumes of mud to the oceans (Box 4.1). Thus actively converging margins are the main donor of mud to sedimentary basins. On the other hand, in stable craton and passive margins, erosion rates are low and residence times are long. Such peneplained, low-relief cratons and passive margins are insignificant or negligible sediment donors, as in much of the territory of Brazil. The interaction of exposure time (residence time), relief and climate can be seen in Fig. 4.1, based on Potter et al. (2005). It can be clearly seen that the greatest contribution of silt and mud to sedimentary basins originates in converging regions with high precipitation (1) and that this contribution decreases gradually according to the scenarios (2), (3) and (4) a minimum in passive regions with little rain.

In conclusion, while climate (rainfall) controls the intensity of denudation, the exposure time of rocks to weathering processes (resident time) is governed by tectonic stability.

4.2.2 Transportation and Deposition

Transportation mechanisms in sorting the detrital sediment load as well as the sedimentation environments and depositional mechanisms where the interrelationships among dissolved oxygen, benthic organisms, sedimentary structures, colour

Box 4.1 Geomorphological Balance in the Andean-Tropical Region
Although the rate of denudation (erosion + mass removal) is high in the Andean-tropical region, conditions of high precipitation and temperature contribute to rapid decomposition, so that in many places the weathering rate favours the formation of relatively mature residual soils. In addition, it is considered that there is a high contribution of mass removal to the denudation. Furthermore, the beds of rivers drag high volumes of debris that originated from mass movements with abundant amount of mud (Montero et al. 2011).

Relative degree of tectonic activity	**Guayanas shield**		**Convergence areas (Andes range)**	
	Prolonged tectonic tranquility that extends the time rocks are exposed to weathering		Active tectonically controlled instability time rocks are exposed to weathering action	
Type of relief	**Very low**	**Low**	**Moderate**	**High**
Weathering profile	High probability of developing a deep and mature profile	Moderate probability of developing a mature profile	Low probability of developing a mature profile; depends on the relative stability of the landscape	Minimum probability of developing a mature profile
Denudation rate	**Low**	**Medium**	**High**	**Very high**
Geomorphological balance	Weak erosion with minimal mass removal	Strong erosion with predominance over mass removal	Probable equilibrium between mass removal and erosion	Maximum contribution of mass removal to denudation

and organic and pyrite content play a fundamental role as they impart properties to the mudrocks that will determine their engineering properties.

4.2.2.1 Transport Mechanisms

As shown by Potter et al. (2005) mudrocks are formed from clays transported in suspension (wash load) and silt-sized particles that are transported as both bed-load and suspension. Energy is supplied mainly by streams, waves, tidal currents, turbidity currents in lakes and oceans, and by ocean currents, including deep ocean contour currents. Rivers and deltas contribute most continentally derived fine detritus and deposit it in lakes and oceans, with contributions from glaciers and mass movements. Below sea level, landslides and slumps are the source of most of the turbidity currents that transport mud and silt into deep water. Coastline position

	RESIDENCE TIME, RELIEF AND RAINFALL	DECREASING CONTRIBUTION OF THE VOLUME OF MUD AND SILT
CONVERGENT MARGINS	**Wet highlands** Limited Resident Time / heavy Rainfall Major donor [1] –large volumes of silt and mud, with expandable and layered clays in suspension plus large volume of diverse chemical species in solution /dissolved sulfates and carbonates.	1
	Dry highlands Limited Resident Time / Scarce Rainfall Minor to moderate donor [2] -small to moderate volume of mud and silt plus suspended and chemical loads of diverse composition. Minor export of eolian clay and silt.	2
CRATONS AND PASSIVE MARGINS	**Wet lowlands** Increased resident time / heavy rainfall Minor donor [3]- small volumes of mud with gibbsite and kaolinite in suspension, plus a restricted chemical species in solution.	3
	Dry lowlands Longer resident / low rainfall Negligible donor [4] except for eolian export of clay and silt of diverse composition. Appreciable export of eolian clay and silt.	4

Fig. 4.1 Relation between relief and climate and its role in the production of different types of sediments (modified from Potter et al. 2005)

changes associated with sea level fluctuations, influence the background turbulence, fauna presence, bioturbation, and oxygenation. Thus, mud accumulates on platforms areas, slopes and the sea bed at different depths, with contributions from mass movements and turbidity currents. Additionally, fines are brought to deep-sea areas by wind, including loess silt. Many of these processes are involved in the generation of different styles of stratification and various structures characteristic of mudrocks the interpretation of which helps to recognize cycles and sequences in the stratification of these rocks.

4.2.2.2 Sedimentation Environments

A sedimentary environment favourable to the formation of mudrocks requires calm waters with minimal energy conditions conductive to the deposition of mud and silt. These conditions occur preferentially in deltaic and lagoon environments, as long as they are sheltered from the waves or in marine environments—*neritic zone* in a band parallel to the coast up to 200 m deep in shallow waters and *bathyal zone* in water 200 m to 2000 m deep.

4.2.2.3 Depositional Mechanisms

The clay and fine silt particles that make up the mud are deposited in the sedimentary basins mainly as aggregates of particles and rarely as individual particles. As aggregates, they are deposited in flocculated form (*flocculation*) or as faecal pellets (*pelletization*). Flocculation occurs through electrostatic attraction and the action of van der Waals forces between clays particles. The excess of negative charges present on the surface of clay particles, tends to cause mutual repulsion such that clays are kept too far apart for aggregation to occur unless there are large concentrations of charged ions in the surrounding water. However in sea water dissolved ions counteract the natural electrostatic repulsion of clays, which cause them to flocculated in which they come into contact and form into aggregated groups of particles. Pelletization comprises a biological process that results in the formation of chains or aggregates of the clay particles together with faecal matter and fossil debris (faecal pellets).

Potter et al. (2005) drew attention to the practical consequences of clays flocculating or not flocculating in the water column. When flocculation occurs the initial fabric of the deposit is open and contains much water and is therefore potentially unstable. In contrast, without flocculation the initial deposit has a more stable sub-parallel structure with less water, showing higher density and shear strength as well as lower propensity to erosion by flowing water (Potter et al. 2005).

Taking into account the slow rate of settling of mud in water, the prevalence of turbulence in nature and the great abundance of mudrocks, Pryor (1975) consider that most mud is deposited as aggregates rather than as individual particles. This high contribution of pelletization is consistent with the great abundance of faecal pellets and related aggregates reported in modern marine and lacustrine muds. Pelletization by organisms together with flocculation are the most efficient processes of transport and deposition of mud in sedimentary basins. On the other hand, Potter et al. (2005) consider that: (1) clay-sized particles settle to the bottom wherever bottom turbulence and currents are minimal regardless of water depth: a lagoon a few meters deep or an offshore basin more than 1000 m deep; (2) the organic matter concentrates in suspension in quiet, low-energy depressions along with mud, because its low density imparts a low fall velocity. This is why there is an inverse relationship between more silt and sand and the less organic matter (Potter et al. 2005).

In the case of the mudrocks, when the aggregated particles that make up the mud reach the bottom of the basin they form a solid matrix with saturated pores. Although in principle the water which saturates the pores has the same composition as that of the upper water column, a lack of water circulation due of confinement from newly sediments can result in very significant changes (Box 4.2).

Box 4.2 Role of Oxygen

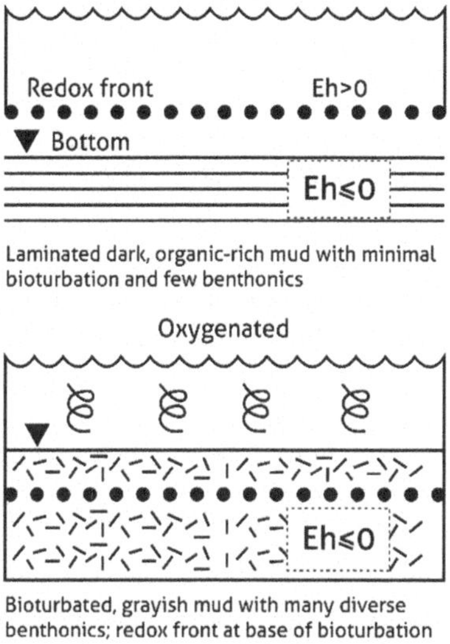

A sediment at the instant of deposition will consist of loose, detrital particles, pellets and other aggregations, crystals, organic fragments, colloidal mud, or mixtures of these substances. Therefore, when a lamina is formed it becomes the interface between the beneath deposited material and sedimentation medium (Krumbein and Sloss 1951). Accordingly, the depositional interface represents an important boundary condition which separates two different physicochemical regions (Krumbein and Sloss 1951).

Benthic fossil content, degree and kinds of bioturbation (agitation of sediments by organisms), colour (of rock) and pyrite content are mainly controlled by the amount of dissolved oxygen in the water column and in the pore water of the mud immediately beneath the sediment–water interface. In turn, bottom turbulence and the amount of organic debris in the mud control the dissolved oxygen content. The main concept on oxygen role is a redox front—the contact between oxygen-rich and oxygen-poor water—which may be present above, below or at the sediment–water interface (Potter et al. 2005).

Accordingly, if the redox front is in water column above the sediment–water interface, macro-invertebrates that would otherwise disrupt the sediment are not present, so lamination is well developed. Dark coloured, laminated, organic rich, muds containing pyrite are frequently formed in such marine environments. When the redox front is below the sediment–water interface, bioturbation by benthic organisms occur, so muds contain small amounts of organic matter and they tend to be lighter in colour. Pyrite may occur in small quantitates or be frequent; however, it will form at greater depth in the sediments. When a local concentration of organic matter in an oxidized mud is presented a corresponding local and isolated redox front around the organic matter will occur (Potter et al. 2005).

The interrelationships among dissolved oxygen, benthic organisms, sedimentary structures, colour and organic and pyrite content are illustrated above (Potter et al. 2005) They are fundamental to the interpretation of mudrocks.

The relationship between dissolved oxygen concentration and the influence of fauna on bioturbation, as well as the character and composition of the mud are listed in Table below. The resulting type of mudrock and its probably geotechnical behaviour in terms of being rock-like or soil-like are also shown.

Oxygen level	Fauna presence	Bioturbation	Character of the mud composition	Mudrock
Anoxic environment	Limited	Poor	Laminated, dark, abundant organic matter and pyrite	**Soil-like type**
Oxic environment	Abundant	Strong	Grey, organic matter and pyrite scarce or absent	**Rock-like type**

4.2.3 *Burial Diagenesis*

Burial diagenesis of mudrocks lead to several changes such as progressive conversion of expandable to not expandable clay minerals, precipitation of cements and loss of porosity, which contribute to the formation of more durable and less prone to slaking and swelling materials.

According to Potter et al. (2005), in the sedimentary basins the burial of the sediments is accompanied by the subsidence of the basin in such a way that the sea progressively increases its level creating enough space of accumulation. As a consequence, the sediments of mud are gradually being buried and with the increase of overburden and temperature with the depth, they compacted progressively until hardening (lithic), becoming mudrock, with physical and chemical characteristics that are reflected in its texture and mineralogy. For convenience, two stages are considered in the sediment burial process: a first stage *shallow burial* occurs very early when the first few meters of sediment have been stacked, and a second stage *deep burial* occurs much later when great thicknesses are accumulated.

In shallow burial there are important changes in the porosity of the sediments and strong biological reactions. In addition, these processes involve exchange between the mud and the water that covers it. In deep burial the compaction rate is slower and linear and some abiotic reactions occur such as the generation of hydrocarbons and chemical reaction such as the transformation of less stable minerals (e.g. Ca-Plagioclase) into more stable (e.g. Na-Plagioclase). Because in this case the processes occur far from the surface, the deposition environment exerts little influence; instead, the burial rate and the geothermal gradient control the events.

The effect of the burial can be visualized comparing the compaction of clays (mudrocks) with the one of sands (Sandstones). The porosity of a freshly deposited mud with a value of 70–80% is reduced very little with depth of burial whereas the porosity of a sandstone can be reduced from 50% to 25% under the same conditions of burial (Fig. 4.2).

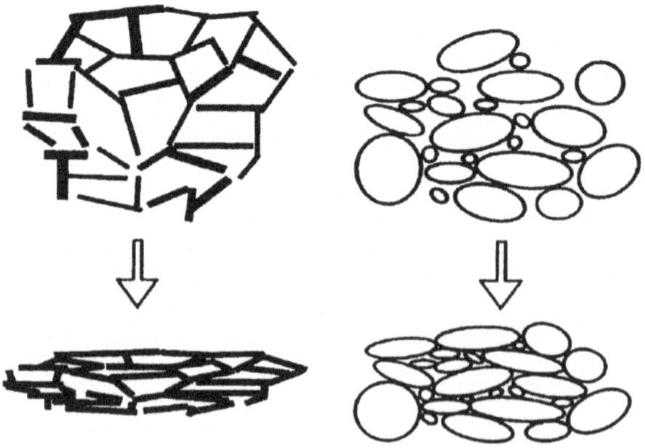

Fig. 4.2 Comparison of compaction between clays and sands (after Potter et al. 2005)

Sands have higher permeabilities than muds regardless of their age. After burial in the sedimentary basins when these deposits are consolidated and lithified, the sandstones retain appreciable but very variable permeabilities, while in the mudrocks the permeability of the matrix is very low as it is mainly secondary through discontinuities.

The contrast is due to the fact that newly deposited mud has a weak structure characterized by an open clay sheet fabric, with some silt, sand, pellets, organic matter and possibly some pyrite generated by bacteria, which tends to collapse and/or deform gradually. On the other hand, the sand fabric, comparatively denser with well-packed grains (depending on the gradation) and abundant quartz and without organic matter, is less prone to collapse by pressure as well as the porosity is reduced by the cement that is formed in the pores. Under the effect of burial in the diagenesis process the clay sheets of smectite and interlayered clays are mostly transformed into chlorite and illite while the quartz grains that form the sandstones maintain their chemical stability.

From the physical and chemical processes that accompany the compaction of the rocks in the natural processes of burial, the first ones produce more significant effects on the characteristics of deformability and strength of the rocks. However, at depth the mineralogical transformations help to fill the voids and bind the particles, so that porosity is reduced in more aged mudrocks (Potter et al. 2005).

Due to the compaction, the mud reduces its volume by expelling a large amount of water upwards and to the sides, whereby the porosity and permeability are substantially reduced with the depth at the same time as the density and the seismic velocity increase.

4.2.4 Uplift, Unloading and Weathering

The schematic geological history of muddy sediment during the burial and after the reduction of the overburden can be investigated using laboratory consolidation tests as it is illustrated in Fig. 4.3 (Cripps and Taylor 1981). Immediately after sedimentation mud has a porosity of between 70% to 90% which corresponds to 50–80% water by weight (Weaver 1989) as represented schematically in Fig. 4.3(A) point *a*. The accumulation of more sediments causes an increase of the overburden pressure and a decrease in water content. The line *a-b-c* in Fig. 4.3(A) represent *normally consolidated* clays where point *b* indicates a load equal to the present overburden pressure. The shear strength of uncemented *normally consolidated* clay is proportional to the existing overburden pressure and, therefore is represented in Fig. 4.3(B) by a straight line passing through the points *a-b-c*.

Unloading the sediment by erosion to the present overburden (point *e*) brings the sediment to an *over-consolidated* condition. Although there is some increase in porosity, the void ratio does not increase to the corresponding to point *b* on the chart. Thus, although the sediment is under the same overburden as at point *b* the water content (and porosity) is much less in the over-consolidated state. As well as

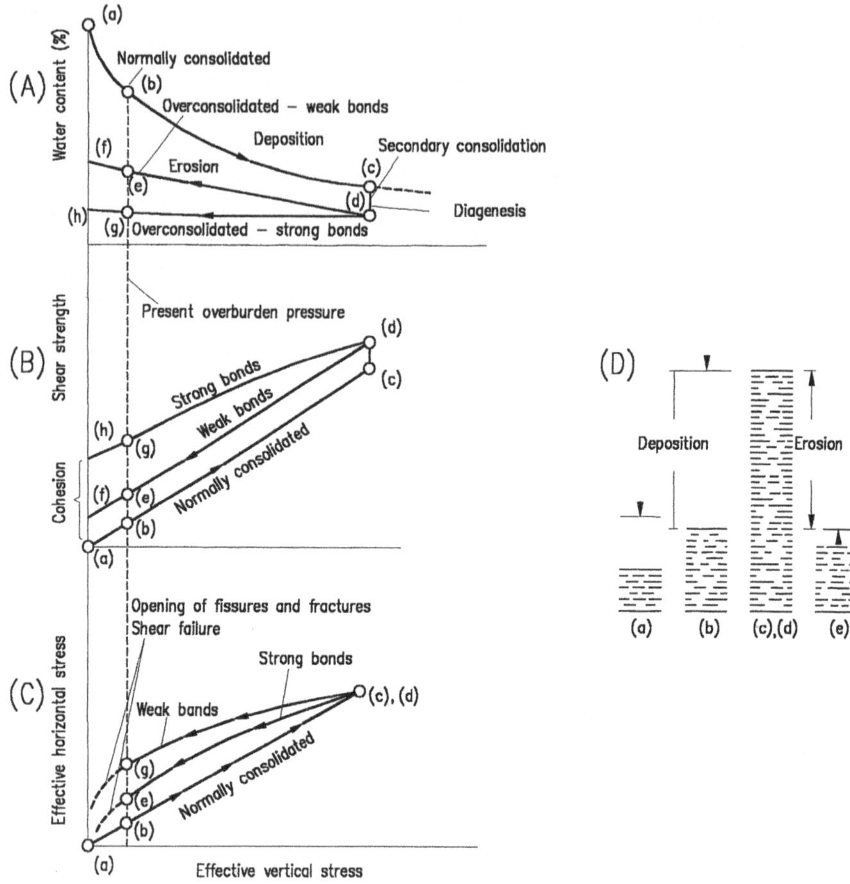

Fig. 4.3 Schematic geological history of a mudrock: (**A**) compaction and unloading, (**B**) effect of compaction and unloading on shear strength and (**C**) variation of horizontal and vertical stresses due to compaction and unloading (after Cripps and Taylor 1981)

this, due to the rearrangement of particles and denser state of the sediment, the shear strength is higher.

If the sediment represented by point *c* remains under the same effective pressure over a long time period a further reduction in water content occurs as a consequence of secondary consolidation. During this period the sediment may be subject to diagenetic changes including recrystallization of particles, adhesion between particles and precipitation of cements. The bonds formed between particles through these processes are termed diagenetic bonds and because of them the sediment becomes stronger, less permeable and more brittle.

On unloading, because over-consolidated clays and mudrocks possess recoverable strain energy stored in compressed flexible flake-shaped clay particles, the materials swell and the water content increases. The extent and timing of this expansion depends on the strength of the diagenetic bonds between particles as is shown by lines *d-e-f* and *d-g-h* in Fig. 4.3(A). The latter rebound curve as compared with the former shows that expansion of clays and mudrocks with strong diagenetic bonds will be restricted. However, as the breakdown of strong bonds is time-dependent, the strain energy will be released progressively thus leading, in the long-term, to softening of over-consolidated clays and mudrocks.

The importance of diagenetic inter-particle bonding in the structure of the material is demonstrated when the horizontal to vertical stress ratio is modelled as schematically illustrated in Fig. 4.3(C). On unloading the changes in vertical stress are larger than those in horizontal stress (as the vertical expansion is greater than the horizontal expansion) due to lateral restrain resulting in high horizontal stress. Removal of overburden leading to the development of stress relief joints coarsely parallel to the ground surface. For strongly bonded types the horizontal expansion is more restricted because the bonding prevents the expansion. As a result, the effective horizontal stress is lower than that developed for weakly bonded types. However, destruction of these bonds in stronger types during weathering processes causes high horizontal stresses to be developed later in the relatively shallow weathered zone leading to mudrock disintegration. Thus, in compaction mudrocks the effects of burial are rapidly lost, while in cemented mudrocks degradation may occur over an extended period.

The weathering processes promote both physical breakdown and chemical alteration result in the fragmentation of the material and disintegration of inter-particle bonds (dissolution of the cements and other mineralogical changes) and lead to the release of latent strain energy in the sediment. The resulting expansion is accompanied by an increase in water content and compressibility, and a decrease in shear strength. The disintegration of the bonds is consequently related to the degree of weathering and thus, an indurated mudrock can eventually achieve the status of a remoulded *normally consolidated* clay in the more weathered stages.

4.3 Engineering Geological Characterization of Mudrocks

4.3.1 Introduction

A first task for the engineering geological characterization of mudrocks is to define the scale that the study is performed. Accordingly, mega, macro and micro scales may be used including different assessments for the features of mudrock behaviour (Fig. 4.4) (Torres and Alarcón 2007).

In rock mass scale—megascale—the main geological, geomorphological and geostructural features are recognized through the application of routine methods and procedures from field reconnaissance including geological mapping, discontinuity characterization, rock mass classification—GSI, RMR, etc. and from site investigation works (geophysical surveys, boreholes and in situ tests).

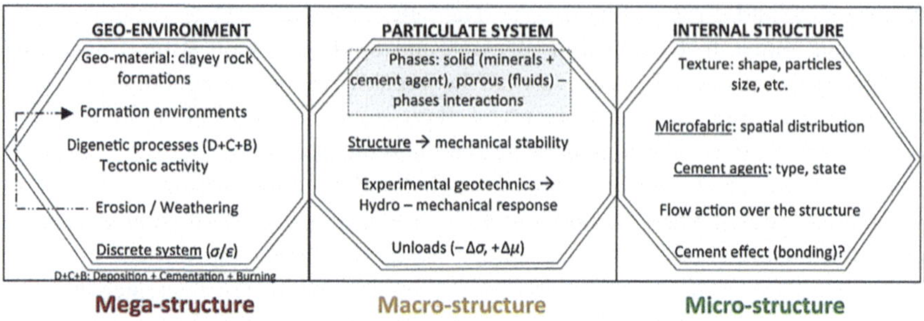

Fig. 4.4 Spatio-time scales relationship and key features for engineering geological characterization of mudrocks (after Torres and Alarcón 2007)

At macroscale samples collected from the rock mass are used for laboratory testing in order to characterize the physical and engineering properties of intact rock. Those results are related should be in line with the physical and geomechanical properties exhibited by the rock mass, but the effects of rock mass structures are not measured in tests.

At micro scale (internal structure) mineral composition and textural features of mudrocks are characterized aiming to predict its geomechanical behaviour (σ—ε) and durability.

Although, the three scales of observation are complementary and are required to provide a complete and accurate understanding of mudrock behaviour, it is difficult to do this for practical situations. Instead rock mass classification systems and physical and numerical modelling approaches are used to derive the engineering response of the rock mass to changes brought about by engineering works, taking into account intrinsic material behaviour of different parts of the systems and the effects that are observed at one scale are related intrinsically with the observed behaviour on the others scales of the study. This can be exemplified in a cut slope (mega-structure) where dipping bedding planes may cause failure. Shear strength of bedding planes may be determined in laboratory testing (macro). The value of strength will depend on rock fabric (micro), which becomes lower due to particle orientation due to progressive failure.

4.3.2 Definition and Terminology

The classification and definition of the terms related to fine-grained sedimentary rocks have always given rise to confusion in the technical literature Potter et al. (1980). This is because many terms, including clay, mud, shale, mudstone, mudrock, claystone, and clay rock, have multiple uses and definitions, and some of them have changed with time Potter et al. (2005).

For example, the term *shale* has been defined (Tourtelot 1960 ; Potter et al. 1980) as the general class name for fine-grained sedimentary rocks. However, because *shale* can also imply that the rock possesses fissility by virtue of being laminated there is misuse of this term. The double significance of the term shale led to Ingram (1954) proposing that *mudrock* be used to designate that class of fine-grained rocks. This was adopted by Dunbar and Rodgers (1957) and Blatt et al. (1980), who attributed to it a significance similar to the term sandstone and limestone.

In the review of the terminology of fine-grained sedimentary rocks Stow (1981) and Stow and Piper (1984) define mudrocks as a rock in which more than 50% of the grains are siliciclastic composition and smaller than 63 μm (silt and/or clay size). The term siliciclastic means formed out of silicate minerals and clastic granular material, and is applied to detrital grains that are mainly constituted of clay minerals and quartz. This definition implies that mudrocks can encompass materials that contain little or no clay minerals, a factor that greatly influences their performance in engineering situations.

The term *mudstone* is advocated by Potter et al. (2005) as the suffix *stone* puts this general class term for fine-grained rocks into same hierarchy as other important sedimentary rock sub families, such as sandstone, limestone and dolostone. Unfortunately, Dunham (1962) refers to fine-grained limestone as mudstone and the term mudrock is preferable as it has a single significance and has increasingly been accepted in engineering geological terminology.

In Table 4.1 a set of common terms used to describe clearly fine-grained sediments and sedimentary rocks are summarized.

Table 4.1 Common terms used to describe fine-grained sediments and sedimentary rocks

Term	Significance
Sand	Material between 0.06 mm and 2 mm in diameter.
Mud	Material smaller than 0.06 mm (silt and clay mixture).
Silt	Material between 0.002 mm and 0.06 mm in diameter.
Clay	Material finer than 0.002 mm.
Mudrock	A general term for the class of fine-grained sedimentary rock that contains at least 50% of siliciclastic grains finer than 0.06 mm.
Siltstone	Indurated rock comprised more than 2/3 of silt-sized particles.
Mudstone	Indurated rock containing between 2/3 to 1/2 of silt-sized particles.
Claystone	Indurated rock comprised more than 1/2 of clay-sized particles.
Argillite	Weakly metamorphosed rock, firmly indurated without slaty cleavage.
Slate	Fine-grained metamorphic rock with slaty cleavage.
Stone	Suffix applied when rock has a layering greater than 1 cm (siltstone, mudstone and claystone).
Shale	Suffix applied when rock has a layering less than 1 cm (siltshale, mudshale and clayshale).
Sandy	Applies when rock is formed by more than 10% of sand sized particles.
Calcareous	Applies when rock contains between 10% and 50% of carbonates.
Fissility	Property of splitting or dividing into thin fragments along closely spaced (<1 cm) planes parallel to the bedding or lamination.

4.3.3 Geological Characterization of Mudrocks

4.3.3.1 Mineral Composition

Mudrocks are composed of more than 50% siliciclastic debris that includes clay minerals, quartz and feldspar they also contain non-detrital minerals, such as carbonates, iron oxides, pyrite and organic matter.

Clay Minerals

Clay minerals generally found in mudrocks are kaolinite, illite, smectite, chlorite and mixed-layer clays (Box 4.3). These minerals are derived mainly by the weathering of rocks and soils forming the landmass and they usually constitute the largest fraction of siliciclastic grains in mudrocks. Clay minerals can undergo minor structural changes during weathering and deposition, although flocculation and the formation of faecal pellets can lead to the presence of silt-sized aggregations of clay minerals. Structural modifications of clay minerals, ruled by temperature, pressure, pore water chemistry and duration of exposure to particular environmental conditions, can occur during burial diagenesis. Such modifications lead to the conversion of smectites into illites, illitization resulting in the disappearance of expandable layers within the structure of the minerals, which promote the increase of the mechanical stability of the rocks concerned. See Czerewko and Cripps (2012).

Quartz

Quartz occurring as angular single silt-sized particles is the main detrital constituent of mudrock which can form 20–30% of the whole rock (Potter et al. 1980). Quartz is presented usually either due to the abrasion and collisions suffered by the coarse quartz grains during the transport process, or from the release of silt-size grains from fine-grained crystalline clay rocks. Quartz may also occur as a diagenetic constituent in mudrock, formed as a by-product of the process of illitization of mixed-layer illite-smectite clays (Blatt 1982).

Feldspar

Feldspars together with quartz constitute almost the entire resistate fraction of mudrocks. Feldspar generally occurs in lesser amounts than quartz. Secondary (diagenetically derived) feldspar may also be present in mudrocks.

Box 4.3 Structure and Composition of the Main Clay Minerals

Structurally clay minerals are composed by alternate silica tetrahedral layers and alumina octahedral layers that form 1:1 sheet type (e.g. kaolinite group) and 2:1 sheet type layers (e.g. groups of illite-smectite and chlorites). Mixed-layered clay minerals correspond to the regular or irregular overlaying of two or more different types of layers (e.g. illite-smectite, chlorite-smectite). Clay minerals have an extensive and complex mineralogy that is a result of the combination of the number and arrangement of sheets, ionic substitution in both tetrahedral and octahedral layers and the type and nature of the interlayered materials.

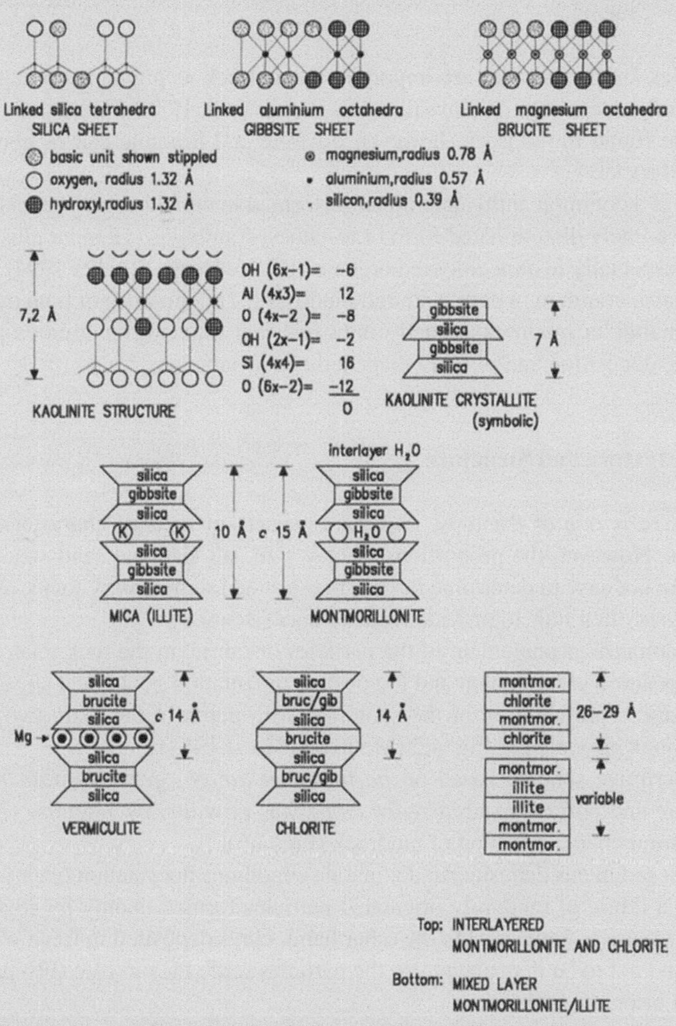

Schematic representation of the structure and composition of the main clays minerals (after Taylor and Cripps 1984)

Carbonates

Calcite, dolomite and siderite are the common carbonate minerals in mudrocks. These minerals can occur, either as cement, or as discrete particles. Calcite is generally the dominant carbonate phase. Its origin is varied as it may occur as particulate material in the form of skeletal debris, as inorganic precipitated microcrystals spread through the rock, or concentrated in nodules or concentrations formed by diagenetic processes (Tucker 1994).

Other Constituents

Iron oxides and hydroxides are important in mudrock as pigmentation agents and occur chiefly as coatings on clays minerals (Potter et al. 1980). Hematite is the main iron oxide found in mudrock; however, goethite and limonite can be abundant in weathered rocks.

Pyrite is a common authigenic mineral (formed in situ during diagenesis) which occurs in a finely disseminated form or as cubes, framboids (globular clusters) and nodules, especially in dark coloured organic-rich mudrocks (Tucker 1994). Organic matter is also common in dark coloured mudrocks. The presence of both pyrite, and organic matter in mudrocks usually indicates that reducing conditions prevailed during the deposition and/or early diagenetic environments.

4.3.3.2 Texture and Structure

Particle size is one of the most important aspects of textural characterization of mudrocks. However, the proportions of clay size, silt size and sand size fractions present are not easy to determine in such fine-grained sedimentary rocks, as particle size analysis often fails to provide realistic data (Box 4.4).

The geometric arrangement of the particles or fabric, in the rock relates closely to the deposition environment and the post-sedimentation geological stress history. A schematic representation of the evolution with depth of the main two types of association is presented in Fig. 4.5 (Moon and Hurst 1984).

A descriptive scheme based on *particle orientation*, *grain-to-grain relations*, *clay fabric* and *carbonate occurrence mode* was provided by Jeremias (2000) for microtextural characterization of mudrock (Fig. 4.6).

In most sediments deposited under marine conditions flocculation of clay particles produces a fabric of randomly orientated particles formed mainly by edge-to-face (EF) inter-particle contacts. On the other hand, clays deposited in fresh water conditions tend not to be flocculated and the particles settle face-to face (FF) producing a parallel orientated fabric.

Compaction produces a general increase in particle orientation with depth. At low overburden pressure (shallow depth) the sediments with EF inter-particle contacts are reorganized into small FF randomly oriented domains (Bennett et al. 1981).

Box 4.4 Grain Size Determination in Mudrocks

Grain size determination is a difficult task in mudrocks. Field criteria such as appearance (hand lens observation) and sense of touch (grittiness, smoothness and plasticity) are used to estimate roughly the relative amounts of silt-sized and clay-sized particles present in mudrocks. In laboratory particle size distributions can be determined by size analysis (sieving and sedimentation) and petrographic and SEM techniques. Size analysis carried out in mudrocks may reflect more the effectiveness of the disaggregation and dispersing techniques applied rather than the original particle assemblage of the sediment (Spears 1980; Blatt et al. 1980; Hawkins and Pinches 1992). Size measurements using petrographic thin-sections are limited to the coarse silt and sand fractions. SEM studies provide very valuable data but only very small volumes of rock are examined and they are not practical for routine analysis. Furthermore, clay minerals may flocculate during sedimentation and deposition as silt size aggregates rather than single particles. In addition, diagenetic processes may result in changes to existing or the formation of new clay minerals, which has the effect of modifying the initial size distribution.

With increasing depth these small domains tend to coalesce to form larger domains, but they do however retain a randomly orientated structure. Sediments with inter-particle contacts of the FF type are subject to layering as the depth increases, producing a structure that is oriented in the direction of the stratification. Notwithstanding these differences other factors such as the bioturbation, the presence of silt and sand size particles, carbonates and organic matter are also important in the initial formation of the fabric and they may influence post-depositional changes in fabric that occur during compaction, Other diagenetic process as mineralogical transformations may also alter the fabric. Thus silty clays are liable to possess a more random fabric than a pure clay.

Fissility and *stratification* are common features in mudrocks. The former is the tendency of the rock to split along laminations or bedding planes while stratification (bedding and lamination) is the result of vertical variations in composition, texture and/or fabric. Fissility tends to form in response to weathering action in which laminar weakness present in the rock are exploited so that splits into a series of thin sheets. The separation between the fissility planes generally decrease with the degree of orientation of clay minerals and as organic matter content increases. On the other hand, the presence of carbonates and of silt and sand-sized quartz grains in the rock will tend to give rise to thicker partings. However, in addition to the predisposition of the sediment to parting along fissility planes, the structure displayed is mostly the result of the stress relief and weathering due to exhumation.

The thickness classes mostly generally used for describing stratification and fissility are presented in Fig. 4.7. In order that no overlap of thickness classes between beds and laminae exists a usually accept arbitrary distinction at 10 mm was

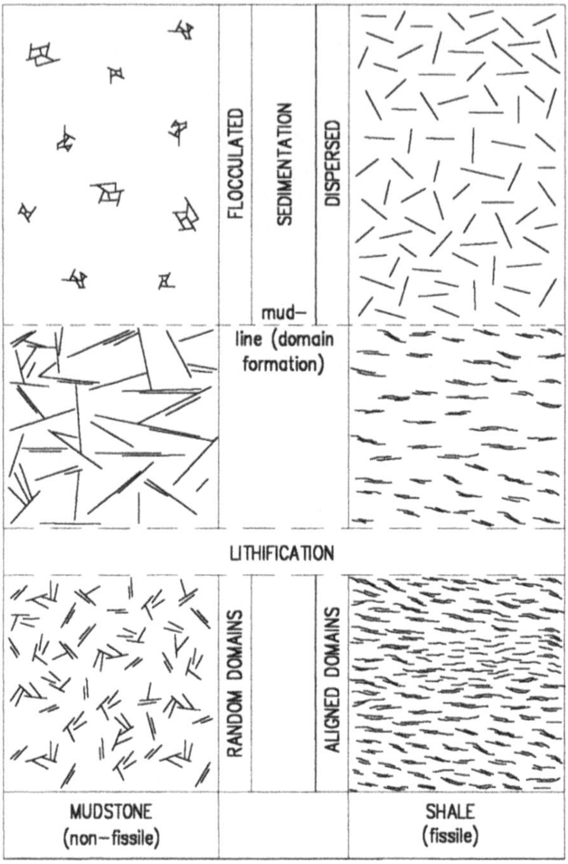

Fig. 4.5 Schematic representation of the evolution of the fabric of respectively originally flocculated and dispersed textured mudrocks with the depth of burial (after Moon and Hurst 1984)

proposed (Ingram 1954; Potter et al. 1980). According to the classification proposed by Campbell (1967) lamina is the lesser megascopic unit identified in a sedimentary sequence, reaching a thickness of 200 mm or more; therefore, thicker classes of laminae overlap the finer classes proposed for beds.

Different types of lamination in terms of variations in fabric, grain size and colour (Fig. 4.8) can be distinguished in mudrocks (Lundegard and Samuels 1980). Fabric lamination is produced by the parallel orientation of clay mineral grains a few microns in size, being the most common type. Grain size lamination is due to the differences in settling rates of different components which usually produces alternating layers of clay mineral and silt-size quartz grain material. In the context of glacial deposits, these layers are known as varves. Colour laminations are produced by alternated layers of different colours, where this may reflect subtle differences in grain composition and staining by iron compounds.

FEATURE	VISUAL DESCRIPTION	
A - Particle orientation	1 - Microlamination absent, the rock constituents have a random orientation.	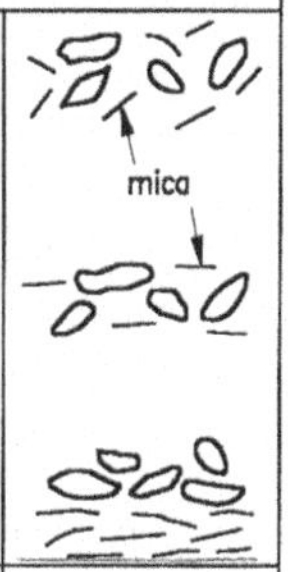
	2 - Microlamination absent, but the platy minerals *e.g.* micas show some preferred orientation.	
	3 - Well-developed lamination or microlamination.	
B - Grain-to--grain relations	1 - Predominantly grain-to-grain relations by long or point contacts, and/or by the development of cementation bonds.	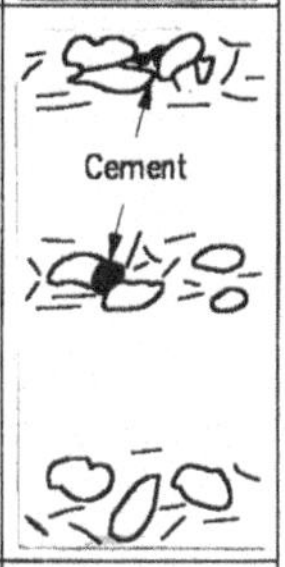
	2 - Occurrence, both of grain-to-grain contacts (1), and of floating grains (3).	
	3 - Predominantly floating grains.	
C - Clay fabric	1 - Mostly clay particle arrangements of EE and EF types, loose structure, high porosity.	
	2 - Clay particle arrangements of EF, FF and low-angle EF types, denser and less porous structure than the former (1).	
	3 - Predominantly clay particle arrangement of the FF type, very dense structure, low porosity.	
D - Carbonate occurrence mode	1 - Carbonate minerals occur mainly as individual grains greater than 5 μm in diameter, or as sparry cement.	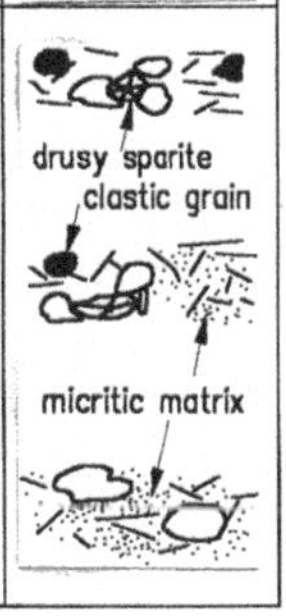
	2 - Carbonate minerals occur in both modes described in (1) and (3).	
	3 - Carbonate minerals essentially occur as a fine-grained micrite matrix.	

Fig. 4.6 Descriptive scheme for microtextural characterization of mudrocks (Jeremias 2000)

4.3.3.3 Colour

Colour is the most obvious feature of mudrocks and for this reason, it is always recorded in descriptions. However, the description of colour can be very subjective and it is therefore recommended that Munsell colour charts are used (Geological

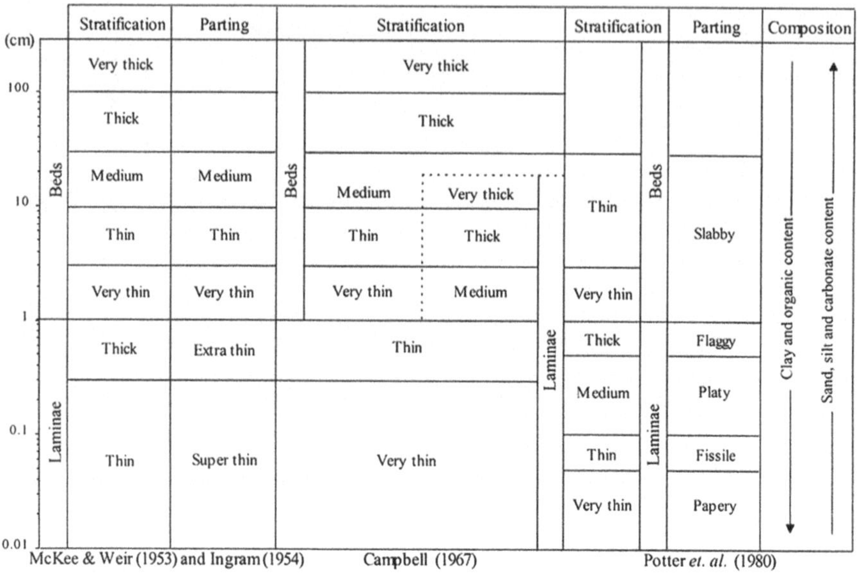

Fig. 4.7 Thickness classes of stratification and fissility (adapted from McKee and Weir 1953; Ingram 1954; Campbell 1967; Potter et al. 1980)

Fig. 4.8 Different types of lamination in mudrocks (after Lundegard and Samuels 1980)

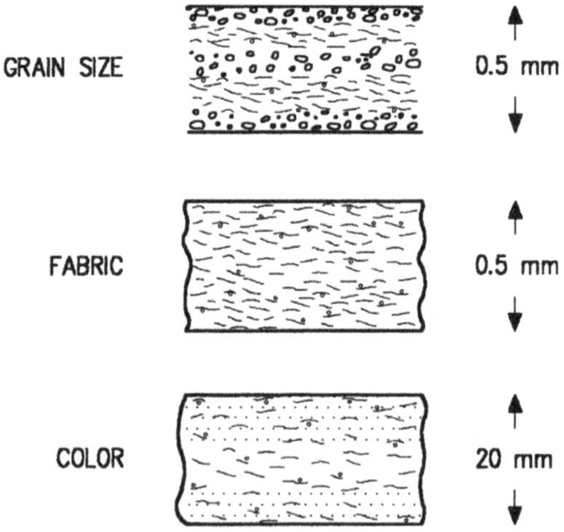

Society of America 1995). The content of organic matter and oxidation state of iron are the main factors which control the colour of mudrocks. The former affects the shades between grey and black, while the latter produces shades of red, purple, brown, yellow and green.

4.3.4 Geotechnical Characterization of Mudrocks

By reputation mudrocks are regarded as poor engineering materials in the construction and mining industries. In comparison with other rock types they are more prone to failure in slopes, embankments and surface mining highwalls. They are also susceptible to rapid loss of strength and disintegration when exposed at the surface. Pyrite is a commonly occurring but very minor and significant constituent of many mudrocks. On exposure to surface weathering conditions it is liable become oxidized which results in acidity and the liberation of sulphate-rich solutions. Besides being responsible for a loss of strength this can cause pollution of surface and ground water, rapid chemical attack on construction materials and large changes in volume. The particular geotechnical features that characterize mudrocks are described hereunder.

4.3.4.1 Plasticity

Plasticity is related to the strength of inter-particle bonds in a clay and also to the type, fabric and amount of clay minerals present (Heley and MacIver 1971). Clay mineral interlayer cation type and concentration and the amount and composition of water, particularly in terms of changes in the ordered molecular layers of water associated with the clay mineral surfaces, also affect plasticity (Gillott 1987). The assessment of plasticity in more indurated mudrock types can be problematic since the materials are difficult to disaggregate without affecting the size distribution of the particles present.

4.3.4.2 Strength/Deformability

The strength and deformability are related to the degree of aggregation, type and arrangement of the mineral grain that make up the mudrock. The density (porosity) and degree of cementation also exert controls, where that latter is particularly changed by weathering processes (Heley and MacIver 1971). These properties are also influenced by micro-scale features such as fissures. Mudrocks, especially, the stress-relieved laminated types, possess anisotropy so that measurements parallel, inclined and perpendicular to the main structure (bedding, lamination, parting, etc.) are required to obtain a full understanding of their performance.

The formation of a rupture surface due to movement in shear leads to the re-orientation of platy clay particles. The fabric of the rock and the strength of inter-particle bonds together with the angle between any plane structure in the rock (stratification, lamination, etc.) and the shearing direction are the main controls over this reorientation. For undisturbed samples, the measured peak shear strength parallel to any lamination or clay mineral preferred orientation will be less than that for a random or flocculated fabric (Gillott 1987).

Residual shear strength is mainly determined by degree of orientation of the platy clay particles, and it also depends upon pore water composition, grain size and mineralogical composition (Mesri and Cepeda-Diaz 1986). Tests performed on pure clay minerals show that the smallest values of residual shear strength were obtained for sodium montmorillonite samples in which the predominant arrangement of the particles is in face-to-face domains such that low inter-particle contact pressures are developed. However, tests performed with the same samples after treatment with carbon tetrachloride which caused aggregation of the platy clay minerals particles into clusters, gave higher results similar to the ones obtained with silt-quartz samples, which demonstrated that residual shear strength is more influenced by the lamellar character of particles than by mineralogy per se.

4.3.4.3 Swelling

Swelling is a characteristic property of mudrocks which is a result of the following features: (a) stress-relief of the geological materials and (b) physico-chemical reactions with water. The latter include hydration of clay minerals and chemical alteration of non-swelling minerals into swelling minerals (ISRM 1983).

Mechanical swelling occurs in response to elastic and time-dependent stress unloading which may be caused by excavation and removal of overburden in the course of engineering works or by natural erosion of the land surface (Taylor and Smith 1986). Since mudrocks are low permeability materials, the balance of moisture in the rocks during swelling can be inhibited due to deficient water percolation within the pore space. Thus, unloading can give rise to the development of negative pore water pressures (suction) in materials that will correspond to an increase in the effective stress acting on the rock. Mechanical swelling and the reabsorption of water causes softening and weakening of the rock and high stresses developed during this process may produce failure of the fabric of the rock, that is, fissuring.

The physico-chemical swelling mechanisms related to the hydration of clay minerals can be inter-crystalline or osmotic and intra-crystalline water absorption mechanisms, which are shown schematically (Fig. 4.9). The term inter-crystalline swelling is used to characterize the water absorption phenomena that occur on the external surfaces of clay crystals and in the voids between crystals. On the other hand, according to Gillott (1987), intra-crystalline expansion is a result of the forces of attraction that link structural sheets being weaker than the attractive forces responsible for the moisture uptake. The water absorbed by this mechanism forms successive monolayers on the surfaces of clay minerals that keep them apart.

The absorption of water by the internal surfaces of clay minerals is influenced by the nature of the interlayer cations. The smectite group together with vermiculite, swelling chlorites and some mixed-layered minerals are the main clay minerals that exhibit intra-crystalline swelling.

Osmotic swelling results from the interaction between the diffuse double water layers possessed by clay minerals particles which is related to the thickness of the double layer. In turn this is controlled by the dielectric constant, type and concentra-

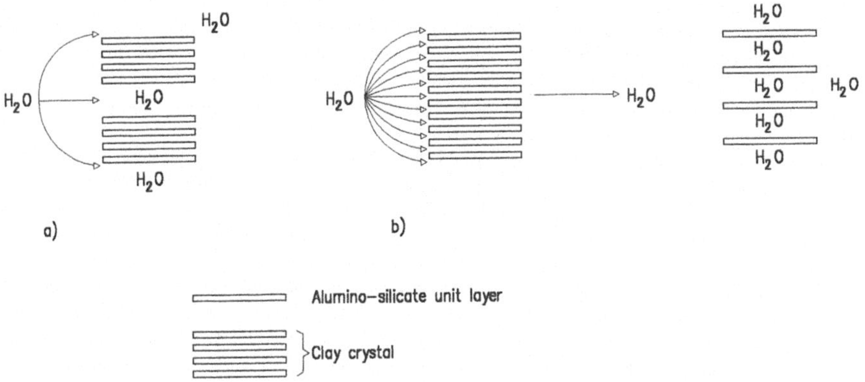

Fig. 4.9 Moisture uptake by clay minerals: (**a**) inter-crystalline absorption of water in a non-expanding lattice; (**b**) intra-crystalline absorption of water in an expanding lattice (after Gillott 1987)

tion of cations and temperature of the pore fluid (Taylor and Smith 1986, Hongxi 1993; Bell et al. 1993).

According to the model proposed by Taylor and Smith (1986), if the space between the clay platelets is less than 15 Å, the atoms in the clay minerals develop attractive forces of the van der Waals type. In these conditions the cations are uniformly distributed between any two clay platelets, and do not form separate diffuse double layers. If the space between the clay particles is greater than 15 Å then electrostatic repulsion forces dominate and separate double layers are formed, thereby increasing the distance between the clay platelets (Fig. 4.10). In the first case, when double layers overlap, there is an excess cation concentration between the clay particles. Equilibrium is restored by free water being drawn into the system from adjacent pores and capillaries. In practical terms, monovalent cations in weak concentrations give rise to more extensive diffuse double layers, while an increase of valence and/or concentration will reduce swelling.

The swelling of mudrocks can also be a result of the chemical alteration of non-expansive minerals in expansive minerals. The hydration of anhydrite leads to the formation of gypsum with a maximum increase of volume of about 60% (ISRM 1983). The oxidation of pyrite, frequently by complex biochemical processes, produces sulphuric acid and hydrated iron sulphates that cause an increase in volume (Penner et al. 1973).

The magnitude of swelling in mudrocks is controlled by mineralogical factors such as nature and relative amount of expandable and non-expandable clay minerals, the extension of cementation, and by textural factors such as compaction, microfissuration and lamination (Ordaz and Argandoña 1981; Kojima et al. 1981). However, textural factors only seem to play the principal role in controlling mudrock swelling behaviour in the absence of expandable clay (Ordaz and Argandoña 1981; Sarman and Shakoor 1990; Sarman et al. 1994). In such mudrocks swelling depends on

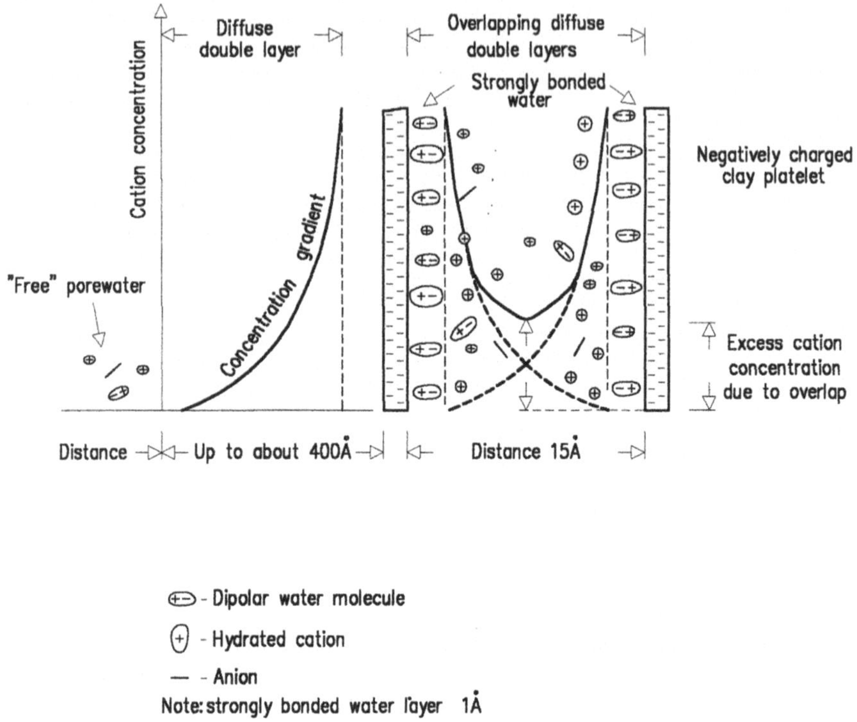

Fig. 4.10 Model proposed to explain osmotic swelling mechanism between two clay platelets (after Taylor and Smith 1986)

porosity defined by the geometry and distribution of the pore-crack system which controls the water suction by capillary action. Porosity seems also to control swelling strain rate and swelling anisotropy behaviour. However, swelling anisotropy of mudrocks may also be due to mineral particle preferred orientation, grain size, lamination, parting and the arrangement of voids.

4.3.4.4 Durability/Slaking

In the context of engineering geology, the term durability may be defined as the resistance to weathering breakdown of the rocks with time after exposure to surface and/or underground environmental conditions (Gamble 1971; Franklin and Chandra 1972). The weathering agents responsible for rock breakdown are variable and may include, in order of importance, wetting and drying, frost action and salt crystallization, leaching, mechanical abrasion, solution and chemical alteration (Franklin and Chandra 1972). In the context of engineering geology time should be related to the *life* duration of the engineering structures. The evaluation of mudrock durability is extremely important, as breakdown may occur on engineering time-scales. Due to rapid degradation the initial properties used in design may be entirely inapplicable to the behaviour of the material during the life of a structure, or even during the construction period. Problems can also arise due to changes in behaviour and disintegration that occurs during sampling operations.

The durability of rock materials is controlled by intrinsic factors inherent to their characteristics and by extrinsic factors that reflect environmental conditions to which rocks are exposed or to which they will be subjected. Climate, topography, hydrological conditions are all important factors which assume a different significance on a case by case basis, according to the circumstances of the specific application. Accordingly, studies and methods that seek to evaluate the durability component of mudrock based on the intrinsic characteristics of these materials, in relation to the most probable weathering mechanisms in the most common applications of these rocks, are preferred.

Research has shown (Russell 1982; Shakoor and Brock 1987; Taylor 1988; Dick and Shakoor 1992; Dick et al. 1994) that mudrock durability is controlled by lithological, mineralogical and structural characteristics, such as, clay content, proportion of expandable clay minerals, cementing agents, rock fabric and presence or absence of microstructures like laminae, slickensides and microfissures. As such it is closely related to swelling capacity.

Mudrock durability is usually evaluated as the resistance to slaking in term of a slake durability parameter. Slaking has been defined as the disintegration of rocks when exposed to air or moisture or to alternate drying and wetting. A more elaborate definition of slaking that focuses on the underlying mechanisms and the rate of breakdown has been proposed by Andrews et al. (1980) and quoted by Perry and Andrews (1982):

> Slaking is the short-term physical disintegration of a geologic material following the removal of confining stresses. Breakage may result either from the establishment or occurrence of sufficient stresses within the material or from the decrease in structural strength. The significance of disintegration rate is dependent on the specific engineering considerations; for definition, short-term may be taken to mean less than several years.

The mode, rate and degree of slaking depend on several factors which include particle size, mineralogy, rock fabric and bedding characteristics (Perry and Andrews 1982). Two modes of slaking of mudrock are reported by Perry and Andrews (1982). In the first one *chip slaking*, fragments that range in thickness from 0.6 to 2.0 mm and in length and width from 2.5 to 15.0 cm, are produced which exhibit resistance to further degradation. In the second mode *slaking*, disintegration into inherent fine-grained particles leads to the complete destruction of the original structure of the rock.

According to Taylor and Cripps (1987) the main mechanisms that influence slaking are as follows: (a) incidence of sedimentary structures and discontinuities, (b) pore air compression, and (c) swelling of expansive clay minerals (described above under the heading of swelling). The important discontinuities in disintegration are the bedding surfaces of depositional or diagenetic origin and structural discontinuities, such as joints, fissures and fractures, and shearing surfaces.

The pore air mechanism is a function of the capillary and suction characteristics of the materials as achieved from the study of breakdown of Coal Measures mudrocks (Taylor and Spears 1970). Mudrock breakdown, as a result of simple static slaking tests, was less when the tests were conducted in vacuum, than when conducted in air. This mechanism operates in such way that during dry periods evaporation from the surfaces promotes suction, which in turn increases shearing resistance of rock fragments. In the case of extreme desiccation, the voids remain filled with air. In these conditions, if the material is rapidly immersed in water, the air in the pores is compressed by the development of capillary pressures in the outer pores, which can cause failure of the rock fabric along the weakest planes.

4.3.5 Laboratory Tests

Observations and descriptions in the field in order to determine the influence of bedding, joints, fissures, faults and other significant discontinuities, geological structure and groundwater conditions in order to assess rock mass behaviour may be complemented by laboratory studies that provide additional information on mineralogical, textural, chemical and geotechnical features of mudrocks (Taylor and Spears 1981). These studies should be carefully planned since many of the laboratory techniques require the use of sophisticated instruments, which are expensive and time consuming to operate. Given the objectives of a specific study the cost *versus* information obtained as a result of the use of the various laboratory techniques available, should always be pondered. The more valuable laboratory techniques for the study of mudrocks are hereunder presented.

4.3.5.1 Mineralogical, Textural, Chemical Characterization

X-Ray Diffraction

X-ray diffraction is the best technique to determine the mineral constituents of mudrocks. This technique can provide both qualitative and semi-quantitative determinations of the mineral species present. For general characterization of a mudrock a whole rock randomly oriented sample is used. For detailed study of the clay minerals, it is necessary to work on 2 µm fraction and performed pretreatments, glycolation and heating. These procedures are required for study of swelling clay minerals and to distinguish kaolinite from chlorite.

Polarizing Microscopy

The polarizing microscope may be useful in the study of the mineralogy of mudrock constituents in the sand-sized and silt-sized fractions as well as in the study of small scale features including cross bedding, particle orientations, micro-laminations and other micro-structures.

Scanning Electron Microscopy

Scanning electron microscopy (SEM) is used to study both mineralogical and textural aspects of mudrocks. Photomicrographs are very useful for the study of textural aspects such as fabric, shape and size of the grains and pore space geometry. Mineral phases may be distinguished using backscattered electron (BSE) images, which provide atomic number contrast. Analytical facilities such as an energy dispersive X-ray analysis system (EDS) can be jointly with SEM enabling the determination the composition of grains. Such data enable the role of particular mineral phases, and association between mineral phases, within the mudrock fabric to be studied.

Porosimetry

Mercury intrusion porosimetry (MIP) and gas intrusion (BET) are very useful in the study of microtextural aspects of mudrocks such as effective porosity, pore distribution, surface porosity and particle size. These features are linked with breakdown processes commonly observed in mudrocks.

Chemical Analyses

Chemical techniques are commonly used in the study of mudrocks. Bulk chemical analysis for both major and trace elements can be obtained using wet chemical, inductively coupled plasma atomic absorption spectrometry (ICP-AAS), atomic absorption (AA) or X-ray fluorescence (XRF) methods. XRF is a rapid procedure that enables analysis of most mudrock components. This enables validation of mineralogical determinations. In terms of assessing weathering effects chemical data have limited value because although mineralogical changes may occur, weathering may result in only small changes in chemical composition.

4.3.5.2 Identification Test

Natural Water Content

The determination of natural water contents very easy to perform and gives valuable information. High natural water contents may indicate the presence of mudrocks with greater amounts of clays minerals, in particular expandable clay minerals. In addition to natural water content, water absorption resulting from the immersion of rock fragments and water adsorption due to exposure specimens to specific moisture conditions are also frequently determined (ISRM 2007a).

Density and Porosity

Several testing procedures for determining density and total and effective porosity are given in ISRM (2007a). Difficulties arise to determine effective porosity when mudrock slake in water, being necessary to use alternative methods such as mercury intrusion porosimetry to measure the pore volume.

Particle Size and Atterberg Limits

The determination of particle size (ISO 2016) and Atterberg limits (ISO 2018) requires the prior disaggregation of the material. The inter-particle bonding characteristics of the mudrocks may not permit disaggregation of the material into individual particles which remain in aggregated or clustered of group of particles. This affects the determination of grain size and measured plasticity, reflecting the values obtained the method and effort used to process (disaggregate) the material.

Methylene Blue Adsorption

The adsorption of methylene blue by clay minerals, using the spot method, makes it possible to measure their hydrophilic surface characteristics and thus evaluates the capacity of these minerals to retain water. The fact that methylene blue is not adsorbed by inert minerals (quartz, feldspar, carbonates, etc.) makes it possible to characterize as routine test the clay fraction of rocks without having to isolate it from rest of the material (NF 1998).

4.3.5.3 Strength and Deformability

Uniaxial Compressive Strength Tests

Specific appropriate testing procedures for uniaxial compressive strength are provided by ISRM (2007b) and ASTM (2014). The requirements for the sample preparation of samples into regular shapes constitute the main drawback of this test for mudrocks. Difficulties arise in forming specimens having suitable dimensions for testing especially in laminated mudrocks. Uniaxial compressive tests can be used for the assessment of strength anisotropy in mudrocks. Once more, coring suitably orientated specimens parallel and perpendicular to the plane of weakness could be the main difficulty.

Tensile Strength Tests

The suggested methods for laboratory testing of tensile strength are given in ISRM (2007c). The application of direct tension to specimens involves laboratory testing difficulties and often diametral compression (Brazilian) tests are used. However, the latter give in general higher values, because diametral testing predetermines the failure surface to specimens.

Point Load Test

The suggested method is given in ISRM (2007d). The point load test is extensively used in indurated rock types but is not suitable for weak or weathered materials. When axial distance of irregular samples is lesser than 25 mm, the inaccuracies increase and the results show low significance. It is recommended in mudrock testing the use of as close as possible equant samples with the loading direction normal to the bedding or parting planes. The index values are related to the uniaxial compressive strength, being advised to determine a correlation factor for each individual mudrock formation.

Schmidt Rebound Test-Hammer

Testing methods for the Schmidt rebound test-hammer are provided by ISRM (2007e). The Schmidt hammer is mainly used for in situ testing but it can also be used in laboratory tests on large diameter core samples and/or on block samples. Correlations converting rebound number to uniaxial compressive strength values have been proposed for mudrocks. However, for an individual mudrock formation, it is advised to determine a site-specific correlation.

National Coal Board Cone Indenter (NCB)

The National Coal Board (NCB) cone intender was developed by Mining Research Development Establishment (MRDE) and the test procedures are given by National Coal Board (1977). The cone indenter is a portable instrument which tests the strength of very small specimens not bigger than 12 × 12 × 6 mm. The test is suitable for testing small chips of fractured or thinly parted mudrocks and does not require elaborate specimen preparation. Correlations have been attempted between uniaxial compressive strength and cone indenter number. However, a site-specific correlation factor should be determined before cone indenter values can be used to predict compressive strength.

4.3.5.4 Swelling

Swelling Strain

Several methods to determine swelling strain are provided by ISRM (2007a). Methods for determination of axial strain and triaxial (three orthogonal axes) swelling strains are proposed. In general, the specimens are prepared in such way that the axial axis or one of the orthogonal axes be normal to the bedding or parting. The latter one has also given information about swelling anisotropy. ISRM (2007f) also provided a test method specially designed for argillaceous rocks to measure both axial and radial free swelling strains on an unconfined disc shaped specimen (although irregular specimens may also be tested). The radial strain is measured with flexible stainless steel band calibrated at 0.1 mm intervals which is attached to the specimen. The main drawback of this method is that the radial swelling strain cannot be monitored during the test. It can only be determined at the end of the test.

Powder swelling test consists of the measurement in a graduated glass cylinder of the difference between the volume of the dry powdered sample and after the addition of distilled water. Powder swelling tests are a useful method of separating mineralogical and textural influences from each other.

Swelling Stress

A variety of swelling stress strain tests is provided in ISRM (2007f). In less indurated mudrocks oedometer test equipment may be used to determine swell pressure and free swell. In tests suggested by ISRM (2007f) a loading device capable of continuous adjustment to keep the volume of the specimen constant is required. An alternative procedure was used successfully to measure the swelling stress of Cretaceous mudrocks using for that an electrical load cell placed in a rigid frame (Jeremias 1993).

For the design and analysis of structures in swelling mudrocks, ISRM (2007f) has suggested a special method for determining the axial swelling stress as function of the axial swelling strain. Using this it is possible to estimate the swelling strain necessary to reduce the swelling stress to a value required for the particular application.

4.3.5.5 Durability

Ageing Tests

Mudrock durability assessment must be related to a specific agent of weathering, which can be obtained by carrying out several ageing tests that simulate the conditions that operate those mechanisms. The most common are cycling wetting, freezing and thawing and drying, however, these tests are very time-consuming and therefore they are mainly used in research studies.

Slake Durability Test

As climatic slaking is the most effective weathering agent it is the one that in routine studies has been included in tests to assess mudrock durability. Susceptibility to slaking may be evaluated by several test methods, however, the slake durability test developed by Franklin and Chandra (1972), recommended by ISRM (2007a) and standardized by ASTM (2016) is by far the most commonly used method and the two cycle slaking durability (I_{d2}) is used in several classifications of mudrock durability. The main drawback of the slake durability test is that it can overestimate the durability of mudrocks if the test samples breakdown mostly in fragments larger than 2 mm.

Static Slake Test

The one-cycle slaking test consists of placing in water an oven-dried rock fragment and observing its behaviour over a period of 24 h. Qualitative indexes have been provided allowing classification of the slaking behaviour (Lutton 1977; Dusseault et al. 1988; Hopkins and Deen 1984; Czerewko and Cripps 2001). The five-cycle slaking test consists of an extension of the one-cycle test that involves five wet–dry cycles in which the material is washed over a 2 mm sieve after each cycle, providing a quantitative index defined as the percentage of material retained on a 2 mm sieve in the end of five cycles.

4.3.5.6 Compaction Tests

The change of grading between uncompacted and compacted material has been defined as a measure of the degradability of the mudrock occurring during the compaction due to breakdown of individual fragments. Earthwork's French practice has been provided two tests to evaluate the mudrock degradability. The first one named *coefficient de fragmentabilité* (FR) consists in the comparison of the grading curves obtained before and after compaction with Proctor normal hammer (NF 2017). The second test named coefficient de dégradabilité (DR) consists in the comparison of the grading curves obtained before and after five wet–dry cycles (NF 1992).

4.4 Mudrock Classifications

4.4.1 Mudrock Geological Classifications

4.4.1.1 Geological Criteria Used for Classification of Mudrocks

Several criteria have been used by different authors for classification of fine-grained sedimentary rocks. Picard (1971) has reviewed those used in North American literature which included texture (particle size), parting (fissility), mineral composition, colour, chemical composition, degree of metamorphism and tectonic association or depositional environment.

Texture is characterized by the particle size of the components of mudrock, that is, mainly by the relative amounts of clay size and silt size fractions present on these rocks. However, the boundary between silt and clay can be taken as 4 μm (1/256 mm) using the Wentworth's scale or as 2 μm like it is defined in geotechnical engineering. The higher silt limit (silt-sand boundary) can also have different values as it is shown in Fig. 4.11.

Structure is characterized by fissility and stratification. On mudrocks fissility can be related either with the preferred alignment of clay particles or with compositional laminae which produces weakness surfaces parallel to the stratification. In spite of fabric alignment may account to produce anisotropy of properties (such as strength) of fresh mudrocks, the development of fissility is very enhanced by superficial weathering processes, being absent in rocks collected at depth. Consequently, as classification factor the utility of fissility is limited to the surface samples (Lundegard and Samuels 1980).

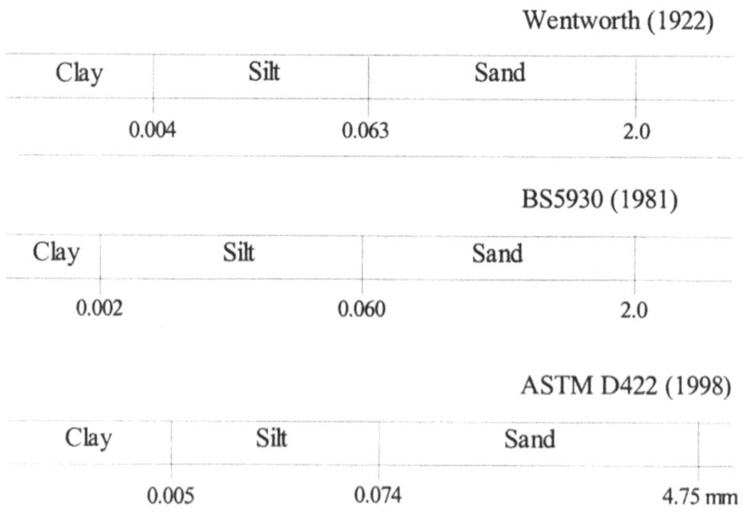

Fig. 4.11 Comparison of the grain size limits defined by the Wentworth (1922), BS (1981) and ASTM (1998) scales

Stratification may be characterized by variations in colour, composition and/or fabric and is subdivided in laminae and beds at the 10 mm boundary. The classes of thickness for each one are given at Fig. 4.7. When beds are presented the suffix *-stone* is attached accordingly to the grain size class (silt, mud or clay) while if laminae occur, the suffix *-shale* is used. Thus, shale means a laminated mudrock.

Accordingly, grain size and stratification seem to be the most helpful properties for describing and classifying mudrocks. Descriptive modifiers such as colour, mineralogy, fossil content, type of fracture and state of induration are commonly attached to the root names proposed in classifications in order to define better the mudrock.

4.4.1.2 Classifications

Several schemes to describe and classify mudrocks have been proposed in the last decades, however, there is no a geological classification for these fine-grained sedimentary rocks widely accepted (Pettijohn 1975; Potter et al. 1980; Lundegard and Samuels 1980; Grainger 1984). Czerewko and Cripps (2012) provided some guidance to describe key features particular to mudrocks (Box 4.5).

Descriptive schemes based on features, such as texture and structure (see Sect. 4.4.1.1), with some genetic significance have been the way chosen to classify these rocks. The mudrock geological classifications most widely accepted are reviewed next.

Classifications based on the grain size and on the splitting (fissility) characteristics of fine-grained sedimentary rocks that contain more than 50% of silt and/or clay proposed by Ingram (Ingram 1953), Folk (1968), Blatt et al. (1980) and Stow (1981) are presented in Table 4.2.

Potter et al. (1980) proposed a classification for mudrock that does not use fissility as classification factor, but the presence or absence of lamination to distinguish between massive and "shale" (laminated) mudrocks. Lundegard and Samuels (1980) proposed a classification very similar to the previous one, but suggested that 10% of laminae is a generally acceptable limit to consider mudrock as a "shale" in rocks with less than 2/3 of silt and recommend the use of the term siltstone for all rocks with a percentage of silt greater than 2/3 Fig. 4.12.

Spears (1980) has proposed a mudrock classification based on quartz percentage. The silt and clay fractions in mudrocks can be estimated from the proportions of quartz and clay minerals present in the rock, as a decrease in grain size is followed by a decrease in the percentage of quartz (Spears 1980).

Weaver (1989) advocated the need of a new term *physil* which include all sheet silicate minerals and has no size mean. In Table 4.3 the classification of fine-grained rocks for both non-indurated and indurated types based on texture and on the physil content is presented. The terms physil and physilitic can be used if the physil content is greater than 50% (physil clay, calcareous physil claystone, etc.) or less than 50% (calcareous physilitic clay, physilitic silt, etc.).

Dick and Shakoor (1992) and Dick et al. (1994) have used a modified form of Potter et al.'s classification, suggesting the placement of the textural division between mudstones and claystones at 50% of clay because this boundary reflects a change in the durability of mudrocks.

Box 4.5 Guide to the Description of Mudrock Features (After Czerewko and Cripps 2012)

Attribute	Descriptive adjectives
Induration	Enables decision on description as soil or rock. If resistant to slaking in water and hard—rock; if susceptible to slaking in water, deformable and "earthy" consistency—soil. Strength depends on moisture state; dry sediment is stronger than wet, and rock strength varies with moisture content; sampling may impair strength.
Strength	Strength is designated based on the degree of induration. For soil, use field consistency values based on manual assessment (e.g. stiff); when shear strength measurements are made, use strength terms (e.g. high strength). For rock, definition based principally on manual field assessment using geological hammer and knife, may be confirmed with UCS measurement: indurated mudrocks range from extremely weak to medium strong; metamudrocks are stronger depending on weathering.
Structure	Standard terms—see structure Fig. 4.7. Include description of lithology and textural interrelationship, as complex features may be present with structured strata such as "thin beds of cross bedded" mudstone.
Colour	Use Munsell colour chart for consistency. Important for correlation; likely environment of formation and indication of likely behaviour of material, that is, red colour—likely formation under oxidizing continental environment. Most important to mudrocks is relationship between colour on the Fe^{3+}/Fe^{2+} ratio. A decrease in this ratio gives an increase in colour from red → green → grey (more Fe^{2+} indicates the presence of pyrite). Organic carbon controls colour: <0.2–0.3%C = light-grey to olive grey; 0.3–0.5%C = mid-grey; >0.5%C = dark-grey to black.
Accessory minerals	Calcareous (slightly to very—based on level of effervescence when assessed with HCl), carbonaceous, dolomitic, ferruginous, glauconitic, gypsiferous, pyritic, micaceous, sideritic, phosphatic.
Rock name	See classification of Fig. 4.12.
Additional information	Presence of fossils—record type (generic such as bivalve and retain for identification), abundance, condition, orientation. Inclusions—nodules (with mineral type and details); gravel, sand, silt partings or pockets, etc.
State of weathering	Alteration seen as distinct discolouration, significant strength reduction to discontinuities and presence of lithorelicts (note orientation). Standard Eurocode 7 (EC7) approach too limited for mudrocks, use specific schemes where available as guides (i.e. Lias and London Clay, Mercia Mudstone).
Fractures	Use EC7 standard terms and procedures. For rock supplement with details such as nature of fragmentation (e.g. conchoidal, hackly, brittle, splintery, slabby, fissile).

Full engineering descriptions of the intrinsic condition and mass properties using descriptive adjectives provided in EC7 (CEN 2004, 2007). This list offers limited guidance for key features particular to mudrocks

Table 4.2 Classifications of fine-grained sedimentary rocks containing more than 50% grains less than 63 μm (silt and/or clay) suggested by Ingram (1953), Folk (1968), Blatt et al. (1980) and Stow (1981)

	Ingram terminology		
	No connotation as to breaking characteristics	Massive	Fissile
No connotation as to relative amounts of silt and clay	Mudrock	Mudstone	Mudshale
Silt predominant over clay	Siltrock	Siltstone	Siltshale
Clay predominant over silt	Clayrock	Claystone	Clayshale
	Folk terminology		
Grain size of mud fraction		Non-fissile	Fissile
>2/3 Silt		Siltstone	Siltshale
Subequal silt and clay		Mudstone	Mudshale
>2/3 Clay		Claystone	Clayshale
	Blatt, Middleton & Murray terminology		
Ideal size definition	Field criteria	Non-fissile	Fissile
>2/3 Silt	Abundant silt visible with hand lens	Siltstone	Siltshale
>1/3 < 2/3 Silt	Feels gritty when chewed	Mudstone	Mudshale
>2/3 Clay	Feels smooth when chewed	Claystone	Clayshale
	Stow terminology		
	Mudrock (>50% siliciclastic, >50% less than 63 μm)		
Basic terms unlithified	Approximated proportions/grain size	Lithified/ non-fissile	Lithified/ fissile
Silt	>2/3 Silt sized (4—63 μm)	Siltstone	Siltshale
Mud	Silt and clay mixture (<63 μm)	Mudstone	Mudshale
Clay	>2/3 Clay sized (<4 μm)	Claystone	Clayshale
Metamorphic terms			
Argillite	Silt and clay mixture	Slightly metamorphosed/ non-fissile	
Slate	Silt and clay mixture	Metamorphosed/fissile	
Textural descriptors	*Approximated proportions*		
Silty	>10% Silt size		
Muddy	>10% Silt size or clay size (applied to non-mudrock sediments)		
Clayey	>10% Clay size		
Sandy, pebbly, etc.	>10% Sand size, pebble size, etc.		
Compositional descriptors	*Approximated proportions*		
Calcareous	>10% $CaCO_3$ (foraminiferal, nannofossil, etc.)		
Siliceous	>10% SiO_2 (diatomaceous, radiolarian, etc.		
Carbonaceous	>1% Organic carbon		
Pyritiferous			
Ferruginous	Commonly used for contents greater than about 1—5%		
Micaceous			
Others			

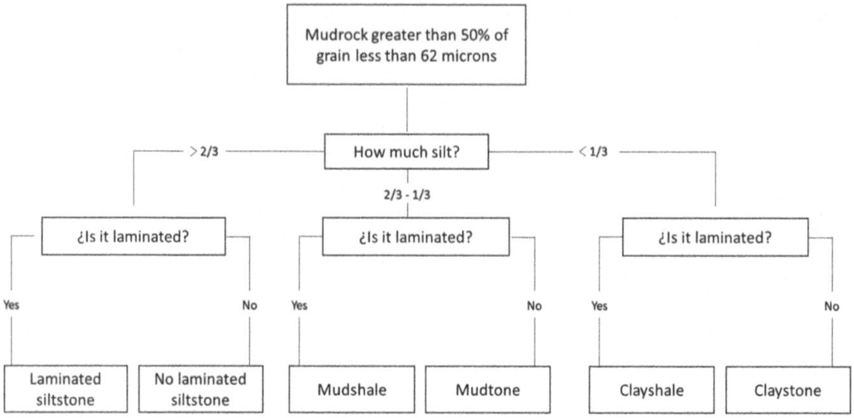

Fig. 4.12 Recommended diagram for the classification of mudrocks according with the scheme proposed by Lundegard and Samuels (1980)

Table 4.3 Classification of fine-grained rocks based on proportion of phyllosilicates (after Weaver 1980, 1989)

Texture	Composition			
	Non-indurated		Indurated	
	>50% Physils	<50% Physils	>50% Physils	<50% Physils
>50% Silt (1/256 to 1/16 mm)	Physil silt	Physilitic silt	Physil siltstone	Physilitic siltstone
>50% Clay (<1/256 mm)	Physil clay	Physilitic clay	Physil claystone	Physilitic claystone

Jeremias (2000) adopted a classification scheme to describe the fine-grained rocks of the north Lisbon area using: (a) the boundaries between claystone and mudstone at 20% quartz+feldspar, mudstone and siltstone at 40% quartz+feldspar and siltstone and sandstone at 60% quartz+feldspar suggested by Spears (1980), Grainger (1984) and Taylor (1988) for quartz alone, (b) a correction for the dilution effect to mudrocks with a carbonate content greater 5% (Spears 1980), (c) the methodology proposed by Hawkins and Pinches (1992) that classifies extremely weathered rocks as soils and not as rocks (Table 4.4).

Czerewko and Cripps (2012) used a quartz content of less than 20% to distinguish claystone and shale, between 20% and 40% to separate silty shale and mudstone, and a percentage of more than 40% to differentiate siltstone. Fissile and non-fissile materials are distinguished based on the fragment shape. Materials are classified as fissile when the ratio of shorted to intermediate dimension (flakiness index) are greater than 2/3 and the ratio of highest to lowest strength are more or equal to 2. These authors advocated the use of the terms fissile mudstone or fissile claystone instead of shale (Czerewko and Cripps 2012).

According to the Eurocode 7 (CEN 2004, 2007) mudrocks are described as rocks constitute by grains of clay minerals, quartz and feldspar less than 0.063 mm in size. These rocks are classified in ISO (2017c) as shown in Table 4.5. Rocks with at least

Table 4.4 Classification adopted to the fine-grained rocks of the north Lisbon area

Quartz+feldspar content[a]	Field criteria	Non-indurated[b]	Indurated	
			Massive	Laminated
>60%	Grains recognized to naked eye or with hand lens	Sand	Sandstone	NA
40–60%	Abundant silt visible with hand lens	Silt	Siltstone	Siltshale
20–40%	Feels gritty when chewed[c]	Mud	Mudstone	Mudshale
<20%	Feels smooth when chewed[c]	Clay	Claystone	Clayshale

NA not applicable
[a]Adapted from Spears (1980), Grainger (1984) and Taylor (1988)
[b]Rocks with more than 10% of carbonates have the term calcareous before the root name
[c]Not recommended in routine investigation

Table 4.5 Classification of mudrocks according to ISO (2017c)

Genetic group		Sedimentary			
		Clastic sedimentary			
Usual structures		Bedded			
Composition		Grains of rock, quartz, feldspars and clay minerals			At least 50% of grains are of carbonate
Predominant grain size (mm)	0.063	Argillaceous or Lutaceous	Mudstone shale: fissile mudstone	*Siltstone*: 50% fine grained particles	Marlstone
	0.002			*Claystone*: 50% very fine particles	

50% carbonate grains are classified as marlstone. However, it is likely that rocks classified as marlstone will contain less carbonate than 50% and this component may be present both as grains and cement. Soil-like mudrocks are identified, described and classified according with ISO (2017a) and ISO (2017b).

In conclusion, the various lithotypes which comprise mudrocks should be distinguished on the basis of texture (grain size) and structure (presence of absence of lamination) using geological classifications of mudrocks such as provided by Lundegard and Samuels (1980) and ISO (2017c), or based on the percentage of the resistate fraction (quartz or quartz+feldspar proportions) determined by laboratorial procedures or field criteria following schemes such as proposed by Jeremias (2000) or Czerewko and Cripps (2012).

4.4.2 Classifications of Mudrocks for Engineering Proposes

4.4.2.1 Engineering Definition of Mudrock

A continuous geotechnical spectrum of materials from loose sands and soft clays to hard rocks exists; however, a main concern in geotechnical practice dealing with mudrocks has been the placement of a soil–rock boundary (Hencher 1993).

From a geological point of view as noted by Cripps and Taylor (1981), there is no distinction between mudrocks and over-consolidated clays and the differentiation between soil-like and rock-like properties are markedly influenced by factors such as the state of induration (bonding and cementation), rebound history and degree of weathering. One of the first approaches to this problem was the distinction between "compacted shales" and "cement shales" proposed by Mead (1936) and followed by Underwood's (1967) classification which recognizes the importance of the soil-like and rock-like character in geotechnical properties of mudrocks. In the first group the materials are the result of consolidation due to weight of overlying sediments, they have no inter-granular cement and are subdivided, according to their grain size characteristics. In the second group the rocks have inter-granular cement which may be calcareous, siliceous, ferrous, gypsiferous, phosphatic, etc. The various lithotypes in this group are classified according to their chemical composition.

Further attempts for a definition of a soil–rock boundary rely commonly on the results of laboratorial tests, usually based on uniaxial compressive strength (UCS) values (Box 4.6). Field strength criteria to recognize soil-like or rock-like mudrocks

Box 4.6 Engineering Criteria for Mudrock Definition

Morgenstern and Eigenbrod's (1974) criteria used the undrained shear strength value of 1.8 MPa, that is, an uniaxial compressive strength (UCS) of 3.6 MPa to differentiate between mudstones and clays. From tests conducted on a variety of materials which included both soil and rock samples (sands and clays; mudstones, limestones, schists, etc.) Rocha (1977) concluded that cohesion and uniaxial compressive strength are the best parameters to distinguish between soils and rocks. Based on the results obtained Rocha (1977) proposed that materials with a cohesion greater than 0.3 MPa and a uniaxial compressive strength larger than 2 MPa are classified as rocks. The latter value was used by ISRM (1981) in BGD (Basic Geotechnical Classification) for the lower limit of rock strength. Rocks were assigned to very weak (UCS 0.6–1.25 MPa) and weak (UCS 1.25–5 MPa) following BS (1981) and IAEG (1981) recommended a UCS value of 1.25 MPa as the lower limit of strength of rocks. De Freitas (1981) suggested the term mudrock should be applied only to argillaceous siliciclastic rocks, whereas Taylor and Spears (1981), considered that at least stiff clays should be included in this group of rocks, because there is no clear distinction between the boundary between soil and rock. According to the method of Morgenstern and Eigenbrod (1974) a compressive strength value of 0.6 MPa (undrained shear strength of 0.3 MPa) was suggested by Grainger (1984) for the lower value of soil-like mudrocks. Eurocode 7 (CEN 2004, 2007) uses the same value for the division between engineering soil and rock, where soils have an undrained shear strength lower than 0.3 MPa (uniaxial compressive strength of 0.6 MPa), and rocks are materials stronger than that value.

are provided in several classifications, but there is no a scheme widely accepted (Box 4.7). In Fig. 4.13 some of the limits suggested for different institutions for the soil–rock boundary are presented and as it is shown no standardization exists for definition of weak rock. Thus, mudrocks with low UCS values are classified as rocks according to some classifications and as soils by others. This can have serious geotechnical and contractual implications in engineering works where these rocks

Box 4.7 Strength Criteria for Mudrocks (After Czerewko and Cripps 2012)

Term	Strength	Description
Strong mudrock	σ_c 50–100 MPa	Can only be scratched by knife or pick end of a geological hammer, and can only be broken with more than one firm hammer blow.
Medium strong mudrock	σ_c 25–50 MPa	Can be deeply scored by a knife or pick end of a geological hammer, and a thin slab can be broken by heavy hand pressure, Specimen is readily fractured with a single firm blow of geological hammer or split with a knife blade. Cannot be peeled with a pocket knife.
Weak mudrock	σ_c 5–25 MPa	Small gravel-sized fragment can be deformed with heavy finger pressure, shallow indentations readily made by firm blow with point of geological hammer. Can be peeled by a pocket knife with difficulty.
Very weak mudrock	σ_c 1–5 MPa	Crumbles onder firm blows with point geological hammer, can be peeledby a pocket knife.
Extremely weak mudrock	σ_c 0.6–1 MPa	Can be indented by thumbnail.
Extremely high strength clay	S_u 0.3–0.6 MPa	Field description will generally be as a "very stiff clay". Crumbles, does not remould, can be indented by thumbnail.
Very high strength clay	S_u 0.15–0.3 MPa	Determine by testing—field description as "very stiff clay".
High strength clay	S_u 0.075–0.15 MPa	
Medium strength clay	S_u 0.075–0.04 MPa	Field description will generally be as a "stiff clay". Crumbles, breaks, remoulds to lump.
Low strength clay	S_u 0.04–0.02 MPa	Field description will generally be as a "firm clay". Cannot be remould, rolls to thread.
Very low strength clay	S_u 0.02–0.01 MPa	Field description will generally be as a "soft clay". Moulds by light finger pressure.
Extremely low strength clay	$S_u < 0.01$ MPa	Field description will generally be as a "very soft clay". Extrudes between fingers.

σ_c Uniaxial compressive strength, S_u Undrained shear strength ($\sigma_c = 2S_u$)

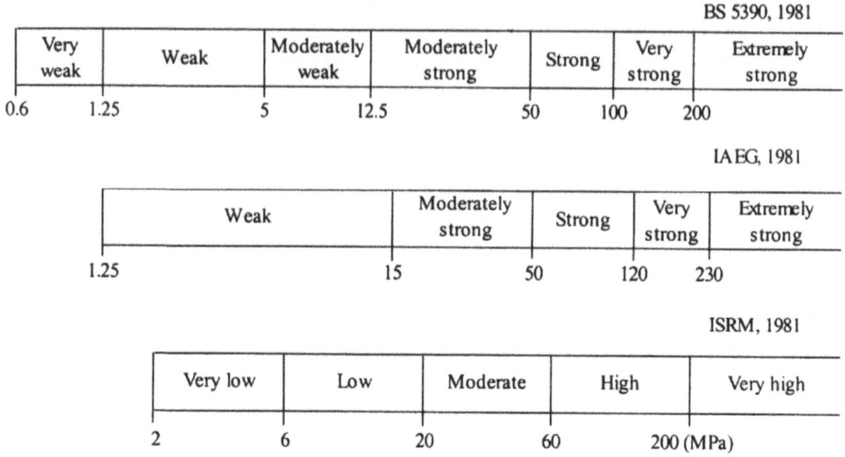

Fig. 4.13 Definition of weak rock according to various standards (adapted from ISRM 1981; IAEG 1981 and BS 1981)

occur. Another problem as noted by Oliveira (1993) is the proper term of weak rock which is also a relative concept in engineering because any material may be considered weak if it fails under loads imposed by the construction of structures.

4.4.2.2 Classifications

In the last decades, mainly due to the increase of engineering construction involving mudrocks, several classification schemes of rock material based on mechanical and engineering properties, were proposed (Deen 1981). The great sensitivity to water of mudrocks determines that durability, namely, the susceptibility to disintegration must be considered in engineering geological classifications (Pinho 2003). Thus, the classification of mudrocks on the basis of these schemes provides criteria on selection of the materials, which can be applied in site selection, design, construction and maintenance of major structures. Table 4.6 presents the soil and rock characteristics used in some of the classifications found in the literature.

The data presented in Table 4.6 show that several classifications use the slake durability test to predict the behaviour of mudrocks. Plasticity obtained from the Atterberg limits is adopted mainly to classify the less indurated types. Strength parameters determined from uniaxial and triaxial compressive tests, or from point load tests are customarily used to distinguish between durable and non-durable mudrocks and to subdivide the harder types.

Deere and Miller (1966) have proposed a general classification chart for intact rock materials based on uniaxial compressive strength and Young's modulus, which differentiated between specific rock types, including mudrocks (Fig. 4.14).

Table 4.6 Soil and rock characteristics used in engineering geological classifications of mudrocks

Soil and rock characteristics	Classifications														
	Deere and Miller (1966)	Gamble (1971)	Deo et al. (1974)	Morgenstern and Eigenbrod (1974)	Lutton (1977)	Olivier (1976)	Franklin (1981)	Grainger (1984)	Rodrigues (1988)	Felix (1987)	Taylor (1988)	Dick et al. (1994)	Venter (1998)	Czerewko and Cripps (2001)	Jeremias (2000)
Mineralogy (from XRD analysis)								X			X				
Anisotropy (Flakiness ratio)								X							
Microfracture frequency index												X			
Porosity									X						
Dry density															X
Grain size								X							
Ansorption water												X			
Moisture absorption														X	
Atterberg limits		X		X			X	X							
Methylene blue adsorption value														X	X
Uniaxial compressive strength	X					X		X		X	X				
Undrained Shear strength				X				X							
Young's modulus (Et or Es)	X														
Point load strength						X	X	X			X				
Cone indenter number								X							
Free swelling strain						X			X	X					

(continued)

Table 4.6 (continued)

Soil and rock characteristics	Classifications														
	Deere and Miller (1966)	Gamble (1971)	Deo et al. (1974)	Morgenstern and Eigenbrod (1974)	Lutton (1977)	Olivier (1976)	Franklin (1981)	Grainger (1984)	Rodrigues (1988)	Felix (1987)	Taylor (1988)	Dick et al. (1994)	Venter (1998)	Czerewko and Cripps (2001)	Jeremias (2000)
Slake durability (evaluated from Jar slake)			X	X										X	
Slake durability (evaluated from slake durability test)		X	X		X		X	X			X	X			X
Rate of slaking				X											
Durability (evaluated from cyclic wet-dry tests[a])			X												
Durability (evaluated from 5 cyclic wet-dry tests)													X		

[a]Tests carried out using a sodium sulphate solution

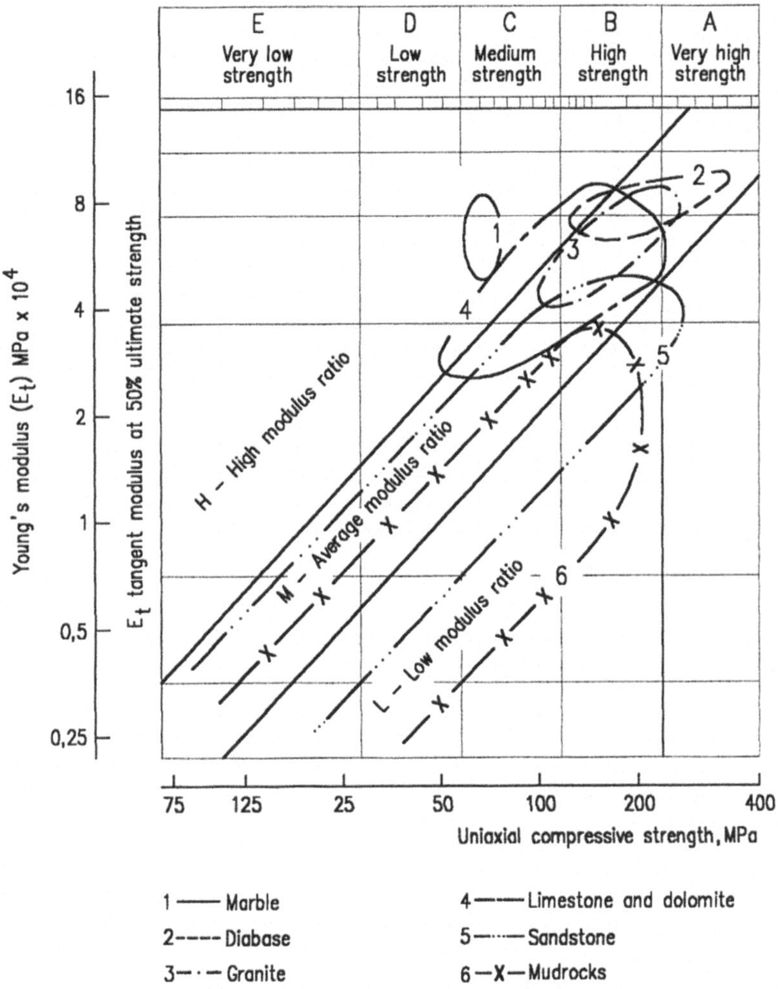

Fig. 4.14 Engineering classification of intact rock showing fields for igneous, sedimentary and metamorphic rocks (after Deere and Miller 1966)

Thcy noted that the envelope for mudrocks extends into the zone of low modulus ratio (modulus/uniaxial compressive strength) which they concluded to be a result of anisotropy ascribable to presence of bedding and lamination. However, the application of this classification to the Karoo mudrocks by Olivier (1976) revealed its inability to differentiate between rocks of high and low durability. This he attributed to the lack of consideration in the classification of swelling properties of the rocks studied.

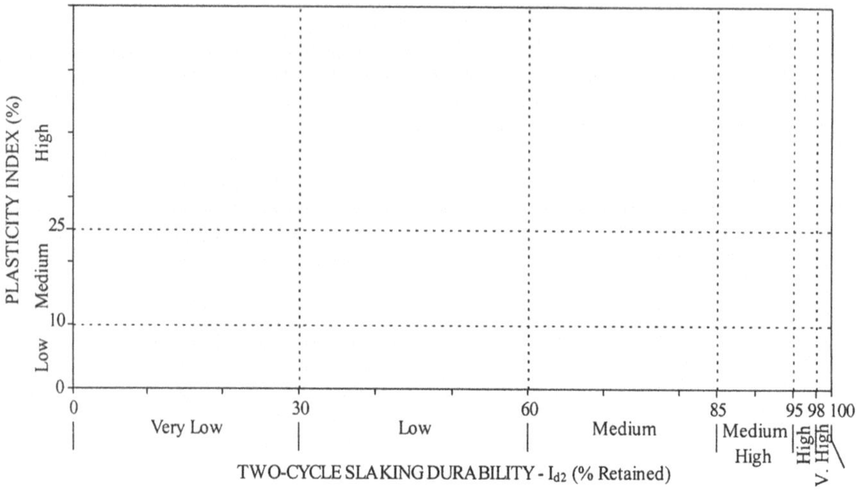

Fig. 4.15 Durability-plasticity chart of Gamble (1971)

Gamble's (1971) durability-plasticity classification of mudrocks recognizes six classes of durability based on plasticity index and two-cycle slake durability. The durability classes are defined in the graph presented in Fig. 4.15. The application of this classification by Olivier (1976) to the Karoo mudrocks in South Africa showed that it is not sensitive to the important distinction between durable and non-durable mudrocks. This is mainly because the rock material disintegrates into fragments that are retained on the 2 mm mesh of the slake durability apparatus and thus over-estimated the durability of the rocks. As noted by Franklin (1981) another problem concerned with the application of this classification to hard mudrocks, is related to disaggregatibility of such rocks in order to determine the Atterberg limits.

Morgenstern and Eigenbrod (1974) proposed an engineering classification for argillaceous materials based on a strength softening test, slaking test, rate of slaking and liquid limit. From the results of the former test they suggested a major differentiation between clays and mudstones on the basis of an undrained shear strength of 1.8 MPa. They suggested also that the liquid limit can be used to predict the amount of slaking and together with the rate of swelling classed the materials in terms of slaking characteristics. Unfortunately, the complete scheme proposed by Morgenstern and Eigenbrod (Morgenstern and Eigenbrod 1974) is too time-consuming to use in most practical situations, tests to determine strength softening being particularly laborious. Accordingly, only that part of the scheme relating to the assessment of slaking characteristics on the basis of liquidity limit and slaking rate is usually adopted (Chapman et al. 1976).

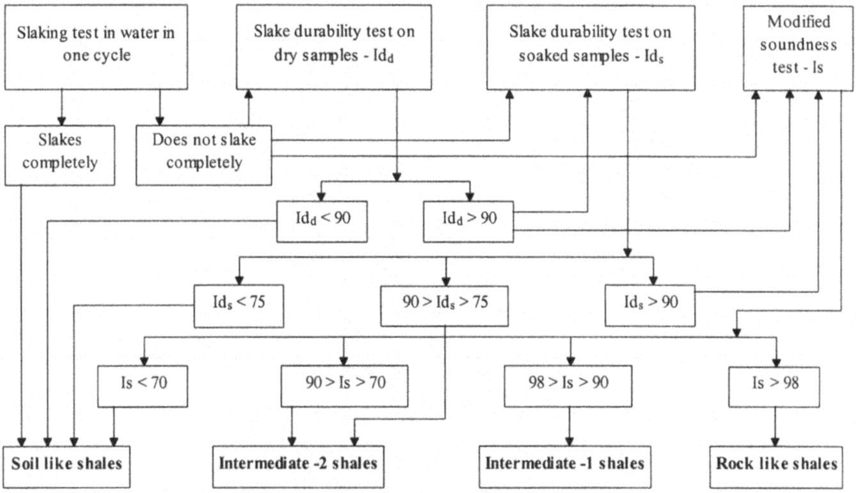

Fig. 4.16 Classification of mudrocks for embankment construction (after Deo et al. 1974)

Wood and Deo (1975) proposed a classification for assessment of mudrock durability for use in highway embankments. This is based on a slake test, slake durability test performed on dry and soaked samples, and a modified sodium sulphate soundness test. According to the flow chart in Fig. 4.16 the rocks were classified in different embankment-use categories from soil-like shales (non-durable) to rock-like shales (durable).

A Geodurability classification was developed by Olivier (1976, 1979) for the siltstones, sandstones and mudstones of Karoo Supergroup in South Africa and is based on uniaxial compressive strength or point load strength, and "Duncan" free swelling coefficient (Fig. 4.17). He proposed six classes to rank the rock materials from very poor to excellent and considered that one of the limitations of his classification is the variation of the index properties selected within rock units or lithological horizons. To overcome this problem, he recommended that large number of samples be tested in order to provide representative values. However, the need for a great number of swelling tests or uniaxial compressive tests in mudrocks remains a problem.

Lutton (1977) carried out an extensive study relating slake-durability, construction lift thickness and embankment performance on mudrocks placed in highway fills in USA. On the basis of this research he proposed the graph presented in Fig. 4.18 where the results of slake durability test can be used to evaluate and predict the behaviour of existing embankments, as well as to define the lift thickness to be used on construction of new embankments. Following this work Strohm et al. (1978) suggested technical guidelines for highway embankment design based on durability characteristics of mudrocks.

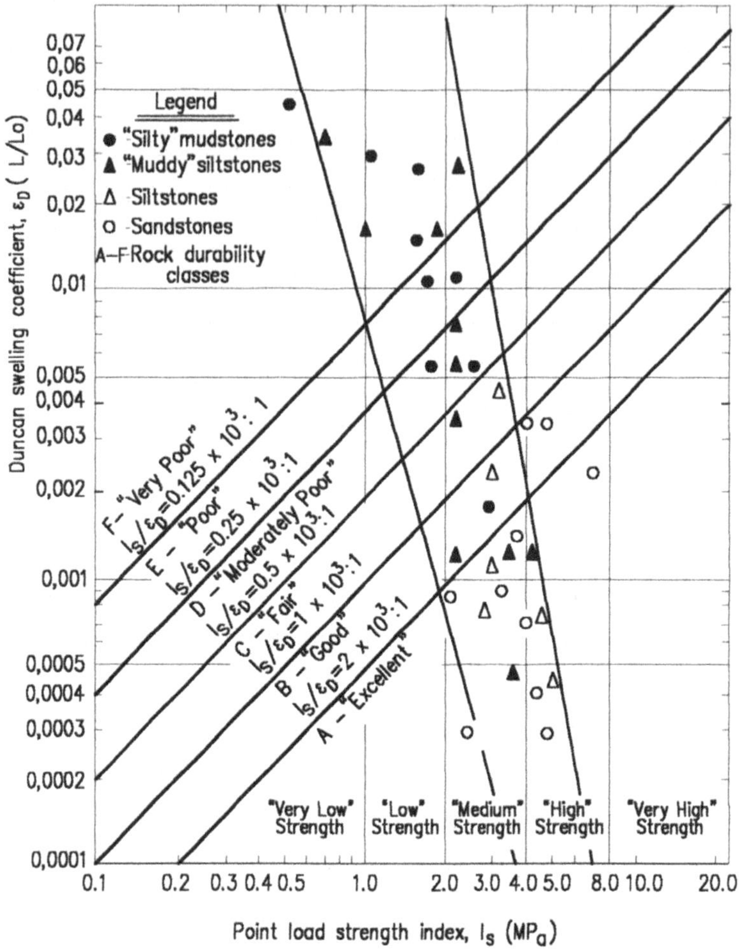

Fig. 4.17 Olivier's geodurability classification developed for strong mudrocks and arenaceous types of the Karoo Supergroup (after Olivier 1976)

Franklin (1981) proposed a mudrock rating system based on slake durability, point load strength and plasticity index. The shale rating chart shown in Fig. 4.19 provides a continuous and quantitative classification of mudrocks. A rating number is assigned according to: (a) its slake durability and plasticity if the mudrock is soil-like and has slake durability index less than 80%, (b) its slake durability and strength (point load strength) if the mudrock is rock-like and has slake durability index (I_{d2}) greater than 80%. Franklin (Franklin 1981) attempted correlations linking the rating number to aspects of engineering performance of mudrocks observed in construction projects such as foundations, embankments and cut slopes, however, as noted by Franklin (1981) care should be taken in using these correlations (e.g. in design) because they are based on limited data.

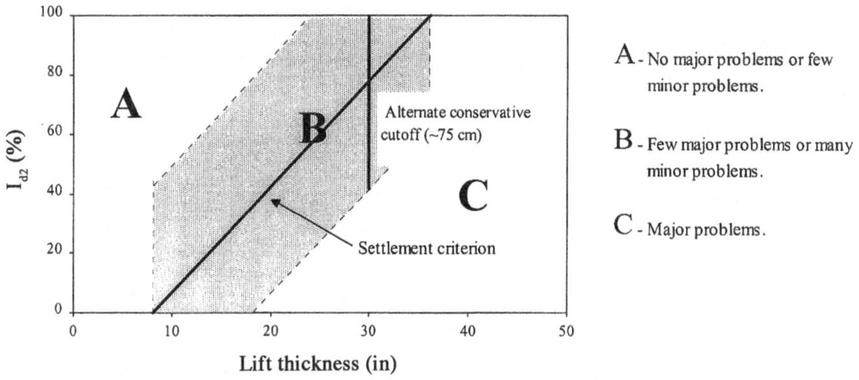

Fig. 4.18 Criterion for evaluating embankment construction on the basis of slaking behaviour of materials (after Lutton 1977)

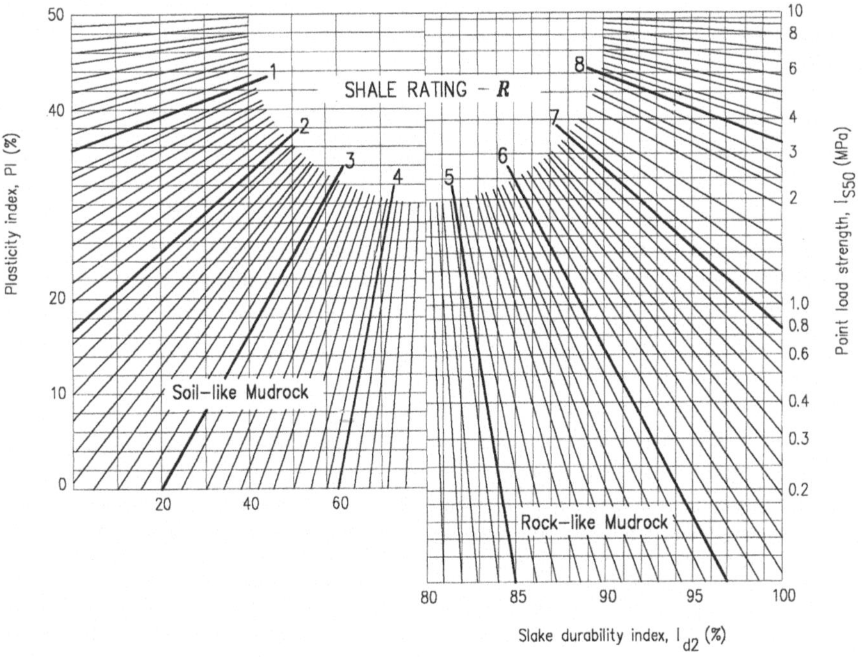

Fig. 4.19 Mudrock rating chart (after Franklin 1981)

Grainger (1984) proposed an engineering classification of mudrocks (Fig. 4.20) based on composition, slake durability, compressive strength and an anisotropy criterion. As noted by Taylor (1988) Grainger's scheme unifies some previous classification suggestions. He used the quartz content method proposed by Spears (1980) to subdivide the more indurated mudrocks and Morgenstern and Eigenbrod's

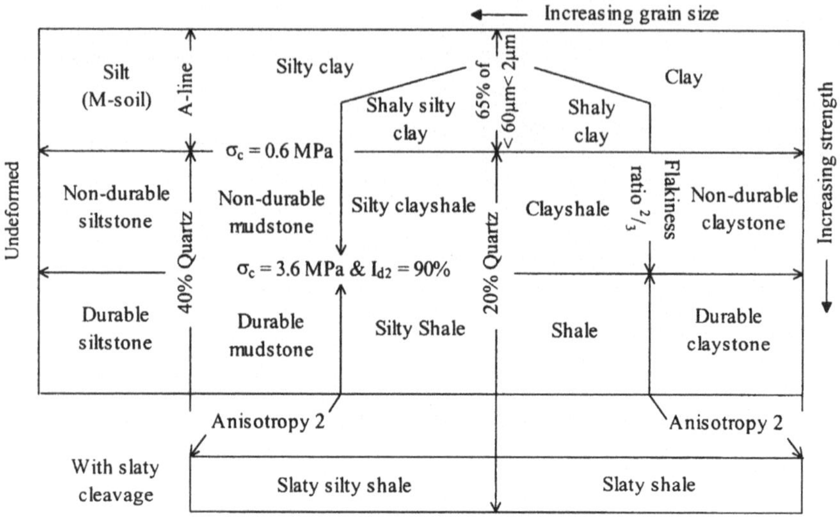

Fig. 4.20 Grainger's (1984) mudrock classification for engineering purposes

(1974) criterion in conjunction with a slake durability (I_{d2}) of 90% to distinguish between durable and non-durable mudrocks. He also suggested that the "shale" lithotypes are distinguished on the basis of the assessment of strength anisotropy. As strength anisotropy is not practical to measure in non-durable mudrocks, Grainger (1984) suggested that a criterion of flakiness ratio (shortest dimension divided by intermediate dimension) of particles produced by natural or artificial fracture should be used to assess the anisotropy character of materials, while for durable mudrocks strength anisotropy is determined by point load or cone indenter testing in orthogonal directions.

Using a wide variety of Portuguese and Algerian carbonate rocks ranging from pure limestone to argillaceous rich carbonate rocks, Rodrigues (1988) proposed a classification based on free swelling strain and effective porosity. The latter parameter was determined by vacuum saturation and buoyancy techniques. Rodrigues (1988) justified the selection of these classification parameters because they evaluate two different main properties of rock materials, respectively, the content of clay minerals and the amount of pore space. Figure 4.21 shows the classification scheme. As noted by Rodrigues (1988), although the limits of the different fields are arbitrary, reflect differences in the behaviour of the rocks in ageing tests and construction works. The use of this type classification is not practical in mudrocks as effective porosity requires mercury intrusion porosimetry which is not suitable for routine classification.

Felix (1987) proposed a classification for sandstones which recognized nine classes of durability based on free swelling strain and the ratio between saturated and dry uniaxial compressive strength. The use of these classification parameters as

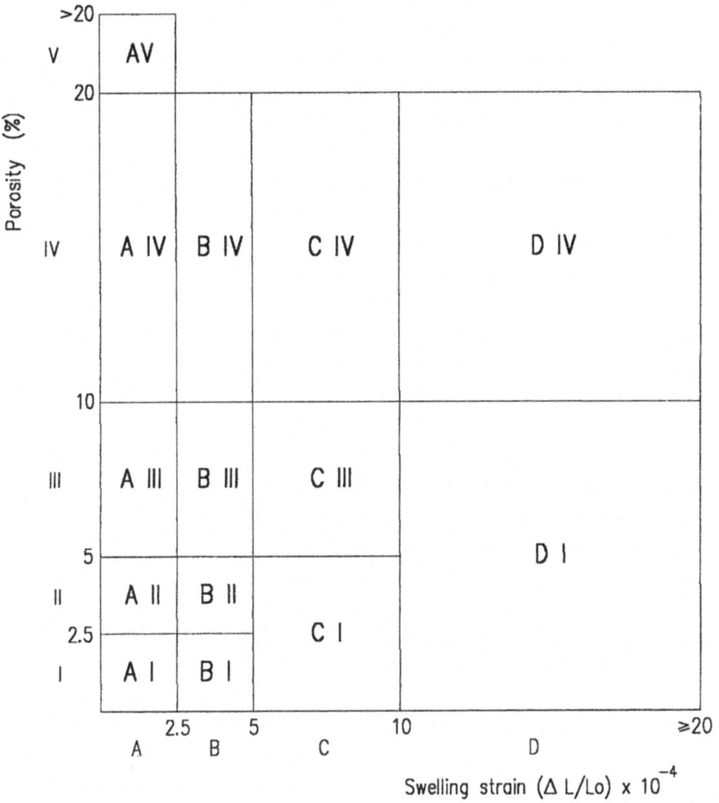

Proposed porosity classes

I - Very low porosity rocks
II - Low porosity rocks
III - Medium porosity rocks
IV - High porosity rocks
V - Very high porosity rocks

Proposed swelling strain classes

A - Very low swelling rocks
B - Low swelling rocks
C - Medium swelling rocks
D - High swelling rocks

Fig. 4.21 Proposed abacus for the geotechnical classification of carbonates rocks (after Rodrigues 1988)

criteria was questioned by Rodrigues and Jeremias (1989) as both parameters are liable to clay minerals in the behaviour of the rocks. Furthermore, with mudrocks it is difficult to obtain specimens suitable for compression testing and to prevent slaking during saturated compression tests. This is especially the case in the softer, less durable materials.

Taylor (1988) developed a classification for British Coal Measures based on composition, uniaxial compressive strength and slake durability. Using the Spears's

Table 4.7 Mudrock durability classification system based on lithological characteristics (after Dick et al. 1994)

Durability	Claystone	Mudstone		Shale	Combined siltstone-siltshale	Argilite
		Slicken sided	I_{mf} (microfractures/cm)	Absorption (%)	Absorption (%)	
High: $I_{d2} > 85\%$	NA	NA	>0.4	< 5.5	< 5	All
Medium: $50 < I_{d2} < 85\%$	NA	NA	0.8 0.4	10 5.5	9.5–5	NA
Low: $I_{d2} < 50\%$	All	All	>0.8	>10	>9.5	NA

Note: *NA* not applicable

(1980) method of quartz content to evaluate the grain size of mudrocks, Taylor (1988) suggested the boundary between siltstones and mudstones of 40% quartz (like Spears 1980) and the boundary between siltstones and sandstones of 60% quartz. Following the Morgenstern and Eigenbrod's (1974) criterion he suggested that durable Coal Measures mudrocks should be distinguished on the basis of a compressive strength of more than 3.6 MPa and a slake durability index achieved on a third cycle (200 rotations per cycle) greater than 60%.

Dick et al. (1994) proposed a mudrock durability classification based on lithological characteristics of a large variety of mudrocks (Table 4.7). These researchers considered that durability behaviour of mudrocks varies accordingly to lithotype and suggested a geological differentiation of mudrocks into claystones, mudstones, siltstones, shales and argillites based on the amount of clay-size material, the presence or absence of laminae, and the state of induration. They also suggested that the assessment of durability of different mudrock types can be achieved from quantitative relationships between slake durability index (I_{d2}) and specific lithological characteristics. According to the results obtained, as indicated in Table 4.7, they concluded that claystones correlate best with the quantity of expandable clay minerals, mudstones with the frequency of microfractures and both siltstones and shales with water absorption value.

Venter (1998) carried out an extensive study of the engineering properties of South Africa mudrocks in order to use these materials in road construction. The 10% FACT (10% fines aggregate crushing value) on dry and soaked samples, and Modified Texas Ball Mill (soaked abrasion test) tests were selected to evaluate general durability. However, as mudrock breakdown is not directly or adequately classified by those tests, Venter (1998) proposed a five cycle wet–dry test and a table with a schematic illustration of range of breakdown behaviours to evaluate this feature.

Czerewko and Cripps (2001) proposed a scheme for prediction of mudrock durability based upon several index tests as modified jar slake test, moisture absorption and the methylene blue value. Data from UK mudrocks were used to developed this methodology scoring each test in order to produce an overall rating between 3 and 9 as well as a classification of durability behaviour in engineering situations is also provided (Table 4.8). They also advocated that modified jar slake test constitute a

Table 4.8 Classification of mudrocks in terms of Rank Total Value (after Czerewko and Cripps 2002)

Rank total value	Sample evaluation	Comment
3	Extremely durable material—not prone to swell or slake	Suitable for engineering appraisal using rock test procedures
4–6	Durable material—may gradually swell and slake	May suffer deleterious effect from rock testing procedures
7–9	Non-durable material—prone to rapid swell and slake	Category of rock and soil—requires non-routine testing approaches

Table 4.9 Evaluation of durability index (DI) from results of index tests (after Jeremias 2000)

Slake durability		Dry density		Methylene blue adsorption value		Durability index	
I_{d2} (%)	Rank value	γ_d (Mg. m^{-3})	Rank value	MBA (g/100 g fines)	Rank value	DI	Total rank
<50	1	<2.19	1	4.1	1	Low	<6
50 85	2	2.19–2.38	2	2.8–4.1	2	Medium	6–7
>85	3	>2.38	3	<2.8	3	High	>7

simple and accurate means to evaluate the slaking potential of mudrocks being capable of distinguishing between low-durability mudrocks and it is adequate for studying the material behaviour on engineering situations as slopes and excavations (Czerewko and Cripps 2001).

Based on previous works of Rodrigues (1988) and Rodrigues and Jeremias (1990) which showed that swelling strain and porosity were two fundamental parameters in the evaluation of the durability of carbonate and greywacke rocks, because they characterize two different features of the materials: the nature and amount of clay minerals and the amount and character of the void space, it was concluded by Jeremias (2000) that these two parameters also control the behaviour of mudrocks. According the mudrock durability classification scheme proposed used the dry density to assess the porosity, the methylene blue adsorption value to evaluate the nature and quantity of clays and the slake durability test for breakdown assessment. Data from Portuguese mudrocks (Lisbon area) were used to score each index test as shown in Table 4.9 and a durability index (DI) was defined as the sum of the rank values determined for the various parameters. As the mudrocks studied are low to medium durability this classification scheme was developed mostly for rocks with those durability ranks. In high-durability mudrocks porosity can be determined accurately by buoyancy techniques using vacuum saturation, and thus is preferable to use a ranking based on porosity values determined directly.

In conclusion, geotechnical classifications of mudrocks are based on the results of index tests and a scheme as the proposed by Grainger (1984) seems to be adequate for evaluate the engineering performance of these rocks as well as the schemes proposed by Jeremias (2000) and Czerewko and Cripps (2001) which evaluates the importance of the nature and amount of clay minerals and of the void space seem also to be appropriate for assessment of mudrock durability.

4.5 Case Studies Related to Mudrocks

Taken into account the widespread of mudrocks in geological record, they are often interested in engineering works. The particular characteristics and behaviour of these rocks are liable to arise problems requiring to avoid them specific work methodologies and measures.

This section includes case studies of commonly engineering situations related to mudrocks, namely, linked to road construction and tunnelling.

4.5.1 Road Construction

4.5.1.1 The Case of A10 Motorway, North of Lisbon, Portugal

The A10 motorway link the Lisbon North Area to Lisbon/Oporto A1 motorway. The A10 motorway was designed with three lanes in each direction; the section under study (Bucelas—Arruda dos Vinhos) has a total length of 8500 m, of which 5275 m correspond to earthworks, 2950 m to viaducts and 275 m to a tunnel (Jeremias and Cravidão 2005).

Engineering Geological Setting

This section of A10 motorway is located on Jurassic sedimentary formations of Arruda dos Vinhos Basin, included in Portuguese western Mesocenozoic border (Fig. 4.22). The motorway alignment cuts an important anticlinal structure

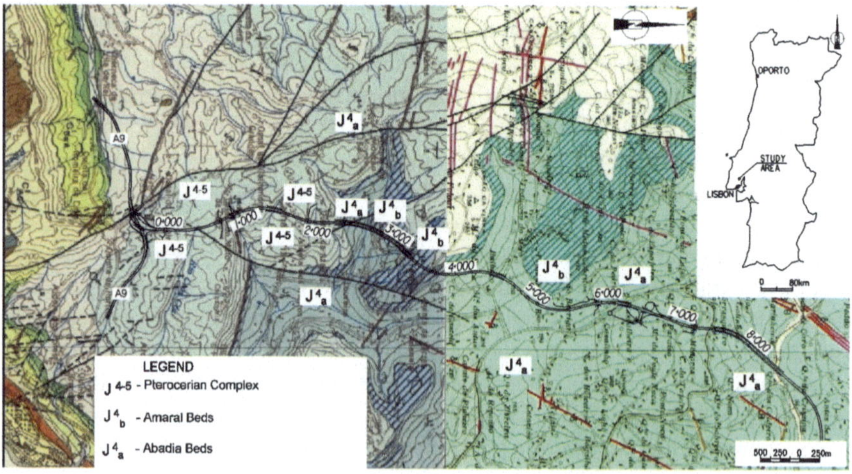

Fig. 4.22 Geological setting of motorway A10 Bucelas—Arruda dos Vinhos section

constituted by Jurassic sediments, outcropping in the core the Abadia Beds, which show large deformations as a result of faulting. In the section the dip of strata is about 10° to 20° towards SE.

From South to North, the motorway alignment intersects firstly the pterocerian marly-limestone complex, constituted by mudrocks with interbedded limestone layers often showing karst features. From the zone of Mato Forte tunnel (after km 3 + 725) up to the end of section occurs the Abadia Beds formation constituted by mudrocks, chiefly grey and greenish micaceous, mudstone, mudshale, siltstone and siltshale with interbedded irregular masses of limestone and sandstone layers.

The Abadia Beds mudrocks are constituted mainly by quartz, feldspar, clay minerals including expandable species (kaolinite, illite, smectite and mixed-layer illite-smectite) and carbonates (calcite and dolomite). The Abadia Beds mudrocks are over-consolidated weak rocks that developed more or less intensely diagenetic bonds, chiefly by carbonate cementation. When exposed at surface the weathering processes promote the disintegration of the inter-particle bonds, which lead to an increase in water content and compressibility, and a decrease in shear strength.

The geomorphological evolution which occurred at the hillsides of the area of the motorway section originated slope debris composed chiefly by fines, which may reach 8 m of thickness, covering almost continuously the Jurassic formations. The debris is constituted by disturbed soils with residual shear strength characteristics. The fine fraction and the differentiated behaviour associated to dry/saturated state of these materials control their geotechnical characteristics. Accordingly, the debris appear to be stiff in dry seasons while when saturated show low shear strength and high deformability characteristics.

Slope failures are relatively frequent after long-period and heavy rainfalls. The presence of an interface almost parallel to the slope surface, between the slope debris and the harder and less permeable mudrock formations constitutes a potential sliding surface due softening and wetting processes associated with the intense and/or frequent groundwater flow over that surface. As a consequence, the evolution of slopes along the motorway alignment is controlled by the occurrence of mudslides and debris flows. Previous slides may be recognized by head scars present in the hillsides of the area under study. In general, the slip surface of these mudslides can occur in the upper layer of mudrock mass, at the interface between the debris and the mudrock formations or pass inside the debris.

Motorway Construction

During the motorway construction a landslide, which reached in its final stages, a 450 m width and 150 m length, occurred at road section at km 5 + 700 (Figs. 4.23 and 4.24). This failure was preceded of the earthworks carried out to construct the motorway platform giving rise in its toe to cuts of about 6 m high with a 1/2 (v:h) geometry. Superficial reconnaissance identified the occurrence of very long fissures with wide apertures over a large area with an average slope angle of about 10°.

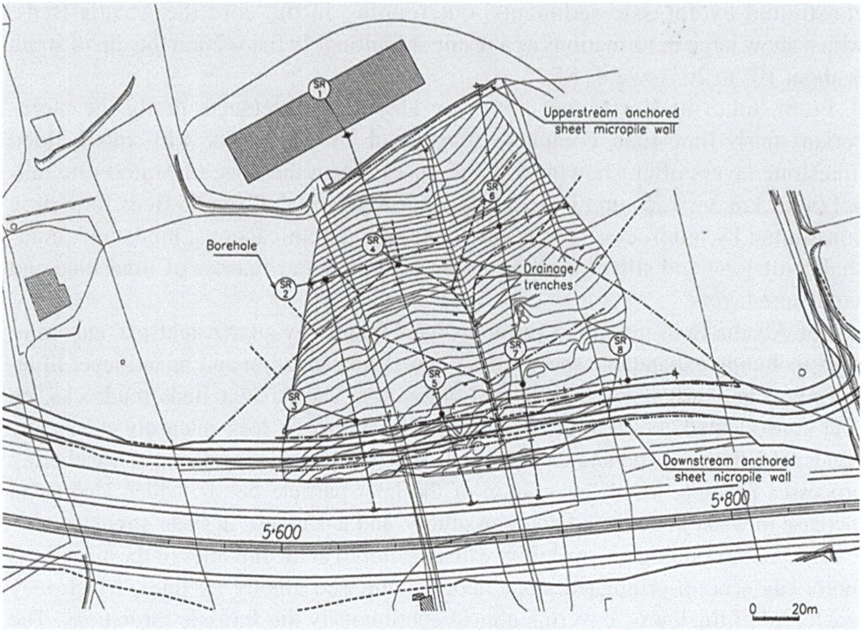

Fig. 4.23 Landslide area map showing the site investigations works carried out as well as the main stabilization measures performed (Jeremias and Cravidão 2005)

Fig. 4.24 Aerial view of the unstable area

At the top of the unstable area there is a platform where an industrial pavilion was installed, and at which toe a small gabion wall was constructed (Jeremias and Cravidão 2005).

A site investigation program was carried out including trial pits and boreholes, which enabled to characterize the upper debris layer and the bedrock composed by mudrocks (Fig. 4.23). This bedrock includes a top layer saturated and very soft that constitutes the interface between the more indurated mudrock formations and the slope debris. Engineering properties from mudrocks of North Lisbon area are presented in Table 4.10. The fieldworks performed confirmed that the local water table was close to the surface.

A global stabilization solution was developed according to the geotechnical constraints identified which included the construction of a continuous drainage rockfill buttress in the upstream area adjacent of the motorway platform as well as the construction of several drainage trenches, which intersected the bedrock (Figs. 4.25, 4.26, 4.27, and 4.28). These trenches played an important role in the stabilization of the landslide area: either through the groundwater drawdown or through the local slope material improvement. In addition to these large scale spatial stabilization measures, two anchored sheet micropile walls founded in the mudrock formations were constructed, located at the toe of the gabion wall existing on the base of the industrial platform slope and upstream of motorway platform close to the service road (Fig. 4.26).

The effective and routine character of the drainage measures and the construction of rockfill buttresses were particularly important for the resolution of slope failures characterized by a mudrock bedrock complex and with non-regular geometry. Moreover, in specific areas close to important infrastructures, anchored sheet micropile walls revealed to be very effective and of routine construction. Over the past years a satisfactory behaviour of the infrastructure was observed (Jeremias and Cravidão 2005).

4.5.1.2 The Case of the Cut Slope Located at EN120, Odemira, Portugal

The rock mass that constitutes the slope under study belongs to the flysch formations of the Baixo Alentejo Flysch Group (BAF), namely, the Formation of Mira of latest Visean to Namurian age. The flysch formations of the BAF which is a stratigraphic unit with an approximate area of 8000 km^2, extends across more than a half of the South Portuguese Zone depositional area (Pinho 2003).

Mudrocks, namely, mudshales/clayshales occur associated to sandstones and related rocks, representing specific depositional environments. The BAF formations are flysch-type deposits constituted, as a rule, by sequences of usually thick greywacke beds, usually of few tens of centimetres thick but sometimes reaching some metres thick, which alternate with thin black to dark-grey shale beds, which include sometimes, intercalations of fine grained conglomerate beds. All the sedimentation of these turbiditic deposits should have been controlled by tectonics. Simultaneously, a low-grade metamorphic event (chlorite zone) should have occurred, associated with the first Hercynian deformation phase, affecting these flysch deposits.

Table 4.10 Mudrock properties from North Lisbon area (Jeremias 2000)

	ρ_d (Mg.m^{-3})	N (%)	MBA (g/100 g fines)	PI (%)	σ_c (MPA)	E_{sec} (GPA)	$CI_{(0.635)}$	ε_s (%)	ε_z (%)	AV (%)	I_{d2}	I_{d5}
Siltstone–Siltshale	2.16–2.37	11.7–20.6	2.6–4.6	9.6–22.5	12.6–23.4	1.17	0.89–1.74	10.1–18.0	1.7–16.7	3.7–29.7	34.3–90.6	4.0–77.6
Mudstone	2.03–2.39	12.4–22.3	3.5–5.8	15.4–27.4	3.2–26.0	1.41	0.51–1.51	13.2–19.4	2.8–11.6	3.5–12.3	4.1–92.3	0.0–82.4
Mudshale	1.93–2.45	12.3–22.7	2.9–6.0	12.9–20.2	1.5–27.6	–	0.78–1.63	12.2–21.2	2.0–9.9	2.7–12.3	14,7–92.7	0.1–82.8

ρ_d dry density, *N* effective porosity, *MBA* methylene blue adsorption value, *PI* plasticity index, σ_c ultimate compressive stress, E_{sec} secant modulus, $CI_{(0.635)}$ cone indenter number measured for steel blade deflection of 0.635 mm, ε_s swelling strain measured on remoulded specimens, ε_z swelling strain measured on intact specimens perpendicular to bedding/lamination, *AV* volumetric strain, I_{d2} and I_{d5} percentages of retained material, respectively, after 400 and 1000 rotations

Fig. 4.25 Detail of a drainage trench

Fig. 4.26 Detail of the downstream anchored sheet micropile wall

In flysch deposits, mudrocks alternate with greywacke sandstone beds, being this alternation of distinct rock beds frequently repeated in a sequence that can reach hundreds of meters to kilometres, in which mudrocks can reach more than two thirds of these continuous turbiditic sequence. The greywacke beds represent the deposition of coarse sands in deep waters due to turbiditic currents from submarine landslides. Between these cataclysmic phenomena occurs the slow deposition of

Fig. 4.27 General view of the section at the end of the slope stabilization works

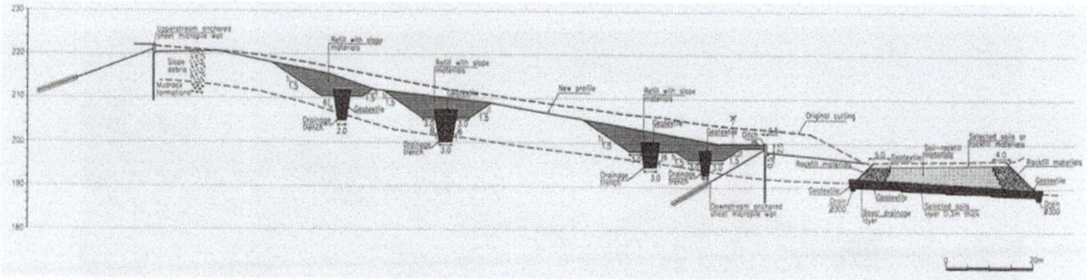

Fig. 4.28 Stabilization solution profile adopted to km 5 + 700 (Jeremias and Cravidão 2005)

clays. Subsequent deformation of these turbiditic formations cause a lot of fractures in the more brittle and higher strength greywackes and numerous folds and shearing in the more ductile and weaker mudrocks.

The general direction of the structures is NW-SE, with vergence for SW. One of the fundamental features of the structural geology of the study area is the presence of large fractures, some deep as indicated by the presence of dolerites in the Messejana-Odemira Fault.

The complex system of discontinuities, which the FBA formations present, has a great influence on the stability of the excavation slopes, especially in the slightly altered rock masses. These are areas of low strength to cutting, especially in the shales, which conditioning the geotechnical behaviour of the rock mass. The excavation slopes in main motorways and the seaside natural cliffs on FBA formations permitted to verify that the planar failure and toppling are the main types of rock failures in these rock masses.

The rock materials of the BAF are heterogeneous and anisotropic, in particular the mudrocks. The heterogeneity and anisotropy revealed in the rocks materials of BAF, are also verified at rock mass scale. The rock masses of BAF are characterized by a great structural complexity, with intense folding and shearing, that affects the flysch formations with alternation of mudrocks and greywackes beds with distinct mechanical properties. As a result of its complex structure the rock masses of FBA present a weathering zone with irregular boundaries, sometimes several metres unlevelled between the fresh to slightly weathered rock mass and the highly to completely weathered rock mass constituted by weak rocks and residual soils.

This case study, fully described in Requetim et al. (2006), is part of a more comprehensive experimental study, about these rock materials, described in previous works (Pinho 2003; Pinho et al. 2004, 2007, 2009, 2015). The main physical, mechanical and weathering properties of the mudrocks and greywackes of FBA were studied, especially the mudrocks ones that behave as soft rocks. The slope under study is located in the town of Odemira (Baixo Alentejo, Portugal), at km 103 of the EN120 motorway, with a slope face direction of N45°W and a dip of 73°SW next to the River Mira (Fig. 4.29). The stability analysis carried out of this steep rock slope, in which small landslides have already occurred, was necessary and important, due not only to its dimensions (24 m high and 104 m long), as well as the fact that the slope is confined by a building area on the top and by a road (EN120), with an intense traffic, on the base, could imply the probable occurrence a major landslide, that would cause serious material damages and eventually, loss of lives. Between this road and the River Mira on SW, there is a slope of approximately 9 m in height that already has stabilization works (gabions).

In order to perform a geotechnical zoning of the rock mass, geological site reconnaissance was carried out, in which lithological and structural data were collected. Thus, scan lines were made observing and recording discontinuities, having the rock mass been divided by sectors along the slope. A set of samples of carbonaceous clayshale, the predominant rock material on the slope under study, with different states of weathering, was collected, in order carry out a laboratory testing programme to determine the fundamental physical properties and index parameters of the rock material.

Fig. 4.29 General view of cut slope located at EN120 motorway in Odemira, Portugal

The results of the structural data obtained in the field and laboratory characterization of the mudrock layers shows that there is a strong influence of the lithological and structural characteristics on the stability of the rock slope. The repeated alternate sequence of mudrock with greywacke sandstone beds influence the water flow in BAF flysch-type deposits, because being the mudrock layers almost impermeable, greywacke sandstone beds tend to conduct the water. So water flow is more confined, resulting higher flow velocities and a bigger probability of rock slides and internal erosion in surface excavations. Besides that, the rock mass is extremely fractured (F5 to F4), and the state of weathering is notorious throughout the rock mass, and very heterogeneous, varying between the slightly weathered state (W2) and the highly weathered state (W4), according to the classification proposed by the International Society of Rock Mechanics. Near the faults, the rock mass is quite friable, behaving like a soil.

The slope is covered with vegetation, including trees and bushes. The carbonaceous clayshales, with some quartz veins, constitute the rock substrate of the slope, presenting a layer of soil cover with about 15 cm of thickness. During the field work and after the occurrence of precipitation, it was observed rock falls along the entire slope, some of blocks of considerable dimensions.

An analysis of the slope stability was made based on the analysis of geometrical and physical characteristics of the main discontinuity sets measured on the slope face. The various types of failure that can occur on a slope are associated with different combinations of discontinuities. Slope stability analysis was performed using the Markland method (Hoek and Bray 1981). The application of this method allowed to conclude that in addition to the probability of rock falls, the most common rock slide is the wedge failure.

The calculation of the factor of safety was accomplished by the method of limit equilibrium, using two-dimensional analysis for plane failure based on the method of wedge geometry which only admits the existence of friction (Hoek and Bray 1981). Thus, because it is not possible to collect samples for laboratory strength tests, due to the difficulty to access the rock mass that constitutes the slope, the value of 25° for the angle of friction was considered in the stability analysis, on the basis of the results obtained in shear strength tests performed on similar materials to those of the case study (Pinho 2003).

The stability analysis provided safety factors of 1.24–1.34 for the wedge failure in most of the slope extension, which indicate a high probability of slope instability.

Considering that it is not possible any change in the slope geometry, due to the local urbanization and the dimension of the blocks does not allow any kind of punctual stabilization, the study allowed to suggest some of the following slope remedial measures:

(a) Drainage of surface water through drainage system (crest and cascade ditches, and a collector or drain to convey the water to outside of the slope);
(b) Regularization of the face of the slope by eliminating the prominent blocks;
(c) Construction of small walls of mortar-walled rockfill, masonry blocks or cemented boulders in crushed areas of the rock mass with abundant fractures and very weathered rock;
(d) Reinforced steel wire mesh, fixed to slope surface with steel cables and expanded bolts or grouted bolts in more jointed rock areas.

4.5.2 *Tunneling: The Case of Kalydona Tunnel, Greece*

The twin tunnel of Kalydona is part of Ionia Highway in western Greece, which connects Antirion with the city of Ioannina. The length of the underground section in the left bore is 1197 m while the right bore is 1190 m, while their distance is 28 m. The maximum overburden is 100 m. The area that the tunnel is excavated has a steep morphology and is characterized by the presence of flysch formations, comprising both sandstones and mudstones interlayered in places.

4.5.2.1 Geological: Geotechnical Conditions

The geological formation of flysch belongs in the Ionian geotectonic zone and is formed by layers of sandstone and mudstone in different proportions, reaching to a thickness of 2000 m. The flysch in the area is composed by (a) thickly bedded sandstones, (b) bedded sandstones with thin mudstone intercalations, (c) mudstone with thin fine grained sandstone intercalations and (d) thinly bedded mudstones. The tunnel is excavated mainly in mudstones, in sandstones interlayered with mudstones and in a short length in sandstones (Istria 2008a). In the present case study, emphasis is given in the section of the tunnel, where mudstones were encountered.

Based on the geotechnical investigation, which comprised the execution of eight boreholes (Istria 2008b), the mudstone formations were discriminated in two geotechnical units (as shown in Fig. 4.30):

- Unit II: the mudstone formation is fresh to slightly weathered and is slightly fractured.
- Unit III: the mudstone formation is fresh and is moderately fractured.

4.5.2.2 Geotechnical Properties

The uniaxial compressive strength of the mudstone presents intense layering with presence of thin sandstone interlayers in places.

The average uniaxial compressive strength, σci, of mudstone is equal to 18.1 MPa, with a range between 7 and 31 MPa. As the mudstone is exposed to weathering, the strength decreases significantly in Unit II. The geotechnical properties of the mudstone are presented in Table 4.11.

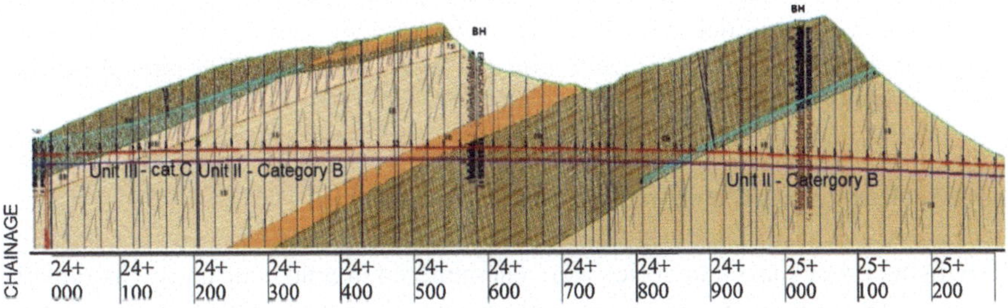

Fig. 4.30 Longitudinal section of Kalydona tunnel (right bore)

Table 4.11 Physical and mechanical properties of mudstone

Property	Mean value	Min value	Max value
Unit weight γ (kN/m³)	26.1	23.8	26.8
Uniaxial compressive strength, σ_c (MPa)	18.1	7.0	31.6
Bulk modulus, E_t (GPa)	23.5	1.0	71.4
Tensile strength (MPa)	3.0	2.2	3.7
Point load strength, I_{s50} (MPa)	1.7	0.03	4.27

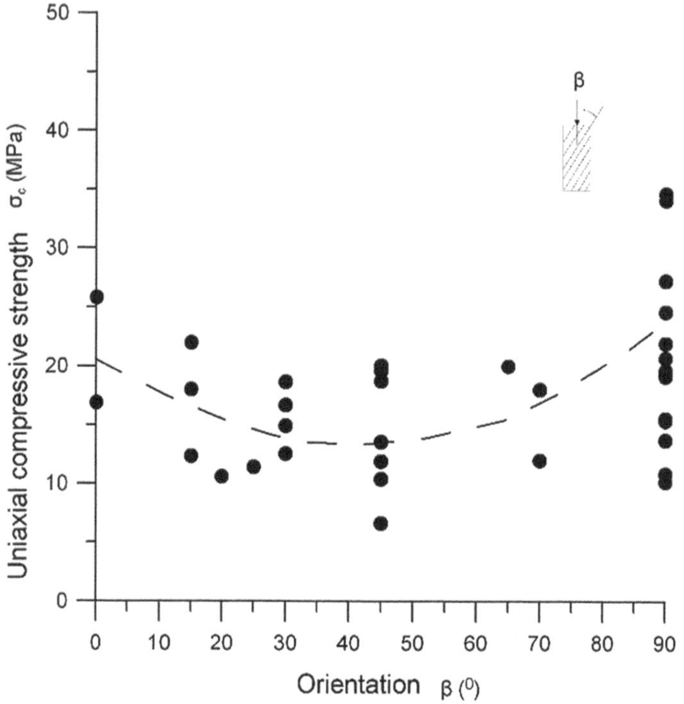

Fig. 4.31 Variation of uniaxial compressive strength due to bedding orientation

The uniaxial compressive strength of the mudstone is strongly affected by the orientation of loading in relation to the bedding planes, thus strength anisotropy should be also taken into account. The uniaxial compressive strength variation with loading orientation has a "U" shape, while the maximum strength is encountered perpendicular and parallel to the bedding planes ($\beta = 90°$ and $\beta = 0°$) and the minimum is at an angle equal to 45° (Saroglou and Steiakakis 2010). The strength variation is presented in Fig. 4.31

4.5.2.3 Rock Mass Categories: Design Parameters

Two distinct rock mass categories were encountered in mudstone along the tunnel, having similar lithology, fracturing and weathering degree. The rock mass category B consists of thickly bedded mudstone, with thickness of 5–15 m and slightly fractures (RQD: 70–100%). The rockmass is intersected by bedding and two discontinuity systems and its behaviour is expected to be very goof with very little to no convergence. It is encountered for a total tunnel length of 740 m. Rock mass category D consists of slightly weathered mudstone with moderate fracturing (RQD: 40–60%). It is encountered for a total length of 100 m. Groundwater presence is not expected in the tunnel as the permeability of the rock mass is very low, while attention should be given to relieve hydraulic pressures. The rock mass categories and the geotechnical parameters, which were used in the tunnel design, are presented in Table 4.12.

4.5.2.4 Tunnel Construction

The proposed support categories were assessed based on the geological–geotechnical discrimination of mudstone behaviour along the tunnel.

More specifically, in unit II it was anticipated that a light support category (cat. A or B) is applied. In unit III, a heavier excavation-support category was designed (cat. C). The heavier support category D was used only in tunnel portal areas, where the anticipated conditions were worst. The support measures as prescribed by the design are given in Table 4.13. The geotechnical units which were encountered were as expected and assessed in the design. The tunnel instabilities, which occurred, were gravitational only, due to the formation of wedges by intersecting joints and bedding planes of mudstone. The convergence was less

Table 4.12 Rock mass categories (Istria 2009; Yiouta-Mitra et al. 2011)

Geotechnical unit	II	III
Rock mass category	B	D
Expected tunnel length of category (%)	63	8
R.Q.D. (%)	70–100	40–60
GSI	55–65	40–50
Rock mass strength, σ_{cm} (MPa)	2–5	2–3
Modulus E_m (GPa)	10–15	3–6
Overburden height (m)	100	50

Table 4.13 Tunnel support measures

Support category	A	B	C	D
Excavation step (m)	3.0–5.0	2.0–3.5	1.5–2.5	1.5–2.0
Shotcrete	10 cm, C20/25 fibre-reinforced 40 kg/m^3	12 cm, C20/25 fibre-reinforced 40 kg/m^3	16 cm, C20/25 fibre-reinforced 40 kg/m^3	18 cm, C20/25 fibre-reinforced 40 kg/m^3
Steel ribs			Lattice girder LG 70-10-30-20	Steel sets HEB 140
Rock bolts	Fully grouted Φ25 S250 3.0 m 2.5 × 2.5 m grid	Fully grouted Φ25 S250 4.0 m 2.0 × 2.0 m grid	Fully grouted Φ25 S250 4.0 m 1.5 × 2.0 m grid	Fully grouted Φ25 S250 4.0 m 1.5 × 1.5 m grid

than expected in all sections, which were excavated. Some minor occurrences of hydraulic pressure were encountered, but they were effectively resolved with drainage holes.

A typical outcrop of the mudstone formation (unit II) at the tunnel face is shown in Fig. 4.32, while a typical temporary support category (cat. C) is presented in Fig. 4.33.

Fig. 4.32 Characteristic outcrop of mudstones at the tunnel face

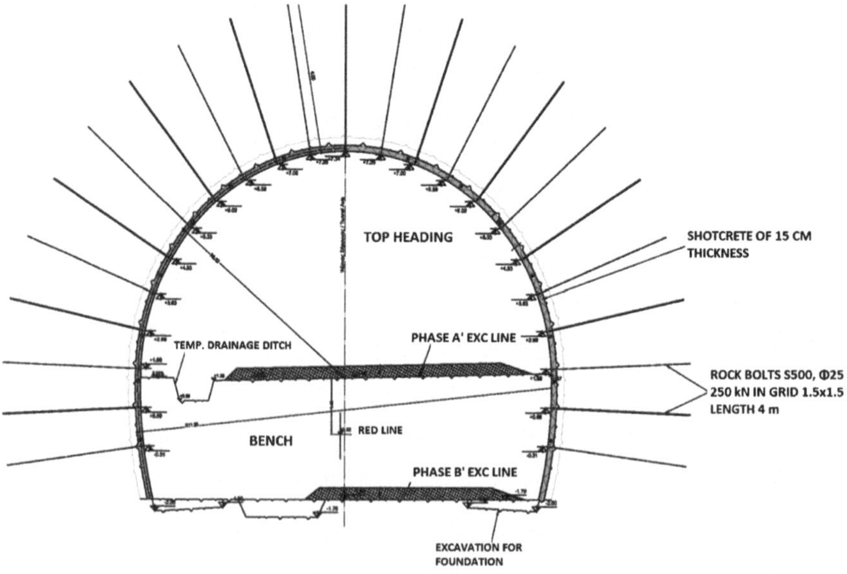

Fig. 4.33 Typical view of temporary support category (cat. C)

The excavation-support measures anticipated during design and applied during construction of the tunnel had minor deviations from those determined in the design. The support category A was applied in greater length than estimated; up to 600 m and category B was applied for a total length of 200 m. Category C was applied for a total length of 100 m, as expected from design.

4.6 Summary

Mudrocks constitute a large part of the geological record and are geographically widely disturbed and because of that they are often interested in engineering works. Mudrocks are the end product of complex geological processes including transport, deposition and burial of muddy sediments and, once they are formed, of exhumation and weathering. Composition and structure are key features of mudrocks which strongly control its geotechnical properties and behaviour under different engineering situations.

Mudrocks are fine to very-fine grained rocks being difficult to describe and characterize, requiring time-consuming and expensive laboratory techniques for detail investigations. However, index tests can provide valuable information concerning the most important engineering properties and long-term behaviour of such rocks.

Terminology associated to mudrock given rise to confusion in technical literature with the same terms being used with different significances. Geological and geotechnical classifications for mudrocks are extensively reviewed in this chapter, as there is no a generally accepted standard for these rocks.

From geotechnical point of view, mudrocks ranging, mostly, from non-indurated soils to moderate strong rocks and a division between compaction mudrocks (soil-like mudrock) and cemented mudrocks (rock-like mudrocks) are generally accepted. Compaction mudrocks formed just by compaction due to burial are compressible and low strength and durability materials, showing a rapid degradation at ground surface. Soil schemes are adequate for the description and characterization of these mudrocks and they behave in most engineering situations as soils. Cemented mudrocks present bonding and cementation to variable extended. They are less compressible and stronger materials ranging from hard soils to moderate strong rocks. These mudrocks may undergo time-depending degradation due to weathering leading to behave as a soil. Slaking resulting from cyclical changes of moisture disintegrated the rock into sand and gravel-sized particles. More durable materials are described and characterized following the rock procedures and in most engineering situations behave as rock.

Under the scope of mudrocks as soft rocks, guidance has been given in this chapter concerning the description, characterization, terminology and classification of these rocks. Finally, various case studies of engineering works involving mudrocks were presented.

Acknowledgements The authors would like to thank Dr. John Cripps of the University of Sheffield for the helpful suggestions and comments that result in significant improvements to this chapter.

References

Andrews DE, Withiam JL, Perry EF, Crouse HL (1980) Environmental effects of slaking of surface mine spoils: Eastern and Central United States. Bureau of Mines, U.S. Department of Interior, Denver, CO, final report, 247p

ASTM D422 (1998) Standard test method for particle-size analysis of soils. ASTM International. www.astm.org

ASTM D7012 (2014) Standard test methods for compressive strength and elastic moduli of intact rock core specimens under varying states of stress and temperatures. ASTM International. www.astm.org

ASTM D4644 (2016) Standard test method for slake durability of shales and other similar weak rocks. ASTM International. www.astm.org

Bell FG, Cripps JC, Culshaw MG, Entwisle DC (1993) Volume changes in weak rocks: prediction and measurement. In: Anagnostopoulos A et al (eds) Geotechnical engineering of hard soils-soft rocks. Balkema, Rotterdam, pp 925–932

Bell FG, Entwisle DC, Culshaw MG (1997) A geotechnical survey of some British Coal Measures mudstones with particular emphasis on durability. Eng Geol 46:115–129

Bennett RH, Bryant WR, Keller GH (1981) Clay fabric of selected submarine sediments: fundamental properties and models. J Sediment Petrol 51:217–232

Blatt H (1982)Sedimentary petrology. Freeman, New York, 514p

Blatt H, Middleton GV, Murray RC (1980) Origin of sedimentary rocks, 2nd ed. Prentice-Hall, New Jersey, 782p

BS 5930 (1981) Code of practice for site investigations

CEN (2004) EN 1997-1:2004. Eurocode 7: geotechnical design part 1: general rules

CEN (2007) EN 1997-2:2007. Eurocode 7: geotechnical design part 2: ground investigation and testing

Campbell CV (1967) Lamina, laminaset, bed and bedset. Sedimentology 8:7–26

Chapman DR, Wood LE, Lovell CW, Sisiliano WJ (1976) A comparative study of shale classification tests and systems. Bull Assoc Eng Geol 13:247–266

Cripps JC, Czerewko MA (2017) The influence of diagenetic and mineralogical factors on the breakdown and geotechnical properties of mudrocks. In: Rutter EH et al (eds) Geomechanical and petrophysical properties of mudrocks. Geological Society, London, Special Publications, p 454

Cripps JC, Taylor RK (1981) The engineering properties of mudrocks. Q J Eng Geol 14:325–346

Czerewko MA, Cripps JC (2001) Assessing the durability of mudrocks using the modified jar slake index test. Q J Eng Geol Hydrogeol 34:153–163

Czerewko MA, Cripps JC (2002) The classification of mudrock durability based on index properties. In: van Rooy JL, Jeremy CA (eds) Engineering geology for developing countries. Balkema, Rotterdam, pp 182–189

Czerewko MA, Cripps JC (2012) Mudrocks, clays and pyrite. In: Burland J et al (eds) ICE manual of geotechnical engineering, vol 36. Thomas Telford, London, pp 481–516

De Freitas MH (1981) Introduction: mudrocks of the United Kingdom. Q J Eng Geol 14:241–242

Deen RC (1981) The need for a scheme for the classification of transitional (shale) materials. Geotech Test J 4:3–10

Deere DU, Miller RP (1966) Engineering classification and index properties for intact rock. Air Force weapons laboratory, Kirtland Air Base, New Mexico, Technical report N. AFWL-TR-65-116. University of Illinois, Urbana, 300p

Deo P, Wood LE, Lovell CW (1974) Use of shale in embankments. National Research Council, Transportation Research Board, Special Report, 148, pp 87–96

Dick JC, Shakoor A (1992) Lithologic controls of mudrock durability. Q J Eng Geol 25:31–46

Dick JC, Shakoor A, Wells N (1994) A geological approach toward developing a mudrock—durability classification system. Can Geotech J 31:17–27

Dunham RJ (1962) Classification of carbonate rocks according to depositional texture. In: Ham WE (ed) Classification of carbonate rocks, Memories of American Association of Petroleum Geologists, vol 1. American Association of Petroleum Geologists, Tulsa, pp 108–121

Dunbar CO, Rodgers J (1957) Principles of stratigraphy. Wiley, New York, 356p

Dusseault MB, Cimolini P, Soderberg H, Scafe DW (1988) Rapid index tests for transitional materials. Geotech Test J 15:281–346

Felix C (1987) Essais et critères de choix pour des grès (molasses) de substitution lors de travaux de restauration. Chantiers/Suisse 18:419–423

Folk RL (1968) Petrology of sedimentary rocks. Hemphill's, Austin, 170p

Franklin JA (1981) A shale rating system and tentative applications to shale performance. Transp Res Rec 790:2–12

Franklin JA, Chandra R (1972) The slake-durability test. Int J Rock Mech Min Sci 9:325–341

Gamble JC (1971) Durability-plasticity classification of shales and other argillaceous rocks. PhD thesis, Univ. Illinois, Urbana-Champaign, 161p

Istria General Consulting Ltd (2008a) Final geological design for Kalydona tunnel, Ch. 24+040 to Ch. 25+250 Section 1 (Report IR.D.T.278.B)

Istria General Consulting Ltd (2008b) Geotechnical and evaluation report for Kalydona tunnel, Ch. 24+037 to Ch. 25+255 (Report IR.D.T.279.B)

Istria General Consulting Ltd (2009) Temporary support design report, Ionia Odos/Kalydona tunnel. (Report IR.D.T.461.C)

Gillott JE (1987) Clay in engineering geology. Elsevier, Amsterdam, p 468

Geological Society of America (1995) Rock colour chart. The Geological Society of America, Boulder. 8th printing

Grainger P (1984) The classification of mudrocks for engineering purposes. Q J Eng Geol 17:381–387

Hawkins AB, Pinches GM (1992) The engineering description of mudrocks. Q J Eng Geol 25:17–30

Heley W, MacIver BN (1971) Engineering properties of clay shales. Development of classifications indexes for clay shales. Technical Report N° S-71-6, vol 1. US Army, Engineering Waterways Experiment Station, Vicksburg, MS, p 89

Hencher SR (1993) Conference summary. In: Cripps JC et al (eds) The engineering geology of weak rock. Balkema, Rotterdam, pp 499–504

Hoek E, Bray JW (1981) Rock slope engineering, 3rd ed. Institution of Mining Metallurgy, London, 358p

Hongxi L (1993) Physico-chemical properties of swelling soft rocks. In: Anagnostopoulos A et al (eds) Geotechnical engineering of hard soils-soft rocks. Balkema, Rotterdam, pp 665–670

Hopkins TC, Deen RC (1984) Identification of shales. Geotech Test J 7:10–18

IAEG (1981) Rock and soil description and classification for engineering geological mapping. Bull Int Assoc Eng Geol 24:235–274

Ingram RL (1953) Fissility of mudrocks. Bull Geol Soc Am 64:869–878

Ingram RL (1954) Terminology for the thickness of stratification and parting units in sedimentary rocks. Bull Geol Soc Am 65:937–938

ISO 14688-1 (2017a) Geotechnical investigation and testing Identification and classification of soil. Part 1: identification and description, 23p

ISO 14688-2 (2017b) Geotechnical investigation and testing. Identification and classification of soil. Part 2: principles for a classification, 11p
ISO 14689 (2017c) Geotechnical investigation and testing. Identification, description and classification of rock, 21p
ISO 17892-4 (2016) Geotechnical investigation and testing Laboratory testing of soil. Part 4. Determination of particle size distribution, 31p
ISO 17892-12 (2018) Geotechnical investigation and testing. Laboratory testing of soil. Part 12. Determination of liquid and plastic limits. 27p
ISRM (1981) Basic geotechnical description of rock masses. Int J Rock Mech Mining Sci Geomechan Abstr 18:8–110
ISRM (1983) Characterization of swelling rock. Pergamon Press, Oxford, 3p
ISRM (2007a) Determining water content, porosity, density, and related properties and swelling and slake-durability index properties. In: Ulusay R, Hudson JA (eds) The complete ISRM suggested methods for rock characterization, testing and monitoring: 1974-2006. ISRM Turkish National Group, Ankara, pp 83–98
ISRM (2007b) Determining uniaxial compressive strength and deformability of rock materials. In: Ulusay R, Hudson JA (eds) The complete ISRM suggested methods for rock characterization, testing and monitoring: 1974-2006. ISRM Turkish National Group, Ankara, pp 151–156
ISRM (2007c) Determining tensile strength of rock materials. In: Ulusay R, Hudson JA (eds) The complete ISRM suggested methods for rock characterization, testing and monitoring: 1974-2006. ISRM Turkish National Group, Ankara, pp 177–184
ISRM (2007d) Determining point load strength. In: Ulusay R, Hudson JA (eds) The complete ISRM suggested methods for rock characterization, testing and monitoring: 1974-2006. ISRM Turkish National Group, Ankara, pp 121–132
ISRM (2007e) Determination of the Schmidt rebound hardness. In: Ulusay R, Hudson JA (eds) The complete ISRM suggested methods for rock characterization, testing and monitoring: 1974-2006. ISRM Turkish National Group, Ankara, pp 107–108
ISRM (2007f) Laboratory testing of swelling rocks. In: Ulusay R, Hudson JA (eds) The complete ISRM suggested methods for rock characterization, testing and monitoring: 1974-2006. ISRM Turkish National Group, Ankara, pp 199–216
Jeremias FT (1993) Determinação das pressões de expansão em rochas argilosas. Geotecnia 68:81–88
Jeremias FT (2000) Geological controls on the engineering properties of mudrocks of the North Lisbon Area. PhD thesis, University of Sheffield, 464p
Jeremias FT, Cravidão J (2005) Slope debris during the construction works of Bucelas Arruda dos Vinhos section of A10 motorway. Proceedings of Géoline 2005, Lyon, 10p
Kojima K, Saito Y, Yokokura M (1981) Quantitative estimation of swelling and slaking characteristics for soft rock mass. In: Proceedings of the international symposium on weak rock, Tokyo, pp 219–223
Krumbein WC, Sloss LL (1951) Stratigraphy and sedimentation. W. H. Freeman and Co, San Francisco, 660p
Lundegard PD, Samuels ND (1980) Field classification of fine-grained sedimentary rocks. J Sediment Petrol 50:781–786
Lutton RJ (1977) Design and construction of compacted shales embankments. vol 3—slaking indexes for design. Federal Highway Administration, US Department of Transportation, Report n° FHWA-RD-77-1, 94p
Mead WJ (1936) Engineering geology of dam sites. Transactions, 2nd international congress on large dams, Washington DC, 4, pp 183–198
Mesri G, Cepeda-Diaz AF (1986) Residual shear strength of clays and shales. Géotechnique 36:269–274
McKee ED, Weir GW (1953) Terminology for stratification and cross-stratification in sedimentary rocks. Bull Geol Soc Am 64:381–389

Montero JM, Torres MC, Palomino CA, Cortés R (2011) Rocas lodosas de la cordillera oriental de Colombia. Boletín Colombiano de Geotecnia, 15, 63p
Moon CF, Hurst CW (1984) Fabric of muds and shales: an overview. In: Stow DAV, Pipper DJW (eds) Fine-grained sediments: deep water processes and facies. Blackwell, Oxford, pp 579–593
Morgenstern NR, Eigenbrod KD (1974) Classification of argillaceous soils and rocks. J Geotech Eng ASCE 100:1137–1156
National Coal Board (1977) NCB cone indenter. Mining research and development establishment handbook N° 5, London, 5p
NF P94-066 (2017) Coefficient de fragmentabilité des matériaux rocheux
NF P94-067 (1992) Coefficient de dégradabilité des matériaux rocheux
NF P94-068 (1998) Détermination de la valeur de blue de méthylène d'un sol ou d'un matériau rocheux par i'essai à la tache
Oliveira R (1993) Weak rock materials. In: Cripps JC et al (eds) The engineering geology of weak rock. Balkema, Rotterdam, pp 5–15
Olivier HJ (1976) Importance of rock durability in the engineering classification of Karoo rock masses for tunnelling. In: Proceedings of the symposium on exploration for rock engineering. Balkema, Johannesburg, pp 137–144
Olivier HJ (1979) A new engineering-geological rock durability classification. Eng Geol 14:255–279
Ordaz J, Argandoña V (1981) Swelling characteristics of some mudrocks from Asturias (Spain). In: Proceedings of the international symposium on weak rock, Tokyo, pp 231–235
Penner E, Eden WJ, Gillott JE (1973) Floor heave due to biochemical weathering of shale. In: Proceedings of the 8th international conference on soil mechanics and foundation engineering, Moscow, pp 151–158
Perry EF, Andrews DE (1982) Slaking modes of geologic materials and their impact on embankment stabilization. Transp Res Rec 873:22–28
Pettijohn FJ (1975) Sedimentary Rocks, 3rd edn. Harper and Row, New York, p 628
Picard MD (1971) Classification of fine-grained sedimentary rocks. J Sediment Petrol 41:179–195
Pinho AB (2003) Geotechnical characterization of weak rock masses—the Baixo Alentejo Flysch Group. PhD thesis, University of Évora, 272 p
Pinho AB, Duarte IMR, Rodrigues-Carvalho JA (2007) The Baixo Alentejo (Portugal) flysch rocks: physical properties and correlations. In: Proceedings of the 11th congress of international society for rock mechanics
Pinho AB, Duarte IMR, Rosa FRS, Rodrigues-Carvalho JA (2015) Influence of anisotropy on the mechanical properties of Baixo Alentejo flysch rocks (Portugal), The 13th ISRM Congress, Montréal, QC, Canada, 10-13 May 2015. In: Proceedings of the 13th congress of international society for rock mechanics
Pinho AB, Rodrigues-Carvalho JA, Gomes CF (2004) State of rock weathering evaluation by index tests. In: Proceedings of the 9th national geotechnical congress, pp 199–208
Pinho AB, Rodrigues-Carvalho JA, Gomes CF, Duarte IMR (2009) Overview of the evaluation of the state of rock weathering by visual inspection. In: Culshaw MG et al (eds) Engineering geology for tomorrow's cities. Geological Society Engineering Geology Special Publications, London, p 22
Potter PE, Maynard JB, Depetris PJ (2005) Mud and mudstone, introduction and overview. Springer-Verlag, Berlin, 297p
Potter PE, Maynard JB, Pryor WA (1980) The sedimentology of shale: a study guide and reference source. Springer Verlag, New York, p 306
Pryor WA (1975) Biogenic sedimentation and alteration of argillaceous sediments in shallow marine environments. Geol Soc Am Bull 86:1244–1254
Requetim L, Pinho AB, Duarte IMR, Lopes L, Valente AR (2006) Stability analysis of an excavation slope in Odemira. In: Proceedings of the 7th national geological congress. University of Évora

Rocha M (1977) Alguns problemas relativos à mecânica das rochas dos materiais de baixa resistência. Memory 491, LNEC, Lisboa

Rodrigues JD (1988) Proposed geotechnical classification of carbonate rocks based on Portuguese and Algerian examples. Eng Geol 25:33–43

Rodrigues JD, Jeremias FT (1989) Contribuição para o estudo das rochas grauvacóides e para a avaliação da sua durabilidade. Internal report 101/89-NP, LNEC, Lisboa

Rodrigues JD, Jeremias FT (1990) Assessment of rock durability through index properties. In: Proceedings of the 6th congress of the international association of engineering geology, pp 3055–3060

Russell DJ (1982) Controls on shale durability: the response of two Ordovician shales in the slake durability test. Can Geotech J 19:1–13

Sarman R, Shakoor A (1990) Prediction of volumetric increase of selected mudrocks. In: Proceedings of the 6th congress of the international association of engineering geology, Amsterdam, pp 459–465

Sarman R, Shakoor A, Palmer DF (1994) A multiple regression approach to predict swelling in mudrocks. Bull Assoc Eng Geol 31:107–121

Saroglou H, Steiakakis C (2010) Experimental analysis of the uniaxial compressive strength of anisotropic layered flysch type rocks and suggestions for use in the GSI classification. In: 6th Hellenic conference of geotechnical and geoenvironmental engineering, Vólos, pp 243–250

Shakoor A, Brock D (1987) Relationship between fissility, composition and engineering properties of selected shales from northeast Ohio. Bull Assoc Eng Geol 24:363–379

Spears DA (1980) Towards a classification of shale. J Geolog Soc 137:125–129

Stow DAV (1981) Fine-grained sediments: terminology. Q J Eng Geol 14:243–244

Stow DAV, Piper DJW (1984) Deep-water fine-grained sediments: history, methodology and terminology. In: Stow DAV, Pipper DJW (eds) Fine-grained sediments: deep water processes and facies. Blackwell, Oxford, pp 3–16

Strohm WE, Bragg GH, Ziegler TW (1978) Design and construction of compacted shale embankments: Volume 5 - Technical guidelines. Federal Administration, US Department of Transportation Research Record, 790:33-41

Taylor RK (1988) Coal measures mudrocks: composition, classification and weathering processes. Q J Eng Geol 21:85–99

Taylor RK, Cripps JC (1984) Mineralogical controls on volume change. In: Attewell PB, Taylor RK (eds) Ground movements and their effects on structures. Surrey University Press, Glasgow, pp 268–302

Taylor RK, Cripps JC (1987) Weathering effects: slopes in mudrocks and over-consolidated clays. In: Anderson MG, Richards KS (eds) Slope stability. Wiley, New York, pp 405–445

Taylor RK, Smith TJ (1986) The engineering geology of clay minerals: swelling, shrinking and mudrock breakdown. Clay Miner 21:235–260

Taylor RK, Spears DA (1970) The breakdown of British coal measures rocks. Int J Rock Mech Min Sci 7:481–501

Taylor RK, Spears DA (1981) Laboratory investigation of mudrocks. Q J Eng Geol 14:291–309

Torres MC, Alarcón A (2007) Some aspects of degradability of argillaceous rocks. In: Estire J, Olalla C (eds) Preservation of natural stone and rock weathering. Taylor & Francis Group, London, pp 67–47

Tourtelot HA (1960) Origin and use of the word "shale". Am J Sci 258-A:335–343

Tucker ME (1994) Sedimentary petrology, 2nd ed. Blackwell, Oxford, 260p

Underwood LB (1967) Classification and identification of shales. J Soil Mech Found Div ASCE 93:97–116

Venter JP (1998) An overview of the engineering properties and the use of mudrock in road construction in South Africa. In: Proceedings of the 8th congress of the international association of engineering geology, pp 2785–2796

Weaver CE (1980) Fine-grained rocks: Shale or physilites. Sedimentary Geology 27:301-313

Weaver CE (1989) Clays, muds and shales. Developments in sedimentology 44. Elsevier, Amsterdam, 819p

Wentworth CK (1922) A scale of grade and class terms for clastic sediments. J Geol 30: 377–392

Wood LE, Deo P (1975) A suggested system for classifying shale materials for embankments. Bull Assoc Eng Geol 12:39–55

Yiouta-Mitra P, Steiakakis C, Merziotis D, Nomikos PP, Sofianos AI (2011) "Ionia Odos: tunnelling through undisturbed anisotropic flysch formation", 1st scientific congress on tunnels and underground structures in South-Eastern Europe, "Using Underground Space", Dubrovnik

Chapter 5
Sandstones in Dam Foundations and Tunnels

Eraldo L. Pastore

5.1 Introduction

Sandstones are usually classified as soft rocks. According to Kanji (2014), these rocks are a critical material since they present a series of problems. They have a behavior intermediate between soil and hard rock and in many cases, they cannot be tested neither in soil mechanics laboratories due to its high resistance, nor in rock mechanics laboratories as they are too soft to be trimmed and tested. Consequently, there are often many difficulties to test such materials. Additionally, the sampling of soft rocks with conventional rotary drillings partially or totally destroys the rock core and sometimes triple barrels may not be suitable.

Sandstones are sedimentary clastic rocks, that is, formed by the accumulation and subsequent diagenesis of mineral grains or rock particles with more than 50% grain sizes ranged between 2 mm (very coarse sand) and 0.06 mm (very fine sand). The main types of sandstone are quartz sandstone, which is the most common, followed by arkose and greywacke.

At the beginning of the formation process, sandstones are deposits of alluvial sand in aeolian or aqueous environment, which by suffering the diagenesis (natural chemical and physical processes) over geologic time become layered rocks with more or less mechanical strength, depending on the time and type of active diagenetic process. The most common diagenetic processes are overburden pressure generated by the own weight of the sedimented layers, cementation of grains formed by the precipitation of chemical elements in the pores, and authigenesis which is the formation of mineral in situ during the diagenetic process.

Diagenetic processes transform sand deposits into sandstone rock, which presents high cohesion between the grains, has greater strength against erosion, piping,

E. L. Pastore (✉)
WRC Consultoria Empresarial, São Paulo, Brazil

M. Kanji et al. (eds.), *Soft Rock Mechanics and Engineering*,
https://doi.org/10.1007/978-3-030-29477-9_5

and liquefaction phenomenon, and is excavated with rippers associated with pop-shooting or explosives, only.

Sand deposits such as recent deposits which have not yet suffered diagenetic processes, or part of old deposits which have not undergone diagenetic processes (did not turn into rock), continue being sands from the engineering point of view, that is, materials without cohesion, with low shear strength, and easily excavated. In Geology, due to classic definitions, sands are also called sandstone which has generated numerous misconceptions regarding the classification and geotechnical/geomechanical properties of these materials in the civil construction works. Sand layers can also be formed from breakage of sandstone rocks exposed to weathering processes acting on the ground surface that disintegrate the grains, leading to a residual material without cohesion and low strength. The layers thus formed are located on the ground surface and are denominated residual sandstone soil.

5.2 Classification of Sandstones

Sandstones are usually classified expeditiously for engineering purposes in four categories according to their degree of coherence, for indirect evaluation of their geo-mechanical properties and resistance to excavation. Table 5.1 shows the classification based on the degree of coherence adopted by the ABGE (Brazilian Association of Engineering Geology), associated with it the corresponding geotechnical properties of each category based on information from innumerous scientific papers published on the subject. In this table, it is worth clarifying that alluvial and aeolian sands with low or without coherence can also be classified as C4 (incoherent sandstone), due to the grain size composition and geotechnical behavior, very close to residual sandstone soil.

As foundation of dams or structures, sandstones are the "rock" with greater range of uniaxial compression resistance (100 MPa > σ_c > 0.4 MPa).

The following classification of sandstones based upon the unconfined compressive strength (σ_c) is proposed in Table 5.2, together with the soft rock classification of International Society of Rock Mechanics (ISRM 1978).

The clear understanding of the different engineering behaviors between these four sandstone categories is of great practical importance in the design where the realistic geological and geotechnical boreholes and outcrops descriptions are essential to the correct identification and evaluation of natural rock and soil mass properties.

In sandstone deposits, not homogeneous layers are very common, occurring as lenses or even sand layers (C4) interspersed with strength sandstone layers (C2/C1), due to nonuniformity in the diagenetic processes at site. In the case of rhythmites, the interspersion of layers of siltstones and mudstones repetitively and alternating with sandstone layers are also quite frequent due to genesis of the deposit.

The existence of sand layers (C4), interspersed with resistant layers (C2/C1), can be easily identified in a natural slope or excavation by the observation of "pits," caused by the internal erosion (*piping*) of these materials when they intercept water flow occurrences (Figs. 5.1 and 5.2).

Table 5.1 Classification, geomechanical properties and behavior on sandstone excavation works

Category	Name	Sandstone features Guidicini et al. (1972)	Typology of sandstone	σ_c (MPa)	SPT (30 blows) recovery (%)	Sonic speed (m/s)	Execution method (at site experience)	
					Kanji et al. (1981)		Maranesi et al. (1983)	Koshima et al. (1983b)
C1	Coherent sandstone	Hard to break by hammer blows, producing sharp-edged fragments Surface difficult to scratch by steel blade Excavation by blasting, only	Strongly cemented with silica, limonite, and/or carbonates	Oliveira and Caruso (1979) Red: 100 White: 49 Partially: 34 Few: 29	– 60–100%	–	By blasting	–
C2	Moderately coherent sandstone	Hard to breaks by hammer blows, producing sharp-edged fragments Easy to scratch by steel blade Excavation by blasting	Low cemented or lithified	Koshima et al. (1983a) >3.6	– 10–60%	–	By blasting	Pop-shooting and CAT D8/9H ripper
C3	Low coherent sandstone	Breaks easily by hammer blow, producing fragments that can be broken by hand Surface easy to scratch by steel blade. Scarify by ripper	Weakly or not cemented, few lithified	Koshima et al. (1983a) 3.6–0.4	SPT (∗) > 60 0–10% (∗)SPT: Standard Penetration Test	2100–3000	Bulldozer blade	Pop-shooting and CAT D8/9H ripper
C4	Incoherent sandstone	Breaking with finger pressure, disaggregating. Can be cut with a steel blade. Friable and be excavated with a bulldozer blade	Without cementing or lithification, friable/sands	Koshima et al. (1983a) <0.4	5 < SPT < 60	800–1400	Bulldozer blade	–

Table 5.2 Proposed classification of sandstone based on the σ_c

Class	Description	σ_c (MPa)	ISRM	σ_c (MPa)
C1	Hard sandstone	>30	Hard rock	>30
C2	Soft sandstone	$30 > \sigma_c > 4$	Soft rock	$30 > \sigma_c > 8$
C3	Very soft sandstone	$4 > \sigma_c > 0.4$	Very to extremely soft rock	$8 > \sigma_c > 0.3$
C4	Friable sandstone (sand)	<0.4	Soil	<0.3

Fig. 5.1 Friable sandstone strata (C4) interspersed into more resistant sandstones with "pits" formed by internal erosion (*piping*). (Photo E.L. Pastore)

5.3 Brazilian Dam Foundations on Sandstone

The Brazilian expertise in building of dams in sandstone foundation presented in this chapter is based on innumerous papers, specially that published by Pastore et al. (2015).

Four large dams and about 32 small hydroelectric plants were built on sandstone foundations in Brazil. Among the small hydro plants, 12 are concrete dams and 6 earth dams on sandstone foundation. Eight of concrete structures have foundations on sandstone rock or soft sandstone rock (less resistant sandstones) and four on sandy soil (sand).

Fig. 5.2 "Pits" formed by the phenomenon of *piping* in friable sandstone (C4). (Photo E.L. Pastore)

5.3.1 Structures on Sandstone Rock Foundations (C1 and C2)

The Porto Primavera navigation lock (Maranesi et al. 1983) and Dona Francisca dam (Coulon and Coelho 1983) structures are of importance due to the amount of investigations made on the behavior and properties of the sandstones. The Porto Primavera lock studies on sandstones were published by Maranesi et al. (1983). Studies regarding the geological and geotechnical aspects, related to the geotechnical behavior on the slopes, were published by Nogueira Jr et al. (1983), and the geomechanical characteristics by Ré et al. (1983).

At the Porto Primavera lock foundation, the following types of sandstone occur: C1 to C2 in the upstream and downstream channels; types C3 to C4 sandstone intercepted with C1 to C2 sandstone in the lock chamber and downstream area of the upstream channel.

In Dona Francisca power plant, concrete structure foundations are comprised of sandstone and siltstone and mudstone interspersed levels. In feasibility phase, the designer was concerned with the structures stability; but during the studies of the bidding design, it was found that, in fact, the geological concerns were referred to the siltstone/mudstone levels due to their low shear strength.

5.3.2 Structures on Sand Soil Foundation (C4 Sandstone)

In concrete structures and earth dams built on sandy soils, geotechnical laboratory and in situ geologic investigations, and performance, must be of concern of designers and owners. A good example in Brazil was recorded in the Curuá-Una dam concrete structures' foundation where site investigation included infiltration tests in boreholes and laboratory triaxial tests. The Porto Primavera and Pedra Redonda hydro projects, are also good examples, where the campaigns were more elaborate, even including liquefaction tests for the study of the behavior of sands in earth dam foundations (Cruz 1996).

5.4 Control Devices

There are innumerous flow control devices for dam foundations that can be used to control the flow rates and *piping* occurrence. The selection of the control device type will depend on the nature of the foundation, the water head, cost, and the construction schedule. The flow control system including sealing and drainage devices for each case must be designed and dimensioned based on seepage analysis (Finite Elements Analysis) which allows for the evaluation of the efficiency of possible solutions. For that, it is essential that the modeling should be based on in situ research data and reliable tests.

5.4.1 Sealing Control Devices

Sealing devices are intended to prevent all or part of the seepage/leakage through foundation, in this case, by reducing the flows and hydraulic gradients to acceptable levels.

- *Concrete structures and earth or rockfill dam foundations on coherent sandstone (C1/C2).*

Seepage/leakage occurs through fractures and bedding planes, and a classical grouting curtain can be designed as sealing control, as in these features the grout penetrates when applied under pressure (Fig. 5.3).

- *Concrete structures and earth or rockfill dam foundations on sandstone (C1/C2) with interception of slim or thick layer friable sandstone (C4).*

Seepage occurs through the sandstone fractures C1/C2 and through the layers of friable sandstone. The use of grouting alternative only reaches the fractures of C1/C2 sandstones and cannot penetrate the pores of the friable sandstone.

Foundation sealing alternatives are designed to intercept the layers of friable sandstone, which can be done with key shear or trenches filled with soil (in the case

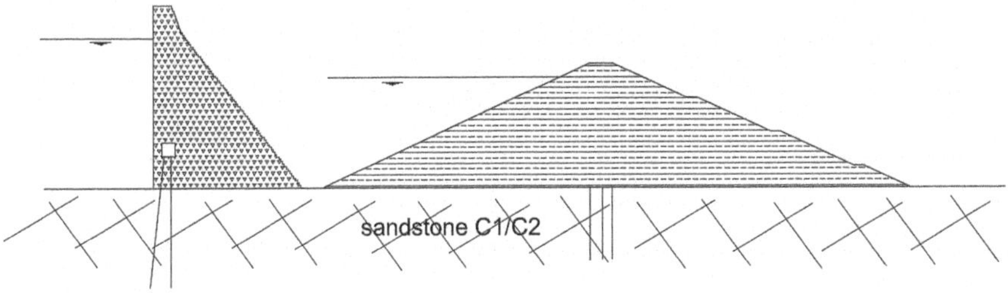

Fig. 5.3 Grouting curtain in coherent sandstone. (Figure E.L. Pastore)

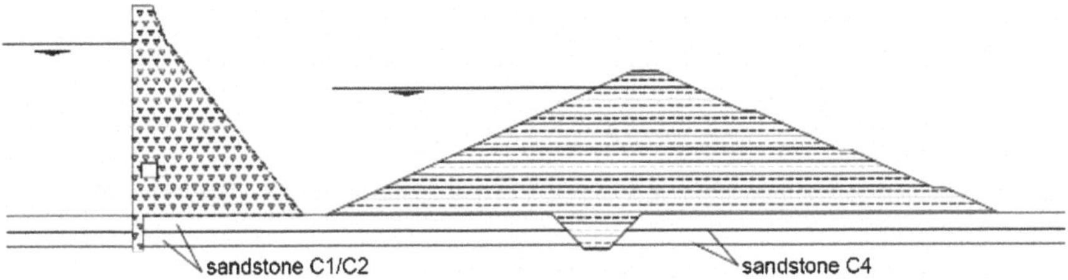

Fig. 5.4 Key shear and trenches in C1/C2 sandstone interbedded with C4 friable sandstones. (Figure E.L. Pastore)

of dams) and concrete (structures) (Fig. 5.4). In deep permeable features, grouting curtains can be used, but as we have seen they are only efficient in fractures of C1/C2 sandstone. If the occurrence of these features also extends to dam abutment foundations, a more radical solution may be required to prevent flow water throughout downstream areas and abutment downstream dam zones. In this case, *Jet-grouting* or even a plastic diaphragm alternative inside abutment foundation may be required.

- *Earth and rockfill dam foundations consisting of thick friable sandstone layer over coherent sandstone.*

Cutoff at a depth of 3–4 m or even reaching the coherent sandstone, combined with an upstream "impervious blanket" may be sufficient. To ensure a good *performance* of the soil blanket, the author recommends applying a high-density polyethylene (HDPE) geomembrane—type blanket over the soil layers (Fig. 5.5).

- *Earth and rockfill dam foundations on medium thickness friable sandstone layer over coherent sandstone.*

Cutoff combined with sealing trench may be sufficient, provided that the trench bottom achieves the coherent sandstone. As in other cases, this treatment should be completed with a downstream drainage system, in the author's opinion (Fig. 5.6).

- *Earth or rockfill small dam foundations on large thickness friable sandstone layer over coherent sandstone.*

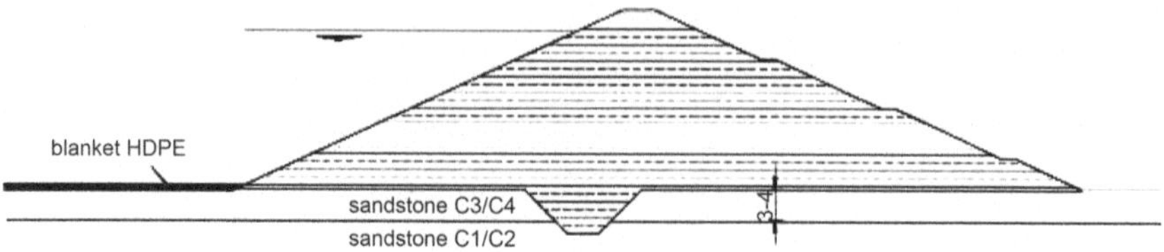

Fig. 5.5 Cutoff and HDPE type blanket (Figure E.L. Pastore)

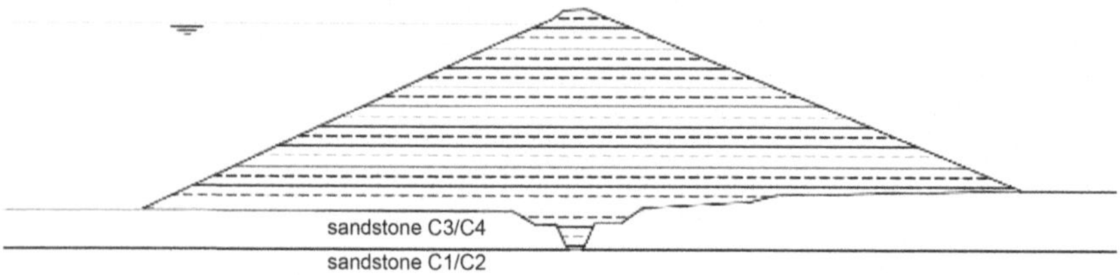

Fig. 5.6 Foundation sealing with trench and cutoff combined (Figure E.L. Pastore)

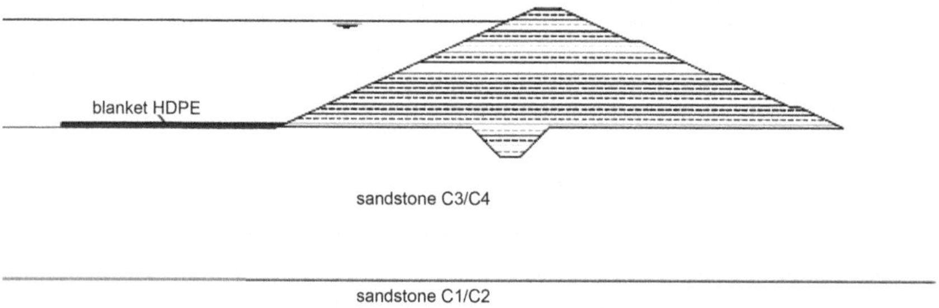

Fig. 5.7 Partial cutoff combined with upstream impervious sealing blanket. (Figure E.L. Pastore)

Partial sealing trench, even not reaching the impermeable rock, together with an "impervious upstream blanket" may be enough for flow control (Fig. 5.7). In this case, the geologic abutment conditions should be deeply investigated.

- *Earth and rockfill large dam foundations on thickness friable sandstone layer over coherent sandstone.*

A flow control sealing trench alternative may require deep and expensive excavations to reach the coherent sandstone. A combination of a partial trapezoidal trench with a vertical trench (*cut-off* type) filled with soil or concrete, reaching the coherent sandstone may be feasible (Fig. 5.8a). The flow control treatment at abutment zones may require a similar solution. The *Jet-grouting* solution aims to build a "wall" formed by aligned adjacent columns, or two rows for best sealing. The *Jet-grouting* alternative must be concerned to related columns alignment and effective flow control. This depends on the penetration and removal of the friable sandstone layers. In these cases, the plastic diaphragm wall is more guaranteed, for flow and leakage control, in the author's view (Fig. 5.8b).

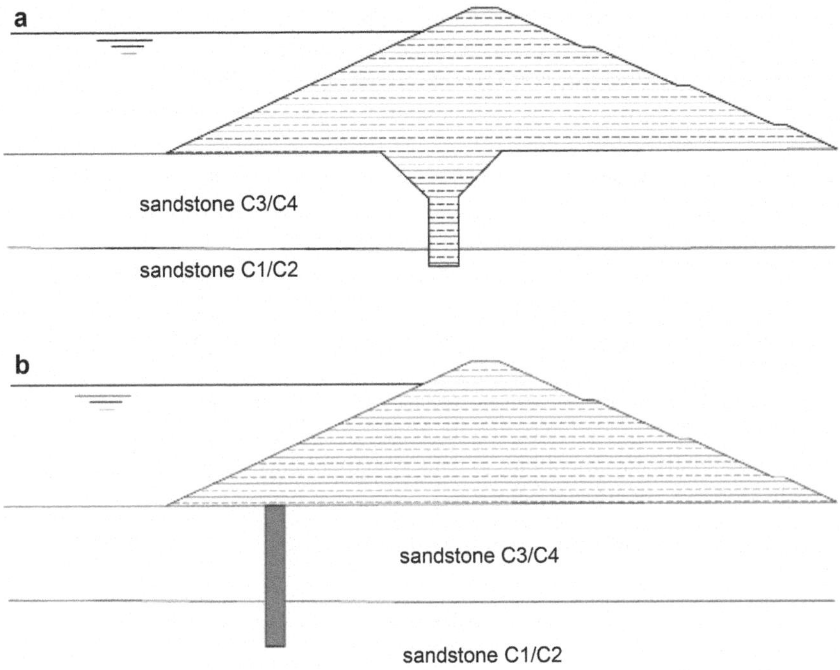

Fig. 5.8 (**a**) Combination of partial trapezoidal and vertical trenches (Figure E.L. Pastore). (**b**) Jet grouting or plastic diaphragm wall (Figure E.L. Pastore)

5.4.2 Drainage Control

The effective leakage and gradients' flow control by the sealing system foundation structures are feasible by combining both the *sealing* with the *drainage* systems, only. Drainage has the function of conducting the seepage/leakage water throughout the foundation of the structures of the dam and abutments safety to downstream areas.

Supplementing the sealing devices mentioned in the previous chapter, the drainage project, including vertical filters, horizontal drains, relief wells, drainage trenches, horizontal drains, and so on, are important protection solutions in the flow control of sandstone foundations.

These sealing and drainage solutions require deep field investigations and local inspections of the foundation and abutment zones often still less known in the design phase. Occurrence of "pits" mentioned in item 2 of this paper may be masked by layers of soil and vegetation, and they are found out when the soil and vegetation are removed during the construction, only.

Inner erosion and *piping* are responsible for most accidents and dam breaks. In sandstones foundation, sealing *cutoffs,* internal drainage systems; relief drains, and inverted filters of granular materials on downstream areas, to reduce the foundation and exit gradients, are applicable solutions which have guaranteed the safety in the post-filling phase.

The design and selection of granular materials of the internal drainage system are part of the protection measures, which cannot be minimized by the designer. Among these concepts, the filter criterion Terzaghi–Bertram (Cruz 1996) consider the retention (and protection from progressive erosion) and permeability. In the selection of materials to be applied in drains, filters, and transition zones, other considerations must be met regarding particle size and potential of segregation, non-cohesive materials, and so on.

The criterion of limiting the maximum diameter (D_{mx}) of the grain or particle to 7.5 cm (3 in.) in order to limit the segregation during handling and spreading operations (sandwich filters, transition soil core, and riprap), should be considered at the design stage and segregation of filter materials must be avoided during the construction.

Well graded materials, with uniformity coefficient ($C_u = D_{60}/D_{10}$) ≥ 25 are not recommended. Well graded materials such as gravel or crushed rock largely graded become potentially incompatible, and in noncompliance with the filter criterion Terzaghi-Bertram, mentioned below:

- To prevent *piping* and protection from the sandstone particle entrainment:

$$\frac{F_{15}}{B_{85}} < 5$$

where B_{85} is the size at which 85% of the total base material (B) particles are smaller and F_{15} the size at which 15% of the total filter material (F) particles are smaller.

Cohesive materials are generally less susceptible to *piping*. Cruz (1996) recommends the value may be above 5, in case there is experience with the material under study, and special filtering tests are performed.

- To meet the drainage condition, one must meet the following condition:

$$\frac{F_{15}}{B_{15}} > 5$$

where B_{15} is the size at which 15% of the total base material (B) particles are smaller.

Both expressions are well known in dam engineering. In addition, drainage sand selected, limiting the percentage of fines <#200 (0.0074 mm) to less than 3% is highly recommended, due to the substantial effect of fine material in drain material permeability.

In steep valleys' abutments, it is most recommended that transition zones should be enlarged in view of occurrence of shear zone and cracks and arching effects in these areas. This recommendation is extended in regions subject to seismic phenomena. One potential phenomenon in sandstones, as well as alluvial soils, is the formation of holes such as those shown in Figs. 5.9 and 5.10, which later with increasing leakage leads to formation of *sinkholes* and the intense inner erosion and failure, if there is no immediate intervention.

Fig. 5.9 Interbedded weathering silt material in sound sandstone foundation (Photo E.L. Pastore)

Fig. 5.10 Erosions in sandstones foundation due to seepage (Photo E.L. Pastore)

Sherard (1979) reports the occurrence of *sinkholes* after grouting services in the earth dam foundations of glacial formation or with similar properties, such as well graded particle size with a maximum diameter (D_{mx}) of gravel or crushed stones, and silts and clays on the "fine side" of the grain size curve.

De Mello (1975) addresses the issue of internal erosion and stability for gap-graded soils and proposes a method, named *Retention Ratio (RR) criterion*. In this method, the grain size curve of the filter material is split into two grain size curves for the portions finer and coarser, respectively. For the two gradation curves RR is calculated from the Terzaghi filter criterion: $RR = F_{15}/B_{85}$.

Sherard (1979) applied the method on a glacial soil (which may also result from alluvial and colluvial soil mixtures or alluvium), divided the soil particle size into two portions at fraction of 1.0 mm. Thus, obtaining two curves: a fine fraction (<1.0 mm), where the value is D_{85} and, a coarse fraction (>1.0 mm) where the obtained value is $D_{15.}$

The example cited by Sherard (1979) showed the noncompliance of the soil with the Filter Criteria, according to the $D_{85/}D_{15}$ < 4–5 analytical method. In practice, if both soils were placed adjacently, the thinner portion could be eroded through the voids of the coarser fraction during the flow, or in an extreme case, the erosion of the fine material through the voids could initiate a process of internal erosion.

In sandstone foundation, where the execution of grouting curtain along the axis under the *cutoff* and on the abutments often shows to be ineffective, *sinkholes* in the downstream foundation zones can occur if a protection "barrier" (diaphragm wall or concrete wall) is not executed.

One of the reasons for the noneffectiveness of grouting curtains is the dissolution of cementation of sandstone by the water flow. As the cementation of sandstone is lost due to chemical reactions between water and the "cement", the sandstone becomes fine sand and easily carried off, if there is a free downstream exit. The erosion process occurs back to front, that is, from downstream to upstream. This phenomenon occurred in a dam in the state of Ceará, Northeast of Brazil, at the spillway foundation, at the discharge channel area and one of the abutments, several years after the end of dam construction (Souza 2013).

5.5 Sandstone Behavior in Tunnels

In Brazil, there are few cases of tunnels built in sandstones mass. More recently in Ceará Water Belt, the following water supply tunnels were excavated in sandstones: Veneza tunnel, 2322 m lengths, Arajara tunnel, 300 m length, Sítio Alto 1 tunnel, 480 m length and Sítio Alto 2 tunnel, 620 m length. The predominant Geological Formations are Missão Velha and Mauriti sandstones. There are no more details recorded about these cases.

A pressurized tunnel about 8000 m length was excavated in metasandstones (C1/C2) with intercalated layers of friable sandstone (C4) in Serra das Agulhas Small Power Plant, Minas Gerais State. Rock burst phenomenon occurred in some stretch where rock cover attempts 250 m, with an estimated horizontal/vertical stress ratio $K = 4$.

The following case (Pastore 2012) concerns a railroad tunnel belonging to TransNordestina Railroad built in a soft sandstone with some intercalated siltstone layers.

5.5.1 Tunnel Location and Geometry

The tunnel is located in the State of Ceará where the railroad passes under the CE-293 highway, near Brejo Santo town. It has 60 m extension, 10 m height, and width of 10.60 and 7.60 m, with maximum coverage of 19 m showing an oval section as defined in the project. Figure 5.11 shows the geometric section of the tunnel while Fig. 5.12 shows the same already concluded.

5.5.2 Geomechanical Characterization of the Sandstone Rock Mass

Geomechanical sandstone rock mass characterization was made based on the examination of the core samples of four rotary drills carried out on site and rocky exposures in the cut slopes of the railroad near the designed tunnel.

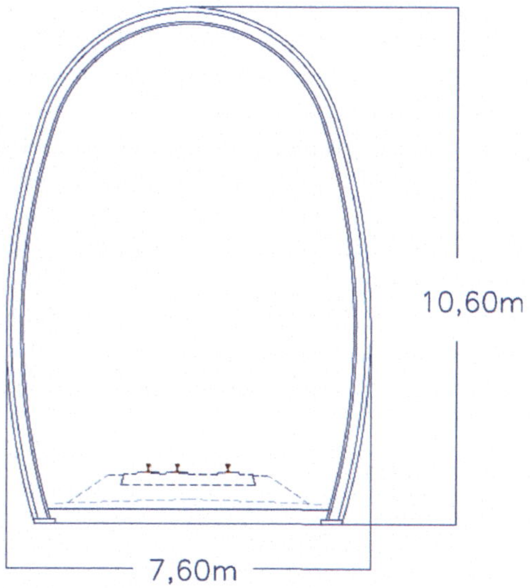

Fig. 5.11 Tunnel geometric section. (Figure E.L. Pastore)

Fig. 5.12 Railroad tunnel view. (Photo E.L. Pastore)

The mass is composed of reddish sandstones with whitish grey siltstones interbedded layers approximately above the elevation 387.00 m and predominance of reddish sandstones in greater depth. According to the geological map of the State of Ceará (CPRM 1982), this package of sediments belongs to the Santana Formation of Araripe Group (Lower Cretaceous).

The sandstones are incoherent (C3/C4) closer to the surface of the ground and more coherent (C2) in greater depth. A layer of friable sandstone (C4) with thickness of 0.3 m was detected in SM 02, at 18.90–19.20 m depth. The sandstone composition is predominantly medium-size, yellow-reddish, with grains composed almost exclusively by quartz.

Subvertical fracture systems have been identified in cutting slope which, along with horizontal sedimentary plans showed great potential for formation of unstable rock blocks, especially on the roof of the tunnel. This fact, however, was not consummated in excavations. The water level was not detected in exploratory surveys and excavations.

The geomechanical sandstone mass properties were estimated through the GSI—Geological Strength Index (Hoek 1994) and assumed the following input data to obtain the shear strength envelope and deformation modulus:

H (depth) = 19 m
$\gamma = 20$ KN/m^3
GSI = 40
σ_c sandstone = 20 MPa

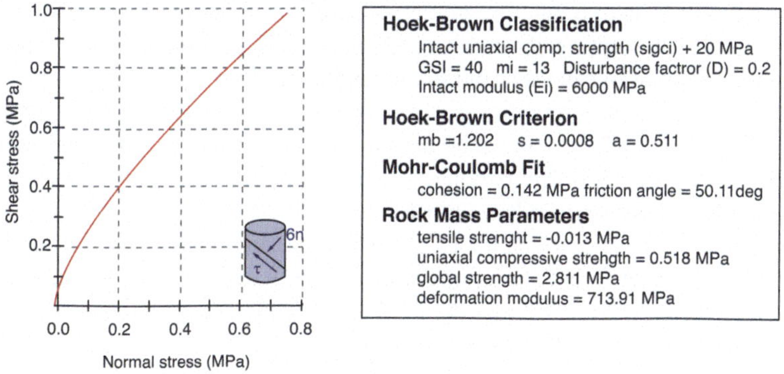

Fig. 5.13 Sandstone rock mass non-linear Hoek and Brown envelope and deformation modulus

E_i sandstone = 6000 MPa
mi and D = 13 and 0.2, respectively

The shear strength envelope was obtained from RocLab software (Fig. 5.13).

The following geomechanical parameters of rock mass were considered in the definition of the best excavation geometry in terms of stress distribution and in tunnel system support design:

mb = 1.202, s = 0.0008 and a = 0.511
E_m ~ 700 MPa (deformation modulus of rock mass)
$\nu = 0.25$

5.5.3 *Definition of the Section Geometry and Design of Support System*

The Phase 6 RocScience software was used for the definition of the section geometry and design of support system, with the following contour conditions:

D = 0.2 (blasting rock mass damage parameter)
K = 0.5
Water level: nonexistent
Height from the floor to the terrain surface: 28 m
Loading stages: 0.2 (first stage), 0.2 (second stage), 0.1 (third stage), 0.5 (last stage). Percent of the total load applied in each stage of excavation
Section geometry: oval
Excavation sequence: top heading and bench
Excavation method: NATM—D&B

The support system applied in numerical analysis consisted essentially of steel sets and shotcrete reinforced with steel fibers in various combinations with variations in the spacing of steel sets and in the thickness of the shotcrete.

The ultimate support system was composed of steel sets spaced every 0.80 m in the whole section of tunnel and shotcrete reinforced with metal fibers in the thickness of 40 cm to be applied in two layers: a primary with 25 cm thick together with the steel sets and other secondary with 15 cm thick.

Figures 5.14, 5.15, 5.16, and 5.17 present the results obtained in terms of displacements in the different stages of excavation with and without the applied support system while Table 5.3 shows the results of displacements, the number of yield elements and the average factors of safety.

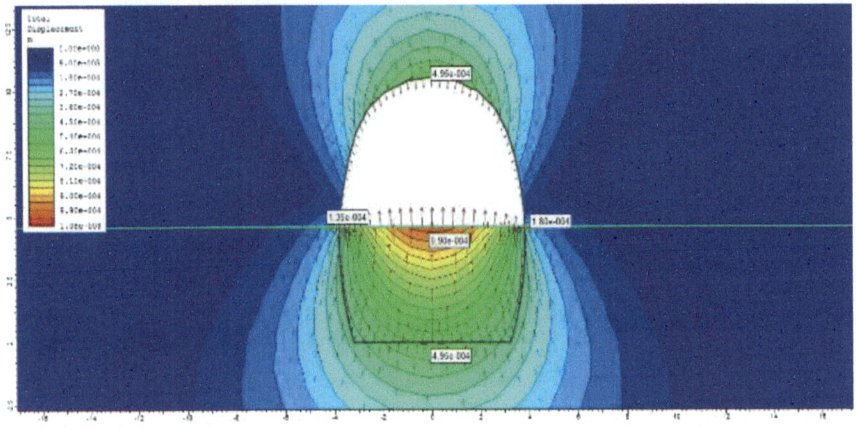

Fig. 5.14 First stage: top heading excavation: displacements

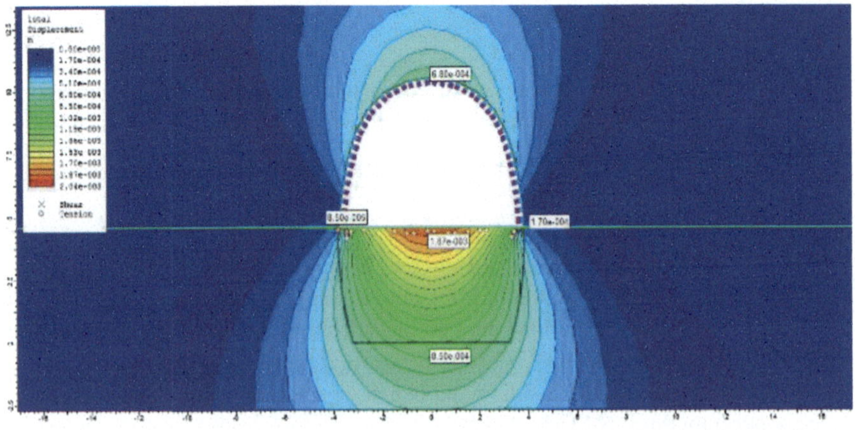

Fig. 5.15 Second stage: top heading with steel sets and shotcrete applied: displacements

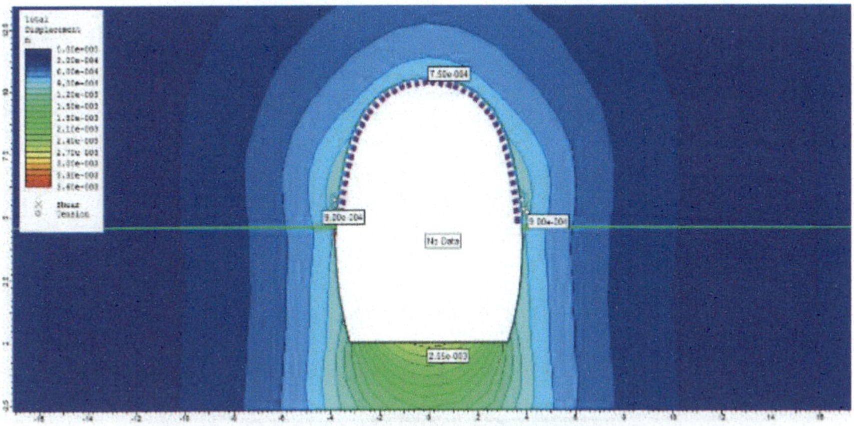

Fig. 5.16 Third stage: bench excavation: displacements

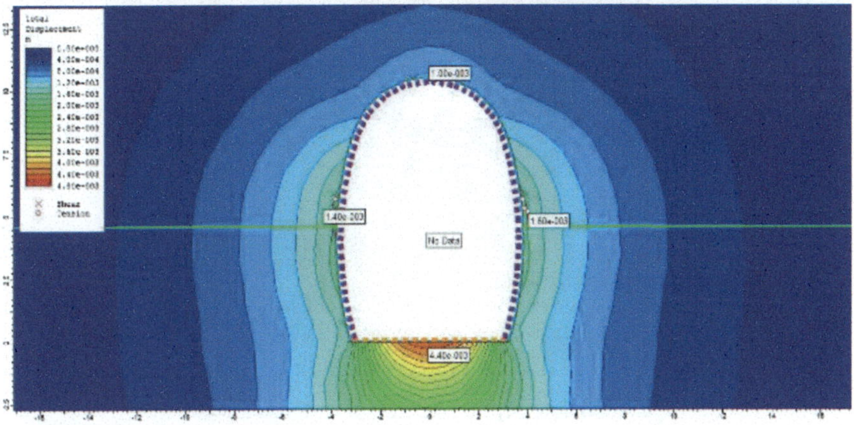

Fig. 5.17 Fourth stage: bench with support applied: displacements

It appears from these results that the support system agrees with the criterion of the executive design which established safety factor >1.3 at the end of construction.

The excavation was carried out carefully, with smooth blasting, and advances of 1.5–2.0 m maximum, in function of the incertitude of the knowledge of the geomechanical behavior of the rock mass in the excavation, due mainly to its diversity and unpredictability of pockets and sand lenses.

During the excavation of the bench, due to the best rock mass geomechanical behavior in relation to the admitted in numerical analysis, and with the ends of the steel sets supported on the walls, the prolongation of these elements was eliminated. So in the bench walls, a 20 cm layer of shotcrete without fibers + steel grid + 20 cm of shotcrete with fibers was applied (Fig. 5.18).

Table 5.3 Results of numerical analyses

Condition	Number of yield elements				Maximum displacement (mm)				Safety factor			
	LW	RW	Roof	Floor	LW	RW	Roof	Floor	LW	RW	Roof	Floor
Top heading without support	4	4	4	4	0.13	0.18	0.49	0.99	1.6	1.6	1.8	2.6
Top heading with support	18	18	18	18	0.68	0.08	0.17	1.87	1.8	2.1	5.0	2.3
Bench without support	17	17	17	17	0.90	0.90	0.75	2.55	1.0	1.0	5.2	1.3
Bench with support	19	19	19	19	1.4	1.6	1.0	4.4	1.3	1.3	4.4	1.3

Notes: *LW* left wall, *RW* right wall

Fig. 5.18 Application of shotcrete and steel grid in bench walls (Photo E.L. Pastore)

5.6 Conclusions

Less cemented, fragile, and low resistant features as well as permeable sand pockets forming preferred paths for water flow and internal erosion can frequently occur in sandstone rock mass. The experience of recent decades shows that the location and dimensions of such features are only possible when the features are exposed, mapped, measured, and tested. It is the only way in which the necessary treatment can be properly specified and executed.

The geologic investigation program should use triple barrels in the surveys to recover 100% of the drilled material and optical profiling of the bores. In addition, the participation of skilled professionals in the description and classification of the drill cores, based primarily on the degree of coherence of the material, is fundamental for the interpretation and preparation of sections and geological models. All these procedures, unfortunately, are not being always followed, which contributes to the increase in the occurrence of geological risk and collapse of structures in sandstones.

References

Coulon FK, Coelho DGW (1983) Strength parameters of sandstones at Dona Francisca dam. Simpósio Sul Brasileiro de Geologia, 1, Porto Alegre (in Portuguese)

CPRM (1982) Geologic map of *Ceará* State. Scale 1:500.000 (in Portuguese)

Cruz PT (1996) *100 Brazilian Dams*. Oficina de Textos, São Paulo, 648p (in Portuguese)

De Mello V (1975) Some lessons from unsuspected, real and fictitious problems in Earth dams engineering in Brazil. In: Sixth conference for Africa on soil mechanics & foundation engineering, South Africa, pp 285–304

Guidicini G, Oliveira AMS, Camargo FP, Kanji M (1972) A method for the preliminary classification of rock massifs. Week of Applied Geology, 4, Theme 3, São Paulo, pp 275–282 (in Portuguese)

Hoek E (1994) Strength of rock and rock masses. ISRM News J 2(2):4–16

ISRM (1978) Suggested methods for the quantitative description of discontinuities in rock masses. Int J Rock Mech Min Sci Geomech Abstr 15:319–368

Kanji M (2014) Engineering works affected by soft rocks. In: ISRM specialized conference CBMR/ABMS/ISRM, Goiania

Kanji N, Vasconcelos ML, Guedes MG (1981) Methodological aspects of geological and geotechnical investigations of Bauru sandstones. In: Brazilian congress of engineering geology, Itapema, 3, vol 2, pp 257–270 (in Portuguese)

Koshima A, Frota R, Lozano MH, Hoenisch JC (1983a) Behavior and geomechanical properties of the Bauru sandstone. In: Symposium on geotechnics of the Upper Paraná Basin, São Paulo, vol IIB, pp 173–192 (in Portuguese)

Koshima A, Vincenzo M, Rezende MA, Bertolucci JC (1983b) Constructive aspects of a channel in soft rock – Bauru sandstone. In: Symposium on geotechnics of the Upper Paraná Basin, São Paulo, vol IIB, pp 193–210 (in Portuguese)

Maranesi DA, Araújo JS, Marques JD, Oliveira CA, Serra Jr E, Vasconcellos ML (1983) Geological and geotechnical aspects of sandstone in the Caiuá Formation in the foundation of the sluice gates and Porto Primavera Dam. In: Symposium of the Upper Paraná Basin, Theme IIB, São Paulo, pp 211–224 (in Portuguese)

Nogueira Jr J, Vasconcelos ML, Tagliavini R (1983) Geotechnical behavior of sandstone Caiuá in slope excavation of the slice gates of Porto Primavera. In: Symposium on geotechnics of the Upper Paraná Basin, São Paulo, vol IIB, pp 243–260 (in Portuguese)

Oliveira J, Caruso LG (1979) Technological characterization of Sedimentary Rock Cemented by iron oxides. In: Regional symposium of geology, Rio Claro, 2, vol 2, pp 23–28 (in Portuguese)

Pastore EL (2012) Design and construction of a railway tunnel in sandstones. In: Third Brazilian congress of tunnels and underground structures. CBT/ABMS. São Paulo (in Portuguese)

Pastore EL, Cruz PT, Freitas MS (2015) Dam foundations and structures in sandstones: nature of massifs and control of flow and erosions. In: XXX Brazilian seminar of >large dams, Foz de Iguaçu (in Portuguese)

Ré G, Taniguchi H, Amaral A (1983) Geomechanical characteristics of the Caiuá sandstone. In: Symposium about Upper Paraná Basin. ABGE/ABMS/CBMR, São Paulo (in Portuguese)
Sherard J (1979) Sinkholes in dams of coarse, broadly graded soils. In: International conference on large dams, 13, India
Souza L (2013) Assessment of behavior of the foundation of a dam in sandstone rock – case study of the Jaburu I dam, Fortaleza (in Portuguese)

Chapter 6
Geomechanical Characterization of Evaporitic Rocks

Mauricio Giambastiani

6.1 Introduction

In general, evaporites are mono-mineral rocks formed by minerals of density <2.3 g/cm^3, Mohs hardness <3, high solubility and easy recrystallization under normal temperature and pressure conditions, low thermal conductivity, low strength, high deformability, and time-dependent behavior (creep, expansion, etc.).

These characteristics strongly influence the behavior of rock masses, formed totally or partially by these rocks, such as subsidence by dissolution, time-dependent deformations in underground works, collapse, sinkhole, and caverns in the subsoil, metal and concrete corrosion, etc.

Due to their particular properties, in the 1980s great efforts were made worldwide to study evaporites as underground repositories of atomic wastes. It was within the framework of this effort that the following conferences were organized: The Mechanical Behavior of Rock Salt in 1981 (Pennsylvania State University—USA) and in 1984 (Federal Institute for Geosciences and Natural Resources—Germany). These meetings were repeated in France (1993), Canada (1996), Romania (1999), culminating with the VII Conference in France (2012).

Research projects were carried out in the field of sulphate rocks (gypsum and anhydrite) by Madsen and Nüesch from ETH (Switzerland) between 1990 and 2000, Giambastiani and Pehovaz Alvarez at the School of Engineering of São Carlos (Brazil) between 2000 and 2010 and those of Alonso, Oldecop, Berdugo, and others from the Polytechnic University of Catalonia (Spain) between 2000 and 2012, and most notably the Technical Workshop on Tunnels in Saline and Expansive rock masses held in Barcelona in 2011.

M. Giambastiani (✉)
Universidad Nacional de La Rioja (UNLaR), La Rioja, Argentina

M. Kanji et al. (eds.), *Soft Rock Mechanics and Engineering*,
https://doi.org/10.1007/978-3-030-29477-9_6

6.2 Geological Aspect of Evaporitic Rocks

Evaporites and their derived rocks are sedimentary deposits of chemical origin characterized by very fast accumulation processes due to the evaporation of saturated waters into salts. The accumulation process requires that the evaporation rate exceeds the rate of water inputs (Spalletti 2017). These rocks can accumulate in marine, marginal, and continental environments. Most have been formed in marine and marginal marine environments and are particularly common in the Middle Cambrian, Permian, Jurassic, and Miocene.

The most common minerals that make up evaporites are halite, anhydrite, and gypsum, but numerous other minerals may appear (Table 6.1).

As evaporation progresses and brine is concentrated, a more or less orderly precipitation of salts follows the sequence: gypsum-anhydrite, halite and potassium salts, and magnesium (sylvite, carnallite, polyhalite, and kieserite). This phenomenon is cyclical and very fast as it can be observed in the stratigraphic columns. To accumulate 4 m of salts requires the evaporation of 300 m of sea water (Spalletti 2017).

The most important sedimentary structures are cross-stratification, ondules and graded. Massive, nodular (chicken wire), laminar (halite-gypsum, halite-anhydrite, salt-clay), enterolytic and drying cracks interrupting lamination are also very common in gypsum and anhydrite deposits.

The diagenesis of evaporites produces changes in mineralogy and structures. One of the most significant processes of diagenesis and exhumation of evaporitic deposits is the anhydrite-gypsum transformation. A large number of gypsiferous surface formations come from the in situ replacement, by hydration of anhydrous subsurface formations (Orti Cabo 2010). The hydration of the anhydritic rock is initiated through the greater permeability layers of the rock, such as stratification, cracks, fractures, and surfaces of nodules, and progresses towards the interior of the anhydrous rock. Several hydration systems are developed:

(a) "By veins," which cross the anhydrous rock in all directions;

Table 6.1 Evaporitic minerals

Minerals		Composition
Chlorides	Halite	$NaCl$
	Silvite	KCl
	Carnallite	$CaMgCl_2{\cdot}6H_2O$
Sulphates	Anhydrite	$CaSO_4$
	Gypsum	$CaSO_4{\cdot}2H_2O$
	Polihalite	$K_2MgCa_2(SO_4)_4{\cdot}2H_2O$
	Kieserite	$MgSO_4{\cdot}H_2O$
	Epsomite	$MgSO_4{\cdot}7H_2O$
Carbonates	Calcite	$CaCO_3$
	Dolomite	$MgCaCO_3$
Borates	Borax	$N_2B_4O_7$

(b) "Concentric," as the hydration fronts move in a centripetal and progressive manner towards the interior of the more impermeable anhydrous masses,
(c) "Massive," in which the gypsum progresses as a front, without leaving structures of its passage through the anhydrous rock.

An important aspect of the anhydrite hydration in the different varieties of secondary gypsum is that the geological process occurs without any modification of the lithofacies indicating that the process usually operates in an isovolumetric mode. That is to say, without swelling or deforming the rock, expelling the generated excess of calcium sulphate solution (Orti Cabo 2010).

Another of the most notable geological aspects in the evolution of evaporites, and which has a great impact on their geomechanical properties, is the diapirism or halokinesis. The term halokinesis refers to the spontaneous movement of a saline mass (typically sodium chloride), which is usually induced by the gravitational instability of the mass itself when it supports a thick coverage of denser sediments. Saline Tectonic refers to saline masses movements driven by structural causes, generally tangential or tensional stresses. The term saline diapirism is reserved for the penetration and rupture of the sedimentary cover phenomenon by ascending saline masses (saline intrusion and eventual extrusion).

The halokinetic behavior of the salt is based on its low density and ease of creep (plastic behavior). Other properties of the salt, such as thermal conductivity, water content, deformability of the crystalline lattice of sodium chloride, and the presence of K-Mg salts accompanying halite, facilitate this process.

The density of the halite mineral is relatively low 2.17 g/cm^3 at room temperature. Contrary to other geological materials, the salt rock loses density in depth as the temperature rises 1.90 g/cm^3 at 801 °C, because of the near absence of porosity in this rock and the expansion of the mineral by the increasing temperature, the volume increase exceeds the shrinkage due to the confining pressure. This behavior of the salt facilitates its ascent by "buoyancy" as, with the progressive burial, the contrast of densities is greater.

Halite and anhydrite have the highest thermal conductivity of sedimentary rocks, which makes saline masses behave as very effective heat conductors, deforming regional thermal gradients and significantly influencing the diagenetic reactions of the host rocks. Small amounts of water (fluid inclusions, intercrystalline water) greatly increase salt flow and creep.

6.3 Index and Geomechanic Properties of Evaporites

There is an intimate dependence between the index properties and the resistance and deformation properties of rocks in general and the evaporites in particular. As rocks are three-phase systems (minerals and pores and microcracks filled or not with water), density, moisture content, degree of saturation, absolute and effective porosity, hardness of the minerals, grain size, microcracks, mineralogy, and so on can influence the strength and deformability of rocks and rock masses.

This chapter will focus on the index and geomechanical properties of the evaporites with an emphasis on sulphate rocks, integrating the experimental results obtained in original research with those found by other authors consulted.

6.3.1 Index Properties

The properties that depend on geological history and that aid in the geotechnical classification of rocky materials, receive the name of index properties. These help to characterize the varied structures, textures, mineralogical components, and associations present in the rocks. These indexes serve to explain certain geomechanical behaviors and are important in some applications (Goodman 1980).

6.3.1.1 Density

Table 6.2 shows the densities of the main evaporite forming minerals (Robertson 1962; Urai et al. 2008).

Table 6.3 shows the density average values of anhydrite and gypsum calculated with the results published by Bell (1981, 1994), Fabre and Dayre (1982), Bilgin (1982), Clark (1966), Rybach (1975), Wheildon et al. (1974), Zierfuss (1969), Bullard and Niblett (1951), Herrin and Clark (1956), Müller and Briegel (1978), Alejano et al. (1999), Howart and Christian-Fear (1997), Pfeifle and Hansen (1998), Giambastiani (2005), Gumusoglu and Ulker (1982), Papadopoulos et al. (1994), Karacan and Yilmaz (2000), Yilmaz and Sendir (2002), Auvray et al. (2004).

Table 6.2 Density of evaporitic minerals

Mineral	Density (g/cm^3)
Halite	2.04–2.16
Gypsum	2.20–2.4
Anhydrite	2.2–2.98
Carnallite	1.57–1.61
Sylvite	1.86–1.99
Polyhalite	2.78–2.79

Table 6.3 Density of anhydrite and gypsum (g/cm^3)

Rock	Average	Stand. dev.	CV%
Anhydrite	2.86	0.09	2.97
Gypsum	2.25	0.12	5.16

Table 6.4 Mohs hardness of evaporitic minerals

Mineral	Mohr hardness
Anhydrite	3–3.5
Gypsum	1.5–2
Thenardite	2.5
Halite	2.5
Sylvite	2
Carnalite	2.5
Borax	2
Polihalite	2.5–3

6.3.1.2 Mohs Hardness of Evaporitic Minerals

The Mohs hardness index arranges the minerals in order according to their resistance to being scratched by different objects, on a scale from 1 to 10. Table 6.4 shows that the main minerals of the evaporites have hardness between 1 and 3.

There is a premise that rocks cannot have greater resistance than that of the major minerals that form them. The low Mohs hardness of the primary minerals of evaporites dictates that their resistance will be from low to intermediate, as it will be seen later in this chapter.

6.3.1.3 Porosity

Porosity is one of the index properties that most influences the mechanical properties of rocks (Goodman 1980). In evaporites the voids correspond to microcracks, presence of clastic matrix and dissolution microcavities. In general, porous rocks are more deformable and less rigid. In porous rocks, pores collapse and nucleation of microcracks are frequent as mechanisms of response to external loads. Table 6.5 shows the average of absolute porosity values of anhydrite and gypsum rocks (Bell 1981, 1994; Fabre and Dayre 1982; Zanbak and Arthur 1986; Müller and Briegel 1978; Giambastiani 2005; Gumusoglu and Ulker 1982; Karacan and Yilmaz 2000; Auvray et al. 2004).

As it can be observed the average absolute porosity of gypsum is greater than that of anhydrite. It is probable that the higher porosity gypsum values are due to the secondary origin of the majority of samples resulting from the exhumation of anhydrite deposits. According to International Association of Engineering Geology (1979) the anhydrites are classified as low porosity and gypsum as low to medium porosity. Figure 6.1 shows the direct relationship between the increase of the uniaxial compression strength and the decrease of the rocks total porosity.

6.3.1.4 Sonic Velocity

The sonic velocity is a very interesting index because it presents an intimate relation with other intrinsic properties of the material such as density, porosity, and presence of fluids. It indirectly measures the integrity of the material and can reasonably

Table 6.5 Absolute porosity of anhydrite and gypsum

Rock	Average	Stand. dev.	CV%
Anhydrite	2.54	1.85	73
Gypsum	5.43	3.31	61

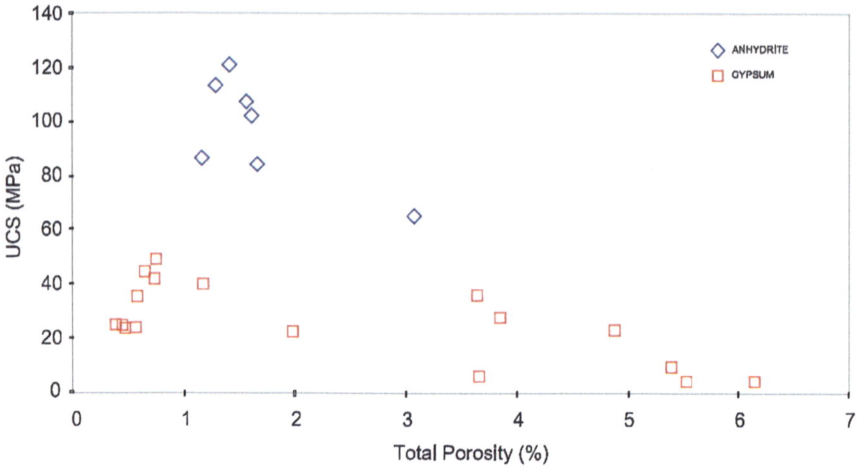

Fig. 6.1 Uniaxial compressive strength vs. total porosity of anhydrites and gypsum rock (Giambastiani 2005)

estimate elastic parameters such as the deformability modulus and the Poisson coefficient for both intact rock and rock masses (Goodman 1980; Barton 2007).

Table 6.6 presents the sonic velocity data collected in the bibliography consulted for sulphatic rocks (Bell 1994; Clark 1966; Schwerdtner et al. 1965; Wheildon et al. 1974; Giambastiani 2005; Karacan and Yilmaz 2000; Auvray et al. 2004).

The higher average sonic velocity of anhydrites (4460 m/s) in relation to gypsum (3720 m/s) is probably due to the fact that the anhydrites are formed by denser minerals and have lower average porosity.

Figures 6.2 and 6.3 show the correlation between sonic velocity vs. density and sonic velocity vs. total porosity of the sulphatic rocks studied by Giambastiani (2005).

It is verified for both rock types, that with decreasing density there is a decrease in sonic velocity (Fig. 6.2) and that there is an inverse relationship between porosity and sonic velocity (Fig. 6.3). Apparently for these rocks the sonic velocity depends more on the crystals packing than on the mineralogy.

Mavko (2017) reports for salt rock a primary velocity (V_p) between 4500 and 5500 m/s, secondary velocity (V_s) between 2500 and 3100 m/s and density between 2.1 and 2.3 g/cm^3. For the anhydrite reports a $V_p = 4000–5500$ m/s, $V_s = 2200–3100$ m/s and density of 2.9–3.0 g/cm^3.

Yan et al. (2014) have made various laboratory measurements on halite salt (Gulf Coast domes) for investigating the temperature and pressure effect on seismic velocities of halite salt (Fig. 6.4). It is observed that the velocities increase with the

Table 6.6 Sonic velocity (m/s) of anhydrite and gypsum

Rock	Average	Stand. dev.	CV%
Anhydrite	4460	920	21
Gypsum	3720	1050	28

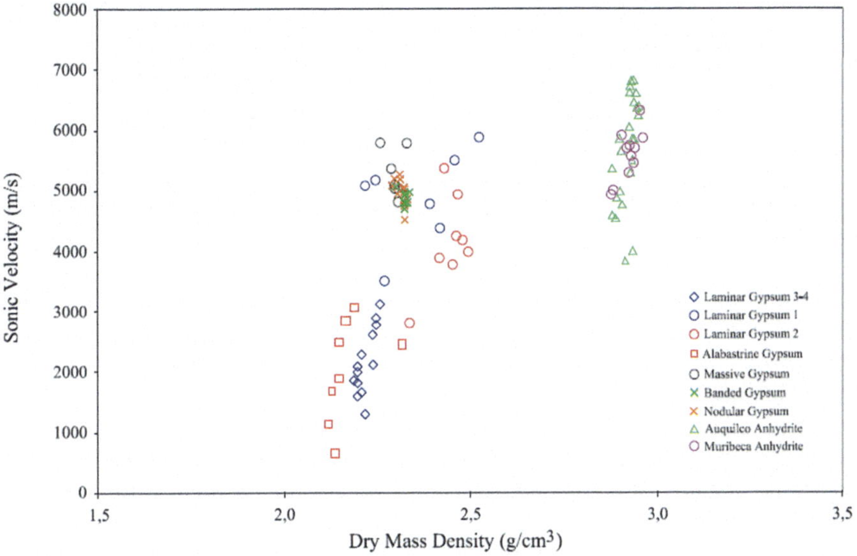

Fig. 6.2 Sonic velocity vs. density of sulphatic rock of Brazil and Argentina (Giambastiani 2005)

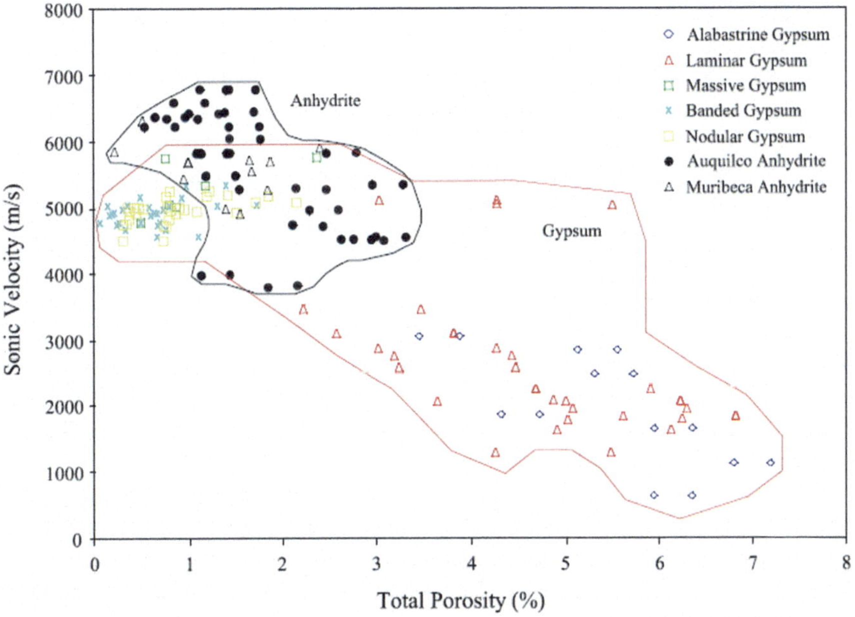

Fig. 6.3 Sonic velocity vs. total porosity sulphatic rock of Brazil and Argentina (Giambastiani 2005)

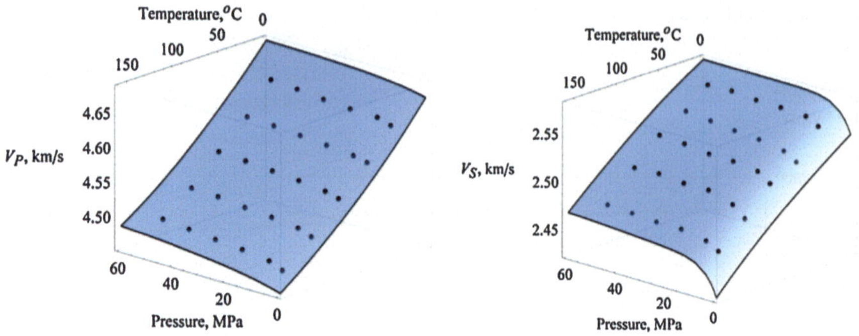

Fig. 6.4 Temperature and pressure effects on P-wave velocity and S wave velocity for halite rock (Yan et al. 2014)

increase of the confining pressure (closing of cracks and pores) at stable temperature and decreases with the increase of the temperature at constant pressure. It is emphasized that the V_p at room temperature is approximately 4650 m/s and $V_s = 2600$ m/s.

Zong et al. (2017) studied samples of halite rocks from Ontario (Canada), Louisiana (USA), and Houston (USA). Table 6.7 summarizes the average values of velocities V_p and V_s and the rocks densities, which are consistent with those presented by other authors.

They also studied the effect of confinement pressure on velocities V_p and V_s, confirming that both increase with increasing confinement (Fig. 6.5).

Tixier and Alger (1970) *apud* Daniels et al. (1980) show the test results on evaporite rocks in the Permian Paradox Member of the Hermosa Formation (EEUU) (Table 6.8).

Daniels et al. (1980) also studied rocks of the same geological formation and concluded that halite rocks have a dry density of approximately 2.05 g/cm^3 and a sonic velocity of approximately 4500 m/s. The anhydrite has a density of 2.95 g/cm^3 and a velocity of about 5000 m/s. Potash minerals (carnallite and sylvite) have very low apparent bulk densities (less than 2.0 g/cm^3) and high velocity (5000 m/s).

6.3.1.5 Point Load Index (Is)

Another great value index to estimate the uniaxial strength of intact rocks is the point load index developed by Broch and Franklin (1972). The Is values for anhydrite and gypsum reported by Bell (1981, 1994) and Papadopoulos et al. (1994) are respectively 3.7 and 1.3 MPa.

The empirical relationship between the uniaxial strength (UCS) and the point load Index (Is) for gypsum tested by Papadopoulos et al. (1994) ranges from 12 to 21, whereas the tests carried out by Bell (1994) indicate that the same ratio varies between 14 and 17 for gypsum and between 26 and 29 for anhydrites. The data

Table 6.7 Velocity and density measured in three sets of salt samples (Zong et al. 2017) with permission of the authors)

Sample location	V_p (m/s)		V_s (m/s)		δ (g/cm^3)
Goderich salt mine, Canada	Direction of symmetric axis	4750	2460		2.15 ± 0.01
	Halfway between two symmetric axes	4440	2920	2470	
Bayoy salt dome, USA	4450 ± 20		2590 ± 30		2.15 ± 0.01
Hockley salt mine, USA	4640 ± 90		2700 ± 50		2.18 ± 0.01

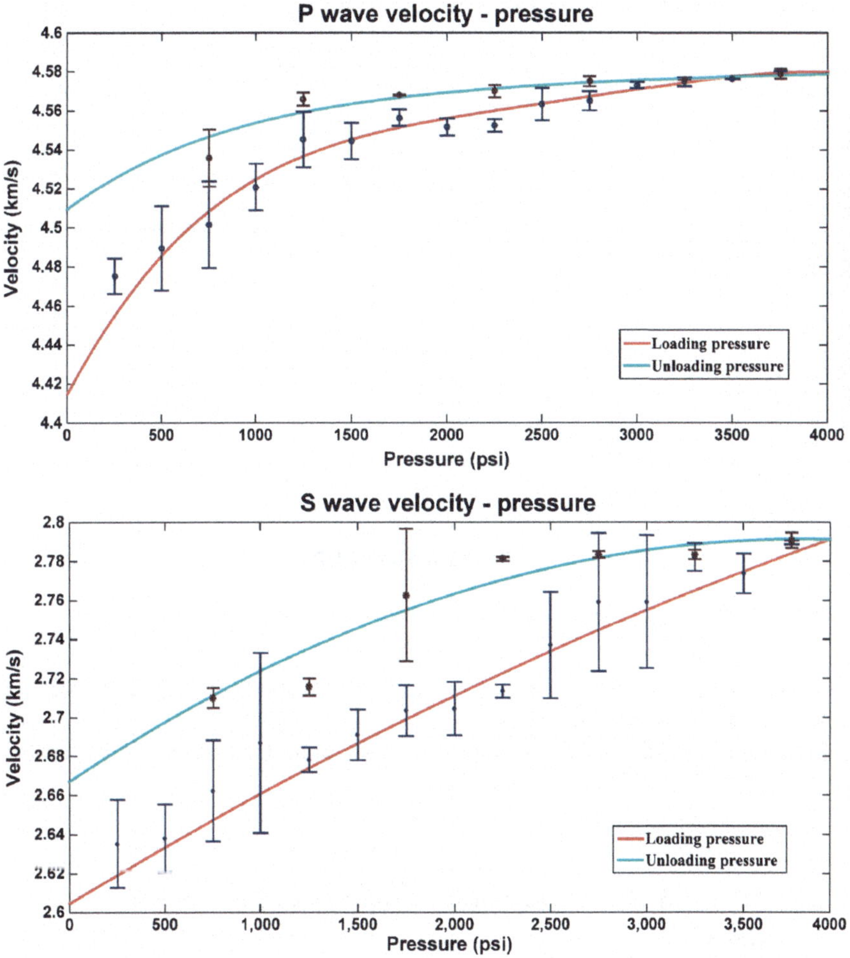

Fig. 6.5 Pressure effects on P-wave velocity and S wave velocity for halite rock (Zong et al. 2017—with permission of the authors)

Table 6.8 Physical properties measured in different evaporites deposits of Paradox member (Tixier and Alger 1970 *apud* Daniels et al. 1980)

Rock	Density (g/cm^3)	Velocity (m/s)	Natural radioactivity	Water content
Halite	2.16	4400–6500	None	Very low
Sylvite	1.99	4600–6500	High	Low
Anhydrite	2.96	4100	None	Very low
Carnallite	1.61	4400–6500	Low	High
Dolomite	2.87	3500–6900	None	Low
Gypsum	2.32	2000–3500	None	Intermediate
Shale	2.2–2.6	2300–4700	High	Interm—high

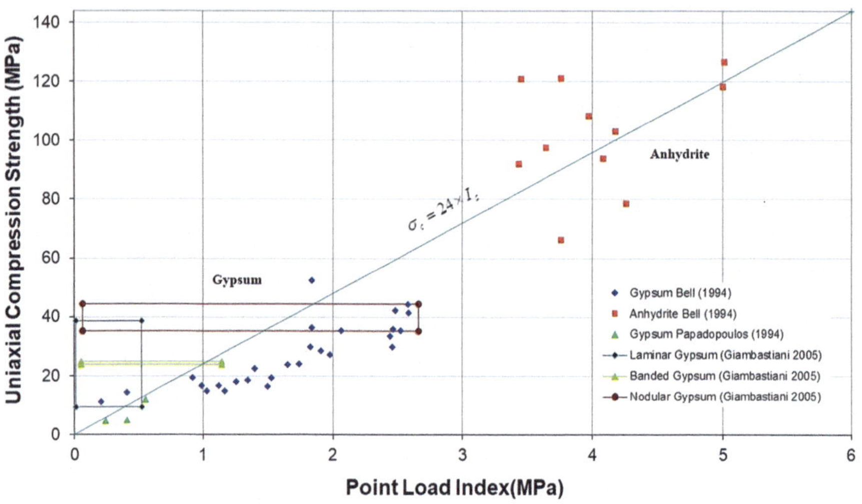

Fig. 6.6 Uniaxial compression strength vs. point load index for sulphatic rocks

published by Papadopoulos et al. (1994), Bell (1994) and Giambastiani (2005) for these rocks are shown in Fig. 6.6.

In literature, no data have been reported on the point load index of other evaporite rocks.

6.3.2 *Strength and Deformation Properties of Evaporitic Rocks*

Evaporitic rocks present very complex behaviors due to the influence of the confining pressure, the temperature and the time. Due to their high plasticity and the ability to deform over time, the complex relationships between dissolution-precipitation phenomena are added.

This complex system is almost impossible to represent with a broad constitutive model that comprehends the complex network of behaviors that evaporites have. Most of the rheological models proposed to describe the thermomechanical behavior of the evaporites are based on the Truesdell and Noll (1965) axiom which states that the constitutive "laws" of a material must provide the history of stress vs. deformation and temperature. This means the execution of innumerable number of laboratory tests, which is clearly impossible to perform and forces specialists to develop approximate constitutive laws. In this way, when adjusting a constitutive law, physical variables measured by the tests (stress, deformation, temperature, and time) and parameters obtained during the data adjustment process must be involved.

Since the properties of the evaporites depend strongly on temperature and time, and the formulation of constitutive models becomes gradually more complex, this chapter will be organized in two sections. The first one dealing with the strength and deformation parameters in the quasi-static conditions and the second one dealing with the time-dependent behavior with the introduction to models of creep behavior of rocks.

6.3.2.1 Strength and Deformation Properties on Quasi-Static Conditions

Conventional geomechanical tests (Brazilian test, uniaxial test, and triaxial test) allow for obtaining a constitutive law, independent of time in quasi-static conditions. This results in a strength envelope, elastic moduli (E and ν), dilatance (ψ), tensile strength (σ_t), uniaxial compression strength (σ_c) and the influence of the confining stress on strength and deformability, and so on.

Rock Salt (Halite Rock)

The salt rock is strongly dependent on mechanical properties and temperature since deformability and work hardening increase and decrease, respectively, with temperature rise. Variations in the confining pressure have relatively little effect upon the yield stress. In contrast to confining pressure, changes in temperature have a marked effect upon the mechanical properties of salt.

Also the stress rate, magnitude of differential stress, grain size, thickness of salt body and % of impurities influence on the mechanical properties.

The plastic deformation of rock salt is related to the translation gliding systems in halite (Prucha 1968). These are functions of the face-centered cubic lattice. Most commonly halite glides by translation on the dodecahedron (110) and cube (001) planes in a [110] direction.

Borchert and Muir (1964) *apud* Prucha (1968) point out that fine-grained rock salt is less easily deformed than coarse-grained rock salt because of the more abundant interlocking and "off-setting" of the glide planes at grain boundaries. Serata and Gloyna (1959) *apud* Prucha (1968) state that fine-grained material flows more

Table 6.9 Strength and deformation properties of salt rock (Halite rocks)

	Brazilian test	Unix. comp. strength	E	E_{din}	G	K		c	
	MPa	MPa	GPa	GPa	GPa	GPa	ν	MPa	ϕ
Average	2.2	28.5	23.9	29.8	10.5	24.4	0.31	5.4	38
Stand. dev.	2.2	8.9	12.6	4.	2.6	5.1	0.05	2.9	7
Min	0.2	13.3	1.5	24.6	6.6	19.3	0.24	2.7	29
Max	7.3	60	55.6	35.2	14.5	37.1	0.37	9	46

readily than coarse-grained material at lower temperatures, but that with increasing temperature the situation is gradually reversed.

Table 6.9 presents the bibliographic compilation of results of short-term rock mechanic tests of rocks formed by halite (Robertson et al. 1958; Robertson 1962; Serata and Gloyna 1959; Gunter and Parker 1961; Fairhurst et al. 1991; Hume and Shakoor 1981; Hansen et al. 1984; Devries et al. 2002; Fuenkajorn and Daemen 1988; Boontongloan 2000; Plookphol 1987; Fuenkajorn and Jandakaew 2003; Phueakphum 2003; Wetchasat 2002; Kenzakoo 2006; Sriapai et al. 2012; Kolano and Flisiak 2013). It is emphasized that the uniaxial strength varies between 13 and 60 MPa (average 28.5 MPa), which places these rocks in the transition zone between soft rocks and intermediate strength rocks. The modulus of deformability varies to 1.5 and 55.6 GPa with an average of 23.9 GPa. The modulus ratio (E/σ_c) varies between 110 and 930 with an average of 840 falling in the high MR field. The cohesion and friction angle are on average 5.4 MPa (from 2.7 to 9 MPa) and 38° (29° to 46°) respectively, under ambient temperature conditions.

The influence of the confinement pressure on the variation of the mechanical parameters of the halite rocks, Hume and Shakoor (1981) emphasize the increase of the density, the modulus of elasticity and the Poisson coefficient with the increase of the confinement (Fig. 6.7).

The effect of temperature on the physical and mechanical properties of halite rocks was studied by Hume and Shakoor (1981), Sriapai et al. (2012), and others. In general the increase in temperature decreases the confined and unconfined strength (Fig. 6.8), the modulus E and ν (Fig. 6.9).

Hume and Shakoor (1981) point out that creep increases with temperature and with confining pressure (Fig. 6.10).

Sulphatic Rock (Anhydrite and Gypsum Rock)

The strength and deformation properties of the sulphatic rocks were studied by Richards (1933), Höfer and Menzel (1964), Handin and Hager Jr (1957), Clark (1966), Grob (1976), Müller and Siemes (1974), Muller and Briegel (1978), Bell (1981, 1994), Fabre and Dayre (1982), Bilgin (1982), Gumusoglu and Ulker (1982), Dean and Johnson (1989), Diehl and Savage (1989), Papadopoulos et al. (1994), Pfeifle and Hansen (1998), Alejano et al. (1999), Karacan and Yilmaz (2000), Al-Harthi (2001), Yilmaz and Sendir (2002), Auvray et al. (2004), Giambastiani (2005), Pehovaz Alvarez (2009), and Sriapai et al. (2012) (Tables 6.10 and 6.11).

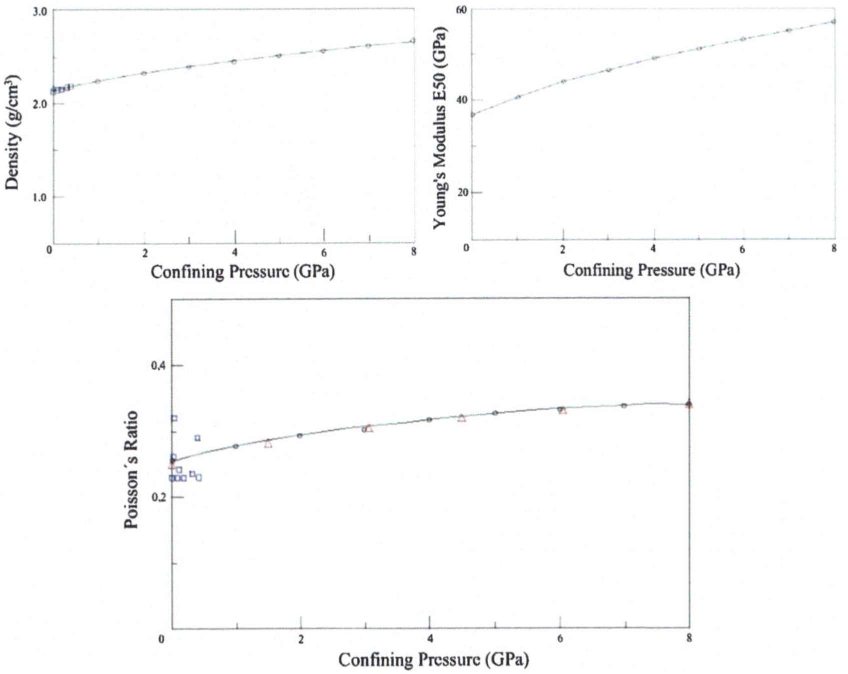

Fig. 6.7 Variation of the density, modulus *E* and *ν* with confining pressure for salt rock. (Modify from Hume and Shakoor 1981)

The anhydrite rocks exhibit average uniaxial compression strength of 75.7 ± 26.7 MPa and tensile strength of 5.4 ± 1.7 MPa, classifying them as medium to high strength rocks.

Gypsum rocks have tensile strength (2.5 ± 1.1 MPa) and uniaxial compression strength (27.8 ± 10.2 MPa) lower than anhydrites and similar to salt rocks, classifying them in the transition of soft and intermediate rocks. The *E* modulus of the anhydrites is twice that of the gypsum, being respectively 50.9 ± 25.1 GPa and 23.1 ± 13.6 GPa. The *ν* for gypsum is 0.27 ± 0.06 and for the anhydrite is 0.30 ± 0.04. Anhydrite rocks are much more resistant than gypsum and halite rocks. This is possibly related to Mohs hardness of the minerals involved, having the anhydrite hardness 3–3.5, the plaster hardness 1.5–2 and halite hardness 2.5.

Handin and Hager Jr (1957), Müller and Siemes (1974), Müller and Briegel (1978), Bell (1981, 1994), Bilgin (1982), Pfeifle and Hansen (1998), Pehovaz Alvarez (2009), Giambastiani (2005), and Karacan and Yilmaz (2000) performed triaxial tests at normal temperature (20–25 °C) to study the influence of confining pressure on the strength and deformability of sulphate rocks. Figure 6.11 shows the envelope of the Mohr-Coulomb criterion obtained from the data compiled from the bibliography for the anhydrite rocks and Fig. 6.12 shows the strength envelope for gypsum rocks.

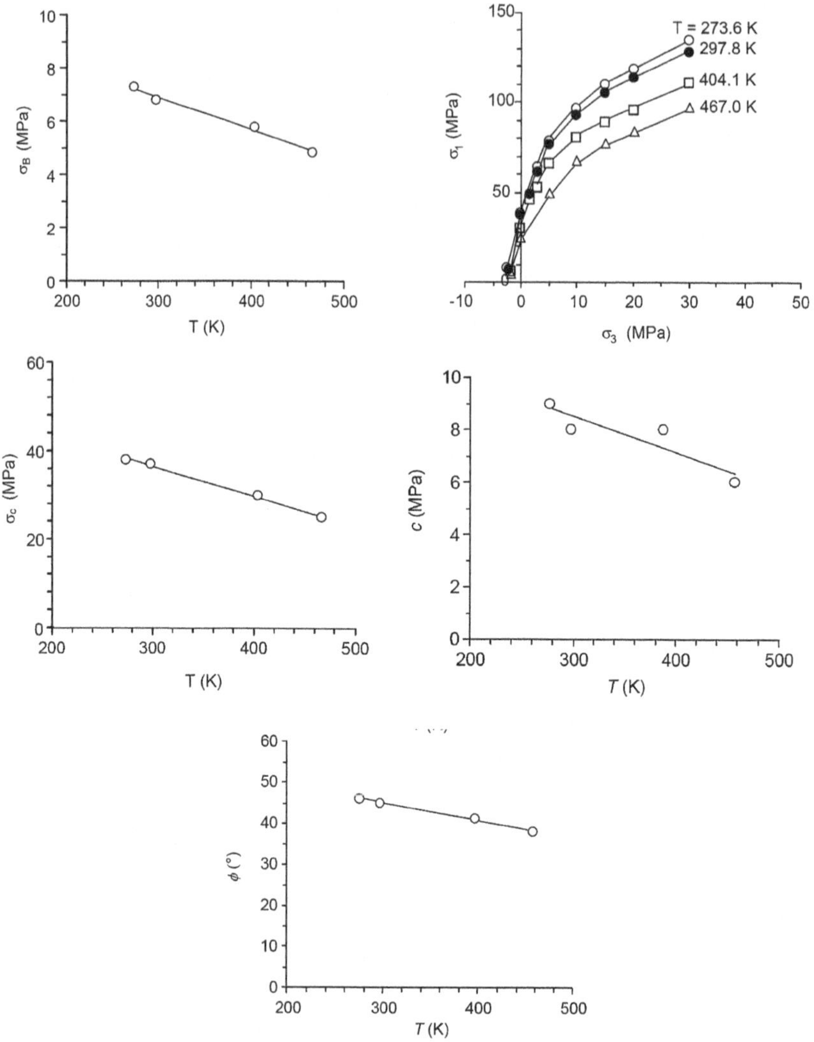

Fig. 6.8 Strength parameters vs. temperature. (From Sriapai et al. (2012)—with permission of the authors)

For anhydritic rocks the strength envelope fits with a cohesion and friction angle of 12.4 MPa and 44° respectively. Figure 6.11 shows that for values of σ_n greater than 50 MPa the line that best adjusts have a cohesion value of approximately 50 MPa and a friction angle of 20°. For values of σ_n less than 50 MPa the line is more inclined and the cohesion changes to approximately 20 MPa and the friction angle reaches 50°.

For gypsum rocks the strength envelope shows that the cohesion and friction angle obtained with a correlation coefficient of 0.93 is respectively 7.2 MPa and 27° (Fig. 6.12).

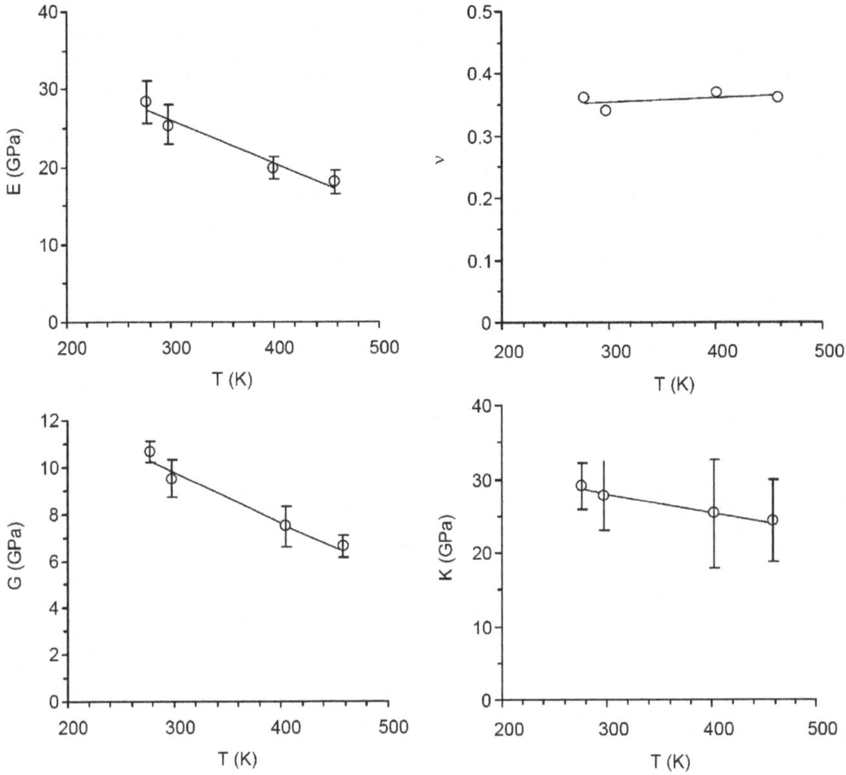

Fig. 6.9 Deformability modulus vs. temperature. (From Sriapai et al. (2012)—with permission of the authors)

The adjustment of the experimental data to the Hoek and Brown rupture criterion gives the following results (Table 6.12).

It is observed that the obtained value of 13.2 is very close to the 12 ± 2 values suggested for these rocks by Hoek et al. (2002). The value $m_i = 7.6$ for gypsum is in the lower limit of the 10 ± 2 reference value proposed for these rocks by Hoek et al. (2002).

Figure 6.13 shows the classification of sulphate rocks according to the Modulus Ratio (Deere and Miller 1966). Anhydrites can be classified as BH, CH, and DH, while those of gypsum can be classified as DH and EH. The halite rocks closely match the gypsum rocks. Both gypsum and halite rocks present some cases with DM, EM, DL, and EL modules, especially related to rocks of higher porosity. It was observed that rocks with porosity less than 2% have MR greater than 500.

Handin and Hager Jr (1957), Clark (1966), Müller and Siemes (1974), and Müller and Briegel (1978) have also investigated the geomechanical behavior of sulphatic rocks at temperatures up to 500 ° C and under different deformation rates (Fig. 6.14).

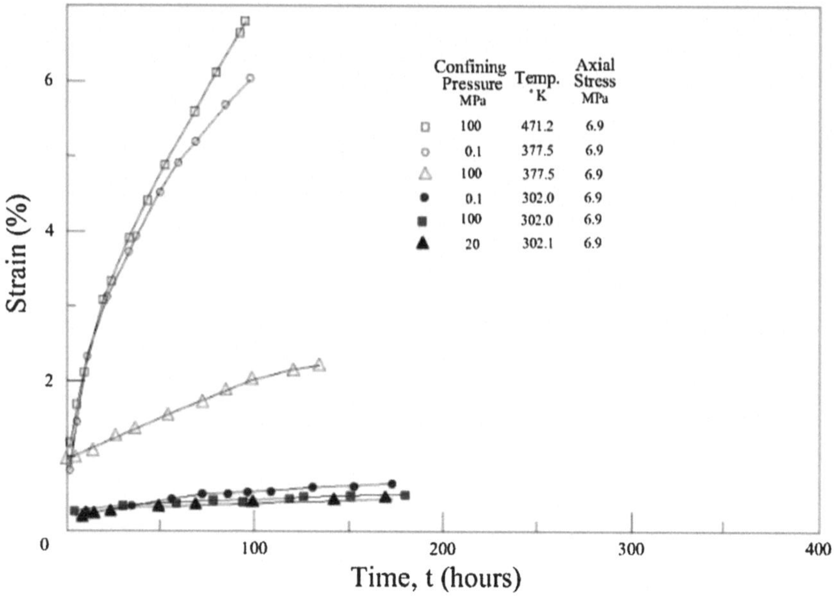

Fig. 6.10 Influence of temperature and confining pressure on Creep behavior. (From Hume and Shakoor (1981))

Table 6.10 Strength and deformational properties of anhydrite

	Brazilian test	Unix. comp. strength	E		c	
	MPa	MPa	GPa	ν	MPa	ϕ
Average	5.4	75.7	50.9	0.3	22.6	34
Stand. dev.	1.7	26.7	25.1	0.04	16.2	4
Min	2.6	40.5	9	0.24	6.4	29
Max	8.2	130	109.5	0.38	43.4	40

Table 6.11 Strength and deformational properties of gypsum

	Brazilian test	Unix. comp. strength	E		c	
	MPa	MPa	GPa	ν	MPa	ϕ
Average	2.5	27.8	23.1	0.27	10.8	24
Stand. dev.	1.1	10.2	13.6	0.06	2.0	2
Min	0.5	4.8	0.8	0.20	0.7	22
Max	5	45	50	0.39	12	27

Giambastiani (2005) used stress-strain curves of the uniaxial and triaxial compression tests of the anhydrites Auquilco to distinguish the initiation thresholds for plastification, onset of dilatancy and rupture following the criteria of Bieniawski (1964) and Martin and Chandler (1994). The normalized plastification threshold to the uniaxial compression strength is 0.7 for alabaster gypsum, 0.3 for laminar gyp-

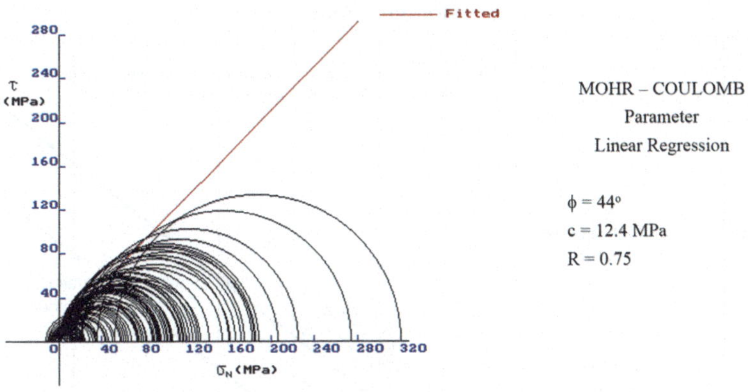

Fig. 6.11 Envelope of the Mohr-Coulomb criteria for anhydritic rocks

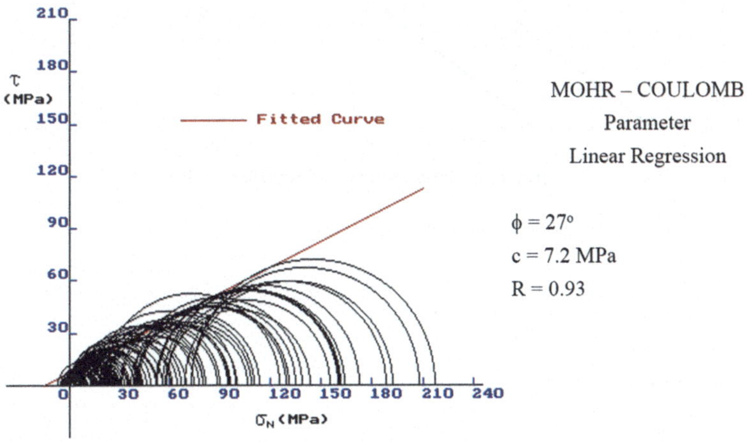

Fig. 6.12 Envelope of the Mohr-Coulomb criteria for gypsum rocks

Table 6.12 Hoek and Brown parameters os anhydrite and gypsum rock (Giambastiani 2005)

	Hoek and Brown parameter				
Rock	σ_{ci} (MPa)	σ_t (MPa)	m_i	s	R
Anhydrite	77.2	5.8	13.2	1	0.71
Gypsum	26.7	3.5	7.6	1	0.91

sum, 0.29 for massive gypsum, 0.23 for banded gypsum, 0.27 for nodular gypsum, 0.26 for Muribeca anhydrite, and 0.44 for Auquilco anhydrite. The envelope that marks the beginning of the normalized dilatancy is 0.73 for alabaster gypsum, 0.67 for laminar gypsum, 0.96 for massive gypsum, 0.92 and 0.85 for banded and nodular gypsum, 0.84 for the Muribeca anhydrite, and 0.89 for the Auquilco anhydrite.

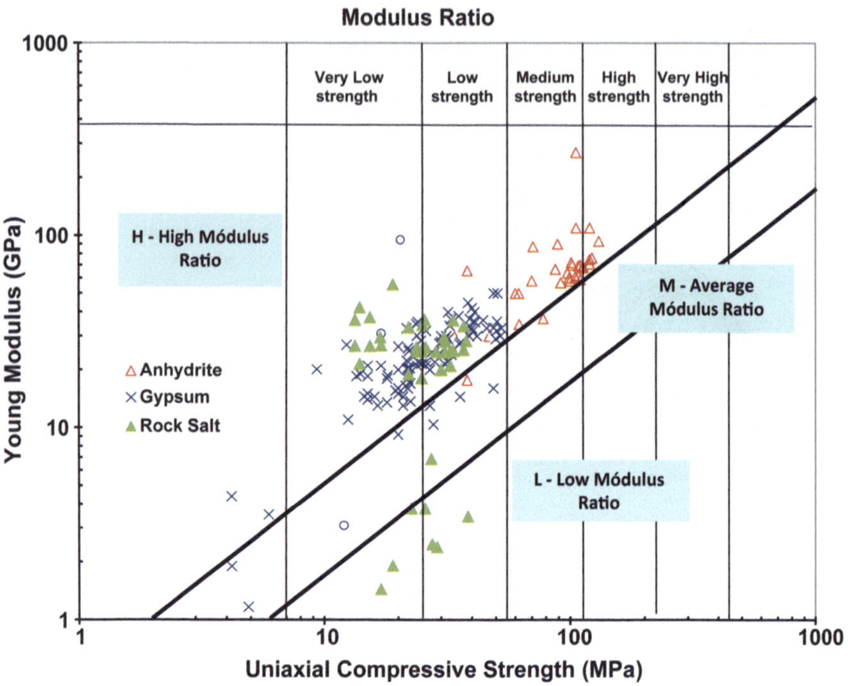

Fig. 6.13 Modulus ratio classification of evaporite rocks

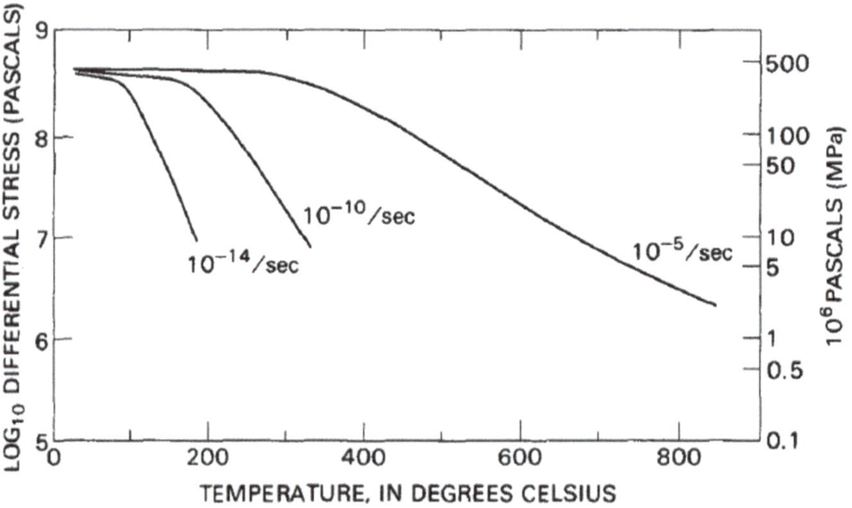

Fig. 6.14 Strength of anhydrite at 0% strain and different strain rates. The confining pressure is 150 MPa. (From Müller and Briegel (1978))

6.3.2.2 Time-Dependent Properties

Most of the geomechanical problems associated with evaporitic rocks are related to time-dependent behavior. The mechanical behavior of salt rock has been studied for many decades by the need to design safe salt mines and to develop improved design and performance assessment predictions for nuclear waste disposal, oil and gas storage, and petroleum drilling (Langer 1981; Pfeifle and Senseny 1982; Hunsche and Hampel 1999; and Cella 2003 among others).

Many research results show that salt rock and other evaporites exhibit pronounced time-dependent deformation or creeps under relatively low stress level and have very low permeability and porosity. Salt rock can creep to a very large strain without fracturing and tends to be self-healing. In recent years, due to these intrinsically properties, salt rock is considered as an ideal material to store natural gas, petroleum, and wastes, especially nuclear wastes.

The rheologic behavior of rocks is influenced by many internal factors, such as mineralogy, impurities, grain size, texture, porosity, and humidity, and external factors such as stress level, confining pressure and temperature (Vutukuri and Lama 1978; Dusseault and Fordham 1993; Hunsche and Hampel 1999; Griggs 1939; Bell 1994; Langer 1981; Cristescu 1989). Many authors have investigated the time-dependent behavior using uniaxial and triaxial tests. Carter et al. (1993) found that the time-dependent properties of salt strongly depend on the temperature. Various so-called deformation mechanism maps have been proposed in order to locate the domain in the stress ± temperature ± strain rate space, where each of the different processes controls the inelastic flow under steady-state conditions (Munson and Wawersik 1993). Hunsche (1994), Chan (1997), Yang et al. (1999), and Cella (2003) reported a large number of uniaxial and triaxial test results and analyzed the confining pressure and deviatory stress effects on the creep strain of rock salt.

There are three possible approaches to analyze the experimental results of creep tests: using empirical "laws," using laws based on rheological models or equations based on the physical mechanisms of deformation.

The empirical laws were the first attempt to obtain a mathematical equation that represented the creep in rocks. These laws are obtained by adjusting a theoretical curve either potential or exponential, to the points of deformation vs. time graph (Langer 1981; Lindner and Brady 1981). Griggs (1939) proposed the following equation for gypsum rocks:

$$\dot{\varepsilon} = a + b\log t + ct$$

These laws have the advantage of their simplicity but do not reflect the tertiary fluency and do not reflect the physical mechanisms involved in the deformation.

Another way to represent creep is through rheological models. Figure 6.15 presents a large variety of rheological models proposed for salt rocks and compiled by Dusseault and Fordham (1993).

Another way of elaborating a creep law is through the study of the physical mechanisms of creep deformation. This approach is highly accepted within the sci-

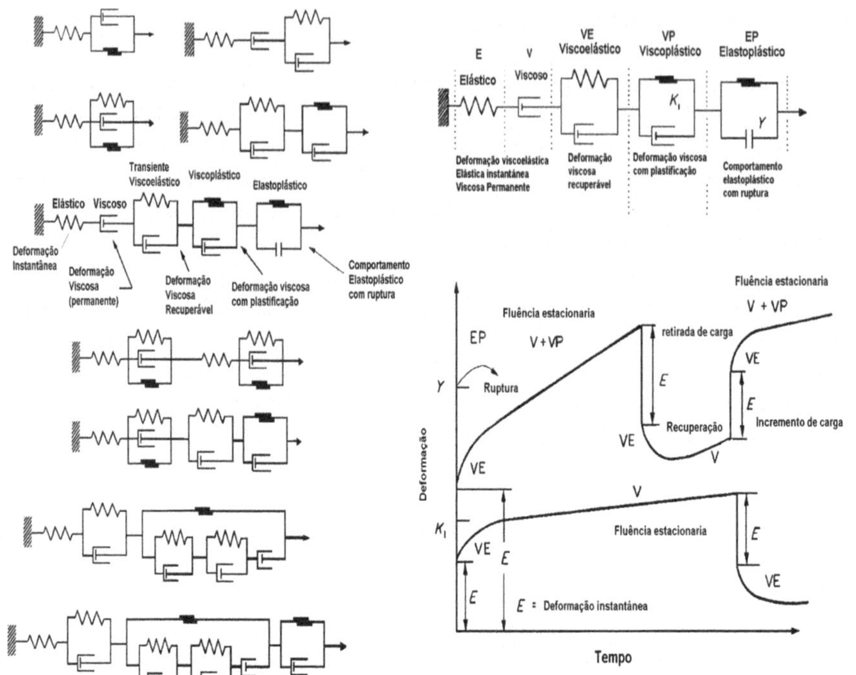

Fig. 6.15 Rheological models for salt rocks (Dusseault and Fordham 1993)

entific community because it allows overcoming the limitations of the previous methods and because it allows incorporating the effect of temperature and of other intrinsic parameters like the mineralogy and the grain size in the constitutive models (Passchier and Trouw 1998).

These phenomenological models are based on the concept of activation energy and assume that intracrystalline deformation mechanisms are predominant in the creep of rocks (Lindner and Brady 1981). Thus each deformation mechanism requires a certain amount of energy to be activated. This activation energy would be related to the kinetics of deformation in an analogous way as the chemical kinetics depends on the temperature.

At the scale of the polycrystalline aggregate, the rocks present the following types of deformation mechanisms: cataclastic flow, plastic flow, intracrystalline or reticular sliding, twinning, recrystallization, grain boundary migration and diffusion.

The cataclastic flow or cataclastic deformation is a type of permanent deformation that occurs at low temperature and whose mechanisms are: intercrystalline dislocation, crushing, fracture and granulation. This deformation necessarily involves increasing the volume (dilatancy). In plastic flow the minerals suffer distortion, twinning and intracrystalline dislocation. Intracrystalline or reticular sliding are movements that are experienced by the crystalline reticles by sliding the molecules from a crystallographic plane at a distance that is multiple of the original distance between the molecules. In this way, the crystal structure is not damaged and there is no increase in volume. This mechanism is very sensitive to temperature increases and to decreases in the rate of deformation but insensible to the confining pressure. Under favorable conditions of temperature and deviatory stress, the intracrystalline slip can be accompanied by recovery or dynamic recrystallization which explains the hardening behavior experienced by most rocks.

The equilibrium between intracrystalline slides and recovery or dynamic recrystallization at high temperatures gives rise to the steady-state flow, which represents the rate of deformation as a function of constant conditions of tension, temperature and microstructure. The semi-empirical flow laws for rocks are based on the Arrhenius equation which considers that the reactions must have certain energy to overcome a barrier in the reaction. This energy barrier is called the activation energy (Q). The Arrhenius equation, which postulates temperature dependence of deformation, arises from the expectation that fluctuations in atomic vibrations can help the applied stress to overcome barriers to the movement of intracrystalline slides. Based on this theory, several authors proposed an equation for steady flow (for high temperature and pressure) by intracrystalline sliding that takes the following form:

$$\dot{\varepsilon} = A \cdot \exp\left(\frac{-Q}{RT}\right) sen\left(B\sigma\right)$$

or

$$\dot{\varepsilon} = A \times \exp\left(\frac{-Q}{RT}\right) \times \sigma^{n} \quad \left(\text{For dislocation climb}\right)$$

where

A = constant related with deformation rate (s^{-1})
$\dot{\varepsilon}$ = steady flow rate
Q = activation energy ($J\ mol^{-1}$)
R = universal gas constant ($J\ mol^{-1}\ K^{-1}$)
T = temperature (K)
B = constant
σ = stress level (MPa)
σ_n = potential law creep for low stress

In plastic deformation by mechanical twinning, the new positions of the crystalline lattice are equally stable where the sliding part is the mirror image of the other adjacent part of the lattice. The mechanical twinning is insensitive to large changes in temperature, hydrostatic pressure, rate of deformation and the magnitude of the deformation. On the contrary, it is sensitive to deviatory stress and grain size.

Cella (2003) performed triaxial tests on Brazilian salt rocks at different temperatures and differential stress. The objective was to obtain the adjustment parameters of the steady state model of these rocks. The dual mechanism deformation equation

(Dusseault et al. 1987) contemplates the change of mechanism that occurs in successive stages of increase of differential effective voltage. The initial mechanism that dominates over a certain range of differential stress loses importance for the next mechanism when approaching a stress level that is determined experimentally. The tests were run with a confinement stress of 10 MPa, differential stress of 6–20 MPa and constant temperature of 86 °C or 359 K. The deformation mechanism map for salt rocks presented by Dusseault and Fordham (1993), indicates that, for the differential stress used in the tests (between 6 and 20 MPa), at a constant temperature of 86 °C, the exponent "*n*" of the stress factor reaches values between 3 and 4.5, in the regime of deformation by glide dislocation in the reticule of halite. For differential stress greater than 10 MPa, the exponent "*n*" was 7.9 which place it in the domain of stable microcracking:

$n = 1$–2 for Diffusion or pressure dissolution
$n = 3$–6 for Glide dislocation
$n = 2.3$–4.7 for Grain boundary sliding
$n = 6$–10 for Microcracking

Cella (2003) observed that even with differential stress from 16 MPa, the samples did not show evolution towards tertiary creep. Possibly this is due to the fact that in the microcracking process, healing of cracks occurs due to the migration of fluids from the fluid inclusions to the planes of microcracks.

Finally, the constitutive equation derived from the tests was:

$$\dot{\varepsilon} = 1.92 \times 10^{-6} \log\left(\frac{\sigma_d}{10}\right)^n$$

where

$n = 4$ ($\sigma_d < 10$ MPa)
$n = 7.94$ ($\sigma_d \geq 10$ MPa)

Uniaxial Creep Test at Room Temperature of Sulphatic Rocks

The rheological properties of the sulphatic rocks were studied by Griggs (1939) for gypsum under unconfined conditions and at room temperature by exploring the effect of the presence of fluids and the strain on the deformation rate. Subsequently, Misra (1962) studied the creep of anhydrites at 15 and 600 °C. Bell (1994) performed uniaxial creep tests on gypsum and anhydrite. Stowe (1985) reported the results for uniaxial and triaxial creep tests on anhydrites of the Salado Formation (USA). Pfeifle and Hansen (1998) compiled the anhydrite creep tests performed for the WIPP project showing that, for confining pressure between 5 and 15 MPa, the secondary creep rates vary between $5 \cdot 10^{-11}$ and $2 \cdot 10^{-6} s^{-1}$.

Giambastiani (2005) performed uniaxial creep tests at room temperature and different stress levels, of sulphatic rocks from Brazil and Argentina (Fig. 6.16).

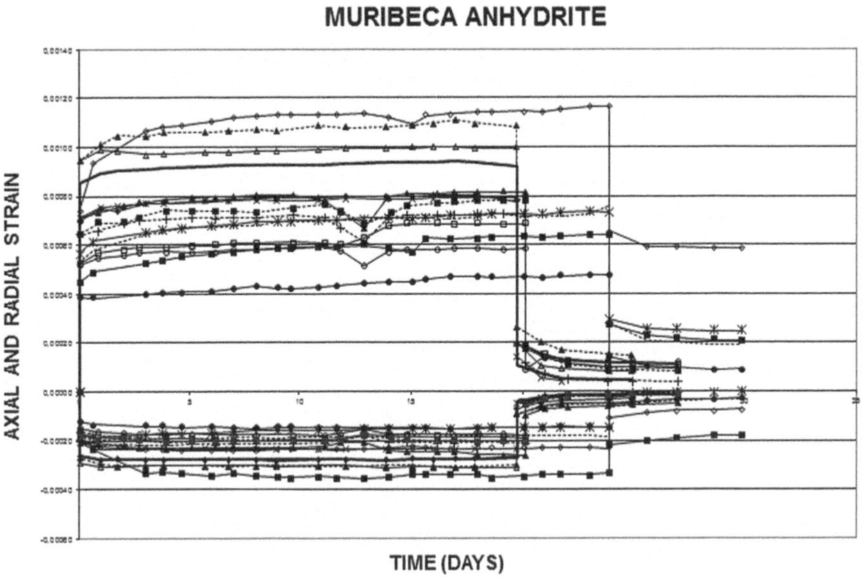

Fig. 6.16 Deformation vs. time curves of the Muribeca anhydrite

The deformation (axial and radial) vs. time curves show an instantaneous elastic deformation, a curved section of the primary or transient creep and a straight section with slight inclination corresponding to the secondary creep where the rate of deformation is constant.

From the deformation vs. time curves the different rheological models were adjusted to characterize the primary and secondary creep phases. The rheological models used were the Maxwell, Boltzmann and Burger models. The Burger, visco-elastic model, was the best fit and its representative equation is:

$$\varepsilon = \frac{2\sigma_1}{9K} + \frac{\sigma_1}{3G_2} + \frac{\sigma_1}{3G_1} - \frac{\sigma_1}{3G_1} e^{\frac{-G_1 t}{\eta_1}} + \frac{\sigma_1}{3\eta_2} t$$

where

K = time-independent volumetric modulus
G_2 = instantaneous shear modulus
G_1 = constant that controls the amount of delayed elasticity
η_1 = constant that controls the rate of delayed elasticity
η_2 = viscous flow rate corresponding to the secondary creep (Goodman 1980)

Table 6.13 presents a summary of the parameters of the Burger model for the rocks studied by Giambastiani (2005).

It was observed that the studied rocks have a threshold of primary creep varying between 1% and 28% of the uniaxial compression strength for the anhydrites and

Table 6.13 Rheological models of anhydrite and gypsum rocks (Giambastiani 2005)

Rock	Rheologic models		Parameter K (MPa)	G_1 (MPa)	G_2 (MPa)	η_1 (MPa.dia)	Deformation rate (s^{-1})
Anhydrite auquilco	20 MPa < σ_o < 70 MPa	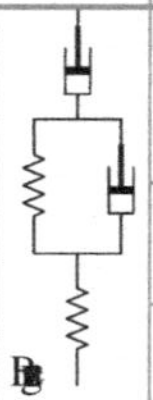	$13{,}152\sigma_o^{0.3200}$	$16{,}333\sigma_o - 14{,}370$	$24{,}137\sigma_o^{0.1177}$	472,309 ± 503,695	$-1.92 \times 10^{-11} + 5.92 \times 10^{-13}$ σ_o
	$\sigma_o > 70$ MPa						$-1.67 \times 10^{-9} + 2.26.10^{-11}\sigma_o$
Anhydrite muribeca	$\sigma_o >$ 15–17 MPa		$16{,}214\sigma_o^{0.3225}$	$7991\sigma_o - 138{,}294$	$7532\sigma_o^{0.3471}$	$4832\sigma_o + 73{,}476$	$1.11 \times 10^{-12}\sigma_o - 3.63 \times 10^{-11}$
				$\sigma_o >$ 15–17 MPa		$\sigma_o >$ 15–17 MPa	for $\sigma_o < 60$ MPa
Gypsum banded	$\sigma_o >$ 1–8 MPa		36,590 ± 15,179	$3267\sigma_o + 3211$	15,012 ± 4888	33,624 ± 21,354	$4.43 \times 10^{-11} \pm 3.84 \times 10^{-11}$
Gypsum nodular	$\sigma_o >$ 4.5–13 MPa		42,340 ± 17,397	$11{,}889\sigma_o - 11{,}2644$	15,067 ± 6594	66,934 ± 42,960	$3.86 \times 10^{-11} \pm 2.16 \times 10^{-11}$
Gypsum laminar	$\sigma_o >$ 4.9–10 MPa		9631 ± 53,712	$8069\sigma_o - 39{,}601$	6165 ± 4565	53,840 ± 97,043	$3.87 \times 10^{-11} \pm 4.97 \times 10^{-11}$
							for $\sigma_o > 10$ MPa

between 4% and 14% for the gypsum. This means the initial viscoelastic deformations would begin to manifest themselves at low stress levels. The parameter η_1, that determines the rate or speed of the delayed elasticity, is not sensitive to the applied stress level. The secondary creep, controlled by the parameter η_2 (viscosity), was analyzed from the plot of strain vs. axial strain rates (Fig. 6.17).

Figure 6.17 shows a more or less clear trend between the stress and the deformation rate, despite a relative dispersion in the gypsum and in the Muribeca anhydrite.

It is also observed that the rate of deformation on both rocks increases as a function of the applied stress. The gypsum differs well from the anhydrites with a steady-state creep starting stress of 6–10 MPa for gypsum and 20–30 MPa for anhydrites. In the anhydrites the trajectory could be divided into two sections of different inclinations whose inflection point would be around 50–60 MPa. These two trajectories would indicate, a priori, that two deformation mechanisms would be acting at different stress levels (Langer 1981; Wawersik 1985; Dusseault and Fordham 1993). The creep tests indicate that the gypsum present greater viscoelastic and irreversible deformations than the anhydrites. The deformation rates in the secondary creep are not substantially different between anhydrites and gypsum ranging from 10^{-5} to 10^{-8} dia^{-1}. The difference resides in the fact that the anhydrites require stress levels between two and three times greater than that of the gypsum to show steady state flow.

Time-Dependent Behavior by Anhydrite-Gypsum Transformation

The time-dependent deformations observed in some underground excavations were attributed to the increase in volume resulting from the anhydrite–gypsum transformation by hydration of the first one. The theory is supported by physical-chemical arguments that show that at the molar level and in an open system, the volume of gypsum is 62% greater than the volume minerals the anhydrite. In nature the diagenesis of the primary gypsum generates anhydrite and the exhumation of this gives origin to the secondary gypsum. There is almost unanimity among the geologists that this process is isovolumetric and not isomolar, the reason why the mineralogical replacement happens without changes in the volume.

Sahores (1962) carried out the first study on the mechanical phenomena associated with the expansion of anhydrite, questioning the traditional concepts about the mechanism of anhydrite transformation and the magnitude of the stresses attributed to the phenomenon (70 MPa). The expansion tests performed on anhydrite samples at the maximum expansion pressure is 1.6 MPa.

Ordoñez et al. (1990) studied the expansion mechanisms of tertiary marls formed by expansive clays and anhydrite. They concluded that the expansion of the marls has two causes: (a) expansion of clays and (b) dissolution of anhydrite and recrystallization as gypsum. The expansion of the massifs was attributed to the precipitation of gypsum in the fractures induced by the excavation. The gypsum crystallization pressure was calculated between 2 and 3 MPa.

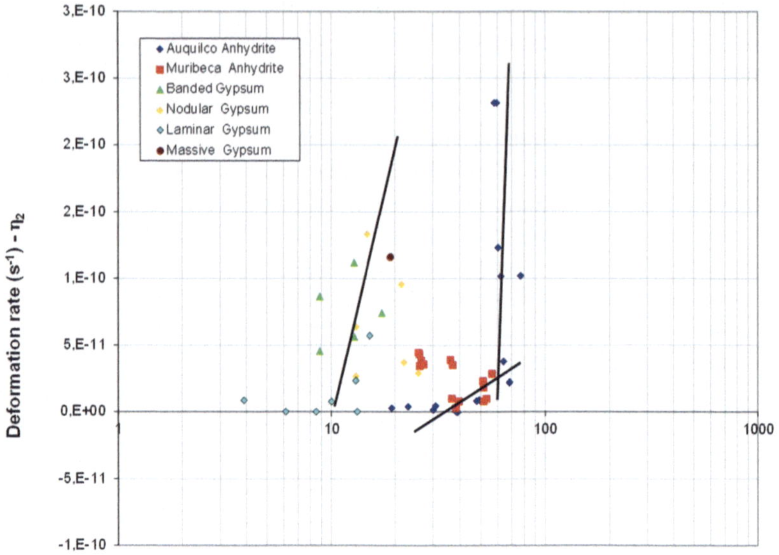

Fig. 6.17 Deformation rate vs. axial stress for sulphatic rocks (Giambastiani 2005)

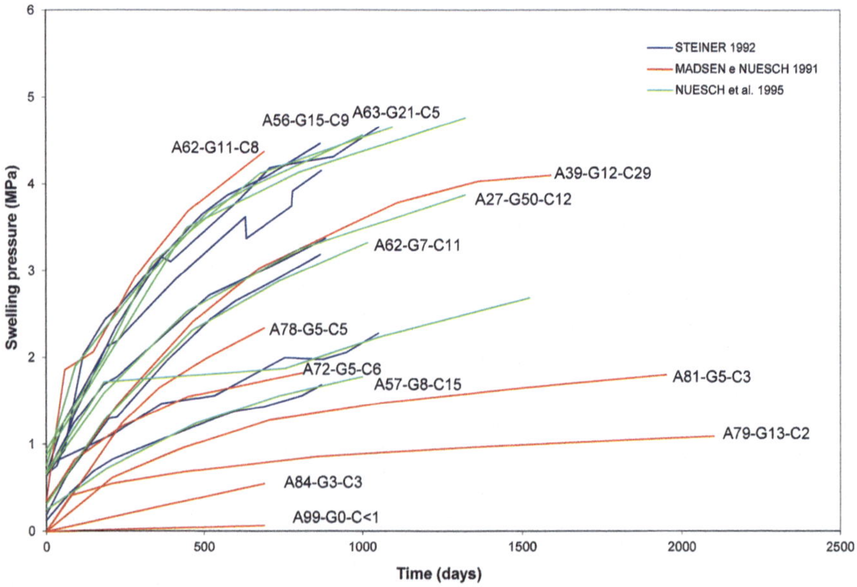

Fig. 6.18 Expansion pressure tests on anhydritic marls. *A* anhydrite, *G* gypsum, *C* clays. (Modified from Steiner (1993), Madsen and Nüesch (1991) and Nüesch et al. (1995))

Figure 6.18 presents the results of tests conducted by Steiner (1993), Madsen and Nüesch (1991) and Nüesch et al. (1995). Expansion pressures range from 0.1 MPa for massive and pure anhydrites to almost 5 MPa for anhydrous marls. The expansion pressure has a medium and long term tendency to stabilization.

Nüesch et al. (1995) observed that 1 cm long gypsum fibers were formed in the contour of the specimen and pseudomorphic growth on precursor anhydrite and gypsum crystals by thin layer separating the claystone. The gypsum grows on the anhydrite in a diffusive way leaving anhydrite residues. The advance of the gypsum front is favored by the presence of layers of clay. Madsen and Nüesch (1991) concluded that the expansion pressures measured were practically zero for rocks with 99% anhydrite and up to 4–4.5 MPa for rocks with 40–60% anhydrite and 8–29% of expansive clays.

Nüesch et al. (1995) studied sulphatic claystone from the Belchen tunnel (Switzerland). The tests showed that the rocks with clays and anhydrite expand more than the pure clay ones. Also, the anhydrite-gypsum transformation occurs by dissolving the anhydrite and gypsum precipitation under super-saturated solution in calcium sulphate conditions. They also observed dissolution and corrosion of anhydrite crystals, precipitation of gypsum in the clay-free layers of sulphates and formation of gypsum crystals in microfractures. Expansion pressures range from 1.75 to 5 MPa.

Oldecop and Alonso (2012) studied the expansion phenomena in the rock massif of the Lilla tunnel (Spain) by laboratory expansion tests. The main conclusion of this analysis is that the dissolution of anhydrite and precipitation of gypsum is the most likely mechanism.

Giambastiani (2005) performed free axial expansion tests on samples of anhydrite and gypsum in order to investigate the physical-chemical mechanisms of the anhydrite-gypsum transformation. Five samples of the Auquilco anhydrite, a Muribeca anhydrite sample and five gypsum samples were tested, one of which was 38% anhydrite. The samples were immersed in both a supersaturated solution of calcium sulphate and distilled water.

Gypsum and anhydrite samples submerged in distilled water for 125 days show a decrease in volume due to the dissolution of anhydrite and gypsum. When distilled water is replaced by the supersaturated solution, the anhydrite samples reverse the negative trend, although the gypsum samples experienced volumetric stability.

The experimental results of the samples submerged in supersaturated solution show that, under laboratory and open system conditions under normal conditions of pressure and temperature, the transformation of the anhydrite is expressed in volume increases due to the formation of an external layer of gypsum crystals, in addition to the formation of a whitish gypsum layer inside the sample. The whitish layer is formed by microcrystalline gypsum of smaller grain size than the anhydrite crystals it replaces. It is possible to observe pseudomorphs of anhydrite completely replaced by gypsum. The replacement is isovolumetric. The process starts with the dissolution of the anhydrite (unstable phase), supersaturation of the solution and deposition of crystalline gypsum. In order for the anhydrite–gypsum transformation process to generate swelling, the existence of a supersaturated solution at the interface is necessary; Otherwise the dissolution process generates a decrease in volume.

The deformation rates calculated for anhydrites vary between 0.6% and 1.1%/year, much lower than the rates reported by Nüesch et al. (1995) but would generate convergence movements compatible with those reported by Steiner (1993).

6.4 Geotechnical Problems in Evaporitic Rocks

There are numerous evidences in the literature of geotechnical problems in evaporitic rocks associated with several phenomena resulting from the properties of these geological materials:

- Dissolution—Karstification (support loss)
- Expansion (anhydrite dissolution-gypsum growth)
- Creep
- Rockburst
- Metal Corrosion
- Concrete damage

Gignoux and Barbier (1955) present several cases of tunnels excavated in anhydrite and gypsum massifs in France with problems of expansion and dissolution (L'Isère bypass tunnel, L'Arc tunnel) and possible rockburst (tunnel of Derivation of Tinée, slopes of the road of Mont-Cenis in Savoia, Tunnels of Derivation of L'Arvan and L'Oued the Lil and in the rail tunnel of Col de Tende.

Brune (1965) cites rockburst events and floor upsets due to the release of strains accumulated in rock mass as a consequence of the expansion of anhydrite in the USA. He also mentions problems with dissolution caves in gypsum layers that give rise to subsidence and collapse.

Calcano and Alzura (1967) include numerous cases of gypsum dissolution in dam's foundations, such as the formation of large cavities in the Hondo, MacMillan, Sandford and Redrock dams in the United States. Problems of gypsum dissolution settlements occurred at San Fernando, Olive Hills, Rattlesnake, and Sandford dams.

Lori and Frosio (1962) presented the experience of 30 years of observations in the diversion gallery of the Idro-Vobarno Hydroelectric Power Plant (Italy) excavated in an anhydrite formation. The lining of the gallery experiments fractures by the continuous floor uplift in the anhydrite section.

Grob (1976) discusses the problem of expansion deformations in the rocks of the Keuper formation in the Belchen Tunnel, consisting of marl with intercalations of gypsum and anhydrite, with expansion pressures obtained in the laboratory varying between 0 and 4 MPa.

Krause (1976) presents problems of inverted arc rising and the formation of dissolution cavities in tunnels of Baden-Württemberg (Germany), in the period 1840–1975, excavated in Triassic rocks of the Keuper formation (anhydrite/gypsum + clays). There are several convergences in tunnels including the Wagemburg tunnels (70 cm in 2 years), Kappelesberg tunnel (4.5 m in 85 years), Sohlstollen tunnel (1 m in 25 years), Hasenberg tunnel, and Engelberg tunnel.

James and Lupton (1978) *apud* Irfan and Özkaya (1981) report problems of dissolution in the foundations of the Mosul (Iraq) and channels in Syria, respectively.

At the Symposium on Geotechnical Problems in the Construction of Tunnels in Rock Masses containing Anhydrite and Gypsum, made in 1975 by the IAEG, where presented different approaches on the problem of expansion by the transformation of anhydrite into gypsum.

Steiner (1993) presents an excellent synthesis on tunnels excavated in expansive rock massifs such as claystones, shales, and clayey anhydrites (Upper Hauenstein Tunnel, Hauenstein Base Tunnel, Belchen Tunnel, Felsenau Gypsum Mine, Elgelberg Experimental Tunnel, Railway Express (Germany), Wagenburg Tunnel (Germany), Heslach Road Tunnel (Germany), and Freudenstein High Speed Rail Tunnel (Germany)).

Maury (1993) cites examples of floor uplift and large convergences of walls in the Genevreuille (1855–1858), Bozberg (1871–1875), Krappelisberg (1878–1880), Ricken (1903–1908), and Hanenstein (1912–1915) tunnels. In some cases the movements were very fast with rates from several millimeters to tens of millimeters per year or very slowly.

Rahn and Davis (1996) referred to the presence of sinkholes and caves in gypsum in the Rapid City area, South Dakota.

Gysel (2002) describes severe stability problems attributed to the anhydrite dissolution phenomena: Punt dal Gall-Ova Spin tunnels and Sçhanf-Ova Spin tunnels from the Engadin Hydroelectric Plant (Switzerland) and the Chixoy Hydroelectric Plant (Guatemala).

From the many cases where squeezing (creep) problems were manifested in soft rock tunnels, only squeezing effects were reported on anhydrite rocks in the Simplon (Switzerland—Italy) and Tauern (Austria) tunnels (Steiner 1993).

Acknowledgments We thank Chairman Prof. Milton Kanji and the vice-chairman Prof, He Manchao for the invitation to make a contribution in this book, and Dr. Paulo Cella for providing a copy of his PhD thesis on salt rocks from Brazil. I would like to thank the anonymous referees for their suggestions, which have led to significant revisions and improvements.

References

Alejano LR, Garcia-Bastante F, Alonso E, Taboada J (1999) Back-analysis of a rockburst in a shallow gypsum room and pillar exploitation. In: Ninth international congress on rock mechanics, Paris, pp 1077–1080

Al-Harthi AA (2001) Environmental impacts of the gypsum mining operation at Maqna area, Tabuk, Saudi Arabia. Environ Geol 41:209–218

Auvray C, Homand F, Hoxha D, Didier C (2004) Influence du temps et de l'hygrometrie sur le comportement du gypse. Rev Fr Geotech 106–107:41–51

Barton N (2007) Rock quality, seismic attenuation and anisotropy. Taylor and Francis Group, London, 721p

Bell FG (1981) Geotechnical properties of some evaporitic rocks. Bull IAEG 24:137–144

Bell FG (1994) Survey of the engineering properties of some anhydrite and gypsum from the north and midlands of England. Eng Geol 38(1–2):1–23

Bieniawski ZT (1964) Mechanism of brittle fracture of rock. Int J Rock Mech Min Sci 4:395–406

Bilgin N (1982) Cuttability of evaporites. Bull IAEG 25:85–95

Boontongloan C (2000) Engineering properties of the evaporitic and clastic rocks of Maha Sarakam Formation, Sakon Nakhon evaporite basin. MS thesis, Asian Institute of Technology, Thailand

Borchert H, Muir RO (1964) Salt deposits: The origin, metamorphism and deformation of evaporites. Van Nostrand Co, London, 338p

Broch E, Franklin JA (1972) The point-load strength test. Int J Rock Mech Min Sci 9:669–497
Brune G (1965) Anhydrite and gypsum problems. Eng Geol 2(1):26–38
Bullard EC, Niblett ER (1951) Terrestrial heat flow in England. Geophys J Int 6(4):222–238
Calcano CEF, Alzura PR (1967) Problems of dissolution of gypsum in some dam sites. In: Bull. of the Venezuelan Society of Soil Mechanics and Foundation Eng., Caracas, Venezuela, pp 1–36
Carter NL, Handin J, Russell JE, Horseman ST (1993) Rheology of rocksalt. J Struct Geol 15(9/10):1257–1271
Cella PR (2003) Desenvolvimento e execução de ensaios triaxiais de fluência estacionária em rochas salina sob altas pressões e temperaturas. Tese (Doutoramento) – Escola Politécnica, Universidade de São Paulo, São Paulo, 2003, 189p
Chan KS (1997) A damage mechanics treatment of creep failure in rock salt. Int J Damage Mech 6:122–152
Clark SP (1966) Handbook of physical constants. Geological Society of America. Memoir 97, The Geological Society of America, Inc., New York, 587p
Cristescu ND (1989) Rock rheology. Kluwer Academic Publishers, Dordrecht, 336p
Daniels JJ, Kite RJ, Scott JH (1980) Geophysical well-log measurements in three drill holes at Salt Valley, Utah. Open-file report 81-36
Dean WE, Johnson KS (1989) Anhydrite deposits of the United States and characteristics of anhydrite importance for storage of radioactive waste. US Geol Survey Bull 1794
Deere DU, Miller RP (1966) Engineering classification and index properties of intact rock, air force laboratory technical report no. AFNL-TR-65-116, Albuquerque, NM
Devries KL, Mellegard KD, Callahan GD (2002) Salt damage criterion proof-of-concept research. Topical report, DE-FC26-00NT41026 prepared for the U.S. Department of Energy, Pennsylvania
Diehl SF, Savage WZ (1989) Section 3. Physical properties of anhydrite. In: Dean WE, Johnson KS (eds) Anhydrite deposits of the United States and characteristics of anhydrite importance for storage of radioactive waste. US Geol Surv Bull 1794, pp 91–132
Dusseault MB, Fordham ChJ (1993) Time-dependent behaviour of rocks. In: Hudson JA (ed) Comprehensive rock engineering: principles, practice and projects, Cap. 6, Vol 3. 119–149 Pergamon Press
Dusseault MB, Rothemburg L, Mraz DZ (1987) The design of openings in saltrock using a multiple mechanism viscoplastic law. In: Proc 28th symp rock mech, ISRM, NARM, Tucson, USA, 1987
Fabre D, Dayre M (1982) Properties Geotechniques de gypses et anhydrites des Alpes de Savoie (France). Bull IAEG 25:91–98
Fairhurst CM, Midea NF, Eston SM, Fernandes AC, Bongiovanni LA (1991) Rock mechanics studies of proposed underground mining of potash in Sergipe, Brazil. In: Seventh ISRM congress, 2–8 Sept, Montreaux, Switzerland, pp 131–134
Fuenkajorn K, Daemen JJK (1988) Boreholes closure in salt. Technical report prepared for the U.S. Nuclear Regulatory Commission, report no. NUREG/CR-5243 RW. University of Arizona
Fuenkajorn K, Jandakaew M (2003) Compressed-air energy storage in salt dome at Borabu district, Thailand: geotechnical aspects. In: Proceedings of the thirty-eighth symposium on engineering geology and geotechnical engineering. University of Reno, Nevada, pp 377–391
Giambastiani M (2005) Comportamiento dependente do tempo de rocas sulfaticas de anhidrita e yeso. Tese de Doutorado. Escola de Engenharia de São Carlos (EESC-USP), 431p
Gignoux M, Barbier R (1955) Géologie des presas et des aménagements hydrauliques. Masson et Cie edit, 343p
Goodman RE (1980) Introduction to rock mechanics. John Wiley and Sons, USA, 478p
Griggs D (1939) Creep of rocks. J Geol 47:225–251
Grob H (1976) Swelling and heave in Swiss tunnels. Bull Intern Assoc Eng Geol 13:55–60
Gumusoglu MC, Ulker R (1982) The investigation of the effect of gypsum on foundation design. Bull IAEG 25:99–105
Gunter BD, Parker FL (1961) The physical properties of rock salt as influenced by gamma rays: Oak Ridge Natl. Lab. Health Physics Div. Rept ORNL-302 7, 68p

Gysel M (2002) Anhydrite dissolution phenomena: three cases histories of anhydrite karst causes by water tunnel operation. Rock Mech Rock Eng 35(1):1–21

Handin J, Hager RV Jr (1957) Experimental deformation of sedimentary rocks under confining pressure: test at room temperature on dry samples. Amer Assoc Petrol Geol Bull 41:1–50

Hansen FD, Mellegard KD, Senseny PE (1984) Elasticity and strength of the natural rock. In: Proceedings of the first conference on the mechanical behavior of salt. Trans Tech Publications, Clausthal-Zellerfeld, pp 71–83

Herrin, H.E.; Clark, S. P. Jr. 1956. Heat flow in west Texas and eastern New Mexico: Geophysics, 21, 4, 1087–1099

Hoek E, Carranza-Torres C, Corkum B (2002) Hoek-Brown failure criterion – 2002 edition. In: Proc NARMS-TAC conference, Toronto, 2002, 1, pp 267–273

Höfer KH, Menzel W (1964) Comparative study of the pillar loads in potash mines established by calculation and by measurement below ground. Int J Rock Mech Min Sci 1:181–198

Howart SM, Christian-Fear T (1997) Porosity, single-phase permeability and capillary pressure data from preliminary laboratory experiments on selected samples from marker bed 139 at the waste isolation pilot plant. Sandia report SAND94-0472/1. Sandia National Laboratories. vols 1–3

Hume HR, Shakoor A (1981) Chapter 3 – Mechanical properties. In: Gevantman (ed) Physical properties data for rock salt. U.S. Dept of Commerce – National Bureau of standard. Monograph 167, 282p

Hunsche U (1994) Uniaxial and triaxial creep and failure test on rock: experiment technique and interpretation. In: Cristecu NS (ed) Visco-plastic behavior of geomaterial. Springer-Verlag, New York

Hunsche U, Hampel A (1999) Rock salt – the mechanical properties of the host rcok material for a radioactive waste repository. Eng Geol 52:271–291

International Association of Engineering Geology (1979) Classification of rocks and soils fore engineering geological mapping. Part 1: Rock and soil materials. Bull Int Assoc Eng Geol 19:364–371

Irfan TY, Özkaya I (1981) Engineering geological mapping of gypsiferous formations, Sivas, Central Eastern Turkey. Bull Int Assoc Eng Geol 24:33–37

James, A.N.; Lupton, A.R.R. 1978. Gypsum and anhydrite in foundations of hydraulic structures. Geotechnique, 28, no3, 249–272

Karacan E, Yilmaz I (2000) Geotechnical evaluation of Miocene gypsum from Sivas (Turkey). Geotech Geol Eng 18:79–90

Kenzakoo T (2006) Relationship between mineralogy and engineering properties of rock salt. MSci thesis, Suranaree University of Technology, 172p

Kolano M, Flisiak D (2013) Comparison of geo-mechanical properties of white rock salt and pink rock salt in Kłodawa salt diaper. Stud Geotech Mech 35(1):119–127

Krause H (1976) Sulphate rocks in Baden-Wurttenberg and their importance to civil engineering. Bull IAEG 13:45–49

Langer M (1981) The rheological behaviour of rock salt. In: Proceed. int. workshop on salt mechanics, Pennsylvania State University, Nov 1981, Aachen, Germany

Lindner EN, Brady BHG (1981) Memory aspects of salt creep. In: Proceedings of the first conference on the mechanics behavior of salt. Trans Tech Publications, Clausthal-Zellerfeld, pp 241–273

Lori A, Frosio G (1962) Treinta años de servicio de la galeria de desviación (Central Hidroelécrica "Idro-Vobarno", que atravieza una formación de anhidrita. I Coloquio Internacional sobre las obras públicas en los terrenos yesiferos. Tema 2°, Comunicación C 2–11, Madrid (España), pp 239–265

Madsen FT, Nüesch R (1991) The swelling behaviour of clay sulfate rocks. 7o Internat. Congress of rock mechanics, Aechen, Germany, 1, vol 285–288

Martin CD, Chandler NA (1994) The progressive fracture of lac du bonnet granite. Int J Rock Mech Min Sci 31(6):643–659

Maury V (1993) An review of tunnel, underground excavation and boreholes collapse mechanism. In: Hudson (ed) Comprehensive rock engineering: principles, practice and projects, cap. 14, vol 4, pp 369–411
Mavko G (2017) Stanford Rock Physics Laboratory https://pangea.stanford.edu/courses/gp262/Notes/8.SeismicVelocity.pdf
Misra AK (1962) An investigation of the time-dependent deformation or "creep" in rocks. PhD thesis, Sheffield University
Müller WH, Briegel U (1978) The rheological behaviour of polycristaline anhydrite. Eclogae Geol Helv 71(2):397–407
Müller P, Siemes H (1974) Strength, ductility and preferred orientation of anhydrite under mantle pressure up to 5 kilobars at temperatures up to 300 °C. Tectonophysics 23(1–2):105–127
Munson DE, Wawersik WR (1993) Constitutive modeling of salt behavior – state of the technolog. In: Proceedings of the seventh international congress of the rock mechanics, vol 3. A.A. Balkema, Netherlands, pp 1797–1810
Nüesch R, Madsen FT, Steiner W (1995) Long time swelling of anhydritic rocks: mineralogical and microstructural evaluation. In: Eighth internat. congress of rock mechanics, Tokyo, pp 133–138
Oldecop L, Alonso E (2012) Modelling the degradation and swelling of clayey rocks bearing calcium-sulphate. Int J Rock Mech Min Sci 54:90–102
Ordoñez S, Soriano A, Garcia Del Cura MA, Esteban F (1990) Swelling mechanism of tertiary anhydritic-dolomitic shales, 6º Congr intern IAEG, pp 1963–1971
Orti Cabo F (2010) Evaporitas: introducción a la sedimentología evaporítica. In: Arche A (ed) Sedimentología del proceso físico a la cuenca sedimentaria. Textos Universitarios 46. Consejo Superior de Investigaciones Científicas, Madrid, 2010, 1287pp
Papadopoulos Z, Kolaiti E, Mourtzas N (1994) The effect of crystal size on geotechnical properties of Neogene gypsum in Crete. Q J Eng Geol 27:267–273
Passchier C, Trouw R (1998) Microtectonics. Springer, New York, 289p
Pehovaz Alvarez I (2009) Estudo de mecanismos de deformação lenta da gipsita bandada da Chapada de Araripe em ensaios de fluência monitorados por emissão acústica. Tese de Doutorado –Escola de Engenharia de São Carlos – EESC-USP, 356p
Pfeifle TW, Hansen FD (1998) Database of mechanical and hydrological properties of WIPP anhydrite derived from laboratory-scale experiments. Contractor report SAND98-1714. Sandia National Laboratories
Pfeifle TW, Senseny PE (1982) Steady-state creep of rock salt in geoenginnering. In: Proceedings of 23rd symposium on rock mechanics, Berkeley, 25–27 Aug 1982. AIME, New York, pp 333–340
Phueakphum D (2003) Compressed-air energy storage in rock salt of the Maha Sarakham Formation. MS thesis, Suranaree University of Technology, Thailand
Plookphol T (1987) Engineering properties of the evaporite in the Khorat Plateau. MS thesis, Asian Institute of Technology, Thailand
Prucha JJ (1968) Salt deformation and decollement in the Firtree point anticline of Central New York. Tectonophysics 6(4):273–299
Rahn PH, Davis AD (1996) Gypsum foundation problems in the Black Hills area, South Dakota. Environ Eng Geosci 2:213–223
Richards TC (1933) On elastic constants of rock with seismic application. Proc Phys Soc 45(246):70–81
Robertson EC (1962) Physical properties of evaporitic minerals. USGS report TEI 821, 90p
Robertson EC, Robie RA, Books KG (1958) Physical properties of salt, anhydrite and Gypsum: Preliminary Report. United States Department of the Interior Geological Survey, Trace Elements Memorandum Report 1048
Rybach L (1975) Thermal problems in the storage of radioactive wastes in anhydrite. Ver Schweiz Petrol Geol Ing Bull 41:100, 1–100,13
Sahores J (1962) Contribution a l'etude des phenomenes mécaniques accompagnant l'hidratation de l'ánhidrite. Thése University Toulouse, Ver. Materiaux de Construction Pub. Tec. 126

Schwerdtner WM, Tou JC, Hertz PB (1965) Elastic properties of single crystals of anhydrite. Can J Earth Sci 2(6):673–683
Serata S, Gloyna EF (1959) Development of design principle for disposal of reactor fuel waste into underground salt cavities: Univ. of Texas, Civil Eng. Dept., Tech. Rept. Contract AT (11·1)-490, 173p
Spalletti LA (2017) Evaporitas. Apuntes de Catedra. Catedra de Sedimentologia, Facultad de Ciencias Naturales y Museo – Universidad Nacional de la Plata (Argentina). http://www.fcnym.unlp.edu.ar/catedras/sedimentologia/pdf/evaporitas.pdf
Sriapai T, Walsri C, Fuenkajorn K (2012) Effect of temperature on compressive and tensile strengths of salt. ScienceAsia 38:166–174
Steiner W (1993) Swelling rocks in tunnels: rock characterization, effect of horizontal stresses and construction procedures. Int J Rock Mech Min Sci Geomech Abstr 30(4):361–380
Stowe RL (1985) Creep test of WIPP (Waste Isolation Pilot Plant) site anhydrite core. Final report. United States
Tixier MP, Alger RP (1970) Log evaluation of nonmetallic mineral deposits. Geophysics 35(1):124–142
Truesdell C, Noll W (1965) The non-linear field theories of mechanics. Handbuch der Physic, III/3. Springer-Verlag, Berlin
Urai JL, Schléder Z, Spiers CJ, Kukla PA (2008) Flow and transport properties of salt rocks. In: Littke R, Bayer U, Gajewski D, Nelskamp S (eds) Dynamics of complex intracontinental basins. Springer, New York
Vutukuri VS, Lama RD (1978) Handbook on mechanical properties of rocks, vol 1–3. Trans Tech Pub, 406p
Wawersik WR (1985) Determination of steady State Creep rates and activation parameters for rock salt. In: Pincus HJ, Hoskins ER (eds) Measurements of rock properties at elevated pressures and temperatures. ASTM STP 869. American Society for Testing and Materials, USA, pp 72–92
Wetchasat K (2002) Assessment of mechanical performance of rock salt formations for nuclear waste repository in northeastern Thailand. MS thesis, School of Geotechnology, Suranaree University of Technology, Thailand
Wheildon J, Evans TR, Girden RW (1974) Thermal conductivity, density and sonic velocity measurements of samples of anhydrite and halite from sites 225 and 227. In: Initial reports of the deep sea drilling projects, vol 23. US Government Printing Office, Washington, pp 909–911
Yan F, Han D-h, Yao Q, Li H (2014) Seismic velocities of halite salt: anisotropy, dispersion, temperature and stress effects. In: SEG Denver 2014 annual meeting, pp 2783–2787
Yang CH, Daemen JJK, Yin J-H (1999) Experimental investigation of creep behavior of salt rock. Int J Rock Mech Min Sci 36:233–242
Yilmaz I, Sendir H (2002) Correlation of Schmidt hardness with unconfined compressive strength and Young's modulus in Gypsum from Sivas (Turkey). Eng Geol 66:211–219
Zanbak C, Arthur RC (1986) Geochemical and engineering aspects of anhydrite/gypsum phase transitions. Bull Assoc Eng Geol 23(4):419–433
Zierfuss H (1969) Heat conductivity of some carbonate rocks and clayed sandstones. Am Assoc Pet Geol Bull 53(2):251–260
Zong J, Stewart RS, Dyaur N, Myers MT (2017) Lab measurements and Gulf of Mexico well log analysis. Geophysics 82(5):1–80

Chapter 7
Site Investigation for Soft Rock Mass

Zhigang Tao, Cuiying Zhou, Luís Ribeiro e Sousa, and Zhao Feifei

7.1 General Considerations

To identify the characteristics of rock mass, including bedding plane, joints, faults and other discontinuities, detailed field investigations are often required. Field investigation method includes geophysical survey and field openings such as trench, gallery, shaft, and borehole (Lau 1999; Ulusay and Hudson 2007; Wyllie and Mah 2010). It is impossible to establish a common guideline for the field investigation program as, in addition to the complexity of the rock mass, each engineering project has its specific characteristics (Wyllie and Mah 2010; Rocha 2013; Sousa et al. 2019).

In the case of soft rock, sampling and sample preparation are usually difficult. Borehole technique can hardly provide rock core in low strength soft rocks, nor sufficient information on the infillings of discontinuity and permeability of the rock formation (ASTM 1999; Ulusay and Hudson 2007; Wyllie and Mah 2010; Poullain 2012). Core samples can be seriously damaged by groundwater, drilling friction and vibration in diamond drilling, as shown in Fig. 7.1 (Kanji 2014).

Sampling and sample preparation are often difficult for weathered rock. Porto granite is medium-grained two-mica granite surrounding part of Metro do Porto. It is considered as soft rock as it belongs to weathered crystalline rock (Begonha and Braga 2002; Babendererde et al. 2006). It is featured by variety of weathering level,

Z. Tao · Z. Feifei
State Key Laboratory for Geomechanics and Deep Underground Engineering (SKL-GDUE), China University of Mining and Technology, Beijing, China

C. Zhou
Zhongshan University, Guangzhou, China
e-mail: zhoucy@mail.sysu.edu.cn

L. Ribeiro e Sousa (✉)
University of Porto, Porto, Portugal

M. Kanji et al. (eds.), *Soft Rock Mechanics and Engineering*,
https://doi.org/10.1007/978-3-030-29477-9_7

Fig. 7.1 Rock core of disaggregated lava flow (Kanji 2014)

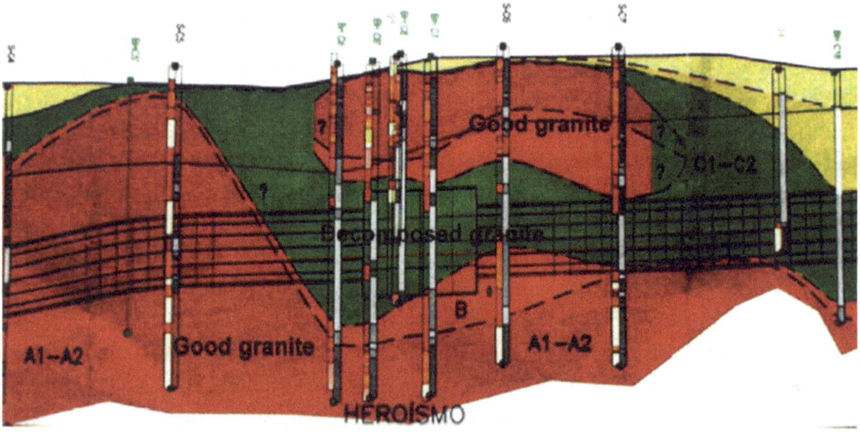

Fig. 7.2 Predicted geology for the Heroísmo station

that is, the Metro do Porto tunnels passed through several weathering levels ranking from fresh granite to residual soil. Thus, the ground behavior varies from strong rock mass to low cohesion granular soil. The granite is less weathered if it is intruded by aplitic or pegmatitic dykes. The weathering depth may reach few tens meters if the rock is affected by stress release in the process of Douro valley deepening. The locations of the weathered rocks are completely irregular and erratic; and the thickness of weathered rock may vary quickly from zero to several meters.

A typical case is Heroísmo station where serious weathered granite is covered by sound granitic rock mass (Fig. 7.2). Figure 7.3 shows borehole samples, left is weathered granite samples that are deeply buried than samples shown in right side, which is drilled from near surface.

Another engineering case of weathered granite is the base rock of Alto Rabagão dam, Portugal, as shown in Fig. 7.4. This hydroelectric scheme includes a solar power generating system. The station collects solar energy using floating photovoltaic cells

Fig. 7.3 Different borehole simples of Porto granite at Metro do Porto (Babendererde et al. 2006)

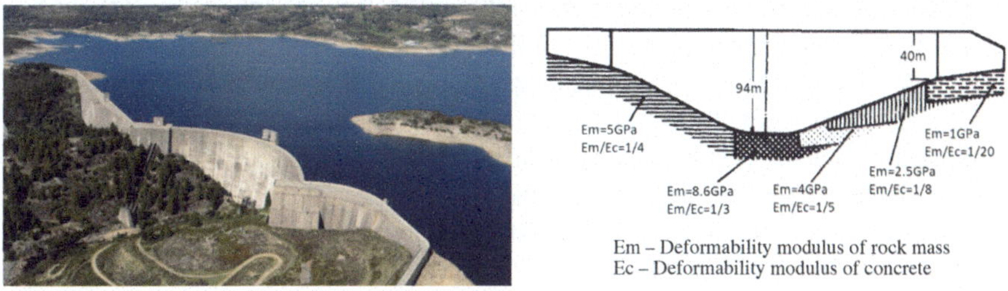

Fig. 7.4 Alto Rabagão dam foundation (Rocha 1975)

during daytime and generates hydroelectric power at night and peak hours. The right abutment of the main concrete arch is founded on highly deformable rock mass, whose deformability modulus varies from 8.6 to 1.0 GPa. The right-most abutment has a deformability modulus about 1.0 GPa, cohesion of 0.4 MPa and friction angle of 52°, with rock uniaxial compressive strength of 2.0 MPa. It belongs to decomposed granite formation where the discontinuity does not affect the deformability and strength of the rock mass significantly. Due to the low strength of the formation, sampling and sample preparation are quite difficult (Rocha 1975, 2013).

Coring includes rock quality observation and measurement of discontinuities. For soft rocks, appropriate preserving measure should be undertaken as some of them are susceptible to degradation. The orientation of discontinuity should be obtained using proper techniques, such as integral sampling method (Rocha 2013) or borehole TV cameras (Sousa 1994; Sousa et al. 1995; Colog Inc. 1995). It can also be obtained through axial scanning of the borehole using TV cameras. Figure 7.5 shows an elliptical

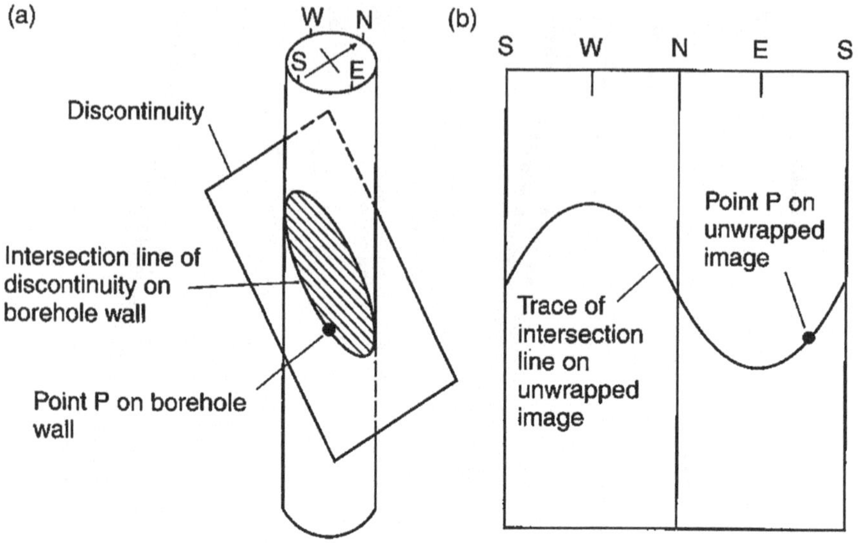

Fig. 7.5 Orientation of a discontinuity in a borehole (Colog Inc. 1995; Wyllie and Mah 2010)

intersection image of a discontinuity in borehole (Fig. 7.5a) and the trace of the intersection line on unwrapped plane (Fig. 7.5b). One Borehole Inspection System (BIS) system, developed at LNEC, Portugal, uses a miniature TV camera operating axial and radial inspections manually. Several applications have been reported at Alto Lindoso hydroelectric scheme and Pracana dam in Portugal, and Cahora Bassa hydroelectric scheme in Mozambique (Sousa 1994; Sousa et al. 1995).

Recently, a new televiewer has been developed using ultrasonic technology (Gaillot et al. 2007; GIF 2009; Zoback 2011; Mount Sopris 2018), which makes it possible to get direct wellbore imaging about discontinuities and faults. This equipment scans the borehole with a centralized rotating ultrasonic transducer generating wellbore image by acoustic reflectance, as shown in Fig. 7.6 (Zoback 2011). Acoustic image logs can reveal natural fracture information, tensile fractures induced by drilling, breakouts, and so on (Fig. 7.7).

Geophysical survey provides a broad view of subsurface conditions, which is recommended in preliminary stages of site investigation programs. Geophysical logging may use artificial sources, such as electrical, nuclear, or acoustic, which promote a disturbance in environment and measure consequent response. It can obtain geotechnical information such as weathering, faults, fracturing, and rock surface fragmentation (Lau 1999; Nickmann et al. 2006; Wyllie and Mah 2010).

Commonly used geophysical methods include electrical methods, electromagnetic methods such as ground penetration radar, acoustic methods such as acoustic televiewer, and seismic wave methods. Among them, the surface refraction is the most commonly used method for evaluation of seismic wave velocities (Wyllie and Mah 2010). Both the particle motion direction and spreading speed of P wave and S wave are different. The velocity of S wave, V_S, is about half of the P wave, V_P. The ratio of V_P/V_S only depends on the Poisson's ratio of the rock. Figure 7.8 shows V_P for different formations.

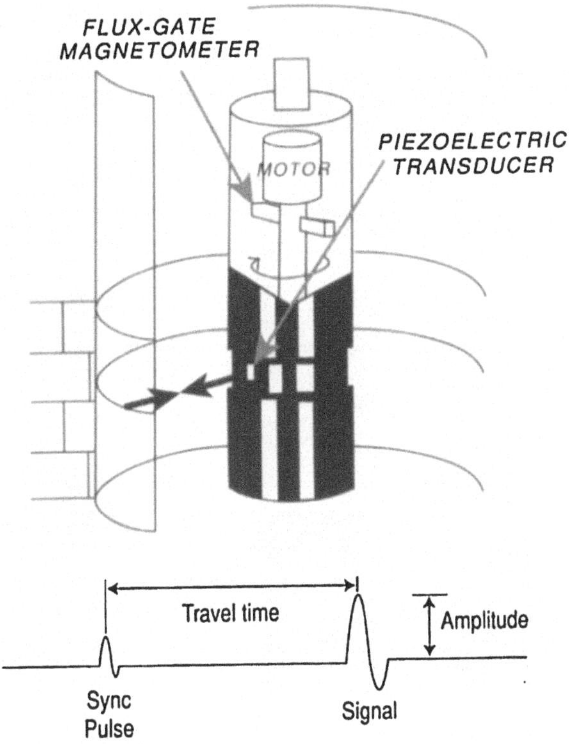

Fig. 7.6 Ultrasonic transducer mounted in a rotating shaft. (Adapted from Zoback 2011)

In mining engineering, large deformation of soft rocks frequently occurs (He 2014; He et al. 2012, 2015). These soft rocks are often characterized by large porosity, high expansibility, poor cementation, high surface fragmentation, and weathering. Figure 7.9 shows roof subsidence in an underground roadway where the surrounding rock deformation and support failure are severely due to high ground stress and swelling of the rock. Figure 7.10 shows the high-steep slope on the west side of the slope of Heilongjiang Open-pit Coal Mine in Inner Mongolia. It can be seen that a weak layer of thickness of 4–12 m exists in No. 6 coal seam. The weak layer consists of clay and muddy clay, whose shear strength reduces greatly when it comes in contact with water. The overlayer of the soft layer is about 45 m height. Therefore, when the weak layer contacts water, it is prone to large deformation and sliding collapse.

Soft rock sampling is an important issue. The strength of soft rock is often low, and soft rock may also be disaggregated or sensitive to moisture contents, causing sampling and sample preparation difficulties. Diamond drilling may destroy the drilling core. Consequently, it is necessarily to develop better sampling and preserving methods for soft rocks.

The SKL-GDUE of China University of Mining and Technology in Beijing and the Zhongshan University in Guangzhou developed new soft rock sampling equipment. This chapter presents details of these sampling equipment. Section 7.2 presents details of the equipment developed at SKL-GDUE and field-testing data. Section 7.3 explains the equipment developed at Zhongshan University.

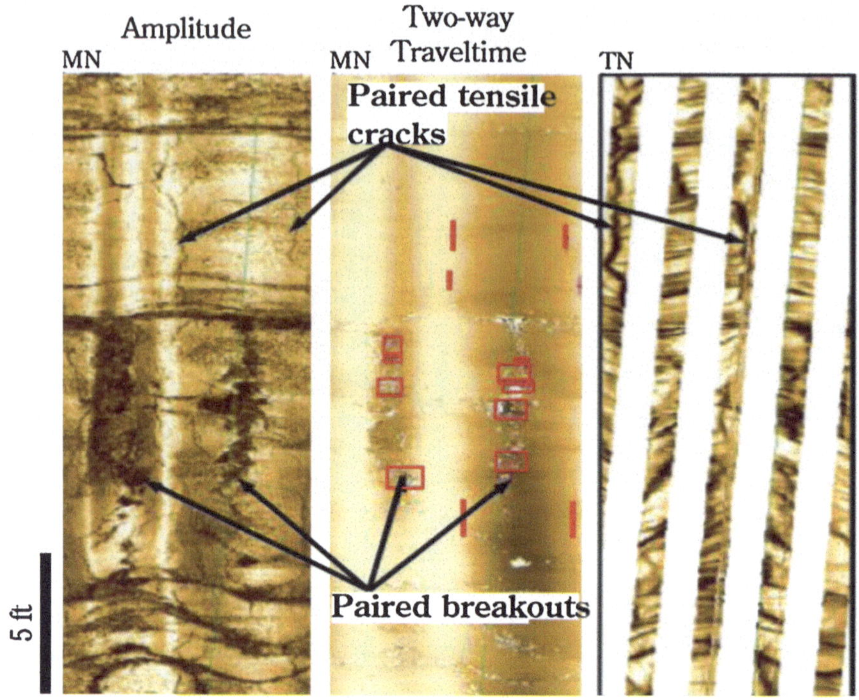

Fig. 7.7 Borehole wall in an acoustic televiewer (Gaillot et al. 2007)

Fig. 7.8 P wave velocity for different rocks (Griffiths and King 1988)

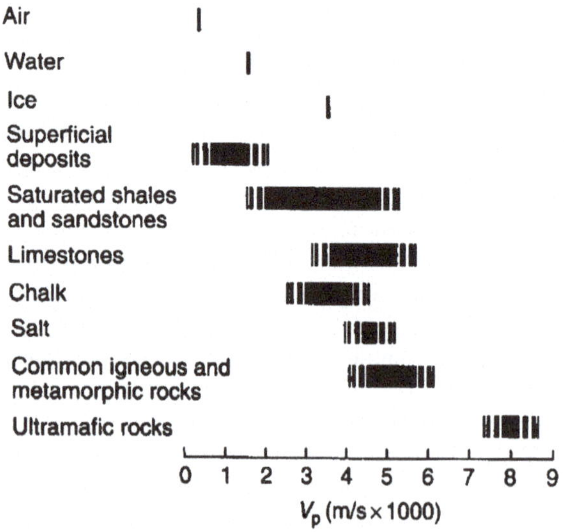

Fig. 7.9 Roof subsidence of the heavy vehicle line roadway (He et al. 2012)

Fig. 7.10 Sedimentary soft rocks in Heidaigou open pit coal mine in Inner Mongolia, China

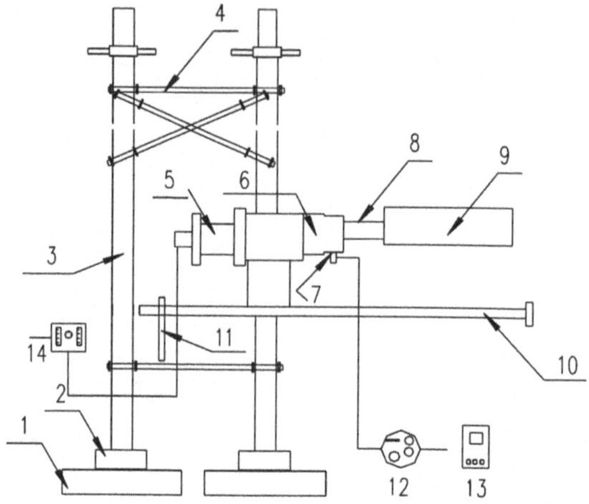

Legend:

1 - base
2 - jack
3 - column
4 - truss
5 - air motor
6 - reducer box
7 - water nozzle
8 - output shaft
9 - single pipe drilling tool
10 - guide
11 - feed lever
12 - water flow monitor
13 - infrared velocimetry
14 - pressure monitor

Fig. 7.11 Diagram of sampling drill for non-swelling samples (He et al. 2018)

7.2 Equipment Developed at SKL-GDUE

7.2.1 Introduction

The SKL-GDUE has developed new equipment for soft rock sampling in the field. The equipment included a field drilling machine, a portable cutting machine and a portable sample box. It can be used for field sample drilling, cutting and polishing, storage, and transportation.

Soft rocks are quite different in the buried depth, strength, and expansibility. To meet different sampling requirements, the developed sampling drilling machine includes non-swelling rock drill and swelling rock drill. Rotary cutting drilling method is adopted in both drilling machines.

The non-swelling sampling drill is mainly composed of the double column truss rig, drilling driven, single pipe drilling tool, and monitoring system. The installation diagram of the whole structure is shown in Fig. 7.11 and a physical drilling rig is shown in Fig. 7.12.

This equipment has four parts. Part I is composed of components 1–4 in Fig. 7.11. The base can stand the whole equipment stably. The column is in the central of the small pillars that are slidable at the top. The bottom jack controls the height of the column, so that upper and lower parts of the equipment can contact with the roof and floor completely. Two columns are cross-connected with each other to keep structure stable during operating. Part II is composed of components 5–11 in Fig. 7.11. The pneumatic motor provides power. The reducer box controls the rotational speed. Drilling speed can be selected according to the lithological characteristics. The lever and guide can advance and retreat the drilling rig in drilling direction. Part III is a single pipe drilling tool, which is represented by 9 in Fig. 7.11. For less swelling rock, sampling is quick and effective. Part IV is monitoring system, which is composed of components 12–14 in Fig. 7.11, including water flow monitor, infrared velocimetry,

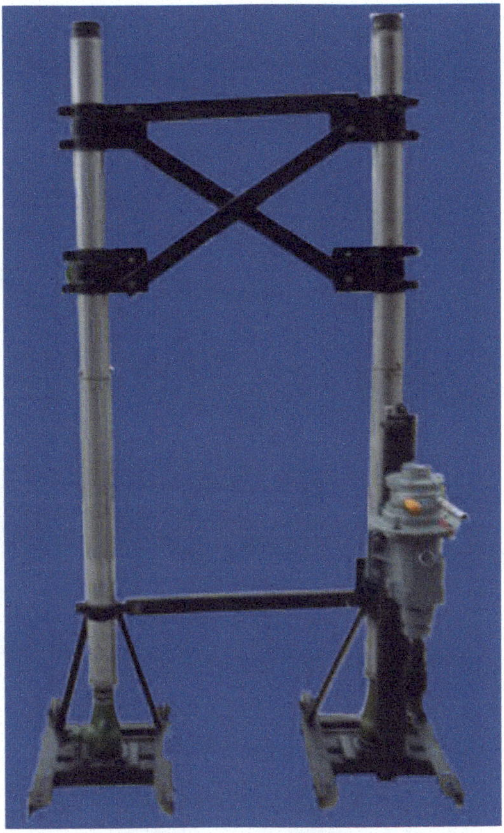

Fig. 7.12 Picture of a sampling drill for non-swelling samples

and pneumatic pressure monitor. It can help to select appropriate water flow, rotation speed, and pneumatic pressure during operating.

The drilling rig is installed on a column, which is made of seamless steel pipe with outer diameter of 80 mm and inner diameter of 70 mm. The column bottom is fixed on the jack. 720° sampling can be achieved by adjusting the driller support to adjust sampling angles, as shown in Fig. 7.13. More details about the sampling drill are provided in the report of He et al. (2018).

7.2.2 *Field Applications*

To test the practicability of the sampling drill, the equipment has been used at a number of sites under different lithological formation, such as Nanfen Iron Ore Mine, Benxi; Shajihai Coal Mine of Shenhua Group; Nantun Coal Mine of Yankuang Group; Jisan Coal Mine of Yankuang Group; Yuncheng Coal Mine; Nanshan Coal Mine of Hebei Coal Group; Hongyang Coal Mine of Shenyang Coal Group; and Chaganzhuoer Mine of Jizhong Energy Resources. Over 800 standard samples were

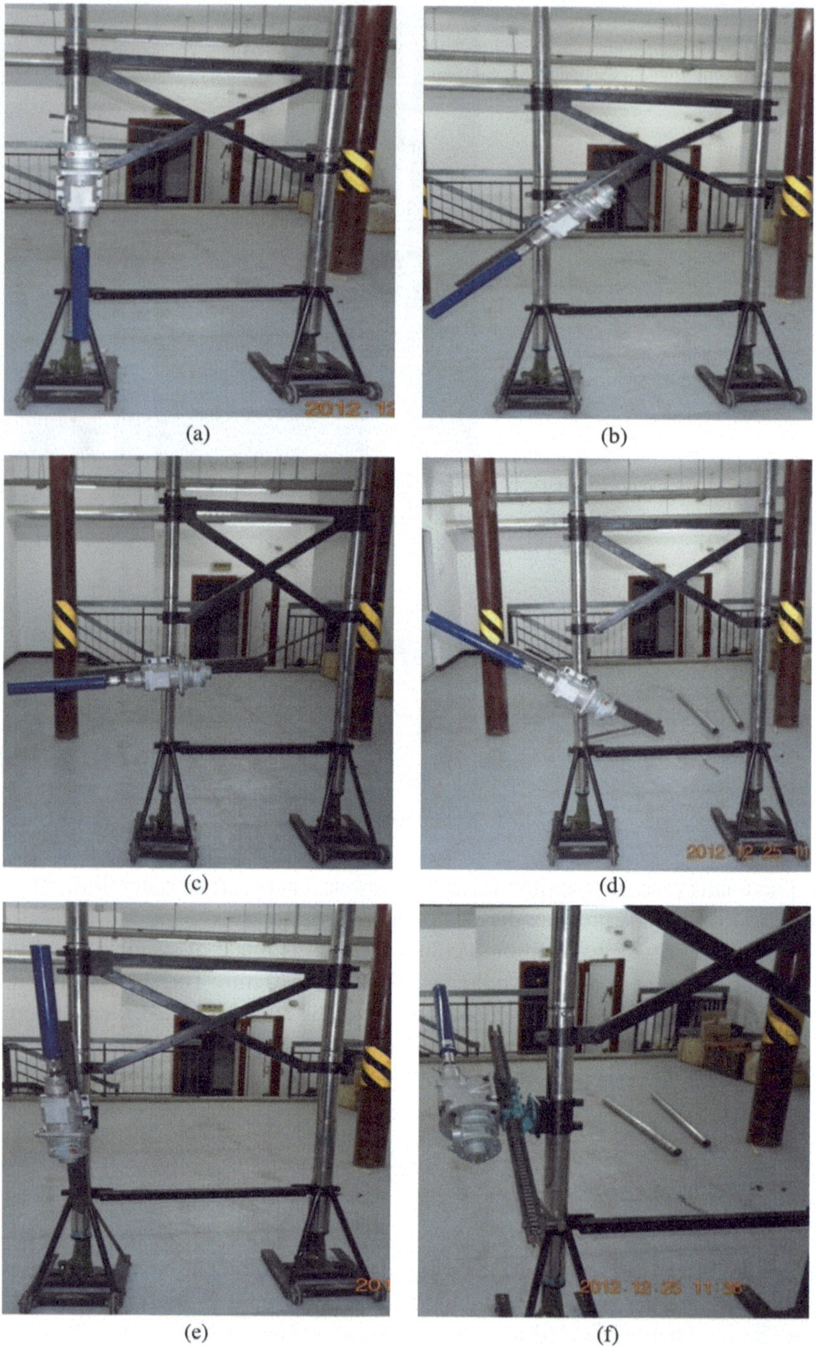

Fig. 7.13 Drilling at different sampling angles (He et al. 2018). (**a**) Vertically downward. (**b**) Downward slant. (**c**) Horizontal direction. (**d**) Upward slant. (**e**) Vertically upward. (**f**) Three-dimensional rotary

Fig. 7.14 Field application of sampling drill (He et al. 2018). (a) Application at Nanfen open iron mine (Liaoning, 2012). (b) Application at Shajihai coal mine (Sinkiang, 2012)

carried out. Figure 7.14 shows field applications at an iron mine and a coal mine. Figure 7.15 shows drilled samples from field.

SKL-GDUE has also developed portable cutting machine for field sampling. After the rock sample is drilled in the field, the cylinder sample can be transported to laboratory for further processing. However, as cutting and polishing have a failure rate, the transported specimens should have a margin for testing. This increases field working load and transportation cost.

To improve sampling efficiency, a portable automatic feeding cutting machine for rock sampling under complex environments has been developed. The schematic diagram of the portable sample cutting machine is shown in Fig. 7.16. It is composed of the sample holding device, sample cutting system, and sample polishing system.

Features of the portable sample cutting machine are the following:

1. The support frame is portable and made of high-strength anticorrosive aluminum. Its three-dimensional structure diagram is shown in Fig. 7.17 and a photograph is shown in Fig. 7.18.
2. Cutting machine and grinding machine are connected with the support frame by a fixing plate, the fixing location can be adjusted according to sample length.
3. The sample is fixed on the clamping device with a V-shaped block and a push rod. The location of clamping device is determined by the ruler.

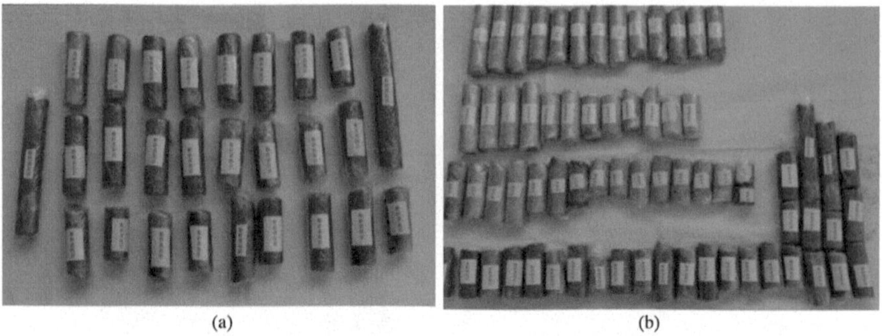

Fig. 7.15 Samples drilled in field (He et al. 2018). (a) Chlorite amphibolite at Nanfen open iron mine. (b) Silt rocks at Shajihai coal mine

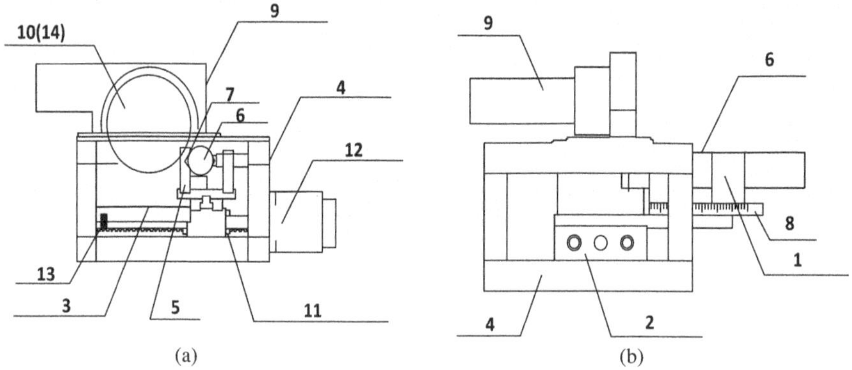

Legend:
1-holding device; 2-supporting block; 3-protective cover; 4-support frame; 5-"V"-shaped located block; 6-sample; 7-rubber gasket; 8-ruler; 9-cutting machine; 10-cutting saw blade; 11-leadscrew; 12-motor; 13-stop valve; 14-polishing plate.!

Fig. 7.16 Schematic of portable sample cutting machine (He et al. 2018). (a) Front view. (b) Left view

4. After the rock sample is clamped, the sample is pushed to cutting position manually.
5. The sample holding device is connected with the leadscrew drive by a sliding block and rolling groove guide. The stepping motor drives the leadscrew drive system to feed and retreat the sample automatically.
6. After cutting, millstone blade can be installed and used to grind the rock sample at required smoothness.

Some standard samples cut by the portable sample cutting machine in the field are shown in Fig. 7.19.

A portable sample box has also been designed taking into considerations of proper measures in sample storage and transportation. It is portable and shockproof. The original state of rock samples can be reserved during transportation. A schematic diagram of the portable sample box is shown in Figs. 7.20 and 7.21.

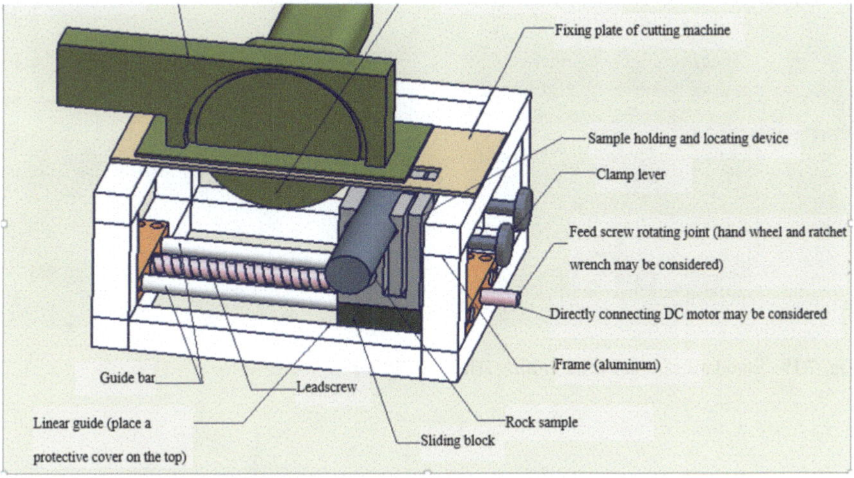

Fig. 7.17 Diagram of the sample cutting machine (He et al. 2018)

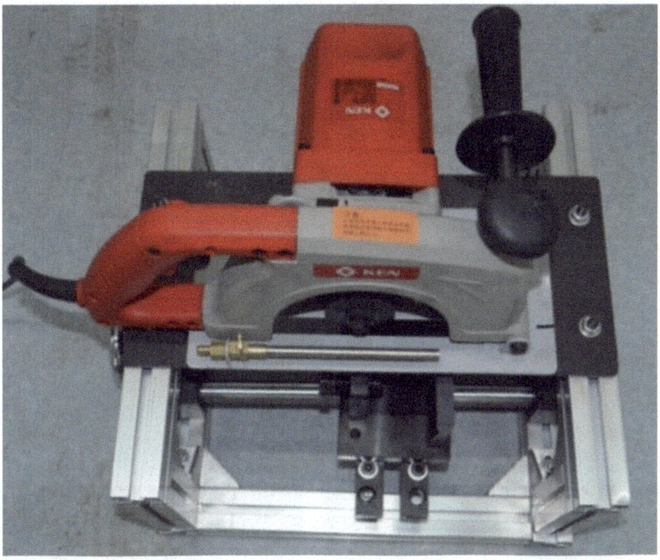

Fig. 7.18 Portable sample cutting machine (He et al. 2018)

7.2.3 Field Sampling Technology

On-site sampling drills, including non-swelling sampling drill and swelling sampling drill, is presented here. To study the influence of water flow on drill rotation speed and pneumatic pressure, an experimental scheme of soft rock drilling at different bedding angles was designed. The experiments analyzed the effect of bedding

Fig. 7.19 Standard samples after cutting (He et al. 2018)

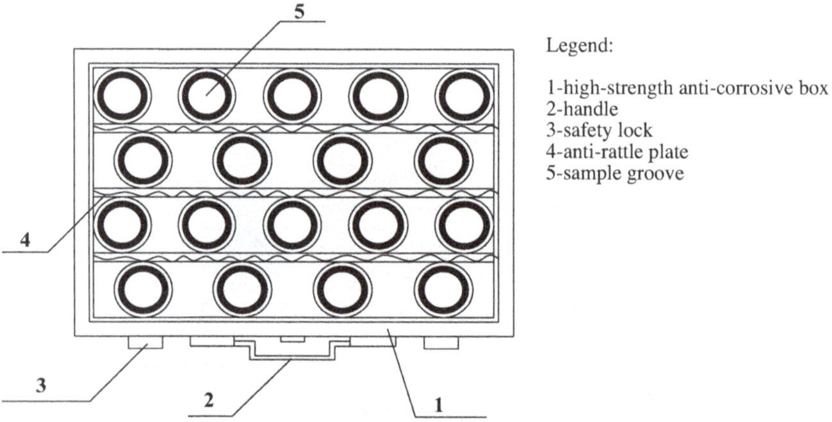

Fig. 7.20 Schematic diagram of structure of portable sample box (He et al. 2018)

Fig. 7.21 The portable sample box (He et al. 2018)

structure on drilling at different bedding angles. Five bedding angles were considered (0°, 30°, 45°, 60°, and 90°). The bedding angle refers to the angle between sample's axial and normal direction of the bedding surface.

Three groups of experiments were adopted:

1. Experiment I: a single pipe drilling tool was adopted. Different bedding angles of the soft rock were drilled while the pneumatic pressure and rotation speed were known. The relation between the water flow and drilling speed was recorded. The optimal water flow at different bedding angles and drilling speed were analyzed.
2. Experiment II: a single pipe drilling tool was adopted. Soft rocks were drilled at different bedding angles under known water flow rate. The relationship between rotation speed and drilling speed is studied. Besides, if the rotation speed was the same, the experiment analyzed the drilling speed at different bedding angles.
3. Experiment III: coring drill was adopted. The soft rock was drilled at different bedding angles at certain pneumatic pressure and rotation speed. The relation between the water flow and drilling speed during double pipe drilling is studied. The experiment analyzed the influence of bedding angle on water flow and drilling speed.

The experiment process was carried out on the site of Shajihai Coal Mine, Sinkiang, as shown in Fig. 7.22.

7.2.4 *Effects of Bedding Structure While Drilling*

7.2.4.1 Experimental Result of Water Flow Rate and Drilling Speed

In Experiment I, the pneumatic pressure and rotation speed were set to be constant as 0.5 MPa and 1300 r/min, respectively. Five bedding angles of 0°, 30°, 45°, 60°, and 90° were adopted for drilling. Water flow rate in the test was adjusted.

Figure 7.23 shows the curves of water flow and the drilling speed at different drilling angles. With the increasing of the water flow, the drilling speed linearly increases at the initial stage. When water flow rate reaches about 7–8 l/min and more, the drilling speed remained about 11 cm/min. That is, when the water flow rate is more than 8 l/min, the drilling speed will not be increased with the increasing of water flow. The best water flow was W_b, which is the minimum water flow when the drilling speed reaches the highest stable speed. When the best water flow was reached, the average drilling speed was 11.3 cm/min.

The combined curves of Experiment I are shown in Fig. 7.24, it can be seen that:

1. With the increasing of water flow, the drilling speed will increase. Water can lubricate drilling and reduce drill bit temperature, which increase the drilling speed;
2. All curves have a similar trend, that is, drilling speed will be increased with the increasing of water flow but with a limit. The water flow refers to the best water flow at the turning point.

Other conclusions were described in the report of He et al. (2018).

(a) (b) (c) (d) (e) (f)

Fig. 7.22 Sampling experimental process (He et al. 2018). (**a**) Water flow monitoring record. (**b**) Pneumatic pressure monitoring. (**c**) Drilling process. (**d**) Drilling of different bedding angles. (**e**) Drilling process. (**f**) Disassemble double pipe and remove sample

7.2.4.2 Experimental Results of Rotation Speed and Drilling Speed

Through Experiment II, the relationship of drilling speed and rotation speed was obtained for different bedding angles of the soft rock. The curves are shown in Fig. 7.25. More results are shown in the report of He et al. (2018).

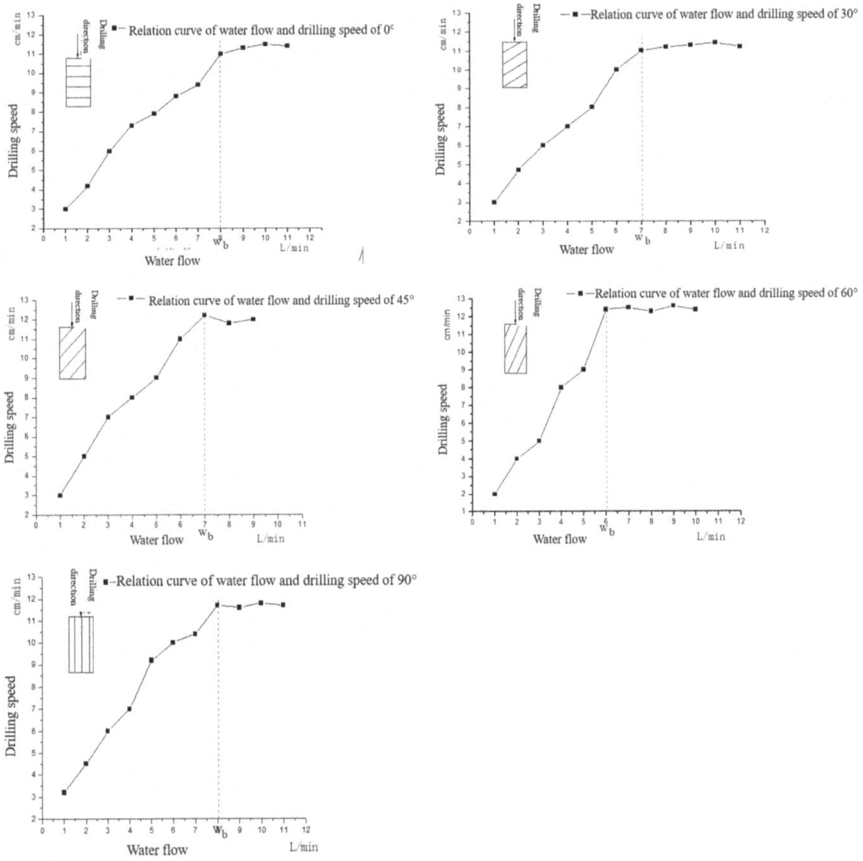

Fig. 7.23 Curves of water flow and the drilling speed for different angles (He et al. 2018). (**a**) Curve of water flow and the drilling speed of 0°. (**b**) Curve of water flow and the drilling speed of 30°. (**c**) Curve of water flow and the drilling speed of 45°. (**d**) Curve of water flow and the drilling speed of 60°. (**e**) Curve of water flow and the drilling speed of 90°

7.2.5 Summary of the Field Sampling System

The feature of the developed on-site sampling system can be summarized as following:

- The sampling system is composed of on-site sampling drill, portable sample cutting machine and portable sample box; the sampling drill is classified into sampling drill of non-swelling sample and sampling drill of swelling sample.
- The sampling drill of non-swelling sample is mainly composed of the double column trussed support, drilling mechanism, single pipe drilling tool and monitoring system. It is light in weight, convenient to move, adjustable and all direction sampling.

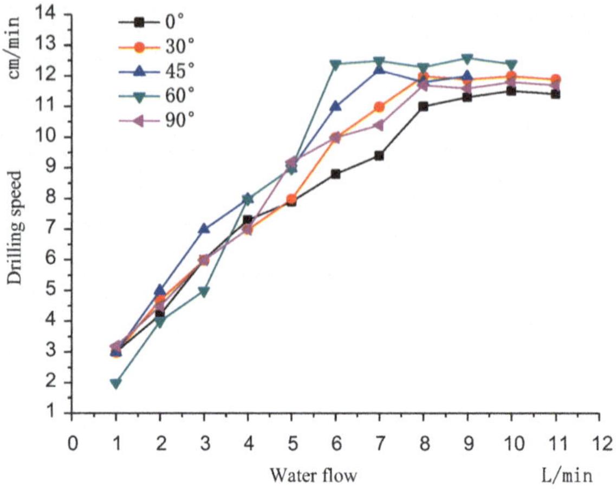

Fig. 7.24 Relation curve of water flow and the drilling speed of different angles (He et al. 2018)

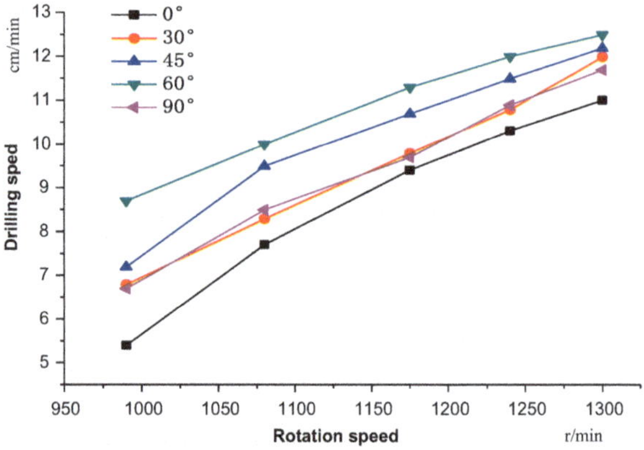

Fig. 7.25 Curves of rotation speed and drilling speed at different angles (He et al. 2018)

- For strong swelling rock sampling, a coring pipe tool was adopted, whose inner pipe does not rotate to protect the sample integrity. The inner pipe is composed of two half pipes that are driven by the rock core pipe along the axis. After two pipes are disassembled, the sample can be taken out.
- The portable sample cutting machine is convenient to carry. The sample can be cut and polished on the site as standard rock testing sample.
- The portable sample box has features of portable and shockproof in transportation.

Through the experiments of sampling and drilling bedding structure effect, the following several points were concluded:

- With the increasing of water flow rate, drilling speed increases but within a limit. The turning point of the flow rate is the best water flow W_b, which is the minimum water flow for the highest stable drilling speed.

7.3 Equipment Developed at Zhongshan University

7.3.1 Introduction

Experimental study of deformation and failure of soft rock is important on the softening mechanism of soft rock. However, due to limitation of test conditions, experiment has been severely restricted. According to the existing soft rock test equipment, the major problems are as follows:

- The soft rock sample preparation equipment often uses water as a coolant. During preparation, the rock sample can easily be disintegrated under water flow, resulting in the failure of sample preparation.
- Most of the pressure chambers in triaxial testing apparatus are enclosed and opaque, which makes it impossible to conduct multi-scale observation on the development of crack, clusters and fracture on the sample during experiment. Some researchers have developed pressure chamber with observation windows, which can only carry out microscopic observations of the development of surface cracks. The observation and measurement of internal and external damage of rock cannot be carried out at the same time, and the process of rock deformation and destruction cannot be connected. It is necessary to develop new technique and test methods to determinate internal and external damage of soft rock simultaneously during experiment.
- The traditional electromagnetic sensor cannot be used in water because of water's conductivity. Therefore, when water is used as the confining medium, it cannot directly measure the distribution of strain (stress).

Therefore, it is significance to develop related sample preparation and triaxial test equipment for soft rock.

7.3.2 Soft Rock Sample Preparation Equipment

Aiming at soft rock sample preparation, a research team at the Zhongshan University has developed soft rock coring equipment, dual-face cutting/grinding rock processing system and soft rock crack prefabrication equipment, which improve the successful rate of soft rock sample preparation. Technical indicators are described in detail in the report of Zhou (2018).

Soft rock sample preparation is often difficulty due to softening and disintegration of soft rock when contacts water. The soft rock sample can be destroyed under cooling water and mechanical disturbance of equipment. A new type of intercooled soft rock coring bit has been developed, which avoids the direct contact between the cooling water and the rock sample during coring.

The intercooled rock coring bit consists of a drill bit with an internal cooling water circulation tubule, a water valve soft plug and a thermally conductive copper ring,

Fig. 7.26 Intercooled soft rock coring bit (Zhou 2018)

which can realize a new coring process to avoid the contact between cooling water and soft rock samples (Fig. 7.26). The soft plug locates at the top of the drill bit, above which is the internal cooling water circulation tubule inlet and outlet. The space under the soft plug is the rock core. Soft plug separates the cooling water and the rock core into upper and lower spaces. The cooling water can only be discharged from the top of the soft plug through the outlet of the tubule. The thermally conductive copper ring bounds outside the drill bit and conducts the heat without cooling water tubule.

The top of drill bit is internal threaded and be installed in commonly used coring machine. A closed space is formed on the top of the soft plug since the cooling water of the vertical coring machine can only enter into the circulation tubule of the drill bit through the water inlet. After flowing through the circulation tubule, the water eventually flows out through the water outlet. Since the incoming cooling water is under pressure, it will jet from the outlet without falling on the rock sample below.

The automatic coring machine is low noisy, durable and easy to operate (Fig. 7.27). It uses an intercooled coring bit of Φ25–200 mm, with the drilling depth up to 400 mm. The coring process avoids direct contact between the cooling water and the soft rock sample, which greatly improves the success rate of soft rock coring. The machine is highly automated, clamping the sample with a special fixture, so that the drilling process can be finished in one button after adjusting the drilling process of high/medium/low speed according to the hardness of the rock specimen.

7.3.3 Dual-Face Cutting/Grinding Rock Processing System

The current rock processing equipment has one function, such as coring, cutting, or grinding. Fixtures of this equipment are normally straight bar, leading to stress concentration in the rock samples, which is may cause crack in the rock samples during the processing, especially for soft rock specimen. To solve this problem, our team has developed a dual-face cutting/grinding rock processing system to cut and grind in one equipment. The fixture system has been redesigned to fit the nature of soft rock.

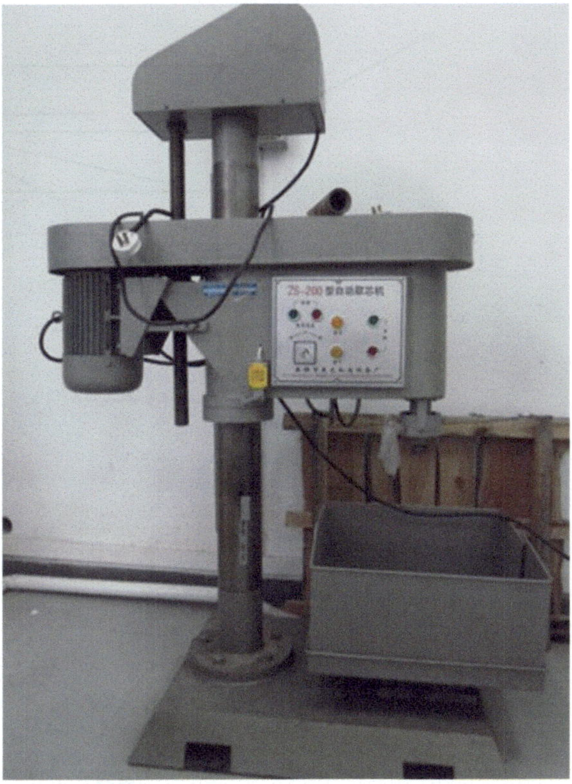

Fig. 7.27 Automatic coring machine (Zhou 2018)

This dual-face cutting/grinding rock processing system consists of base, elevator, workbenches, cutting/grinding motor system, and fixture system (Fig. 7.28). The elevator consists of the main part, pulleys, and knob. The workbench consists of platform, basket, slide rail, groove, and ancillary components. Cutting/grinding motor system consists of two motor systems distributed symmetrically. Each of them consists of base, regulator frame, regulator knob, regulator lever, regulator connection block, grinding stone, motor, blade, fixed knob, and so on. The fixture system consists of base, lower clamp rod, fixing frame, threaded rod, bolt, and upper clamp rod.

The workbench contained soft rock sample can ascend and descend. When the workbench descends to the cutting position, the cutting/grinding motor system on both sides of the workbench moves towards the sample. A high-speed rotating blade cut both ends of the sample to two parallel surfaces. Thereafter, the workbench ascends to the polish position. Through a back-and-forth movement of the workbench, the two end surfaces of the sample are polished by high speed grinding stone. In this way, the cutting and grinding can be accomplished by just holding the sample once, which reduce the sample clamping times improving the success rate of sample processing.

As red-layer soft rock is easily broken in prefabricating, an ultrafine twist-drill-bit-drip cutting method is designed. Compared with the cutting method of laser,

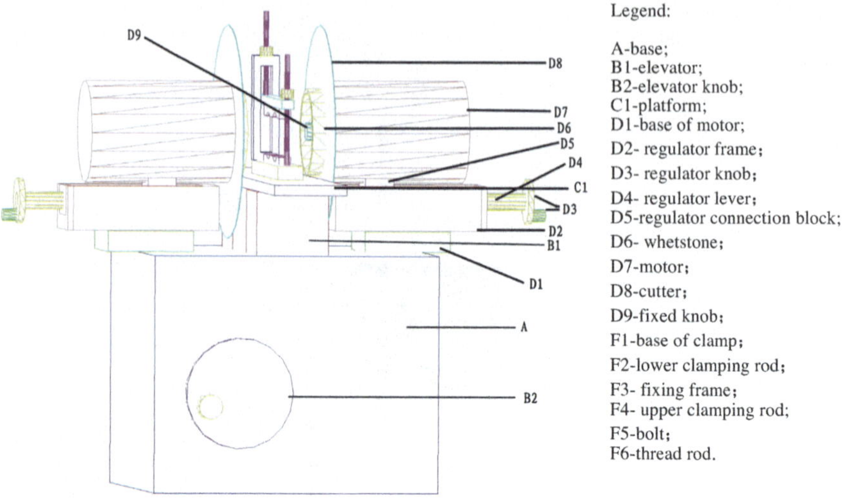

Fig. 7.28 Diagram of the dual-face cutting/grinding rock processing system (Zhou 2018)

Fig. 7.29 Soft rock crack prefabricated equipment (Zhou 2018)

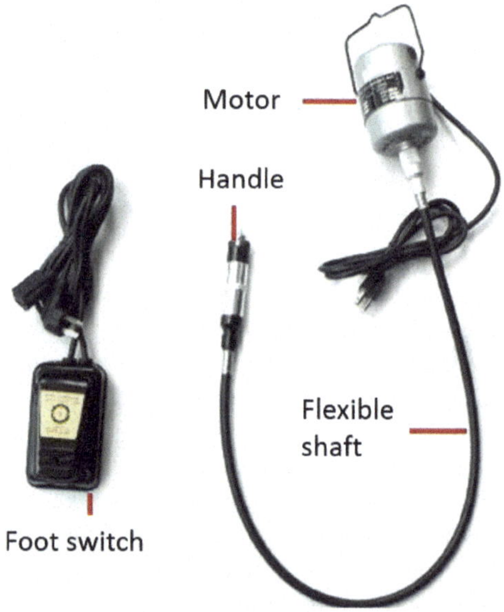

ultrasonic drilling machine and high-pressure water jet cutting machine, the equipment has less vibration, less water consumption, and simple controlling.

Soft rock crack prefabricated equipment consists of foot switch, motor, flexible shaft and handle, as shown in Fig. 7.29. The motor is connected to the foot switch with the power cable, and the speed of the motor is adjustable by the foot switch. The handle is connected with the motor through the flexible shaft, and the output power of the motor is transmitted to the drill bit on the handle through the flexible shaft. High-power mute motor is used in this device to reduce the noise. Flexible shaft is stainless steel hose band for anti-corrosion and durable, and the handle can be easily replaced; 500–25,000 stepless speed-change can be achieved.

7.3.4 Water–Stress Coupling Meso-mechanics Test System

Conventional rock triaxial apparatus uses oil to provide confining pressure and requires that the rock sample must be protected by an envelope. It is impossible to simulate softening conditions in certain practical engineering (taking into account water–stress coupling), and cannot observe the soft rock in water–stress coupling.

Therefore, developing a test equipment capable of simulating the deformation and failure process of soft rock in the real engineering environment was important to study the water–rock coupling.

The Zhongshan University has developed "TAW-100 water–stress coupling soft rock meso-mechanics servo triaxial test system" by improving the loading system, pressure chamber and microscopic observation part of the traditional rock triaxial testing machine. More details are given in chapter entitled "Methodology for Evaluating Mechanical Properties of Soft Rock Masses by Laboratory and In Situ Testing" (Sousa et al. 2019), and in the report of Zhou (2018).

References

ASTM (1999). Standard practice for rock core drilling and sampling of rock for site investigation. American Standard Designation D 2113-99, p 20

Babendererde S, Hoek E, Marinos P, Cardoso AS (2006) Chapter 3: Geological risk in the use of TBMs in heterogeneous rock masses – the case of "Metro do Porto". In: Matos, Sousa, Kleberger, Pinto (eds) Geotechnical risks in rock tunnels. Taylor & Francis Group, London, pp 41–52

Begonha A, Braga MS (2002) Weathering of the Oporto granite geotechnical and physical properties. Catena 49:57–76

Colog Inc. (1995) Image processing system (BIP). Golder CO, and Raax Co. Ltd., Australia

Gaillot P, Brewer T, Pezard P, Yeh E (2007) Borehole imaging tools – principles and applications. Scientific Drilling, no. 5, September

GIF (2009) Borehole explorations. Brochure, Ettlingen

Griffiths DH, King RF (1988) Applied geophysics for geologists and engineers, 2nd edn. Pergamon Press, Oxford

He M (2014) Latest progress of soft rock mechanics and engineering in China. J Rock Mech Geotech Eng 6:165–179

He M, Chen X, Zhang G, Sousa LR (2012) Large deformations analysis in deep coal mines in China. In: Sousa, Vargas, Fernandes, Azevedo (eds) Innovative numerical modeling. Taylor & Francis, London, pp 333–353

He M, Sousa RL, Müller A, Vargas E Jr, Sousa LR, Chen X (2015) Analysis of excessive deformations in tunnels for safety evaluation. Tunn Undergr Sp Technol 45:190–202

He M, Tao Z, Wang X (2018) In: China University of Mining and Technology (ed) Site investigation of soft rock masses. Report, SKL-GDUE, Beijing, p 52
Kanji M (2014) Critical issues in soft rocks. J Rock Mech Geotech Eng 6(2014):186–195
Lau KC (1999) A review of downhole geophysical methods for ground investigation. Geo Report no. 9, Hong Kong, p 67
Mount Sopris (2018) Teviewers, Mount Sopris Instruments, Brochure. Denver
Nickmann M, Spaun G, Thuro K (2006) Engineering geological classification of weak rocks. Geological Society of London, London, IAEG paper no. 492, p 9
Poullain J (2012) Drilling and sampling of soil and rock. PHD Center, Fairfax, p 22
Rocha M (1975) Some problems regarding rock mechanics of low strength materials. V. Pan-American congress on soil mechanics and foundation engineering, Buenos Aires, pp 489–514 (in Portuguese)
Rocha M (2013) Rock mechanics. LNEC, special edition, Lisbon, p 410 (in Portuguese)
Sousa LR (1994). Observation of the underground works of Cahora-Bassa. Complementary observation plan. LNEC report 146/94, Lisbon (in Portuguese)
Sousa LR, Leitão NS, Vicente D (1995) Use in geotechnical works of TV cameras for prospecting boreholes. In: Fifth Portuguese geotechnical congress, Coimbra (in Portuguese)
Sousa LR, Sousa RL, Zhou C, Karam K (2019) Methodology for evaluating mechanical properties of soft rock masses by laboratory and in situ testing 6(3)
Ulusay R, Hudson J (2007) The complete ISRM suggested methods for rock characterization. Testing and monitoring. ISRM Turkish National Group, Ankara, p 628
Wyllie D, Mah C (2010) Rock slope engineering, 4th edn. Taylor & Francis Group, London, p 431
Zhou C (2018) Study and development of sample preparation and triaxial test equipment for soft rocks. Report. Zhongshan University, Guanzhou, p 37
Zoback MD (2011) Reservoir geomechanics. Cambridge University Press, Cambridge, p 449

Chapter 8
Evaluation of Geomechanical Properties of Soft Rock Masses by Laboratory and In Situ Testing

Luís Ribeiro e Sousa, Rita Leal e Sousa, Zhou Cuiying, and Karim Karam

8.1 Introduction

In recent years, the evaluation of geomechanical parameters in rock masses and particularly in soft rock masses has gone through significant improvements. This is partly due to improved measuring equipment and partly due to better numerical techniques. New instruments and equipment for in situ and laboratory tests allow for a more accurate evaluation of the properties of soft rock masses. Advancements in data mining (DM) techniques allow for better tools for decision making. The combination of improved instrumentation and more powerful numerical techniques allow for a better characterization of the geomechanical parameters of rock masses.

Due to the variability of rock formations, and the expense both in time and cost of obtaining subsurface information, there is a large degree of uncertainty associated with the evaluation of geotechnical properties. This is made even more uncertain given the complex geological processes involved and the inherent difficulties in geomechanical characterization (ASCE 1996; Sousa et al. 1997, 2010; Yufin et al. 2007; Miranda 2007; Miranda et al. 2009). As a result, the evaluation of geomechanical parameters is often carried out through in situ and laboratory tests along with the application of empirical methodologies (Bieniawski 1989; Barton 2000; Hoek 2007a, b; Miranda et al. 2018). For deformability characterization, in situ tests

L. R. e. Sousa (✉)
Tongj University, Shanghai, China

China University of Mining & Technology, Beijing, China

R. L. e. Sousa
Stevens Institute of Technology, Hoboken, New Jersey, USA

Z. Cuiying
Zhongshan University, Guanzhou, China

K. Karam
Sarooj Construction Company, Muscat, Oman

M. Kanji et al. (eds.), *Soft Rock Mechanics and Engineering*,
https://doi.org/10.1007/978-3-030-29477-9_8

are normally carried out by applying a load in a certain way and measuring the corresponding deformations of the rock mass. For strength characterization, tests include shear and sliding tests which are often performed in low strength surfaces. These strength tests are expensive and strength parameters are often inferred indirectly by, for example, the Hoek and Brown (H-B) strength criteria normally associated with the GSI empirical system (Hoek 2007a; Hoek et al. 2008; Hoek and Marinos 2009). Laboratory tests are only relevant to a small rock volume and consequently it is necessary to perform a considerable number of tests in order to characterize the variability in geomechanical parameters, even if much effort is put in obtaining representative samples. Laboratory tests such as the determination of uniaxial compressive strength (UCS), point load and discontinuities tests are nonetheless important for empirical methodologies. Based on experience, it can be said that it is necessary to obtain direct geomechanical information from the site in study and it is insufficient to interpolate and extrapolate data from other sites.

Soft rocks exhibit unfavorable behavior related to low strength, high deformability, fast weathering, as well as others (Kanji 2014; He 2014). As such, sometimes testing equipment needs to be modified to accommodate for this. Soft rocks can be sedimentary rocks and weathered igneous and metamorphic rocks, or so-called residual rocks (Rocha 1975; Kanji 2014). The deformability moduli of soft rock masses, even those for residual formations can be considerably higher than those for soil formations. Lower bound of the deformability moduli of soft rock masses can be in the range of 0.4 GPa. For soils, lower values of cohesion can be in the range of 0.3 MPa, with friction angle of the same order of that for soft rocks (Rocha 1975).

For design purposes shear strength parameters are often selected rather than determined. The selection of deformability and strength parameters requires mainly sound engineering judgment, experience and on the use of empirical systems (ASCE 1996; Wyllie 1999; Sousa et al. 2010). The selection of design shear strength parameters in soft rocks is dependent on the particular site characteristics which include geological structures, planes of weakness, discontinuities, amongst others. In the soft rocks, discontinuities tend not to be as significant as for hard rocks. Failure envelopes for upper and lower bounds of shear strength can be determined for different types of potential failure surfaces, for example, in intact rock, with clean discontinuities and with filled discontinuities, and low strength surfaces. Technical Engineering and Design Guides issued by the US Army Corps of Engineers describe the appropriate selection procedures (ASCE 1996). Using the H-B (Hoek & Brown) criterion, Serrano and Olalla (2007) demonstrate the applicability of this criterion and the identification of applicable parameters based on the theory of plasticity. The rock mass is assumed to behave as a continuum and expressions are based in the theory of elasticity. Therefore, the selection of design parameters involves the selection of Poisson's ratio and deformability modulus. For almost all rock masses, Poisson's ratio varies between 0.1 and 0.35, and as a rule, lower values correspond to poorer quality rock masses. The selection of an adequate deformability modulus is important in order to make reliable predictions of deformations (Sousa et al. 2010).

Rock masses are in general described as heterogeneous and discontinuous media and their mechanical properties depend on both the rock material and the discontinuity sets. In weak rock masses, the influence of discontinuities on the mechanical properties of the rock mass is not very significant. The evaluation of mechanical characteristics is influenced by the dimensions of the tested mass. In general, for deformability evaluation, the rock mass can be divided into zones, where each zone is considered as homogeneous based on its degree of weathering and discontinuity network. A reduced number of in situ large scale tests can then be performed in each zone. Small scale in situ tests in soft rocks should be performed in larger numbers for each zone. It is also important to perform laboratory tests for both deformability and strength in these type of rocks, and for the application of empirical systems, since the mechanical properties of the soft rock mass depend strongly of the properties of the rocks (Sousa et al. 1997).

The traditional empirical systems such as like RMR, Q and GSI systems (Bieniawski 1989; Barton 2000; Hoek 2007a; Kanji 2014) are also used in soft rocks. Strength is normally obtained by the H-B criterion using the GSI system (Hoek et al. 2005; Hoek 2007a). The use of the Q system is also recommended for the development of a new strength criterion (Barton 2013). There is, however some room for improvements to be made using artificial intelligence techniques, data mining (DM) techniques in particular. For example, a new empirical system was developed for heterogeneous volcanic rocks and validated using DM (Miranda et al. 2018).

Heterogeneities in rock masses are of great importance when evaluating mechanical properties of soft rocks, such as conglomerates, and residual rock masses. This is the case for the residual granite formations of Porto (Miranda et al. 2014), where the behavior of the rock mass is very unpredictable. As such, adequate measures need to be taken when planning and constructing engineering systems such as tunnels, for example continuous characterization of the tunnel face, real time monitoring to avoid accidents (Fig. 8.1), (Miranda 2007; Sousa et al. 2010; Sousa and Einstein 2012). A case study illustrating the characterization of a conglomerate for a fill dam in Japan will be presented in Sect. 8.2.4.

When dealing with rock masses, an important issue is the occurrence of surfaces with low strength. These can lead to significant consequences such as the accident

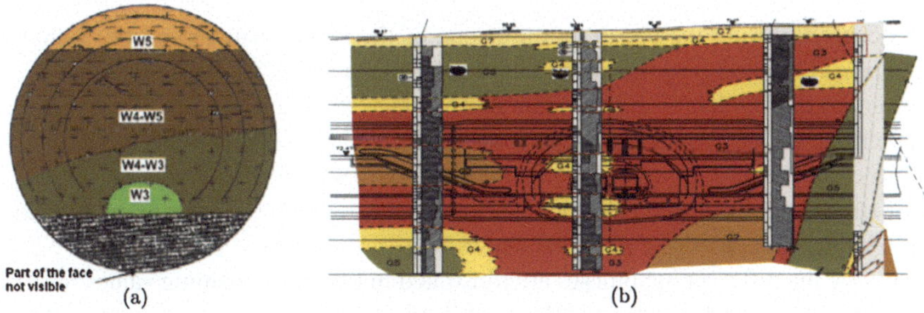

Fig. 8.1 Mixed face conditions found during the construction of the Metro do Porto: (**a**) View of a TBM tunnel face with different weathering degrees; (**b**) Cross-section of Bolhão underground station (Babendererde et al. 2006; Miranda et al. 2014)

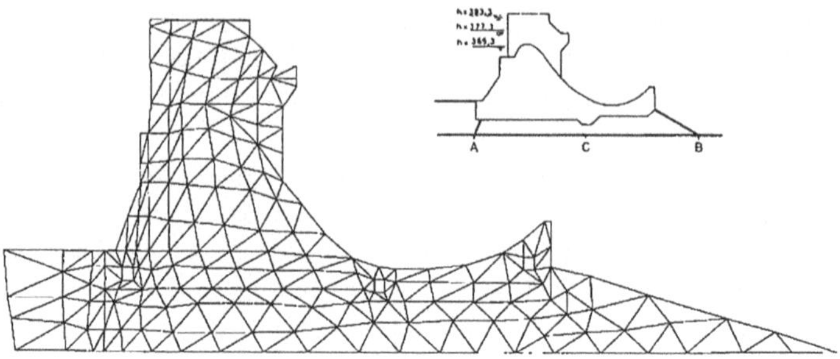

Fig. 8.2 Numerical model used for the study of Água Vermelha foundation. (Adapted from Pedro et al. 1975)

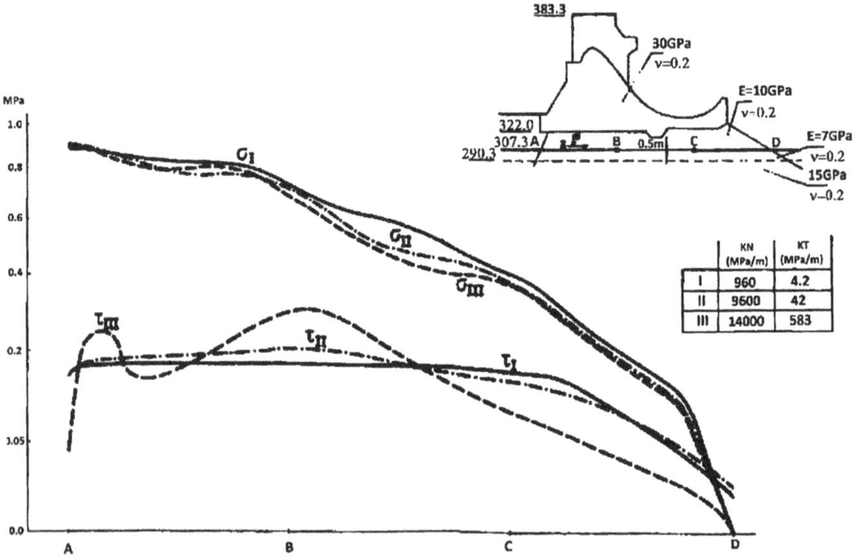

Fig. 8.3 Numerical results for the Água Vermelha foundation (Pedro et al. 1975; Rocha 1975)

that occurred during the construction of one of the surge chambers of the Cahora Bassa Hydroelectric Scheme in Mozambique (Rocha 1978; Sousa 2006, 2010).

Another case is that of the foundation of Água Vermelha dam in Brazil, in which an extensive basaltic circular sub-horizontal surface with thickness of 50 cm occurred (Pedro et al. 1975; Rocha 1975; Pacheco et al. 2017). Different finite element (FE) calculations were performed in order to analyze the structural behavior of the foundation for three hypotheses. Figure 8.2 shows the FE model used (Pedro et al. 1975). Results for the different hypotheses are illustrated in Fig. 8.3, assuming values for the normal and tangential stiffnesses KN and KT, respectively. The low strength surface was assumed to exhibit elastoplastic behavior with zero cohesion and a friction angle of 28°. Failure occurred in the extreme downstream zone for hypotheses I and II. The distribution of tangential stresses tends to be uniform when the deformability of the low strength increases. It is possible to conclude that the risk of progressive rupture decreases with increasing deformability of the surface (Rocha 1975).

The chapter discusses in situ and laboratory tests used for deformability and strength evaluation and presents results in soft rock masses case studies. It is not the intention in this chapter to deal with all the laboratory and in situ tests for soft rock masses, but rather to establish a methodology to characterize the mechanical properties of these type of formations. Results of extensive testing on a very heterogeneous conglomerate rock mass at the site of a dam in Japan are presented.

8.2 Methodology for Evaluating Deformability

8.2.1 Deformability Evaluation

The evaluation of the deformability parameters of rock masses is essential for different applications in Civil and Mining Engineering. For many reasons, such as the heterogeneous and discontinuous nature of rock masses, anisotropy and others, the evaluation of the parameters is largely influenced by the dimensions of the tested volumes (Sousa et al. 1997; Rocha 1975, 2013).

The deformability characteristics of a rock mass can be determined by means of in situ and laboratory deformability tests, in which a load is applied in a specific manner, and the corresponding rock mass deformations are measured. To assess these characteristics, several types of tests are conducted on the rock and on discontinuities, which make it possible to obtain an acceptable representation of the rock mass behavior during and after the execution of projects. In some cases, rock masses are only homogeneous at very large scales and it becomes not economically feasible to perform representative tests. An insufficient test volume causes problems of scale effects (Rodrigues and Graça 1983; Simon and Deng 2009). The basic scale effects in deformability evaluation can be explained by the theory of statistics as follows (Grossmann 1993): whenever a rock mass deforms, the global deformation in a certain direction is the sum of a large number of small aleatory individual deformations in the same direction, of the different constitutive elements of the rock mass. Thus, from the laws of statistics, the global deformation has a normal distribution, whose mean value is independent of the number of summed individual deformations. Therefore, if the tests performed are chosen randomly, the experimental results should always present the same mean deformability and a standard deviation proportional to the square root of a significant length of the tested volume.

From these considerations, the deformability results obtained from small-scale tests are much more variable than those from large-scale tests. Thus, a sufficient number of tests should be performed in order to compensate for this variability. The tests for determining the deformability of rock masses must be performed prior to design. During construction, it is possible to obtain the response of the entire engineering structure by monitoring and comparing the observed results with the predictions by numerical models. The reliability of the models depends fundamentally on the parameters that govern the behavior of the engineering system. Thus, it is important to use system identification techniques in order to re-evaluate the deformability parameters of the rock mass by taking into account the results obtained from monitoring and previous field tests (Castro and Sousa 1995).

Table 8.1 illustrates typical values for the deformability for soft rocks as well as shear strength values and UCS (Unconfined Compressive Strength). For low deformability rocks, deformability moduli can be considerably higher than those for soils. Table 8.2 shows the results of deformability moduli for soft rocks and corresponding rock mass values, for different types of formations in Spain, Angola, Taiwan, Greece, Iran, and Japan (Rocha 1975; Sousa et al. 1997; Hoek 2007a).

In the evaluation of deformability, it is important to consider anisotropy that exists in rock masses. In soft rock masses the anisotropy is mainly related with the rock anisotropy that influences the rock mass properties. Tests as shown in Table 8.2 were conducted parallel and normal to the schistosity. The relationship between tests perpendicular and parallel to schistosity varied between 0.3 and 0.9. In residual rocks, anisotropy also exists due to weathering effects (Rocha 1975). Therefore, deformability of soft rock masses should consider time effect (He et al. 2015b). Heterogeneity is also an important aspect to be considered, since all soft rock masses with low strength presents in general these characteristics. This is true for sedimentary rocks and for residual rock formations, which is the case for instance for a conglomerate formation from Yulin caves in China (Fig. 8.4) or for residual rock granite formations in Porto, Portugal (Fig. 8.5), (Medley 1994; Sousa et al. 2015).

Table 8.1 Deformability and strength range values for soft rocks (Rocha 1975)

Type of rock	E (GPa)	Shear strength		UCS (MPa)
		c (MPa)	φ (°)	
Claystone and siltstone				
Low	0.4–3.0	0.5–3.0	30–35	2–12
Average	3.0–8.0	3.0–6.0	35–40	12–25
High	8.0–30.0	6.0–12.0	40–55	25–80
Sandstone and conglomerates				
Low	0.5–4.0	0.5–3.0	30–40	2–12
Average	4.0–10.0	3.0–8.0	40–50	12–40
High	10.0–60.0	8.0–16.0	50–65	40–150
Limestones				
Clay	0.5–5.0	0.5–4.0	30–40	2–20
Shale	1.0–10.0	1.0–6.0	30–40	4–30
Sound	20.0–100.0	10.0–40.0	40–50	40–250
Schists				
Decomposed	0.4–2.0	0.4–2.0	30–35	1.5–8
Weathered	2.0–15.0	2.0–12.0	35–40	8–40
Sound	15.0–80.0	10.0–20.0	40–65	40–200

Table 8.2 Deformability in soft rocks and rock masses

Formation	Location	Region	E (GPa) Rock (E_r)	Rock mass (E_{RM})	$\alpha = E_{RM}/E_r$	Comments
Schist	Cedillo	Spain	90	40	0.44	Parallel to schistosity
Schist	Cedillo	Spain	40	12	0.30	Perpendicular to schistosity
Schist	Alcantara	Spain	140	5	0.04	Parallel to schistosity
Grés	Cambambe	Angola	65	8.6	0.13	Rocha (1964)
Conglomerate	Aviaki	Greece	60	6	0.10	Rocha (1964)
Siltito	Aviaki	Greece	15	1.5	0.10	Rocha (1964)
Argilito	Karum	Iran	11.5	7	0.61	Rocha (1964)
Limestone	Karum	Iran	50	7.5	0.15	Rocha (1964)
Sandstone	Mingtan	Taiwan	22.3	4.2	0.19	Perpendicular to schistosity
Sandstone	Mingtan	Taiwan	22.3	3.7	0.17	Parallel to schistosity
Siltstone	Mingtan	Taiwan	10.6	3.3	0.31	Perpendicular to schistosity
Silstone	Mingtan	Taiwan	10.6	5.7	0.56	Parallel to schistosity
Interbedded sandstone and siltstone	Mingtan	Taiwan	12.8	2.8	0.22	Perpendicular to schistosity
Interbedded sanstone and siltstone	Mingtan	Taiwan	12.8	3.0	0.23	Parallel to schistosity
Conglomerate	Site B	Japan	0.23	0.74	3.23	Sousa et al. (1997)
Conglomerate	Site C	Japan	0.22	0.50	2.27	Sousa et al. (1997)

For engineering purposes, it is useful to define a modulus reduction factor α, which represents the ratio of deformability modulus between rock mass and a smaller element of the rock material, as presented in Table 8.2. The two cases for conglomerates exhibit reduction factors greater than others due to the way the tests were performed in these heterogeneous formations.

Figure 8.6 represents the modulus reduction factor vs. the RMR coefficient. The correlation is based on values referred in the publication of Sousa et al. (2010). In accordance to the derived equation and considering that soft rocks can vary with RMR values between 40 and 10, the modulus reduction factor varies, respectively between 0.25 and 0.10, which can be considered in accordance to what is presented in Table 8.2.

In the following sections, in situ deformability tests are described which include borehole tests, plate load tests and flat jack tests, as well as laboratory triaxial tests including rockburst tests. A special reference is also made for the opening of galler-

Fig. 8.4 Conglomerates at Yulin caves (Sousa et al. 2015)

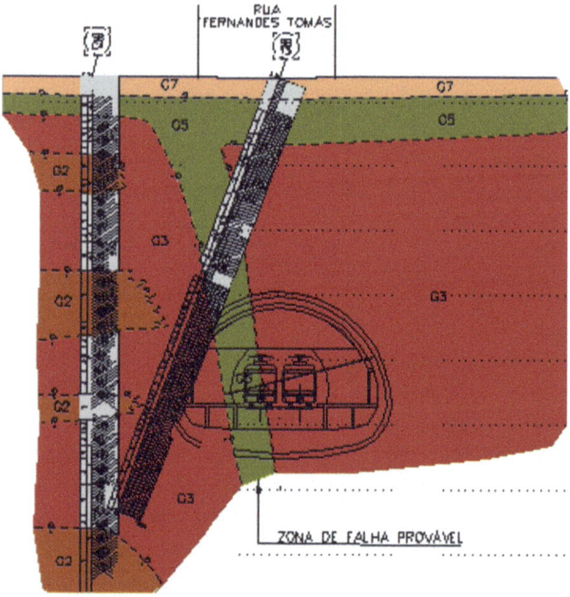

Fig. 8.5 Geological cross-section of Bolhão station (Miranda et al. 2014)

ies and in situ hydromechanical water loading tests. Sections are closed with reinforced concrete plugs and measuring instruments are installed (Ulusay and Hudson 2007; Rocha 2013). A section is dedicated to a deformability investigation on a heterogeneous soft rock mass.

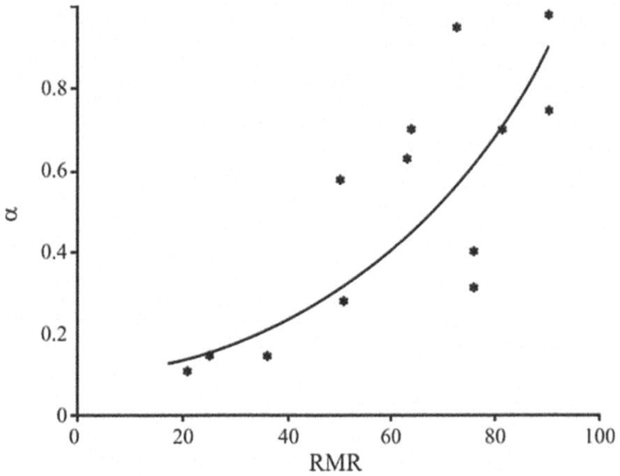

Fig. 8.6 Modulus reduction factor vs. RMR (Sousa et al. 2010)

Table 8.3 Suggestions about the use of large scale tests for deformability evaluation

Situation	E (GPa)	Large scale tests
A	$E \geq 10$	Advisable
B	$5 \leq E < 10$	Necessary
C	$0.1 \leq E < 5$	Necessary with high precision

The general methodology, as mentioned in the Sect. 8.1, is to section the rock mass into different zones which are assumed homogenous (Graça 1983; Sousa et al. 1997). In each zone, several borehole tests can be performed to investigate the rock mass and deliver samples used for deformability tests, dilatometer or pressuremeter tests. The distribution of the tests can be of two distinct types: they can either be performed at random locations using a large number of tests in order to estimate mean values such as mean deformability; or they may be performed at in locations where the deformability modulus is expected to be lower (Sousa et al. 1997). To interpret the results, the criterion used for the location of the boreholes must be considered. In the case where a large number of tests are performed, there can be three distinct situations as shown in Table 8.3. For situation C, which corresponds to soft rock masses it is necessary to perform large-scale deformability tests with high precision, involving a representative test volume of the rock mass. In the same tested volume borehole tests should be performed, and the results obtained from both methods should be compared and correlated. With established correlations for each zone established, the results of other small scale borehole tests can then be corrected. Of course a large number of laboratory tests should be performed in each zone as they will be used for the application of empirical systems and also to compare with the in situ tests. The different types of deformability and strength tests, both in situ and in the laboratory, are described in Table 8.4.

Table 8.4 In situ and laboratory tests for deformability and strength (adapted from Sousa et al. 2010)

Purpose of tests	In situ tests	Laboratory tests
Deformability	Geophysical tests Dilatometer/pressuremeter LFJ and SFJ Borehole jacking Chamber pressure	Uniaxial compression Triaxial compression Swelling Creep
Strength	Direct shear Rock pressuremeter Uniaxial compression Borehole jacking	Uniaxial compression Direct shear Triaxial compression Direct tension Brazilian Point load

8.2.2 In Situ Deformability Tests

8.2.2.1 Borehole Tests

In situ deformability tests performed on a small volume are usually executed in boreholes, since the tests are relatively economical to be performed and in general the boreholes have already been drilled before in order to survey the rock mass. Moreover, preparation of the borehole walls in the test stretch is not required.

Borehole deformability tests can be grouped into the following two main types: (1) the pressure load is applied by means of a flexible membrane completely attached to the borehole walls with a rotationally symmetric pressure (referred to as dilatometers); and (2) the pressure load is applied by means of rigid plates attached to two arcs of the borehole's circumference (referred to as borehole jacks).

In the first type of test, rock pressuremeters are used to measure global volumetric deformation (Birid 2014, 2015). These tests are generally used for soil and weak rock and have accuracy limitations, because they measure volumes and not displacements. The use of rock pressuremeters for deep foundation design is very important, particularly for rock foundations in skyscrapers (Failmezger et al. 2005; Sousa et al. 2010). Other equipment is sometimes used and includes dilatometers with direct measuring devices in which several radial deformations can be obtained (Ulusay and Hudson 2007).

The second type of test corresponds to a more complex theoretical loading situation, such as in the case of the Goodman jack. The greater robustness of this equipment in comparison with the dilatometer is why it is still in use (Ulusay and Hudson 2007; Slope Indicator 2010), (Fig. 8.7).

The most suitable borehole tests for rock mass deformability characterization are dilatometer tests. The LNEC BHD (BoreHole Dilatometer) dilatometer is an old and reliable equipment (Graça 1983; Rocha 1974, 2013). It consists of a steel body, enveloped by a rubber jacket, which transmits pressure to the borehole walls. It was designed to carry out deformability tests for rock masses in NX (76 mm diameter) boreholes, operating under normal conditions up to 150 m deep (Fig. 8.8). The pressure

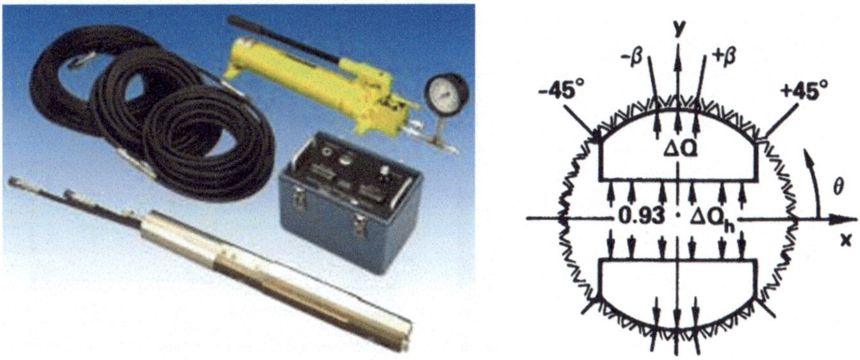

Fig. 8.7 Stiff dilatometer tests (Slope Indicator 2010)

Fig. 8.8 Borehole dilatometer BHD (Rocha 2013)

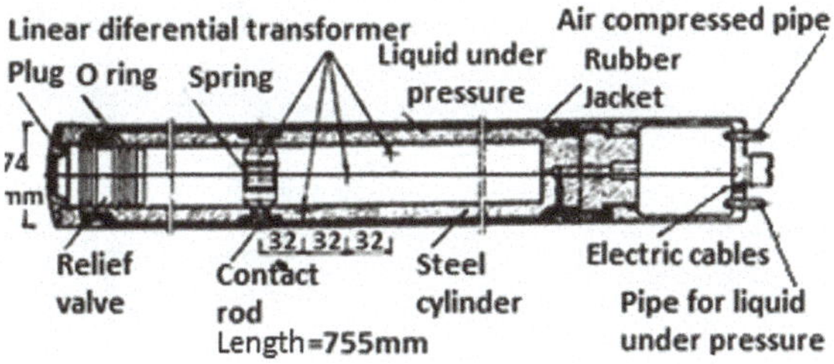

is obtained by pumping water, which can reach 20 MPa. The deformation is measured along four diameters, 45° apart, with the help of four pairs of sensors, which are connected to differential transducers. The deformability modulus, also called dilatometer modulus, is evaluated by assuming a continuous, homogeneous, isotropic and linear elastic medium, in accordance to the theory of elasticity for a borehole of infinite length with a uniform pressure acting on its wall (Graça 1983). These tests are easy to perform, but the tested volumes are small, usually between 0.5 and 1 m^3, and may not be representative of the rock mass. In soft rock masses, the results from dilatometer tests are more representative of rock mass behavior.

Other equipment is also available, such as the dilatometer probe presented by SolExperts (2016) and shown in Fig. 8.9. The specifications include measurements in depths up to 1400 m with borehole diameters of 96, 101, 122 and 146 mm, and maximum pressures up to 20 MPa (SolExperts 2016). Figure 8.10 shows a typical diagram pressure-deformation obtained in a point of a borehole.

Interpretation of these tests is difficult due to variations in the behavior of the rock mass during tests. In foundations, the initial stresses are small, and the tensile circumferential stresses induced by the applied pressure are usually higher than the initial circumferential stresses in the borehole walls. Thus, the deformability modulus obtained from the elasticity formula is not usually the true elastic modulus, but of an equivalent isotropic and homogeneous medium. Several assumptions can be made for the BHD dilatometer depending on the initial stresses (Pinto 1981; Sousa et al. 1997).

Fig. 8.9 Dilatometer probe SolExperts (2016)

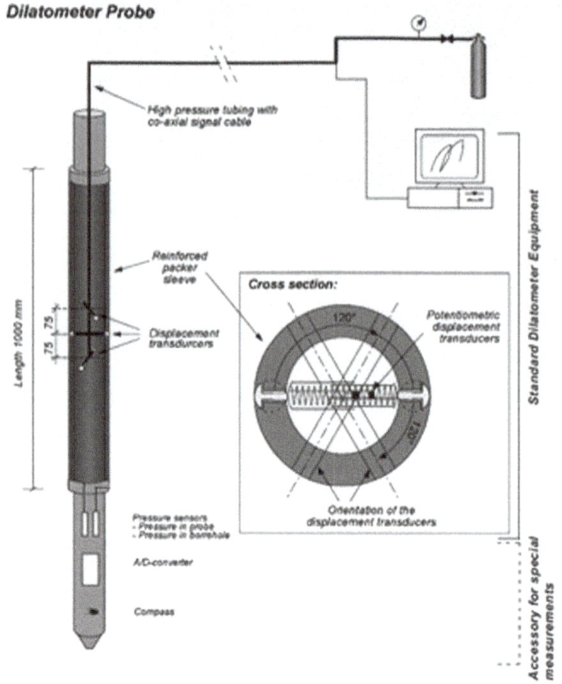

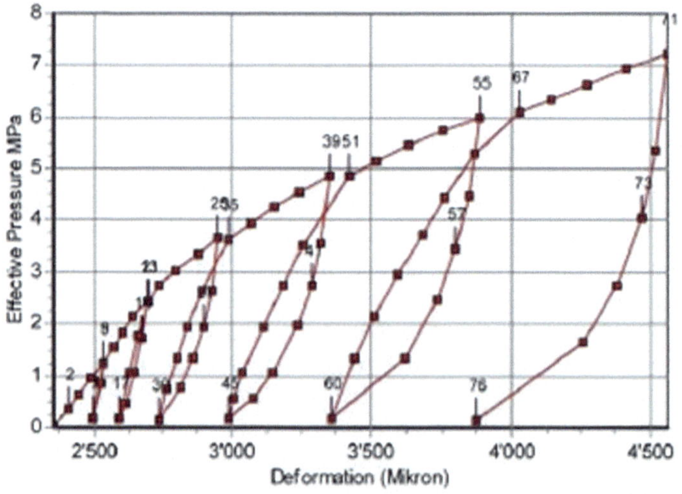

Fig. 8.10 Diagram pressure vs. deformation in the dilatometer (SolExperts 2016)

In soft rocks, it is necessary to have a rigorous characterization of the deformability of the rock masses. In dilatometer tests results, the moduli of deformability better represent reality. This is confirmed by the results in Fig. 8.11 where correlations between the ratio of large deformability and dilatometer tests are correlated

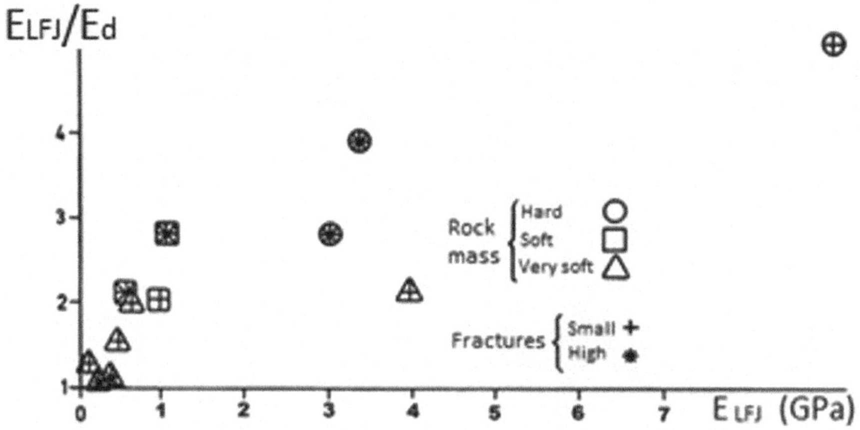

Fig. 8.11 Relationship between dilatometer deformability tests and large deformability tests (Graça 1983)

with the modulus of deformability of large deformability tests obtained by LFJ tests (Graça 1983). Figure 8.11 shows that the relation E_{LFJ}/E_{dil} can reach the value 5 for a rock mass with excellent quality, which can decrease to values near 1 for soft rock masses.

8.2.2.2 Plate Load Tests

In plate load tests (PLT), the load is applied on an existing surface and the response of the undisturbed rock mass is mixed with a superficial zone that is disturbed by the superficial stress release. To avoid this, sometimes only the deformation occurring at a certain distance from the load surface is measured, but measured deformations are usually smaller and therefore the results cannot be very accurate.

In situ PLTs should be based on a jack test (River Bureau 1986; Ulusay and Hudson 2007). The plate test using surficial loading is mainly performed in small tunnels or test adits. Two areas about 1 m in diameter are loaded using jacks. A typical test set up is shown in Fig. 8.12. Another scheme is shown in Fig. 8.13, for a double-plate bearing test with two hydraulic jacks applied to the walls of a gallery with a loading area of about 1 m^2 (Rocha 2013).

If the load is applied by a rigid-disk type plate to obtain a uniform displacement behavior, the modulus of deformability for the rock mass E_{MR} is given by Eq. (8.1):

$$E_{MR} = \frac{(1-\vartheta^2)(F_2 - F_1)}{2a(W_2 - W_1)} \tag{8.1}$$

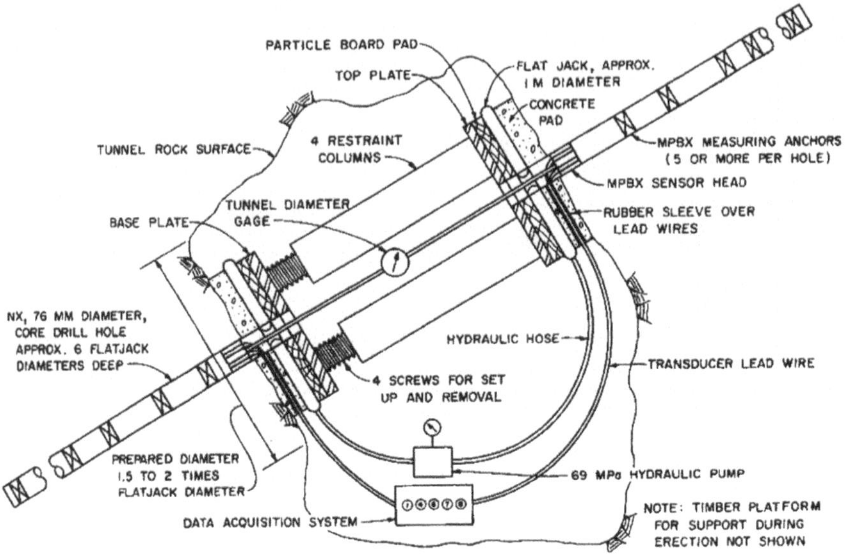

Fig. 8.12 Uniaxial jack test (Ulusay and Hudson 2007)

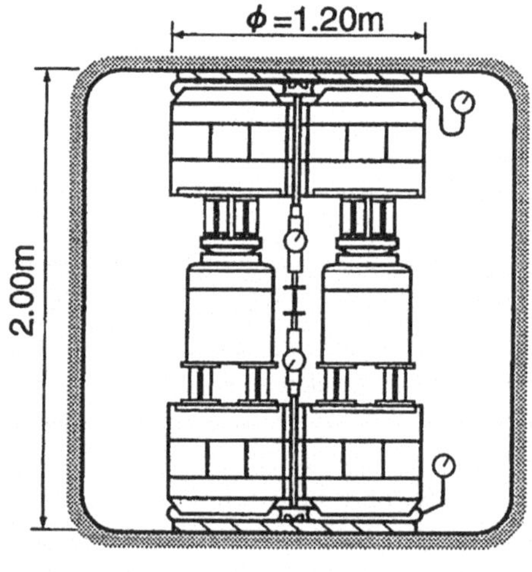

Fig. 8.13 Plate loading test arrangement used at LNEC (Sousa et al. 1997)

If the load is applied by a flexible flat-jack to obtain a uniform distribution load, E_{MR} is given by:

$$E_{MR} = \frac{2\left(1-\vartheta^2\right)\left(r_2 - r_1\right)\left(p_2 - p_1\right)}{W_2 - W_1} \tag{8.2}$$

In Eqs. (8.1) and (8.2), a is the radius of the rigid disk, ν Poisson ratio, F_1 and F_2 the loads of two points in the load vs. displacement curve, and W_1 and W_2 the displacement values corresponding to the pairs of forces and pressures F_1 and p_1, and F_2 and p_2, respectively.

There are indicated two types of PLTs as shown in Fig. 8.14. Two areas diametrically opposite in the test adit are loaded simultaneously, for example using flat jacks positioned across the test drift, and the rock displacements are measured in bore-

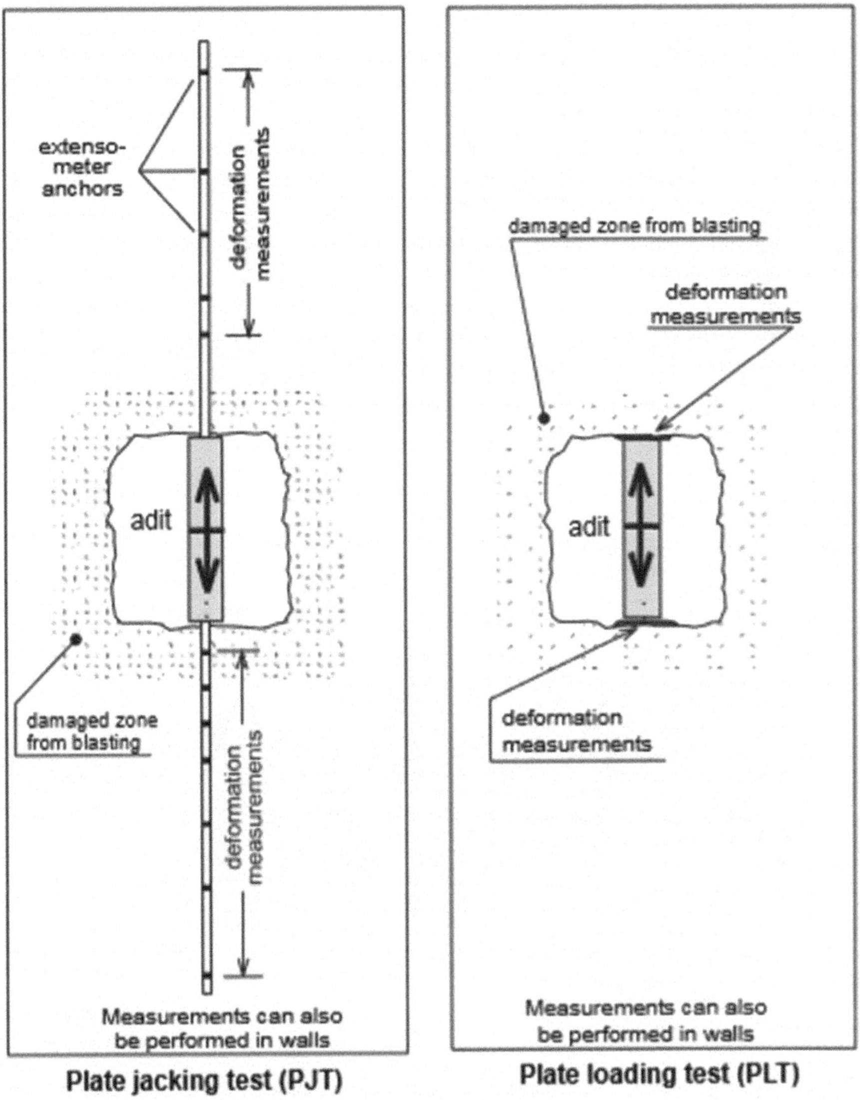

Fig. 8.14 Methods for PLT tests (Palmstrom and Singh 2001)

holes behind each loaded area (Palmstrom and Singh 2001). Alternatively (Fig. 8.14) the PLT measures the displacements at the loading surface of the rock, while the previous case records the displacements in drill holes beyond the loading assembly.

Mingtan Pump Station, in Taiwan

PLT tests were used at the Mingtan Pump Station, in Taiwan (Rodrigues et al. 1978; Hoek 2007a). The power cavern complex is located in sandstone, sandstone with siltstone interbeds and several siltstone beds belonging to the Waichecheng series. The sandstones are fine grained to conglomeratic and sometimes quarzitic. In general, they are strong to very strong although they are slightly to moderately weathered. Locally, softer zones of highly weathered material was encountered. The siltstones are moderately strong and almost always sheared. Occasionally, massive sandstone beds occurred with a thickness of up to 7 m. The geology of the Mingtan powerhouse is presented in Fig. 8.15. The general appearance of the rock mass in an exploration adit is shown in Fig. 8.16. Figure 8.17 shows the in situ PLT's.

Laboratory and in situ tests were carried out in the 1970s for the Mingtan Projects and the detailed design of the Mingtan project started in 1982 (Hoek 2007a). The rock mass in the powerhouse was classified using the RMR and Q systems as: Jointed sandstone—class 2; Bedded sandstone—class 3; Fault and shear zones—classes 4 and 5, as shown in Table 8.5. The in situ deformation modulus values for the rock mass are listed in Table 8.6.

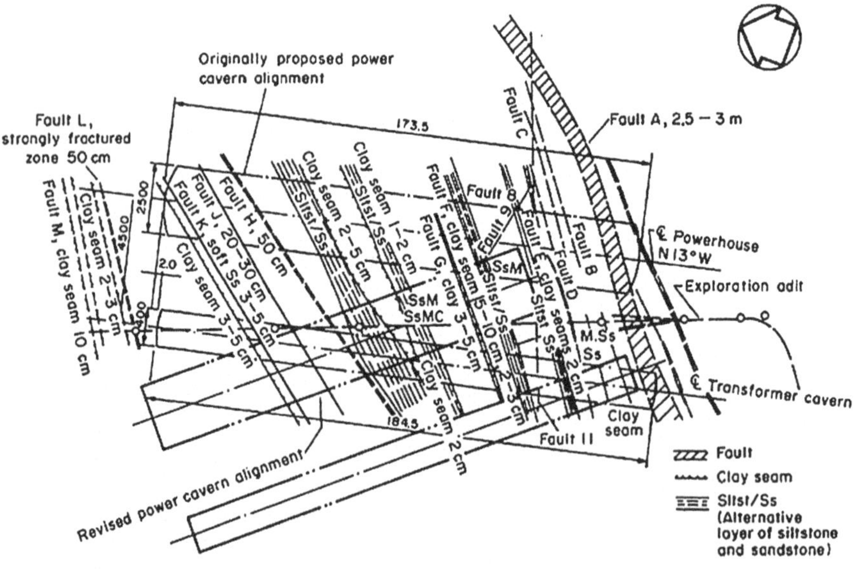

Fig. 8.15 Geological plan of the Mingtan underground powerhouse complex (Hoek 2007a)

Fig. 8.16 Sandstone and siltstone sequence in a adit (Hoek 2007a)

Fig. 8.17 Plate load tests at New Tienlun hydroelectric project (Hoek 2007a)

Table 8.5 Rock mass classifications for the powerhouse area (Hoek 2007a)

Rock type	RMR	Q	Rock quality
Jointed sandstone	63–75	12–39	Good
Bedded sandstone	50–60	7–13	Fair to good
Fault and shear zones	10–33	0.1–11	Very poor to poor

Table 8.6 Deformation modulus of the rock mass from PLT tests

Rock type	Comments	PLT tests *E* (GPa)
Sandstone	Normal to bedding	3.2–5.1
	Parallel to bedding	2.3–5.0
Interbedded sandstone and silstone	Normal to bedding	2.8
	Parallel to bedding	3.0

8.2.2.3 Flat Jack Tests

In Large Flat Jack (LFJ) tests, a load is applied on the walls of a slot and the deformability of large volumes of rock masses in undisturbed conditions is assessed. The authors believe that LFJ tests are the most appropriate large scale tests for evaluating in situ the rock mass deformability. An LFJ test developed by LNEC is widely used by this institution in numerous studies in soft rocks (Pinto 1981). The advantage of this type of test in comparison with other slot tests is that the jacks can apply the pressure directly on the rock surface. The LNEC flat jacks contain four deformeters which measure the variation of the slot opening. The jacks can be placed in slots opened side by side and tested either simultaneously or individually, which allows one to conduct the test on a volume large enough to be representative of the rock mass. Figure 8.18 shows an arrangement of three flat jacks placed side by side.

The tests are usually interpreted using the Theory of Elasticity for a half-space, with a distributed normal load:

$$E_{LFJ} = \left(1 + \nu^2\right) K \frac{\Delta p}{\delta} \tag{8.3}$$

where E_{LFJ} is the deformability modulus obtained with these tests, ν is the Poisson ratio, K a constant to be evaluated, Δp the pressure change and δ the deformation caused by Δp. Constant K may be calculated by numerical methods and it is a function of the number of LFJ tests, the chosen deformeters and the depth of a crack that often develops in the plane of the slots during the tests (Pinto 1983; Rocha 2013).

An alternative LFJ arrangement is shown in Fig. 8.19 where in which each slot contains two independent jacks about half of the size of the old jacks and by a central borehole with 100 mm in diameter. The deformeters are placed in three independent measuring columns, installed in a central borehole and in two boreholes with 76 mm in diameter.

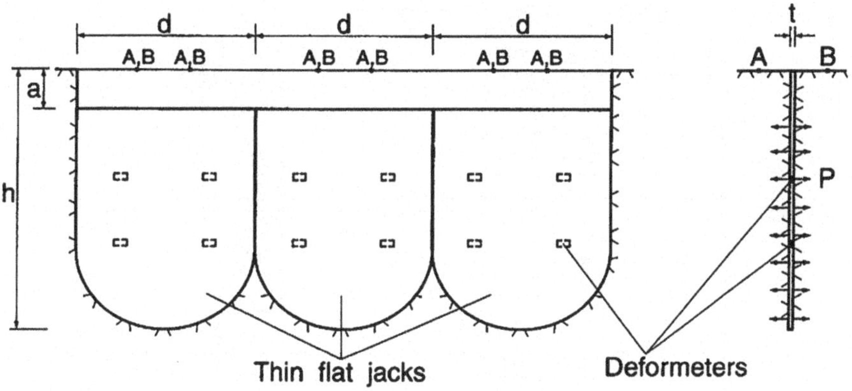

Fig. 8.18 Large flat arrangement with three slots

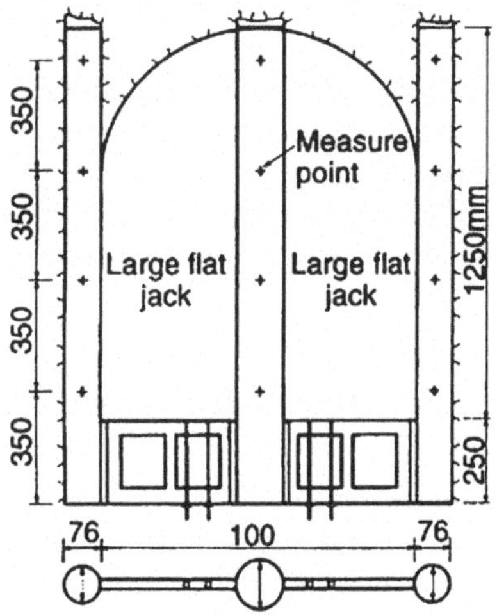

Fig. 8.19 New large flat jack setup (Sousa et al. 1997)

Another flat jack technique developed at LNEC is the Small Flat Jack (SFJ) using small scale jacks which are used mainly for stress measurements (Fig. 8.20), (Pinto and Graça 1983). However it can also be used to obtain the deformability of the rock mass assuming isotropic or anisotropic media (Martins and Sousa 1989). The following formula was obtained using results of a 3D finite element model (FEM):

$$E_{SFJ} = K \frac{\Delta p}{\delta} \tag{8.4}$$

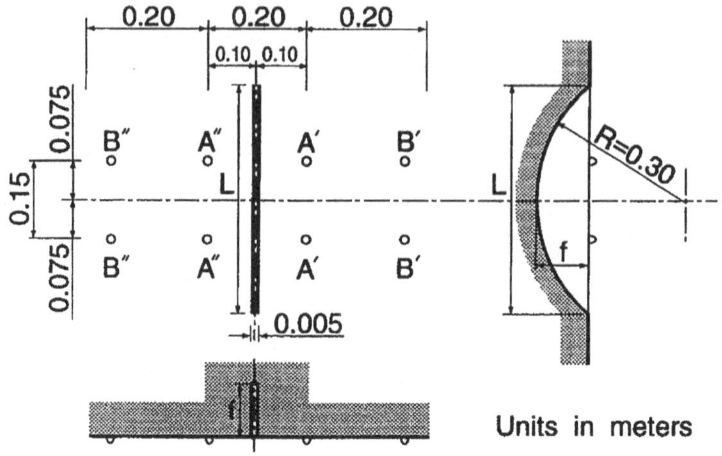

Fig. 8.20 SFJ test. Geometry of a slot (Pinto and Graça 1983)

where E_{SFJ} is the deformability modulus, K a constant obtained through the FEM model, Δp the cancellation pressure and δ measured displacements between pairs of points.

An important application of the use of LFJ methods was at Mingtan Pumped Station, in Taiwan (Rodrigues et al. 1978; Hoek 2007a), presented in Sect. 8.2.2.2.

8.2.3 Triaxial Techniques and Procedures in Laboratory

High performance True Triaxial Testing (TTT) of rocks can simulate the in situ stress environment with three principal stresses with the goal to characterize the deformation and strength behavior of rocks. The state of the art of TTTs was discussed and reviewed during the Workshop on TTT of rocks held in Beijing (Kwasniewski et al. 2013). Some of the work has been published in Kwasniewski (2013), Lade (2013), Li et al. (2011), Mogi (2013), and Tshibangu and Descamps (2013).

In soft rocks Alexeev et al. (2013) presented experimental results of deformation and fracture of coals under true triaxial compression at various stress states. Lu et al. (2012) presented an experimental study of wellbore deformation in a deep claystone formation by using a large-scale physical true triaxial simulating equipment. Using cylindrical samples of Kimachi sandstone, Fujii et al. (2013) performed tests where confining pressures were applied through a liquid and the axial stress was applied by solid pistons as minimum principal stress. Tarasov (2012) demonstrated that hard rocks can exhibit dramatic embrittlement at a certain range of confining pressures, and proposes a special shear rupture to explain the phenomenon.

A special water–stress coupling meso-mechanics test system was developed at Zhongshan University, China (Zhou 2018). It uses conventional rock triaxial apparatus with oil to provide confining pressure and requires that the rock sample be protected by an envelope. In practice, it is very difficult to simulate softening conditions of soft rock (taking into account water-stress coupling), and it is difficult cannot meet the requirements of observing meso-characteristics on the surface of soft rock under water-stress coupling. This was the motivation to develop the apparatus to simulate the deformation of and failure process of soft rock in the real environment. The so-called "TAW-100 water-stress coupling soft rock meso-mechanics servo triaxial test system" was created improving the loading system, pressure chamber and microscopic observation part of traditional rock triaxial testing machines (Zhou 2018). The design of the new system is shown in Fig. 8.21, where A is the pressure chamber; B is the loading system; C is the measuring system; D is the servo control system; E is the microscopic observation system; F is the computer system; G are the rock samples. The loading system consists of an axial pressure system and a two-direction confining pressure servo loading system. The system can adjust the axial pressure and radial confining pressure on rock samples with different time intervals according to different experimental programs in order to make it possible to simulate the pressure situations for soft rock under various construction programs, and provide the function of filling and compressing the pressure chamber with water or oil. The axial loading system is illustrated in Fig. 8.22. The two-direction confining pressure servo-loading system is presented in Fig. 8.23. It consists of a rigid support, ball screw, servo-motor, water and oil reservoirs, piston hydraulic sensor, etc.

More details of this equipment can be found in the report of Zhou (2018).

A modified true triaxial test system to for TTT of rocks was developed at SKL-GDUE, of the China University of Mining and Technology, of Beijing (He et al. 2013) and is shown in Fig. 8.24. This innovative equipment has a single face unload-

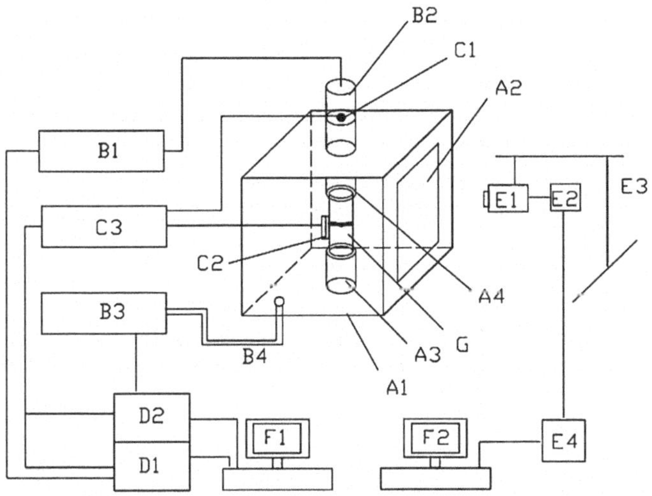

Fig. 8.21 Design of a new test system

Fig. 8.22 Axial pressure loading system

ing device that allow one to perform rockburst tests in laboratory (He and Sousa 2014; He et al. 2015a). Figure 8.25 describes the dropping test used for unloading a face which is performed through a change of a piston movement.

Since its development, a large number of rockburst tests were performed and the results of the tests collected, and stored in a database which was analyzed. This included studies from several countries including China, Italy, Canada and Iran. DM techniques were applied and predictive models were developed for the rockburst maximum stress (σ_{RB}) and for a rockburst risk index (I_{RB}). All the tests were of strainburst type. A rockburst critical H_e was calculated by the following expression:

$$H_e = 18.52\sigma_{RB} \tag{8.5}$$

where σ_{RB} is the rockburst maximum stress obtained in the test and H the depth where the sample was collected. The index I_{RB} was proposed and calculated from the formula (He et al. 2015a):

$$I_{RB} = \frac{H}{H_e} = 0.054\frac{H}{\sigma_{RB}} \tag{8.6}$$

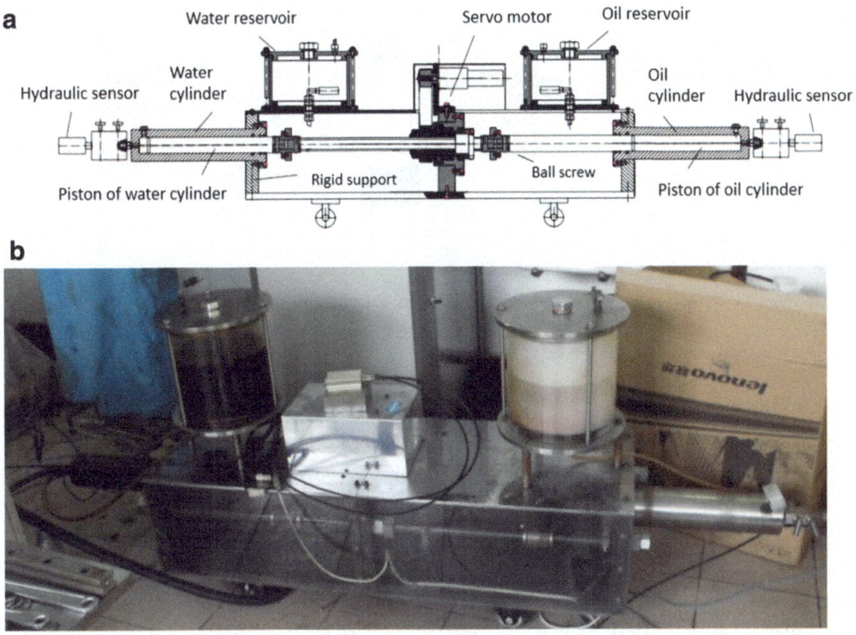

Fig. 8.23 Two-direction servo confining pressure loading system. (**a**) Design diagram of two-direction servo confining pressure loading system. (**b**) Photo of two-direction servo confining pressure loading system physical

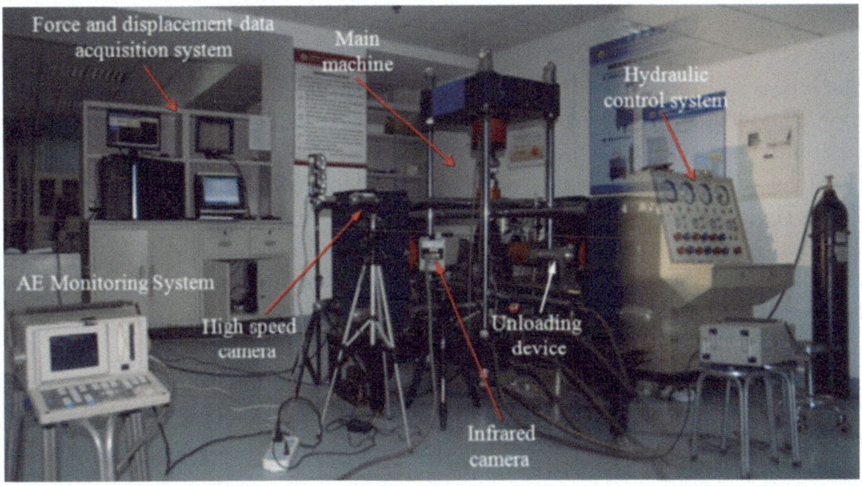

Fig. 8.24 Rockburst testing system (He 2006)

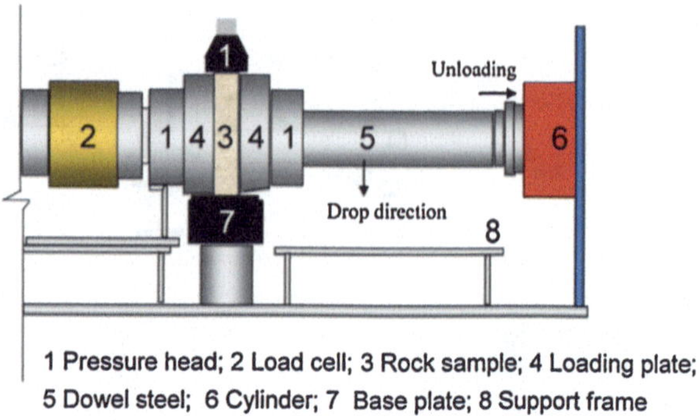

Fig. 8.25 Illustration of the dropping system for load bar and loading plate (He 2006)

Table 8.7 Statistical parameter values of soft rocks obtained through rockburst tests

Parameters	Schist	Coal	Dolomite	Limestone	Mudstone	Sandstone	Shale	Slate
UCS (MPa)	–	11.9	–	24.1	11.5	83.4	8.0	58.3
E (GPa)	–	2.4	–	9.7	2.0	24.5	3.0	13.5
ν	–	0.28	–	0.24	0.37	0.24	0.37	–
H (m)	1000	507	250	3375	910	854	500	500
H_e (m)	1413	352	2315	1278	306	1878	343	1587
σ_{RB} (MPa)	76.3	19.0	125.0	69.0	16.5	101.4	18.5	85.7
I_{RB}	0.81	1.65	1.11	3.49	3.02	0.62	1.54	0.34

Some of most important statistical characteristics concerning the database are presented in Table 8.7 for soft rocks (schist, coal, dolomite, limestone, mudstone, sandstone, shale, and slate). The DM models made it possible to predict the parameters in Table 8.7 with high accuracy using data from the rock mass and a specific project. The DM modeling techniques comprised MR—Multiple Regression, ANN—Artificial Neural Networks, and SVM—Support Vector Machines.

A classification in accordance with the rockburst index was established for the possibility of occurrence of rockburst depending on the value of I_{RB}. For $I_{RB} < 0.6$ is classified as Low; $0.6 < I_{RB} \leq 1.2$ is classified as Moderate; $1.2 < I_{RB} \leq 2.0$ is classified as High; and $I_{RB} > 2.0$ is classified as Very High. Figure 8.26 shows the relation between I_{RB} and σ_{RB}.

Among the DM algorithms used only MR provides an equation relating the output and the input variables. The modeling and evaluation is discussed in He et al. (2015a). The equations are:

$$\sigma_{RB} = 9.132 - 0.013H + 0.381UCS + 0.364E + 1.211\sigma_{h1} - 0.069\sigma_{h2} + 0.365\sigma_V \quad (8.7)$$

$$I_{RB} = 1.432 + 8.035 10^{-4} H - 8.429 10^{-4} UCS - 0.009E - 0.007\sigma_{RB} - 0.074K \quad (8.8)$$

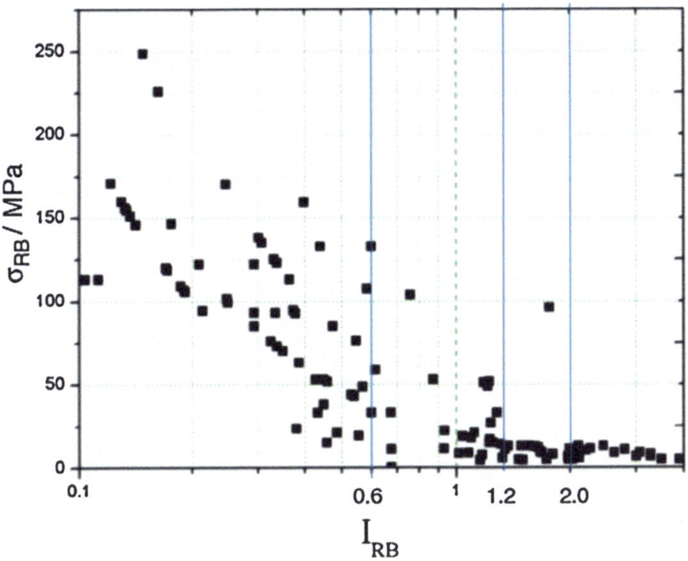

Fig. 8.26 Distribution of I_{RB} vs. σ_{RB} (He et al. 2015a)

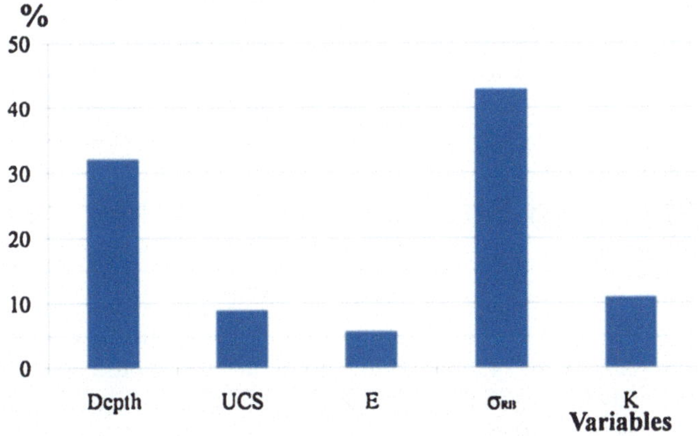

Fig. 8.27 Importance of variables for predicting I_{RB} using ANN model (He et al. 2015a)

In Eqs. (8.7) and (8.8), σ_{h1} and σ_{h2} are, respectively, the horizontal in situ stresses in a perpendicular and in the face to be unloaded, and K is ratio between the average in situ horizontal stresses and the in situ vertical stress due to overburden. Figure 8.27 shows the relative importance of the variables for predicting I_{RB} in the more accurate ANN model.

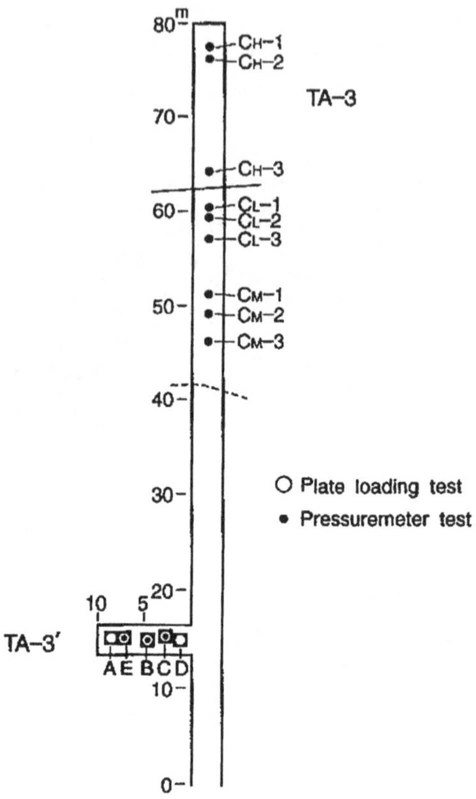

Fig. 8.28 Galleries TA-3 and TA-3′. Location of the tests

8.2.4 Deformability Investigation of Soft Rock Mass at a Dam Site in Japan

In this section the evaluation of the deformability parameters is discussed in the context of the foundations of a dam in the north of Japan (Sousa et al. 1997). The rock mass is highly heterogeneous and consisted of lapilli tuff and conglomerates.

A main gallery about 80 m long was excavated as shown in Fig. 8.28, and several in situ and laboratory tests were performed on rock samples. In situ PLTs were performed on a small gallery designated TA-3′ perpendicular to the main gallery (Fig. 8.29). The PLTs were performed considering rigid plates with 30 and 60 cm in diameter, mainly in the conglomerate formation. Nine in situ pressuremeter tests were performed in the locations indicated in Fig. 8.28, namely C_H-1, C_H-2, C_H-3, C_L-1, C_L-2, C_L-3, C_M-1, C_M-2, and C_M-3. Some laboratory tests were carried out that included uniaxial compression tests and measurement of the velocities of waves and specific gravity.

PLT tests were performed at five locations in the gallery TA3′ designated by A (No. 1), B (No. 2), C (No. 3), D (No. 3′), and E (No. 4). The loading tests were

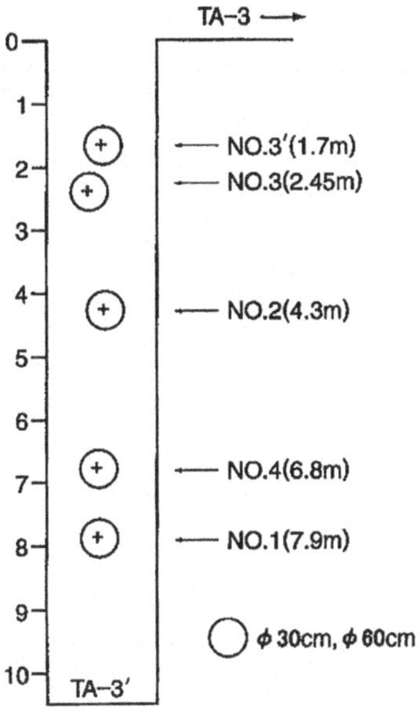

Fig. 8.29 Detail of the location of plate load tests at the gallery TA-3′

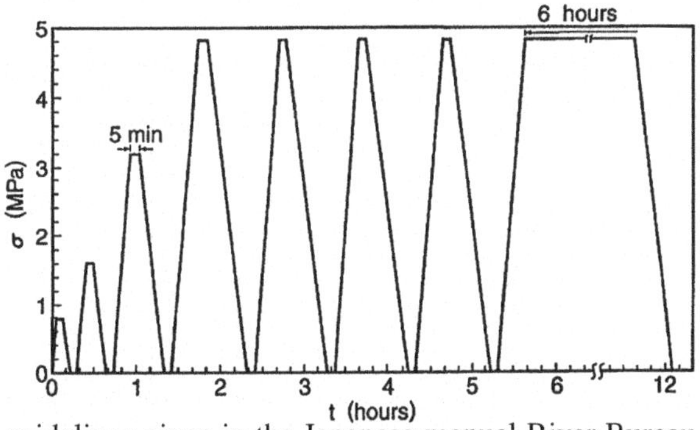

Fig. 8.30 Loading pattern for the plate load tests

performed according to the guidelines given in the Japanese manual River Bureau (1986). The loading pattern was based on four incremental steps of 0.8, 1.6, 3.2, and 4.8 MPa for four cycles repeated for the last increment of load as indicated in Fig. 8.30. A final endurance loading cycle was carried out for 6 h. The coefficient of deformation *D*, tangential coefficient E_t and secant coefficient of elasticity E_s were determined in accordance with the example of load-displacement curves presented in Fig. 8.31.

Figure 8.32 is a schematic of the bedrock observed in the exploratory TA-3′, which was excavated along the right bank of a river in the conglomerate layer (Sousa et al. 1997). The conglomerate is pale dun-greenish dun and its heterogeneous bedrock consists of gravel and matrix. The bottom of the exploratory gallery is relatively dry and the ground water level was about 40–80 cm below the bottom. The gravel is made up of lapilli tuff, coarse-grained tuff and mudstone. The main gravel size ranged from 5 to 50 cm, but large gravel measuring 2 m in size was

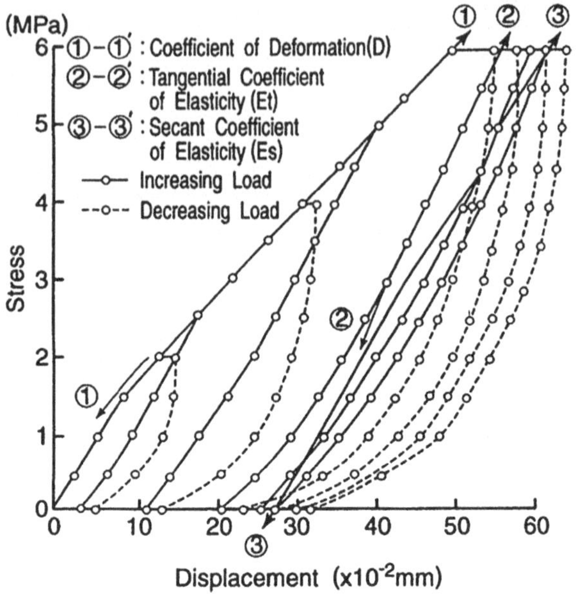

Fig. 8.31 Example of load vs. displacement curves

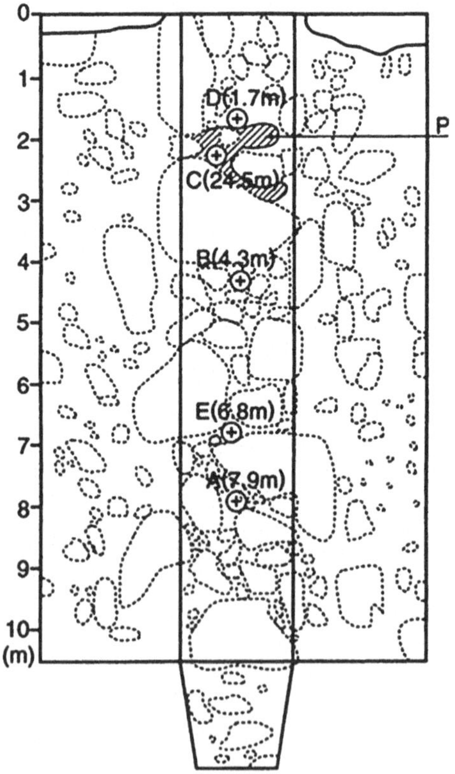

Fig. 8.32 Heterogeneity involved by PLT tests

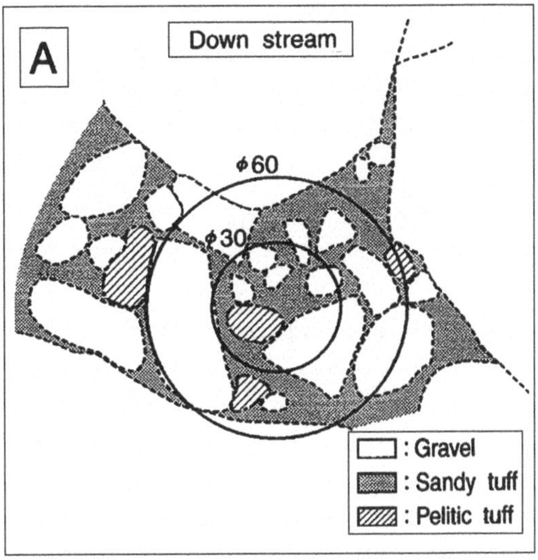

Fig. 8.33 Sketch of the test surface A

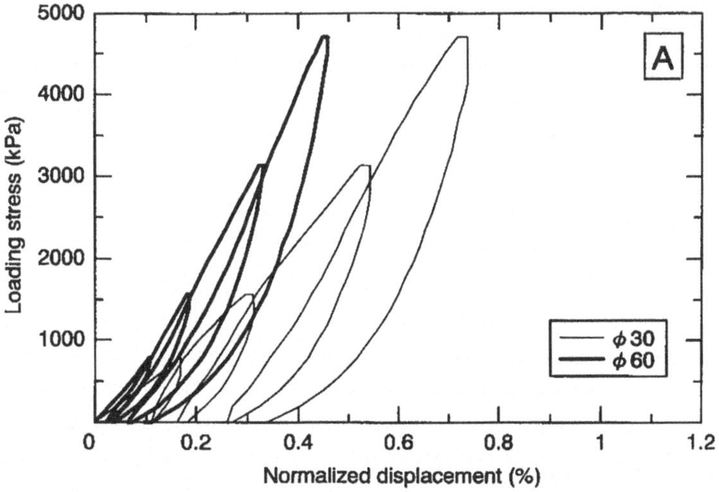

Fig. 8.34 Diagram of loading stress vs. displacement for the PLT tests at A

found in places. The matrix is soft and consisted of sandy tuff and politic tuff, and has a low degree of consolidation.

The test site consists of gravel of various sizes and a small-scale matrix that lies in between. A detailed sketch of the rock formation in the location was made for each site location. Figure 8.33 presents a detailed sketch for the test surface A. In each test site a PLT with 30 cm diameter was first performed, followed by a PLT with 60 cm diameter. The tests were performed in areas where the matrix and gravel were in an approximate equal proportion.

Figure 8.34 shows typical results for the relationship between loading stresses and displacements at site A. The normalized displacements represent the average

Table 8.8 PLT test results

	30 cm plate			60 cm plate		
Position	D (MPa)	E_t (MPa)	E_s (MPa)	D (MPa)	E_t (MPa)	E_s (MPa)
A	578	1014	818	886	1307	906
B	413	808	639	736	1159	789
C	47	439	410	487	953	641
D	634	1119	852	914	1413	931
E	736	1225	1019	1243	1667	1511

Table 8.9 Pressuremeter test results

Test	Class	D (MPa)	E_t (MPa)
C_L-1	C_L	30	210
C_L-2	C_L	350	890
C_L-3	C_L	60	300
C_M-1	C_M	1260	1890
C_M-2	C_M	2160	6900
C_M-3	C_M	1730	2770
C_H-1	C_H	1870	13,000
C_H-2	C_H	2890	5200
C_H-3	C_H	2590	8650
B	C_L	220	1240
C	C_L	130	540
E	C_L	690	1200

values of the four points on the stiff plate as a percentage, obtained by dividing the displacements by the diameter of the plate. The results show that the slope of the tangent of the curves increases as displacements of the plate increases, which means that the rock formation gets harder with increasing stresses.

Tests conducted at another site, C, showed extremely large deformations for the PLT's with 30 cm diameter, but the PLT with diameter 60 cm did not show such large deformations. This happened because the plate probably was on the matrix, where the gravel area ratio is different to the other tests.

Table 8.8 summarizes the PLT results. In order to calculate the coefficient of deformability D, the range of loading stress was from 2 MPa (test B), to 2.2 MPa (tests C, D and E), to 2.4 MPa (test A) to 4.8 MPa (for all tests). E_t was calculated for a lower pressure between 1.0 and 2.8 MPa and for a higher stress between 4.0 and 4.8 MPa. E_s was determined for the range of stresses between 0 and 4.8 MPa.

Several pressuremeter tests were carried out for a 66 mm diameter borehole. Six tests were performed on lapilli tuff (classes CM and CH) in gallery TA-3′ and six in the conglomerate rock mass, three being in the same location of plate load tests B, C and E as indicated in Fig. 8.28. Three additional tests were performed in class CL in TA-3. Some results are shown in Table 8.9.

Several laboratory tests were performed in order to evaluate the deformability modulus and the uniaxial compressive strength (Sousa et al. 1997). A summary of

Table 8.10 Deformability moduli at a heterogeneous conglomerate for different tests

Test	Deformation coefficients	Maximum (MPa)	Mean (MPa)	Minimum (MPa)
PLT (60 cm)	D	1243	853	487
	E_t	1667	1300	953
PLT (30 cm)	D	736	482	47
	E_t	1225	939	439
Pressuremeter	D	690	247	30
	E_t	1240	730	210
Lab. Uniaxial	E	304	199	98

the test results are shown in Table 8.10. For the uniaxial tests, a mean value of 199 MPa was obtained for the elasticity moduli, with a maximum of 304 MPa and a minimum of 98 MPa. For the uniaxial compressive strength a mean value of 1.4 MPa was obtained. The deformability moduli obtained using laboratory tests were generally lower than the values obtained with the in situ tests.

An overall analysis of the results for the different type of tests (plate load, pressuremeter and laboratory uniaxial tests) in the same type of rock mass, namely conglomerate of class C_L, was made. The results suggested that coefficients D and E_t decrease with the reduction in test volume. This is probably due to the fact that the tests were performed in locations with lower deformability coefficients which were chosen based on safety considerations (see Fig. 8.32). When the in situ PLTs were conducted on larger volumes, as with the 60 cm diameter rigid plate, the influence of more rigid blocks was more significant and the rock mass was observed to be stiffer. The comparative analysis of the results shows that the in situ plate tests with small diameters cannot represent behavior of the rock mass well. The mean deformability values obtained through large scale PLTs are closer to the real mean deformability value of the rock mass in the tested zone.

Figures 8.35 and 8.36 compare deformability coefficient D and tangential coefficient E_t. Figure 8.35 shows the values of the in situ deformability coefficients as: 30 cm plate load tests (D_{30}) against pressuremeter tests (D_d); 60 cm plate load tests (D_{60}) against pressuremeter tests; plate load tests D_{60} against plate load tests D_{30}. Using the least square method the following gradients were obtained: $D_{30} = 1.12D_d$; $D_{60} = 2.00D_d$; and $D_{60} = 1.61D_{30}$. Figure 8.36 compares tangential coefficients E_t. The relations were as follows: $E_{t30} = 1.12E_{td}$; $E_{t60} = 2.00E_{td}$; and $E_{t60} = 1.61E_{t30}$.

In conclusion, because of the heterogeneous nature of the rock mass, the deformability parameters are very much influenced by the dimension of the tested volumes and by the methodology followed. There is a significant difference in the deformability parameters that are obtained by PLTs of different diameters. This has real practical and important implications, specifically that if the PLT of diameter 30 cm results are used, then the Japanese regulations would not allow the dam to be constructed, whereas with the PLT of 60 cm results the permissions could be obtained to build the dam (Sousa et al. 1997).

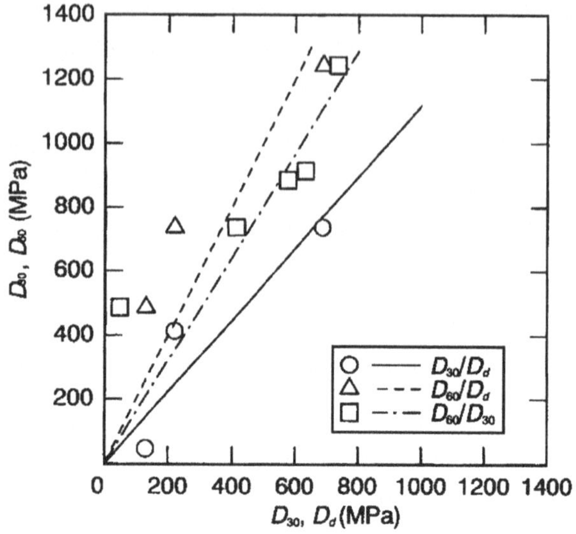

Fig. 8.35 Comparison between in situ deformability coefficients

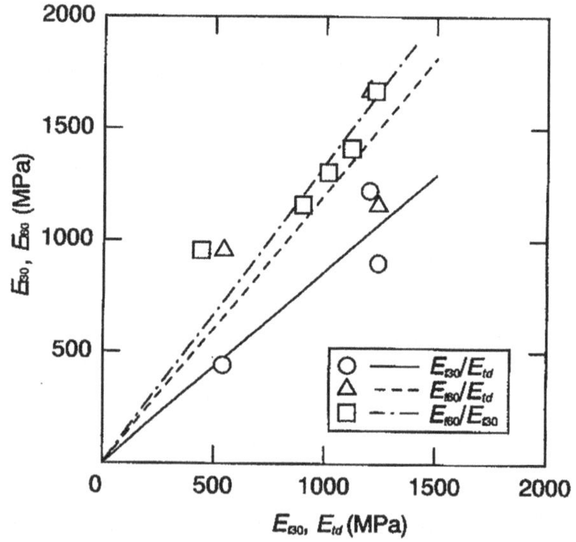

Fig. 8.36 Comparison between in situ tangential coefficients of elasticity E_t

8.3 Methodology for Evaluating Strength

8.3.1 Strength Evaluation

Rock mass strength parameters can be determined using large scale in situ and laboratory tests for intact rock and discontinuities. The major in situ tests are sliding or shearing along discontinuities, in the fault filling materials and along other low strength surfaces and at rock mass/concrete interfaces. Laboratory tests for intact rock strength evaluation are shown in Table 8.4 (ASCE 1996; Rocha 2013).

In general, the unconfined compressive strength (UCS) of soft rocks varies between 2 and 20 MPa, and fiction angles normally range between 30° and 45° and cohesion values are typically not lower than 0.4 MPa (Rocha 1975). In addition, the shear strength of soft rock masses is less influenced by discontinuities than for hard rock, and the influence diminishes as the strength of the rock decreases. In the case of continuous fractures or continuous surfaces of low strength, each case should be considered individually as in the case of Água Vermelha presented in Sect. 8.1 (Rocha 1975; Pedro et al. 1975). There are cases where it is reasonable to adopt intact rock strength for the strength of the rock mass as was the case for Alto Rabagão dam, in Portugal, and for São Simão in Brazil (Rocha 1974).

Empirical systems provide good guidelines to estimate rock strength parameters for a given failure criterion. The GSI (Geological Strength Index) system was specifically developed to estimate rock mass strength parameters (Hoek 2007a). The system uses a qualitative description of two fundamental parameters of the rock mass: its structure, and the condition of its discontinuities. This system has also been used for evaluation of heterogeneous rock masses in the Porto Metro and tunnels in Greece that are in difficult rock mass conditions like flysch (Hoek et al. 2005; Babendererde et al. 2006).

Usually, the calculation of the GSI value is based on correlations with modified forms of the RMR and Q indices, taking into consideration the influence of groundwater and orientation of discontinuities (Hoek 2007a). Other approaches, defined by several authors, can be used for the GSI evaluation (Miranda 2003, 2007). Based on experimental data and theoretical fracture mechanics, the H-B criterion for rock masses is given by:

$$\sigma_1^{'} = \sigma_3^{'} + \sigma_c \left(m_b \frac{\sigma_3}{\sigma_c} + s \right)^a \tag{8.9}$$

where σ'_1 and σ'_3 are, respectively, the maximum and minimum effective principal stresses, m_b is a reduced value of the m_i parameter which is a constant of the intact rock, and s and a are parameters that depend on characteristics of the rock formation. Serrano and Olalla (2007) extended this failure criterion to 3D in order to consider the intermediate principal stress in the failure strength of rock masses.

The H-B criterion has seen developments introduced, as well as limitations (Douglas 2002; Carter et al. 2007; Carvalho et al. 2007). Once the value of GSI is determined, the parameters of the H-B criterion can be calculated through the following equations:

$$m_b = m_i \exp\left(\frac{GSI - 100}{28 - 14D} \right)$$

$$s = \exp\left(\frac{GSI - 100}{9 - 3D} \right) \tag{8.10}$$

$$a = \frac{1}{2} + \frac{1}{6}\left(\frac{GSI - 100}{9 - 3D}\right)$$

where D is a parameter developed for the underground works of the Porto Metro that depends on the disturbance to which the rock mass formation was subjected from blasting and stress relaxation (Hoek et al. 2002).

Guidelines were established for estimating D where $D = 0$ for excellent quality controlled blasting or excavations by TBMs and $D = 1$ for very large open pit mine slopes (Hoek 2007a). There is a drawback in this formulation since the excavation process affects a damaged zone near the excavation and it is not an intrinsic characteristic of the rock mass. As such it only represents the disturbed zone near the excavation. For GSI > 25, m_b can also be calculated through the expression:

$$m_b = m_i s^{1/3} \tag{8.11}$$

For many cases of rock masses and for certain geotechnical software, it is convenient to use the equivalent cohesion c' and friction angle φ' to the H-B criterion. The range of stresses should be within $\sigma'_{1,RM} < \sigma_3 \leq \sigma'_{3RM}$. The value σ'_{3RM} can be determined for each specific case using.

$$\frac{\sigma'_{3max}}{\sigma'_{cm}} = 0.47\left(\frac{\sigma'_{cm}}{\gamma H}\right)^{-0.94} \tag{8.12}$$

For the shear strength of rock discontinuities the Q system can be used (Barton and Choubey 1977; Barton 2016). The failure criterion is expressed by:

$$\tau = \sigma_n \tan\left(JRC\,log_{10}\left(\frac{JCS}{\sigma_n}\right) + \varphi_r\right) \tag{8.13}$$

where τ, σ_n, φ_r, JCS, and JRC are shear strength, normal stress, residual friction angle, joint compressive strength and joint roughness coefficient, respectively.

In the case of Bakhtiari dam site this empirical criterion was used with success for estimating the shear criteria of the discontinuities based on a statistical analysis (Sanei et al. 2015).

8.3.2 Use of Data Mining DM Techniques

Data Mining (DM) techniques have been used in many fields and recently in geotechnical engineering in different applications (Miranda 2007; Miranda and Sousa 2012; Sousa et al. 2012, 2017; Miranda et al. 2018). They are adequate as an advanced technique for analyzing large and complex databases with geotechnical information within the framework of an overall process of Knowledge Discovery in Databases (KDD). KDD processes have been carried out in the context of rock mechanics using the geotechnical information of two hydroelectric schemes built in Portugal on mainly granite rock formations. The main goal was to find new models to evaluate strength and deformability parameters (namely friction angle, cohesion and deformability modulus) as well as the RMR index. Databases of geotechnical data were assembled and DM techniques were used to analyze and extract new and useful knowledge. The procedure allowed one to develop new, simple, and reliable models for geomechanical characterization of rock masses using different sets of input data, and which can be applied in different situations depending on availability of information (Miranda et al. 2009, 2011; Miranda and Sousa 2012).

With a database obtained from Venda Nova II hydroelectric scheme, the Mohr-Coulomb parameters derived from the H-B criterion were computed. For poorer rock mass conditions, the peak and residual parameters can be considered similar because a perfectly plastic post-peak behavior can be assumed. Prediction models for φ' and c' were developed. Figure 8.37 shows a plot of the most important parameters in the prediction of φ' (Miranda and Sousa 2012). There is a large number of variables that are significantly related to the prediction of φ' and several show similar importance. Having said this, the most important variables are: (1) UCS which is expected since it is a measure of strength; (2) the ratio between Jw/SFR; and (3) the Q index (with logarithmic transformation) as well as other variables related to the Q system. This was somewhat less expected because the Q system is normally used only for classification purposes and not for the calculation of strength parameters since it considers the rock mass as a continuum despite the Jr/Ja ratio being a strength index for joints. The Q index has shown to be complete and can be used for the prediction of geomechanical parameters. This was confirmed in Barton (2013).

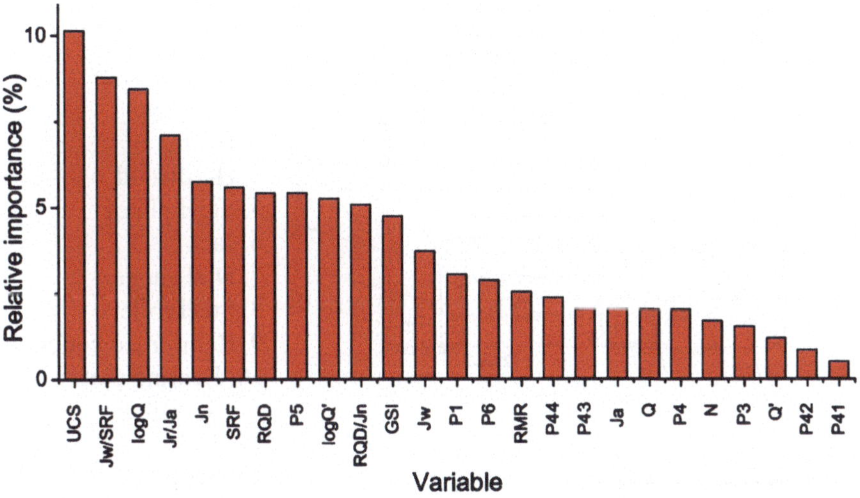

Fig. 8.37 Relative importance of the attributes for the prediction of friction angle (Miranda and Sousa 2012)

As an alternative classification system for the RMR system, the HRMR system has been developed. It adapts to the level of knowledge of the rock mass providing a classification with different accuracy levels, using a database including sound and very weathered granite formations. It is based on a large database of cases and has been validated in case studies. It allows one to visualize the surface and underground structures vividly (Miranda et al. 2014).

8.3.3 In Situ Shear Tests

Direct in situ shear tests can be conducted in discontinuities containing an infilling that has a critical influence on sliding stability. A typical set up of the direct shear equipment is shown in Fig. 8.38 (Wyllie 1999).

When studying dam sites, direct shear tests can be used on concrete-rock surfaces and special continuum surfaces with low strength such as in Água Vermelha dam (Rocha 2013).

In situ tests performed at LNEC for different dam sites studies were carried out as illustrated in Fig. 8.39, where a square-section of 70 cm was used (Rocha 2013). In order to obtain a more regular distribution of normal stresses, a shear force was applied by means of an inclined jack. The axes of the jacks pass through the center of the volume to be tested. This layout was advantageous for tests inside galleries. Once the vertical force was applied, which enabled the deformability of the rock

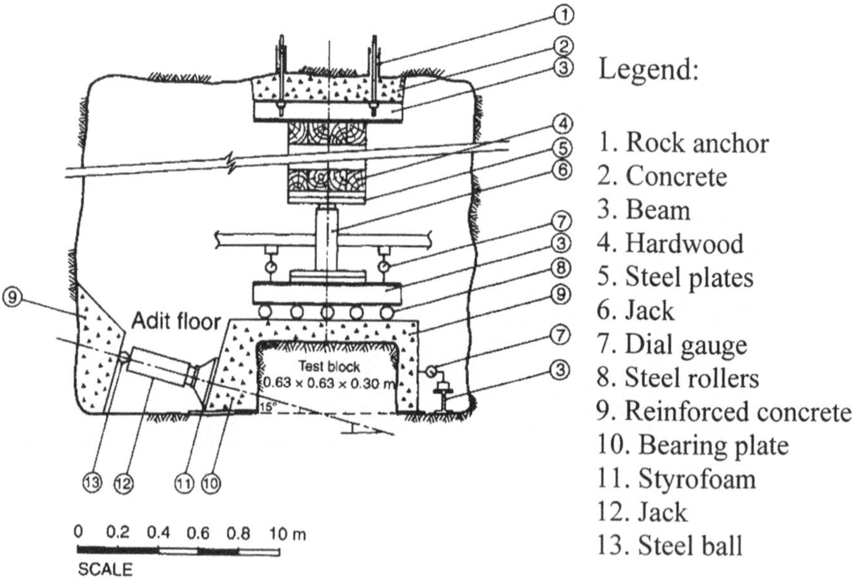

Fig. 8.38 Typical setup of an in situ shear test (Wyllie 1999)

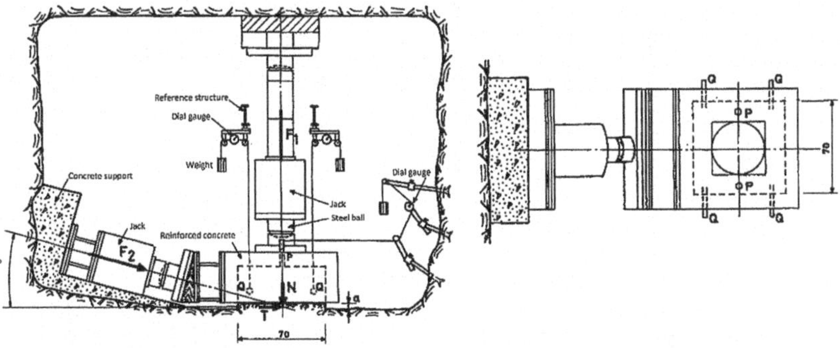

Fig. 8.39 In situ LNEC shear test (Rocha 2013)

Table 8.11 Shear strength of rocks tested in situ (adapted from Rocha 1964)

Rock formation	Type of test	Cohesion (MPa)	Friction angle (°)	Scheme	No. tests
Shale	Rock normal to schistosity	0.2	69	Bemposta (Portugal)	9
	Concrete-rock parallel to schistosity	0.2	60		5
	Concrete-rock parallel to discontinuities	0.2	63		3
Shale	Rock normal to schistosity	1.8	59	Valdecañas (Spain)	7
	Concrete-rock parallel to schistosity	0.4	62		3
Shale	Rock parallel to schistosity	0.4	59	Miranda (Portugal)	4
	Rock normal to schistosity	0.6	64		10
	Concrete-rock parallel to schistosity	0.4	62		8
	Concrete-rock normal to schistosity	0.7	60		8
Shale	Rock normal to schistosity	0.1	70	Alcantara (Spain)	16
	Concrete-rock parallel and normal to schistosity	0.1	56		28
Shale	Rock parallel to schistosity	0.2	50	Cambambe (Angola)	4
Sandstone	Rock parallel to schistosity	0.1	60	Cambambe (Angola)	4
	Concrete-rock parallel to schistosity	0.2	53		4

mass or of the discontinuity surface to be determined, the inclined force was applied by steps. Results of tests conducted by LNEC at several soft rock dam foundations are presented in Table 8.11. The rock are shale and sandstone, in dam sites in Portugal, Spain and Angola. Values of cohesion between 0.1 and 1.8 MPa were obtained, with an average of 0.4 MPa and friction angles between 53° and 70°, with an average of 60°.

Table 8.12 In situ shear tests on bedding planes at the Bakhtiari dam site (Sanei et al. 2015)

c_p (MPa)	φ_p (°)	φ_r (°)	JCS	JRC	σ_n (MPa)	τ_p (MPa)	τ_r (MPa)
0.36	31.9	31.7	27	6.4	1.05	0.94	0.93
0.31	34.6	34.5	20	8.7	1.55	1.33	1.31
0.10	33.0	32.3	26	7.1	0.89	0.60	0.59

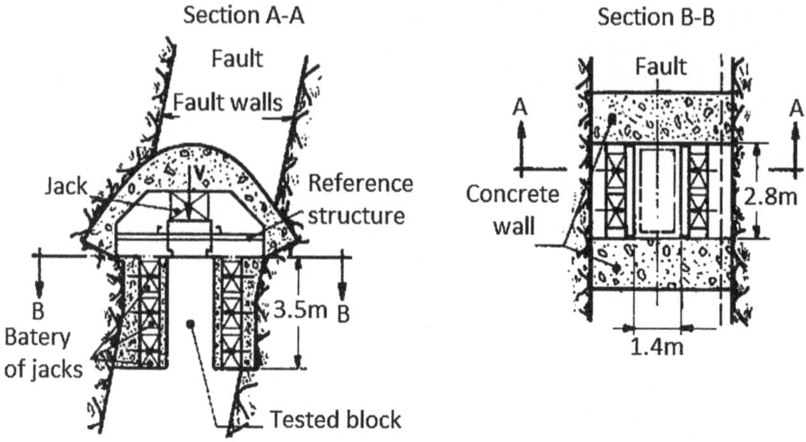

Fig. 8.40 Tests at Kurobe IV in Japan—large blocks were tested in this dam foundation (Rocha 2013)

In situ tests were recently performed at Bakhtiari dam site in Iran. This is the site of a hydroelectric power plant which includes the design and construction of a 315 m high, double curvature concrete dam and an underground powerhouse with a nominal output of 1500 MW (Sanei et al. 2015). Limestones layers constitute the foundation of the dam and of the powerhouse. In situ direct shear tests along rock discontinuities were performed. Three in situ tests on bedding planes were also performed on blocks with 70 × 70 × 35 cm^3 in a gallery at Bakhtiari dam site. The rate of shear displacement ranged from 0.1 mm/min to 0.5 mm/min. The normal force was applied by hydraulic jacks and the shear force via a pressure plate. In the tests the normal stress varied from 0.59 to 7.03 MPa. The mechanical properties of the bedding planes obtained are given in Table 8.12, including cohesion and peak and residual friction angles. The parameters JCS and JRC of the Q system (Barton 2016) are also given as well as the applied pressures during tests (normal and peak and residual values).

In situ shear tests were performed at the Kurobe dam foundation site in Japan in basaltic formations (Rocha 2013). These tests were on a large scale, and are illustrated in Fig. 8.40. Tests were performed on a section of size 2.5 × 3.5 m^2. Tangential forces were successively applied in two perpendicular directions.

More recent equipment to determine the shear strength of rock masses in boreholes include the rock borehole shear test (RBST) apparatus, which is used in a

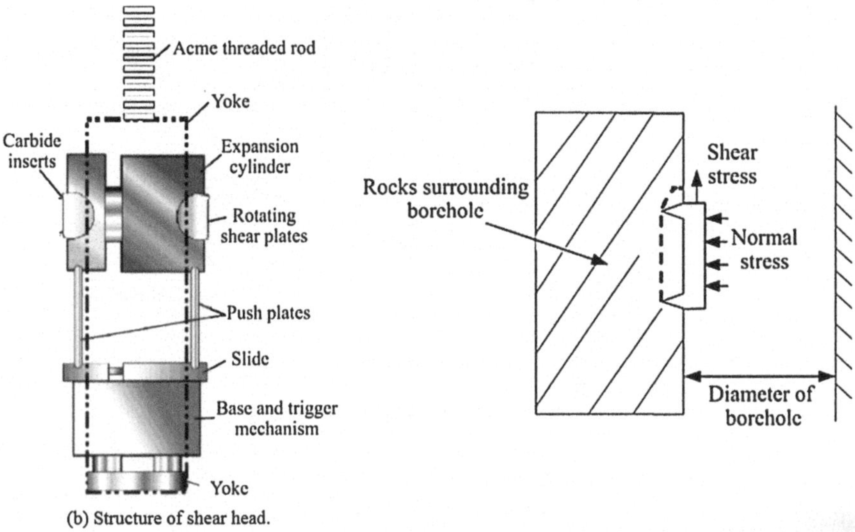

Fig. 8.41 Rock borehole shear tests (Yufei et al. 2012)

76 mm diameter borehole as shown in Fig. 8.41. This has found wide use in USA, Japan, Korea, and China (Zhao et al. 2012). This equipment was applied at the Xiangjiaba Hydropower Station which is one of the cascade power stations on the Jinsha River, in China. Due to the complicated geological conditions of the dam foundation, several groups of RBSTs tests were conducted on the black mudstone in the dam foundation. Forty three groups of shear strengths of black mudstone samples were obtained from RBSTs, and the shear strength parameters (c' and φ') were calculated. The average values of the internal friction coefficient and cohesion were 0.47 (about 25°) and 0.8 MPa, respectively.

8.3.4 Laboratory Tests of Discontinuities

It is common to determine the shear strength of discontinuities by using direct shear tests on core samples containing joint surfaces or bedding planes. Figure 8.42 presents a schematic shear box test with a rock sample under constant normal force (ASTM 2002). Shear tests are generally carried out with a constant normal force or stress. Dilatation can be inhibited by the surrounding rock mass and the initial normal stresses may increase with shear displacements. Different shear modes are illustrated in Fig. 8.43. In (a) and (c), the normal force is controlled, whereas in (b) and (d) the normal displacements are controlled.

Figure 8.44 shows the results of a test performed on a fractured basalt sample. It shows different types of behavior namely, tangential stresses vs. tangential displacements for different normal stresses; normal stresses vs. normal displacements; the

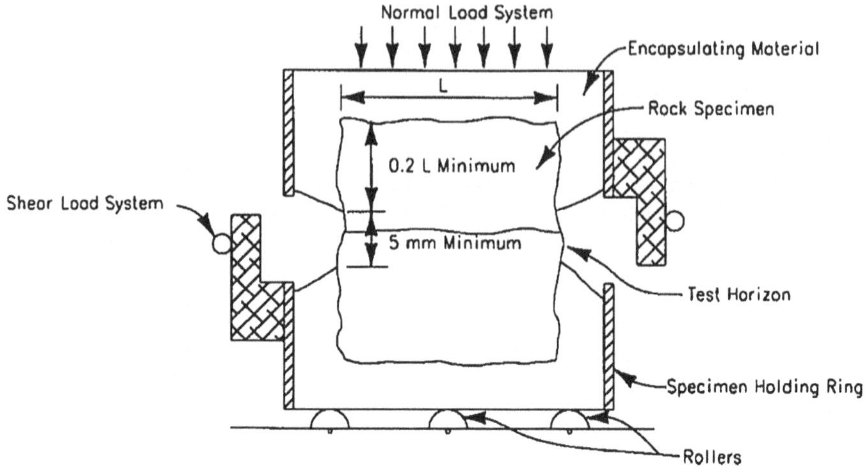

Fig. 8.42 Direct shear box with a specimen (ASTM 2002)

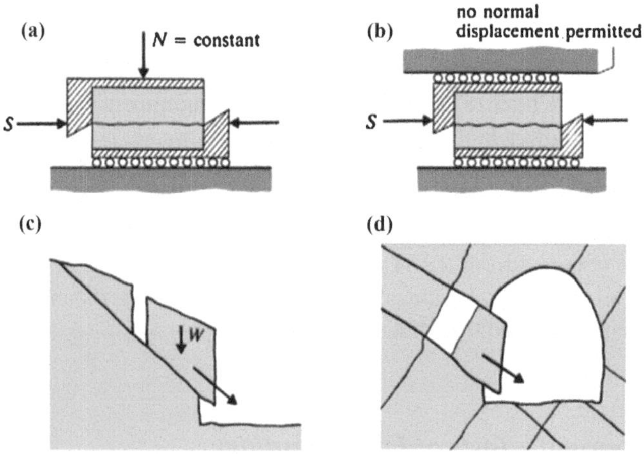

Fig. 8.43 Different situations of shear modes (Brady and Brown 2005)

dilatancy behavior of the discontinuity during the tests; and the peak and residual Coulomb criteria.

Other equipment has been developed by several laboratories and entities (Rocha 1974; Brady and Brown 2005; Wyllie and Mah 2010; Sanei et al. 2015). Figure 8.45 illustrates the classic equipment developed at LNEC (Rocha 2013). Figure 8.46 shows a special shear test developed at the University of Porto that allows one to conduct shear tests at normal stresses while allowing one to impose small deformations in the direction of the plane of the discontinuity. Figure 8.47 shows a portable device developed at the Imperial College that allows one to conduct tests in laboratory and in the field.

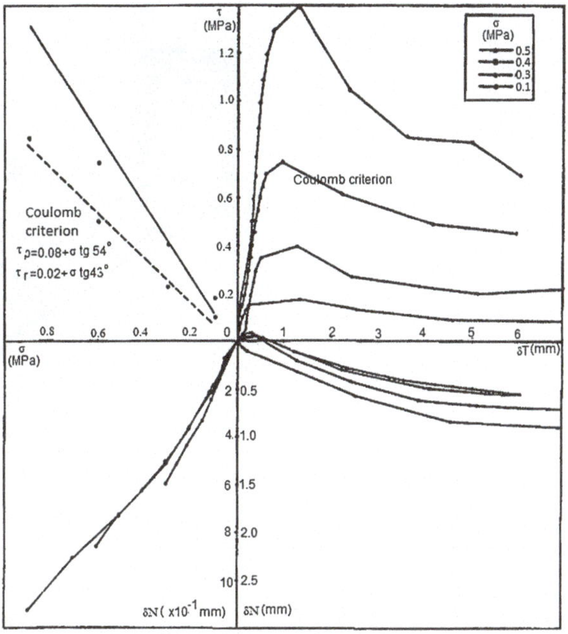

Fig. 8.44 Laboratory shear test results in a fractured basalt sample

Fig. 8.45 Discontinuity shear tests at LNEC

Fig. 8.46 Discontinuity shear tests at University of Porto

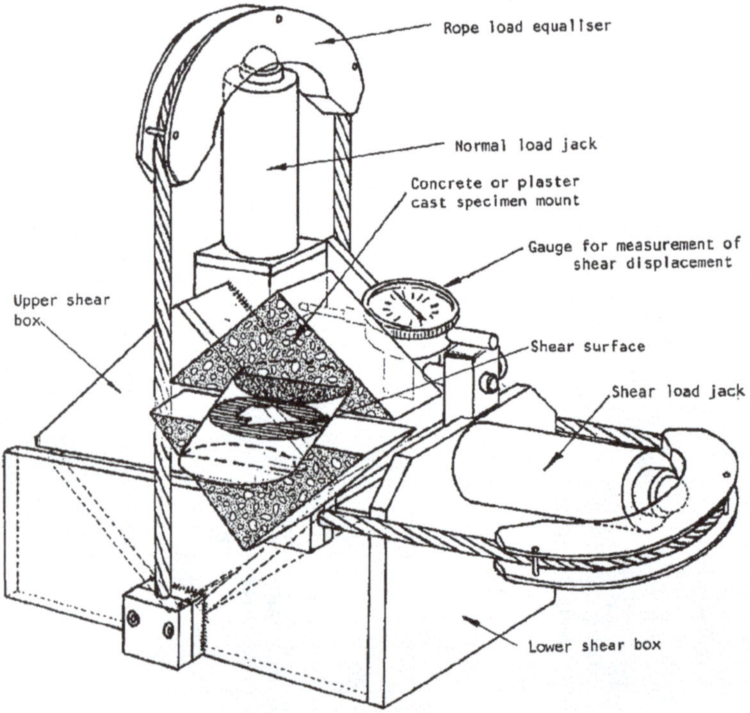

Fig. 8.47 Portable in situ test developed at Imperial College (Brady and Brown 2005)

Direct shear laboratory tests were performed on limestone samples from the Bakhtiari dam site in Iran (Sanei et al. 2015) which was discussed in the previous section. The purpose of these tests was to evaluate peak and residual shear strengths by applying normal and shear loads following the ISRM methods (Ulusay and Hudson 2007). The samples ranged from 5.4 to 14.8 cm in length in the direction of

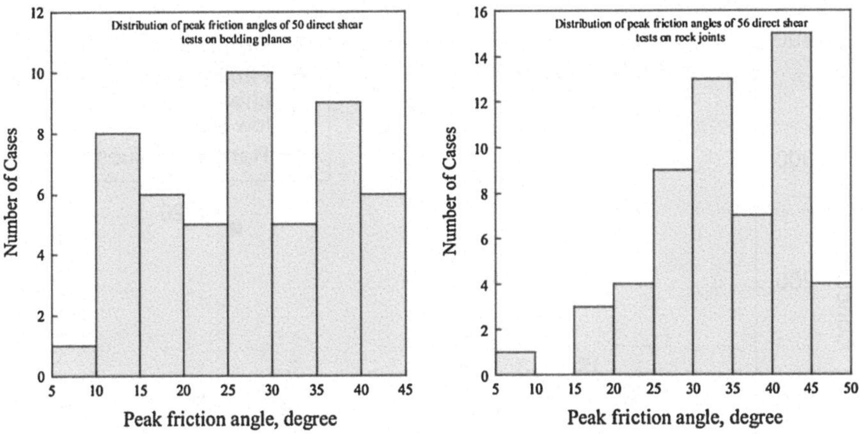

Fig. 8.48 Peak friction angles of discontinuities in limestones (Sanei et al. 2015)

shear. The normal stresses ranged between 0.47 and 3.2 MPa. The shear force was applied continuously while controlling the rate of shear displacement. In addition, the joint roughness coefficient JRC and joint compressive strength were evaluated. About 50 tests were conducted on bedding planes and 56 tests on rock joints. The distribution of peak and residual shear strengths are shown in Fig. 8.48.

In soft rocks, particularly weathered rocks, discontinuities are occupied with filling material. Consequently, strength of the joints is affected by the nature of the filling materials. Often, material like clay is present, and it is essential to investigate if these filler materials are continuous or if blocks are in direct contact (Rocha 1964; Wyllie and Mah 2010). Figure 8.49 shows the results of direct shear tests carried out in mainly soft rocks. These types of rocks are grouped depending on infilling materials (Wyllie and Mah 2010).

Other shear tests such as triaxial tests and torsion tests can be conducted. These are addressed in Wei et al. (2015). For the complete ISRM suggested methods for rock characterization and testing reference is made to Ulusay and Hudson (2007).

8.4 Conclusion

This chapter presents various methods to evaluate the geomechanical properties of rock masses, with emphasis on soft rocks. In general, there is a large degree of uncertainty associated with the evaluation of the deformability and strength properties of soft rock masses. The selection of appropriate values of these parameters requires a combination of in situ and laboratory tests as well as engineering judgment. The use of techniques based on artificial intelligence, particularly those based on Data Mining, have contributed to developing new geomechanical models for the evaluation and characterization of geomechanical properties of soft rock masses.

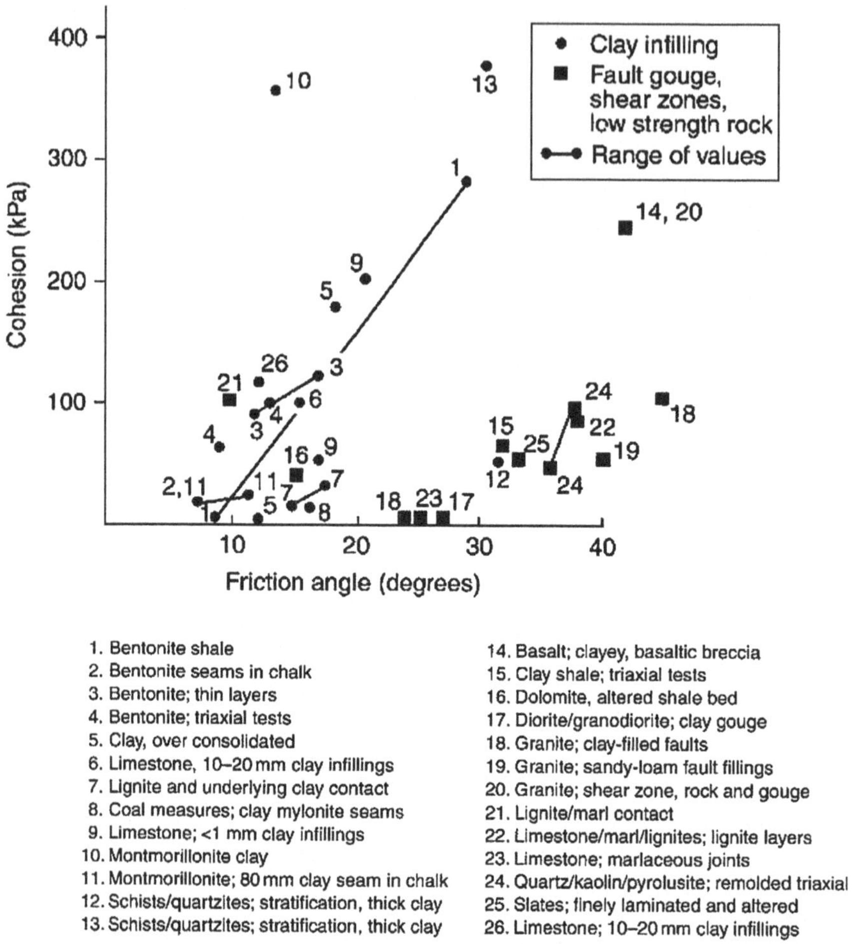

Fig. 8.49 Shear strength of filled discontinuities (Wyllie and Mah 2010)

Common in situ and laboratory tests were described and the results of different tests were illustrated through several case studies. In general, refining predictions depends on better testing equipment, both in situ and in the laboratory, as well as the use of refined numerical models for interpretation. A case study of a dam in Japan with foundations on a heterogeneous conglomerate formation was described. In these formations, the geomechanical parameters are heavily influenced by the dimension of the tested volumes and of the methodology followed. The case illustrated that sometimes this can affect whether the construction of an engineering structure can proceed or not.

References

Alexeev AD, Revva VN, Molodetski AV (2013) Stress state effect on the mechanical behavior of coals under true triaxial compression conditions, chapter 21. In: Kwasniewski M, Li X, Takahashi M (eds) True triaxial testing of rocks. Taylor & Francis, London, pp 281–291
ASCE (1996) Rock foundations (technical engineering and design guides as adapted from US Army Corps of Engineers, no. 16). American Society of Civil Engineers, New York, p 129
ASTM (2002) Standard test method for performing laboratory direct shear strength test of rock specimens under constant normal force. ASTM D5607-02
Babendererde S, Hoek E, Marinos P, Cardoso S (2006) Characterization of granite and the underground construction in metro do Porto. In: Matos AC, Sousa LR, Kleberger J, Pinto PL (eds) Geotechnical risk in rock tunnels. Taylor & Francis, London, pp 41–51
Barton N (2000) TBM tunneling in jointed and faulted rock. Balkema, Rotterdam, p 172
Barton N (2013) Shear strength criteria for rock, rock joints, rockfill and rock masses: problems and some solutions. J. Rock Mech Geotech Eng 5(2013):249–261
Barton N (2016) Non-linear shear strength descriptions are still needed in petroleum geomechanics, despite 50 years of linearity. 50th US Rock Mechanics Symposium, ARMA Symposium, 16-252, Texas, p 14
Barton N, Choubey V (1977) The shear strength of rock joints in theory and practice. J Rock Mech 10:1–54
Bieniawski ZT (1989) Engineering rock mass classifications. Wiley, New York, p 251
Birid K (2014) Comparative study of rock mass deformation modulus using different approaches. 8th Asian Rock Mechanics Symposium, Sapporo, pp 553–563
Birid K (2015) Interpretation of pressuremeter tests in rocks. ISP7 Pressio Conference, pp 289–299
Brady B, Brown ET (2005) Rock mechanics in underground engineering. Kluwer Academic Publishers, New York, p 645
Carter T, Diederichs M, Carvalho J (2007) An unified procedure for Hoek-Brown prediction of strength and post yield behaviour for rock masses at the extreme ends of the rock competency scale. Proc. 11th ISRM Congress, Lisbon, pp 161–164
Carvalho J, Carter T, Diederichs M (2007) An approach for prediction of strength and post yield behaviour for rock masses of low intact strength. Proc. 1st Canada USA Rock Symposium, Vancouver, p 8
Castro AT, Sousa LR (1995) Interpretation of the monitored behavior of a large underground powerhouse using back-analysis techniques. Int. Conf. on engineering mechanics today, Hanoi
Douglas K (2002) The shear strength of rock masses. PhD Thesis, School of Civil and Environmental Engineering, University of New South Wales, Sydney, p 284
Failmezger R, Zdinak A, Darden J, Fahs R (2005) 50 years of pressiometers, vol 1. Press of ENPC/LCPC, Paris, pp 495–503
Fujii Y, Takahashi N, Takahashi T, Takemura T, Park H (2013) Fractographical analyses of the failure surfaces from triaxial extension tests on Kimachi sandstone, chapter 25. In: Kwasniewski M, Li X, Takahashi M (eds) True triaxial testing of rocks. Taylor & Francis, London, pp 323–329
Graça JC (1983) Deformability – BHD method. In: Recent developments in rock mechanics. LNEC, Lisbon, pp 29–58 (in Portuguese)
Grossmann N (1993) New developments in the in-situ determination of rock mass parameters. Course on dam foundations in rock masses. LNEC, Lisbon
He M (2006) Rockburst disasters in coal mine. Glob Geol 9(2):121–123
He M (2014) Latest progress of soft rock mechanics and engineering in China. J Rock Mech Geotech Eng 6:165–179
He M, Sousa LR (2014) Experiments on rock burst and its control. AusRock 2014: Third Australasian ground control in mining conference, Sydney, pp 19–31
He M, Jia XN, Gong W, Liu G, Zhao F (2013) A modified true triaxial test system that allows a specimen to be unloaded on one surface, chapter 19. In: Kwasniewski M, Li X, Takahashi M (eds) True triaxial testing of rocks. Taylor & Francis, London, pp 251–266

He M, Sousa LR, Miranda T, Zhu G (2015a) Rockburst laboratory tests database: application of data mining techniques. J Eng Geol Geol Geotech Hazard 185(2015):116–130
He M, Sousa RL, Muller A, Vargas JRE, Sousa LR, Chen X (2015b) Analysis of excessive deformations in tunnels for safety evaluation. J Tunnel Undergr Space Technol 45(2015):190–202
Hoek E (2007a) Practical rock engineering. www.rocscience.com
Hoek E (2007b) Big trends in bad rock. Therzaghi lecture. ASCE J Geotech Geoenviron Eng 127(9):726–740
Hoek E, Marinos P (2009) Tunneling in overstressed rock. Symp. EUROCK 2009, Dabrovnik, p 13
Hoek E, Carranza-Torres C, Corkum B (2002) The Hoek-Brown failure criterion – 2002 edition. Proc. 5th north American rock mechanics Sym. And 17th tunneling Assn of Canada conf. NARMS-TAC, Toronto, pp 267–271
Hoek E, Marinos PG, Marinos VP (2005) Characterization and engineering properties of tectonically undisturbed but lithologically varied sedimentary rock masses. Int J Rock Mech Min Sci 42(2005):277–285
Hoek E, Torres C, Diederichs M, Corkum B (2008) Integration of geotechnical and structural design in tunneling. 56th Annual geotechnical engineering conference, Minneapolis, p 53
Kanji MA (2014) Critical issues in soft rocks. J Rock Mech Geotech Eng 6:186–195
Kwasniewski M (2013) Mechanical behavior of rocks under true triaxial compression conditions – a review, chapter 8. In: Kwasniewski M, Li X, Takahashi M (eds) True triaxial testing of rocks. Taylor & Francis, London, pp 99–138
Kwasniewski M, Li X, Takahashi M (2013) True triaxial testing of rocks. CRC Press, Taylor & Francis Group, London, p 367
Lade PV (2013) Estimating the parameters for a three-dimensional failure criterion for rocks from a single test, chapter 15. In: Kwasniewski M, Li X, Takahashi M (eds) True triaxial testing of rocks. Taylor & Francis, London, pp 213–222
Li X, Shi L, Bai B, Li Q, Xu D, Feng X (2011) True-triaxial techniques for rocks – state of the art and future perspectives, chapter 1. In: Kwasniewski M, Li X, Takahashi M (eds) True triaxial testing of rocks. Taylor & Francis, London, pp 3–18
Lu Y, Chen M, Jin Y, Yang P, Yuan J, Fan K (2012) Experimental study of wellbore deformation in a deep claystone formation, chapter 22. In: Kwasniewski M, Li X, Takahashi M (eds) True triaxial testing of rocks. Taylor & Francis, London, pp 293–300
Martins CS, Sousa LR (1989) Recent advances in the interpretation of the small flat jack method. ISRM Congress, Montreal
Medley EW (1994) Engineering characterization of melanges and similar block-in-matrix rocks (bimrocks). PhD thesis, Department of Civil Engineering, University of California at Berkley
Miranda T (2003) Contribution to obtaining geomechanical parameters for the modeling of underground works in granite rock masses. MSc Thesis, University of Minho, Guimarães, p 186 (in Portuguese)
Miranda T (2007) Geomechanical parameters evaluation in underground structures. Artificial Intelligence, Bayesian probabilities and inverse methods. PhD Thesis, University of Minho, Guimarães, p 291
Miranda T, Sousa LR (2012) Application of data mining techniques for the development of geomechanical characterization models for rock masses. In: Sousa LR, Vargas E, Fernandes MM, Azevedo R (eds) Innovative numerical modeling in geomechanics. Taylor & Francis, London, pp 245–264
Miranda T, Correia AG, Sousa LR (2009) Bayesian methodology for updating geomechanical parameters and uncertainty quantification. Int J Rock Mech Min Sci 46(7):1144–1153
Miranda T, Sousa LR, Correia A (2011) Development of models for geomechanical characterization using Data Mining techniques applied to a database gathered in underground structures. 45th US Rock Mechanics Symposium, San Francisco, 10 in CD-ROM
Miranda T, Sousa LR, Tinoco J (2014) Updating of the hierarchical rock mass rating (HRMR) system and a new subsystem developed for weathered granite formations. J Min Sci Technol 24:769–775

Miranda T, Sousa LR, Gomes AT, Tinoco J, Ferreira C (2018) Volcanic rocks geomechanical characterization by using empiric systems. J Rock Mech Geotech Eng 10:138–150
Mogi K (2013) How I developed a true triaxial rock testing machine, chapter 9. In: Kwasniewski M, Li X, Takahashi M (eds) True triaxial testing of rocks. Taylor & Francis, London, pp 139–157
Pacheco F, Caxito F, Moraes L, Marangoni Y, Santos R, Soares A (2017) Basaltic ring structures of the Serra Geral formation at the southern region, Triângulo Mineiro, Água Verrmelha region, Brazil. J Volcanol Geotherm Res 355:136–148
Palmstrom A, Singh R (2001) The deformation modulus of rock masses – comparisons between in-situ tests and indirect estimates. J Tunnel Undergr Space Technol 16(3):115–131
Pedro JO, Sousa LR, Teles M, Ramos JM (1975) Study of the Água Vermelha dam by the finite element method. LNEC Report, Lisbon (in Portuguese)
Pinto JL (1981) Determination of the deformability modulus of weak rock masses by means of large flat jacks (LFJ). ISRM Symp. on weak rocks, Tokyo
Pinto JL (1983) Deformability – LFJ method. In: Recent developments in rock mechanics. LNEC, Lisbon, pp 3–27 (in Portuguese)
Pinto JL, Graça C (1983) State of stress – SFJ method. In: Recent developments in rock mechanics. LNEC, Lisbon, pp 41–58 (in Portuguese)
River Bureau (1986) Manual for river works in Japan. Design of dams. Ministry of Construction, Tokyo
Rocha M (1964) Mechanical behavior of rock foundations in concrete dams. Proc 8th Congress on large dams, Edinburgh, vol 1, pp 785–831
Rocha M (1974) Present possibilities of studying foundations of concrete dams. 3rd ISRM Congress, Denver, pp 879–897
Rocha M (1975) Some problems regarding rock mechanics of low strength materials. V Pan-American Congress on Soil Mechanics and Foundation Engineering, Buenos Aires, pp 489–514 (in Portuguese)
Rocha M (1978) Analysis and design of the foundation of concrete dams. ISRM Symp. on rock mechanics applied to dam foundations, Rio de Janeiro, vol 3, pp 3.11–3.70
Rocha M (2013) Rock mechanics, special edn. LNEC, Lisbon, p 445 (in Portuguese)
Rodrigues FP, Graça JC (1983) Heterogeneity. In: Recent developments in rock mechanics. LNEC, Lisbon, pp 123–159 (in Portuguese)
Rodrigues FP, Grossman NF, Rodrigues LF (1978) Rock mechanics tests of the Mingtan pumped storage project. LNEC Report, Lisbon, 2 Vol
Sanei M, Faramarzi L, Fahimifar A, Goli S, Mehinrad A, Rahmati A (2015) Shear strength of discontinuities in sedimentary rock masses based on direct shear tests. Int J Rock Mech Min Sci 75(2015):119–131
Serrano A, Olalla C (2007) Bearing capacity of shallow and deep foundations in rock with the Hoek and Brown failure criteria. Proc. 11th ISRM Congress, Lisbon, vol 3, pp 1379–1392
Simon R, Deng D (2009) Estimation of scale effects in intact rock using dilatometer tests results. Geohalifax2009, pp 481–488
Slope Indicator (2010) Goodman jack. Slope Indicator, Washington, DC, p 23
SolExperts (2016) Dilatometer measurements. Brochure. SolExperts, Monchaltorf, p 2
Sousa LR (2006) Learning with accidents and damage associated to underground works. In: Matos AC, Sousa LR, Kleberger J, Pinto PL (eds) Geotechnical risks in rock tunnels. Taylor & Francis, London, pp 7–39
Sousa RL (2010) Risk analysis for tunneling projects. PhD Thesis, MIT, Cambridge, p 589
Sousa RL, Einstein H (2012) Risk analysis during tunnel construction using Bayesian networks: Porto metro case study. Tunnel Undergr Space Technol 27(2012):86–100
Sousa LR, Nakamura A, Yoshida H, Yamaguchi Y, Kawasaki M, Satoh H (1997) Evaluation of the deformability of rock masses for dam foundations. Analysis of deformability investigation results of heterogeneous bedrock. Technical Memorandum of PWRI, no. 3514, Tsukuba City, 45p
Sousa LR, Chapman D, Miranda T (2010) Deep rock foundations of skyscrapers. J Soils Rocks 33(1):3–22

Sousa LR, Miranda T, Roggenthen W, Sousa RL (2012) Models for geomechanical characterization of the rock mass formations at DUSEL using data mining techniques, US Rock Mechanics Symposium, Chicago, 14p (CD-ROM)
Sousa LR, Wang X, Guo Q, Dias D, Yuan P, Sousa RL (2015) Stability and risk analysis of ancient cavities in historical areas. The case of Yulin caves. China Int Symp on Scientific problems and long-term preservation of large-scale ancient underground engineering, Longyou, pp 411–418
Sousa LR, Miranda T, Sousa RL, Tinoco J (2017) The use of data mining techniques in rockburst assessment. J Eng 3(2017):552–558
Tarasov B (2012) Superbrittle failure regime of rocks at conventional triaxial compression, chapter 27. In: Kwasniewski M, Li X, Takahashi M (eds) True triaxial testing of rocks. Taylor & Francis, London, pp 343–350
Tshibangu JP, Descamps F (2013) The FPMs (UMons-Belgium) device for investigating the mechanical behavior of materials subjected to true triaxial compression, chapter 4. In: Kwasniewski M, Li X, Takahashi M (eds) True triaxial testing of rocks. Taylor & Francis, London, pp 51–60
Ulusay R, Hudson J (2007) The complete ISRM suggested method for rock characterization, testing and monitoring: 1974–2006. ISRM Turkish National Group, Ankara, Turkey, p 613
Wei Y, Fu W, Nie D (2015) Nonlinearity of rock joint shear strength. Strength Mater 47(1):205–212
Wyllie D (1999) Foundations on rock. E & FN Spon, London, p 401
Wyllie D, Mah CW (2010) Rock slope engineering. Taylor & Francis, London, p 431
Yufei Z, Wang X, Zhang X, Jia Z, Zeng X, Zhang H (2012) Rock borehole shear tests in dam foundation of Xiangjiaba hydropower station. J Rock Mech Geotech Eng 4(4):360–366
Yufin S, Lamonina E, Postolskaya O (2007) Estimating of strength and deformation parameters of jointed rock masses. 5th Int. Work. on applications of computational mechanics in geotechnical engineering, Guimarães, pp 3–15
Zhao Y, Wang XG, Zhang XH, Jia Z, Zeng X, Zhang H (2012) Rock borehole shear tests in dam foundation of Xiangjiaba hydropower station. J Rock Mech Geotech Eng 4(4):360–366
Zhou C (2018) Study and development of sample preparation and triaxial test equipment for soft rocks. Report. Zhongshan University, Guangzhou, p 37

Chapter 9
Interaction Between Water and Soft Rocks

Na Zhang

9.1 Introduction

Groundwater may induce various kinds of geological disasters. The degradation of surrounding rock due to interaction with water has become increasingly prominent in deep underground mining engineering; it can lead to surrounding rock collapse, roof falling, floor heaving and support failure of the deep roadways. It was found that the mechanical properties of the rock samples were weakened with increasing of water contents (Hadizadeh and Law 1991; Taibi et al. 2009; Török and Vásárhelyi 2010; Vásárhelyi and Ván 2006; Yilmaz 2010). In particular, the strength of clayey rocks reduces significantly due to strong water absorption and swelling capacity of the clay minerals in the rock. Therefore, experimental research on water absorption characteristics of soft rock and strength degradation after absorbing water is significant for understanding the mechanisms of roadway deformation and also on the decision-making of the support design for deep coal mines roadways.

In rock engineering practice, attentions have been paid on the rock-water interaction and subsequent strength degradation. Characteristics of water absorption for some kinds of rocks were studied by means of sorption and desorption isotherms (Charrière and Behra 2010; Enke et al. 2010; Gruszkiewicz et al. 2001; Shang et al. 1994; van Olphen 1965). Additionally, most research in this area studied the influence of water and chemical solutions on physical, chemical, and mechanical properties of the rock (Delage et al. 1996; Dunn and Hudec 1966; Feng et al. 2001, 2004a, b; Feucht and Logan 1990; Hadizadeh and Law 1991; Heggheim et al. 2005; Karfakis and Akram 1993; Li et al. 2003; Madejová et al. 2002; Montes et al. 2003; Newman 1983; Post and Tullis 1998; Risnes et al. 2003; Rutter 1974; Zielinski et al.

N. Zhang (✉)
State Key Laboratory for GeoMechanics and Deep Underground Engineering, Beijing, China

School of Mechanics, Architecture and Civil Engineering, China University of Mining and Technology, Beijing, China

M. Kanji et al. (eds.), *Soft Rock Mechanics and Engineering*,
https://doi.org/10.1007/978-3-030-29477-9_9

1982). However, few experimental studies have been performed to study the process of water absorption of the deeply buried soft rock in coal mines. Moreover, it is important to note that the method that specimens were completely immersed in water was commonly adopted in water absorption tests in literature. However, this method can hardly simulate the real water absorption process of the surrounding rock in underground excavation.

In this chapter, a testing instrument to simulate the on-site procedure of water absorption of soft rock was designed by the authors, in which, the water absorption was conducted by one surface of a rock sample under different water pressures. Water absorption experiments were performed to investigate their water absorption characteristics of shale samples collected from a deep underground coal mine in China. The experimental instrument and methodology employed to test the water absorption behavior of soft rock is introduced for the first time. The mineralogical composition and pore structure were also investigated by means of X-ray diffraction analysis, mercury intrusion porosimetry, and scanning electronic microscopic technique. Furthermore, strength degradation of shale samples was quantitatively determined.

9.2 Water Absorption Tests of Soft Rocks

9.2.1 Liquid-Water Absorption Test

9.2.1.1 Samples

Shale samples were collected from the Daqiang coal mine at Shenyang, Liaoning province of China. All samples were located at the depth of 1283–1305 m. Collected specimens were wrapped with ziplock bags and then sealed immediately with wax after being transported to ground to maintain their freshness. After transport to the laboratory, all samples were processed into cylindrical rock cores of $\phi 55$ (diameter) × 110 (height) mm. The physical properties and geological descriptions of the rock samples are listed in Table 9.1. Basic physical parameters were determined according to Standard for Tests Method of Engineering Rock Masses (GB/T50266-1999) (MOC 1999).

9.2.1.2 Liquid-Water Absorption Experiments

Equipment

In deep mining engineering, after excavation the confining pressure and hydrogeological environment of surrounding rock change. The possibility of water absorption of the soft surrounding rock in a moist environment increases causing remarkable strength decrease and large deformation. A computer-automated water

Table 9.1 Basic physical parameters and geological descriptions of the shale samples

No.	Weight (g)	Density (g/cm³)	Moisture content (%)	Geological description
S-1	654.65	2.41	0.084	Grey, dense, obvious bedding
S-2	631.62	2.31	0.040	Grey, dense, obvious bedding, full cover by white-grey filler
S-3	614.82	2.24	0.010	Grey, dense, obvious bedding, a small amount of grey filler
S-4	653.84	2.38	0.339	Grey, dense, obvious bedding, grey filler
S-5	672.16	2.46	0.453	Grey, dense, obvious bedding, brown and grey-white filler
S-6	607.83	2.26	0.122	Grey, dense, obvious bedding
S-7	638.22	2.34	0.366	Grey, dense, obvious bedding
S-8	606.68	2.22	0.021	Grey, dense, obvious bedding, grey-white filler
S-9	611.23	2.27	0.093	Grey, dense, obvious bedding, brown and grey-white filler
S-10	653.10	2.44	0.283	Grey and yellowish-brown, dense, coarse surface

absorption tester was specially designed to investigate the water absorption of rock samples by simulating water absorption processes in real construction environment. As shown in Fig. 9.1, the water absorption tester has two experimental modes: (1) water absorption without water pressure that simulates the water-absorption process under capillarity effect; and (2) water absorption under pressure which simulates the water-absorption process due to groundwater seepage. In mode 2, a certain water pressure (1 m water head in this case) can be loaded on the top surface of the rock sample. Water absorption experiments, without water pressure and under water pressure, can be carried out at the same time using this experimental system.

The tester is composed of the following three main parts (Fig. 9.1):

(a) Main test chamber: it is constituted by two sets of polymer water containers standing on electronic balances placed on two rows of brackets, two sets of water pipes and rock samples placed in small alcoves. Rock samples are connected to water containers through water pipes. One end of the rock sample is in contact with water through a device made of hard polymer attached to one end of the water pipe.
(b) Weighing subsystem: it includes two sets of electronic balance (accuracy = ±0.02 g), which are placed respectively on two brackets located in the main chamber. The balances are connected to a computer via data cable. The weighing data can be transmitted on the basis of this section.
(c) Data acquisition subsystem: it is constituted by a high-performance computer, in which program modules are installed to allow real time data transmission and storage. Moreover, the weight variations of the water containers due to water absorption of rock samples can be monitored by real-time curves displayed on the window of the program interface.

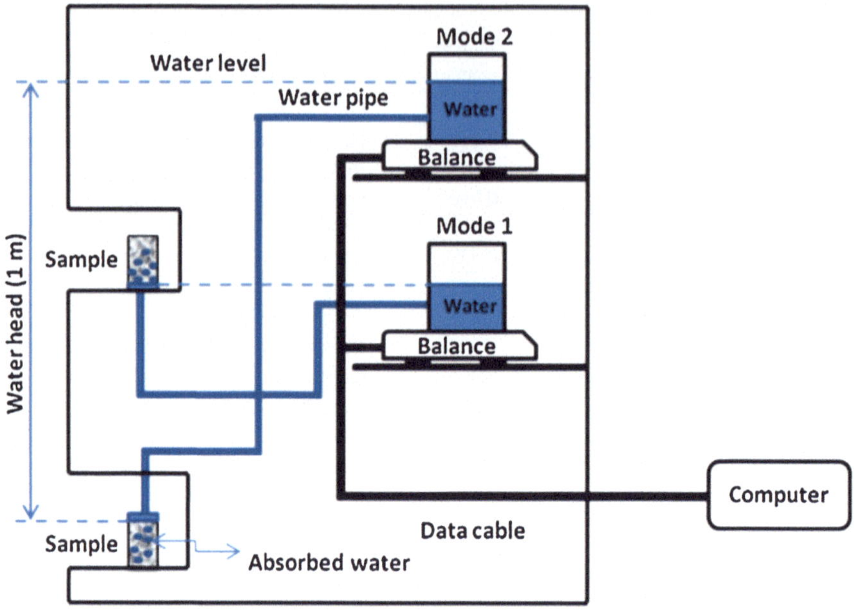

Fig. 9.1 Schematic diagram of the computer-automated water absorption tester for soft rock

Experiments

Water absorption experiments were performed on two groups of shale samples (five samples for each group) under different water pressures (with and without water pressure) at room temperature (~24 °C) for 250 h to obtain their water absorption curves.

9.2.2 *Gaseous-Water Absorption Test*

9.2.2.1 Samples

The soft rock samples used in this study were collected from the Daqiang coal mine which is located in Kangping county of Shenyang city, Liaoning Province of China. All samples were located at the depth of 1283–1359 m. Investigated samples included five mudstones, four coarse sandstones and four fine sandstones.

Collected specimens were wrapped with ziplock bags and then sealed immediately with wax after transported to the ground to maintain their original state. After carried to the laboratory, all samples were processed into cylindrical cores with a uniform size of $\Phi 55 \times 110$ mm using rock drilling, cutting and grinding machines. Numbered samples were then dried in a vacuum drying oven for 1–2 days. Finally, all the dried soft rock samples were kept in desiccators before being used. Table 9.2 lists the basic geological parameters of the 13 clayey soft rock samples investigated in this study.

Table 9.2 Geological description of soft rock samples selected from Daqiang coal mine in China

No.	Lithology	Depth (m)	Geological description
N-1	Mudstone	1304	Grey brown, dense, obvious bedding
N-2	Mudstone	1284	Grey brown, dense, obvious bedding
N-3	Mudstone	1283	Grey brown, dense, obvious bedding, with brown filler
N-4	Mudstone	1304	Grey brown, dense, obvious bedding, with white filler
N-5	Mudstone	1305	Grey brown, dense, obvious bedding, with white filler
CS-1	Coarse sandstone	1354	Grey, dense
CS-2	Coarse sandstone	1354	Grey, dense, less obvious joint
CS-3	Coarse sandstone	1354	Grey, dense, less obvious joint
CS-4	Coarse sandstone	1354	Grey, dense
XS-1	Fine sandstone	1359	Grey brown, dense
XS-2	Fine sandstone	1359	Grey brown, dense
XS-3	Fine sandstone	1355	Dark grey, dense, with few red-brown filler
XS-4	Fine sandstone	1355	Brown, dense

9.2.2.2 Gaseous-Water Absorption Experiments

Equipment

An experimental system was specially developed to measure the water vapor sorption of rock samples. The schematic diagram of the apparatus is shown in Fig. 9.2. The system consists of the following three main parts:

(a) Main test chamber: this section is constituted by a programmable constant temperature and humidity tester with capacity of 320 L. Testing room of the tester can accommodate up to six rock samples. Rock samples are placed on the top surfaces of plastic bars standing on the trays of electronic balances of the testing room.
(b) Weighing subsystem: this section is constituted by six electronic balances (accuracy = ±0.02 g), which are placed on a platform under the main chamber. The balances are connected to a computer via data cable. The weighing data can be transmitted on the basis of this section.
(c) Data collection and analysis subsystem: this section is comprised by a high-performance computer, in which program modules are installed to allow real time data transmission and storage. Moreover, the weight variations of the rock samples can be monitored by the real-time curves displayed on the window of the program interface.

Experiments

Water vapor absorption experiments were carried out on the soft rock samples using the experimental system described in the previous chapter (Fig. 9.2). Soft rock samples were placed in the testing room of the experimental system under relative

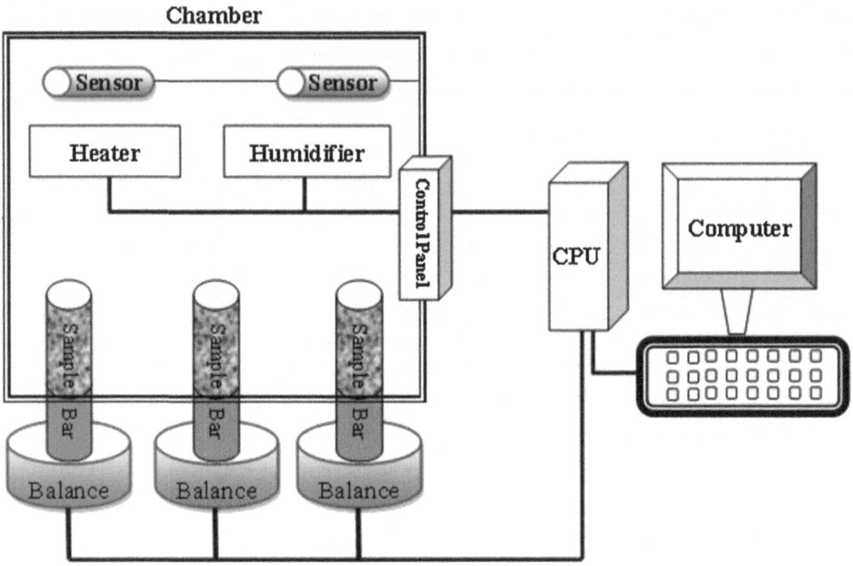

Fig. 9.2 Schematic diagram of the experimental system for water vapor absorption of deep soft rocks

humidity of 95% and temperature of 25 °C for a period of time to observe weight changes of rocks due to vapor absorption.

9.3 Water Absorption Characteristics of Soft Rocks

9.3.1 Liquid-Water Absorption Curve

Changes of water absorption with time for the two groups of shale specimens under different water pressures are presented in Fig. 9.3. It can be seen that the amount of water absorbed by each sample increases with time while the absorption rate decreases with time. Moreover, the water absorption curves varied significantly among samples in spite of the same trend. The water absorptivity differed in all samples due to their differences in physicochemical parameters.

9.3.2 Gaseous-Water Absorption Curve

Changes of water vapor absorption with time for three groups of soft rock samples are presented in Fig. 9.4. It can be seen that the amount of water absorbed by each sample increases with time while the absorption rate decreases with time. Moreover,

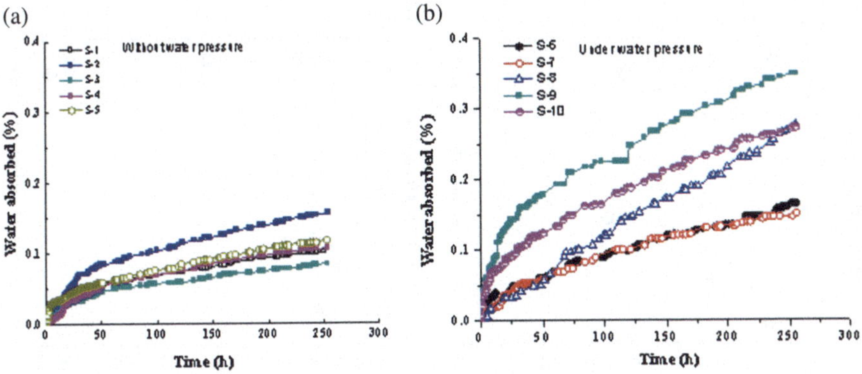

Fig. 9.3 Changes of the water absorption with time for the shale samples without water pressure (a) and under water pressure (b)

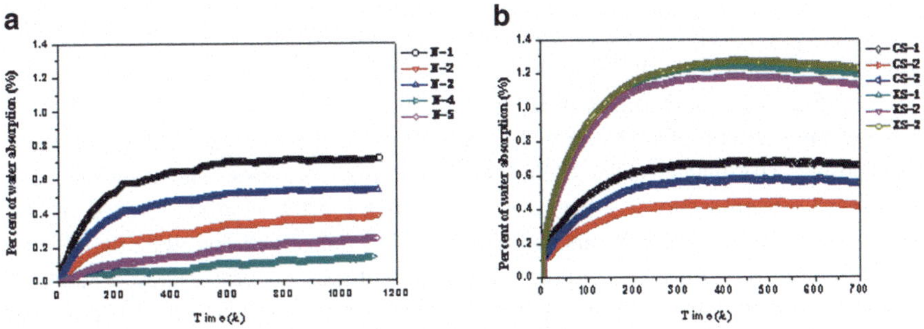

Fig. 9.4 Water vapor absorption curves of the deep soft rock samples. (a) Mudstone. (b) Coarse sandstone and fine sandstone

despite keeping a same trend, the water absorption curves varied significantly among sample groups. It is evident that sandstones have relatively stronger absorptivity than coarse sandstones and mudstones. The variance of water absorptivity of the soft rock samples can be attributed to their difference in physicochemical parameters.

9.4 Effect of Clay Minerals on Water Absorption of Soft Rocks

Clay minerals have long been regarded as a key factor influencing water absorption of rocks. The correlation relationships between water absorptivity and clay minerals for both mudstone and sandstone samples are shown in Fig. 9.5. Compare to mudstone samples, sandstone samples with relatively higher clay contents had stronger

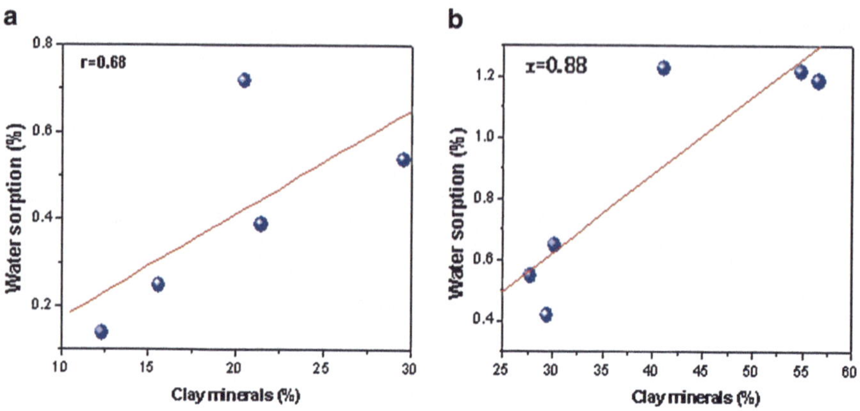

Fig. 9.5 Correlation relationships between water vapor absorption and clay minerals of deep soft rock. (**a**) Mudstone. (**b**) Sandstone

water absorption ability. It is inferred that the higher the content of the clay minerals, the larger water absorption of the soft rocks.

Clay mineral contents is one of the most important parameters affecting water absorption performance (capacity and rate) of clayey soft rocks. Clay minerals have a strong adsorption capacity because of the extra electric charges formed on the particle surface by isomorphous substitution (He et al. 2009; He and Zhao 2012). When interacting with water, mineral grains in soft rock generally show a series of interfacial phenomena such as double electric layer on the surface of grains and ionic exchange. These physicochemical actions directly influence the hygroscopic property of soft rock. Generally speaking, the interfacial phenomena of coarse mineral grains with small specific surface area and lower surface electric charges are much weaker than that of fine mineral grains with large specific surface area and higher surface electric charges. Since the sizes of clay mineral grains in the soft rock are usually far smaller than that of coarse mineral grains such as quartz and feldspar, clay mineral contents become the most critical parameter which can determine the water absorption behavior of soft rocks.

The analytical results of X-ray diffraction for all minerals and clay minerals in all samples are listed in Table 9.3. As Table 9.3 displays, the content of clay minerals in all samples were relatively low and ranged from 7.2% (S-1) to 17.0% (S-4). In addition, illite was the only dominant clay mineral component in all samples.

In general, it was noticed that the samples with higher clay contents showed large water absorption capacity. Based on statistical analyses, positive correlation relationships were found between water absorption capacity and clay mineral contents for both groups of investigated shale samples under different water pressure (Table 9.4). These results indicate that the clay minerals, specifically illite in this case, play an important role in water absorption process of the shale samples with or without water pressure. However, the mechanism affecting water absorption characteristics of the rock samples cannot be well explained only by analyzing the

Table 9.3 Analytical results of X-ray diffraction for all rock specimens

No.	All minerals (%)								Clay minerals (%)	
	Quartz	Potash Feldspar	Plagioclase	Calcite	Dolomite	Pyrite	Analcite	Clay mineral	Illite	Kaolinite
S-1	13.7	2.9	9.0	/	67.2	/	/	7.2	7.1	0.1
S-2	10.6	7.2	15.0	25.4	24	3.2	/	14.6	14.6	/
S-3	19.1	14	16.8	1.6	27.5	7.1	/	13.9	13.6	0.3
S-4	15.2	8.1	5.9	/	53.8	/	/	17.0	16.3	0.7
S-5	14.1	4.3	10.0	/	53.8	2.8	/	15.0	15.0	/
S-6	13.9	5.3	10.9	19.1	40.3	/	/	10.5	10.5	/
S-7	12.5	3.6	8.5	0.7	64.1	/	/	10.6	10.6	/
S-8	22.1	7.8	17.5	/	39.1	/	/	13.5	13.5	/
S-9	12.5	6.7	10.2	31.0	23.6	4.6	/	11.5	11.5	/
S-10	18.8	1.4	13.2	/	49.3	3.5	1.1	12.9	12.9	/

Table 9.4 Results of the correlation analyses between water absorptivity and clay mineral for the shale samples

Experimental mode	Correlation coefficient (r)	Significant (p)
Without water pressure	0.69	0.20
Under water pressure	0.80	0.10

clay mineral contents. Other microscopic characteristics of the rock samples are discussed in the following sections.

9.5 Effect of Pore Structure on Water Absorption of Soft Rocks

Pore structure is another crucial physical parameter influencing water absorption of the rock. Both the amount and the structure of micropores in the rock are important factors for water absorption of rock. The pore size distribution curves for the shale samples are shown in Fig. 9.6. It can be seen that the curves of pore size distribution have similar pattern while the total number and size distribution of micropores differed slightly among different samples. It is evident that rock heterogeneity resulted in the differences in microscopic pore structures of the investigated samples with the same lithology and formation.

Detailed data of porosity and pore size distribution of the shale samples obtained through the mercury intrusion porosimetry analyses are presented in Table 9.5. It is noted that all the shale samples have relatively low porosity ranging from 1.82% to 4.14%. For all the shale samples except S-10, pores with diameter larger than 0.2 μm were most abundant and among them pores with diameter larger than 10 μm were the overwhelming majority. Porosity ($d > 0.2$ μm) is also referred to the effective porosity based on the fact that the film thickness of adsorptive water on the surface of pore is

Fig. 9.6 Pore size distribution curves for the shale samples

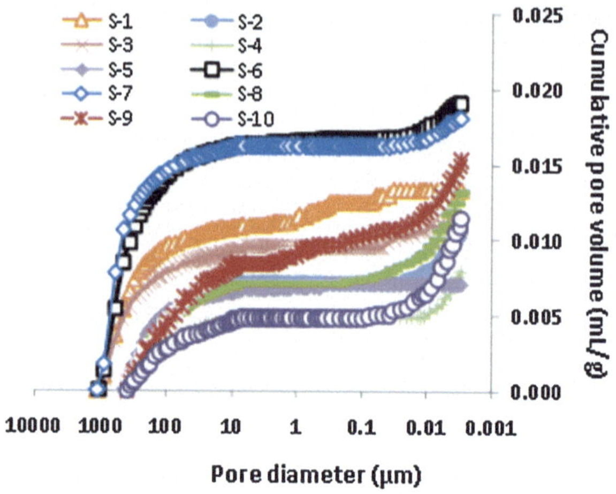

Table 9.5 Results of the mercury porosimetry analyses for shale specimens

	Porosity (%)		Cumulative pore volumes (mL g^{-1})		% of total pore volume	
No.	$d > 0$ μm	$d > 0.2$ μm	$d > 0$ μm	$d > 0.2$ μm	$d > 0.2$ μm	$d > 10$ μm
S-1	3.1458	2.9345	0.0134	0.0125	93.21	80.79
S-2	2.4203	1.6211	0.0106	0.0071	67.51	65.90
S-3	2.4002	2.0919	0.0109	0.0095	86.88	84.25
S-4	1.8448	1.1442	0.0079	0.0049	62.25	59.32
S-5	1.8239	1.8239	0.0071	0.0071	100.00	92.05
S-6	4.1408	3.6205	0.0191	0.0167	87.14	84.42
S-7	3.9248	3.5128	0.0181	0.0162	89.51	88.78
S-8	2.8444	1.5184	0.0133	0.0071	53.83	52.23
S-9	3.3289	2.1322	0.0153	0.0098	63.98	53.25
S-10	2.6909	1.1232	0.0115	0.0048	41.18	40.00

generally more than 0.1 μm. If the diameter of pore in rock is less than 0.2 μm, water cannot easily flow within pores under normal condition (Qin and Li 2004). Normally, it is expected that a rock sample with large effective porosity will absorb more water. Nevertheless, negative correlations were found between water absorptivity and effective porosity for the two groups of shale samples under different water pressures (Table 9.6). The reasons for these negative relationships could be explained from the following perspectives. As shown in Table 9.5, the effective porosities of the investigated samples are relatively small and in a narrow range of 1.12–3.62%. In this situation, the significance of the total amount of micropore contained in rock reduces, while the structural characteristic of micropore, including complexity and connectivity, becomes more important. As a result, the microstructural characteristics of pores in the investigated samples need to be further analyzed.

Table 9.6 Results of the correlation analyses between effective porosity and water absorptivity for the shale samples

Experimental mode	*r*	Sig.
Without water pressure	−0.77	0.15
Under water pressure	−0.60	0.30

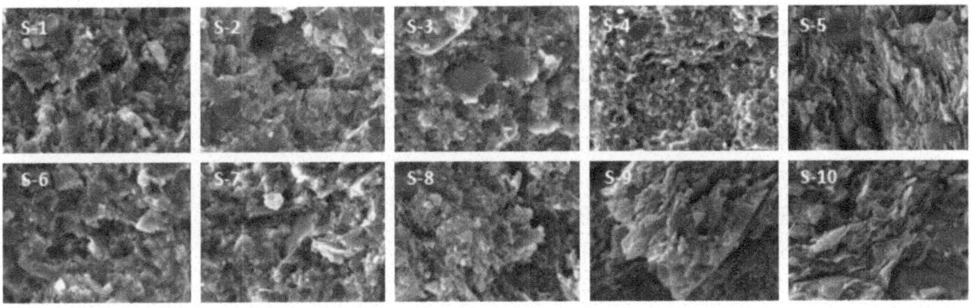

Fig. 9.7 Micropatterns and characteristics of clay minerals in shale samples

The microstructural characteristics of the pores in rock are mainly determined by the distribution and the occurrence of minerals contained in the rock. The contact and connection modes between clay mineral crystals as well as their configuration in internal space of the rock directly determine micropore shapes and in turn the water absorption characteristics of the rock. Microstructural images of SEM analyses for rock samples are shown in Fig. 9.7. It is found that the single crystal state of illite mainly present lamellose and laminated patterns and the congregated state present complicated multilayer support structures. The staggered distribution of illite in intergranular pores forms a large number of micropores. The formed micropore structures mostly consist of triangle and anomalous polygons, in which capillary water readily forms at the corner of pore reducing the effective radius of the pore and the water absorption of rock samples. This explains the observed relative low water absorptivity of the two groups of shale samples (Table 9.7).

9.6 Effect of Water Pressure on Water Absorption of Soft Rocks

The average values of water absorbed by shale samples under two different experimental modes (with or without water pressure) are compared in Fig. 9.8. It is found that the water absorption under water pressure is greater than that of no water pressure by more than two times. As a result, it is suggested that water pressure can significantly increase the water absorption capacity of rock samples.

Table 9.7 Results of uniaxial compressive tests on the shale specimens with different water content

Specimen no.	Water content (%)	UCS (MPa)	Modulus of elasticity (GPa)	Poisson's ratio
S-1	0.175	73.30	18.81	0.26
S-2	0.207	69.71	16.68	0.20
S-3	0.106	71.35	10.86	0.24
S-4	0.482	62.63	14.69	0.18
S-5	0.596	61.59	14.13	0.23
S-6	0.292	76.16	13.35	0.15
S-7	0.518	70.21	11.94	0.18
S-8	0.701	73.10	8.70	0.11
S-9	0.393	68.50	14.50	0.14
S-10	0.528	71.71	15.91	0.29
DS	0	77.70	7.70	0.17
NS	0.101	76.10	13.40	0.16
SS	1.349	60.90	12.70	0.24

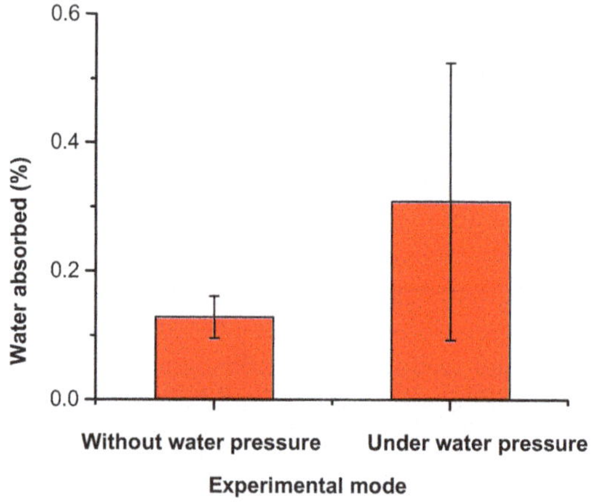

Fig. 9.8 Comparison between the average values of water absorbed by shale samples under two different experimental modes

9.7 Water-Induced Strength Degradation Effect of Soft Rocks

9.7.1 Determination of Mechanical Properties of Hydrated Soft Rock Samples

Uniaxial compression tests were performed on two control dry samples (CS-4 and XS-4) and nine hydrated samples using a self-developed microcomputer servo testing machine (XTR01). Two mudstone samples (N-1 and N-3) were eliminated because of their damage during water absorption experiment. The Suggested Methods for Determining the Uniaxial Compressive Strength and Deformability of Rock Materials were employed for the determination of UCS and Young's modulus.

It is worth mentioning that the Young's modulus on rock samples were determined by their tangent modulus (E_t) measured at stress levels of 50% of UCS due to the complex nature of the investigated soft rock samples.

9.7.2 Mechanical Properties of Hydrated Soft Rock Samples

Results of the UCS tests on the soft rock samples are shown in Table 9.8. Correlation analysis between UCS, Modulus of Elasticity and water content (Figs. 9.8 and 9.9) demonstrated that both of them tend to decrease with the increase of their water contents. The results suggested that with increasing water content induced by vapor sorption the soft rock strength tended to decrease, and meanwhile it was more prone to deformation. Consequently, it is evident that vapor sorption leads to reduced rock strength and increased vulnerability of deformation.

Table 9.8 Basic physical and mechanical parameters of the rock

Specimen no.	Lithology	Water content (%)	UCS (MPa)	Modulus of elasticity (GPa)
N-2	Mudstone	0.39	55.82	16.89
N-4		0.14	61.69	19.66
N-5		0.25	62.33	19.44
CS-1	Coarse sandstone	0.65	61.60	5.1
CS-2		0.42	112.0	12.9
CS-3		0.55	90.95	12.2
CS-4		0	89.50	14.3
XS-1	Fine sandstone	1.19	49.41	10.2
XS-2		1.23	105.05	13.1
XS-3		1.22	15.70	5.6
XS-4		0	106.15	16.8

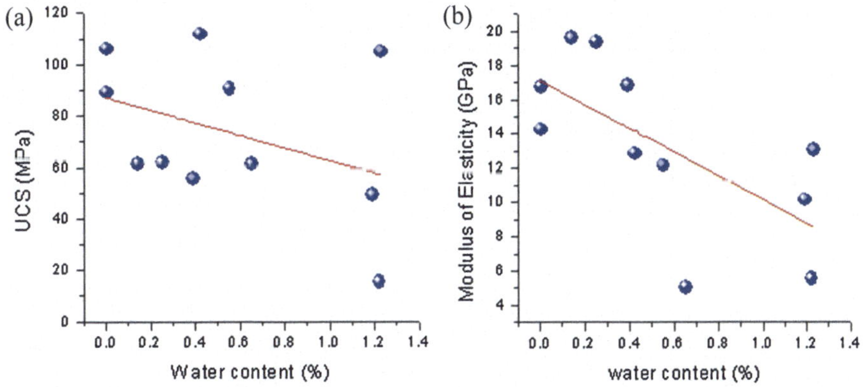

Fig. 9.9 The relationship between water ratio and the uniaxial compressive strength (**a**) and Young's modulus (**b**) of the soft rock samples

Acknowledgement This work was performed under the guidance of Prof. Manchao He at the State Key Laboratory for GeoMechanics and Deep Underground Engineering of Being, China University of Mining & Technology. Special thanks are given to Prof. He for his valuable directions and supports.

References

Charrière D, Behra P (2010) Water sorption on coals. J Colloid Interface Sci 344(2):460–467

Delage P, Schroeder C, Cui YJ (1996) Subsidence and capillary effects in chalks. Dans EUROCK '96, Prediction and performance on rock mechanics and rock engineering - ISRM International symposium, Torino, France, pp 1291–1298

Dunn JR, Hudec PP (1966) Water, clay and rock soundness. Ohio J Sci 66(2):153–168

Enke D, Ruckriem M, Schreiber A, Adolphs J (2010) Water vapour sorption on water absorption and hydrophobic nanoporous materials. Appl Surf Sci 256(17):5482–5485

Feng X, Chen S, Li S (2001) Effects of water chemistry on microcracking and compressive strength of granite. Int J Rock Mech Min Sci 38(4):557–568

Feng X-T, Chen S, Zhou H (2004a) Real-time computerized tomography (CT) experiments on sandstone damage evolution during triaxial compression with chemical corrosion. Int J Rock Mech Min Sci 41(2):181–192

Feng XT, Li SJ, Chen SL (2004b) Effect of water chemical corrosion on strength and cracking characteristics of rocks-a review. Key Eng Mater 261–263:1355–1360

Feucht LJ, Logan JM (1990) Effects of chemically active solutions on shearing behavior of a sandstone. Tectonophysics 175(1–3):159–176

Gruszkiewicz MS, Horita J, Simonson JM, Mesmer RE, Hulen JB (2001) Water adsorption at high temperature on core samples from The Geysers geothermal field, California, USA. Geothermics 30(2–3):269–302

Hadizadeh J, Law RD (1991) Water-weakening of sandstone and quartzite deformed at various stress and strain rates. Int J Rock Mech Min Sci Geomech Abstr 28(5):431–439

He MC, Zhao J (2012) Effects of Mg, Ca, and Fe(II) doping on the Kaolinite (001) surface with H_2O adsorption. Clay Clay Miner 60(3):330–337

He MC, Fang ZJ, Zhang P (2009) Theoretical studies on the extrinsic defects of montmorillonite in soft rock. Mod Phys Lett B 23(25):1–9

Heggheim T, Madland MV, Risnes R, Austad T (2005) A chemical induced enhanced weakening of chalk by seawater. J Pet Sci Eng 46(3):171–184

Karfakis MG, Akram M (1993) Effects of chemical solutions on rock fracturing. Int J Rock Mech Min Sci Geomech Abstr 30(7):1253–1259

Li N, Zhu Y, Su B, Gunter S (2003) A chemical damage model of sandstone in acid solution. Int J Rock Mech Min Sci 40(2):243–249

Madejová J, Janek M, Komadel P, Herbert HJ, Moog HC (2002) FTIR analyses of water in MX-80 bentonite compacted from high salinary salt solution systems. Appl Clay Sci 20(6):255–271

MOC (1999) Standard for tests method of engineering rock masses. GB/T 50266-1999. Chinese Plan Publishing House, Beijing

Montes HG, Duplay J, Martinez L, Geraud Y, Rousset-Tournier B (2003) Influence of interlayer cations on the water sorption and swelling-shrinkage of MX80 bentonite. Appl Clay Sci 23(5–6):309–321

Newman GH (1983) The effect of water chemistry on the laboratory compression and permeability characteristics of some north sea chalks. J Pet Technol 35(5):976–980

Post A, Tullis J (1998) The rate of water penetration in experimentally deformed quartzite: implications for hydrolytic weakening. Tectonophysics 295(1–2):117–137

Qin J, Li A (2004) Petrophysics. China University of Petroleum Press, Shandong, p 344p

Risnes R, Haghighi H, Korsnes RI, Natvik O (2003) Chalk-fluid interactions with glycol and brines. Tectonophysics 370(1–4):213–226

Rutter EH (1974) The influence of temperature, strain rate and interstitial water in the experimental deformation of calcite rocks. Tectonophysics 22(3–4):311–334

Shang S, Horne RN, Ramey HJ (1994) Water vapor adsorption on geothermal reservoir rocks. Geothermics 24(4):523–540

Taibi S, Duperret A, Fleureau J-M (2009) The effect of suction on the hydro-mechanical behaviour of chalk rocks. Eng Geol 106(1–2):40–50

Török Á, Vásárhelyi B (2010) The influence of fabric and water content on selected rock mechanical parameters of travertine, examples from Hungary. Eng Geol 115(3–4):237–245

van Olphen H (1965) Thermodynamics of interlayer adsorption of water in clays. I. Sodium vermiculite. J Colloid Sci 20(8):822–837

Vásárhelyi B, Ván P (2006) Influence of water content on the strength of rock. Eng Geol 84(1–2):70–74

Yilmaz I (2010) Influence of water content on the strength and deformability of gypsum. Int J Rock Mech Min Sci 47(2):342–347

Zielinski RA, Peterman ZE, Stuckless JS, Rosholt JN, Nkomo IT (1982) The chemical and isotopic record of rock-water interaction in the Sherman Granite, Wyoming and Colorado. Contrib Mineral Petrol 78(3):209–219

Chapter 10
Weathering of Rocks in Brazil

Eduardo Antonio Gomes Marques, Eurípedes do Amaral Vargas Jr, and Marcio Fernandes Leão

10.1 Introduction

Weathering is an important aspect of rock behavior especially in tropical regions, where the process is accelerated by temperature variation over a year and the considerable amount of water (from rain). All types of rocks can present considerable change in their mechanical behavior due to physical and chemical weathering. It is therefore important to characterize the influence of weathering on the mechanical behavior of weathering materials, from fresh rock to soil, including its transitional material, to provide useful geotechnical data for design of civil and mining works.

Several authors working in various places worldwide have studied the effects of weathering on geomechanical properties of rocks under different climate conditions. On this context, the works by Ruxton and Berry (1957), Deere and Patton (1971), Dearman (1974, 1976), IAEG (1981), Beavis (1985), Lee and De Freitas (1988), Dobereiner and Porto (1990), Dobereiner et al. (1993), Gupta and Rao (2000), Arel and Önalp (2004), Basu and Aydin (2004), Marques et al. (2005, 2010), Marques and Williams (2015a, b), Marques et al. (2017), and Leão et al. (2016) can be cited as those which, by studying weathering profiles developed on various rock types, have tried to characterize the variations in mechanical properties (Marques et al. 2017).

The study of the influence of weathering and weatherability is closely related to the development of weathering profiles, characterized by several transitional materials between rock and soil, phenomena very common on humid tropical regions as those existing both in Brazil. The occurrence of weathering profiles on geotechnical works involving cut slopes, foundations, underground and open excavations on civil

E. A. G. Marques (✉) · M. F. Leão
Universidade Federal de Viçosa, Viçosa, Brazil

E. d. A. VargasJr
Pontifícia Universidade Católica do Rio de Janeiro, Rio de Janeiro, Brazil

M. Kanji et al. (eds.), *Soft Rock Mechanics and Engineering*,
https://doi.org/10.1007/978-3-030-29477-9_10

and mining works is of fundamental importance for its stability, as the weathering imposes different materials with different mechanical properties to the design (Marques 1998; Marques et al. 2010). In addition to that, the presence of structural discontinuities which, besides favoring water percolation and weathering, also induces variations on mechanical behavior and control stability because of its geometry and strength properties. Thus, a complex and important mechanism for rock and soil failure is provided and has commonly been responsible for several mass movements on roads, foundations and mining cut slopes.

Brazil has a very complex and diverse geology, as can be seen in the map shown on Fig. 10.1. The most common rock types are as follows:

- Sedimentary rock types, such as claystones, siltstones, shales, and limestones, on approximately 50% of the country's total area, on sedimentary basins, as shown on Fig. 10.1.

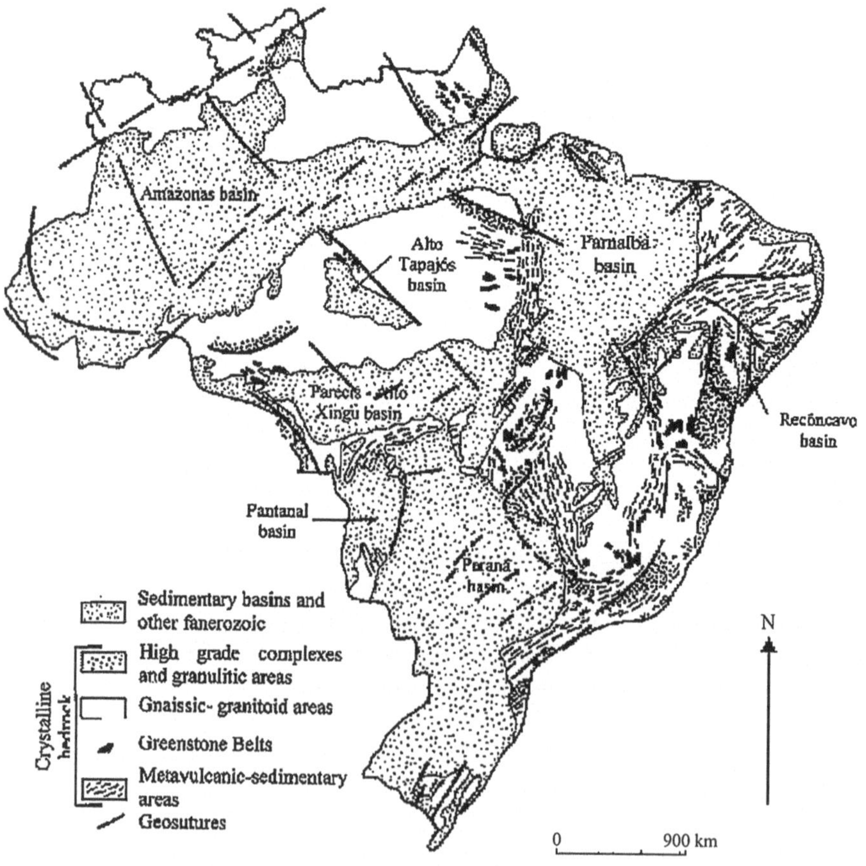

Fig. 10.1 Geological map of Brazil (Marques et al. 2005), showing its varied geology

- Metamorphic rocks, such as gneisses (several different types), amphibolites, migmatites, schists, phyllites, quartzites and marble, occupying gneissic-granitoid and high-grade granulitic complexes, and metavulcanic-sedimentary areas.
- Igneous rocks, such as basalts (Paraná basin), granites, diorites, and sienites occupying gneissic-granitoid areas and part of metavulcanic-sedimentary areas.

Weathering has been object of attention in more recent decades in Brazil, as this comprises an important mechanism controlling rock behavior on these areas. Presence of such profiles is a critical factor in the stability of cut slopes, foundations and underground excavations where the weathering produces materials with extremely variable behavior (reducing strength and increasing deformability and permeability). Further difficulty is provided by the presence of structural discontinuities that controls underground water movement and, hence, chemical weathering. The works by Menezes Filho (1993), Barroso (1993), Marques (1998), Marques et al. (2010), and Leão et al. (2016), should be pointed for metamorphic rocks. Regarding sedimentary rocks there are only few papers, the work of Marques et al. (2005) being one of the most recent. For igneous rock, Basu et al. (2009) have presented a study on evaluation of mechanical behavior on several weathering granite grades.

Another important aspect of considering weathering in design of civil and mining engineering in Brazil is that it can commonly reach several hundreds of meters in depth. Several engineering problems in Brazil are commonly related to weathering and among these problems the following deserves to be highlighted:

- Stability of mining and civil cut slopes (Fig. 10.2a, b).
- Instability of natural slopes (Fig. 10.2c).
- Underground excavation problems (Fig. 10.2c).

In the present chapter the main aspects that controls weathering in Brazil are presented, including their typical morphologies. The intention is to present general characteristics imposed by weathering on a tropical area and their influence on physical, chemical, mineralogical, and mechanical properties of the materials found in weathering profiles, focusing on sound rock to transitional rocky materials.

10.2 Main Aspects Controlling Weathering in Brazil

As already mentioned, weathering is a complex process. Many factors can control its development, but the most important ones are:

- *Rock type and presence of structures that can form discontinuities*—Each rock type has a typical mineralogy, which reacts differently to weathering and both macro and micromineralogy are important to the process. On Fig. 10.3a, b it is shown two different litotypes (phyllite and shale) presenting pyrites that can easily weather to secondary minerals, such as melanetrite and halotrichite. Fractures

Fig. 10.2 Examples of weathering influence on civil and mining engineering: (**a**) An example of road slope failure along a stress relief joint (BR040 Highway, Rio de Janeiro state); (**b**) A weathered level developed along foliation in an itabirite from Alegria Mine (Ouro Preto, Minas Gerais state); (**c**) Failure in a natural slope over residential area during a major disaster on Rio de Janeiro state mountain region; (**d**) Weathering on a road tunnel at Petrópolis (Rio de Janeiro state) has imposed the need of extra support for tunnel excavation

(Fig. 10.3c, d), joints and faults can influence weathering, as these structures are preferential paths for water circulation. Directional structures and/or textures such as foliation, cleavage, and schistosity (Fig. 10.2b), can also influence weathering especially in the presence of discontinuities parallel to them.

- *Natural slope (topography)*—On more inclined slopes, run off transport or landslides (Fig. 10.2a, d) of weathered material downhill, continuously exposes sound rock to weathering attack, in a way that weathered material has low thickness. On less inclined slopes, infiltration prevails over run off and so there is more water to attack rocks and weathering thickness can reach hundreds of meters.
- *Climate (weather)*—Weathering is climate-dependent process as the amount of water and daily and seasonal differences on temperature act as a catalyst for chemical reactions that can affect mineral changes. So on tropical and subtropical climates, commonly found in Brazil, the weathering process is accelerated and this is the main reason for the high weathering thicknesses found in the

Fig. 10.3 Examples of influence of rock mineralogy and rock structures on its weathering. On (**a**) weathering of pyrite in a basic intrusive rock (Santiago et al. 2005) and on (**b**) pyrite crystals on a shale from Amazon River basin (Marques et al. 2005). On (**c**) is shown a residual soil filling a stress relief joint at a cut slope on gneiss. On (**d**) weathering along foliation in a phyllite of iron Quadrangle, Brazil

country. On more dry and cold climates, weathering is mainly due to physical weathering with lower mineral (chemical) change.

- *Extent of weathering process*—Time necessary for chemical decomposition of sound rock considerably varies due to climate type, topography and rock composition and structure. The longer the process can act, the higher will be weathering. So it can be thousand of years for the development of residual soils in dry and cold regions, while in humid and hot regions this process can occur in an engineering time.

Based, on that, the main particularity that makes the development of weathering in Brazil different from other countries, especially those on northern region of the planet, is climate. Similar climate conditions can be found in all southern hemisphere countries—Brazil, South Africa, Angola, Indonesia, Vietnam, Malaysia, some parts of Australia, and so on, and weathering results on similar weathering profiles for similar controlling factors and for the same lithotypes.

Depending on the rock genetic type, some structures can give rise to discontinuities and so control water flow. Sedimentary rocks can present sedimentary structures such as bedding and lamination, while metamorphic rocks can present foliation and cleavage. All rock types can present joints and faults. As all these structures, if defining discontinuities, increase specific surface exposed to water attack, its presence is also an important weathering control, as it controls the amount of chemical weathering. Other physical processes are very important for the development of weathering, such as:

- *Stress relief*—This process is common to all genetic rock types and result in joints with high aperture and with lower spacing closer to surface (Fig. 10.2a).
- *Wetting and drying cycling*—Occurs on both sedimentary rocks and metamorphic rocks of sedimentary origin mainly, due to the presence of clay minerals. It can cause disaggregation of rocks because of the expansion of clay minerals and mixed-layers due to water adsorption (Fig. 10.4).
- *Heating and cooling*—Another process that is common to all kinds of rocks. As some regions in Brazil can present very high daily and seasonal temperature variation, causing rocks to expand and shrink, resulting in rock fatigue, so creating and propaganting existing discontinuities.
- *Erosion effect of water and wind*—Erosion is more efficient on sedimentary rocks, but not exclusively. The effect of water and wind erosion can create different types of discontinuities such as fractures and joints.

Fig. 10.4 Black shale from Ilhas Group and undifferentiated Candeias Formation, from Recôncavo sedimentary basin, Bahia state, Northeast Brazil. It can be seen the effects of disaggregation (slaking) of rock, due to its explosion to weathering in a cut-slope

- *Mechanical excavation effect*—This is a much more localized process when compared to the other previously shown but it can be very important mainly in underground excavations (deep wells, tunneling, and mining). The higher the residual stresses within a rock mass the higher is the stress relief effect. So this type of effect is much more noticeable on igneous (plutonic) and metamorphic rocks, but also occurs on sedimentary rocks.

Many rock-forming minerals are relatively stable under surface conditions, but some can be particularly reactive, such as pyrite, marcasite, pyrrhotite, calcite, dolomite, biotite, garnet (group), olivine, halite, anorthite (calcium feldspar), and others. So in the engineering time scale, the most important chemical weathering processes, as suggested by Taylor (1988), and confirmed for weathering of rocks in Brazil are:

- *Oxidation* of sulphide minerals resulting in new formed minerals and (Fig. 10.5a).
- *Dissolution* of carbonaceous cements and salt minerals.
- *Hydration and other water-dependent effects* such as clay mineral and anhydrite hydration (Fig. 10.5b), ionic sorption, ion exchange, osmosis, and water adsorption.

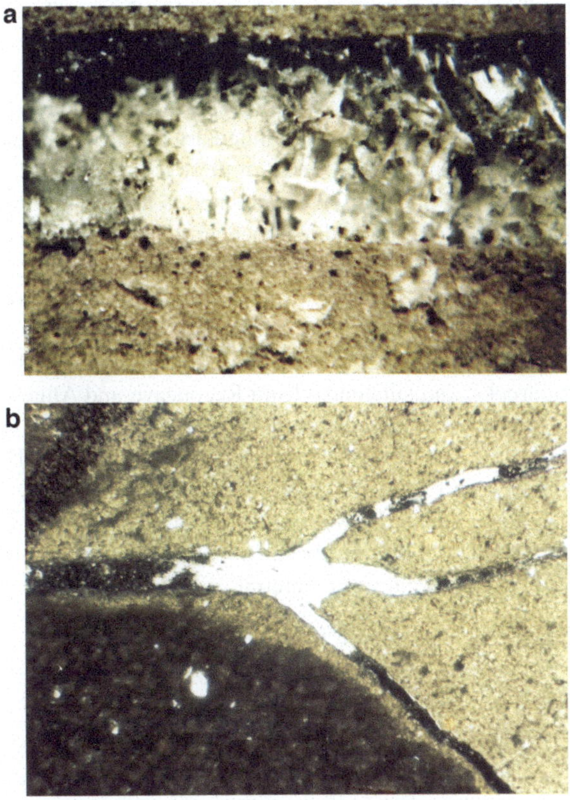

Fig. 10.5 On (**a**) new minerals crystals (Melanterite) formed due to weathering of pyrite in Amazon shale (Marques et al. 2005). On (**b**) the dark green area is an expanded (higher relief) clay mineral area due to its humidification by water, observed in a thin section (Marques et al. 2005) from the shale shown in Fig. 10.4

Table 10.1 Secondary minerals formed by some mineralogical change due to chemical weathering (Taylor 1988)

Original mineral	Secondary (new formed) mineral	Volume increase (%)
Pyrite (FeS_2)	Jarosite [$KFe^{3+}{}_3(OH)_6(SO_4)_2$]	115
	Melanterite ($FeSO_4 \cdot 7H_2O$)	536
	Ferrous Sulfate Anhydrous [FeO_4S]	350
Calcite ($CaCO_3$)	Gypso [$Ca(SO_4) \cdot 2H_2O$]	103
	Bassanite [$2Ca(SO_4) \cdot H_2O$]	189
Illite [$KAl_2Si_3O_8(OH)_2$]	Jarosite [$KFe^{3+}{}_3(OH)_6(SO_4)_2$]	10
	Alunite [$KAl_3(SO_4)_2(OH)_6$]	8

In the sequence, some specific aspects of weathering of sedimentary and metamorphic rocks are presented. Studies on igneous rocks are current under development for the authors in Brazil but no data is presently available.

10.2.1 Weathering of Sedimentary Rocks

On the weathering of sedimentary rocks all three chemical processes described previously—oxidation, dissolution and hydration, and other water-dependent effects can occur, but the last two are the most common, as clay minerals and carbonatic rocks are more common than rocks with sulphide minerals. In Table 10.1, the main mineralogical changes on sedimentary rocks are presented. The same reactions can be easily found in weathering of these rocks in Brazil.

Another important process related to the weathering of sedimentary rocks is the influence of hydration due to penetration of water front into rock matrix, driven by capillarity forces or gas condensation. This process can cause entrapment and pressurization of air in pores, which can be sufficient to promote microfracturing, causing physical disintegration.

Marques et al. (2005) have developed studies on the durability of rocks (shales, mudrocks, and siltstones) from Recôncavo and Amazon River sedimentary basins. All those rocks, when exposed on cut slopes present a rapid disintegration, which commonly causes many geotechnical problems. The weathering mechanisms involved into its degradation where hydration of clay minerals and air entrapment into pores, caused by wetting-drying cycles. These processes have lead to the development of microfracturing and expansion causing loss of strength and high deformability of the weathered material.

10.2.2 Weathering of Metamorphic Rocks

Garnets, biotite, and feldspars are common minerals in Brazilian metamorphic rocks. Garnet and biotite can release considerable amounts of iron due to its weathering, which is leached through microfractures and micropores and deposits on those voids, acting like a cement and so resulting in an increase in strength on moderately weathered (W3 and W3/W4 classes, according to ISRM, 2015) layers of rock mass. Anorthite, on the other hand, rapidly weather to kaolin, reducing rock strength and increasing its deformability.

Works by Barroso (1993), Menezes Filho (1993) and Marques (1998) present detailed studied the influence of weathering on physical and geomechanical properties of three high grade metamorphic rocks (different type of gneisses—Leptinite, Augen gneiss, and Kinzigite, respectively). Both rock matrix and rock masses characteristics were described to form a complete picture of weathering influence on those rocks. By analyzing hand samples and thin sections those authors have determined that fracturing and microfracturing (especially transgranular fractures) present a substantial increase due to weathering. This physical process increases porosity and influence its mechanical behavior. Also, it allows water to access inner portions of rock matrix, leading to an increase in chemical weathering. Figure 10.6 shows some aspects of those studies and weathering influence on physical and mechanical properties (Marques et al. 2010). Table 10.2 presents a detailed description of microscopic (both mineralogical and structural) changes observed on those gneisses due to weathering.

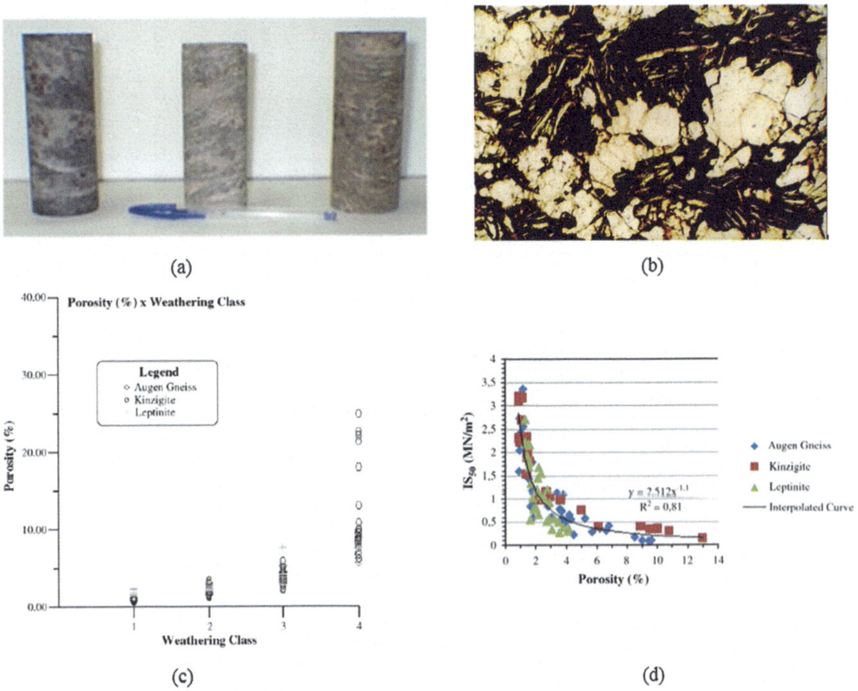

Fig. 10.6 (**a**) Effects of weathering on rock matrix of kinzigite from Rio de Janeiro (Marques et al. 2010). (**b**) Example of weathering effects on micro scale (Marques 1998). Influence of weathering on porosity (**c**) and relation between porosity and point load strength (**d**) for gneisses from Rio de Janeiro (Marques et al. 2010)

Table 10.2 Main mineralogical and microfracturing changes on gneisses from Rio de Janeiro due to weathering (Marques et al. 2010)

Weathering class	Augen Gneiss	Leptinite	Kinzigite
Sound rock (Class 1)	Plagioclases are sound—no signs of alteration—presenting typical white color. Quartz crystals are fractured and elongated. Biotites show a sound aspect, but few are slightly altered.	All minerals are sound. Only few biotite grains show more pale colors at its boundaries.	Plagioclases are lightly transformed into sericite and with few fractures open and oxidized. Biotites show a light discoloration on grain edges. Garnets and Sillimanite have sealed fractures without any oxidation sign. Cordierite presents few open fractures.
Slightly weathered (Class 2)	Great part of plagioclases are still sound. However, some of them show incipient alteration. Quartz is very fractured and elongated. Biotites are slightly discolored.	Plagioclases are lightly transformed into sericite. Biotites with discolored edges are oxidized along cleavage planes. Garnet with intergranular fractures is lightly weathered.	Some plagioclase crystals were transformed into sericite. Several open transgranular fractures are oxidized or filled with argillaceous material. Biotites with discolored edges and a few oxidized cleavage planes. Garnet with trans and intergranular fractures, lightly weathered. Cordierite with fractures filled with iron oxides.
Moderately weathered rock (Class 3)	Plagioclase crystals are opaque, showing evident caulinization. Quartz is very fractured, elongated and contouring phacoidal grains. Biotites are mainly oxidized, showing incipient exfoliation and iron deposit on their cleavage.	Sericite formation from plagioclase and feldspar surfaces is intensified at this stage. Biotites undergo an exfoliation process related to the opening of their cleavage. Garnet grains show a conspicuous of iron oxide along its internal fractures. There is a predominance of transgranular microfractures.	Plagioclase transformed into sericite is surrounded by argillaceous material and present several open, oxidized, trans and intragranular fractures. Biotites presents discolored edges, incipient exfoliation and oxidized cleavage. Garnet with transgranular open fractures and intragranular fractures filled with iron oxide. Cordierite with edges and fractures very oxidized. Argillaceous material more common than in Level W2.
Highly weathered rock (Class 4)	Plagioclases are mostly opaque and caulinized. Quartz are intensely fractured and oxidized. Biotites are oxidized and highly exfoliated.	At this weathering stage almost all plagioclase grains are transformed into clay minerals, especially kaolinite. Biotites are completely exfoliated. Garnets are oxidized.	Plagioclase occurs as rare, very fractured and oxidized crystals. Biotite occurs as a few sound, very fractured and oxidized crystals. Quartz presents oxidized fractures. Garnets crystals are intensely fractured. Argillaceous material occupies great portions of thin sections.

Santiago (2008), by studying laboratory weatherability of metabasic intrusive rocks (amphibolites) for Iron Quadrangle (Minas Gerais state), have also determined the influence of weathering on physical properties and parameters by using cycling tests such as natural cycling and water-oven cycling. Also, the author has analyzed mineralogical changes along the process.

10.3 Typical Morphology of Brazilian Weathering Profiles

Morphology of weathering profiles is an important aspect of weathering because it has a direct relation to geotechnical behavior of rock masses.

In the early stages of weathering studies, several authors (Moye 1955; Ruxton and Berry 1957; Dearman 1976; Dearman et al. 1978; IAEG 1981) divided noticeably different zones into the rock mass, based on three parameters: soil/rock ratio, rock discoloration degree, and presence of original structures, with no or few attention to rock matrix changes. However, because the created models did not completely express field behavior and of some confusion in the use of the method, the philosophy of the study changed and more attention to rock matrix was introduced (Beavis 1985; Barroso 1993; Menezes Filho 1993; Marques 1998; Gupta and Rao 2000; Arel and Önalp 2004; Marques et al. 2010). Recent studies have proved that a complete understand of weathering effects on rock masses can only be accessed if both approaches are used (Fig. 10.7).

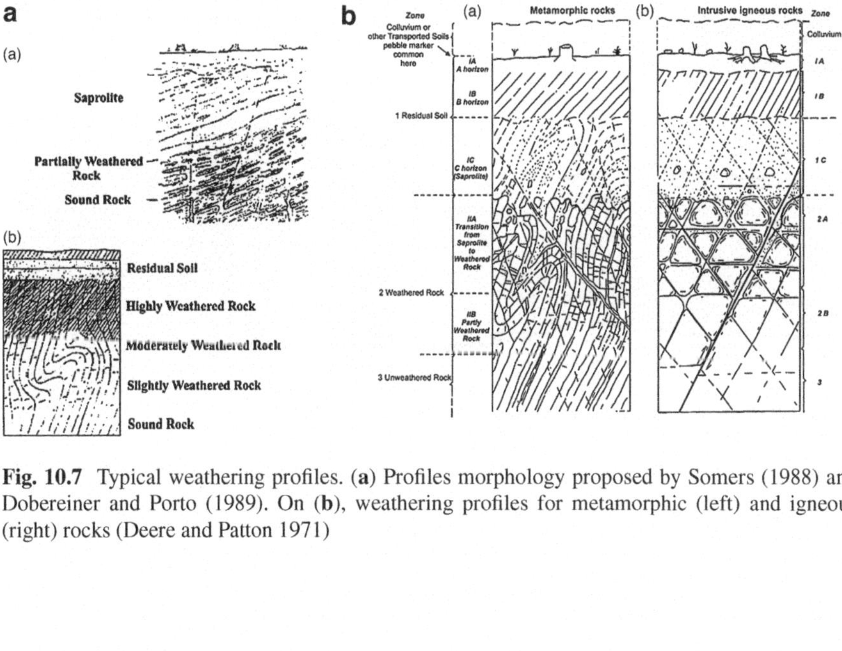

Fig. 10.7 Typical weathering profiles. (**a**) Profiles morphology proposed by Somers (1988) and Dobereiner and Porto (1989). On (**b**), weathering profiles for metamorphic (left) and igneous (right) rocks (Deere and Patton 1971)

According to Price (2009) four basic types of mass weathering (weathering profiles) can be identified:

- *Uniform weathering*—its also called sequential weathering profile, as a gradual and sequential decrease of weathering grade can be observed with depth (Fig. 10.8a).
- *Corestone weathering*—it is characterized by the presence of rounded and almost fresh rock blocks surrounded by decomposed rock or soil. It is typical of igneous rocks, both plutonic and volcanic (Fig. 10.8b).
- *Complex weathering*—an irregular profile due to contrasting layers weatherability and the structural complexity (presence of several geological structures such as joints, fractures, faults and folds). It is very common in metamorphic rocks such as schists and gneisses. Commonly stress relief joints also plays an important role in the morphology as these structures commonly present higher apertures, so allowing water to flow (Fig. 10.8c).
- *Solution weathering*—a specific weathering pattern related to carbonatic rocks, where fractures and bedding planes become open by dissolution and can evolve to karstic forms. This process can also occur on saline rocks (Fig. 10.8d).

Fig. 10.8 Examples of typical weathering profiles (**a**) Marques (1992); (**b**, **c**) Marques et al. (2010)

10.3.1 Typical Weathering Profile Morphology for Brazilian Sedimentary Rocks

According to Dobereiner et al. (1993), typical morphology of weathering profiles for Brazilian sedimentary rocks can be divided in two types, as shown on Fig. 10.9a. Weathering profiles for clayey and silty rocks usually present a superficial layer of tablet-like material with variable (0.2–1.5 m) thickness. The presence of this layer is due to wetting and drying cycling caused by moisture variations related to water table fluctuation or rainfall. Also, changes in air humidity can cause the same process. In cut slopes where clay and silt-rich layers are exposed, this process is very common and occurs in a period of days to months. Whenever the thickness of the layer increases, it can result in landslides, which exposes sound rock and the processes is allowed to restart. This typical morphology can be modified by the presence of interbedded sandstone layers with argillaceous rocks (shales, claystones and siltstones), as shown on Fig. 10.9b. On theses profiles, the same tablet-like layer can occur on claystones and siltstones, letting sandstones layers in balance, which can break, originating rock-falls.

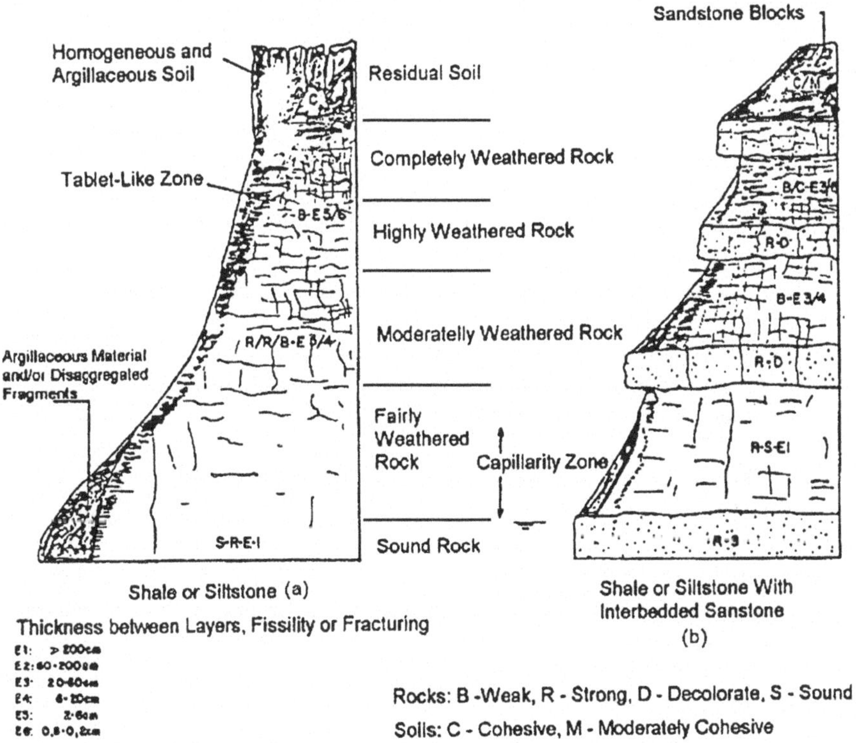

Fig. 10.9 Typical weathering profiles for sedimentary rocks (Dobereiner et al. 1993)

10.3.2 Typical Weathering Profile Morphology for Brazilian Metamorphic Rocks

The main geotechnical problems associated with metamorphic rocks weathering profiles are related to their structural complexity, to its irregular soil–rock contact surface and to their anisotropy (Dobereiner 1989). Some authors (Lee and Freitas 1988; Dobereiner and Porto 1990) have described difficulties when trying to fit the weathering profiles of metamorphic rocks studied by them into previous classifications. These structural complexities formed by the presence of several different types of tectonic structures—folds, foliation, faults, fractures, are fundamental for the weathering process of metamorphic rocks, as they can compose discontinuities through which water can flow and, therefore, promote chemical weathering and control the speed of weathering. Several heterogeneous morphological aspects such as sharp soil–rock contacts, the presence of a weathering front and the influence of subhorizontal stress-relief joints also contribute to make weathering profiles of metamorphic rocks even more complex.

On the weathering profiles of metamorphic rocks in Brazil, a remarkable observation is the importance of structures. Besides tectonic structures, which are important for all types of metamorphic rocks, stress relief joints play an especially role for the development of weathering profiles into high-grade metamorphic rocks, such as observed on Rio de Janeiro gneisses by Barroso (1993), Menezes Filho (1993), Marques (1998), and Marques et al. (2010) (Fig. 10.10d).

Another important aspect of weathering of these rocks is slope inclination, as observed by Marques (1998) for kinzigites from Rio de Janeiro. On more steep areas, stress relief joints mainly control weathering, promoting zoning with a clear tendency of reduction of weathering with depth. This behavior is influenced by fracture density and direction and is also important for the development and for the morphology of these profiles. On more fractured zones, weathering is much more complex, with several sharp and structurally controlled contacts between different weathering classes (Fig. 10.8c).

Some porphyroblastic gneisses can present a weathering profile showing blocks involved in a soil matrix, in a similar way of granites weathering profiles described for several authors, as Ruxton and Berry (1957). This, although, is not common in Brazil.

In a similar way, foliation also is important for the development of weathering, as it can have discontinuity planes parallel to it. If foliation is parallel or subparallel to surface, contacts are abrupt, as can be seen for phyllites from Iron Quadrangle (Minas Gerais state, Brazil) on Fig. 10.10a, b. When foliation is perpendicular to surface, the weathering profile is similar to the one presented on Fig. 10.10c. Weathering on discontinuities developed along foliation also creates sharp contacts between different weathering rock mass classes, as can be observed on Fig. 10.10b.

Fig. 10.10 Different aspects of weathering profiles of metamorphic rocks from Brazil. In (**a**) (from Leão et al. 2016) and (**b**) (from Carvalho et al. 2017) influence of penetrative foliation on weathering of phyllites from Iron Quadrangle, Minas Gerais state. On (**c**) a sequential weathering profile described by Marques et al. (2010); and on (**d**) aspects of weathering of gneisses from Rio de Janeiro (Marques et al. 2010)

10.4 Physical Parameters

Physical parameters of the sedimentary, igneous and metamorphic rocks can basically be evaluated by means of the specific density and the porosity. These two physical properties, especifically, are very important in the verification of the processes of weather degradation that the rocks can suffer. When unweathered, the rocks present ranges of typical values according to their origin.

In sedimentary rocks, the specific density may vary depending on the rock being clayey, the presence and proportion of clasts, existence of heavy minerals in their composition, besides the degree of lithification and cementation and type of cement. When these rocks are of chemical origin there is a greater influence of the presence of clay minerals, iron hydroxide, and bituminous substances.

In general, the geological characteristics directly influence the physical properties, where rocks with equal absolute porosity values may have different permeability, strength, and deformability values. Thus, density and porosity are directly related to the mineral composition of the rock, interstitial structure, silica content of the rock, and the presence of lamellar minerals (mica).

In metamorphic rocks, the factors that can influence the physical properties are fewer, however, they are of complex prediction if compared the sedimentary and igneous rocks. Basically, these properties vary according to the mineral composition, metamorphism class, and tectonic activity in the region, resulting in broad bands of specific density, in which the ones of sedimentary origin present smaller specific mass when compared to those of igneous origin. The porosity in these rocks is practically of fissural origin and rarely exceeds 1–2%. When porosity tends to higher values there is almost always a relationship with microcracking, reaching 10% in serpentinites. The formation of discontinuities in these rocks is dependent on the depth and conditions of deformation (stiffness, hardness) of the nesting rocks, besides the weathering action, for example.

The effect of weathering on sedimentary, igneous, and metamorphic rocks is distinct due to the intrinsic characteristics of these rocks. However, in general, weathering promotes the degradation of the physical properties, with reduction of the specific densities and increase of the porosity and capacity of water absorption as the class of rock weathering advances.

For this reason, the effect of weathering on physical characteristics of rocks can be evaluated through its index properties, in order to correspond to physical peculiarities that directly reflect the mineralogical composition and the voids present in the rock, in order to discern and quantify the rocky matrix (Leão et al. 2016). The index properties are obtained in samples of intact rock which, although not representing the properties of the rock mass, can help to classify it primarily (Azevedo and Marques 2002). Among them, specific mass and porosity are the most representative of the effects of weathering onto geotechnical parameters of altered rocks, considering that the increase of the specific mass and the reduction of porosity correspond, as a rule, to an increase of the resistance and a decrease of the deformability of rocks (Pinheiro 2002).

Not only, but mainly, in rock masses composed by soft foliated rocks, it is common the presence of discontinuities that develop parallel to the foliation. These interruptions of the continuity of the mineral formation and, as a result, of the rock matrix generate voids related to the deformation and rupture of the rocks. The amount of voids can be evaluated by porosity, that is, the ratio between the void volume of a rock sample and its total volume. These voids, pores or fissures, are not necessarily interconnected and may be completely closed (Hawkes and Mellor 1970), making it difficult to be measured. Moreover, primary porosity (pore volume between rock fragments) and secondary porosity (fracture and subsequent alteration of the rock) of the massif can be defined. In high grade metamorphic rocks (such as gneisses), porosity is a good property to evaluate the effects of weathering (Barroso 1993). Other important index properties, in soft silty-clayey rocks (Lashkaripour and Passaris 1995), are water absorption capacity and moisture content. According to Dobereiner (1984), the increase of these parameters represents a significant decrease on rock strength. The absorption of water, obtained by the "Quick absorption technique" (ISRM), makes it possible to evaluate the voids index and the altered state (Martin 1986). In rocks of a soft nature, disintegration may occur during the saturation process, requiring the use of the durability test (Pinho 2003) or different test procedures (Marques et al. 2017).

Figure 10.11 shows typical dry density values for sedimentary rocks (a), igneous (b) and metamorphic (c) and saturated specific mass for metamorphic rocks (d) from several countries, including Brazil (identified as BR in the rocks, according to the classes of change (ISRM 2015). In general, weathering promotes the reduction of the specific dry mass for all the rocks presented.

Brazilian shales and siltstones show, in general, lower values of dry density compared to other sedimentary rocks, probably due to the meteoric conditions to which these rocks were exposed. Comparing the sedimentary rocks (Fig. 10.11a), sandstones present greater dispersion for values of specific dry mass for the same class of alteration when compared to tuffs, limestones, and greywackes. In general, sandstones present good resistance to weathering, due to the high amount of quartz and other resistant minerals, but in clastic sandstones a differential alteration can occur due to the origin of the clast, the structural arrangement of the grains and to the presence and type of cement in the rock. It is also worth noting that the selection of suitable methods for extraction, preparation and decision on representative tests of

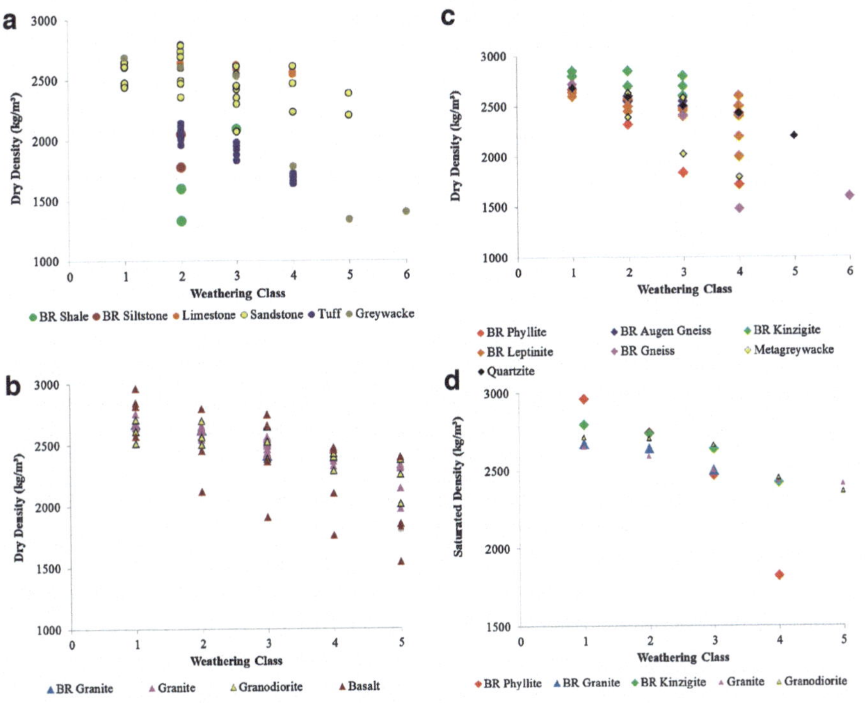

Fig. 10.11 Variation of dry density by weathering in sedimentary rocks (**a**), igneous (**b**) and metamorphic (**c**) and, of the specific mass saturated in metamorphic rocks (**d**). Brazilian rocks are identified by BR and are highlighted in the charts

the physical properties of sedimentary rocks can directly influence the results obtained (Kanji 2014) because these rocks are naturally soft when sound.

However, for the igneous lithotypes (Fig. 10.11b), there is a lower dispersion of the dry density values, with an evident tendency of reduction of this property values with weathering, both for Brazilian granites and for other igneous rocks. In general, there is a reduction of 10% of the dry density for granites when rock changes from class 2 to 3.

In coarser metamorphic rocks (Fig. 10.11c) dry density values are higher when compared to finer rocks (quartzite and phyllite). Despite of an evident tendency of reduction of dry density with weathering, there is a certain dispersion of the results for rocks of lower metamorphic grade as phyllite. For higher-grade metamorphic rocks, as in the case of augen gneisses and leptinites, this behavior can be observed mainly from weathering class 3, probably due to their intrinsic characteristics, such as mineralogy and texture.

In the case of saturated density, the trend is similar, both for Brazilian metamorphic rocks and others from different countries (Anagnostopoulos et al. 1993). It is worth noting the abrupt reduction of saturated density values for phyllites from classes 3 to 4, with a reduction of about 73%, since these rocks are easily disaggregated during the saturation process.

The effect of weathering on porosity promotes changes in the pore size distribution, geometry, connectivity, filling and formation of new pores, and can create systems of great complexity, depending on the rock, leading to variations in permeability and other mechanical properties (Tugrul 2014). Together with mineralogical composition and texture, porosity controls susceptibility of rocks to physical weathering (Hudec 1998).

By analyzing Fig. 10.12 it is possible to note that the porosity and the water absorption capacity increase with the advancement of weathering, but not to the same extent, since the water absorption capacity is dependent on the connectivity of the pores, for example. As weathering progresses, there is an increase in microcracks and voids in the rock.

Figure 10.12a, e shows naturally lower porosity and water absorption capacity for igneous (Fig. 10.12b, e) and metamorphic rocks (Fig. 10.12c, f) when compared to sedimentary rocks, even when undergoing weathering. More significant variations in porosity can be observed in samples of greywacke, due to the alteration of the clasts of the rock.

In low porosity rocks, this characteristic is due to cracking and, to a lesser degree, to intergranular cracking. Igneous rocks present a difference on porosity and water absorption capacity due to the type of magmatic rock. On effusive rocks, such as basalts, porosity is mainly related to gas and water vapor bubbles created during the cooling of the lava. In the cataclastic and weathered zones magmatic rocks develop higher porosity from fracturing, reaching between 10% and 20%, while on massive rock matrix porosity does not exceed 1–3%. Brazilian granites show an increase of up to 8% and 4% for porosity and water absorption capacity, respectively, from Class 2 to 3. Granodiorites show a larger increase in porosity than granites, with less than 1% for class 1, reaching up to 15% for class IV; similar behavior is observed in

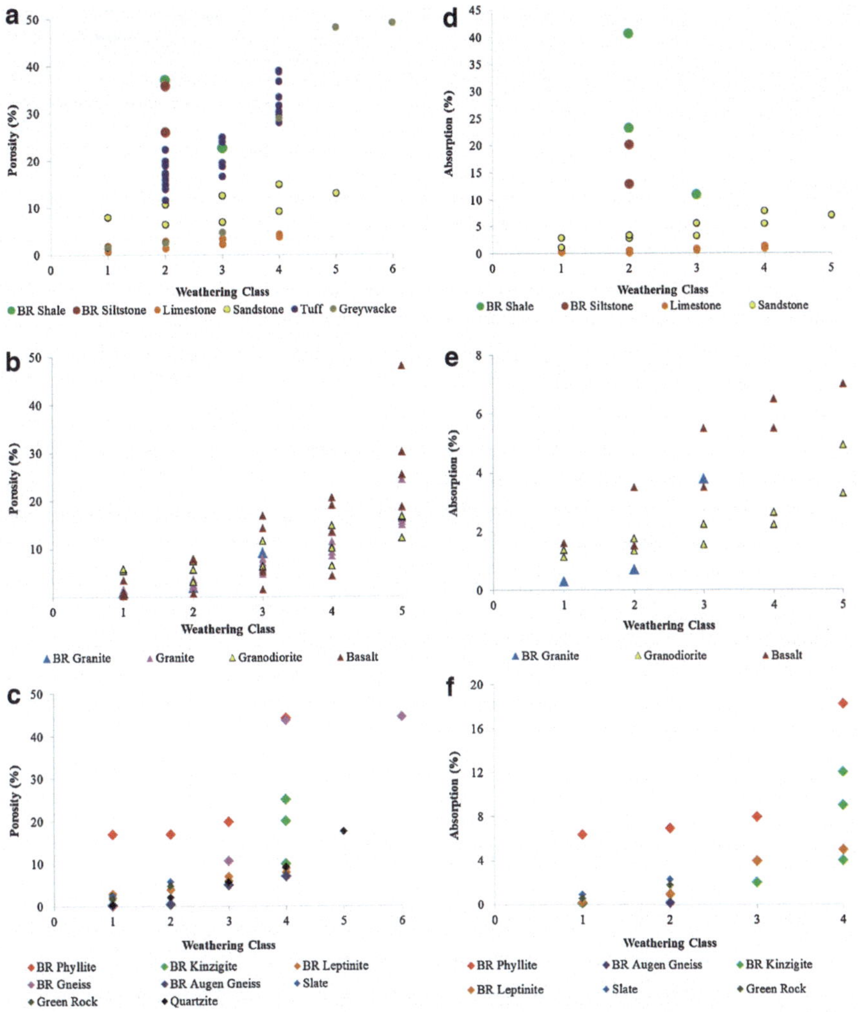

Fig. 10.12 Variation of porosity and water absorption capacity with weathering in sedimentary rocks (**a** and **d**), igneous (**b** and **e**), and metamorphic (**c** and **f**). Brazilian rocks are identified by BR and highlighted in the charts

the water absorption capacity (Ilcidari et al. 2013). Note that the porosity of the basalts is less than 1% for class 1 reaching almost 50% for class 5, due to the increase in microcracks within the grains and matrix due to the weathering (Tugrul 2014).

It is observed that metamorphic rocks present a similar behavior to the one described for igneous rocks, whose porosity and water absorption capacity is mainly function of the increase of rock fracturing, facilitated by the existence of structures such as bundles, lineaments, and foliations that, due to the advance of weathering, tend to propagate along discontinuity planes. In naturally soft and clayey rocks,

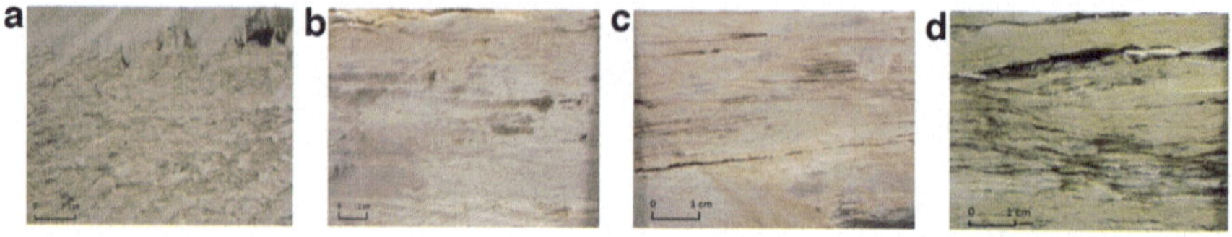

Fig. 10.13 Orthogonal foliations for the weathering classes 1 (**a**), 2 (**b**), 3 (**c**), and 4 (**d**). Aperture and persistence increase with weathering (Leão et al. 2016)

such as phyllites, these values tend to be higher when the rock is sound, compared to other coarse-grained and even some fine-grained metamorphic rocks, such as slates, due to the existence of penetrative foliation planes that facilitate the process of change, as shown in Fig. 10.13.

Values of microfracturing index (I_{fr}) on a Brazilian phyllite studied by Leão et al. (2016) have increased considerably from class 2 to class 4, but decreases from Class 4 to Class 4/5. This result is attributed to physical weathering because this process occurs primarily in rock layers (up to Class 4) of the weathering profile. Additionally, in more weathered zones (Classes 4/5 and 5), chemical weathering frees more iron oxide, which then precipitates along preexisting fractures, commonly sealing them. The micropetrographic index (I_{mp}) for these rocks showed an increase from Class 3 to Class 4/5 and little differentiation between Classes 4 and 4/5 indicates that chemical weathering occurred primarily during weathering of rock material and that almost none occurred during soil formation. These results are similar to the ones found for Australian phyllites.

10.5 Chemical and Mineralogical Characteristics

Amongst the endogenous factors, mineralogical composition of the rock is considered as the main conditioning factor, since each mineral presents a distinct response to alteration (Goldich 1938). Another condition refers to the granulometry, which influences the resistance of rock to alteration due to the mineral specific surface. Aspects such as parent rock texture, recrystallization of grains, presence of geological structures and directional textures (foliation, cleavage, etc.) also facilitate weathering.

Petrographic analysis is a widely used method to evaluate and quantify, microscopically, the presence of (1) microstructures; (2) mineralogy composition, which influences the roughness surface (Barton 1973); and (3) texture. It can also provide useful estimation of porosity of the rocks (Goodman 1989). It should be noted that the first three factors are the main characteristics controlling the intensity of weathering (Hudec 1998), with reverse changes in the crystallization of rock minerals (Goodman 1993). In some of the results presented on this chapter, the use of quantitative indices can be correlated with physical and mechanical characteristics (Ceryan et al. 1998). This method has the advantage of being a direct analysis, with some precision in the identification of important aspects of the rock matrix and the influence of weathering, with limitations on the identification of clay minerals, and have been widely used by several authors (Rodrigues et al. 1978; Russel 1982; Marques and Vargas Jr. 1994).

In rocks as thin as phyllites, in which mineralogy cannot be determined by macroscopic evaluations, the petrographic analysis allows the identification of the effects of weathering in the rock. In samples of seritic phyllites from Batatal Formation (Minas Gerais, Brazil), as the rock changes from class 1 to class 4, the percentage of muscovite (high birefringence) decreases, and the percentage of sericite increases. The distinction between these two minerals occurs on the altered edges (sericita) of the muscovite, where the birefringence colors are noticeably lower than those of the muscovite. In smaller quantities, there is a relative increase of quartz grains in the most altered classes, as this mineral is resistant to chemical intemperism processes. Due to the opening of the foliation plans with the progression of weathering, the percolation of mineralized fluids occurs, filling the discontinuities with minerals, as observed in Fig. 10.14.

In the weathering of igneous rocks, kaolinization is very common, especially in granitic rocks. Minerals like quartz, tourmaline, and muscovite remain almost unchanged. However, feldspars are easily dissolved by the presence of water, which percolates throughout fractures and micro-fractures. The effect of chemical weathering promotes the dissolution of feldspar, creating cavities and forming kaolin and smectite, among other clay minerals. Thus, X-ray diffraction (XRD) techniques can be used, and present advantages in relation to usual petrography descriptions for rocks with very fine granulometry. This technique is very useful for the estimation of clay minerals and mineralogy formation due to weathering (Gidigasu 1971, 1974). This can influence the degree of leaching and laterization, cation exchange, and hygroscopic moisture, as shown in Fig. 10.15. It can be noted that peaks of muscovite occur at all levels and it is worth noting that sericite shows similar signs, which would mark the mineral alteration. However it is not possible to distinguish the proportion between these two minerals. The kaolinite occurs at peaks of 12.2° and is more evident in the alteration classes 3 and 4, indicating that the intemperic processes are more active at these levels. 2:1 clay minerals appear in discrete peaks around 6° from class 1 to class 4, in small proportions (Leão et al. 2016).

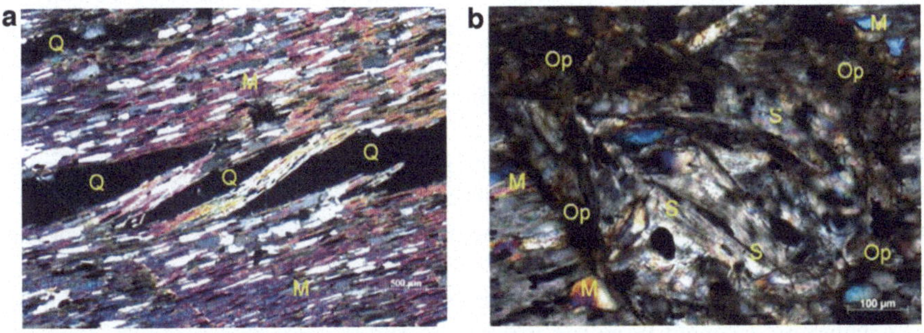

Fig. 10.14 Muscovite bands (M) interspersed with quartz grains (Q) in phyllite class 1 (**a**). In (**b**) sericite formation process (low birefringence) in the muscovite bands, with filling of micro-cracks by oxides and opaque (Op) (Leão et al. 2016)

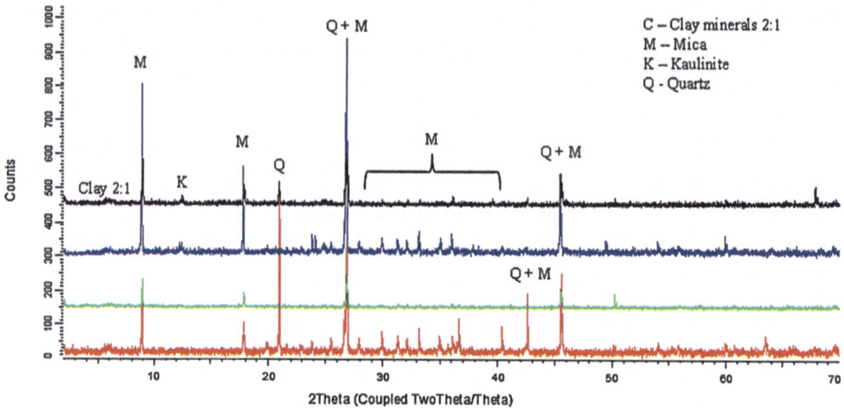

Fig. 10.15 XRD analysis in phytochemical samples of class 1 (red), 2 (green), 3 (blue) and 4 (black). It is possible to observe that the progression of weathering leads to the development of 2:1 expansive minerals (Leão et al. 2016)

The identification of the presence of expansive clay minerals, especially in rocks that have foliation planes and low mechanical resistance, facilitates the understanding of compression and relaxation effects of these planes (Bhasin et al. 1995). It is also worth mentioning that clay minerals associated to discontinuities and moisture variation are also responsible for chemical weathering of rocks (Taylor and Smith 1986; Gökçeoglu and Aksoy 2000).

Geochemical analysis methods seek to quantitatively estimate the mobility/immobility of elements by the action of weathering, leaching, as well as loss/gain of source material (Chadwick et al. 1990). Among the major intemporal processes responsible for the loss/redistribution of these elements are (1) the dissolution of primary minerals, (2) the formation of secondary minerals, and (3) the reduction, transport and exchange of ions (Thanachit et al. 2005).

Analyses by X-ray fluorescence (FRX) allow a qualitative and quantitative evaluation of the concentration of chemical elements present in rock samples. The use of this technique for the understanding of alteration processes is very valid, since intemperic processes denote the mobility of certain chemical elements, especially in supergenic environments (Licht 1998). The behavior of the chemical elements in supergenic conditions, as in the QF, is governed by the pH, oxidation potential (Eh), granulometry, mineral dissolution, hydrological regime, and chemical properties of the elements (Van Der Weijden and Van Der Weijden 1995). Some elements such as titanium and iron present low mobility and great possibility of reprecipitation (motivated by processes of dissolution of primary minerals). Aluminum also obeys this principle, forming secondary minerals such as kaolinite and gibbsite (Holanda and Bueno 2010).

It is common to use mineralogical indexes to characterize stages of evolution of weathering, to quantify matrix weathering and to serve as a guide or reference for geomechanical properties of rocks (Brimhall and Dietrich 1986). Initially, mineralogical indexes were proposed for granites based on the relationships between stable and unstable minerals (Lumb 1962; Dearman and Irfan 1978), later applied to other genetic types. More recently, proposals for mineral alteration indices based on more sophisticated laboratory techniques have been presented, such as electron microscopy (Hu et al. 2014).

Considering the geochemical changes in the protoliths and their products of weathering, especially in climates with water surplus and high annual average temperatures, several researchers proposed chemical indices based on molecular relations of the major elements, very useful to discuss the evolution of a profile of weathering. It is worth noting that the use of these indices differs in the treatment of homogeneous weathering profiles (Sutton and Maynard 1992) and heterogeneous weathering profiles (Ciampone et al. 1992), in which, in the latter case, the presence of discontinuities can generate levels of chemical composition and mineralogical properties.

In some of these indices, presented in Table 10.3, relationships between immobile (slightly soluble) and mobile elements (concentration reduces as weathering and leaching advance) are assumed. (a) There are indices that are based on the percentage of total silica; (b) however, they can be very questionable, as they consider silica in the form of quartz. Indexes based on chemical change consider certain elements as immobilized; (c) aluminum is of low mobility and remains constant during weathering in the weathering-potential index (WPI) and product index (PI) cases, both proposed by Reiche (1943). The chemical index alteration (CIA) is currently one of the most used chemical indices for evaluating weathering. Proposed by Nesbitt and Young (1982), the CIA uses the molecular ratio between larger mobile and immobile elements and functions as an index of the intensity of weathering. There are still indices that consider the relationships previously commented, but relate their concentration when healthy and weathered (d). It is worth noting that aluminum, abundant in the composition of many minerals, can be representative as an evaluation element in chemical indices, as it presents greater mobility, as a result of environments with pH below 4.5 and in the presence of organic acids (Gardner 1980, 1992).

The use of these indices may be interesting to evaluate the effect of chemical weathering on physical weathering. Figure 10.16 presents results of chemical indices in four classes of alteration of a phyllite (W1—Class 1 to W4—Class 4) of the Quadrilátero Ferrífero (QF, southeast Brazil).

The chemical indices ba, ba1, ba2, and ba3 showed reduction of values with the evolution of weathering. In terms of silica content, the indices (Silica/R_2O_3 and R) showed a variation and progression according to the more subtle weathering, showing (Silica/R_2O_3, PI, R, and STI) the non-chemical alteration of quartz in this intemperic evolution. Based on these results it was concluded that a SiO_2 concentration occurs in class 4 materials, reflecting the formation of clay minerals. The R index itself, which is highly recommended for the evaluation of silica loss with alumina, in igneous and metamorphic rocks, due to the increase in weathering, was not very significant for the phyllites studied. Considering the indices that evaluate the chemical alteration of the rock (CIA and CIW), the representativity was low, which was already expected as these indexes are mostly used for evaluating alteration in rocks

Table 10.3 Chemical indexes based on molecular relations of mobilizable/immobilizable elements (a), based on silicon content (b), based on chemical change (c), and based onnormalized indices (d)

Chemical index	Author
(a)	
$ba = \frac{K_2O + Na_2O + CaO}{Al_2O_3}$ $ba1 = \frac{K_2O + Na_2O}{Al_2O_3}$ $ba2 = \frac{CaO + MgO}{Al_2O_3}$ $ba3 = \frac{K_2O + Na_2O + MgO}{Al_2O_3}$	Harrassowitz (1926)
$Base : Al = \frac{K_2O + Na_2O + CaO + MgO}{Al_2O_3}$ $Base : R_2O_3 = \frac{K_2O + Na_2O + CaO + MgO}{Al_2O_3 + Fe_2O_3 + TiO_2}$	Colman (1982)
$b1 = \frac{Al_2O_3}{TiO_2}$	Rocha Filho et al. (1985)
$WR = \frac{CaO + MgO + Na_2O}{ZrO_2}$	Chittleborough (1991)
(b)	
$SF = \frac{SiO_2}{Fe_2O_3}$ $\frac{\text{Silicium}}{R_2O_3} = \frac{SiO_2}{Al_2O_3 + Fe_2O_3 + TiO_2}$	Jenny (1941)
$PI = \frac{SiO_2}{SiO_2 + Al_2O_3 + Fe_2O_3 + FeO + TiO_2}$	Reiche (1943)
$R = \frac{SiO_2}{Al_2O_3}$	Ruxton (1986)
$STI = \frac{\frac{SiO_2}{TiO_2}}{\frac{SiO_2}{TiO_2} + \frac{SiO_2}{Al_2O_3} + \frac{Al_2O_3}{TiO_2}}$	De Jayawardena and Izawa (1994)

(continued)

Chemical index	Author
$(WI-1/WI-2)=\dfrac{\dfrac{SiO_2+CaO}{Fe_2O_3+TiO_2}}{\dfrac{SiO_2+CaO}{Al_2O_3+Fe_2O_3+TiO_2}}$	Darmody et al. (2005)
(c)	
$SF=\dfrac{SiO_2}{Fe_2O_3}$ $\dfrac{\text{Silicio}}{R_2O_3}=\dfrac{SiO_2}{Al_2O_3+Fe_2O_3+TiO_2}$	Jenny (1941)
$PI=\dfrac{SiO_2}{SiO_2+Al_2O_3+Fe_2O_3+FeO+TiO_2}$	Reiche (1943)
$R=\dfrac{SiO_2}{Al_2O_3}$	Ruxton (1986)
$STI=\dfrac{\dfrac{SiO_2}{TiO_2}}{\dfrac{SiO_2}{TiO_2}+\dfrac{SiO_2}{Al_2O_3}+\dfrac{Al_2O_3}{TiO_2}}$	De Jayawardena and Izawa (1994)
$(WI-1/WI-2)=\dfrac{\dfrac{SiO_2+CaO}{Fe_2O_3+TiO_2}}{\dfrac{SiO_2+CaO}{Al_2O_3+Fe_2O_3+TiO_2}}$	Darmody et al. (2005)
(d)	
$B=\dfrac{b_{\text{weathered}}}{b_{\text{sound}}}$	Harrassowitz (1926)
$LCH\,\text{Factor}=\dfrac{I_{\text{weathered}}}{I_{\text{sound}}},\ \text{when}\ I=\dfrac{K_2O+Na_2O}{SiO_2}$	Jenny (1941)
$WI=\dfrac{WPI_{\text{weathered}}}{WPI_{\text{sound}}},\ \text{when}$ $WPI=\dfrac{K_2O+Na_2O+CaO\cdot H_2O}{SiO_2+Al_2O_3+Fe_2O_3+TiO_2+CaO+MgO+Na_2O+K_2O}$	Short (1961) (WI), Reiche (1943) (WPI)
$K=\dfrac{I_{\text{weathered}}}{X_{\text{sound}}},\ \text{where}$ $I=\dfrac{SiO_2}{Al_2O_3}\quad X=\dfrac{K_2O+Na_2O+CaO}{Al_2O_3}$	Rocha Filho et al. (1985)
$\text{Imob}=\dfrac{I_{\text{sound}}-I_{\text{weathered}}}{I_{\text{sound}}},\ \text{when}$ $I=K_2O+Na_2O+CaO$	Irfan (1996)

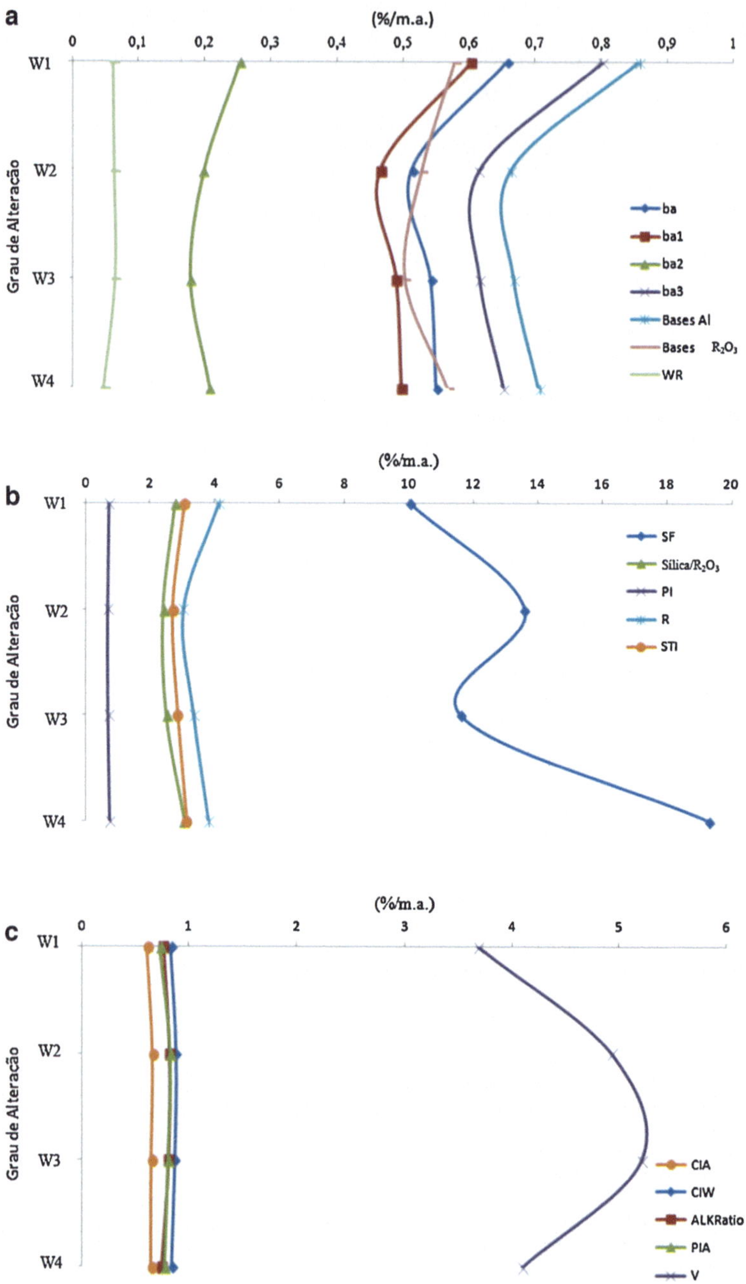

Fig. 10.16 In (**a**), chemical index mobilizable/immobilizable relation (ba, ba1, ba2, and ba3—Harrassowitz 1926; Bases R_2O_3 and Al—Colman 1982 and WR—Chittleborough 1991). In (**b**), silica contents (SF and Sílica/R_2O_3—Jenny 1941; PI—Reiche 1943; R—Ruxton 1986; STI—Jayawardena and Iazama 1994). In (**c**), indexes based on chemical change (CIA—Nesbitt and Young 1982; CIW—Harnois 1988; ALK Ratio—Harnois and Moore 1988; PIA—Fedo et al. 1995; V—Vogt 1972)

rich in feldspar, mineral that was not found in the phyllites under study. However, as these indices take into account elements such as Al and K, present in the chemical formulae of the main minerals of the rock (muscovite and sericite), a slight tendency is observed, originated from the alteration processes that occur in these minerals (Leão et al. 2016).

10.6 Mechanical Properties

In ideal elastic conditions, mechanical properties of geological materials, whether soils, rocks or their intermediate state (altered rocks), by static (stress–strain relations) and dynamic (elastic wave velocity) experiments should result in the same outcome. However, rock masses are not ideal elastic, mainly due to factors such as heterogeneity.

The study of the behavior of rock masses composed by soft rocks and the weathering action on them is still subject of discussion. Their weathered portions may present remnant aspects of the parent rock and the geotechnical-geomechanical parameters of these materials are difficult to obtain in situ (Leão 2015) and in the laboratory. Rocks, when weathered become soft and even when they are naturally soft, they may present problems related to sampling, preparation and selection of tests that can represent its properties variation (Kanji 2014); and representativity and extrapolation in numerical models (Agliardi et al. 2001). Due to their typical brittle behavior, equipment and samplers used for testing hard rocks are not adequate for use on soft rocks and, on the other hand, these rocks are too resistant when subjected to the conditions of ordinary experiments in soil mechanics (Kanji 2014).

There are no internationally accepted standards for the preparation of specimens in altered rocks. Some authors state that the existing standards for hard rocks are, in some ways, very strict (Pells and Ferry 1983; Chiu et al. 1983). In particular, in relation to the height–diameter ratio for uniaxial and triaxial compression tests, the 2:1 height–diameter ratio is acceptable for altered rocks (Chiu et al. 1983), although based on few data. For uniaxial compression, the ISRM (Brown 1981) recommends a ratio of 2.5:1, not excluding soft rocks. Another problem is the possibility that the materials contain expansive properties, compromising the specimens during drying processes. The re-saturation of the samples can generate desagregation and disintegration (Pinho 2003). Finally, in disaggregated rocks, such as sandstones with little cementing, total or partial disintegration of the sample during vacuum sealing may occur, and it may be necessary to adopt alternative saturation techniques, such as progressive saturation (Marques et al. 2017).

In fact, performing tests on weathered rocks, soft and weakly weathered rocks should follow certain precautions without neglecting the standards of execution, and each exception should be analyzed, discussed and included with common sense (Leão 2017).

10.6.1 Strength Characteristics

The point load test is a simple, fast and inexpensive test in which samples can be of different size and shape, as well as easy to reproduce due to equipment portability (ISRM 2015). This test is highly recommended for the characterization of the rocks and may be related to the uniaxial compression strength (ISRM 2015). In clayey rocks, such as claystones and siltstones, with a strength of less than 25 MPa, this test is not recommended due to the penetration of the conical tips in the sample (Hawkins and Pinches 1992). If to be used, a correction in the results will be necessary (ISRM 2015). Even so, the test is a commonly used for the determination of the compressive strength, due to the difficulties in the preparation/machining of the samples.

By comparing distinct rock groups, the strength values may also be very different. Igneous rocks and some varieties of quartzite and sandstones have the highest compressive strengths when compared to other rocks. Sound igneous rocks can reach compression strength of up to 414 MPa (basalts), also driven by the crystalline arrangement and presence of porphyries of reduced porosity. High-grade metamorphic rocks, such as amphibolites, are also included in this category.

The effect of weathering promotes a reduction in the resistance of the rocks, motivated by changes in their microfabric, which controls their resistance. This effect can be observed in Fig. 10.17 for sedimentary rocks (a and d), igneous rocks (b and e) and metamorphic rocks (c and f). In the case of metamorphic rocks in Brazil, results of tests in samples are presented in the orthogonal (O) and parallel (P) directions to the foliation of the rock.

In general it can be noted that the increase of weathering promotes the reduction of both the uniaxial compressive and the point load index strength. Comparing results obtained in tests carried out in samples oriented in the orthogonal direction to the rock foliation, one notices that the values are larger when compared to the ones obtained in tests of samples oriented in the parallel direction to foliation. The texture, mainly concerning the size of the grains and the presence of cementing substances, is the main responsible for the resistance. This can be observed when comparing igneous and high-grade metamorphic rocks to sedimentary and low-grade metamorphic rocks. However, for a specific weathering class, the values of uniaxial compressive strength are almost equivalent (classes 4 and 5). On sedimentary rocks, the presence and type of cementing materials can increase the mechanical strength of the rock and minimize the loss of resistance with weathering, as observed in limestones and sandstones (Fig. 10.17a). Compared with clayey rocks (mudstones and tuffs) or the one with clay-like materials between main minerals (as on marls) the uniaxial compression strength can reduce drastically.

In sound igneous rocks the grains and mineral aggregates promote high resistance and stability. Thus, the range of values for compressive strength varies in broad bands, being dependent on the degree of crystallization and mineralogy of the rocks. For sound materials the reduction of mechanical resistance can be influenced by grain size and by the presence of lamellar minerals (mica) and reduction of the amount of quartz. When these rocks undergo weathering, reduction of the strength

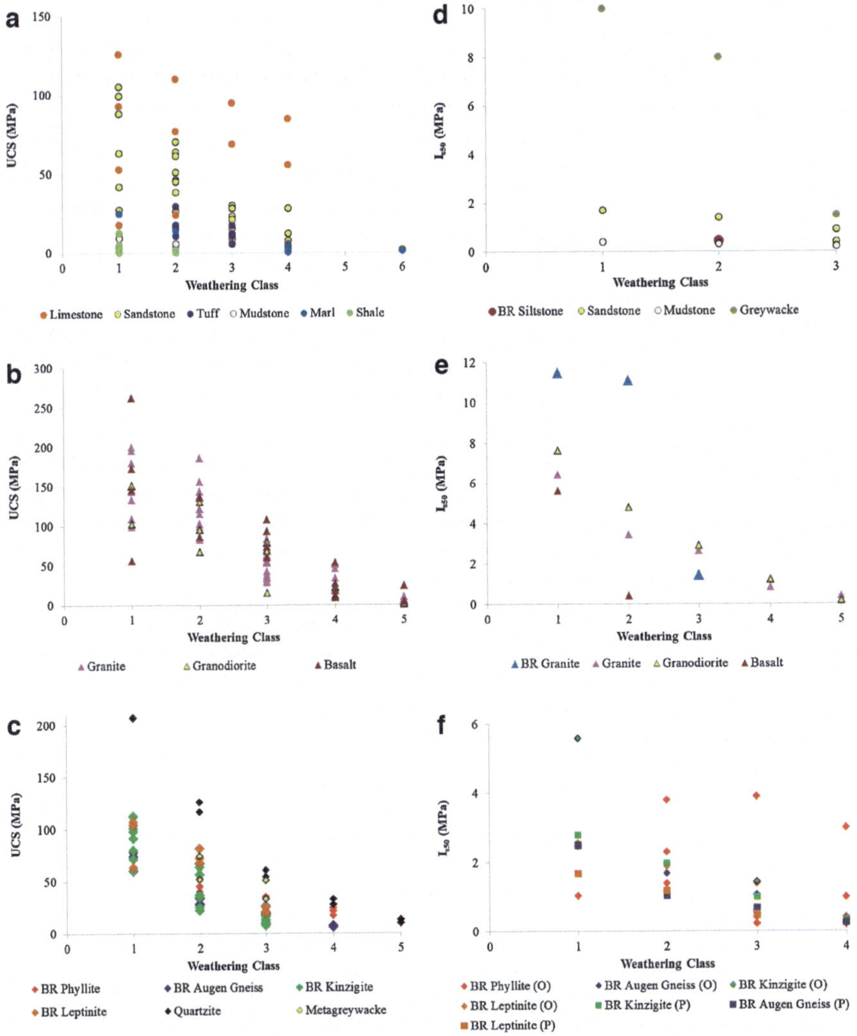

Fig. 10.17 Variation of resistance to simple compression and resistance to point load compression with weathering in sedimentary rocks (**a** and **d**), igneous rocks (**b** and **e**) and metamorphic rocks (**c** and **f**) in the orthogonal (O) direction and parallel (P) to foliation. Rocks from Brazil are identified by BR and highlighted in the charts

occurs according to the intrinsic properties of rocks. Thus, rocks such as granite and granodiorite present greater resistance when compared to basalts (Fig. 10.17b). In general, weathering promotes the development of microfractures, which may be responsible for stress relief in mineral grains (Dobereiner et al. 1993). Due to weathering rock microstructure becomes noticeably more fractured compared to sound materials.

For metamorphic rocks (Fig. 10.17c) strength is mainly dependent on the mineralogical composition of the rock. The presence of planar minerals (such as chlorite, biotite and mica) develops low resistance and alterability planes. Comparing the results presented in Fig. 10.17c, it can be observed that rocks with higher crystallinity (leptinites, augen gneisses and kinzigites) exhibit greater resistance to uniaxial compression when compared to rocks of low metamorphic degree, such as phyllites. Rocks that possess foliation are more susceptible to the presence of clay minerals in the planes and as in igneous rocks, the presence of these minerals reduces the resistance of the rock.

It should also be noted that for all rocks there is an influence of water on the mechanical properties. The water tends to reduce surface energy and crystalline arrangements, reducing the rock's resistance and increasing its deformability, which can be observed progressively in the classes of change. In this case, there may still be a chemical influence on the process, considering that quartz is a mineral with high resistance to chemical degradation (Gupta and Rao 2000). In particular, the clayey sedimentary rocks are very sensitive to variations of moisture content, directly affecting the mechanical properties. The structural arrangement of clayey rocks is far inferior in terms of resistance when compared to the structural arrangement observed in magmatic and metamorphic rocks.

Another issue that influences the resistance of metamorphic rocks is anisotropy (I_A), as shown in Fig. 10.18. This figure shows that for high-grade metamorphic rocks (gneisses) form Rio de Janeiro, there is a significant variability inside the same class of alteration. In general, the larger resistance under compressive stresses corresponds to the forces applied perpendicular to the foliation.

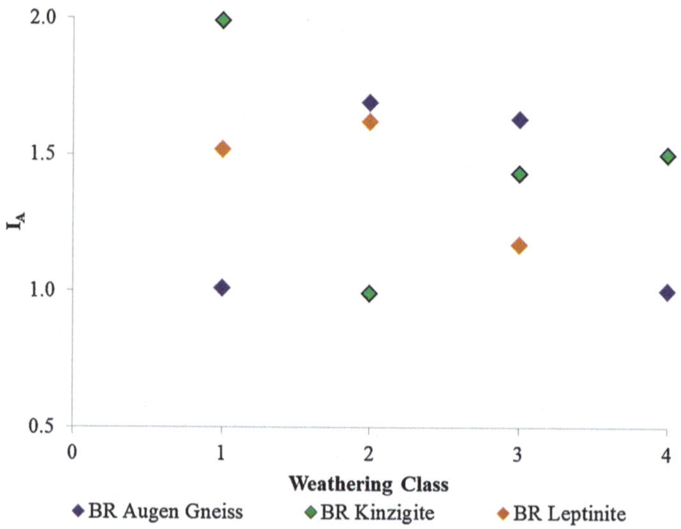

Fig. 10.18 I_A values for Brazilian metamorphic rocks

According to Fig. 10.18, the comparison between the I_A values for metamorphic rocks in Brazil shows a tendency of reduction of rock resistance and anisotropy, with the increase in weathering, whose values for rocks are very similar.

Another important and expeditious test used in the evaluation of rock weathering is the Schmidt hammer or sclerometer test. Due to its simplicity, it allows an in situ measurement without the need of any sample preparation. The presence of roughness on the tested surfaces and the proximity of discontinuities can generate dispersion in the values. Moreover, it is not very representative in very soft rocks (for R—"Rebound," lower than 10) (Pinho 2003). The Schmidt sclerometer, according to Brown (1981), as well as the point load test (PLT) has a reduced application for rocks of very low strength; however, for some soft rocks with some strength, such as cemented sandstones, good correlations can be achieved with the simple compressive strength. Leão and Marques (2016) have obtained good results for the evaluation of the weathering in phyllites by using sclerometry.

Disintegration of the rock in smaller portions due to the alternation of cycles of wetting and drying can occur during weathering. In this context, several methods are suggested as an index test for determining weatherability of rocks, such as slake durability test, cycling tests, and Soxhlet extractor. These tests accelerate rock decomposition more quickly than would occur in nature.

The slake durability test (ISRM 2015) is strongly recommended for soft rocks (Russel 1982; Ojima and Rodrigues 1983; Lee and Freitas 1988; Dick and Shakoor 1990, 1992; Lana 2014), due to its simplicity and speed. The durability index (I_d) is the percentage of dry rock that is retained in a metallic net drum after 1 or 2 complete cycles (I_{d1} or I_{d2}), based in which the rock can be classified according to the criteria proposed by Franklin and Chandra (1972). It is worth mentioning that Gökçeoglu and Aksoy, (2000) state that two cycles are not enough to evaluate the durability of soft rocks, when compared to other durability tests, and three or more cycles may be necessary.

Although in situ and laboratory conditions are distinct, these durability tests show good results for identifying alteration processes. The lack of information for correlation between the results of durability tests and other tests (mechanical, chemical and physical), as well as the lack of clear understanding of the phenomenon of disaggregation (Akai 1997), limit the applicability of these tests.

Figure 10.19 presents weathering assessments using Schmidt sclerometry and slake durability test for sedimentary (a and d), igneous (b and e) and metamorphic (c and f) rocks.

The results of sclerometry (Fig. 10.19a–c) show that, in sedimentary rocks, the dispersion of results is much lower when compared to igneous and metamorphic rocks, but for all related lithotypes there is a reduction of this property with the increase of weathering. Some aspects of the rocks tested may influence *R* values such as the degree of rock polishing/machining, the presence of foliation planes, the existence of loose surfaces in the execution of the test, penetration of the tip in highly altered and friable materials, and rock heterogeneity. When performed in the laboratory, the stability of the sample should also be guaranteed (Katz et al. 2000).

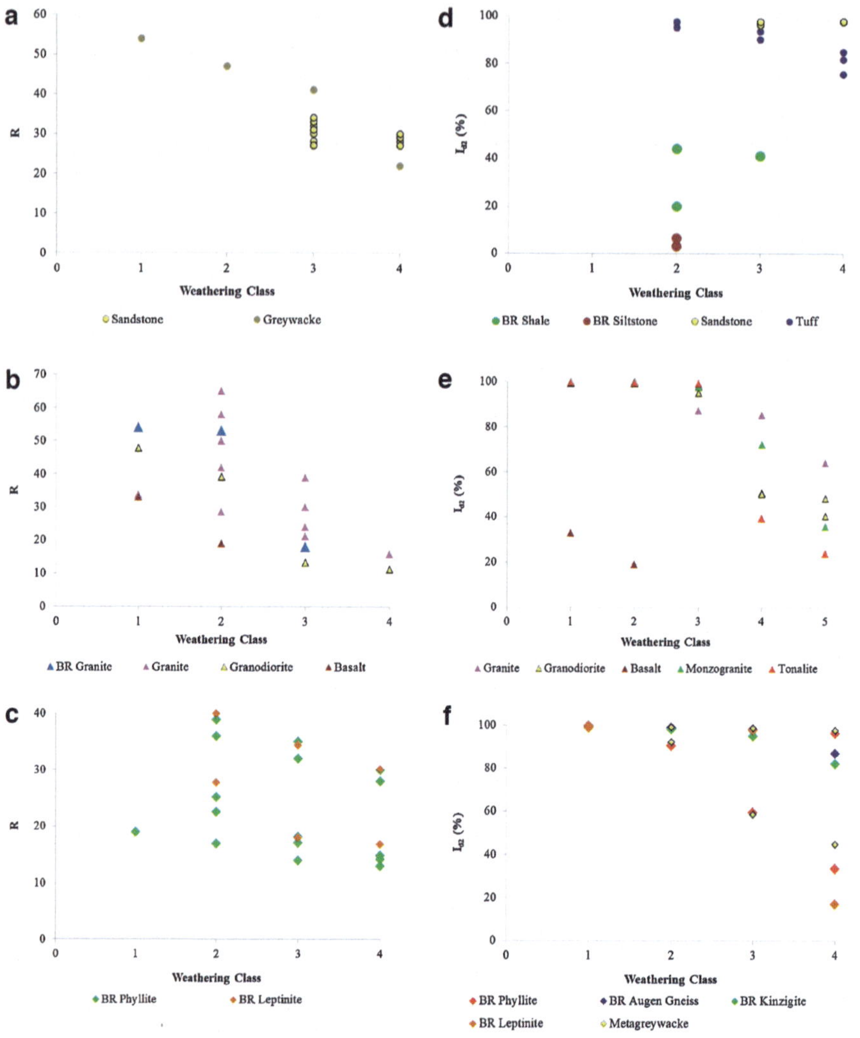

Fig. 10.19 Variability of sclerometric index and durability index with weathering in sedimentary rocks (**a** and **d**), igneous (**b** and **e**) and metamorphic (**c** and **f**). Brazilian rocks are identified by BR and highlighted in the charts

Mechanical anisotropy can be evaluated by performing the test in different directions in relation to the rock structures (foliation planes and bundles, for example), and it should be emphasized that in non-horizontal tests the influence of gravity occurs (Basu and Aydin 2004; Aydin and Basu 2005). Thus, tests performed in the orthogonal direction to the structures present higher values than tests carried out in the parallel direction.

In hard rocks, such as granites, it is recommended to repeat multiple impacts on the same point (Aydin and Duzgoren-aydin 2002). For soft and very altered rocks this procedure modifies the characteristics of the application surface and is not suitable (Leão 2017). Greco and Sorriso-Valvo (2005) performed many measurements of sclerometry, considering the influence of the discontinuity characteristics of the rock mass. For the four groups of rocks studied: (1) plutonic rocks, (2) low to medium metamorphic rocks (phyllites and argillites), (3) augen gneisses and (4) medium to high metamorphic grade (shales, gneisses, amphibolites and migmatites), only the crystalline rocks showed low influence of the discontinuities, being more dependent of these characteristics than the foliated rocks. In this case, it is recommended to carry out the test on a grid drawn on a regular surface of the rock, with application points 5–10 cm apart, according to the conditions found.

The durability is mainly related to the mineralogical composition of rocks, texture and the nature of the fluids that are in contact with the rock during the test. This can be observed from the results for sedimentary rocks (Fig. 10.19d), igneous (Fig. 10.19e) and metamorphic rocks (Fig. 10.19f), where the former tend to show less dispersion than two other genetic lithotypes, but the reduction of I_{d2} with weathering is clear.

It is worth mentioning that, in clayey rocks such as siltstones, claystones and shales, as well as some types of phyllites and slates, characteristics such as contact surface, presence of discontinuities, moisture content, past erosive processes and genetic rock type can directly influence durability (Gautam and Shakoor 2013), as can be observed on I_{d2} values obtained for phyllites compared to coarse metamorphic rocks (Fig. 10.19f).

10.6.2 Deformability Characteristics

The elastic properties in the rocks vary in a wide and direction dependent manner, the value of the modulus of elasticity depends on the direction in which the measurement is made. In general, the modulus of elasticity of the rocks is smaller when the measurement is made perpendicular to the structures present in the rock (stratification, lineage, foliation, for example). In addition, weathering also influences this property, in that it reduces the stiffness of the samples.

The elastic modulus is an important parameter in the practice of geotechnical engineering, and can be obtained from the measurement of the velocity of P waves. Characteristics such as the decrease of calcite in calcareous rocks can exponentially decrease the modulus of elasticity, being a parameter that is important for the evaluation of weathering in these rocks, as shown in Fig. 10.20. The weathering action promotes the development of clay minerals, the decomposition of the texture and the crystalline arrangements, as well as the microfracture, causing a reduction in the modulus of elasticity of the rocks (Momeni et al. 2017).

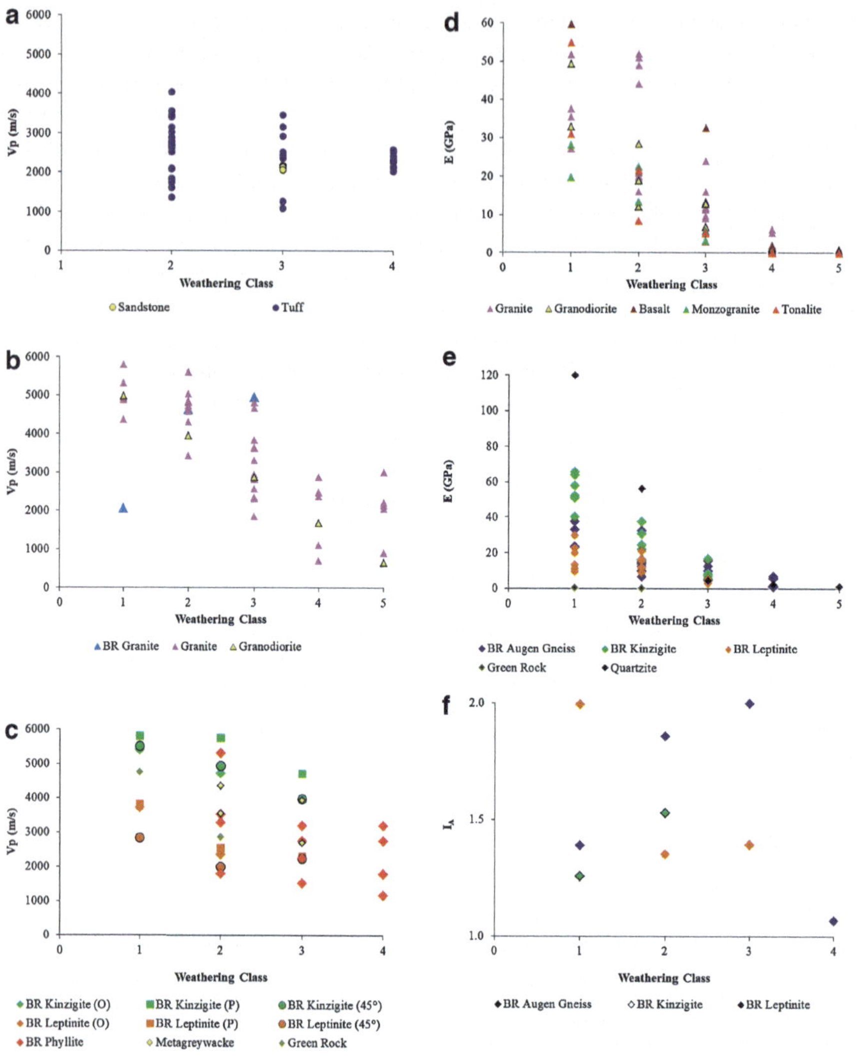

Fig. 10.20 Variation of the velocity of seismic waves and modulus of elasticity with weathering in sedimentary rocks (**a** and **d**), igneous (**b**) and metamorphic (**c** and **e**). Figure (**f**) presents the I_A for the deformability modulus for metamophic rocks. Rocks from Brazil rocks are identified by BR and highlighted in the charts

This effect can be observed in Fig. 10.20 for sedimentary rocks (a and d), igneous (b) and metamorphic (c and e) rocks. In the case of metamorphic rocks in Brazil, results of tests performed in samples are presented in the orthogonal direction (O), with 45° angle, and parallel (P) to the rock foliation. Figure 10.19f presents the I_A for the deformability modulus for metamorphic rocks.

By observing the results presented in Figs. 10.20a–c it can be seen that the values of Vp for all the rocks, reduce as the weathering progresses. The degree of entanglement of the mineral grains, when low, can influence in the composition of the rock primary minerals like quartz and feldspar, considered of low deformability. However, when this bonding is well developed, the values of strength and stiffness tend to be high. As weathering is responsible for the reorganization of the textured rock patterns, in addition to the alteration of primary minerals, the moduli of elasticity (Figs. 10.20d, e) tend to be smaller the larger the change in weathering class is. The development of microfractures and clay minerals also contributes to this effect, causing significant loss of rock elasticity. It should also be noted that the saturation of the sample also causes a reduction in the modulus of elasticity and Vp waves. Considering the direction of propagation of Vp waves due to banding, stratification, and foliations, it is possible to encounter anisotropy in the elastic properties. In this case, measurements performed in the direction perpendicular to stratifications, bundles, and foliations are larger when compared with measurements carried out parallel to these structures. The I_A values presented for metamorphic rocks (Fig. 10.20f) show that sound rocks tend to be more anisotropic when compared to more weathered stages of the same rock.

10.7 Recent Studies

Recently, the authors and their research group have been working on physical, chemical, and mechanical characterization of low-grade igneous, sedimentary, and metamorphic rocks in the southeast region of Brazil. In this context, some sandstone, limestone, and granite weathering profiles are currently under study.

For the study of influence of weathering on igneous rock, several outcrops of granitic rocks were selected, considering the scarcity of data in the scientific literature on complete weathering profiles developed in these rocks, under tropical climates. Several types of calcareous rocks were selected and comprise calcilutites, limestones and siltstones, whose outcrops exposed on road cuts in the State of Minas Gerais (Southeast Brazil). At the same time, the research group is continuing studying low-grade metamorphic rocks, mainly phyllites. The main purpose of all those studies currently under development is to understand the influence of weathering on physical, chemical, mineralogical, and mechanical properties of both the intact rock and the rock mass along the weathering profile and trying to correlate these properties.

It is expected that this study can determine the effect of weathering on the aforementioned properties, through evaluations and tests to be performed, both in intact rock and discontinuities, in situ and in the laboratory. In addition, it is expected to propose a simple and practical classification criterion suitable for different rock geneses (igneous, sedimentary and metamorphic) under tropical climate. Also, the study has the purpose of elucidating the following aspects: (1) morphological characterization of typical weathering profiles for different lithotypes; (2) field and/or laboratory characterization of the properties of the main discontinuities, and (3) identification of the main mechanisms (physical and chemical) responsible for the development of the weathering.

References

Agliardi F, Crosta G, Zanchi A (2001) Structural constraints on deep-seated slope deformation kinematics. Eng Geol 59:83–102

Akai K (1997) Testing methods for indurated soils and soft rocks—Interim report. ISSMFE—TC 22—Technical Commmittee on Indurated Soils and Soft Rocks. Geotechnical Engineering of hard soils—soft rocks. In: Int. Symp. ISSMFE/IAEG/ISRM, Anais… Athens, Greece, vol 3, pp 1707–1737

Anagnostopoulos A, Schlosser F, Kalteziotis N, Frank R (1993) Geotechnical engineering of hard soils—soft rocks, vols 1 and 2. In: Proceedings of an International Symposium under the auspices of the International Society for Soil Mechanics and Foundation Engineering (ISSMFE), the International Association of Engineering Geology (IAEG) and the International Society for Rock Mechanics (ISRM), Athens, Greece

Arel E, Önalp A (2004) Diagnosis of the transition from rock to soil in a granodiorite. J Geotech Geoenviron Eng ASCE 130:968–974

Aydin A, Basu A (2005) The Schmidt hammer in rock material characterization. Eng Geol 81:1–14

Aydin A, Duzgoren-aydin NS (2002) Indices for scaling and predicting weathering-induced changes in rock properties. Environ Eng Geosci 8:121–135

Azevedo I, Marques EAG (2002) Introduction to Rock Mechanics. Viçosa:UFV. p. 363

Basu A, Aydin A (2004) A method for normalization of Schmidt hammer rebound values. Int. J. Rock Mech. Min. Sci. 41:1211–1214

Basu A, Celestino TB, Bortolucci AA (2009) Evaluation of rock mechanical behaviors under uniaxial compression with reference to assessed weathering grades. Rock Mech Rock Eng 42:73–93

Bhasin R, Barton N, Grimstad E, Chryssanthakis P (1995) Engineering geological characterization of low strength anisotropic rocks in the Himalayan region for assessment of tunnel support. Eng Geol 40:169–193

Barroso EV (1993) Estudo das características geológicas e do comportamento geotécnico de um perfil de intemperismo em leptinito. UFRJ: Dissertação (Mestrado em Geologia), UFRJ, Rio de Janeiro

Barton N (1973) Review of a new shear-strength criterion for rock joints. Eng Geol 7:287–332

Beavis FC (1985) Rockweathering. Engineering Geology. Blackwell Scientific,Melbourne

Brimhall GH, Dietrich WE (1986) Constitutive mass balance relations between chemical composition, volume, density, porosity, and strain in metasomatic hydrochemkai systems: results on weathering and pedogenesis. Geochim Cosmochim Acta 51:567–587

Brown RT (1981) Rock characterization testing and monitoring. ISRM Suggested Method. Pergamon Press, Oxford. 211 p

Carvalho TRR, Marques EAG, Leão MF, Illambwetsi AM (2017) Perfis de Intemperismode filito sob clima tropical, na região do Quadrilátero Ferrífero, Minas Gerais, Brasil. In: 16o. Congresso Nacional de Geotecnia, Açores, Portugal. Sociedade Portuguesa de Geotecnia, Lisbon, p 7 (In Portuguese)

Ceryan S, Zorlu K, Gokceoglu C, Temel A (1998) The use of cation packing index for characterizing the weathering degree of granitic rocks. Eng Geol 98:60–74

Chadwick AO, Brimhall GH, Hendricks DM (1990) From a black to a gray box a mass balance interpretation of pedogenesis. Geomorphology 3:369–390

Chittleborough DJ (1991) Indices of weathering for soils and paleosols fonned on silicate rocks. Aust J Earth Sci 38(1):15–20
Chiu HK, Johnston IW, Donald IB (1983) Appropriate techniques for triaxial testing of satured soft rocks. Int J Rock Mech Min Sci Geomech Abstr 20:107–120
Ciampone MA, McVEY DE, Gerke TL, Briggs WD, Zhang Y, Maynard JB, Huff WD (1992) Non-systematic weathering profile in the Blue Ridge Mountains, role of geochemical variables. Geol Soc Am Abstr Prog 24(7):214
Colman SM (1982) Chemical weathering of basalts and andesites: evidence from weathering rinds. US Geol. Surv. Prof. Pap., no. 1246, 51 p
Darmody RG, Thorn CE, Allen CE (2005) Chemical weathering and boulder mantles, Kärkevagge, Swedish Lapland. Geomorphology 67:159–170
De Jayawardena US, Izawa E (1994) A new chemical index of weathering for metamorphic silicate rocks in tropical regions: a study from Sri Lanka. Eng Geol 36:303–310
Dearman WR (1974) Weathering classification in the characterization of rock for engineering purposes in British practice. Bull Int Assoc Eng Geol 9:33–42
Dearman WR (1976) Weathering classification in the characterization of rock: a revision. Bull Int Assoc Eng Geol 13:123–127
Dearman WR, Baynes FJ, Irfan TY (1978) Engineering grading of weathered granite. Eng Geol 12:345–374
Deere DU, Patton FD (1971) Slope stability in residual soils. In: Pan. Conf. Soil Mech. Found. Eng., 4. Puerto Rico, Proceedings… Puerto Rico, ISSMFE, p. 87–170
Dick JC, Shakoor A (1990) The effects of lithologic characteristics on mudrock durability. In: 6 Int. Congress, Int. Assoc. engineering Geology, Anais… Amsterdam, vol 4, pp 3061–3066
Dick JC, Shakoor A (1992) Lithological controls of mudrock durability. Q J Eng Geol Lond 25:31–46
Dobereiner L (1984) Engineering geology of weak sandstones. PhD Thesis, Imperial College of Science and Technology, University of London, 471 p
Dobereiner L (1989) Constructions problems related to excavation on soft rocks. Proceedings 12th Int. Cong. Soil Mech. Found. Eng., General Report. ISSMFE, Rio de Janeiro, p. 4
Dobereiner L, Durvile JL, Restituito J (1993) Weathering of the Massiac Gneiss (Massif Central, France). Bull Int Assoc Eng Geol 47:79–96
Dobereiner L, Porto CG (1990) Considerations on the weathering of gneissic rocks. Eng. Group Meeting on the Geol. of Weak Rock, 26th Annual Conf. of B.G.S., Leeds, 1990. Proceedings... Leeds, B.G.S, pp. 228–241
Fedo CM, Nesbitt HW, Young GM (1995) Unraveling the effects of potassium metasomatism in sedimentary rocks and paleosols, with implications for paleoweathering conditions and provenance. Geology 23:921–924
Franklin JA, Chandra A (1972) The slake durability test. Int J Rock Mech Min Sci Geomech Abstr 9:325–341
Gardner LR (1980) Mobilization of AL and Ti during weathering - isovolumetric geochemical evidence. Chem Geol 30:151–165
Gardner LR (1992) Long-term isovolumetric leaching of aluminum from rocks during weathering: implications for the genesis of saprolite. Catena 19:521–537
Gautam TP, Shakoor A (2013) Slaking behavior of clay-bearing rocks during a one-year exposure to natural climatic conditions. Eng Geol 166:17–25
Gidigasu MD (1971) The importance of soil genesis in the engineering classification of Ghana soils. Eng Geol 5:117–161
Gidigasu MD (1974) Degree of weathering in the identification of laterite materials for engineering purposes: a review. Eng Geol 8:213–266
Gökçeoglu C, Aksoy H (2000) New approaches to the characterization of claybearing, densely jointed and weak rock masses. Eng Geol 58:1–23
Goldich SS (1938) Stone line profiles: importance in geochemical exploration. J Geochem Explor 30:35–61

Goodman RE (1989) Introduction to rock mechanics, 2nd edn. Wiley., 562 p, New York

Goodman RE (1993) Engineering geology. Rock in engineering construction. Wiley, New York. 412 p

Greco R, Sorriso-Valvo M (2005) Relationships between joint apparent separation, Schmidt hammer rebound value, and distance to faults, in rocky outcrops, Calabria, Southern Italy. Eng Geol 78:309–320

Gupta AS, Rao KS (2000) Weathering effects on the strength and deformational behaviour of crystalline rocks under uniaxial compression state. Eng Geol 56:257–274

Harnois L (1988) The CIW index: a new chemical index of weathering. Sediment Geol 55:319–322

Harnois L, Moore JM (1988) Geochemistry and origin of the ore chimmey formation, a transported paleoregolith in the Greenville Province of Souteastern Ontario, Canada. Chem Geol 69:267–289

Harrassowitz HL (1926) Material und Versuch erdgeschichtlicher Auswertung, vol 4. Fortschritte der Geologie und Paläontologie, Berlin, p 14

Hawkes I, Mellor M (1970) Uniaxial testing in rock mechanics laboratories. Eng Geol 4:177–285

Hawkins AB, Pinches GM (1992) The engineering description of mudrocks. Q J Eng Geol Lond 25(1):17–30

Heidari M, Momeni AA, Naseri F (2013) New weathering classifications for granitic rocks based on geomechanical parameters. Eng Geol 166:65–73

Holanda CEF, Bueno GT (2010) Comportamento de elementos químicos em ambiente supergênico e pedogênese – Parque Municipal das Mangabeiras, Quadrilátero Ferrífero (MG). Rev. de Geografia. Recife: UFPE – DCG/NAPA, vol especial VIII SINAGEO, n. 3

Hu R, Oyediran IA, Gao W, Zhang X, Li HL (2014) "Plagioclase solution degree index": a new index to evaluate the weathering degree of granite. Bull Eng Geol Environ 73:589–594

Hudec P (1998) Rock properties and physical processes of rapid weathering and deterioration. In: Moore DP, Hungr O (eds) Proceedings of 8th International congress of IAEG, vol 1. Rotterdam, Balkema, pp 335–341

IAEG (1981) Rock and soil description and classification for engineering geological mapping. Bull Int Assoc Eng Geol 24:235–274

ISRM – International Society of Rock Mechanics (2015) In: Ulusay R (ed) The ISRM suggested methods for rock characterization, testing and monitoring: 2007–2014, 5th edn. Springer, New York

Jenny H (1941) Factors of soil formation. McGraw-Hill., 281 p, New York

Kanji MA (2014) Critical issues in soft rocks. J Rock Mech Geotech Eng 6:186–195

Katz O, Reches Z, Roegiers J-C (2000) Evaluation of mechanical rock properties using a schmidt hammer. Int J Rock Mech Min Sci 37:723–728

Lana MS (2014) Numerical modeling of failure mechanisms in phyllite mine slopes in Brazil. Int J Min Sci Technol 24(6):77–789

Lashkaripour GRE, Passaris EKS (1995) Correlations between index parameters and mechanical properties of shales. In: 8 Int. congress rock mechanics, ISRM, Anais… Tokyo, vol 1, pp 257–261

Leão MF (2015) Análise tensão-deformação de uma barragem de concreto gravidade em solo residual preponderantemente anisotrópico. Dissertação (Mestrado em Engenharia Civil), Departamento de Engenharia Civil, UERJ, Rio de Janeiro

Leão MF, Marques EAG (2016) Morphology and geotechnical characterization of a phyllite weathering profile developed under tropical climate. In: 5th International conference on geotechnical and geophysical site characterisation, Anais…Queensland

Leão MF (2017) Geomechanical behavior of phyllite weathering front in the region of Quadrilátero Ferrífero. Tesis (Geology), Departament of Geology, UFRJ, Rio de Janeiro

Leão MF, Polivanov H, Barroso EV, Marques EAG, Vargas EA Jr (2016) Weathering of metapelites from the Quadrilátero Ferrífero mineral province, southeastern Brazil. Bull Eng Geol Environ. https://doi.org/10.1007/s10064-017-1036-1

Lee SG, Freitas MH (1988) Quantitative definition of highly weathered granite using the slake durability test. Géotechnique 38:635–640

Licht OAB (1998) Prospecção Geoquímica: princípios, técnicas e métodos. CPRM., 35 p, Rio de Janeiro

Lumb P (1962) The properties of decomposed granite. Géotechnique 12(3):226–243

Marques EAG (1992) Study of Alteration and Alterability of Some Shales and Siltites of Recôncavo Sedimentar Basin – Bahia. Dissertation (Geology). Departament of Geology, UFRJ, Rio de Janeiro. (in portuguese)

Marques EAG, Vargas EA Jr (1994) Alteration studies of some shales and siltstones of the Recôncavo sedimentary basin - Northeast Brazil. In: 7 International congress of the International Association of Engineering Geology, 1994, Lisboa. Anais.... A. A. Balkema, Rotterdam, pp 1–8

Marques EAG (1998) Geomechanical Characterization of a Weathering Profile in Kinzigite. Departament of Geology, UFRJ, Rio de Janeiro. (in portuguese)

Marques EAG, Vargas EA Jr, Antunes FS (2005) A study of the durability of some shales, mudrocks and silstones form Brazil. Geotech Geol Eng 25:321–348. https://doi.org/10.1007/s10706-004-1605-5

Marques EAG, Barroso EV, Menezes Filho AP, Vargas EA Jr (2010) Weathering zones on metamorphic rocks from Rio de Janeiro – physical, mineralogical and geomechanical characterization. Eng Geol 111:1–18. https://doi.org/10.1016/j.enggeo.2009.11.001

Marques EAG, Williams DJ (2015a) Weathering Profiles of Bunya Phyllite in Southwest Brisbane - a Geotechnical Approach. In: 12 Australia New Zealand Conference on Geomechanics, 2015. Wellington. Anais... Wellington: Changing the Face of the Earth, v. 1, p. 1–8

Marques EAG, Williams DJ (2015b) Weathering Profiles of Some Sandstones from Sunshine Coast, Australia Morphological and Geotechnical Approach. In: 49th United States Rock Mechanics / Geomechanics Symposium, 2015b, San Francisco. ARMA 2015. San Francisco: American Rock Mechanics Association, pp 1–8

Marques EAG, Williams DJ, Rodrigues IA, Leão MF (2017) Effects of weathering on characteristics of rocks in a subtropical climate: weathering morphology, in situ laboratory and mineralogical characterization. Environ. Earth Sci. 76:602–619

Martin RP (1986) Use of index tests for engineering assessment of weathered rocks. In: 5 Int. Congress, Int. Assoc. Engineering Geology, Anais… Buenos Aires, 5th, pp 433–450

Menezes Filho AP (1993) Geological–geotechnical aspects of an Augen gneiss weathering profile. M.Sc. Thesis, Department of Geology, Federal University of Rio de Janeiro, Brazil 229p. (in portuguese)

Momeni A, Hashemi SS, Khanlari GR, Heidari M (2017) The effect of weathering on durability and deformability properties of granitoid rocks. Bull Int Assoc Eng Geol 76:1037–1049

Moye DG (1955) Engineering geology for the Snowy Montain scheme. J Inst Eng 27:287–298

Nesbitt HW, Young GM (1982) Early proterozoic climates and plate motions inferred from major element chemistry of lutites. Nature 279:715–717

Ojima LM, Rodrigues JD (1983) Weathering of phyllite in Morgavel Tunnel. In: Int. IAEG Symp. on engineering geology and underground construction, Anais… Lisboa, vol 1, pp I1–I14

Pells OJN, Ferry MJ (1983) Needles stringency in sample preparation standards for laboratory testing of weak rocks. In: 5 Int. Congress rock mechanics, ISRM, Anais… Melbourne, vol 1, pp A203–A207

Pinheiro AL (2002) Análise de rupturas em taludes no morro do Curral, Ouro Preto, Minas Gerais. Dissertação (Mestrado em Engenharia Mineral). UFOP, Ouro Preto, MG, 116 p

Pinho AB (2003) Caracterização Geotécnica de Maciços Rochosos de Baixa Resistência – o Flysch do Baixo Alentejo. Évora: Tese (Doutorado Geologia), Universidade de Évora, Évora

Price DG (2009) In: de Feitas MH (ed) Engineering geology, principles and practice. Springer-Verlag, Berlin, p 442

Reiche P (1943) Graphic representation of chemical weathering. J Sediment Petrol 13:58–68

Rodrigues FP, Grossman NF, Rodrigues LF (1978) Rock mechanics tests of the Mingtan pumped storage project. LNEC Report, Lisbon, 2 Vol.

Rocha Filho P, Antunes FS, Falcão MFG (1985) Quantitative influence of the weathering upon the mechanical properties of a Young gneiss residual soil. In: First Int. Conf. Geomech. Trop. Lateritic Saprolitic Soils Brasilia, v. 1, p. 281–294, Anais… Brasil

Russel DJ (1982) Controls on shale durability: the response of two Ordovician shales in the slake durability test. Can J 19:1–3

Ruxton BP, Berry L (1957) Weathering of granite and associated erosional features in Hong Kong. Bull Geol Assoc Amer 68:1263–1292

Ruxton BP (1986) Measures of the degree of chemical weathering of rocks. J Geol 76:518–527

Santiago LOR, Marques EAG, Costa TAV (2005) Characterization of the influence of weathering and aletrability on the mechanical strength of some rocks of the Quadrilátero Ferrífero. In: 11° CBGE, Florianópolis. p. 1845–1858. (in portuguese)

Santiago LOR (2008) Mineralogical, Mechanical and Alterability Characterization of Intrusive Basic Rocks and Sericitic Phyllites of the Quadrilátero Ferrífero Region. Tesis (Civil Engineering) – UFV, Viçosa. (in portuguese)

Short NM (1961) Geochemical Variations in Four Residual Soils. J Geol 69(5):534–571

Somers GF (1988) Foundation problems of residual soils. In: International Conference on Engineering Problems of Residual Soils, Beijing, China, 1988. Proceedings…, pp 154–171

Sutton SJ, Maynard JB (1992) Multiple alteration events in the history of a sub-huronian regolith at lauzon bay, Ontario. Can J Earth Sci 29(3):432–445

Taylor RK, Smith TJ (1986) The engineering geology of clay minerals: swelling, shrinking and mudrock breakdown. In: Proc. Conference Clay Minerals, Anais… Durham, vol 21, pp 235–260

Taylor RK (1988) Coal measures mudrocks: composition, classification and weathering processes, vol 21. Quartely Journal Engineering Geology, London, pp 85–99

Thanachit S, Suddhiprakarn A, Kheoruenromne I, Gilkes R (2005) The geochemistry of soils on a catena on basalt at Khon Buri, northeast Thailand. Geoderma 135:81–96

Tugrul A (2014) The effect of weathering on pore geometry and compressive strength of selected rock types from Turkey. Eng Geol 75:215–227

Van Der Weijden CH, Van Der Weijden R (1995) Mobility of major, minor and some redox-sensitive trace elements and rare-earth elements during weathering of four granitoids in central Portugal. Chem Geol 125:149–167

Vogt T (1972) Sulitjelmafeltets geologi og petrografi. Norges Geologiske Undersokelse 121(3–4):1–560 (in Norwegian, with English abstract)

Chapter 11
Weathering, Erosion, and Susceptibility to Weathering

H. Robert G. K. Hack

11.1 Introduction

Weathering is the change of ground materials under influence of the Earth atmosphere, hydrosphere, cryosphere, biosphere, and by nuclear radiation, mostly causing hard ground to become soft ground. Landslides and instable cut slopes are the most obvious results of weathering but also soil and rock in tunnels may weather and jeopardize tunnel stability. Completely buried ground below foundations of surface structures may weather due to percolating groundwater and air causing a gradual degradation of the geotechnical quality of the ground under a foundation. Under some conditions, weathering may have a reverse effect and cause an increase in ground strength when forming *hard* layers.

Weathering in most grounds is normally a fairly slow process taking long (geological) times to weather a large volume of ground, but it should be realized that small changes of very small quantities of material of a ground by weathering happening in a relatively short time span already may jeopardize an engineering construction. For example, the rock on both sides of a joint plane after excavation being exposed to a new environment may weather by a depth of tens of a millimeter in a couple of days, months, or years after excavation. Then the shear strength of the joint plane may be significantly reduced due to weathering of the asperities on the joint plane and forming a thin layer of low shear strength infill material (e.g., clay) in the joint. Such reduction in shear strength is often enough to allow sliding of a rock block where that would not have been the case along the unweathered joint plane.

Weathering also allows *loss of structure* of ground. The geotechnical quality of ground is to a certain extent due to a tight structure of particles and blocks of ground

H. R. G. K. Hack (✉)
Engineering Geology, ESA, Faculty of Geo-Information Science and Earth Observation (ITC), University of Twente, Enschede, The Netherlands
e-mail: h.r.g.k.hack@utwente.nl

M. Kanji et al. (eds.), *Soft Rock Mechanics and Engineering*,
https://doi.org/10.1007/978-3-030-29477-9_11

material. The tightness of the structure is reduced if weathering causes removal of part of the material or reduces the strength of particles or blocks. The reduction in tightness, among others, allows displacements in the ground, reduction of stresses in the ground, reduction of shear strength between particles and blocks, and hence the overall geotechnical quality of ground.

Thus, weathering has a major influence on the engineering properties of the ground (Anon. 1995; Fookes 1997; Hack 1996; Hencher 2015; ISO 14689-1:2017 2017; Miščević and Vlastelica 2014; Price et al. 2009; Tating et al. 2013). Therefore, weathering and the change of geotechnical properties with time during the lifetime of an engineering construction have to be incorporated in the design of any construction on or below the Earth surface.

Weathering is often assumed to be restricted to the Earth surface, but active weathering may take place deep under the surface, for example, around faults with percolating groundwater down to thousands of meters deep, and at surface weathered material may have moved down into the Earth crust by tectonic and sedimentary processes. Moreover, weathered material may be a relic of weathering under a past climate or environment that has changed since long (Harris et al. 1996; Olesen et al. 2007). Hence, any weathered material can be encountered anywhere at the surface or in the subsurface of the Earth.

11.1.1 Definitions

Weathering in this chapter denotes the chemical, physical and nuclear change of ground due to contact between the ground and the Earth atmosphere, hydrosphere, cryosphere, and biosphere, and due to nuclear radiation, but without any substantial transport of the weathering products. The influence of lightning striking the ground and the influence of bush or underground fires on ground are included in weathering too. Man-action is also of importance as it is often the reason for new or intensified weathering of ground. Some literature identifies a fourth process; biological weathering, but in this chapter biological influence is included in physical and chemical weathering as the processes and mechanisms involved in biological weathering are the same as in physical and chemical weathering. Erosion denotes the transport of weathered material by surface or subsurface processes to another location, i.e., transport by (ground) water, ice, air, or other agents. In this chapter fragmentation without transport is part of the weathering process. Radiation and emission of particles from the Earth itself, space, or from man-made nuclear materials are collectively denoted *nuclear radiation*.

The locations on surface and in the subsurface where weathering takes place are also the locations where often stress relief, intact ground creep, and groundmass creep occur. These effects are virtually always impossible to differentiate from effects caused by weathering. Moreover, the effects and weathering interact, for example, stress relief causes displacement of discontinuities that increases permeability, which allows more groundwater to percolate through the ground that intensifies weathering. Vice versa, removal of material due to weathering and erosion allows ground to destress, increases permeability, and so on. Therefore, in this chapter effects of stress relief, intact ground creep, and groundmass creep are included in the definition of weathering as proposed by Price (1995).

Ground materials are diverse and may be gases, fluids, solids (i.e., minerals, grains, and aggregates of grains or minerals), and any mixture of these and include man-made ground, such as fills and waste dump materials. Ground is commonly differentiated into soil and rock; soil being an aggregate of loose or weakly bonded particles, and rock consisting of particles cemented or locked together, giving rock a tensile strength. Soil and rock are, by some, differentiated based on a compressive strength difference mostly with soil being weaker than 1 MPa and rock being stronger.

A differentiation is made between *intact* and *discontinuous* ground, that is, ground without respectively with distinct planes of mechanical weakness (*mechanical discontinuities*). Faults, joints, bedding planes, fractures, schistosity, foliation, and so on can be mechanical discontinuities. *Incipient discontinuities* or *integral discontinuities* are discontinuities that are not yet developed into mechanical discontinuities, that is, the geotechnical properties of the discontinuity are not yet different from those of the surrounding intact ground. Weathering and changes in stress can be causes for incipient discontinuities to become mechanical, for example, weathering causes incipient bedding planes to become mechanical discontinuities. A groundmass consists of (blocks of) intact ground with discontinuities, if present. In this chapter the terms *intact ground* and *groundmass* are used in which ground can be soil, rock, fill, waste, or any mix. Discontinuities denote mechanical discontinuities, except where indicated otherwise. *Banking* is a term sometimes used to denote layers of intact ground bounded by mechanical discontinuities. Discontinuities are mostly belonging to a *set* or *family* of discontinuities in which the discontinuities have the same origin and broadly the same geotechnical characteristics.

11.1.2 Weathering Terminology

The terminology on weathering and the resulting weathering products is somewhat confusing. In most literature *regolith* is used to denote the zone below the Earth surface in which material is affected by or is the result of weathering (e.g., weathered rock), loose soils such as fluvial and eolian deposits for which also the term *transported regolith* may be used, and cemented soils such as hard layers and crusts. Other literature also includes in regolith volcanic deposits as lava flows and volcanic ash, organic material, groundwater-deposited salts, and groundwater itself. The term *saprolite* may be used for the complete sequence from slightly weathered bedrock to residual soil of a chemically weathered groundmass in a tropical environment (Massey et al. 1989), or only for the completely weathered parts of the bedrock (Wilson 2004). Regolith and saprolite may be used together with the complete weathering sequence denoted regolith but with a subdivision between a top layer named *residual soil* and the gradual change to bedrock below named saprolite

(Singh and Huat 2004). *Laterite* is used in the literature to describe the *hard layer* of iron-cemented residual soil resulting from weathering in a tropical environment (Blight 1989), but by others may be used for any red-colored residual soil formed in a tropical environment. In this chapter only the ISO 14689-1:2017 (2017) standard terminology will be used and other terminology avoided as the definitions are ambiguous and rather vague with regard to engineering properties.

11.2 Weathering

Weathering is part of the normal cycle of ground materials on Earth. Rocks are moved up to surface by tectonic forces or extruded by volcanic action. The rocks at surface are weathered due to exposure to the atmosphere, hydrosphere, cryosphere, biosphere, and nuclear radiation that cause disintegration, changes in mineralogy, dissolution, and rock to become soil (Fig. 11.1). The soil may be eroded by wind, rivers, and glaciers, transported, and finally may end up in lakes or in sea where it

Fig. 11.1 Exposure of physically and chemically weathered granodiorite (an intrusive, igneous rock comparable to granite); zones of fresh, hard, and strong granodiorite (bluish-grey colored) occur in between zones with granodiorite fallen apart in smaller blocks and zones completely decomposed into loose soil material (brownish colored). Note the general increase of decay upwards to the original topographical surface (vegetated) (Road T710, Falset—Gratallops, Catalunya, Spain; photo courtesy of W. Verwaal 2002)

becomes sediment. The sediment is buried by subsequent deposits and under an increasing overburden load consolidated, compacted, possibly cemented, heated, melted and coagulated, and/or changed in mineralogy, and becomes rock again. On the other hand, weathering products such as dissolved minerals may be transported by (ground) water and precipitate at other locations and become cement in between particles and discontinuities, leading to increased strength of intact ground and discontinuities, and may result in hard layers or crusts (Sect. 11.8).

11.2.1 Physical Weathering

Physical weathering results in the breakdown of groundmasses into progressively smaller fragments without changing its components. The main processes in the breakup of ground are differential expansion due to temperature variations, wetting and drying, freezing and thawing, pressures of water and ice in pores between particles in intact ground and in discontinuities, (re)crystallization pressures, hydration, swelling and shrinkage of minerals due to water absorption (e.g., some clay minerals), growth of plant and tree roots in ground and discontinuities, and animal action. Lightning striking the ground and fires (bush and underground) may also cause reduction of fragments (Allison and Bristow 1999; Knight and Grab 2014).

11.2.2 Chemical Weathering

Chemical weathering designates the reactions between minerals and weathering agents and results in decomposition, new forming, and solution of minerals (Fig. 11.2). Presence of water, groundwater, or air with dissolved or vapor of chemical agents is of major importance as these react with the ground and groundmass material. Some minerals are more susceptible to chemical reactions with weathering agents than other depending on crystal structures and chemical characteristics. For example, silicate minerals based on lattices of a repeating grid of silicon and oxygen weather very slowly if the silicon-oxygen groups are firmly bond directly to each other (e.g., quartz), but weather fast when the oxygen atoms are bond to intermediate atoms as magnesium or iron (e.g., olivine). Other minerals are easily soluble, easily react with oxygen, or with weak carbonic acid (hydrolysis). Temperature is important, as chemical reactions are faster under higher temperatures. Therefore, chemical weathering is faster in tropical areas with a warm and humid environment than in an arid polar or mountain environment with low temperatures and little humidity. Biotic influences, induced by living organisms such as plants, bacteria, and worms, and the influence of chemicals exerted by living organisms and dissolved in (ground) water are included in chemical weathering. Chemical weathering may also occur due to man-action, such as polluted groundwater, released acids in mining (Fig. 11.3), fertilizer dissolved in groundwater, and polluted air and rain (e.g., *acid rain*).

Fig. 11.2 Road cut slopes excavated in gypsum-cemented siltstone about 5 years after excavation. The slopes were excavated as a plane surface with a slope angle of about 60°. The slopes are instable and eroded due to solution of gypsum after excavation. The north side is less affected than the south side, likely due to more direct sunlight on the north side and the pre-failing wind direction. The protrusive banks in the south side and greyish areas in the north side are layers mainly consisting of gypsum with little silt that are slightly more resistant to solution and erosion (Road C44 near Vandellòs, Catalunya, Spain)

Fig. 11.3 Acid groundwater causing deterioration of rock mass and steel support in a mine tunnel. Oxidation of metal sulfides by oxygen-rich percolating groundwater caused forming of sulfates and acid water. The blue and white stained areas are sulfates precipitated from seepage water (Gold, silver, copper, and lead mine Rammelsberg, Goslar, Harz, Germany; photo courtesy S. Carelsen 2010)

11.2.3 Nuclear Weathering

The Earth and any other object in space are subject to a bombardment with different types of radiation consisting of galactic and solar waves and particles from space that may change atom and mineral structures, and disintegrate minerals, ground and groundmasses (Pieters and Noble 2016; Sim et al. 2017). The effect is known as *space weathering* and the effects are evident on planets and objects in space without atmosphere and magnetic field. On Earth, the Earth atmosphere and the Earth magnetic field stop most radiation from space reaching the Earth surface. Radiation emitted by radioactive materials in the ground of the Earth itself and by man-made nuclear materials have similar effects. The effects of nuclear weathering at or in the Earth surface are likely very small or negligible for engineering compared to weathering by the Earth atmosphere, hydrosphere, cryosphere, and biosphere, except probably for intense radiation by natural nuclear fission and man-made nuclear materials over long periods of times, such as in radioactive waste repositories (Sect. 11.4.4).

Fig. 11.4 Decrease of spacing between mechanical discontinuities (*banking* spacing) nearer to the original topographical surface due to weathering of incipient bedding plane discontinuities into mechanical discontinuities. Note the very thin layer of weathered material and topsoil derived by weathering from the limestone rock mass that consists of fairly pure calcareous material (cut slope along road T322-Montroig to Colldejou, 1.3 km SE of Colldejou, Falset, Spain; photo courtesy W. Verwaal 2001)

11.2.4 Development and Weathering of Discontinuities

Discontinuities in groundmasses weather as well. The spacing between mechanical discontinuities at surface is mostly smaller; thus, the block size because incipient discontinuities develop into mechanical due to weathering (Fig. 11.4). The shear strength of discontinuities may reduce due to weathering of the discontinuity wall, solution effects, or forming or change of infill material. For example, in weathered calcareous rock masses containing some clay, it is often found that the walls of the bedding planes are slightly weathered and that a thin clay infill is found in the discontinuities. The infill is the leftover of the weathered rock material.

Mechanical discontinuity sets may develop sequential (Tating et al. 2015). At the onset of weathering first some of the incipient bedding planes develop into mechanical discontinuities, followed at later stages in the weathering process by other sets of discontinuities with different origin. Also, the persistence of discontinuities may extend per set such that in less weathered groundmasses virtually no persistent discontinuities are present, developing over one set being persistent to finally all discontinuities in all sets being persistent in more severely weathered masses. The properties of discontinuities, for example, shear strength, may also be different because of the different duration of weathering since the change from incipient to mechanical discontinuities.

11.3 Lithology and Types of Weathered Groundmasses

Weathering is to a considerable extent depending on the lithology, texture, and structure of the groundmass and groundmass materials. The number of different materials and groundmasses is virtually infinite and therefore also the effects of weathering on groundmasses. However, it is possible to broadly group different groundmasses affected by weathering that show more-or-less comparable effects of weathering.

11.3.1 Standard Weathering Profile

Throughout the literature and in the standards for geotechnical work mostly a sketch of a typical weathering profile is included, however such profiles are *idealized* and do not apply to all weathering profiles. Figure 11.5 shows the typical weathering profile for a weathered gneiss and schist groundmass in a tropical climate.

11.3.2 Sandstone, Intrusive and Metamorphic Groundmasses

Sandstones, many intrusives, and many metamorphic groundmasses show a weathering profile similar to those shown in Figs. 11.5 and 11.6. The weathering starts from and via the discontinuities due to percolating groundwater and air in the discontinuities. The discontinuities change progressively from incipient to mechanical discontinuities (Sect. 11.2.4), become more permeable, thus allow more groundwater and air to percolate, and so allow more and more weathering to occur. Discontinuities become wider, that is, the aperture increases, because intact ground changes into decomposed intact ground which forms the soil-type discontinuity infill material. Coarser grained, igneous, and metamorphic rocks, and sometimes sandstones may form core stones with a different state of weathering than the surrounding more weathered material, and various sets of discontinuities may have different states of weathering and resulting characteristics due to differences in duration of weathering.

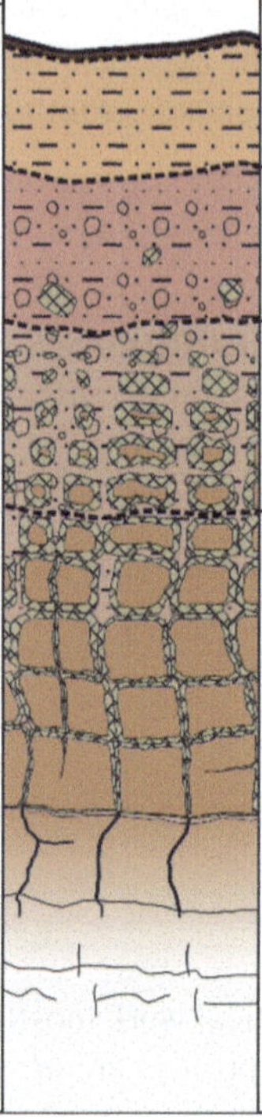

Indicative depth below surface (m)	Rock mass weathering grade	Description (as rock mass for fresh to moderately weathered, and as soil mass for highly weathered to topsoil)
0 0.2	Topsoil	Dark brown, black, sandy clayey SILT with organic matter and man-made material
0.9 - 1.2	Residual soil	Reddish brown to light brown, low sphericity, very angular, sandy clayey SILT of high plasticity
2 - 3	Completely	Reddish brown to red, low sphericity, very angular, clayey silty sandy GRAVEL to gravelly sandy clayey SILT of high plasticity, relict discontinuities (foliation and joints), with in lower part occasionally cobble- and boulder-sized completely decomposed core stones with white kaolin
3.5 - 4.5	Highly	Reddish brown, low sphericity, very angular, clayey silty sandy GRAVEL, relict discontinuities (foliation and joints), with cobble- and boulder-sized completely and partially decomposed core stones consisting of mainly mica and feldspar with white kaolin
5.5 - 8	Moderately Slightly	Light brown, light grey, white, medium- to coarse-sized minerals with phenocrysts, very weak to very strong, schistose and foliated, SCHIST and GNEISS in part migmatized, upwards increasingly weathered leading to core stones surrounded with zone of decomposed intact rock material (hatched), discontinuities discolored (bottom) to filled with completely decomposed and altered intact rock (top), white kaolin in intact rock and discontinuities, porphyroclasts (garnet) becoming markedly weathered (weaker) starting from moderately weathered
20	Fresh (un-weathered)	Light grey, white, medium- to coarse-sized minerals with phenocrysts, moderately strong to very strong, schistose and foliated, SCHIST and GNEISS in part migmatized

Fig. 11.5 Example of a weathering profile in a groundmass consisting of weathered gneiss and schist in a tropical climate (Weathering grades follow Table 11.2 for rock masses). (Modified from Tran et al. 2019)

Fig. 11.6 Weathered cut slope in gneiss and schist in a tropical climate (Weathering grades follow Table 11.2 for rock masses) (Yên Bái City, Vietnam; photo courtesy D. Alkema, 2010)

11.3.3 Calcareous Groundmasses

Weathered calcareous groundmasses may show a weathering profile as shown in Fig. 11.5, but generally the amount of weathered material is less as the carbonate for a part or completely dissolves in groundwater and vanishes from the environment. Therefore, completely weathered and residual soil layers formed by weathering of calcareous rock masses are mostly very thin because only the non-dissolvable material (e.g., clay) remains, if the calcareous rock mass contains any non-dissolvable minerals. A very pure calcareous groundmass, that is, consisting nearly completely of calcareous minerals, does not produce any layers of completely weathered material and residual soil (Fig. 11.4), and also no infill material in discontinuities. Only the spacing of discontinuities becomes smaller with increasing state of weathering as incipient discontinuities change into mechanical, resulting in smaller block sizes.

11.3.4 Varying Susceptibility to Weathering in One Groundmass

A groundmass with different lithologies with different susceptibilities to weathering may develop a weathering profile in which the less susceptible layers are weathered less than other layers. Consequently, the state of weathering will vary depending on the lithology (Fig. 11.7).

Fig. 11.7 Different lithologies with different susceptibility to weathering result in groundmasses with layers with different states of weathering. The grey, light-brown sandstone is more resistant to weathering than the dark-brown, black shale that is already completely weathered and partially eroded within 5 years after excavation giving the inserts in the groundmass (Flysch deposit of sandstone layers mainly consisting of quartz grains, alternating with shale; Kota Kinabalu, Malaysia, 2011)

11.3.5 Very Thin Bedded Groundmasses

Thin-bedded fine-grained sedimentary materials may weather very rapidly directly after exposure due to the influence of atmosphere, hydrosphere, and cryosphere or when stresses change (Huisman 2006; Miščević and Vlastelica 2014). Figure 11.8 shows calcareous sedimentary deposits with many incipient bedding planes with very small spacing probably also including some small quantities of organic material. When the groundmass is exposed the mas destresses and becomes subject to temperature and humidity variations. Incipient sedimentary bedding planes become then mechanical discontinuities and rapidly within a couple of years, the groundmass falls apart.

11.3.6 Disking in Borehole Cores: Weathering of Borehole Cores

A similar feature as with thin-bedded sedimentary rocks may occur in borehole cores. Newly formed mechanical discontinuities split the borehole core in *disks* when arriving at surface. The process is described by the term *disking* or *discing*.

Fig. 11.8 Thin-bedded sedimentary mudstone and claystone with angular more-or-less equidimensional rock blocks of some 20 cm fall apart in rounded very small flaky particles of millimeter size due to stress relief and weathering within a couple of years after exposure in a warm temperate to semiarid Mediterranean climate. The insert photo shows layers of bluish-greenish mudstone and claystone alternating with pinkish dolomite layers (Length hammer about 30 cm) (Hostal Slope, along N420 road, km 846, Falset, Spain; large photo courtesy M. Huisman 2004)

Various processes and mechanisms sometimes acting together are responsible. Mud- and claystones and shales show a disking effect in which closely spaced incipient bedding planes become mechanical discontinuities due to stress relief and temperature and humidity changes. Presence of swelling minerals is also of influence (Ohta 2001). Disking may be time dependent and cores directly after drilling may show little disking but this increases with time (Hoek and Brown 2018) which is very similar to the weathering as described in Sect. 11.3.5. The effect of disking is mostly more severe for borehole cores from larger depth (Stacey 1982). Core disking may also be due to the existence of (highly anisotropic) in situ horizontal stresses in relation to the tensile strength of the material even if the material in situ is unfractured and not bedded, for example, disking of granite without any form of incipient bedding, layering, or fracturing (Lim and Martin 2010).

11.3.7 Groundmasses Containing Easily Dissolvable Minerals: Gypsum and Salt

Anhydrite, gypsum, and salts are generally very easily dissolving in (ground) water. This causes that the complete groundmass with these minerals or part of it vanishes with the water and when the soluble minerals forms the cement between the particles in the groundmass the groundmass develops into a loose uncemented soil (Fig. 11.2) (Huisman et al. 2006). Salt dissolves virtually instantly after excavation at contact with water, that is, first rain, while anhydrite and gypsum may dissolve a in a couple of years in a temperate or humid climate. Therefore, exposed groundmasses with anhydrite, gypsum or salts exist in arid climates only.

11.3.8 Karstic Groundmasses

Karst is weathering by solution in carbonate, anhydrite, gypsum, or salt containing groundmasses where part of the groundmass is dissolved and removed by groundwater (Pesendorfer and Loew 2004; Schmitz and Schroeder 2009). The process starts with dissolving material along discontinuities due to percolating groundwater resulting in small cavities and holes along discontinuities. In strong rock masses with not too fast dissolving minerals (e.g., calcareous minerals), these may grow into caves, sometimes of a size allowing underground streams and rivers, such as the river Lesse in Belgium (Caves Han-sur-Lesse 2018). In weaker groundmasses solution pipes often infilled with materials from above, may have penetrated deeply into the calcareous groundmass (Price et al. 2009).

11.3.9 Volcanic Groundmasses

Lava and ash are formed relatively rapidly as these cool very fast resulting in small minerals of which many are unstable under surface conditions (Yokota and Iwamatsu 2000). Some coagulated lavas, as for example, some basalts, have a very low permeability that prevents minerals to be in contact with weathering agents as water, but ash has a high permeability, and thus many options for contact between unstable ash minerals and agents, and weathering of ash is very fast. Minerals in volcanic ashes disintegrate quite rapidly, dissolve, or weather into clay minerals (James et al. 2000; Yokota and Iwamatsu 2000). Weathering of volcanic deposits may be 30 times faster than weathering of granite or shale under the same conditions (Dosseto et al. 2012). The rapid weathering produces the very fertile agricultural soils in volcanic areas.

11.3.10 Groundmasses with Sulfides: Oxidation of Sulfides

Weathering by oxidation of sulfides such as pyrite is influenced by bacteria, air, and oxygen-rich water that cause the sulfides to decompose into an oxide and sulfuric acid. The oxide minerals are mostly less strong or have a different volume than the original sulfide minerals and therefore, reduce the mechanical strength parameters of the ground. The acid may be aggressive to the groundmass, engineering structures, and the environment (Fig. 11.3) (Brattli and Broch 1995; Oyama and Chigira 2000). Sulfur and sulfides such as pyrite, may spontaneously combust if in contact with oxygen-rich air, leading to underground fires if igniting nearby combustible materials (Sect. 11.3.11) (Wu et al. 2006).

11.3.11 Organic Groundmasses: Coal and Peat Fires and Peat Weathering

Combustion and fires of coal and peat are also a form of weathering (Kuenzer and Stracher 2012). These may occur in coal seams and peat layers that are in contact with oxygen-rich air. Combustion may start at surface exposures and can run underground for kilometers if faults or subsidence cracks allow fresh air to reach the coal or peat and eject exhaust gasses. Poor mining practices, uncontrolled exploitation, or natural spontaneous combustion of coal may be the onset of coal fires and bush fires may be the onset of peat fires. Oxidation of sulfur and sulfides present in the coal or surrounding layers may start or accelerate the fire (Sect. 11.3.10) (Didier et al. 2008). Underground coal fires are notoriously difficult to stop and sometimes may burn for many tens to thousands of years (Krajick 2005). The burning process converts the coal and peat in gasses with a large volume reduction of the original seams and layers. In the organic layers but also in surrounding non-organic layers minerals may change or fall apart due to the heat and heating may fracture intact ground, leading to a geotechnically less strong ground. The inverse may also happen if minerals in surrounding layers are cooked or baked such that the layers become stronger. The ground remaining after the fire may have cavities partially or completely filled by collapsed overlying ground, open discontinuities, and major subsidence at surface can occur. The layers between the layer the fire was in to surface may have irregular boundaries and subsidence cracks.

Weathering of peat is due to water or air with oxygen percolating through peat layers that cause (bio) oxidation and disintegration of organic material in the peat. The peat undergoes a large reduction in volume and land subsidence is the result, such as in Venice, Italy (Gambolati et al. 2005), Ireland, Finland, The Netherlands, and many other countries.

11.3.12 Poisonous Fluids and Gasses

Particular geological conditions may cause weathering that forms (highly poisonous) arsenic groundwater, for example, in parts of India and Bangladesh, China, South and North America, and Eastern Europe (BGS Arsenic 2001; Ghosh and Singh 2009; VanDerwerker et al. 2018; Welch et al. 2000). Although the origin of the arsenic and the process how it ends up in the groundwater are not known in all detail in all areas, it is likely that in Bangladesh and India, the arsenic stems from coal seams, micas, pyrite-bearing shale, or gold deposits, that are weathered, eroded, transported, and became soil deposits at another location. These later deposits are in contact with the groundwater in which the arsenic is released. Possibly anaerobic conditions, bacterial or man-action, such as contamination of the groundwater with fertilizer, may be of influence in releasing the arsenic from the soil deposits into the groundwater. Another very dangerous weathering process is the release of deadly quantities of hydrogen sulfide gas from reactions between methane released by coal and bacterial sulphate reduction of anhydrite (Alborz Service Tunnel, Iran; Wenner and Wannenmacher 2009).

11.3.13 Man-Made Groundmasses: Fills and Waste Dumps

Man-made groundmasses such as dumped or dredged fills for land reclamation weather if the fill or dump materials are prone to weathering. Municipal solid waste dumps may weather due to decay of materials, often forming gas and fluids and leaving cavities in the mass (Machado et al. 2008). Groundwater below and around waste dumps may become polluted by a whole series of chemicals of which many are unfriendly for the environment, poisonous, aggressive for engineering structures, or may react with surrounding groundmasses.

11.4 Weathering Rate and Depth of Weathering

The rate of weathering, that is, the weathering per time unit, is highly variable and strongly depends on the type of groundmass, environment, climate, and local circumstances, such as erosion, the accessibility of the groundmass for (ground) water and air, and dissolved minerals in (ground) water and vapor in air. Nuclear radiation may also have an influence on weathering rates. Under some conditions weathering may have a significant impact on engineering structures within a few years whereas in other no influence is observed in centuries (Cabria 2015; Hack et al. 2003; Huisman et al. 2006; Tating et al. 2013; Tran et al. 2019). Consequently, the same holds for the weathering depth into the groundmass. In humid and warm environments, in situ weathering from surface may go down to tens and often more than 100 m below surface (Fookes 1997; Lumb 1983; Qi et al. 2009). In dry climates, however, the in situ weathered zone is often just a few decimeters or meters deep. The depth of the weathered zone is less where weathered material is removed by erosion (Sect. 11.4.2) or by solution into (ground) water.

11.4.1 Environment and Climate

The environment and climate have a major influence on rate and depth of weathering. In a tropical humid climate chemical weathering is dominant and minerals fall apart very rapidly under influence of chemical reactions. In more temperate climates physical weathering becomes dominant, whereas in arid polar or dry mountain climates physical weathering will be virtually the sole mechanism of weathering (Lamp et al. 2017). Further important factors are the amount of sunlight on the groundmass, wind force, and prevailing wind direction. The sunlight and wind allow for fast drying of the ground after rainfall and hence reduce chemical reactions that involve water and thus reduce the chemical weathering rates. Larger differences in temperature between day and night and over the seasons and alternating wetting and drying due to cyclic rainfall increase physical weathering. Soluble (e.g., carbonate, anhydrite, gypsum, and salts) materials fall apart or dissolve faster in a wet climate than in a drier climate.

11.4.2 Erosion

Erosion predominantly occurs at the Earth surface and is generally less relevant in underground works. However, underground water flows, including water leaking from sewage and water mains, may transport soil or infill materials. Erosion by itself may lead also to effects similar as those in weathering, for example, the saltation of sand that reduces particle size of sand grains and creating dust particles in wind (Shao et al. 1993). Grinding of rock blocks over the bedrock in rivers and glaciers reducing block size of a moving block, and also fracturing, loosening, and unlashing rock blocks and particles from the bedrock (*plucking*) by moving water, ice, or wind are other examples (Anderson and Anderson 2010; Singh et al. 2011).

Weathered materials often form a good insulation of the underlying groundmass from the influence of the atmosphere, hydrosphere, cryosphere, and often also biosphere. This decelerates weathering, slows further weathering in depth, and when the layer of weathered material is thick enough effectively stops further weathering. Inversely, erosion causes the insulation to be removed, exposing the groundmass to the environment, accelerate weathering, and allow further progressive weathering in depth of new fresh ground. Erosion thus increases the rate of weathering of the underlying material (Huisman et al. 2011; Tating et al. 2019).

11.4.3 Accessibility of Groundmass for Weathering Agents

Weathering of soil-type material mostly progresses through intact material and discontinuities, if present. The weathering agents, such as water and air, percolate through the pores and channels between pores in intact ground, and through discontinuities. The permeability of intact rock-type material is normally quite low and therefore weathering of rock masses mostly starts from the discontinuities through which the weathering agents circulate and develops further into the intact rock material from the discontinuities. Causing, for example, a gradual decrease in strength of intact rock towards the discontinuity. In many groundmasses discontinuities give thus the access to the groundmass for weathering agents, and the number of permeable discontinuities determines for a considerable extent the rate of weathering in rock masses. Discontinuities, such as faults, with percolating groundwater may exist at any depth below surface and groundmasses at large depth of thousands of meters may be subject to active in situ weathering (Katongo 2005).

Man-made influences such as the damage caused by the tools and means used for making an excavation may allow faster and deeper weathering. The excavation method may create fractures, widen existing discontinuities, and change incipient into mechanical discontinuities, collectively denoted *backbreak*, that allow water and air to infiltrate easier and deeper into the groundmass (Fig. 11.9) (Hack et al. 2003).

Generally, more backbreak is caused when higher energy levels are used in shorter time spans, that is, blasting gives a high energy peak with many backbreak fractures, while scouring by a river creates an exposure virtually without any backbreak as energy is applied over many hundreds or thousands of years. Table 11.1 illustrates the damaging influence of excavation methods in use for surface and underground excavations with quantitative factors for the damage. The factors are correction factors applied to groundmass properties.

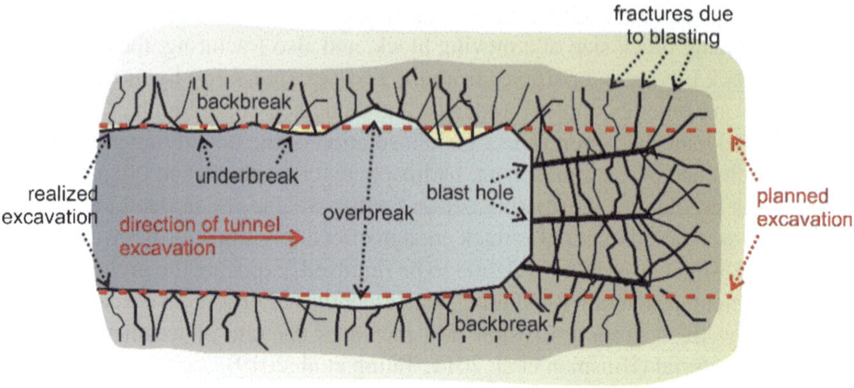

Fig. 11.9 Discontinuities in a groundmass due to blasting and present as backbreak in the walls, roof, and floor of a tunnel

Table 11.1 Excavation damage ("backbreak") factors for groundmasses

SSPC[a] (slope)		FRHI[b] (slope)		Hoek - Brown[c,d] (GSI disturbance factor)				MRMR[e] (underground mining)		MBR[f] (underground mining)	
Excavation	Factor	Excavation	Rating	Excavation (slope)	D	Excavation (tunnel)	D	Excavation	Factor	Excavation	Factor
Natural/hand-made[g]	1.00	Smooth excavation	−1	In some softer rocks excavation can be carried out by ripping and dozing and the degree of damage to the slope is less (mechanical excavation)	0.7	Mechanical or hand excavation in poor quality rock masses (no blasting) results in minimal disturbance to the surrounding rock mass	0	Boring	1.00	Boring	1.00
Pneumatic/hydraulic hammer[h]	0.76	Regular cut	3								
		Manual cut	4								
Controlled blasting	0.99	Controlled blasting	1			Excellent quality controlled blasting or excavation by Tunnel Boring Machine results in minimal disturbance to the confined rock mass surrounding a tunnel	0	Controlled blasting	0.97	Controlled blasting	0.94–0.97
				Small scale blasting in civil engineering slopes results in modest rock mass damage, particularly if controlled blasting is used. However, stress relief results in some disturbance: Good blasting	0.7						
Blasting with result: Good	0.77							Good blasting	0.94	Good blasting	0.90–0.94
Blasting with result: Open discontinuities	0.75	Regular blasting	5			Where squeezing problems result in significant floor heave, disturbance can be severe unless a temporary invert is placed: With invert	0				
Blasting with result: Dislodged blocks	0.72			Small scale blasting (as above): Poor blasting	1.0						
Blasting with result: Fractured intact rock	0.67					Where squeezing problems (as above): No invert	0.5				
Blasting with result: Crushed intact rock	0.62	Poor blasting	8	Very large open pit mine slopes suffer significant disturbance due to heavy production blasting and also due to stress relief from overburden removal (production blasting)	1.0	Very poor quality blasting in a hard rock tunnel results in severe local damage, extending 2 or 3 m, in the surrounding rock mass	0.8	Poor blasting	0.80	Poor blasting	0.90–0.80

Notes: SSPC, MRMR and MBR factors range from 1.00 for negligible damage to 0.62 respectively 0.80 for maximum damage, FRHI is expressed as point rating from −1 for no damage to 8 for maximum damage, GSI disturbance factor ranges from 0 for undisturbed to 1.0 for maximum disturbance

[a]Hack et al. (2003)

[b]Singh (2004)

[c]The description of D is referenced with example photographs of excavation damage

[d]Rocscience (2011) and Hoek and Brown (2018)

[e]Laubscher and Jakubec (2001)

[f]Cummings et al. (1984)

[g]Care should be taken that discontinuities due to stress relief are not considered excavation damage

[h]This value is based on hammer sizes up to 5 m length with a diameter of 0.2 m

11.4.4 Nuclear Radiation Effects

Some research has been done on the influence of nuclear radiation on groundmasses on Earth. In facilities for long-term storage of radioactive nuclear waste, degradation of the surrounding groundmass, brines, and groundwater is a subject of limited research (Lainé et al. 2017; Lumpkin et al. 2014; Soppe and Prij 1994). Changes of atoms and minerals under influence of radiation is also occurring in natural nuclear reactions (Oklo reactor; Bracke et al. 2001; Meshik 2009). Weathering rates due to nuclear radiation are likely very low compared to the rates under other influences but may be of importance for stability of rooms, tunnels, and shafts guaranteeing access to underground radioactive-waste repositories over very long time spans of tens of thousands of years.

11.4.5 Local Environment and Engineering Influence

Weathering is strongly dependent on local conditions, such as surface run-off water, snow and ice, vegetation and animals, chemicals in groundwater, salt-containing water and air, and man-made action. Water run-off over and through the near surface of an exposure may erode weathered surface material and expose new fresh ground to the environment progressing weathering of the fresh ground. Surface run-off water will have a larger influence on a slope in a groundmass with a small block size than with a larger block size because smaller blocks are more easily flushed away by run-off water. Steeper slopes are less affected than less steep slopes because the water runs over faster and water less easily penetrates the groundmass. Snow and ice may block seepage from discontinuities where discontinuities are outcropping at an exposure, which may lead to water pressures in the pores and discontinuities, causing intact material to break, existing discontinuities to widen, and incipient discontinuities to become mechanical. In situ in the groundmass, frozen water that expands because ice has a larger volume than water, may cause the same changes.

Roots of trees and other vegetation may grow into the ground and discontinuities resulting in fractures in intact ground, changing incipient into mechanical discontinuities, root wedging that dislodges blocks, opening of discontinuities, and subsequently access for (more) water and air and thus more and faster weathering. Vegetation is normally not present in underground excavations; however, small plants may start growing in lighted tunnels. Actions by other living species such as fungus, bacteria, and larger animals as worms and other burying animals, may influence weathering in surface but also in underground excavations by for example, making burrows that allow water and air to penetrate the groundmass more easily. Vegetation and animals may also excrete chemicals that attack the ground material causing degradation.

Ground and surface water may be polluted such as water seeping from sewages, landfills, waste dumps, ore mines and tailings, or be polluted with fertilizer. Natural groundwater in some grounds may also be aggressive, for example, acid water corroding the tunnels of the London Underground (Rainey and Rosenbaum 1989) or water around the tunnel for the Niagara Tunnel project, Canada (Hughes et al. 2007). Air pollution has a negative effect on all ground, and also on natural and non-natural construction and building materials by reactions between the minerals in the ground or construction materials and the chemicals in the polluted air (e.g., *acid rain*) (Doehne and Price 2010). This causes chemical weathering but may also cause physical weathering due to precipitation and crystallization in pores and discontinuities of chemicals from polluted air or newly formed chemicals.

Salt-water and salt containing vapor may penetrate the groundmass and discontinuities (Cabria 2015; Mottershead 1989; Rodriguez-Navarro et al. 1999; Ruiz-Agudo et al. 2007; Wellman and Wilson 1965). The salt may directly react with ground material and cause degradation of the ground material, or salt may precipitate in pores and discontinuities and expansion of salt crystals causes breaking of intact ground, widen existing discontinuities, and change incipient into mechanical discontinuities. The salt may originate from salty capillary groundwater, polluted air, or seawater spray. Salt in the atmosphere from seawater spray may be present up to many tens of kilometers from the coast.

11.5 Description of Weathered Groundmasses

Weathering is gradational and to be able to use gradational features of weathering in engineering normally the weathered groundmass is characterized and divided into different layers with similar weathering characteristics defined by a range of characteristics or parameters significant for engineering applications (Dearman 1995). Note that this only describes the weathering as is at the time of characterization and does not include the future changes due to weathering. Future weathering is discussed in Sect. 11.7.

11.5.1 Weathering Features Important in Engineering

Each type of groundmass weathering has a specific impact on engineering. Uniformly weathered layers following, for example, Fig. 11.5, give the least difficulties with forecasting the geotechnical properties in the subsurface, while properties of irregular layers with different states of weathering or layers resulting from different susceptibilities to weathering are far more difficult to forecast in detail. Often the division in states of weathering for engineering purposes is based on the following characteristics (Price 1995):

- Effectively unweathered. Weathering has not influenced the groundmass to an extent that weathering must be considered in design of engineering works.
- Significantly weathered. The design should incorporate some limited influence of weathering on engineering properties, for example, weathering of discontinuities may have reduced shear strength of discontinuities.
- Severely weathered. Weathering must be considered otherwise the engineering applications will not be stable or functional.
- For rock masses: sufficient of the rock material in a rock mass is decayed to the geotechnical condition of a soil to make the mass behave as a soil-type mass geotechnically.

11.5.2 Standard Schemes for Weathering Description

The above is used to define standard descriptions for the division of a weathered groundmass in different layers with different states of weathering, and thus with different impact on engineering design. Many standard schemes have been proposed, some fairly simple, such as in Table 11.2, or following more elaborate schemes as in Fig. 11.10. A simple scheme is easy to use but may not always be suitable. The more elaborate schemes apply better to various groundmasses and materials, but may be (too) complex and may be misunderstood (Price et al. 2009).

11.6 Quantitative Influence of Weathering on Geotechnical Properties

Table 11.3 shows examples of properties of intact ground and groundmasses indicating how ground properties change with increasing grade of weathering. The table shows the considerable changes in material properties as a consequence of weathering.

Quantification of the grades of weathering in terms of the reduction of geotechnical properties of rock masses is shown in Fig. 11.11. The graph is based on data from various authors and from different rock types and rock masses. The influence of weathering is quite clear in the decrease of intact rock strength over the complete sequence from fresh to completely weathered rock masses and for the decrease in discontinuity spacing and condition of discontinuities (determining the shear strength) down to moderately weathered rock masses. The influence of weathering on spacing and condition of discontinuities de-accelerate or invert from moderately to highly weathered which may be attributed to cementation processes in discontinuities often happening in higher grades of weathering.

Quantification of grades of weathering in terms of the reduction of geotechnical properties of a groundmass have been done by various authors (Bieniawski 1989;

Table 11.2 Weathering description and characterization of intact rock and rock mass (ISO 14689-1:2017 2017)

Intact rock		Rock mass		
Term	Description	Term	Description	Grade
Fresh	No visible sign of weathering/alteration of the rock material.	Fresh	No visible sign of rock material weathering; perhaps slight discoloration on major discontinuity surfaces.	0
Discolored	The color of the original fresh rock material is changed and is evidence of weathering/alteration. The degree of change from the original color should be indicated. If the color change is confined to particular mineral constituents, this should be mentioned.	Slightly weathered	Discoloration indicates weathering of rock material and discontinuity surfaces.	1
		Moderately weathered	Less than half of the rock material is decomposed or disintegrated. Fresh or discolored rock is present either as a continuous framework or as core stones.	2
Disintegrated	The rock material is broken up by physical weathering, so that bonding between grains is lost and the rock is weathered/altered towards the condition of a soil in which the original material fabric is still intact. The rock material is friable but the mineral grains are not decomposed.	Highly weathered	More than half of the rock material is decomposed or disintegrated. Fresh or discolored rock is present either as a discontinuous framework or as core stones.	3
		Completely weathered	All rock material is decomposed and/or disintegrated to soil. The original mass structure is still largely intact.	4
Decomposed	The rock material is weathered by the chemical alteration of the mineral grains to the condition of a soil in which the original material fabric is still intact; some or all of the mineral grains are decomposed.	Residual soil	All rock material is converted to soil. The mass structure and material fabric are destroyed. There is a large change in volume, but the soil has not been significantly transported.	5

Hack and Price 1997). Table 11.4 gives an example in which the factors are based on the weathering at surface of a wide range of rock masses such as limestone, sandstone, shale, granodiorite, and slate in the Mediterranean climate of northeast Spain. The factors in the table are multiplied with the geotechnical property to give the weathered property in a particular grade of weathering.

APPROACH 2 rock is moderately strong or stronger in fresh state UNIFORM MATERIALS		
grade	classifier	description
I	fresh	Unchanged from original state
II	slightly weathered	Slight discoloration; slight weakening
III	moderately weathered	Considerably weakened, penetrative discoloration; large pieces cannot be broken by hand
IV	highly weathered	Large pieces can be broken by hand; does not readily disintegrate (slake) when dry sample immersed in water
V	completely weathered	Considerably weakened; slakes in water; original texture apparent
VI	residual soil	Soil derived by in-situ weathering but having lost retaining original texture and fabric

APPROACH 3 heterogeneous masses (mixture of relatively strong and weak material) HETEROGENEOUS MASSES		
zone	description[(2)]	typical characteristics
1	100% grades I-III	Behaves as rock; apply rock mechanics principles to mass assessment and design
2	> 90% grades III < 10% grades IV-VI	Weak materials along discontinuities; shear strength stiffness and permeability affected
3	50 - 90% grades I-III 10 - 50% grades IV-VI	Rock framework still locked and controls strength and stiffness; matrix controls permeability
4	30 - 50% grades I-III 50 - 70% grades IV-VI	Rock framework contributes to strength; matrix or weathering product control stiffness and permeability
5	< 30% grades I-III 70 - 100% grades IV-VI	Weak grades will control behavior. Corestones may be significant for investigation and construction
6	100% grades IV-VI	May behave as soil although relict fabric may still be significant

APPROACH 4 (moderately weak or weaker in fresh state) CLASSIFICATION INCORPORATES MATERIAL AND MASS FEATURES		
class	classifier	description
A	unweathered	Original strength, color, fracture spacing
B	partially weathered	Slightly reduced strength, slightly closer fracture spacing, weathering penetrating in from fractures, brown oxidation
C	distinctly weathered	Further weakened, much closer fracture spacing, gray reduction
D	de-structured	Greatly weakened, mottled, litho-relicts in matrix becoming weakened and disordered, bedding disturbed
E	residual or reworked	Matrix with occasional altered random or apparent litho-relicts, bedding destroyed. Classed as reworked when foreign inclusions are present as a result of transportation

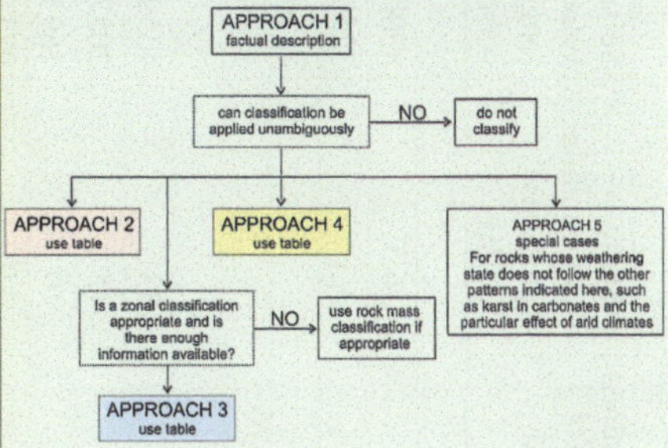

Fig. 11.10 Description state of weathering (BS 5930:1999 1999)

11.7 Susceptibility to Weathering

To guarantee the safe and sound design for the whole lifetime of an engineering structure it is important to know what the geotechnical properties of the groundmass are going to be at the end of the lifetime, that is, *What is the susceptibility to weathering of the groundmass?* Comparing the condition of the groundmass in similar exposures but with different excavation dates is the most common method to establish susceptibility to weathering. Preferably the exposures should be on short distance from each other and from the construction site. The weathering processes should be the same and be the same as those going to act around and influence the future engineering structure; hence, geomorphological and environmental setting should be the same.

Published quantitative data on future changes in properties due to weathering and the rate of weathering for engineering purposes are only sparsely known. Laboratory studies are not very reliable for forecasting in situ weathering rates as these depend on the local circumstances and environment, and the tests are limited

Table 11.3 Examples of differences in engineering properties due to weathering

Grade[a]	In-situ unit weight (kN/m^3)	Porosity (%)	Permeability (cm/s)	Unconfined compressive strength (MPa)	Unconfined tensile strength (MPa)	Static deformation modulus (GPa)	Seismic velocity longitudinal wave (m/s)	Shear wave (m/s)	Schmidt hammer number	Rock mass friction (°)	Rock mass cohesion (kPa)
Dolorite[b]											
0–1	28.0	0.4		170	45	16.5	4500		64		
2	27.6	0.5		122	27	3.3	3250		53		
3	27.0	1		71	13		2150		45		
4	26.2	3.2		41	7		1600		25		
Granodiorite[c]											
0	26.3	1.5		138		33	4359	2567		47	17
1	25.9	4.6		79		15	2057	1693		46	16
2	25.4	1.9		41		10	1693	1111		38	14
3	24.0	5.7		32		4.9	973			17	8
4	19.8	24		0.1		0.008				6	3
5	14.7	44									
Basalt[d]											
0	26.1[e]	2	1×10^{-9}	110	9	58					
2	25.7[e]	4	1×10^{-8}	75	7	48					
3	23.0[e]	10	5×10^{-8}	30	3	23					
4	21.6[e]	36	1×10^{-5}	8		10					
5	16.5[e]	45	1×10^{-4}								
Sandstone[f]											
0	25.0[g]			101							
1	26.3[g]			58							
2	23.6[g]			21							
3	23.8[g]			8							

Grade[a]	In-situ unit weight (kN/m^3)	Porosity (%)	Permeability (cm/s)	Unconfined compressive strength (MPa)	Unconfined tensile strength (MPa)	Static deformation modulus (GPa)	Seismic velocity longitudinal wave (m/s)	Shear wave (m/s)	Schmidt hammer number	Rock mass friction (°)	Rock mass cohesion (kPa)
Gneiss and schist[h,i]											
4	16.8	53	3×10^{-5}	0.36						32	5
5	16.3	54	1×10^{-5}	0.29						21	22

[a]Grade follows the classification in Table 11.2 for rock masses
[b]Dolerite at Stirling Castle, UK (Price 2000)
[c]Granodiorite data from Krank (1980), except rock mass friction and cohesion. Granodiorite rock mass friction and cohesion from slope back analysis in Granodiorite in the Falset area, Spain, from Hack (1998)
[d]Tuğrul (2004)
[e]Dry density
[f]Tating et al. (2013)
[g]Values may be influenced by precipitation of iron in particular weathering grades
[h]Values for 2 years after excavation
[i]Tran et al. (2019)

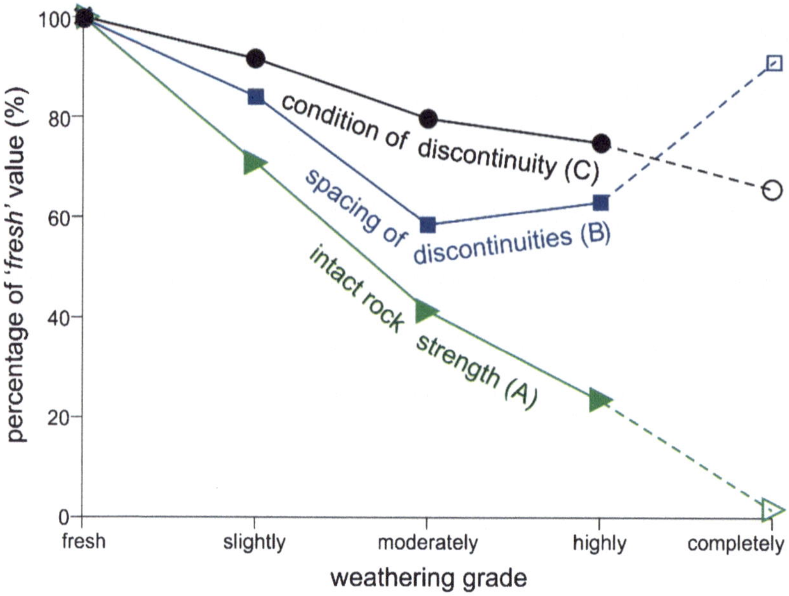

Fig. 11.11 Influence of weathering on intact rock and rock mass properties. Notes: Data averaged after normalization with values for fresh equal 100%. Standard deviation around 15–23% (percent point) for slightly through highly weathered; data for completely weathered are few and average not reliable. Weathering grade refers to rock mass weathering following Table 11.2. *Spacing of discontinuities* based on rock block size and form following Taylor (1980) in Hack et al. (2003) or on discontinuity spacing. *Condition of discontinuity* (determining the shear strength) following sliding criterion (Hack et al. 2003) or friction and cohesion properties for discontinuities. Data: A: 1, 5, 6, 7, and 10; B: 2, 3, 4, and 5; C: 5, 8, and 9. (1) Begonha and Sequeira Braga (2002), (2) Ehlen (1999), (3) Ehlen (2002), (4) GCO (1990), (5) Hack and Price (1997), (6) Marques et al. (2010), (7) Pickles (2005), (8) Reißmüller (1997), (9) Snee (2008), (10) Tuğrul (2004)

Table 11.4 Adjustment factors (WE) for different geotechnical properties of a rock mass (Hack and Price 1997; Hack et al. 2003)

Grade[a]	Description	Intact rock strength	Overall spacing of discontinuities	Overall condition of discontinuities	Rock mass friction	Rock mass cohesion
0	Fresh	1.00	1.00	1.00	1.00	1.00
1	Slightly	0.88	0.93	1.00	0.95	0.96
2	Moderately	0.70	0.89	0.99	0.90	0.91
3	Highly	0.35	0.63	0.89	0.59	0.64
4[b]	Completely	0.02	0.55	0.80	0.31	0.38

[a]Grade follows the classification in Table 11.2 for rock masses
[b]"Completely weathered" is assessed in granodiorite only

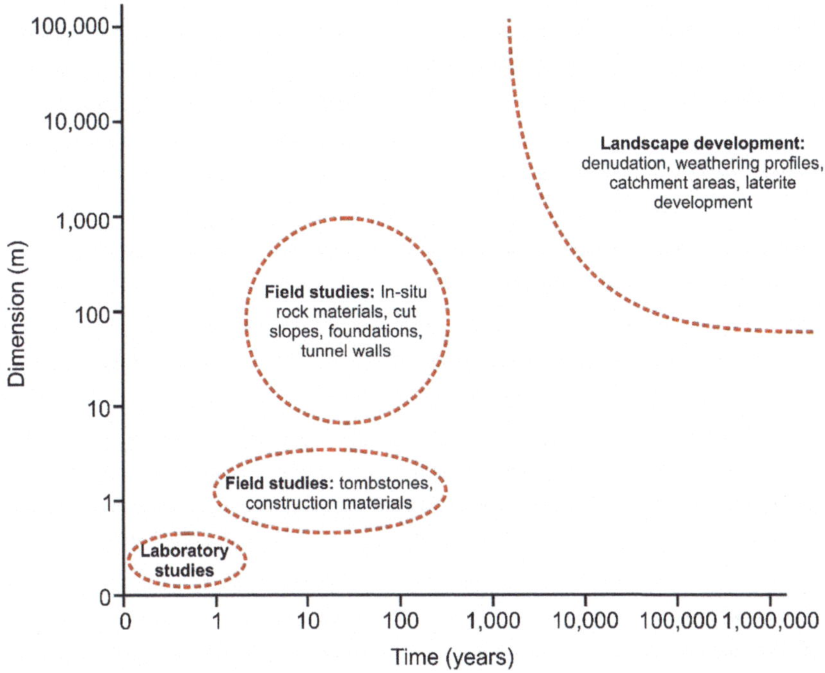

Fig. 11.12 Research to weathering and erosion as function of space and time. (Modified from Huisman 2006)

in groundmass volume and time (Sect. 11.9.1). How groundmasses deteriorate over long (geological) periods over large areas (landscape development) is reasonably well investigated by geological, geomorphological, denudation, and soil forming studies (Sect. 11.7.1). The influence of weathering on intact rock used as building material or gravestones has been extensively studied in situ and rates for loss of material due to weathering have been established by many researchers (*tomb- or gravestone geology*, Sect. 11.7.2). However, few studies are published for 50–100-year time spans over areas from tens up to a couple of hundreds of meters, the typical engineering lifetime and size (Fig. 11.12). This is understandable, as in situ testing is virtually impossible while local variations and inhomogeneity make geological methodologies complicated and unreliable. Some rock mass classification systems have factors that quantitatively assess the future weathering (Sect. 11.7.3) and some recent studies to weathering rates are presented in Sect. 11.7.4.

11.7.1 Loss of Material, Denudation Studies Over Long (Geological) Time Spans

Loss of material due to weathering resulting in denudation over relatively large areas and long (geological) time spans is extensively studied as it may reveal data over past climates and CO_2 presence in the atmosphere, hydrosphere, and cryosphere

Table 11.5 Examples of denudation rates

Lithology	Area	Denudation rate[a] (mm/year)	Present-day climate[b]	
Granitic[c]	Chile	0.02–0.07	Bsk/Csa	Semiarid, Mediterranean
Granitic[d]	Boulder, USA	0.022	Dfc/Dfd	Boreal, mountain
Granulites, migmatites, gneisses, schists, phyllites, granitoids, and alluvial beds[e]	Ambato range, Sierras Pampeanas, Argentina	0.038–0.12	Bsk/Bsh/Cwb	Arid/warm temperate
Mica schist, phyllite[f]	Sierra de las Estancias, Betic Cordillera, Spain	0.034	Bsk	Semiarid, mountain
Mica schist, phyllite[f]	Sierra de los Filabres, Betic Cordillera, Spain	0.054	Csa/Csb/Bsk	Humid to semiarid, mountain
Mica schist[f]	Sierra Cabrera, Betic Cordillera, Spain	0.164	Bsk/Bwh	Semiarid, mountain
Granitic, carbonate and quartz-bearing metasedimentary rocks[g]	Ganges, Northern Himalayas	0.5	Cwa/Cwb	Humid; warm temperate, mountain
Quaternary sediments[g]	Ganges, main stem	0.17	Cwa	Humid, warm temperate
Quaternary sediments[g]	Ganges, southern tributaries	0.03	Cwa/Aw	Humid, warm temperate/tropical
Basalt[h]	Paraná, Brazil	0.006	Csb	Humid, warm temperate

[a]Rates based on ^{10}Be cosmogenic radionuclide (CRN) analysis, if reported
[b]Climate according Kottek and Rubel (2017)
[c]Vázquez et al. (2016)
[d]Dethier and Lazarus (2006)
[e]Nóbile et al. (2017)
[f]Schoonejans et al. (2016) and Vanacker et al. (2014)
[g]Rahaman et al. (2017)
[h]Da Conceição et al. (2015)

(Ahnert 1994; Lebedeva et al. 2010). Denudation is mostly established by measuring the differences in quantities of chemical elements in rivers and streams flowing into and out of an area. If more leaves than flows into the area, the differences are a measure for the loss of material. Denudation over large areas and large time spans is dependent on active tectonic uplift and mountain forming, vegetation, and influences by man, such as land use and (de)forestation. It should be realized that the climate and environment may have undergone major changes during the periods over which the denudation rates are established. Table 11.5 lists various denudation rates for different lithologies under different present-day climates.

11.7.2 Loss of Material, Tombstone Geology Studies Over Short Time Spans

Loss of material of various intact rock types over short time spans and tested on relatively small samples is done mainly on building and construction stones (Doehne and Price 2010; Fookes et al. 1988; Morgan 2016; Selby 1993; Winkler 1986). The amount of intact rock material lost under influence of weathering in a temperate climate on a forested slope in Japan came to 1.3%/year for tuff material, 0.1 for limestone, 0.025 for crystalline schist, and 0.01%/year for granite. The samples were exposed directly at the Earth surface (Matsukura and Hirose 2000). Trudgill et al. (2001) measured the loss of material of exterior building stones of St. Paul's Cathedral in London which consist mainly out of limestone, over a period of 20 years and measured a reduction of 0.01–0.07 mm/year. The results are influenced by air pollution (i.e., SO_2) in London that decreased over the measuring period. Tombstone geology has also been used to establish changes in environment, climate, and air pollution (Meierding 1993). Feddema and Meierding (1987) report values of 0.001–0.067 mm/year for carbonate building stones in areas with varying quantities of air pollution, and Meierding (1993) established weathering rates of over 0.03 mm/year for carbonate rocks in heavily air-polluted areas in the USA. The striking similarity in order of magnitude between rates for loss of material of small building stones over short time spans and loss of material over large areas over long time spans is remarkable (Sect. 11.7.1).

11.7.3 Laubscher's Geotechnical Susceptibility to Weathering Factors

Weathering rates for rock masses in underground excavations and its influence on rock mass geotechnical properties and classification ratings are shown in Table 11.6 (Laubscher and Jakubec 2001). The percentages in Table 11.6 are multiplication factors which are multiplied with the rock mass ratings calculated following Laubscher and Jakubec (2001); for example, the rock mass rating is reduced to 54% of the original rating without susceptibility to weathering if a rock mass is expected to weather from fresh to completely weathered within a half year, whereas it is reduced to 62% if that happens in more than 4 years. It should be noted that conditions in underground excavations are considerably different and in general with less variation than the conditions influencing weathering at surface.

11.7.4 Geotechnical Rate of Weathering

The influence of weathering on geotechnical properties over time spans from 50 to 100 years is thought to be expressed by a logarithmic decrease of properties with time (Colman 1981; Hachinohe et al. 2000; Huisman 2006; Ruxton 1968; Selby 1980; Tating et al. 2013; Utili and Crosta 2011), for example (Huisman 2006):

Table 11.6 Adjustment values for susceptibility to weathering for classification of stability of underground excavations in mining (modified from Laubscher and Jakubec 2001)

Expected grade of future weathering[a]	Description	Percentage adjustment for a rock mass weathered from fresh to indicated grade; after ½ year	1 year	2 year	3 year	≥4 year
		%				
0	Fresh	100	100	100	100	100
1	Slightly	88	90	92	94	96
2	Moderately	82	84	86	88	90
3	Highly	70	72	74	76	78
4	Completely	54	56	58	60	62
5	Residual soil	30	32	34	36	38

Notes: The adjustment is applied to the rating for the stability of the underground excavation of Laubscher's rock mass classification to predict the future stability
[a]The grades of rock mass weathering follow Table 11.2 for rock masses

$$\text{Property}_t = \text{Property}_{\text{initial}} - R_{\text{property}} \log_{10}\left(1+t\right) \tag{11.1}$$

in which property$_t$ is the value of a particular geotechnical property at time t, property$_{\text{initial}}$ is the value of the property initially at time of exposure, that is, at $t = 0$, R_{property} is the *weathering rate* which is property, material, and environment dependent, and t is the time in years (Fig. 11.13). This relation describes the change in time of a property over the full weathering range from fresh groundmass to residual soil. Huisman (2006) incorporated the WE (weathering) factors of Table 11.4 in Eq. (11.1), and established the weathering rates (R_{WE}) for different groundmasses in the Mediterranean climate of Spain (Table 11.7). Table 11.7 lists also the dynamic weathering rates and total decrease of property values for various groundmasses after 30 years of exposure in various climates. Table 11.7 clearly shows the large influence of different climates on the rate of weathering of geotechnical properties and the influence of differences in bedding spacing and presence of soluble materials.

11.8 "Hard" Layers and Crusts

Weathering normally weakens the groundmass, but may also cause forming of various *hard* layers that are harder and stronger than the surrounding groundmass (Bland and Rolls 1998; Winkler 2014). Examples are *duricrust*, *hardpan*, *calcrete*, *laterite*, and *caprock*[1] in dredging (Chesworth 2008). In geotechnical works, these

[1]Note that caprock has multiple meanings: a *caprock* in geomorphology is a mostly stronger layer that is more resistant to weathering and erosion overlying weaker and less resistant layers. *Caprock* may be used also for impermeable layers overlying a salt body protecting the salt from solution by rain and groundwater, and in the petroleum industry *caprock* in used for an impermeable layer capping an oil or gas reservoir.

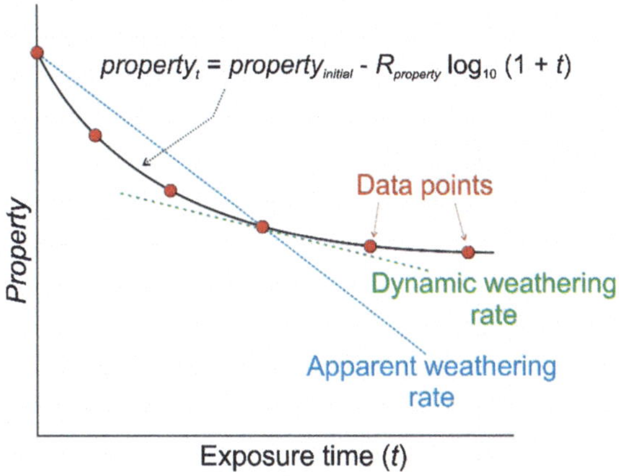

Fig. 11.13 Property vs. exposure time. (Modified from Huisman 2006)

are often simply identified by *hard layer*, *cemented horizon*, or *crust*. Hard layer development is mostly due to precipitation of in groundwater-dissolved minerals, such as salts, carbonates, and iron compounds. Chemical reactions between minerals in different groundwater flows, chemical reactions between groundwater-dissolved minerals and ground materials, and evaporation of water cause precipitation. Solution of minerals due to weathering at another location may be the source of the in groundwater-dissolved minerals, but these may also stem from other sources, such as vegetation or influence of humans (e.g., fertilizer). Hard layer development may be at surface due to, for example, evaporating groundwater but can also be below surface or at or below the sea floor. Existence and location of hard layers are often difficult to predict. The processes involved depend on many local and regional variables that may change over short distance both horizontally and vertically leading to often highly irregular hard layers (Khalifa et al. 2009; Vervoort and De Wit 1997). Hard layers may also occur at locations where these would not be formed at present but are a relic from the (geological) past. Hard layers are common in arid and tropical environments but occur also in Mediterranean and temperate climates.

Most common hard layers are the following. A *duricrust* is a strong and hard layer at or near the surface formed by the accumulation of precipitated minerals by evaporating capillary or seepage water. The thickness is normally a couple of centimeters. A *hardpan* is similar but may also refer to an impermeable, strong layer in a soil mass just below the surface. The soil particles are bond by silica or iron compounds (iron oxides) with calcium carbonate. Hardpan is used also for hard layers formed due to compaction by heavy traffic or cementation by chemicals originating from polluted groundwater. Hard surfaces for temporary roads in arid climates are made in the same way by compaction of the surface layer with simultaneously spraying with salt water. The water evaporates and the precipitated salt cements the soil particles. The intact rock strength of duricrusts and hardpans ranges from material that is easy to crumble by hand to strong rock. *Laterites* are usually residual soils in a tropical environment cemented by iron compounds (oxides and hydroxides) (Blight 1989) and the strength may also range from easy to crumble to strong

Table 11.7 Weathering rate examples

Lithology[a]	Property[b]	Initial value	R_{WE}[c] (1/log [year])	Dynamic weathering rate (at 30 years after exposure as percentage of initial value)[d] (% point/year)	Decrease in 30 years after exposure (as percentage from initial value)[d] (%)	Indicative timespan (year)	Climate[e]	
Clay-containing limestone (bedding spacing <0.1 m)[f]	WE	1	0.052	−0.073	7.8	40	Csa	Mediterranean
Clay-containing limestone (bedding spacing >0.1 m)[f]	WE	1	0.042	−0.059	6.3	40		
Limestone (medium-thick bedded)[f]	WE	1	0.067	−0.094	10	40		
Calcareous shale (clay-/mudstone)[f]	WE	1	0.169	−0.24	25	40		
Gypsum-cemented siltstone[f]	WE	1	0.325	−0.46	48	40		
Gypsum (beds consisting of gypsum)[f]	WE	1	0.133	−0.19	20	40		
Sandstone[g]	IRS	105 MPa	34	−0.45	48	30	Am	Tropical
Gneiss/schist[h,i]	Index prop[j]			−1.15	35	35	Aw	
Gneiss/schist[h,i]	Strength[k]			−1.14	34	35	Aw	
Tuffaceous sandstone[l]	R_s	863 N/mm		−0.34	60	3000	csb	Temperate
Mudstone[l]	R_s	235 N/mm		−0.17	21	3000		

[a]Data (f, g, i) from cut slopes, (l) from natural terraces
[b]WE: weathering factor from SSPC (Hack et al. 2003); IRS: Intact Rock Strength, R_{S0}: Penetration strength based on needle penetration hardness

[c]R_{WE}: (Weathering rate) follows Eq. 11.1
[d]Dynamic weathering rate and total decrease follow logarithmic relation for data from (f, g), linear relation for (i), and exponential relation for (l)
[e]Climate according Kottek and Rubel (2017)
[f]Huisman et al. (2006)
[g]Tating et al. (2013)
[h]Completely weathered material and residual soil only
[i]Tran et al. (2019)
[j]Various index properties of soil, such as Unit Weight, porosity, and saturated conductivity
[k]Various shear and unconfined strength properties
[l]Hachinohe et al. (2000)

rock (see also Sect. 11.1.2). In dredging the term *caprock* is used for stronger cemented layers with intact rock strength values of more than 10–20 MPa in a weaker groundmass. Caprocks are generally layers of conglomerate, breccia, or sand cemented by calcareous, siliceous, or ferruginous material, normally between 10 and 50 cm thick, and may occur as multiple closely spaced layers at or near the sea floor (Verhoef 1997). *Calcretes* are comparable to caprocks formed with calcareous cement but are formed on-shore near surface (Alonso-Zarza and Wright 2010). Calcretes may be meters thick, irregular, laterally extend over hundreds of meters, and be very strong with intact rock strength values exceeding 100 MPa.

11.9 Tests to Establish the State of Weathering and Susceptibility to Weathering

The state of weathering as is can be estimated in laboratory and in situ field tests by comparing test values for the weathered ground material or groundmass to the same but unweathered material or mass. This may give an indication how weathering has influenced the material or mass. Susceptibility to weathering can be established to a certain extent in laboratory tests, however, the long time span in reality has to be simulated within a time span suitable for laboratory testing. For example, freeze–thaw tests with cycling between freezing and thawing conditions within days to simulate seasons. Chemical and physical processes in the ground such as diffusion, may be accelerated in time by using centrifuges. Sample size is limited and effectively restricted to testing of disturbed intact ground material only. Whether the samples and simulated conditions in laboratory tests are representative of reality is often questionable. There are no tests available for determining the susceptibility to weathering of groundmasses in situ over periods up to 50–100 years. This is obvious considering the necessary dimensions of the volume to be tested and the time span over which tests have to be executed.

11.9.1 Laboratory Testing

Laboratory ultrasonic velocity measurements may give an idea about the state of weathering as is of a piece of intact ground by referencing to the measured ultrasonic velocities in a piece of the same but unweathered ground (ASTM D2845-08 2008; Chawre 2018). Higher velocities indicate less weathering and vice versa. It should be noted that there is no direct relation between the state of weathering and ultrasonic velocity. Also, the ultrasonic velocity does not give information on the susceptibility to weathering.

Climate chambers are used to simulate the influence on ground of a changing environment, for example, day–night temperatures, changing seasons, freezing and thawing, and regular wetting and drying due to rainfall (ASTM D5312/D5312M-12 2013; ASTM D5313/D5313M-12 2013; Barros De Oliveira Frascá and Yamamoto 2006). These may be combined with centrifuges (Tristancho et al. 2012). Humidity cells are used to simulate weathering of solids for among other weathering of and chemical changes in mine waste material (ASTM D5744-18 2018). The influence of salt, for example, from sea spray, can be tested by regularly spraying samples with water with dissolved salts (ASTM D5240/D5240M-12e1 2013). Crystallization tests determine the resistance of intact ground to crystallization processes of, for example, salt in pores in rock (BS EN 12370:1999 1999).

Slaking tests are often used to indicate the susceptibility to weathering of intact ground material, for example, the slake durability and the Los Angeles abrasion tests; the later mostly used for determining the durability of toughness and abrasion resistance of aggregate for road pavement (ASTM C131/C131M-14 2014; ASTM D4644-16 2016; Dick and Shakoor 1995; Franklin and Chandra 1972; Hack 1998; Hack and Huisman 2002; Nicholson 2000; Nickmann et al. 2006; Ulusay and Hudson 2007). The tests submerge material in water in a rotating drum and the volume of material that falls apart in a particular time span is a measure for the durability.

Dropping a block of rock from a certain height to investigate how the intact rock fractures under impact (*Drop Test*, CIRIA 2007), and cyclic stressing–destressing tests (Lagasse et al. 2006) may be useful for establishing intact rock integrity. These tests may indicate indirectly the ease with which intact rock fractures due to weathering or how easy incipient discontinuities change into mechanical discontinuities.

11.9.2 In Situ Testing

Indirectly an indication on the state of weathering of a groundmass may be derived from seismic wave characteristics such as wave velocity or wave amplitude (Table 11.3). Measured seismic wave velocities are higher if the wave travels through fresh ground and slower through weathered masses, partially because the measured waves may have travelled around weathered parts of the mass and thus have a longer ray path. The wave amplitude is a function of among others, the absorption of energy in the ground which is higher in weathered than fresh unweathered ground. The wave velocity and amplitude should be correlated to states of weathering established on borehole cores. Seismic waves do not directly give information on the susceptibility to weathering but may give depth of weathering and the thickness of weathered layers. The depth of weathering may in turn give an idea on the rate of weathering as larger depth of weathering in the subsurface often also implies a higher rate. Other geophysical methods that indirectly give an idea about the state of and susceptibility to weathering are resistivity and electromagnetic measurements as these react to the presence of clay that in many grounds will be more present in weathered than unweathered parts of the ground.

11.10 Conclusion

Weathering is a process that governs many engineering applications on and in the Earth. It often transforms originally sound ground into soft ground. Quantities of weathered material do not need to be large as small volumes of ground weathered in a brief time span drastically can change geotechnical properties. Weathering is the reason for many constructions and other engineering applications in which ground is used, to become a disaster. Tests to determine the susceptibility to weathering representative for realistic volumes of groundmass do not exist and published data on time–weathering–degradation relations for groundmasses are few. Therefore, forecasting the influence of weathering on geotechnical properties has to be done by the design engineer based on experience and a priori knowledge without much or no hard data at all. Many engineers do not realize the importance of weathering or are hesitant in taking decisions based on own expertise and a priori knowledge alone. They may assess the state of weathering as is, may mention the existence of future or susceptibility to weathering in general terms in the reporting, but do nothing to implement the consequences in design and construction. Fortunately, the safety factor used in civil engineering instigates excess in design to accommodate for uncertainties in the construction. One of these is ground and future behavior of ground, and therefore not all constructions fail even if susceptibility to weathering is not but should have been taken into account.

References

Ahnert F (1994) Equilibrium, scale and inheritance in geomorphology. Geomorphology 11(2):125–140

Allison RJ, Bristow GE (1999) The effects of fire on rock weathering: some further considerations of laboratory experimental simulation. Earth Surf Process Landf 24(8):707–713

Alonso-Zarza AM, Wright VP (2010) Calcretes. In: Alonso-Zarza AM, Tanner LH (eds) Carbonates in continental settings; facies, environments, and processes, vol 61, 1st edn. Elsevier, Amsterdam, pp 225–267

Anderson RS, Anderson SP (2010) Geomorphology: the mechanics and chemistry of landscapes. Cambridge University Press, Cambridge

Anon (1995) The description and classificatioin of weathered rocks for engineering purposes. Q J Eng Geol Hydrogeol 28(3):207–242

ASTM C131/C131M-14 (2014) Standard test method for resistance to degradation of small-size coarse aggregate by abrasion and impact in the Los Angeles machine. ASTM International, West Conshohocken, PA

ASTM D2845-08 (2008) Standard test method for laboratory determination of pulse velocities and ultrasonic elastic constants of rock (Withdrawn 2017). ASTM International, West Conshohocken, PA

ASTM D4644-16 (2016) Standard test method for slake durability of shales and other similar weak rocks. ASTM International, West Conshohocken, PA

ASTM D5240/D5240M-12e1 (2013) Standard test method for evaluation of durability of rock for erosion control using sodium sulfate or magnesium sulfate. ASTM International, West Conshohocken, PA

ASTM D5312/D5312M-12 (2013) Standard test method for evaluation of durability of rock for erosion control under freezing and thawing conditions. ASTM International, West Conshohocken, PA

ASTM D5313/D5313M-12 (2013) Standard test method for evaluation of durability of rock for erosion control under wetting and drying conditions. ASTM International, West Conshohocken, PA

ASTM D5744-18 (2018) Standard test method for laboratory weathering of solid materials using a humidity cell. ASTM International, West Conshohocken, PA

Barros De Oliveira Frascá MH, Yamamoto JK (2006) Ageing tests for dimension stone - experimental studies of granitic rocks from Brazil; Paper no. 224. In: Culshaw MG, Reeves HJ, Jefferson I, Spink TW (eds) 10th International Congress of the International Association for Engineering Geology and the Environment IAEG; Engineering geology for tomorrow's cities, Nottingham, UK, 6–10 Sept 2006. Geological Society of London, London, p 9

Begonha A, Sequeira Braga MA (2002) Weathering of the Oporto granite: geotechnical and physical properties. Catena 49(1–2):57–76

BGS Arsenic (2001) Arsenic contamination of groundwater in Bangladesh: Final report (BGS Technical Report WC/00/19), vol 2. BGS, Keyworth, UK

Bieniawski ZT (1989) Engineering rock mass classifications: a complete manual for engineers and geologists in mining, civil, and petroleum engineering. Wiley, New York

Bland W, Rolls D (1998) Weathering: an introduction to the scientific principles. Arnold Publishers, London

Blight GE (1989) Design assessment of saprolites and laterites; Invited lecture; General report/ Discussion session 6. In: Publications Committee of the XII ICSMFE (ed) 12th International Conference on soil mechanics and foundation engineering, Rio de Janiero, 13–18 Aug 1989. Balkema, Rotterdam, pp 2477–2484

Bracke G, Salah S, Gauthier-Lafaye F (2001) Weathering process at the natural fission reactor of Bangombé. Environ Geol 40(4):403–408

Brattli B, Broch E (1995) Stability problems in water tunnels caused by expandable minerals. Swelling pressure measurements and mineralogical analysis. Eng Geol 39(3–4):151–169

BS 5930:1999 (1999) Code of practice for site investigations. British Standards Institution, London

BS EN 12370:1999 (1999) Natural stone test methods - determination of resistance to salt crystallisation. British Standards Institution, London

Cabria XA (2015) Effects of weathering in the rock and rock mass properties and the influence of salts in the coastal roadcuts in Saint Vincent and Dominica. University of Twente, Enschede, Netherlands

Caves Han-sur-Lesse (2018) Caves of Han-sur-Lesse. http://www.grotte-de-han.be/en/the-cave-of-han. Accessed 24 Mar 2018

Chawre B (2018) Correlations between ultrasonic pulse wave velocities and rock properties of quartz-mica schist. J Rock Mech Geotech Eng 10(3):594–602

Chesworth WE (2008) Encyclopedia of soil science, Encyclopedia of earth sciences series, 2nd edn. Springer, Dordrecht, The Netherlands

CIRIA (2007) The rock manual. The use of rock in hydraulic engineering, 2nd edn. CIRIA; CUR; CETMEF, C683, London

Colman SM (1981) Rock-weathering rates as functions of time. Quat Res 15(3):250–264

Cummings RA, Kendorski FS, Bieniawski ZT (1984) Caving rock mass classification and support estimation. Engineers International Inc., Chicago

Da Conceição FT, Dos Santos CM, De Souza Sardinha D, Navarro GRB, Godoy LH (2015) Chemical weathering rate, denudation rate, and atmospheric and soil CO_2 consumption of Paraná flood basalts in São Paulo State, Brazil. Geomorphology 233:41–51
Dearman WR (1995) Description and classification of weathered rocks for engineering purposes: the background to the BS5930:1981 proposals. Q J Eng Geol Hydrogeol 28(3):267–276
Dethier DP, Lazarus ED (2006) Geomorphic inferences from regolith thickness, chemical denudation and CRN erosion rates near the glacial limit, Boulder Creek catchment and vicinity, Colorado. Geomorphology 75(3):384–399
Dick JC, Shakoor A (1995) Characterizing durability of mudrocks for slope stability purposes. In: Haneberg WC, Anderson SA (eds) Clay and shale slope instability; reviews in engineering geology, vol 10. The Geological Society of America, Boulder, CO, pp 121–130
Didier C, Van der Merwe N, Betournay M, Mainz M, Kotyrba A, Aydan Ö, Josien J-P, Song W-K (2008) Mine closure and post-mining management; International state-of-the-art (ISRM) ISfRM
Doehne E, Price CA (2010) Stone conservation: an overview of current research, 2nd edn. Getty Conservation Institute, Los Angeles, CA
Dosseto A, Buss HL, Suresh PO (2012) Rapid regolith formation over volcanic bedrock and implications for landscape evolution. Earth Planet Sci Lett 337–338:47–55
Ehlen J (1999) Fracture characteristics in weathered granites. Geomorphology 31(1–4):29–45
Ehlen J (2002) Some effects of weathering on joints in granitic rocks. Catena 49(1–2):91–109
Feddema JJ, Meierding TC (1987) Marble weathering and air pollution in Philadelphia. Atmos Environ 21(1):143–157
Fookes PG (1997) Tropical residual soils; A Geological Society Engineering Group Working Party revised report. Geological Society; Professional handbooks. The Geological Society, London
Fookes PG, Gourley CS, Ohikere C (1988) Rock weathering in engineering time. Q J Eng Geol Hydrogeol 21:33–57
Franklin JA, Chandra R (1972) The slake-durability test. Int J Rock Mech Min Sci Geomech Abstr 9(3):325–328
Gambolati G, Putti M, Teatini P, Camporese M, Ferraris S, Stori GG, Nicoletti V, Silvestri S, Rizzetto F, Tosi L (2005) Peat land oxidation enhances subsidence in the Venice watershed. EOS Trans Am Geophys Union 86(23):217–220
GCO (1990) Foundation properties of marble and other rocks in the Yuen Long-Tuen Mun Area, vol 2/90, Hong Kong
Ghosh NC, Singh RD (2009) Groundwater arsenic contamination in India: vulnerability and scope for remedy. In: 5th Asian Regional Conference of INCID, Special session on groundwater, New Delhi, India, 9–11 Dec 2009. Indian National Committee on Irrigation and Drainage (INCID); International Commission on Irrigation and Drainage (ICID), New Delhi, p 24
Hachinohe S, Hiraki N, Suzuki T (2000) Rates of weathering and temporal changes in strength of bedrock of marine terraces in Boso Peninsula, Japan. Eng Geol 55(1–2):29–43
Hack HRGK (1996) Slope stability probability classification (SSPC). ITC/Technical University, Delft
Hack HRGK (1998) Slope stability probability classification; SSPC; 2nd version. University of Technology Delft; International Institute for Aerospace Survey and Earth Sciences; ITC, Delft, Enschede
Hack HRGK, Huisman M (2002) Estimating the intact rock strength of a rock mass by simple means. In: Van Rooy JL, Jermy CA (eds) 9th Congress of the International Association for Engineering Geology and the Environment (IAEG); Engineering geology for developing countries, Durban, South Africa, 16–20 Sept 2002. IAEG & South African Institute for Engineering and Environmental Geologists (SAIEG), Houghton, South Africa, pp 1971–1977
Hack HRGK, Price DG (1997) Quantification of weathering. In: Marinos PG, Koukis GC, Tsiambaos GC, Stournaras GC (eds) Proceedings engineering geology and the environment, Athens, 23–27 Jun 1997. Balkema, Taylor & Francis Group, Rotterdam, pp 145–150

Hack HRGK, Price DG, Rengers N (2003) A new approach to rock slope stability - a probability classification (SSPC). Bull Eng Geol Environ 62(2):167–184

Harris CS, Hart MB, Varley PM, Warren CD (1996) Engineering geology of the channel tunnel. Thomas Telford Ltd, London

Hencher SR (2015) Practical rock mechanics. CRC, Taylor & Francis Group, Boca Raton, FL

Hoek E, Brown ET (2018) The Hoek-Brown failure criterion and GSI – 2018 edition. J Rock Mech Geotech Eng 11:445–463

Hughes M, Bonapace P, Rigbey S, Charalambu H (2007) An innovative approach to tunneling in the swelling Queenston Formation of Southern Ontario. In: Traylor MT, Townsend JW (eds) Rapid excavation and tunneling conference; RETC 2007, Toronto, Canada, 10–13 Jun 2007. Society of Mining, Metallurgy and Exploration (SME), Littleton, CO, pp 901–912

Huisman M (2006) Assessment of rock mass decay in artificial slopes. University Delft; ITC, Delft; Enschede

Huisman M, Hack HRGK, Nieuwenhuis JD (2006) Predicting rock mass decay in engineering lifetimes: the influence of slope aspect and climate. Environ Eng Geosci 12(1):39–51

Huisman M, Nieuwenhuis JD, Hack HRGK (2011) Numerical modelling of combined erosion and weathering of slopes in weak rock. Earth Surf Process Landf 36(13):1705–1714

ISO 14689-1:2017 (2017) Geotechnical investigation and testing; Identification, description and classification of rock. International Organization for Standardization, Geneva, Switzerland

James P, Chester D, Duncan A (2000) Volcanic soils: their nature and significance for archaeology. In: McGuire WJ, Griffiths DR, Hancock PL, Stewart IS (eds) The archaeology of geological catastrophes, vol 172. Geological Society of London, London, pp 317–338

Katongo C (2005) Ground conditions and support systems at 1 shaft, Konkola mine, Chililabombwe, Zambia. In: The Third Southern African Conference on Base Metals: Southern Africa's response to changing global base metals market dynamics, Kitwe, Zambia, 26–29 Jun 2005. The South African Institute of Mining and Metallurgy, Johannesburg, pp 253–280

Khalifa MA, Kumon F, Yoshida K (2009) Calcareous duricrust, Al Qasim Province, Saudi Arabia: occurrence and origin. Quat Int 209(1–2):163–174

Knight J, Grab SW (2014) Lightning as a geomorphic agent on mountain summits: evidence from southern Africa. Geomorphology 204:61–70

Kottek M, Rubel F (2017) World maps of Köppen-Geiger climate classification; version March 2017. Veterinärmedizinische Universität Wien; Climate Change & Infectious Diseases. http://koeppen-geiger.vu-wien.ac.at/present.htm. Accessed 10 Jan 2019

Krajick K (2005) Fire in the hole; Raging in mines from Pennsylvania to China, coal fires threaten towns, poison air and water, and add to global warming. Smithsonian Magazine May (25 April 2019)

Krank KD (1980) The effects of weathering on the engineering properties of Sierra Nevada granodiorites. University of Nevada, Reeno

Kuenzer C, Stracher GB (2012) Geomorphology of coal seam fires. Geomorphology 138(1):209–222

Lagasse PF, Clopper PE, Zevenbergen LW, Ruff JF (2006) Riprap design criteria, recommended specifications, and quality control; Report 568. NCHRP, Washington, DC

Lainé M, Balan E, Allard T, Paineau E, Jeunesse P, Mostafavi M, Robert JL, Le Caër S (2017) Reaction mechanisms in swelling clays under ionizing radiation: influence of the water amount and of the nature of the clay mineral. RSC Adv 7(1):526–534

Lamp JL, Marchant DR, Mackay SL, Head JW (2017) Thermal stress weathering and the spalling of Antarctic rocks. J Geophys Res Earth 122(1):3–24

Laubscher DH, Jakubec J (2001) The MRMR rock mass classification for jointed rock masses. In: Hustrulid WA, Bullock RL (eds) Underground mining methods: engineering fundamentals and international case studies. Society for Mining, Metallurgy & Exploration, Inc. (SME), Littleton, CO, pp 475–481

Lebedeva MI, Fletcher RC, Brantley SL (2010) A mathematical model for steady-state regolith production at constant erosion rate. Earth Surf Process Landf 35(5):508–524

Lim SS, Martin CD (2010) Core disking and its relationship with stress magnitude for Lac du Bonnet granite. Int J Rock Mech Min Sci 47(2):254–264

Lumb P (1983) Engineering properties of fresh and decomposed igneous rocks from Hong Kong. Eng Geol 19(2):81–94

Lumpkin GR, Gao Y, Gieré R, Williams CT, Mariano AN, Geisler T (2014) The role of Th-U minerals in assessing the performance of nuclear waste forms. Mineral Mag 78(5):1071–1095

Machado SL, Vilar OM, Carvalho MF (2008) Constitutive model for long term municipal solid waste mechanical behavior. Comput Geotech 35(5):775–790

Marques EAG, Barroso EV, Menezes Filho AP, Vargas Jr EA (2010) Weathering zones on metamorphic rocks from Rio de Janeiro—physical, mineralogical and geomechanical characterization. Eng Geol 111(1–4):1–18

Massey JB, Irfan TY, Cipullo A (1989) The characterization of granitic saprolitic soils. In: Publications Committee of the XII ICSMFE (ed) Proceedings of the 12th International Conference on soil mechanics and foundation engineering, Rio de Janeiro, Brasil, 13–18 Aug 1989. International Society for Soil Mechanics and Geotechnical Engineering (ISSMGE); A.A. Balkema, Rotterdam, pp 533–542

Matsukura Y, Hirose T (2000) Five year measurements of rock tablet weathering on a forested hillslope in a humid temperate region. Eng Geol 55(1):69–76

Meierding TC (1993) Marble tombstone weathering and air pollution in North America. Ann Assoc Am Geogr 83(4):568–588

Meshik AP (2009) The workings of an ancient nuclear reactor. Scientific American January (19 April 2019)

Miščević P, Vlastelica G (2014) Impact of weathering on slope stability in soft rock mass. J Rock Mech Geotech Eng 6(3):240–250

Morgan N (2016) Gravestone geology. Geol Today 32(4):154–159

Mottershead DN (1989) Rates and patterns of bedrock denudation by coastal salt spray weathering: a seven-year record. Earth Surf Process Landf 14(5):383–398

Nicholson DT (2000) Deterioration of excavated rockslopes: mechanisms, morphology and assessment. University of Leeds, Leeds, UK

Nickmann M, Spaun G, Thuro K (2006) Engineering geological classification of weak rocks; Paper no. 492. In: Culshaw MG, Reeves HJ, Jefferson I, Spink TW (eds) 10th International Congress of the International Association for Engineering Geology and the Environment IAEG; Engineering geology for tomorrow's cities, Nottingham, UK, 6–10 Sept 2006. Geological Society of London, London, p 9

Nóbile JC, Martini MA, Dávila FM (2017) Cosmogenic 10Be denudation rates and geomorphometric analysis in the Ambato range (28°–29°S), Sierras Pampeanas, Argentina. Quat Int 438:80–91

Ohta T (2001) Core disking and "rockburst" in soft tuffaceous rock masses at Iwate tunnel. Q Rep RTRI 42(3):130–135

Olesen O, Dehls JF, Ebbing J, Henriksen H, Kihle O, Lundin E (2007) Aeromagnetic mapping of deep-weathered fracture zones in the Oslo Region – a new tool for improved planning of tunnels. Nor J Geol 87(1/2):253–267

Oyama T, Chigira M (2000) Weathering rate of mudstone and tuff on old unlined tunnel walls. Eng Geol 55(1–2):15–27

Pesendorfer M, Loew S (2004) Hydrogeologic exploration during excavation of the Lötschberg base tunnel (AlpTransit Switzerland). In: Hack HRGK, Azzam R, Charlier R (eds) Engineering geology for infrastructure planning in Europe; a European perspective, Lecture notes in earth sciences, vol 104. Springer, Berlin, pp 347–358

Pickles A (2005) Rock mass classification for pile foundations. In: The characterization of rock masses for engineering purposes, City University, Hong Kong, 25 Jun 2005. The Geological Society, Hong Kong Regional Group, Hong Kong, p 36 slides

Pieters CM, Noble SK (2016) Space weathering on airless bodies. J Geophys Res Planets 121(10):1865–1884
Price DG (1995) A suggested method for the classification of rock mass weathering by a ratings system. Q J Eng Geol Hydrogeol 26(1):69–76
Price DG (2000) Dolerite once exposed at Stirling Castle, Scotland (personal communication), Delft
Price DG, De Freitas MH, Hack HRGK, Higginbottom IE, Knill JL, Maurenbrecher M (2009) Engineering geology; principles and practice. Springer, Berlin
Qi S, Yue ZQ, Wu F, Chang Z (2009) Deep weathering of a group of thick argillaceous limestone rocks near Three Gorges Reservoir, Central China. Int J Rock Mech Min Sci 46(5):929–939
Rahaman W, Wittmann H, von Blanckenburg F (2017) Denudation rates and the degree of chemical weathering in the Ganga River basin from ratios of meteoric cosmogenic 10Be to stable 9Be. Earth Planet Sci Lett 469:156–169
Rainey TP, Rosenbaum MS (1989) The adverse influence of geology and groundwater on the behaviour of London Underground railway tunnels near Old Street Station. Proc Geol Assoc 100(1):123–134
Reißmüller M (1997) Rottachtales zwischen Bodenschneid, Stolzenbert und Siebligrat sowie Geotechnische Eigenschaften verwitterter Kössener Mergel. Diploma thesis, Technical University of Munich, Munich, Germany. p 128
Rocscience (2011) Disturbance factor; rock mass strength analysis using the generalized Hoek-Brown failure criterion. Rocscience Inc. http://www.rocscience.com. Accessed 14 Oct 2013
Rodriguez-Navarro C, Doehne E, Sebastian E (1999) Origins of honeycomb weathering: the role of salts and wind. GSA Bull 111(8):1250–1255
Ruiz-Agudo E, Putnis CV, Rodríguez-Navarro YC (2007) Role of chemical weathering in salt decay of ornamental stone. In: Proc. MACLA 7, XXVII Reunión De La Sociedad Española De Mineralogía, Jaén, Spain, 11–14 Sept 2007, p 29
Ruxton BP (1968) Measures of the degree of chemical weathering of rocks. J Geol 76(5):518–527
Schmitz R, Schroeder C (2009) Urban site investigation in the Belgian karst belt; Paper 801. In: Culshaw MG, Reeves HJ, Jefferson I, Spink TW (eds) 10th International congress International Association of Engineering Geology and The Environment (IAEG2006); Engineering geology for tomorrow's cities, Nottingham, UK, 6–10 Sept 2006. Geological Society of London, London, p 10
Schoonejans J, Vanacker V, Opfergelt S, Ameijeiras-Mariño Y, Christl M (2016) Kinetically limited weathering at low denudation rates in semiarid climatic conditions. J Geophys Res Earth 121(2):336–350
Selby MJ (1980) A rock mass strength classification for geomorphic purposes: with tests from Antarctica and New Zealand. Z Geomorphol 24(1):31–51
Selby MJ (1993) Hillslope materials and processes, 2nd edn. Oxford University Press, Oxford, UK
Shao Y, Raupach MR, Findlater PA (1993) Effect of saltation bombardment on the entrainment of dust by wind. J Geophys Res Atmos 98(D7):12719–12726
Sim CK, Kim SS, Lucey PG, Garrick-Bethell I, Choi Y-J (2017) Asymmetric space weathering on lunar crater walls. Geophys Res Lett 44(22):11273–211281
Singh A (2004) FRHI-a system to evaluate and mitigate rock fall hazard in stable rock excavations. J Inst Eng India 85:62–75
Singh H, Huat BBK (2004) Origin, formation and occurrence of tropical residual soils. In: Huat BBK, Gue SS, Ali FH (eds) Tropical residual soils engineering, 1st edn. CRC Press, London, pp 1–34
Singh VP, Singh P, Haritashya UK (2011) Encyclopedia of snow, ice and glaciers. Springer Netherlands, Dordrecht, The Netherlands
Snee C (2008) Engineering geology and cavern design for New York City. In: Roach MF, Kritzer MR, Ofiara D, Townsend BF (eds) 9th North American Tunnelling, NAT 2008, San Francisco, 8–11 Jun 2008. Society for Mining, Metallurgy & Exploration, Littleton, CO, pp 364–372

Soppe WJ, Prij J (1994) Radiation damage in a rock salt nuclear waste repository. Nucl Technol 107(3):243–253

Stacey TR (1982) Contribution to the mechanism of core discing. J South Afr Inst Min Metall 82(9):269–274

Tating FF, Hack HRGK, Jetten VG (2013) Engineering aspects and time effects of rapid deterioration of sandstone in the tropical environment of Sabah, Malaysia. Eng Geol 159:20–30

Tating FF, Hack HRGK, Jetten VG (2015) Weathering effects on discontinuity properties in sandstone in a tropical environment: case study at Kota Kinabalu, Sabah Malaysia. Bull Eng Geol Environ 74(2):427–441

Tating FF, Hack HRGK, Jetten VG (2019) Influence of weathering-induced iron precipitation on properties of sandstone in a tropical environment. Q J Eng Geol Hydrogeol 52(1):46–60

Taylor HW (1980) A geomechanics classification applied to mining problems in the Shabanie and King Chrysotile asbestos mines, Rhodesia. University of Rhodesia, Harare, Zimbabwe

Tran TV, Alkema D, Hack HRGK (2019) Weathering and deterioration of geotechnical properties in time of groundmasses in a tropical climate. Eng Geol 260:105221

Tristancho J, Caicedo B, Thorel L, Obregón N (2012) Climatic chamber with centrifuge to simulate different weather conditions. Geotech Test J 35(1):159–171

Trudgill ST, Viles HA, Inkpen R, Moses C, Gosling W, Yates T, Collier P, Smith DI, Cooke RU (2001) Twenty-year weathering remeasurements at St Paul's Cathedral, London. Earth Surf Process Landf 26(10):1129–1142

Tuğrul A (2004) The effect of weathering on pore geometry and compressive strength of selected rock types from Turkey. Eng Geol 75(3–4):215–227

Ulusay R, Hudson JA (eds) (2007) The blue book; The complete ISRM suggested methods for rock characterization, testing and monitoring: 1974–2006. Commission on Testing Methods ISRM, International Society for Rock Mechanics (ISRM), Turkish National Group, Ankara, Turkey

Utili S, Crosta GB (2011) Modeling the evolution of natural cliffs subject to weathering: 1. Limit analysis approach. J Geophys Res Earth 116(F1)

Vanacker V, Bellin N, Molina A, Kubik PW (2014) Erosion regulation as a function of human disturbances to vegetation cover: a conceptual model. Landsc Ecol 29(2):293–309

VanDerwerker T, Zhang L, Ling E, Benham B, Schreiber M (2018) Evaluating geologic sources of arsenic in well water in Virginia (USA). Int J Environ Res Public Health 15(4):787

Vázquez M, Ramírez S, Morata D, Reich M, Braun J-J, Carretier S (2016) Regolith production and chemical weathering of granitic rocks in central Chile. Chem Geol 446:87–98

Verhoef PNW (1997) Implications for the site investigation of rock dredging projects. Taylor & Francis, Rotterdam

Vervoort A, De Wit K (1997) Correlation between dredgeability and mechanical properties of rock. Eng Geol 47(3):259–267

Welch AH, Westjohn DB, Helsel DR, Wanty RB (2000) Arsenic in ground water of the United States: occurrence and geochemistry. Groundwater 38(4):589–604

Wellman HW, Wilson AT (1965) Salt weathering, a neglected geological erosive agent in coastal and arid environments. Nature 205:1097

Wenner D, Wannenmacher H (2009) Alborz service tunnel in Iran: TBM tunnelling in difficult ground conditions and its solutions. In: Proceedings of the 1st Regional and 8th Iranian tunneling conference, Tehran, Iran, 18–20 May 2009. Iranian Tunnelling Association (IRTA) & Tarbiat Modares University, Tehran, Iran, pp 342–353

Wilson MJ (2004) Weathering of the primary rock-forming minerals: processes, products and rates. Clay Miner 39(3):233–266

Winkler EM (1986) The measurement of weathering rates of stone structures: a geologist's view. APT Bull 18(4):65–70

Winkler E (2014) Stone in architecture; properties, durability. Springer, Berlin

Wu C, Xia C, Li Z (2006) Safety assessment system for evaluating spontaneous combustion of sulfide ores in mining stope. In: Huang P, Wang Y, Li S, Zheng C, Mao Z (eds) Progress in safety science and technology, Proceedings international symposium on safety science and technology (2006 ISSST), Changsha, China, 24–27 Oct 2006. Beijing, China, China Occupational Safety and Health Association, Beijing Institute of Technology, Science Press, pp 1599–1603
Yokota S, Iwamatsu A (2000) Weathering distribution in a steep slope of soft pyroclastic rocks as an indicator of slope instability. Eng Geol 55(1–2):57–68

Chapter 12
Degradation Processes in Civil Engineering Slopes in Soft Rocks

Predrag Miščević, Nataša Štambuk Cvitanović, and Goran Vlastelica

12.1 Introduction

The International Society for Rock Mechanics (ISRM) in 1981 defined soft rock as "rock with an uniaxial compressive strength in the range of 0.25 MPa to 25 MPa", which can be classified as "extremely weak" to "weak". However, this definition does not take into account the problems with weathering and deterioration of these rocks during the engineering time period. Since a proper definition of soft rock is not commonly acknowledged, a potential impediment is usually encountered in academic exchanges. Therefore, in theoretical research and engineering applications, the concept should indeed cover a variety of definitions of common laws and the essential characteristics of soft rocks and reflect their basic performances. For example, based on previous studies of the soft rock concept, He (2014) proposed a concept in association with geological soft rock and engineering soft rock. **Geological soft rock refers** to rocks characterised by low strength, large porosity, poor cementation, broken surface, and strong weathering-dependence, all of which basically contain swelling and loose clayey minerals and/or loose, soft, weak layers. The **concept proposed by ISRM is within the scope of geological soft rock**. **Engineering soft rock** refers to the rocks that can produce significant plastic deformation under the usual engineering loads. The concept of engineering soft rock emphasizes the importance of strength characteristics and engineering forces. Among other engineering tasks in soft rock mechanics, the problem with the soft rock definition concept can be considered through the stability of slopes in soft rock formations.

Surface degradation processes and landslides are quite frequent on slopes excavated or naturally formed in soft rock formations (marls, siltstones, mudstones, shales, claystone, etc.). These processes affect the safety and increase the

P. Miščević (✉) · N. Š. Cvitanović · G. Vlastelica
Faculty of Civil Engineering, Architecture and Geodesy, University of Split, Split, Croatia
e-mail: predrag.miscevic@gradst.hr; nstambuk@gradst.hr; goran.vlastelica@gradst.hr

M. Kanji et al. (eds.), *Soft Rock Mechanics and Engineering*,
https://doi.org/10.1007/978-3-030-29477-9_12

maintenance cost of facilities situated at the bottom of these slopes, while also lowering the stability of facilities located at the top of such slopes. The excavation work in these mostly clayey rocks has to be conducted either by blasting or by means of heavy jack hammers. In other words, in the excavation phase, soft rock can be treated as "standard" rock mass. However, newly excavated slopes are susceptible to rapid weathering and, within several months to several years, that is, within the engineering period of time, the rock deterioration process starts both on the slope surface and within the rock mass. These processes can also be observed on many natural slopes formed in soft rock formations.

Examples of impact of weathering processes on slope stability at unprotected cuts can be found on many locations in engineering practice. In Fig. 12.1, an example of a cut slope with a developed process of marl degradation on the slope surface can be seen, as in addition to detached fragments at the bottom of the slope. Landslides on natural cliffs are quite frequent near town Split (Croatia) in flysch formations along the coastline (Fig. 12.2). The sliding process is continuous, as the material deposited at the bottom of the slope is carried away by the sea, and thus, the space is liberated for the next "sliding phase".

Other examples of the landslides in soft rock formation are shown in Fig. 12.3 from locations in Brazil (Quadrilátero Ferrífero region) formed in schist and sericitic phyllite rock formations (Lana 2016). An interesting point in Fig. 12.3 is the failure mode. Circular failures occurred in highly weathered rocks with strength varying from 0.25 to 5 MPa (Fig. 12.3a). In contrast, in a moderately weathered soft rock, in the range 5–25 MPa, structural failure mode occurred, as shown in Fig. 12.3b. Although schists and phyllites present an anisotropic behaviour due to foliation discontinuities, weathering and strength are often the determining factors of failure modes in slopes.

The presented situations are examples of hazards resulting from weathering processes. These hazards can be a threat to human lives, delay construction deadlines, result in closedown of vital roadways, etc.

Fig. 12.1 Marl degradation on the cut slope surface at the Adriatic Coastal Road in Croatia. Location: Podstrana, Croatia. (Miščević and Vlastelica 2012)

Fig. 12.2 Landslide at the sea coast on the natural cliff in flysch formation. Location: Split, Croatia

Fig. 12.3 (**a**) Sliding in sericitic phyllite rock formation, presenting circular failure. (**b**) Sericitic phyllite rock formation, showing wedge failure. Location: Brazil. (Lana 2016)

The weathering of soft rocks has been studied in various fields, including geology, engineering geology, mineralogy, soil and rock mechanics, and geomorphology. However, the relationship between the weathering and landslides or rockfalls is still not well understood. This chapter will discuss the influence of the weathering process on the deterioration of the soft rock and the possible consequences of the deterioration processes on the rock slope stability. The main consequences of the deterioration are constant slope surface erosion and reduction of the material strength inside the rock mass slope. Results of the strength reduction are landslides and rockfalls on the natural or artificial slopes in soft rock masses.

12.2 Weathering of Soft Rocks

According to Fookes et al. (1988), the term weathering of construction materials within engineering time is defined as "the degradation or deterioration of naturally occurring construction materials under the direct influence of the atmosphere, hydrosphere and, the activities of man, within an engineering time scale". The time factor considered is on an engineering time scale, that is, tens of years. Causes of soft rock weathering have been studied by many authors (for example: Alonso et al. 2010; Gökçeoğlu et al. 2000; Görög and Török 2007; Martinez-Bofill et al. 2004; Miščević 1998a, b, etc.). Weathering properties of soft rocks and their behaviour during exposure to external influences are dominantly controlled by their mineralogical composition, preconsolidation history, composition of binding material in their structure, level of cementation, and rock texture. External conditions controlling weathering are climate, topography, chemistry of weathering solutions, hydrology, and vegetation (Huisman et al. 2006).

In the analysis of soft rock weathering, it is very important to recognise that weathering of soft rock occurs in the engineering time of scale when freshly excavated rock is exposed to weathering factors, that is, over a period from a few months to tens of years. For other types of rocks, weathering is correlated with the geological time scale, meaning hundreds and thousands of years.

Weathering affects rock material and the discontinuities in rock mass. Weathering will affect rock masses from the slope surface and discontinuities into the material, and because of this, the material strength of discontinuity walls will be reduced compared to that of the adjacent rock material further away from the discontinuities. In the beginning stages of weathering, the walls will be weakened to a depth less than the height of asperities, steps, or other roughness components of the discontinuity, and the ultimate shear strength will be governed still by the unweathered material strength that needs to be overcome before asperities can shear through and large displacement can occur. On the slope surface, bedding joints are at most exposed to external influences. These are the weakened surfaces in which the detachment of fragments is most likely to occur, and are at the same time most susceptible to external influences, that is, for example to the action of water that penetrates into the rock joints.

Water plays a crucial role in the weathering of these dominantly clayey rocks. This is conducted through the processes of drying and wetting, freezing and thawing (Martinez-Bofill et al. 2004; Yavuz et al. 2006; Sadisun et al. 2005; Pineda et al. 2014; Erguler and Ulusay 2009), and also through various chemical processes that include water. This influence is manifested as the decomposition of binding material from the clayey rock structure and as the disintegration of material into smaller fragments. In other words, this material is simultaneously affected by both physical and chemical weathering processes. An example of simultaneous physical and chemical disintegration of marl sample in contact with water is presented in Fig. 12.4.

Cracking is shown as an example of physical degradation, and dissolution from the surface as an example of chemical weathering. Deterioration caused by the weathering can be described with two forms of degradation process:

1. One form is disintegration in smaller parts with the development of the cracks system (Fig. 12.5a), Zhang et al. 2012. In samples with or without any cracks on the surface, under the weathering process, new cracks develop and all parallel

Fig. 12.4 Simultaneous physical and chemical disintegration of marl sample in contact with water (1—cracking; 2—dissolution)

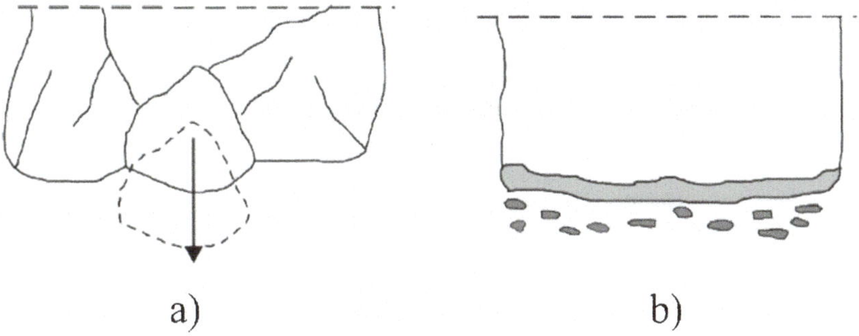

Fig. 12.5 Forms of degradation processes. (**a**) Disintegration in smaller parts with the development of the cracks system; (b) exfoliation from the surface toward the inside (Miščević 1998b; Ciantia et al. 2015)

cracks lengthen. The process usually develops near the surface, but can spread throughout the sample and break the sample apart into smaller pieces.

2. The second form of degradation can be described as the exfoliation from the surface towards the inside (Fig. 12.5b) (resulting eventually in a mass of small angular fragments and flat slivers that could readily be scraped loose by hand). Exfoliation is usually a result of the solubility of clay minerals, but it can also be a result of all weathering processes acting on the surface of sample.

Material on the free slope surface can simultaneously undergo both of the previously described processes. As a result, the material is broken into smaller parts with a larger surface area, and the process of degradation is accelerated.

The above assertions are also confirmed by typical landslide behaviour at cut slopes realised in soft rock formations, that is, the sliding/material fall most often occurs after a period of abundant precipitation, especially if temperatures at the soil surface fall below freezing point during this period. In addition to the influence of water on the weathering process (and hence on the reduction in strength), the following water-related effects can also be observed: soft rocks affected by weathering process have the lowest strength in wet condition; due to weathering of the material, the water

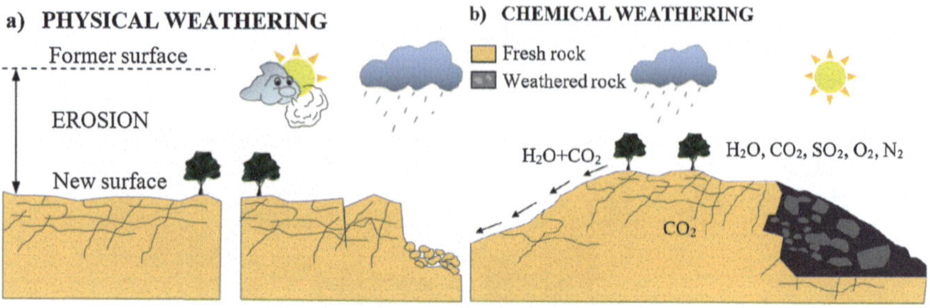

Physical weathering process	*Pressure applied (MPa)*
Freezing	200
Crystallization	2-20
Hydration of salts	100
Clay expansion	2

PROCESSES:
- Differential thermal expansion and insulation
- Wet-dry expansion
- Freeze-thaw action
- Wind and rain action
- Cristalization & expansion

EFFECTS:
- Unloading
- Joints formed
- Incipient fractures opened
- Intergranular and rock mass disintegration

Chemical weathering

PROCESSES:
- Chemical alteration
- Volume change
- Textural change

EFFECTS:
- Alteration of minerals:
  Feldspar+CO_2+H_2O→Clays+Silicas+Cations
  Pyrite+O_2+H_20→Ion Oxide-Hydroxide+Acid solutions
- Dissolution of limestones:
  $CaCO_3+H_2O+CO_2 \rightarrow Ca(H_2CO_3)_2$

Fig. 12.6 (**a**) Process of physical weathering; (**b**) Process of chemical weathering. (Adapted from Ciantia and Castellanza 2016)

penetrates the joints and the pressure builds up within such joints. In addition, the behaviour observed in nature shows that the influence of water pressure on the shear strength along joints/surfaces should be taken into account in stability analyses.

The general influence of physical and chemical weathering on cut slopes is shown schematically in Fig. 12.6 (adapted from Ciantia and Castellanza 2016).

12.2.1 Physical Weathering Processes

The physical weathering is manifested in the breakup of the material due to extension of existing joints or the development of new joints throughout the material (fracturing). Basically, the result of physical weathering is a fragmentation of rock without the change in rock material characteristics.

The breakthrough of joints is a result of the following processes.

12.2.1.1 Wetting and Drying

Due to wetting and drying, rock material swells and shrinks because of a varying quantity of water present in the material, or varying quantities of water bonded to or incorporated in minerals. The associated volume changes lead to a weakening of the material by the repeated tensile and compressive stresses, thus eventually leading to fracturing.

12.2.1.2 Slaking

Slaking is a process associated with wetting and drying. The sudden absorption of water inside dry joints results in the development of pressure inside rock joints, which lead to the extension (deepening) of joints. Slaking is closely associated with the process of suction.

12.2.1.3 Suction

Wetting and drying cycles controlled by atmospheric changes are the main cause of suction changes, associated with the rock physical degradation, leading to the loss of mechanical properties (strength, cohesion—loss of cementation or bonding and stiffness). A possible explanation for the mechanism of degradation (Alonso et al. 2010) of marl fragments is schematically presented in Fig. 12.7 assuming isotropic behaviour.

When wetting occurs, the particle boundary is first wetted and a suction gradient is created inside the rock fragment. This suction gradient induces water transfer and reduces in time until it reaches a zero value when full saturation is attained. As long as there are suction gradients, differential swelling deformations will be developed inside the rock fragment. The geometry, stratification, and confinement of the fragment restrain the swelling displacements and lead to tensile and

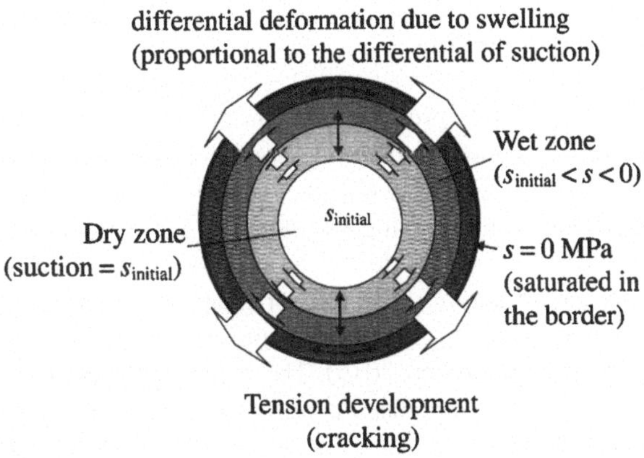

Fig. 12.7 Degradation mechanism explained by differential swelling displacements caused by suction gradients inside the rock (Alonso et al. 2010)

shear stresses, which eventually result in cracking and loss of structure. A similar mechanism can explain the degradation caused by drying (shrinkage instead of swelling deformations).

12.2.1.4 Insolation

Differential rates and amounts of thermal expansion and contraction in rock materials cause tensile stresses within the rock and rock mass. Thus, diurnal temperature changes will lead to tensile stresses and consequently to fracturing. Insolation is also associated with the drying process.

12.2.1.5 Freeze and Thaw

When water freezes into ice, a volume increase occurs of up to 9%. Therefore, freezing of water in pores, fractures, discontinuities, or other openings in a rock material or mass, causes tensile stresses that may be high enough to fracture soft rock.

12.2.1.6 Crystallisation

The volume increase during crystallisation of salts may be up to 5%.

12.2.1.7 Stress Release

Release of stresses, provoked by the removal of material during excavation (relaxation) or erosion from the slope surface, causes development of new joints (listric joints).

12.2.1.8 Swelling

Swelling causes pressure that can disintegrate the rock, and it can develop in the presence of water only. In mineralogical composition, most soft rock contains minerals susceptible to swelling. Some of chemical processes result in materials with increased volume compared with the input ingredients. Increased volume of the material formed inside the joints also causes tensile stresses that may be high enough to fracture soft rock. For example, due to the mineralogical composition of the marl, when it is in contact with water, ingredients from marl form gypsum at the joint walls (Hawkins 2015). The volume of gypsum is up to 98% greater than that of the input components exuded from marl, and pressure is therefore created in joints where the gypsum forming process is under way. This pressure enlarges the

Fig. 12.8 Gypsum growth on the surface of marl sample (white zone marked with an arrow) and developed new joint as a result of pressure caused with swelling in process of gypsum forming

existing joints and initiates formation of new ones. By separating a material from the marl structure in the gypsum forming process, the porosity of this soft rock increases, and hence, the depth of water absorption also increases. This speeds up the process and increases the depth of its influence. An example of gypsum growth on the surface of marl sample is shown in Fig. 12.8.

12.2.2 *Chemical Weathering Processes*

Chemical weathering is a descriptive term for a range of chemical processes that may include a phase change (dissolution into ionic components), or a change of mineral type and structure. Some chemical processes induce mechanical disintegration of soft rocks (cracking), but basically chemical weathering changes the material properties. The development of new joints speeds up physical weathering and enables deeper penetration of chemical weathering effects. Typical weathering types are (Huisman et al. 2006; Oldecop and Alonso 2012):

- Dissolution (release of ions after a reaction of minerals with water).
- Oxidation (loss of electrons from elements, resulting in oxides, or hydroxides if water is present).
- Hydrolysis (during hydrolysis acids react with silicate minerals, releasing silica and metal cations).
- Hydration (involves the addition of water molecules to molecules, usually ferric oxides).
- Ion exchange (cations present in a solution in contact with clay minerals are exchanged with cations present in the clay lattices. This leads to a change in the type of clay mineral and the material properties).
- Chelation (occurs in silicate minerals, and describes the bonding of metal ions to ring-structured organics).

12.2.3 Weathering Intensity

Weathering intensity refers to the degree of weathering or deterioration at the time of observation, while time-dependent weathering intensity (or intensity rate) is the amount of weathering intensity change per unit time (Huisman et al. 2006). In general, factors controlling weathering intensity are previously explained (factors controlling physical and chemical weathering processes). Additional factors controlling weathering intensity on slopes are exposure of slope, orientation of discontinuities in relation to orientation of slope, slope angle, height of slope, method of excavation for artificial slopes, and implemented drainage measures.

Assessment of weathering intensity is usually done by descriptive methods or by index properties. Table 12.1 is an example of descriptive method proposed in the suggested methods for the quantitative description of discontinuities in rock masses (adapted from ISRM 1978; Ulusay and Hudson 2007). These methods are very susceptible to the observer subjectivity, but generally, assessments made by different observers are in very close range of the classification. Materials may sometimes well be somewhere between two classes. The major problem with these widely used descriptive methods is that they are developed for all types of rocks. This means that weathering classifications consider the geological time scale, and not the engineering time scale. For soft rock masses, the engineering time scale (from few months to tens of years) is more important because all weathering effects can occur in the short time after a construction from soft rock is built or a slope is excavated in a soft rock mass formation.

Index properties used for estimation of weathering intensity usually are some kind of strength estimation by simple methods (pocket knife, geological hammer, Schmidt hammer) or tests such as the slake durability test and loss slake index (Miščević and Vlastelica 2011; Bryson et al. 2012; Vlastelica et al. 2018a). However, although strength estimates could be useful for engineering purposes, they do not usually give good correlations with the weathering intensity. On the other side, slake durability index and loss slake index can be useful to predict weathering behaviour of soft rock in a qualitative mode but cannot be used to predict the behaviour of the soft rock under natural conditions. Index properties tests can have severe

Table 12.1 Weathering degree (adapted from Ulusay and Hudson 2007)

Grade	Description
W1	Fresh, no visible sign of material weathering.
W2	Slightly weathered. Rock strength partially affected by weathering.
W3	Moderately weathered. Rock strength affected by weathering.
W4	Highly weathered. Rock strength highly affected by weathering.
W5	Completed weathered. Rock disintegrates after immersion in water.
W6	Residual soil. All rock material is decomposed or disintegrated to soil.

limitations in predicting rock mass performance, especially when discontinuities influence weathering rock mass behaviour.

Weathering intensity rate is very important information for the maintenance of the slopes cut in the soft rock formation, because weathering induces material deterioration and degradation of soft rock strength. With the reduction of a shear strength of material on slopes, the stability of slopes decreases over time. With the deterioration of the material on a slope surface, problems with surface erosion occur. Time functions for weathering intensity are commonly empirical and based on experimental results. The time scale of weathering intensity rate in soft rocks is usually equal or even less than the time in which the construction is in use. It is therefore only important to examine whether the soft rock on the slope is liable to weathering in the engineering time scale. If the problems with weathering effects can be foreseen, then cut slopes should be shaped and protected with the intention to prevent weathering processes. For natural slopes, the same information about weathering intensity rate is essential for the decision about stabilisation measures.

An example of weathering intensity rate is shown in Fig. 12.9 for different soft rock samples (Gautam and Shakoor 2013).

Weathering penetration degree—The specificity of soft rock formation is that weathering penetrates inside to the rock forming weathered zone. However, the weathering intensity on the slope surface is not the same as on the end of weathered zone under surface. The weathering penetration degree generally depends on the same factors controlling the weathering intensity and on the additional factors con-

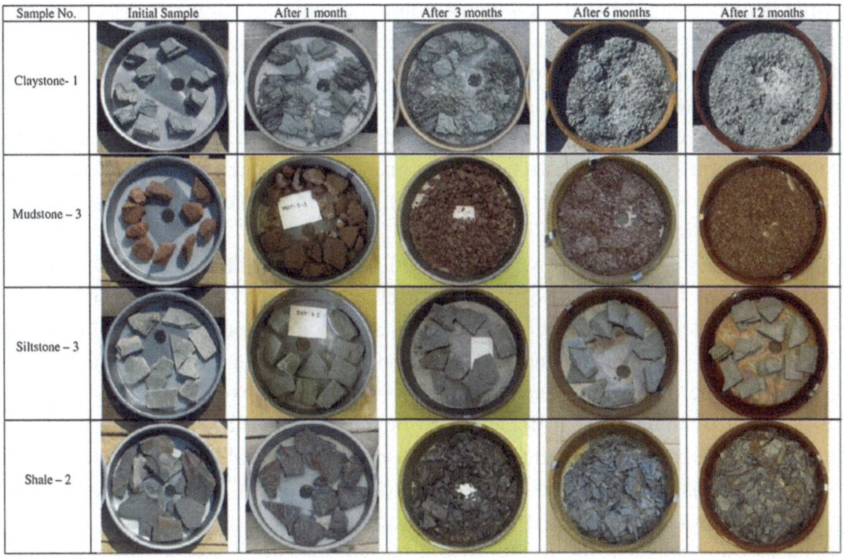

Fig. 12.9 Examples of different weathering rates of some soft rocks (Gautam and Shakoor 2013)—initial samples and samples after 1, 3, 6, and 12 months of exposure to natural climatic conditions

trolling weathering intensity on slopes. The increase rate of the weathered zone usually decreases with an increasing thickness of that same weathered zone.

12.2.4 Time-Dependent Slope Stability

Analyses of slopes cut in flysch formation (mainly formed from marl and marly layers) are presented as demonstrative examples for the problem of the time-dependent slope stability. Other than the fact that marl strength decreases with the time of exposure to atmospheric influences, the depth in which the layer is affected by such influences (weathering penetration degree) is also significant for the analysis of the stability of cut slopes formed in marl. It was established by observing the behaviour of such cut slopes that the influence of water (drying and wetting) is operated in two ways.

First, the process develops on the exposed surface of the rock, which results in constant "ravelling" of the material from the surface (Figs. 12.1 and 12.11). Transported by gravitation and precipitation, the degraded material accumulates at the bottom of the slope, where it eventually disintegrates into a material that can be classified as soil (clayey silt) rather than as rock. The depth of this influence depends on the proportion of carbonate and clayey components in marl, orientation of bedding joints and secondary joints with respect to the slope, and slope inclination or the "rate" at which the disintegrated material is removed from the surface. The total depth of the surface layer that detaches from the surface with the passage of time is also dependant on the meteorological conditions, that is, on the number of dry and rainy periods, the quantity and intensity of precipitation, the exposure of slope to the influence of sun, and the change in air temperature in the area where the slope is located.

Second, the weathering process also spreads deeper down into the rock mass and through the joint system that can easily be penetrated by water. An example of a cut slope on which wet zones near the joints can be seen, while the basic material is "dry", is shown in Fig. 12.10.

In this example, yellow brown deposits on the surface of the bedding joints and secondary joints indicate that seepage occurred even prior to the excavation.

Both above described forms of weathering are documented with several pictures of the slope cut in flysch formations (location town Split in Croatia), which were taken during the observation period of 14 years after the excavation (Fig. 12.11). The quantity of material that detached from the slope surface can be noticed as material accumulated at the bottom of the slope. This material had been removed on several occasions during the observation period. The "depth" of the surface part of the slope that was eliminated from the surface through weathering can easily be noticed as the change in the length of the harder sandstone layers "overhanging" from the slope surface. This sandstone had been initially excavated to the same levels as the surrounding layers, but was not eliminated with weathering. The observations made on similar slopes have shown that the minimum depth of the material

Fig. 12.10 Water seepage through joints formed in flysch

disintegrated from the surface of cut slopes in the Dalmatia (Croatia) region amounts to around 5 cm annually.

The sliding and weathering caused by seepage through joints can be seen in the given example at the left side of the slope. Several bigger blocks, formed at the joints through which the water penetrated, have over time been affected by sliding. The arrow in Fig. 12.11g shows the position of the last rock fall. In 2004 and 2010, the concrete protective structure was extended to protect the man-built facility situated above the cut slope, as the foundations of this facility were endangered by weathering and erosion of slope material underneath the foundations. It can be seen that the excavation was conducted very favourably with respect to bedding joints, that is, it was made perpendicular to bedding joints.

Based on data presented in previous sections, it can be concluded that the stability of slopes cut in marl should not be considered only from the standpoint of material strength immediately after the excavation, or only from the aspect of position of bedding joints and other joints with respect to the cut slope position (Šestanović et al. 1994). The analysis should include the factor of time in which the strength of this material will be reduced with weathering, and the factor of weathering depth should also be taken into account. This, in fact, defines the issue of durability of slopes cut in soft rocks. If the slope is not adequately protected so as to prevent weathering, the resistance of cut slope to weathering will be reduced over time.

A problem is also the weathering that occurs along the joints and through the basic material (marl) in between the joints, which results from water seepage. Joints are quite numerous in flysch formations. The following joint types can be differentiated: bedding joints located at a relatively small distance from one another (thinly layered form), secondary joints resulting from layer bending caused by geological processes such as folding and heaving, and listric joints that occur most frequently as a result of material relaxation after excavation and partly as a result of the excavation method used. In addition, the weathering results in the extension of the existing joints and in the creation of new joints that penetrate into the basic material. This is

Fig. 12.11 Progress of weathering process on slope cut in flysch formation, in period of 14 years. (**a**) First year following the excavation (2003). (**b**) First phase of protective structure completed (2004). (**c**) Seepage through secondary joints (2006). (**d**) Situation in 2007. (**e**) Second phase of protective structure completed (2010). (**f**) Situation in 2011. (**g**) Rock fall at cut slope (2012). (**h**) Situation in 2016

why the rule that favourable excavation with respect to dominant bedding joints ensures good stability is not fully applicable in these formations. This is additionally confirmed by slope sliding examples shown in photographs presented in this chapter. In practical terms, a favourable orientation of the side slope surface does not actually exist as, in addition to along bedding joints, the sliding can occur in any side slope position, and also along secondary joints, random joints, or joints formed by weathering, and through the basic material (marl). The fact that the water seeps through these joints is proven by plant roots, which follow the spreading of such secondary joints.

To prevent long-term instability of cut slopes excavated in marl, it is of utmost significance to block the development of the weathering process. To this end, it is important to prevent the influence of factors that enable weathering, both on the slope surface and in the interior of the slope. Factors that dominantly influence the development of weathering are cyclic processes of drying/wetting, heating/cooling, freezing/thawing and so on. They result in a whole array of secondary processes that contribute to the development of both physical and chemical forms of weathering (swelling, dissolution, gypsum formation, slaking, etc.). In order to block the influence of these processes, the slope surface must be "sealed" and the water runoff over the slope should be controlled. The rock can be sealed in two ways:

1. Prevent removal of degraded material from the slope surface. The degraded material kept on the surface lessens the penetration of the above mentioned influences along the depth. If vegetation cover develops on such surface, the layer will become "reinforced" by roots. This can easily be achieved by making slopes with a relatively low inclination. The natural inclination of degraded material deposited at the slope bottom ranges from 31° to 38° for marls, which means that it cannot be removed by precipitation. In addition, it is certain that vegetation cover will over time naturally develop on such a material.
2. Place surface protection that will prevent the development of these processes toward the interior of the slope (geosynthetics, vegetation cover, sprayed concrete, etc.).

It is therefore important to note that the surface behind the top of the slope must also be treated so as to prevent the penetration of water from higher areas into the slope zone. The weathering along the joints can be generated by water seeping through the joints, because of the dry and rainy seasons (drying–wetting). The weathering is usually not caused by water that does not emerge on the surface of the slope, and in cases when ground water table oscillations are not significant.

The above measures are not necessary for cut slopes that can be proven to be stable for shear strength parameters corresponding to weathered marl (marly clay, as the final product of weathering of marly materials). Owing to the occurrence of secondary joints and the development of weathering along such joints, we are not able to claim with confidence that the slope that has been cut "favourably" will in fact be stable with respect to bedding joints.

Therefore, cut slopes excavated in marl layers cannot be considered as permanently stable if the development of weathering is not blocked. Temporary stability on untreated slopes is possible in periods ranging from several dozens of days to several years, but it is just a question of time when the weathering process will lead to the loss of strength and ultimately to the sliding and detachment of the slope material.

12.3 Deterioration of Soft Rock Slopes

Deterioration (or "decay") of soft rocks, when analysing its impact on slope stability, is a serious problem that is important for infrastructure works and the safe construction of cut slopes (Calcaterra and Mario Parise 2010). Moreover, understanding of this problem and related processes is necessary for the proper prediction of the future behaviour of rock masses in slopes and therefore for the sustainable development in civil and environmental engineering.

Deterioration implies several fundamental processes, time-related and generally slow, acting on a slope, artificial or natural as well, which may cause a significant decrease in rock mass strength and slope stability. These primary processes are *relaxation*, *weathering*, and *erosion*, or in other words, deterioration occurs as a result of their combined action. Depending on the balance between the primary processes, deterioration usually progressively leads to alteration and disintegration of materials and rock masses (although the rock mass quality in some cases may be locally improved by weathering related processes, for example, formation of crust on the rock surface that is stronger than the material beneath it).

After the excavation and initial stress release (primary relaxation), deterioration or decay of a rock mass in geotechnical terms can be described as a process of seeking equilibrium between weathering and erosion as well as on-going stress and strain redistribution (secondary relaxation). For the duration of such a process, which is the rule rather than the exception in artificial slopes, three main situations can be recognised (Huisman 2006):

- Imbalance favouring erosion (weathering delimited).
- Equilibrium between weathering and erosion.
- Imbalance favouring weathering (erosion delimited).

To provide safe construction and predicted engineering lifetime of cut slopes, as well as to limit future economic and environmental risks, assessment of dominant deterioration processes/situations and affected geotechnical units in time, with the rate of decay of rock mass quality, should already be included in the geotechnical design phase (not only current, but also future geotechnical properties are important). However, that is not always possible since the process of deterioration of soft rocks slopes can be difficult to quantify and its mechanisms are still not recognised and understood well.

12.3.1 Relaxation

Excavations cause changes in stress along the slope (stress release or unloading; elastic relaxation and expansive recovery) so that stress distribution may become anisotropic, for example, on the slope face, the horizontal stresses are equal to zero. As the horizontal stress is removed as a result of the excavation, joints tend to undergo dilation. This may further result in the formation of a potential sliding block. Redistribution of stress and strain can be one of the causes of reduced joint roughness (broken "teeth" of the joint as a consequence of joint wall decay caused with weathering processes). In general, the fractured rock mass tends to expand along joints as a result of the stress relief and the removal of lateral confining pressure.

With zero horizontal stress on the cut surface and an unchanged value of vertical stress, shear stress can reach failure condition even in short periods after the excavation when weathering effects are not significant (Fig. 12.12).

Given these considerations, relaxation can be described as the redistribution of stress and strain following excavation (i.e., lowering of the confining stress level), leading to loss of structural integrity. Redistribution of stress and strain is a continuing process from *primary relaxation* as a direct result of excavation itself (volume increase), through deterioration of a rock mass due to the effect of weathering and erosion to *secondary relaxation* (creep).

Primary relaxation with the volume increase effect occurs directly after excavation, during the period of days to months (Huisman 2006), resulting in slightly opened discontinuities that allow access to water. Additional simultaneous effects (some, not necessarily all), linked to each other, are the decrease in normal stress on discontinuity walls and thus a decrease in shear resistance (Fig. 12.12), increased porosity and permeability due to the formation of microcracks, changed paths of water, increased aperture and hydraulic conductivity of the discontinuities, more intense water penetration and circulation, fluctuations in moisture content in the unsaturated zone, shear displacements along discontinuities resulting in breaking of

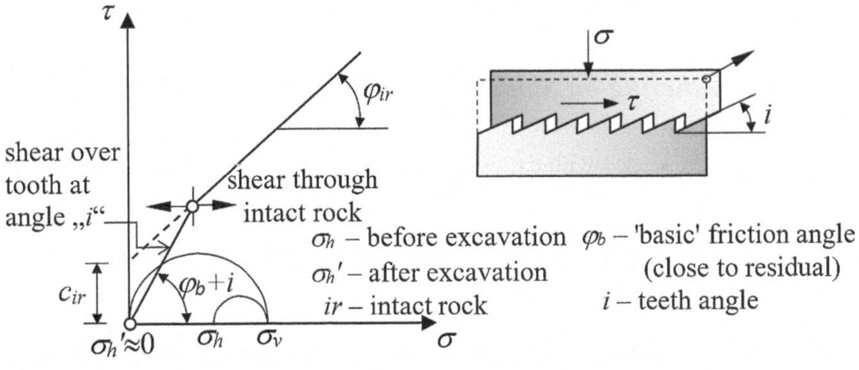

Fig. 12.12 Stress modification after overburden removal in comparison with the bilinear failure criterion of a discontinuity. (Adapted from Wittke 1990; Roje-Bonacci et al. 2009)

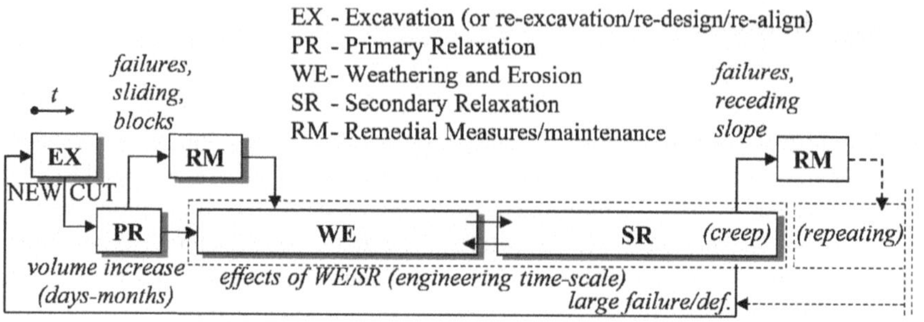

Fig. 12.13 Relationships between primary and secondary relaxation and other processes

deteriorated asperities, decreased shear strength along joints, fracturing of intact rock, and outward displacements. Through the mentioned processes, primary relaxation affects and enhances further weathering and erosion. Due to the physical and chemical weathering and—erosion process—ongoing changes in rock mass strength, density, volume, mineral composition, slope angle, and morphology, the so-called secondary relaxation will occur as continuing long-term process of stress–strain redistribution, often producing an effect known as creep.

The effects of the secondary relaxation are more or less the same as described for primary relaxation. Directly caused by weathering and erosion, secondary relaxation in return affects the intensity and the rate of further weathering and erosion in a slope, and it is very difficult to quantify and isolate the influence of secondary relaxation from other effects.

Primary relaxation shortly after excavation may have a more direct and larger effects on slope stability (failures, sliding) than secondary relaxation over a long time period (one to several magnitudes longer), but continuous secondary relaxation can affect large volumes of the rock mass (receding slope) with effects that are hard to predict, failures as well, and require costly remedial measures, re-excavation/re-design, or, in most cases, permanent maintenance. These relationships are described schematically in Fig. 12.13.

12.3.2 Weathering on Slopes

Over time, on the slopes with exposed surface, the processes occur that are more and more pronounced with the aging of slope. This "aging" as a process of change and transformation related to the climate and environment is called *weathering*, and it includes many effects of physically and chemically weakening of both the intact material and the rock mass as a whole.

From the geophysical point of view, this is the process of weakening the structure of the rocks on the surface, thus accelerating erosion and development of surface forms (landscape). Thus, the influence of the hydrological and environmental processes is being realised at all scales of space and time. In geological time, the decay of minerals leads to long-term deposition of carbon dioxide, which affects the development of the global climate. Therefore, weathering occurs as a result of certain climate, but in the longer time scale, it affects and changes the same climate through the resulting products (CO_2). From the previous discussion, it is evident that this is a key process that controls the morphological and ecological processes, and it has a direct impact on social issues (importance of soil/rock resource to the society). In general, the study of weathering is multidisciplinary, interdisciplinary and transdisciplinary, and requires coupling physical, chemical, and biological processes over a range of spatial and temporal scales and involves a variety of scientific and engineering disciplines. The "critical zone" of interactive physical, chemical, and biological processes, where the intensity of weathering is the most obvious, extends from the lowest levels of underground water to ground surface including vegetation cover. Variables such as the slope orientation and angle, climate, hydrology, tectonics, biological communities, and anthropogenic impacts, control the area affected by weathering, the amount of weathering and the place and the amount of disposal of degraded materials.

Regarding the engineering significance and target aspects of sustainability, durability, and slope stability, weathering is particularly important for soft rocks, which are subject to weathering in the time scale of the duration of human life, that is, they reduce the durability and can be studied within an engineering time scale. The question is how can we improve our ability to predict the process of weathering and its influence on the durability and future stability of the slope, and this is further discussed in Sect. 12.2.3 using the terms of *weathering intensity* and *weathering intensity rate*.

The result of weathering process is constant degradation of natural or artificial slopes. Rock profiles that result from long-term or short-term weathering can be highly inhomogeneous, complex, and irregular (Fig. 12.11), and that may lead to hazardous situations and create serious problems to the geotechnical engineer (rockfalls, plane or wedge sliding, remedial works, and similar).

Weathering affects rock masses in slopes from the slope surface and surfaces of discontinuities into a material or rock mass. The terms that describe this in the literature are *weathering penetration depth* (depth of advancement of weathering front into a slope) and *weathering penetration rate* (rate of advancement of the weathering front into a slope). Here, it should be noted that the water is an essential driving force of the weathering processes (exposure to water and initial opening of joints). Especially in soft rocks, most of these processes depend on the change in the water content, so that the weathering process mainly develops when a material is exposed to the wetting–drying changes.

There are many effects of weathering manifested in the alteration of rock material and rock mass (physical or chemical): an increase in aperture and the frequency of discontinuities; increased porosity, void volume, permeability, and flow rate; microfracturing; decrease in block size and intact rock strength; mineral alteration; weakening of intergranular bonds; reduced material strength and roughness of discontinuity walls; reduction of shear/tensile strength along the discontinuities and cohesion in the filling, and others (Fig. 12.14). Among them, the influence of weathering on discontinuities present in the rock mass is particularly important for the

Fig. 12.14 Degradation effects of weathering (example of marl as soft rock)

slope stability. For example, shear strength along the discontinuity (and thus the stability) will significantly reduce when the weathering penetration depth exceeds the discontinuity roughness.

Factors or conditions that control weathering in artificial slopes are of different nature:

- *Internal*: properties of rock/soil material and rock mass (porosity, permeability, composition of materials—fabric, physical, and chemical stabilities of coexisting mineral phases and their mutual interrelation, surface properties, inner strain, and discontinuities).
- *External*: those related to the environment (climate, topography, chemistry of weathering agents, hydrology, and vegetation).
- *Geotechnical*: design parameters (slope orientation, angle, height, excavation method, and drainage).

Thus, according to the current cognitions, rock mass deterioration results from interaction of internal material features with external conditions and biogenic activity and is further influenced by rock mass characteristics, geotechnical factors, and boundary conditions. In contrast, the durability of rock material is a complex function of the mentioned internal material features.

12.3.3 Erosion

Together with relaxation and weathering, erosion is a primary process in the deterioration of soft rock slopes, which refers to the removal and transport of previously weathered material by water, gravity, or wind, that consequently destroys the rock mass structure. In comparison with weathering, there is a strong distinction: erosion is the process of material transport, while weathering is a process of material change.

Surface erosion is a natural continuous process that acts on the products of weathering, removing them from their original location, and thus exposing "fresh" material to weathering (former and new surface, Fig. 12.6a). It can be said that erosion sets a boundary condition for rock mass weathering (Huisman 2006), and contributes to the overall process of deterioration on a larger scale of rock mass (while weathering acts on a relatively small scale, on in situ materials). Although it affects all types of soil and rock mass, erosion is more evident in the rocks with a higher frequency of discontinuities and small blocks (either weak in a fresh, that is, unweathered state, or weakened by weathering), and in unconsolidated soils. Eroded material may be subjected to further weathering during erosion.

The term *erodibility* signifies the susceptibility of particular soil or rock masses to erosion, and the properties that determine erodibility (aggregation, shear strength, hydrological conditions, and others) generally strongly depend on climatic factors (such as rainfall distribution, daily temperature changes, and frost) with systematic seasonal variations. Therefore, erodibility is a dynamic variable, where a small variation in the parameters of material and mass properties and composition can change it on shorter time scales, for example, between and during separate events such as rainstorms.

In certain conditions of climate, rock mass properties, and the geometry of the slope, erosion may cause substantially increased incidence of slope failures. Joints and fissures in their systems in rock mass can be either temporarily closed or partly opened. In open joints, the water circulates in the direction of gravity (in closed joints water flow is capillary), forming small channels, where the erosion effect of the water becomes more intensive with the degree of slope inclination (Roje-Bonacci 1996). Failure can be caused by additional external forces, as the hydrostatic pressure in surface joints connected to the potential sliding surface and change in the stress state (pore pressure) due to the effect of hydrostatic pressure. As shear strength within the rock mass is smaller along joint systems, it is of importance in slope stability problems, for example, landslides and slips on high cut-offs. Slips of this type, where joints reach the potential sliding surface and wide are so that water can circulate through them by gravity, occur during heavy rainfall.

The other examples where erosion may be the primary deterioration process (smaller influence of weathering) are *erosion by sheet wash (sheet erosion)*, *rill (and interrill) erosion*, and *gully erosion*. If soil is transported by overland flow, that is, if *wet erosion* occurs, a certain critical shear stress on the soil surface needs to be exceeded before erosion could occur. In theoretical studies, erosion is determined by multiplying exceeded shear stress and "erodibility" as the proportionality constant representing the soil erosion susceptibility. Using such simplified models with average values of critical shear stress and erodibility leads to an overestimation or underestimation of soil loss by erosion, and the variations in these parameters should also be taken into account. Quantitative evaluation of erosion (mean erosion and/or specific annual degradation) is possible on the basis of known (measured) data on the amount and mode of distribution of deposited sediments, if they are not removed by activities such as the regular maintenance of roads.

In exposed slopes of the rock types such as hard marls, the erosion mechanism includes two stages: exposure of the surface to the impact of precipitation, and then mobilisation of surface disturbance as the raindrops are lifting soil particles into suspension ready for transport (Kollios 1993). As more and more particles are picked up, a dense suspension with erosive power is formed, and due to variations in surface morphology, flow is channelled and local ravines formed. The velocity of the suspension and its erosive capacity increases as it moves downwards. Therefore, the key to reducing erosion is to slow down the water run-off velocity (geosynthetics as possible intervention). According to the well-known Chezy–Manning equation for open channel flow:

$$V = \frac{1.49}{n} \cdot R_{\mathrm{h}}^{2/3} \cdot S^{1/2} \tag{12.1}$$

where V is average run-off velocity, R_{h} is hydraulic radius corresponding to the area versus wetted perimeter, S is slope or hydraulic gradient, and n is roughness coefficient; the other possible measure is to increase the surface roughness coefficient (vegetation).

Among others factors, some factors of relevance regarding the rate of erosion on slopes are the occurrence of heavy rainfall (trigger effect, increased rates of erosion), slope angle and length of the slope, presence of cracks and macropores (infiltration conditions), surface roughness/vegetation, discharge over the slope, and moisture content of slope material. As a result of changing moisture conditions, erodibility for a given slope angle can change significantly over short time scales during rainstorms. Together with the "static" moisture conditions, seepage is also an important factor that influences erosion, where cracks provide preferential pathways for water infiltration and transport of solutes. Rill and gully formation are enhanced if seepage occurs. Also, if hydraulic impedance is present near the slope surface, erodibility can increase.

By observing the process of natural slope degradation in flysch formations near the sea cost (middle Dalmatia, Croatia), it is determined that material from the bottom of the slope is constantly removed by *wave erosion*, with the consequence of recession of the coastal cliffs at a rate 3–18 cm/year (Vlastelica et al. 2017). This represents a very high recession and erosion rate, if the denudation classification (Kossev 1990) given in Table 12.2 is taken for comparison.

Spatial and temporal variations in rock mass properties and climatic factors, those that change erodibility, in combination with human activity during slope design and construction (slope location, orientation and angle; method of excavation), influence the *development of erosion profiles* in slopes. This may result in the

Table 12.2 Classification according to denudation velocity

Denudation class	Denudation velocity h (mm/year)
Very low (1)	<0.50
Low (2)	0.50–1.00
Average (3)	1.00–2.50
High (4)	2.50–5.00
Very high (5)	>5.00

occurrence of *differential erosion* and weathering, and consequently in hazardous effects of erosion. Namely, erosion of particles formed by weathering (flaky particles and similar product of weathering, Fig. 12.11) leads to a decrease in the slope angle in exposed soft rock layers. Such previous erosion of underlying soft rock layers (especially when unfavourable joint systems are present in the rock mass) undercuts otherwise strong and durable limestone, dolomite or sandstone banks or blocks, which usually results in toppling of layers or rockfall.

Steep artificial cut in soft rocks, a lot steeper than the inclinations of natural slopes in such materials, leads to rapid erosion, and therefore to the sliding and receding of the slope and the undercutting of the otherwise resistant rocks types. Within the long-term continuous process of slope development and deterioration of soft rock slopes, the short-term episodic events influenced by rapid erosion and gravity—sliding, undercutting, rockfall, toppling, etc., will generate some scatter in the observation data.

12.4 Slope Stability in Engineering Timescale

For any analysis of change in geometry of the cut, the depth to which the soft rock mass is affected by weathering has an important role. Since the processes of weathering and erosion occur simultaneously, their combination causes further decomposition of rock mass on the face of the cut, with additional relieving effect through the redistribution stresses. The relative ratio of erosion and weathering has a significant impact on the development of the slope surface of cuts.

Although erosion is most commonly associated with a weakened and weathered soil or rock, it is also present on solid (unweathered) surfaces with lower intensity. Comparing the intensity of erosion and weathering in soft rock mass, one can notice three fundamental relationships of these processes, as discussed previously.

The potential erosion and weathering relationships depend upon the weatherability of rock immediately after excavation, and therefore, depending on their relationship, five basic situations can be noticed, as shown in Table 12.3.

With regard to the type of material and climatic features of the area of research, slope degradation models based on weathering can be used only when the process is in the range of balance of erosion and weathering to an imbalance in favour of weathering, that is, the erosion always takes place on the weathered material.

For example, by observing cuts in flysch rock mass in the Dalmatian region, the Mediterranean climate combined weathering and erosion effect is mainly found to be imbalanced in favour of erosion (Vlastelica 2015).

When steep cuts are observed, it is noticed that weathering is developed on the exposed surface, and erosion is present at a constant rate, which is always lesser than the observed weathering rate. Transported by gravity and precipitation, weathered material builds up at the base of the slope, where it eventually additionally transforms into a soil-like material (Roje-Bonacci 1998). This process has more impact on the cost of maintaining the facility than on global stability. The accumu-

Table 12.3 Combination of weathering and erosion of slopes (Huisman 2006)

	Description	Weathered material present from previous step	Thickness of weathered layer	Weathering intensity on the slope surface
Imbalance in favour of erosion	Only erosion	No	0	Constant
		Yes	Decreases due to erosion, after that 0	Decreases due to erosion, after that constant
	Some weathering, erosion penetrates faster than weathering can develop	No	0	Constant
		Yes	Decreases due to erosion, after that 0	Decreases due to erosion, after that constant
Balance between erosion and weathering	Erosion and weathering penetrate at the equal rate	No	0	Constant
		Yes	Constant, larger than 0	Constant
Imbalance relations in favour of weathering	Some erosion, weathering penetrates into the depth of cut faster than weathering	No	Increasing	Increasing
		Yes	Increasing	Increasing
	Only weathering, no erosion	No	Increasing	Increasing
		Yes	Increasing	Increasing

lated material at the bottom of the slope is usually required by owners to be removed regularly, and therefore, the slope is "moving away" from the original position. In time, this can cause additional instabilities, such as rockfalls or planar sliding on the secondary joint systems.

It was observed that the effect of weathering on the surface of the slope can be developed to a depth of several centimetres to 10 cm (Miščević and Vlastelica 2014). The depth depends on the share of carbonate and clay marl components, position of primary and secondary cracks in relation to the position of the slope, and the angle of the slope. The total thickness of the surface layer depends also on meteorological conditions: number of dry and rainy seasons, quantity and intensity of rainfall, exposure to the sun, and variation in air temperature.

12.4.1 Slope Degradation Models

12.4.1.1 Fisher–Lehmann Model

This model is a combination of Fisher's (1866) mathematical description of degradation of abandoned, initially vertical cut in chalk deposits without the accumulation of eroded material at the bottom of the slope, and Lehmann's (1933) generalised model, which introduces the possibility of initial sloping curvature and accumulation of eroded material at the base of the slope (Fig. 12.15).

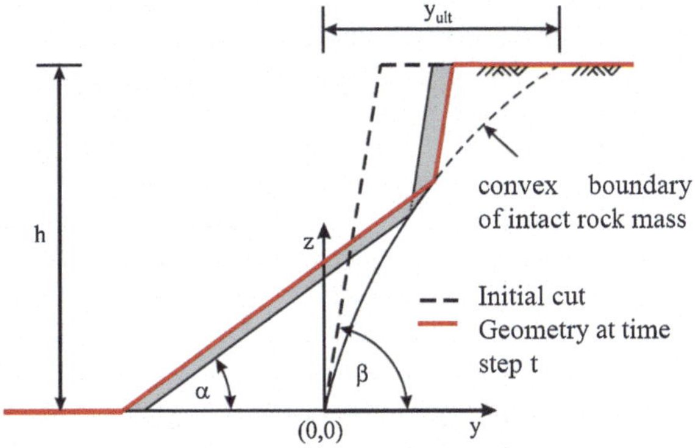

Fig. 12.15 Fisher–Lehmann model (Lehmann 1933)

Change in the morphology of the cut can be described by the following (Eq. 12.2):

$$y(t)=\begin{cases} h\cdot ctg\beta & \text{for } t=0 \\ h\cdot ctg\beta + R_{y,s} & \text{if } 0<t<\left(y_{ult}-h\cdot ctg\beta\right)/R_{y,s} \\ y_{ult} & \text{if } t\geq t_{ult}=\left(y_{ult}-h\cdot ctg\beta\right)/R_{y,s} \end{cases} \tag{12.2}$$

where h is the height of the cut, β is slope angle, and $R_{y,s}$ is annual erosion rate.

According to Hutchinson (1998), erosion of the cut is not necessarily a linear process in time. In an example of a 1.75 m height cut, based on intermittent observations over a period of 15 years, he concluded that the process was non-linear (Fig. 12.16).

Usually, it is possible to notice high initial erosion of cuts (Fig. 12.17); however, this effect should perhaps better be attributed to the initial relaxation of rock mass and local instability depending on the condition and quality of excavation.

Taking this into account, instead of the parameter $R_{y,s}$, a nonlinear function $R_{y,s} = R_{y,s}(t)$ can be introduced or the linear criterion described in (Eq. 12.2) can replace the bilinear criterion (Vlastelica 2015):

$$y(t)=\begin{cases} h\cdot ctg\beta & \text{for } t=0 \\ h\cdot ctg\beta + n\cdot R_{y,s} & \text{if } 0<t<t_1 \\ h\cdot ctg\beta + R_{y,s} & \text{if } t_1<t<\left(y_{ult}-h\cdot ctg\beta\right)/R_{y,s} \\ y_{ult} & \text{if } t\geq t_{ult}=\left(y_{ult}\quad h\cdot ctg\beta\right)/R_{y,s} \end{cases} \tag{12.3}$$

where n is the coefficient of initial relaxation of the cut; t_1 is the time within the effects of initial relaxation of the cut manifests. For example, in the Hutchinson model (Fig. 12.16), these parameters would be $n = 10$ and $t_1 = 3$ years.

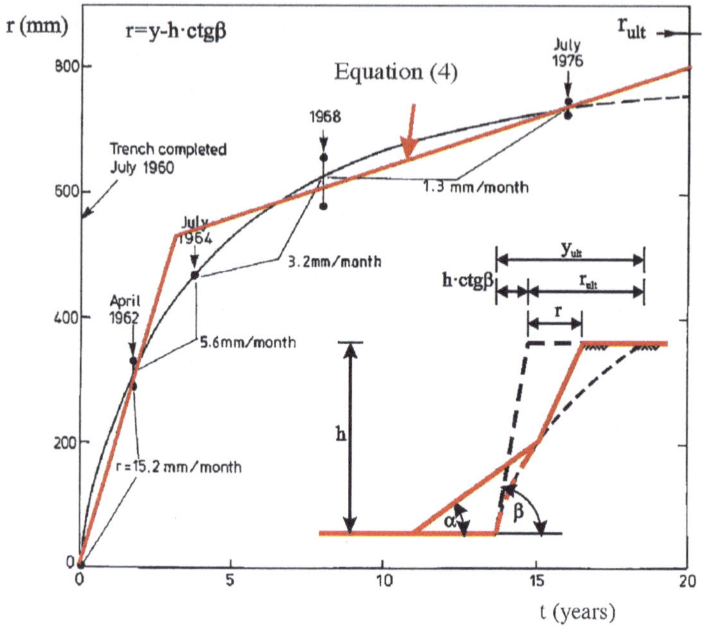

Fig. 12.16 Yearly erosion of "Overton Down" cut, UK (Hutchinson 1998; modified by Vlastelica 2015)

Fig. 12.17 An example of a quick initial change of the profile of the cut, Brnik (Vlastelica 2015). (**a**) Immediately after excavation (July 2013). (**b**) Situation after 7 months (February 2014)

12.4.1.2 Bakker–Le Heux Model

The basic assumptions of the Bakker–Le Heux (1946) model are similar to that of the Fisher–Lehmann, except for the way the free face of the slope face is eroded.

This model assumes uneven wear of the face of the cut (Fig. 12.18), according to which the erosion front rotates around the bottom of the cut. Change in the morphology of the cut can be described by introducing the following:

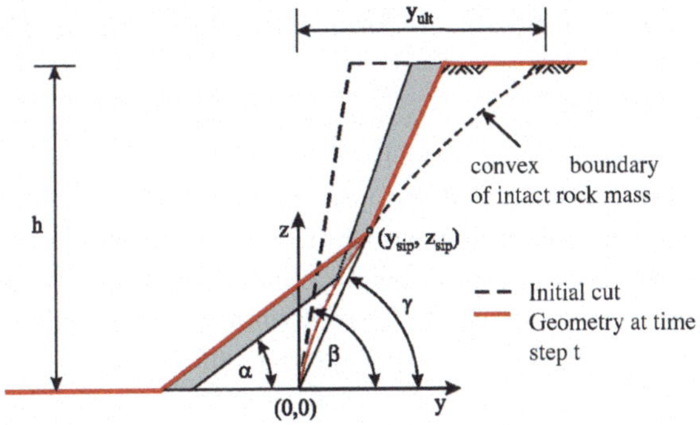

Fig. 12.18 Bakker–Le Heux model (Bakker and Le Heux 1946)

$$\gamma(t)=\begin{cases}\beta & \text{for } t=0\\ \beta-R_{\gamma,s}\cdot t & \text{if } 0<t<(\beta-\alpha)/R_{\gamma,s}\\ \alpha & \text{if } t\geq t_{ult}=(\beta-\alpha)/R_{\gamma,s}\end{cases} \tag{12.4}$$

where γ is the slope angle at selected time step; $R_{\gamma,s}$ is the average annual change of slope angle due to erosion.

By introducing time dependency t, it is possible to determine the position for the top of the scree at a given moment (Huisman et al. 2011):

$$\begin{pmatrix} y_{\text{sip}} \\ z_{\text{sip}} \end{pmatrix}=\begin{pmatrix} \dfrac{z_{\text{sip}}}{tg(\gamma(t))} \\ h\cdot\sqrt{\dfrac{1}{1-2c}\cdot\left(e^{\frac{1-2c}{c-1}\ln\left(\frac{a\cdot tg(\gamma(t))-1}{(a-b)\cdot tg(\gamma(t))}\right)}-1\right)} \end{pmatrix} \tag{12.5}$$

Also, the form of the slope:

$$y(t)=\begin{cases} y_{\text{sip}}-a\cdot(z_{\text{sip}}-z) & za\ 0\leq z\leq z_{\text{sip}}\\ z\cdot ctg(\gamma(t)) & za\ z_{\text{sip}}\leq z\leq h\end{cases} \tag{12.6}$$

Although these models are independent, in some cases, combination of both is possibly more accurate. By comparative analysis, Vlastelica (2015) found that after initial relaxation of the cut, the Fisher–Lehmann model best describes erosion at selected pilot locations in the Dalmatia region. For this reason, the Bakker–Le Heux model is preferably combined with a Fisher–Lehmann model for the simulation of initial relaxation of steep cuts.

12.4.2 Change in Shear Strength Due to Degradation

Repeated drying and wetting has a dominant influence on the development of a series of physical and chemical weathering processes that degrade soft rocks. A question arises whether, following the introduction of processes that cause degradation, it is possible to measure the trend of changes in strength. The strength of rocks, as well as all known classification of soft rocks, is usually displayed by uniaxial compressive strength (UCS) determined in a hydraulic press on cylindrically shaped samples. However, this procedure requires a set of monolithic specimens, which is usually difficult to obtain when sampling a soft rock material. Chiselling and sanding of sample sides would produce a more ideal specimen; however, these techniques are not practical on samples subjected to weathering; therefore, it is necessary to find an alternative (Miščević and Vlastelica 2009).

An alternative way of quantifying the strength of the material is to define its shear strength, usually by Mohr–Coulomb shear strength criteria. However, in this case, conventional laboratory equipment is not adapted to the whole range of appearance and friability of soft rock, which makes it necessary to devise new equipment specifically for this type of material or to adapt the existing equipment to testing requirements (Miščević and Vlastelica 2010; Buocz et al. 2010; Yin et al. 2016). Since the use of existing equipment is more rational from an economic point of view, it is proposed to simplify the experiment by using a portable direct shear apparatus for rocks. When slightly modified from the standardised procedure, the apparatus would make it possible to test the soft rock in both intact and degraded state of the samples subjected to laboratory simulated weathering. For marginal cases, direct shear apparatus for soil is used; however, it is necessary to clearly distinguish in which cases is this appropriate (Roje-Bonacci 1998; Vivoda Prodan et al. 2016). Based on these observation and series of testing on marls from Dalmatian region, Vlastelica et al. (2016a) proposed guidelines for the application of laboratory-simulated weathering on soft rocks, by using standard laboratory equipment (Table 12.4).

12.5 Monitoring of Soft Rock Slopes

Each new insight into the processes that cause the instability of slopes in soft rock, such as erosion or rockfalls, can help mitigate the possible consequences. The development of new technologies in the field of geodesy, TLS (terrestrial laser scanning, sometimes called LiDAR—Light Detection and Ranging) and GB-InSAR (Ground Based Interferometric Synthetic Aperture Radar) opens new paths for monitoring changes in the slope surface and the material from which it was created. These remote sensing technologies can be used on a global level to cover wide areas (InSAR is used for a couple of decades on the spaceborne and airborne platform), but also at the local level as terrestrial or ground-based instruments. This is very

Table 12.4 Guidelines for the application of laboratory-simulated weathering on soft rocks

	The applicability of the portable direct shear apparatus		Laboratory-simulated weathering procedure[c]	
	Unweathered material	Laboratory weathered material	Recommended number of cycles	Remark
Type I	Yes	Yes	>8	–
Type II	Yes	Yes	Do 8 (4[a])	For a larger number of cycles, it is recommended to use the device for direct shear of soil
Type III	Yes	Limited	1[b]	It is recommended to always use the device for direct shear of soil

[a]For Type II marl bordering with Type III (generally with smaller $CaCO_3$ content or LSI ≈ 0.5) maximum of 4 cycles of laboratory-simulated weathering is recommended
[b]For direct shear of intact rock use the direct shear apparatus for rock. Testing of degraded material is usually not possible and usage of direct shear apparatus for soil is recommended
[c]When selecting the proper procedure, it is always recommended to use information about $CaCO_3$ content, absorption, I_{d4} and LSI index of the rock. If it therefore can be concluded that it is a borderline sample between two categories, it is recommended to use a method of the higher category, and if it is necessary correct the procedure during the testing

important for monitoring of cuts and steep slopes, where usually it is not possible to acquire data without properly positioned ground based units.

Although GB-InSAR provides displacement measurements of up to a few square kilometres with sub-millimetre precision and high temporal frequency of acquisition (Ferrigno et al. 2017), resolution is still very rough (from couple to dozen meters, at ranges from 800 to 4000 m). It is therefore not suitable for small-scale problems, such as those occurring in soft rock strata or flysch deposits.

On the other hand, TLS provides sub-centimetre accuracy (Kordić 2014) with resolution of measurement in centimetres (at ranges of a couple hundred meters), which can provide very usable data for engineering geologists and geotechnicians, such as very precise slope morphology, prefailure deformation detection when long-term monitoring methodology is implemented (in certain types of rock masses that are more deformable, such as flysch), and information about the spatial distribution of rock-type properties (when additional information about the intensity of the return laser beam is considered). Lately, RPAS (Remotely Piloted Aircraft System) is used to collect images, which are then processed by Pix4D mapper (structure from motion technology) to generate a point cloud as with TLS, and this method is getting more precise and cost effective (Török et al. 2018).

Spatial distribution of material properties in flysch such as steep slopes can be very important, especially for detection of weak layers, which are more prone to erosion (causing differential erosion), or as a method of detecting faults and discontinuities (Fig. 12.19).

When a sufficient number of checkpoints are referenced in the global coordinate system, then the whole cloud of points can be oriented in the same system, and

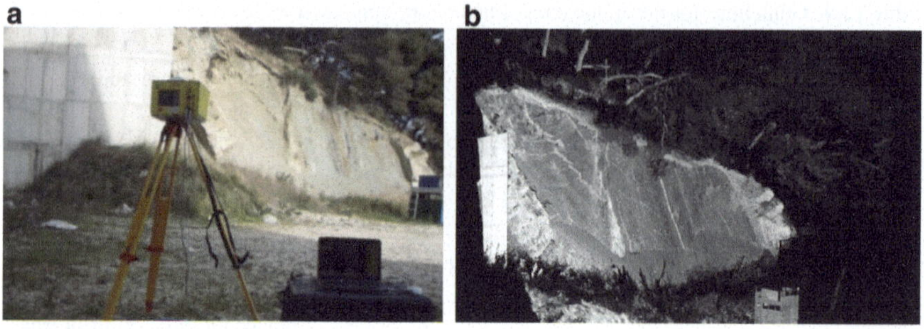

Fig. 12.19 (**a**) Typical cut in flysch rock mass; (**b**) Result of laser scanning is a point cloud with approximately 4 million points. On each point, intensity or return beam is added as grayscale, thus numerically distinguishing points of same reflective properties (Vlastelica et al. 2016b)

information about slope and discontinuities orientation can also be collected. The points can be further specified with colour, that is, RGB component can be associated, when scanners have an integrated and calibrated digital camera.

The use of TLS in this work is presented solely from a basic user's perspective. For more information about the principles of measurement, technology, and instrument performance, readers are referred to additional literature (Kordić 2014; Petrie and Toth 2008; Teza et al. 2007).

12.5.1 Methodology for Detecting Changes in Slope Morphology

The change in morphology of cuts is carried out by comparing the point clouds from different epochs using the following methodology (Vlastelica et al. 2016b):

- Obtaining a reference point cloud (Combine multiple point clouds in case of more scanning positions. Preferably georeferencing is done to determine spatial orientation of a cut).
- Creating a Triangle Irregular Network (TIN) model of the surface of the cut—reference surface (S_0).
- Obtaining new point cloud after a certain period (PC_1, PC_2, …, PC_n).
- Preparing the data for alignment between the multiple epochs.
- Comparing of the PC_i with the S_0.
- Calculating the difference for each different epoch. Distance of each point in PC_i and S_0 is calculated using the "Data vs. reference comparison", wherein the direction of comparison is determined by the direction of the vector perpendicular to the reference plane P_0.
- Creating cross section for comparison or directly comparing point cloud with S_0 in three dimensions.

Preparation of data for comparison (fourth bullet point) between different epochs of measurements is defined by the alignment matrix of the new point cloud with the reference surface. Preferably, the alignment matrix is defined by using a fixed object in the environment (buildings near of cuts or geological members, which are not subject to weathering in engineering time scale), in the following four steps:

- Identification of the stable part of the cut or an object in vicinity of the cut.
- Removal of the part of the point cloud where the changes take place (detachment and deposition) and any unwanted measurements (vegetation, moving objects, etc.).
- Alignment of the chosen stable part using Iterative Closest Point (ICP) algorithm, which defines the alignment matrix of the point cloud.
- Using the alignment matrix on the original point cloud.

Data collected by TLS can be analysed for the whole surface of the cut to give us a three-dimensional insight into the behaviour of the erosion process. Additional identification of the geological members of cuts, through photos and/or intensity of the returned laser beam, is a basis for further analysis of other phenomena, such as landslides, rockfalls, and toppling. Some results can be attributed to the measurement error or errors in methodological approach; therefore, they must be carefully assessed (Vlastelica et al. 2014).

12.5.2 *Presentation of Results as Input to Degradation Models*

Comparisons of different epochs can be displayed three-dimensionally as a field of differences in the face of cuts or two-dimensionally through selected representative cross sections (Fig. 12.20).

In three dimensions, the distances are shown through a field of values (Fig. 12.21).

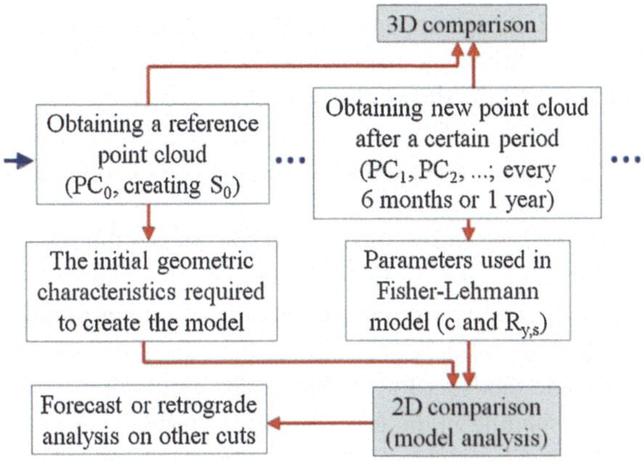

Fig. 12.20 Flowchart presenting data-obtaining procedure and presentation of results (Vlastelica et al. 2016b)

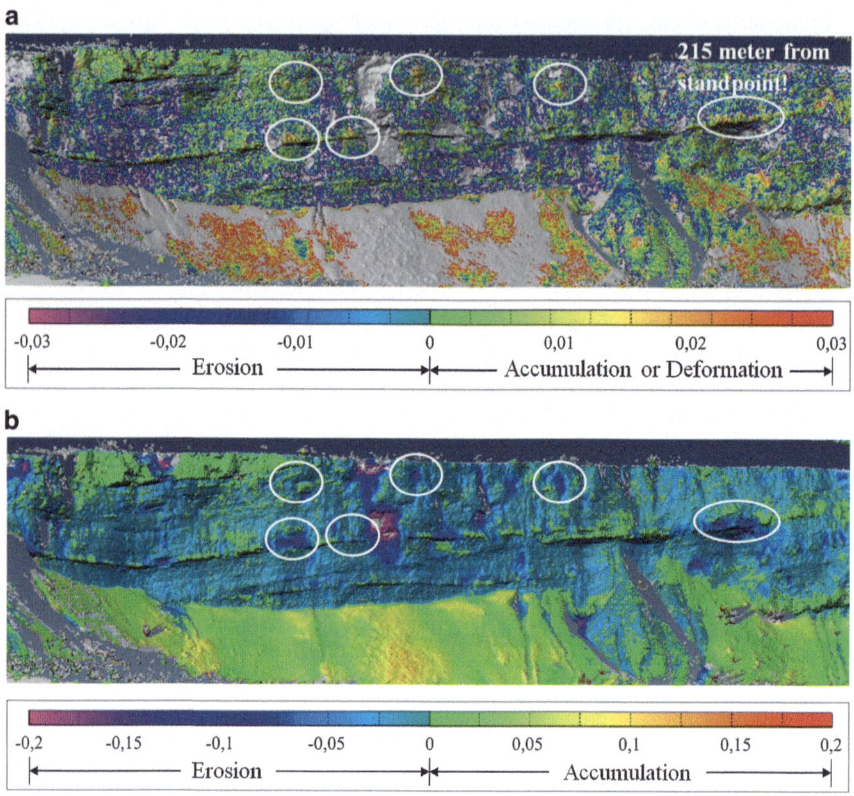

Fig. 12.21 Example of comparison of point cloud PC_i with the reference surface S_0 with prefailure deformation detected (Vlastelica 2015). (**a**) Comparison 3.4.2014–27.6.2014 (in meters). (**b**) Comparison 3.4.2014–8.10.2014 (in meters)

Negative values indicate a lack of material in a given epoch (erosion), or separation of larger rock fragments. Positive values indicate the accumulation of eroded material in the form of talus at the base of the cut or larger blocks due to rockfall (Fig. 12.21b). The positive shift towards the instrument, if it is in the open face of cuts, may indicate a displacement that preceded the rockfall (Fig. 12.21).

Figure 12.22 shows the representative cross sections as a result of observation during a period of 2 years (7.3.2012–10.3.2014), while Fig. 12.22b shows the result of numerical calculation based on the Fisher–Lehmann model of erosion of the cut for the same time period.

Field observation on this and the other pilot sites reveals that, after the initial relaxation of cuts, the Fisher–Lehmann model correctly describes the change in the profile of cuts in soft rocks that occurs alone or in the Dalmatian flysch sequence.

After the models are calibrated by field observations, it is possible to provide retrograde analysis for existing cuts, or to forecast the time needed for complete slope reshaping in the form of talus (Vlastelica et al. 2018b).

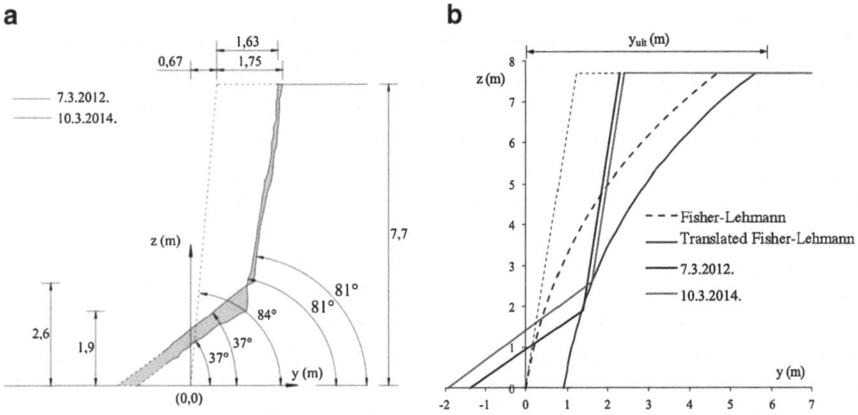

Fig. 12.22 (**a**) Results of observations carried out on pilot location "Žnjan 1" as a typical cross section of the epoch PC_0 (7.3.2014) and PC_4 (10.3.2014). (**b**) Fisher–Lehmann model based on the results of observations carried out on same pilot location (Vlastelica et al. 2016b)

12.6 Conclusions

Newly excavated slopes are susceptible to rapid weathering and, within several months to several years, that is, within the engineering period of time, the rock deterioration process starts both on the slope surface and within the inside of the rock mass. These processes can also be observed on many natural slopes formed in soft rock formations. The main consequences of the deterioration are constant slope surface erosion and reduction of the material strength inside the rock mass slope. Results of the strength reduction are landslides and rockfalls on the natural or artificial slopes in the soft rock masses.

The deterioration is caused with the weathering processes. Weathering properties of soft rocks, and their behaviour during exposure to external influences, are dominantly controlled by their mineralogical composition, preconsolidation history, composition of binding material in their structure, level of cementation, and rock texture. External conditions controlling the weathering are climate, topography, chemistry of weathering solutions, hydrology, and vegetation. This material is simultaneously affected by both physical and chemical weathering processes. Weathering affects not only the rock material, but also and the discontinuities present in the rock mass. Weathering will affect the rock masses from the slope surface and discontinuities into the material, and because of this, the material strength of discontinuity walls will be reduced compared to that of the adjacent rock material further away from the discontinuities.

Weathering intensity refers to the degree of weathering or deterioration at the time of observation, while time-dependent weathering intensity (or intensity rate) is the amount of weathering intensity change per unit time. In general, factors controlling weathering intensity are factors controlling physical and chemical weathering processes. Additional factors controlling weathering intensity on slopes are the exposure of slope, orientation of discontinuities in relation to the orientation of slope, slope angle, height of slope, method of excavation for artificial slopes, and implemented drainage measures.

Weathering intensity rate is very important information for the maintenance of the slopes cut in the soft rock formation, because weathering induces material deterioration and degradation of soft rock strength. With the reduction of the shear strength of material on slopes, the stability of slopes decreases over time. With the deterioration of the material on a slope, surface problems with surface erosion occur. Deterioration implies several fundamental processes, time-related and generally slow, acting on a slope, artificial or natural as well, which may cause a significant decrease in the rock mass strength and slope stability. These primary processes are relaxation, weathering, and erosion, or in other words, deterioration occurs as a result of their combined action. Depending on the balance between the primary processes, deterioration usually progressively leads to the alteration and disintegration of materials and rock masses (although the rock mass quality in some cases may be locally improved by weathering-related processes, for example, formation of crust on the rock surface that is stronger than the material beneath it).

In order to block the influence of weathering processes, the slope surface must be "sealed" and the water runoff over the slope should be controlled. The rock can be sealed in two ways. One is to prevent the removal of degraded material from the slope surface. The degraded material kept on the surface lessens penetration of the above mentioned influences along the depth. If vegetation cover develops on such a surface, the layer will become "reinforced" by roots. The other is to place surface protection that will prevent the development of deterioration toward the interior of the slope (geosynthetics, vegetation cover, sprayed concrete, etc.).

With regard to the type of material and climatic features of the area of research, slope degradation models based on weathering can be used only when the process is in the range of balance of erosion and weathering to an imbalance in favour of weathering, that is, the erosion always takes place on the weathered material. When steep cuts are observed, it is noticed that weathering is being developed on the exposed surface, and erosion is present at a constant rate, which is always less than the observed weathering rate. Transported by gravity, precipitation, weathered material builds up at the base of the slope, where it eventually additionally transforms into a soil-like material. This process has more impact on the cost of maintaining the facility than to global stability. Accumulated material at the bottom of the slope is usually required by owners to be removed regularly; therefore, the slope is "moving away" from the original position. This can cause additional instabilities, such as rockfall or planar sliding on the secondary joint systems.

A question arises whether, following the introduction of processes that cause degradation, it is possible to measure the trend of changes in strength. The strength of rocks, as well as all known classification of soft rocks, is usually displayed by uniaxial compressive strength (UCS) determined for cylindrically shaped samples. However, this procedure requires a set of monolithic specimens, which are usually difficult to obtain when sampling a soft rock material. Some authors suggest a method of chiselling and sanding of the sample sides, which would produce a more ideal specimen. However, these techniques are of no practical use when using samples subjected to weathering; consequently, this method is not suitable for the analysis of changes in strength, and it is necessary to look for an alternative.

Each new insight into the processes that cause the instability of slopes in soft rock, such as erosion or rockfalls, can help to mitigate the possible consequences. The development of new technologies in the field of geodesy, TLS (terrestrial laser scanning) and GB-InSAR (ground-based interferometric synthetic aperture radar) opens new paths for monitoring changes in a slope surface and the material from which it was created. These remote sensing technologies can be used on global level to cover wide areas, but also at the local level as terrestrial or ground-based instruments. This is very important for monitoring of cuts and steep slopes, because it is usually virtually impossible to acquire data without properly positioned ground-based units.

References

Alonso EE, Pineda JA, Cardoso R (2010) Degradation of marls; two case studies from the Iberian Peninsula. In: Calcaterra D, Parise M (eds) Weathering as a predisposing factor to slope movements, Engineering geology Special publications, vol 23. Geological Society, London, pp 47–75

Bakker JP, Le Heux JWN (1946) Projective-geometric treatment of O. Lehmann's theory of the transformation of steep mountain slopes. Proc Koninklijke Nederlandse Akademie van Wetenschappen (KNAW) 49(5):533–547

Bryson LS, Gomez-Gutierrez IC, Hopkins TC (2012) Development of a new loss slake durability index for compacted shales. Eng Geol 139–140:66–75

Buocz I, Rozgonyi-Boissinot N, Görög P, Török Á (2010) Laboratory determination of direct shear strength of granitoid rocks; examples from the host rock of the nuclear waste storage facility of Bátaapáti (Hungary). Cent Eur Geol 53(4):405–417

Calcaterra D, Mario Parise M (2010) Weathering as a predisposing factor to slope movements: an introduction, Engineering geology Special publications, vol 23. Geological Society, London, pp 1–4

Ciantia MO, Castellanza R (2016) Modelling weathering effects on the mechanical behaviour of rocks. Eur J Environ Civil Eng 20(9):1054–1082

Ciantia MO, Castellanza R, Crosta GB, Hueckel T (2015) Effects of mineral suspension and dissolution on strength and compressibility of soft carbonate rocks. Eng Geol 184:1–18

Erguler ZA, Ulusay R (2009) Water-induced variations in mechanical properties of clay-bearing rocks. Int J Rock Mech Min Sci 46:355–370

Ferrigno F, Gigli G, Fanti R, Intrieri E, Casagli N (2017) GB-InSAR monitoring and observational method for landslide emergency management: the Montaguto earthflow (AV, Italy). Nat Hazards Earth Syst Sci 17:845–860. https://doi.org/10.5194/nhess-17-845-2017

Fisher O (1866) On the disintegration of a chalk cliff. Geol Mag 3:354–356

Fookes PD, Gourley CS, Ohikere C (1988) Rock weathering in engineering time. Q J Eng Geol 21:33–57

Gautam TP, Shakoor A (2013) Slaking behavior of clay-bearing rocks during a one-year exposure to natural climatic conditions. Eng Geol 166:17–25

Gökçeoğlu C, Ulusay R, Sönmez H (2000) Factors affecting the durability of selected weak and clay-bearing rocks from Turkey, with particular emphasis on the influence of the number of drying and wetting cycles. Eng Geol 57:215–237

Görög P, Török Á (2007) Slope stability assessment of weathered clay by using field data and computer modelling: a case study from Budapest. Nat Hazards Earth Syst Sci 7:417–422. https://doi.org/10.5194/nhess-7-417-2007
Hawkins AB (2015) Splitting of mudrocks/shales by gypsum growth. In: ISRM Congress 2015 Proceedings - Int'l symposium on rock mechanics
He M (2014) Latest progress of soft rock mechanics and engineering in China. J Rock Mech Geotech Eng 6(3):165–179
Huisman M (2006) Assessment of rock mass decay in artificial slopes. Ph.D. thesis, Delft University of Technology and University of Amsterdam
Huisman M, Hack HRGK, Nieuwenhuis JD (2006) Predicting rock mass decay in engineering lifetimes: the influence of slope aspect and climate. Environ Eng Geosci 12(1):39–51
Huisman M, Nieuwenhuis JD, Hack HRGK (2011) Numerical modelling of combined erosion and weathering of slopes in weak rock. Earth Surf Process Landf 36:1705–1714
Hutchinson JN (1998) A small-scale field check on the Fisher-Lehmann and Bakker-Le Heux cliff degradation models. Earth Surf Process Landf 23:913–926
ISRM (1978) Suggested methods for the quantitative description of discontinuities in rock mass. Int J Rock Mech Min Sci Geol Abstr 15(6):319–368
Kollios A (1993) Geosynthetic design for erosion control of hard marls. In: Anagnostopoulos A et al (eds) Proceedings of an international symposium Geotechnical engineering of hard soils – soft rocks, vol 2. Balkema, Rotterdam, pp 1285–1288
Kordić B (2014) Development of three-dimensional terrestrial laser scanning method for determining and analyzing of landslide surface movements. PhD thesis, Faculty of Geodesy, University of Zagreb, Zagreb
Kossev N (1990) Destruction of rock slopes as a result of deterioration. In: Sixth Int. congress of engineering geology, Amsterdam, Netherlands
Lana MS (2016) Some reflections about engineering behavior of schists and phyllites in Brazil. In: II Specialized Conference on soft rocks, Cartagena, Colombia
Lehmann O (1933) Morphologische Theorie der Verwitterung von Steinschlagwänden. Vierteljahrsschr Naturforsch Gesellschaft Zurich 78:83–126
Martinez-Bofill J, Corominas J, Soler A (2004) Behaviour of the weak rock cut slopes and their characerization using the results of the slake durability test. In: Proc Engineering geology for infrastructure planning in Europe – a European perspective. Springer, New York, pp 405–413
Miščević P (1998a) Effect of drying and wetting on mechanical characteristics of Eocene flysch marl. In: Marić B, Lisac Z, Szavits-Nossan A (eds) Proc. XIth Danube Europian conf. on soil mech. and geotech. eng., Poreč, Croatia, pp 737–741
Miščević P (1998b) The investigation of weathering process in Eocene flysch. In: Evangelista A, Picarelli L (eds) Proc. Second Int. Sym. on hard soils-soft rocks, Naples, Italy, pp 267–272
Miščević P, Vlastelica G (2009) Shear strength of weathered soft rock – proposal of test method additions. In: Proc. Reg. Sym. on Rock Eng. in Diff. Gr. Cond. – Eurock 2009, Cavtat, Croatia. CRC Press/Balkema, Leiden, pp 303–308
Miščević P, Vlastelica G (2010) Shear strength of artificially weathered marl. In: Proc. Reg. Sym. on Rock Mechanics in Civil and Environmental Engineering – Eurock 2010, Lausanne, Switzerland. CRC Press/Balkema, Leiden, pp 119–122
Miščević P, Vlastelica G (2011) Durability characterization of marls from the region of Dalmatia, Croatia. Geotech Geol Eng 29(5):771–781
Miščević P, Vlastelica G (2012) Time-dependant stability of slopes excavated in marl. Gradevinar 64(6):451–461
Miščević P, Vlastelica G (2014) Impact of weathering on slope stability in soft rock mass. J Rock Mech Geotech Eng 6(3):240–250
Oldecop L, Alonso E (2012) Modelling the degradation and swelling of clayey rocks bearing calcium-sulphate. Int J Rock Mech Min Sci 54:90–102
Petrie G, Toth CK (2008) Introduction to laser ranging, profiling and scanning. In: Shan J, Toth CK (eds) Topographic laser ranging and scanning: principles and processing. CRC Press/Taylor & Francis, London, pp 1–28

Pineda JA, Alonso EE, Romero E (2014) Environmental degradation of claystones. Geotechnique 64(1):64–82
Roje-Bonacci T (1996) The landslides on old high cut-offs in the Mediterranean karst. In: Senneset K (ed) Landslides. Balkema, Rotterdam, pp 1163–1168
Roje-Bonacci T (1998) Parameter changes after weathering of soft rock. In: Proc. Sym. on hard soils-soft rocks, Naples, Italy, 12–14 oct 1998, pp 799–804
Roje-Bonacci T, Miščević P, Števanić D (2009) Rock-slides on road cuttings in the Dinaric karst of Croatia: processes and factors. Environ Geol 58:359–369
Sadisun IA, Shimada H, Ichinose M, Matsui K (2005) Study on the physical disintegration characteristics of Subang claystone subjected to a modified slaking index test. Geotech Geol Eng 23:199–218
Šestanović S, Štambuk N, Samardžija I (1994) Control of the Stability and Protection of Cut Slopes in Flysch. Geolog Croat 47/1:139–148
Teza G, Galgaro A, Zaltron N, Genevois R (2007) Terrestrial laser scanner to detect landslide displacement fields: a new approach. Int J Remote Sens 28:3425–3446
Török Á, Barsi Á, Bögöly G, Lovas T, Somogyi A, Görög P (2018) Slope stability and rockfall assessment of volcanic tuffs using RPAS with 2-D FEM slope modelling. Nat Hazards Earth Syst Sci 18(2). https://doi.org/10.5194/nhess-18-583-2018
Ulusay R, Hudson JA (2007) The complete ISRM suggested methods for rock characterization testing and monitoring: 1974–2006. Comission on Testing Methods, ISRM, Lisbon
Vivoda Prodan M, Mileusnić M, Mihalić Arbanas S, Arbanas Ž (2016) Influence of weathering processes on the shear strength of siltstones from a flysch rock mass along the northern Adriatic coast of Croatia. Bull Eng Geol Environ 76:695–711. https://doi.org/10.1007/s10064-016-0881-7
Vlastelica G (2015) The influence of weathering on durability of cuts in soft rock mass. PhD thesis, Faculty of Civil Engineering, Architecture and Geodesy, University of Split, Split
Vlastelica G, Miščević P, Fukuoka H (2014) Rockfall monitoring by terrestrial laser scanning - case study of the Rock Cliff at Duće, Croatia. In: Mihalić S, Arbanas Ž (eds) Landslide and flood hazard assessment, Proc. of the first regional symposium on landslides in the Adriatic-Balkan region. Faculty of Mining, Geology and Petroleum Engineering, University of Zagreb and FCE, University of Rijeka, Zagreb, pp 51–55
Vlastelica G, Miščević P, Pavić N (2016a) Testing the shear strength of soft rock at different stages of laboratory simulated weathering. Gradevinar 68(12):955–966
Vlastelica G, Miščević P, Fukuoka H (2016b) Monitoring of vertical cuts in soft rock mass, defining erosion rates and modelling time-dependent geometrical development of the slope. In: Ulusay R et al (eds) Rock mechanics and rock engineering: from the past to the future. Taylor & Francis Group, London, pp 1249–1254
Vlastelica G, Pikelj K, Kordić B (2017) Erosional processes acting on coastal cliffs in the Split urban zone, Croatia. In: Proc. of the fourth Coastal and Maritime Mediterranean Conference, Split, Croatia, pp 79–84. https://doi.org/10.5150/cmcm.2017.015
Vlastelica G, Miščević P, Štambuk Cvitanović N (2018a) Durability of soft rocks in Eocene flysch formation (Dalmatia, Croatia). Eng Geol 245:207–217
Vlastelica G, Miščević P, Štambuk Cvitanović N, Glibota A (2018b) Geomechanical aspects of remediation of quarries in the flysch: case study of abandoned quarry in Majdan. In: Litvinenko V (ed) Proceedings: Geomechanics and geodynamics of rock masses. Taylor & Francis Group, London, pp 1585–1590
Wittke W (1990) Rock mechanics. Theory and applications with case histories. Springer, Berlin
Yavuz H, Altindag R, Sarac S, Ugur I, Sengun N (2006) Estimating the index properties of deteriorated carbonate rocks due to freeze–thaw and thermal shock weathering. Int J Rock Mech Min Sci 43:767–775
Yin Y, Zhang BY, Zhang JH, Suna GL (2016) Effect of densification on shear strength behavior of argillaceous siltstone subjected to variations in weathering-related physical and mechanical conditions. Eng Geol 208:63–68
Zhang BY, Zhang JH, Sun GL (2012) Particle breakage of argillaceous siltstone subjected to stresses and weathering. Eng Geol 137–138:21–28

Chapter 13
Mining Slopes in Weathered and Weak Rocks

Paulo Cella, Luiz Castro, and Trevor Carter

13.1 Introduction

Weak (or soft) rocks have been defined as constituting an intermediate stage between hard (or stiff) soils and hard rocks. Dependent on mineral composition, some weak rocks have potential to degrade within a period of weeks to years when subjected to moisture content change. Such loss of strength in these weak rocks is not reversible, whereas for many soils the strength loss can be reversed in the short term by changes in water content or loading (Nickmann et al. 2006).

The International Society of Rock Mechanics (ISRM 1981) field assessment of rock strength classifies weak rocks as extremely weak (R0, uniaxial compressive strength, UCS: 0.25–1 MPa), very weak (R1, UCS: 1–5 MPa) and weak R2 (UCS: 5–25 MPa).

It is commonly accepted in the technical community that R0 materials (UCS <1 MPa) should be treated as soils, since the R0 class overlaps with very stiff (S5, UCS in the range of 0.25–0.5 MPa) to hard (S6, 0.5–0.75 MPa) soils. In addition, it is also accepted that from the middle to upper R2 strength range, these materials can be assumed to behave as a rock mass in which the role played by discontinuities becomes dominant. The problem becomes on how to handle the so-called R1 to R2 materials, that is, for materials with UCS in the range of 1 MPa to 10–15 MPa, which present behaviour that transitions from stiff/hard soil to rock. The term "weak rocks" will herein be primarily used to represent R1 and R2 materials.

This chapter describes some of the wealth of experience in mining pit slopes in South America within altered and weathered weak rocks including:

P. Cella (✉)
BVP Engenharia, Belo Horizonte, Brazil
e-mail: paulo.cella@bvpengenharia.com.br

L. Castro · T. Carter
Golder Associates, Toronto, ON, Canada

M. Kanji et al. (eds.), *Soft Rock Mechanics and Engineering*,
https://doi.org/10.1007/978-3-030-29477-9_13

- Approach used to assess intact rock strength.
- Adaptive approach to distinguish the materials within the weathering profile and assist with the creation of the Engineering Geology Model and the Weak Rock Mass Structure Model.
- Discussion on adjustments required to use a familiar rock mass classification for weak rock masses focusing on a connection to strength rather than a behavioural (stable–not stable) approach.
- Proposition of a transition function for the estimation of the weak rock mass strength that bridges the gap between the linear Mohr–Coulomb strength criterion for soil and the non-linear Hoek–Brown strength criterion for rock.
- Presentation of examples of design and performance of pit slopes excavated in weathered and altered rock masses.

13.2 Background

Experience gained in slope designs in Brazil and elsewhere has indicated that within the strength range of R1 to R2, the weak rock mass behaviour varies significantly and would therefore be better captured by subdividing this range into R1− (UCS in the range of 1–3 MPa), R1+ (3–5 MPa), R2− (5–10 MPa) and R2+ (10–25 MPa).

Weak rocks can be found in many mines around the world, ranging from the hydrothermally altered rocks of the Andes (e.g. porphyry copper deposits), through the weathered rocks (or saprolites/saprock) of tropical areas, to the soft iron ore deposits of Brazil and Africa, which feature rocks leached of silica from thinly bedded/banded units. Both alteration and weathering processes can degrade the parent intact rock strength and joint surface conditions. This reduces interlocking of rock blocks and changes the character and competence of the rock mass fabric, that is, the rock mass becoming loosened (Carter and Marinos 2014). The broad nature of bonding, which binds the grains together, and variety of geological backgrounds present a challenge to practitioners for classifying and estimating the strengths of these weak rock masses for open pit or underground mine designs. Table 13.1 outlines some observed similarities and differences between altered and weathered weak rock masses.

13.2.1 Hydrothermally Altered Rocks

Figure 13.1 illustrates the typical halo of alteration seen in gold and in porphyry copper deposits. Examples of hydrothermal alteration processes that can reduce rock strength are (with clay minerals indicated between brackets): propylitic (e.g. montmorillonite and illite), potassic (feldspar and biotite), and moderate argillic (illite, smectite, montmorillonite, kaolinite) to highly (or advanced) argillic (pyrophilitic, dickite, kaolinite).

Table 13.1 General descriptions of rock mass characteristics and behaviour in altered and/or weathered weak rock masses

Hydrothermal alteration processes	Weathering processes
Alteration typically develops through deep geological processes: hydrothermal alteration, metamorphic alteration, etc.	Weathering typically develops through mechanical and chemical decomposition processes.
Hydrothermal alteration can either reduce or increase the strength and stiffness of the original rock. For instance, argillic alteration typically reduces rock strength whereas silicification normally increases it. The amount of strength increase or reduction depends on the degree and type of alteration and the type of clay minerals.	Strength and stiffness of the underlying parent rock tends to decrease with increasing degree of weathering. Weathering generally causes the breakdown of the minerals and blocks, changing the rock mass fabric.
Occurrence of quartz veins within a moderately altered rock mass can increase inter-block shear strength thereby allowing excavation of steeper inter-ramp slope angles. Quartz veinlets in porphyry rocks for example can also increase the intra-block strength, yielding higher intact rock strengths.	Laterization is a process that develops from surface to depth and tends to locally increase rock strength.[a] Laterization occurs through the removal of silica and precipitation of iron by leaching from near surface, thus forming a hard iron-rich cap [labelled "canga" or laterite (pisolitic or ferricrete)]. Saprolite is the term used to define the zone of active leaching. Saprolite occurring immediately beneath the laterite can be of extremely weak strength and/or saturated, causing slope instability.
A wide halo of alteration can exist around sub-vertical shear hosted orebodies (Fig. 13.1) extending from great depth to surface (i.e. covering hundreds of metres of vertical extent). Laterally, alteration can progress horizontally for up to tens to hundreds of metres from a sub-vertical ore zone.	Weathering generally progresses to less than 100 m depth below surface. Weathering can render a very uneven bedrock as it penetrates deeper along sub-vertical faults, foliation and less resistant mineralogy to form a saw-toothed bedrock profile.
The occurrence of swelling clays (e.g. smectite and montmorillonite) further accelerate the degradation of rock strength when the surface is exposed to the atmosphere and/or to water.	Strength and slope performance in saprolite can vary with matrix composition. For example, saprolite from quartz-rich rocks tends to be stronger than that from ultramafic ones.
Metamorphic alteration of limestone creates skarn with variable strengths, (e.g. pyroxenes/garnet (R2) or magnetite (R4)).	Saprolite often contains relict structural fabric remaining from original rock mass discontinuities.
Through the steatization process, alteration of ultramafic rocks can result in the formation of talc, chlorite and other soft platy minerals with the potential to reduce shear strength when evident along discontinuities.	Weathering can also lead to coatings of clays on discontinuities, which cause a reduction in their shear strength. In some cases, saprolites can be collapsible, due to their internal arrangement and porosity.

(continued)

Table 13.1 (continued)

Hydrothermal alteration processes	Weathering processes
Slope performance is impacted by surface water run-off and groundwater. Surface water can cause erosion. Groundwater can impact slope performance through perched or localized seepage, as well as through increase in pore-pressure on relict structures and/or behind a hydraulic barrier formed by a layer with low hydraulic conductivity (e.g. meta sediments or schists) behind the slope face.	
For saprolite or highly altered rock masses, wider geotechnical berms are required to break uninterrupted inter-ramp slope heights at vertical intervals that vary from about 40 to 80 m.	
Variation of rock colour usually provides a good visual clue of the degree of alteration or weathering in crystalline rocks, while leaching from iron or silica bearing rocks is poorly described in changing colours.	

[a]Various leaching and induration processes can lead to the development of hard "crusts" in weathered materials. Desiccation and lateritization processes are examples of reversible and irreversible change processes, respectively

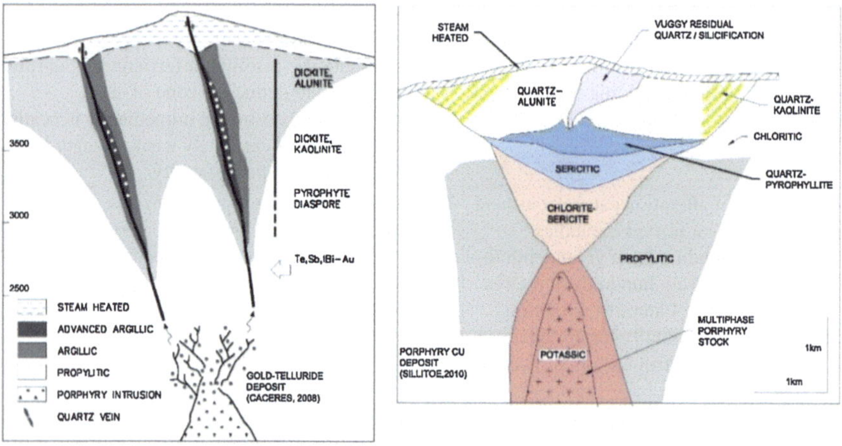

Fig. 13.1 Schematic representation of typical hydrothermal alteration in gold and in porphyry copper deposits

Hydrothermally altered rock masses tend to be associated with tectonic zones resulting in highly fractured rock, which makes the fracture frequency and matrix condition (crushed-fine grained, shattered, fragmented-centimetric fragments of rock) useful parameters to assist in their classification.

The use of specialized drilling fluids tends to improve core recovery and thus parameters, such as RQD and fracture frequency, included in most rock mass classifications. Even with improved drilling methods, the quality of core obtained from weak altered rocks can be variable. Data collected from altered core is sometimes similar to that obtained from competent weathered rocks, that is, good recovery, high RQD, low fracture frequency, but very weak to weak intact strength. In many other situations, core recovery is poor and RQD is barely definable since no easy distinction can be made between natural fractures and those caused by the drilling process. This results in high fracture frequency and strengths varying from R2 to R0 being typically described for most moderate to highly altered rocks (Castro et al. 2013).

13.2.2 Weathered Rocks

Weathered rocks have been termed from surface to depth as: highly to completely weathered or residual soil (R0/W6), saprolite (R1 to R2/W4 to W5, very weak to weak) and saprock (R3/W3, moderately strong and weathered, using ISRM, 1981 classifications), as illustrated on Fig. 13.2. The saprolite will be used herein to describe the saprolitic profile or saprolitic soils that result from the weathering of

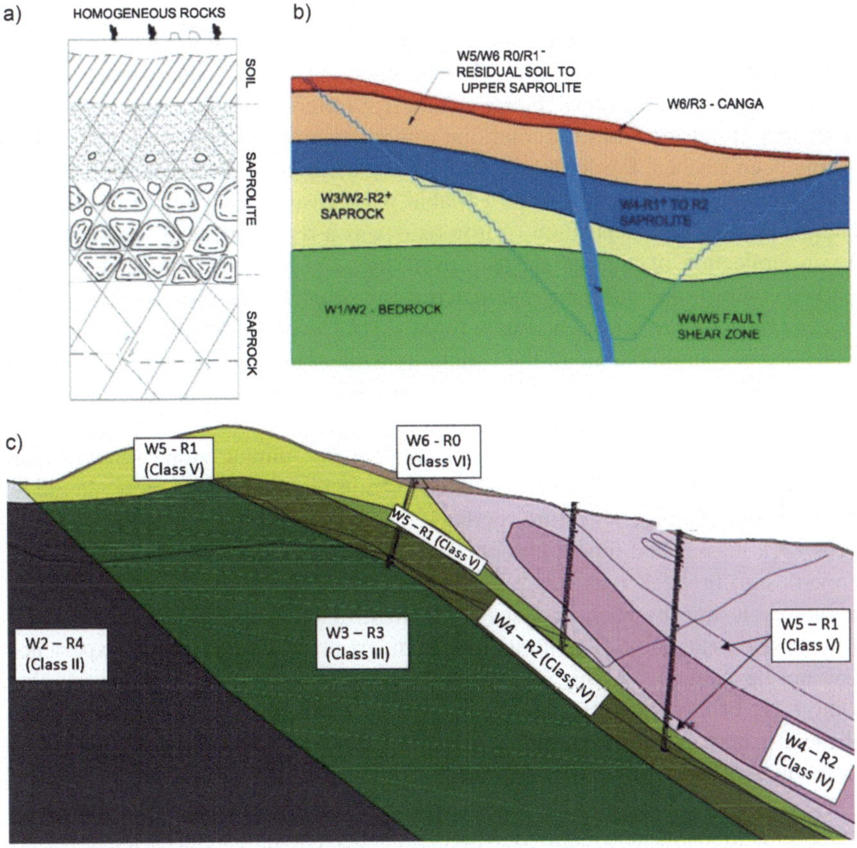

Fig. 13.2 Schematic representation of typical weathering profile: (**a**) after Deere and Patton (1971), (**b**) in a tropical mine environment and (**c**) with the intercalation of completely/highly weathered and moderately weathered (inclined) layers in anisotropic rock mass

the parent rock. Core recovery of saprolite is usually good, with core loss typically occurring in zones where the saprolite is less cohesive and consists of fine sand or silt with little clay. Slope performance in saprolite, however, can be highly variable as different saprolites can be affected by significant strength anisotropy due to differential weathering along foliation or bedding planes, or effects of intercalation of stronger layers in weak masses. Differential weathering may develop where prominent foliation occurs, creating an alternation of R0, R1 and R2 rocks, as illustrated on Fig. 13.2c. Additionally, degradation can be ongoing as saprolite strength can be reduced with exposure time due to either its large deformation with the deepening of the pit (potential for tension cracks to develop) or because of alternating periods of wetting and drying (i.e. heavy rains and dry seasons).

13.2.3 Soft Iron Ore Deposits

In some soft iron ore deposits, supergene alteration of the host rocks (e.g. poor itabirites) causes leaching of silica, while weathering changes primary minerals into clays in crystalline host rocks (Fietze et al. 2013; Jeffcoat et al. 2013). In banded (or laminated) iron formations the condition is more complex, with most of or all silica having been leached. Furthermore, oxidation of the iron minerals can form martite and hematite, which can result in a friable mass easily damaged during drilling. There can also be re-deposition of iron oxides as a replacement of leached silica. Again, collection of appropriate data from core can be very difficult due to the marked contrasts and differences in material competence. Attempts have been made to adapt the GSI classification system for "flysch", proposed by Marinos and Hoek (2001), to account for some of the variability evident in these laminated/banded or inter-fingered rocks (often phyllites and quartzites). However, more work is still required to classify soft iron ore materials.

Despite poor drill core recovery for weak rocks, mining slopes greater than 200–300 m in height have been excavated even in highly to moderately altered or weathered conditions, reflecting, perhaps, lack of a sound relationship between weak rock masses and proper classification systems. Thus, current practice for pit slope design in weak rock relies strongly on previous experience and calibrated laboratory testing by back-analyses of failures, rather than a stricter behavioural rock mass classification. The limitation of this practice is that previous experience may not be available or truly representative of the conditions in question—unless some form of descriptive framework is also provided. Even when this is accomplished, it may still require significant assumptions to be made regarding design values to account for this inability to correctly classify weak materials.

Currently, most existing geomechanical classification systems do not fully address the fundamentals required to predict the behaviour of weak rock masses. As such, it is recommended that unaltered current classification systems should not be applied whenever a given rock mass falls into the range of R0 to R1− and should instead be treated as hard soils. For R1+ and R2− rock masses (UCS in the range of

3–10 MPa), it is considered that adjustments to these classifications are required to better reflect the true response of the rock condition to shearing under saturated conditions. Recommendations are provided in Sect. 13.5 for an appropriate path forward.

13.3 Characterization of Intact Rock Strength

13.3.1 Quick Field Estimate of Rock Strength

The method of field assessment to estimate intact rock strength (or hardness), proposed by the ISRM (1981), constitutes a useful starting point for characterizing weak rocks. However, it does not extend sufficiently into the low end for adequate definition. Accordingly, Table 13.2 presents some suggested additional tactile and visual characterization procedures, including finger pressure and scratching with both thumbnail and pocket knife, and response to geological hammer blows. These quick procedures can be applied in the field to further subdivide the ISRM (1981) description and help better define the subclasses R1−, R1+, R2− and R2+.

Table 13.2 Modified ISRM field strength classification of intact rock strength (after BVP 2012; Cella 2017)

Adapted from ISRM										
					Field assessment					
Strength		Description	UCS (MPa)	I_{S50} (MPa)	Hammer	p-knife scratch	p-knife blow	p-knife blade scratch	Nail scratch	Finger pressure
R6		Extremely strong	>250	>10	Chipped	No	No	No	No	No
R5		Very strong	250-100	<-10	breaks with difficulty after several blows	No	No	No	No	
R4		Strong	100-50	2-4	Breaks after several blows	Superficial scratch	-	No	No	No
R3		Moderately strong	50-25	1-2	Breaks with difficulty after one blow			Very difficult	No	No
R2	R2+ (concrete)	Less strong	25-10	-	Breaks after one blow	Scratches	Small point	Difficult	No	No
	R2- (soil cement)	Soft	10-5		Fragments with one blow		Large point	Produces powder	Superficial scratch	No
R1	R1+ (Brick)	Very soft superior	5-3	-		Deep scratch		Easy produces powder in great amount	Scratches	Breaks locally
	R1- (saprolite)	Very soft inferior	3-1		Crumbles	Cuts apart	Penetrates	Peels	Penetrates	Break corners
R0		Extremely soft	1.0-2.5	-			Penetrates	-	Cuts	Crumbles

Consistence field test method

1-Reaction to the geologist + hammer blow
2- Reaction to scratch with pocket knife
3-Reaction to blow with pocket knife
4-Reaction to scratch with pocket knife blade
5- Reaction to scratch with finger nail
6-Reaction to fingers pressure

Reaction index

High
Fair
Low

Point load strength testing has been found useful for identifying the boundary to R2+, that is, when the UCS is greater than about 10 MPa. The lower practical bound of point load strength test (I_{s50}) is found to be about 0.7 MPa (assuming UCS $\cong 20I_{s50}$) for most typical weak rocks hosting iron deposits, such as schist, phyllite and metabasic rocks.

Other aspects, which contribute to the intact rock strength and its potential to deteriorate when exposed in a slope, and can be described during core logging are: texture (comprising the particle size distribution and its variation with depth), composition, type and amount of clay minerals, anisotropy (foliation/bedding/rock mass structure), degree of alteration, erodibility, intensity and strength of micro-defects (e.g. veinlets and fissures in intact rock); and porosity.

13.3.2 Weathering and/or Alteration Indices

Table 13.3 lists the ISRM (1981) scale of weathering (adapted from Martin and Hencher 1986). This scale proves a useful index measure for weak weathered rock description as it provides a qualitative measure of the degree of weathering of the original rock material and describes modifications in minerals and rock mass structure as compared to the sound, uniform condition. Because of this, the same format of table can be applied to altered rocks by replacing the "W" with "A", for instance, highly weathered W4 ≅ highly altered A4. Table 13.3 also includes comments on the use of the GSI (Geological Strength Index) classification system for these materials (Carter and Marinos 2014).

Geotechnical domains in weathered rocks, as illustrated on Fig. 13.2, should be defined using a combination of weathering degree (Table 13.3) and field strength estimates (Table 13.2).

13.3.3 Laboratory Strength Parameters

The heterogeneous nature of the soil-rock transition creates difficulties for sampling, moulding, consolidation-saturation, as well as for interpretation of laboratory test results, based on a classical soil mechanics framework.

Whenever achievable, undisturbed blocks of soft rocks should be extracted from the bottom of test pits, trenches, road cuts or excavated open pit benches for characterization and strength testing. Particle size distribution, Atterberg limits, porosity and slake-durability tests have been used to characterize R0 to R2− materials. Direct shear tests on specimens extracted from undisturbed blocks can provide a good indication of the shear strength of these materials. Spectrometers (identify the type of clay), X-ray diffraction (XRD, clay mineralogy), free swell (e.g. using ISRM, 1981 procedure) and ethylene glycol tests have been used to evaluate the behaviour of swelling clays.

Table 13.3 Weathering classification (modified after ISRM 1981)

Term	Grade scale	General description	Fracture condition	GSI notes (after Carter and Marinos 2014)
Residual soil	W6	All rock material is converted to soil. The original rock mass structure and material fabric are destroyed. There is large change in volume, but the soil has not been significantly transported. Discolouration evident throughout.	N/A	N/A (advise soil mechanics testing).
Completely weathered	W5	All rock material is decomposed to soil. The original mass structure is commonly preserved. Discolouration extent: throughout.	Filled with alteration minerals	For W5–R0 to R1−, N.A. (consider soil testing with adjustments in saturation and strain rate). W5–R1+ constitutes where GSI is marginally applicable. The structure has been severely disturbed and interlocking between the fragments has been lost. Clayey-sandy zones follow the original structure and rock fragments are not interlocked. Joint condition compared to the matrix strength is Very Poor.
Highly weathered	W4	>50% of the rock matrix is disintegrated to a soil. Fresh or discoloured rock is present either as a discontinuous framework or as corestones. Discolouration extent: throughout.	Filled with alteration minerals	The structure has been highly disturbed and the interlocking between the fragments has been highly loosened. Clayey and sandy products are filling all the discontinuities. Joint condition is Very Poor. The GSI shifts down and right in the chart.
Moderately weathered	W3	<50% of the rock material is decomposed and/or disintegrated to a soil. Fresh or discoloured rock is present either as a discontinuous framework or as corestones. Visible texture of the host rock still preserved. Surface planes are weathered (oxidized or carbonate filling) even when breaking the "intact rock". Discolouration extent: >20% of fracture spacing on both sides of fracture.	Discoloured, may contain thick filling	Interlocking between the fragments has been considerably loosened. Weathering coatings and fragments are filling principal discontinuities and other joints. Joint condition is Poor. The GSI shifts to the poorer structure (e.g. from Very Blocky to Blocky/Disturbed and to the right in the chart).
Slightly weathered	W2	Rock is discoloured along discontinuities (usually oxidized). Less than 5% of rock mass altered.	Discoloured, may contain thin filling	The structure is not changed but the quality of the discontinuity surfaces (shift to the right in the GSI graph). The GSI is reduced to Fair conditions.
Fresh (sound rock)	W1	No visible sign of rock material weathering. No discolouration.	Closed or discoloured	Fresh rocks are generally massive (Intact to Very Blocky). Joint Condition is Very Good (very rough) to Good (rough). Blocks and Surfaces are strongly interlocking. Rock mass may be even more fractured but only in depth where weathering has not been favoured (i.e. along a fault zone). On surface, a fractured rock mass is rarely fresh.

Applicable mainly for crystalline rocks profiles

Uniaxial compression (UCS) and triaxial compression tests can be carried out in materials that are R1+ to R2 or higher strength. However, in foliated or schistose altered/weathered rocks (e.g. phyllite, schist-serpentinites), the influence of the foliation or schistosity causes not only strength anisotropy but also the rock specimen to become too flaky for its cutting or for unconfined laboratory testing. In this case, attempts are made to carry out triaxial tests by moulding the specimen for a more stable anisotropy orientation.

13.3.3.1 Porosity

Porosity represents the amount of void space in the material skeleton. UCS has been found to decrease with increasing porosity in sedimentary rocks (Fig. 13.3a), with the low porosity and high strength being the result of a dense arrangement of grains and/or cementing agents filling the void space between grains. The trend remains well-defined for moderately weathered crystalline rocks. However, for highly weathered conditions, the variation of UCS with porosity is not well identified, perhaps because of the increasing clay infilling, making testing almost insensitive to discriminate for medium porosity (see Fig. 13.3b) from higher and even more for porosities of 15–20% (not represented on Fig. 13.3b, note the asymptotic trend below a UCS of 10 MPa).

13.3.3.2 Triaxial Compressive Strength Tests

Triaxial testing of weak rocks can be challenging. Often, specimen saturation cannot be achieved properly in a condition that works well for soils. This issue can be further compounded by the presence of laterite cementing (increase in the material stiffness) and/or by the presence of mica, aligned nearly at right angles to the specimen axis. In general, for a soil sample that is 90% saturated with percolation and under atmospheric pressure, a theoretical backpressure of almost 500 kPa is

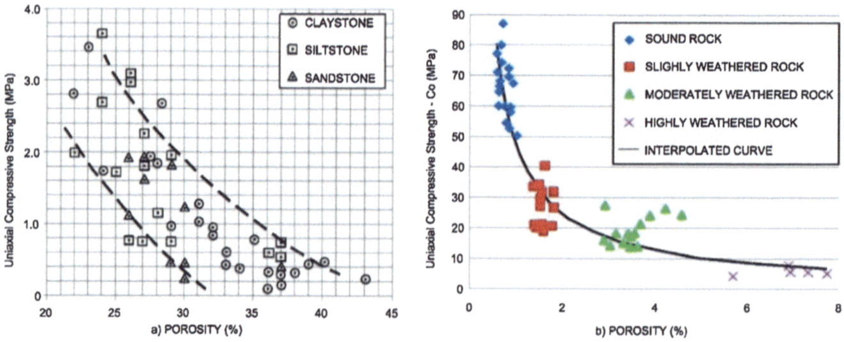

Fig. 13.3 Variation of the UCS with increase in porosity (after, (**a**) Klein 2001—sedimentary rocks and (**b**) Marques et al. 2010—crystalline rocks)

expected to ensure full saturation (Lowe and Johnson 1960). For a soft rock, an appropriate workable backpressure would have to be increased to almost double (i.e. from 800 kPa to more than 1000 kPa). Advances in the future, however, are seen as possible as the latest improvements in triaxial test equipment have extended their capability to apply confining pressures up to 2.5 MPa, which will soon increase the strength database for weak rocks.

Loss of Cohesion in Weak Rocks

Cohesion of weathered materials is related with the bonding between both grains and mineral aggregates and can reach high values when compared to soils. In the R0–R2+ range (UCS = 0.25–25 MPa), cohesion may vary by up to two orders of magnitude. Comparatively, hard rocks generally have cohesion varying by less than one order of magnitude.

The higher the degree or intensity of hydrothermal alteration or weathering the more the cohesion loss will be. For instance, laboratory rock tests were carried out to investigate the altered phyllites at the Simandou Iron Ore deposit in Guinea, where leaching of the silica (quartz) caused a progressive weakening of the phyllites from Phyllite Compact (PHC), Weak Phylite (PHW), through Very Weak Phyllite (PHV) to Extremely Weak Phyllite (PHS). The test results indicated a substantial reduction of the cohesion component and only a marginal reduction in the internal friction angle with the increase in alteration from PHW (R2) to PHS (R0) (Fig. 13.4) (Fietze et al. 2013).

The scatter observed in cohesion values is related to not only the heterogeneities in weathering, but also due to the testing conditions with possibly partial saturation of harder specimens and/or non-equalization of excess pore-pressure throughout the specimen volume occurring in shearing. Moreover, the higher the test strain rate is, the higher the estimated effective strength will be (Sheahan et al. 1996).

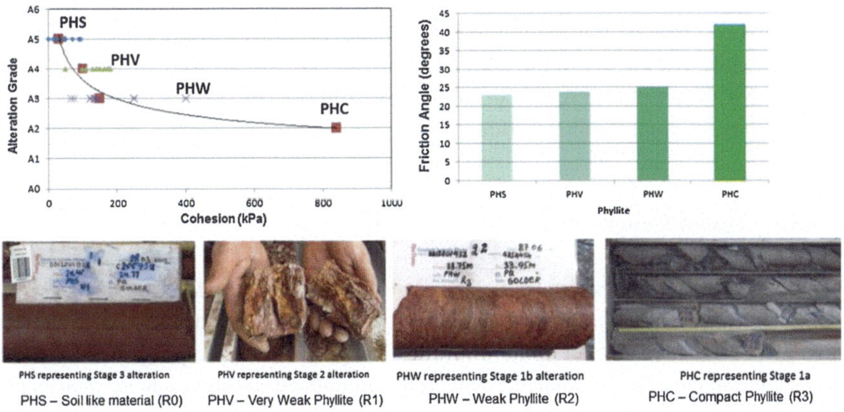

Fig. 13.4 Laboratory rock tests on phyllites at Simandou project (after Fietze et al. 2013)

Table 13.4 Cohesion and friction angles mainly for saprolites from phyllites and schists from near 70 laboratory triaxial CIU tests (after Cella 2017)

	Cohesion (kPa)			ϕ (°)			Confidence interval—*P*—95%			
							Cohesion (kPa)		ϕ (°)	
Lithology	Avg.	S.D.	Median	Avg.	S.D.	Median	Min.	Max.	Min.	Max.
Dolomitic phyllite	64.9	45.8	50.0	24.7	6.4	22	42.5	87.4	21.5	27.8
Undefined phyllite	53.8	41.1	46.0	27.4	6.4	29	34.3	73.3	24.4	30.4
Sericitic phyllite	36.7	38.1	17.0	29.2	7.2	28	17.5	56	25.6	32.8
Schists	61.2	63.8	40.0	28.4	8.5	26.6	33.9	88.5	24.8	32

For example, a sample of micaceous quartzite from the Cercadinho Formation (Brazil), classified as R1+ with slightly weathered mica, turned out to exhibit cohesion values three times higher from an undrained triaxial compressive test (saturated sample) than when determined from a drained test, probably due to unsuitable shear rate which was based on standard soil testing. Anisotropy associated with the relative orientation of mica plates in this situation, were suspected as influencing behaviour, making full saturation difficult and, at the same time, preventing pore-pressures from equalizing during shearing. The same complications have been observed when testing schists and phyllites, again likely because of the presence of fine to very fine-grained and not completely weathered, although comminuted, mica plates.

As a result of the aforementioned testing constraints, the scatter in strength parameters is relatively wide for saprolites derived from these rocks, as obtained from laboratory tests mainly for R1 materials and summarized in Table 13.4.

For friable hematites, a material containing no mica, no significant difference in cohesion is noticeable between undrained and drained tests, in which negative pore pressures and dilation, respectively, have been shown to prevail (BVP 2009).

Anisotropy is another factor that affects intact rock strength of some lithotypes, to a greater or lesser degree, depending on foliation orientation and the scale to be considered. An inaccurate description of the sample and the random orientation of the stress applied to rock foliation can also lead to a wide scatter in triaxial test and direct shear test results. It is recommended that careful testing of samples with foliation set at different angles to the core axis be utilized in order to obtain the classic "U"-shaped behaviour curve for strength of anisotropic intact materials, although this is not easily obtained in practice as a large number of specimens are required.

13.3.3.3 Use of Geochemistry

Geochemical analyses, commonly carried out to define ore content, have also been applied to define weathering or alteration profiles. For instance, Durango and Gavinia (1987) used nickel (Ni), iron (Fe), magnesium oxide (MgO) and silica (SiO_2) contents in a nickel lateritic deposit to evaluate laterization and serpentinization processes along the weathering profile, as a basis for classifying the rock mass into geotechnical domains for canga, laterite, residual soil, saprolite, saprock and fresh bedrock (Fig. 13.5).

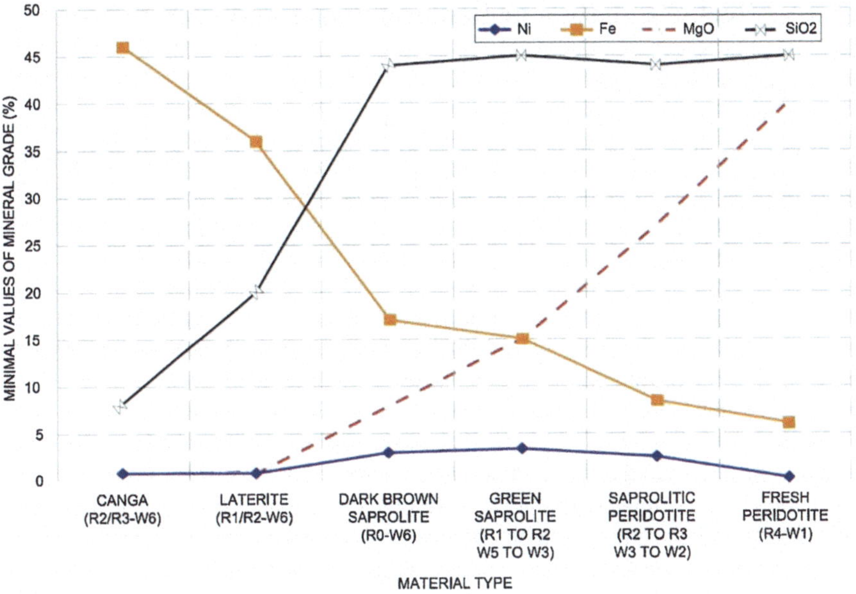

Fig. 13.5 Variation of mineral grade for a weathering profile in a nickel deposit. (Adapted from Durango and Gavinia 1987)

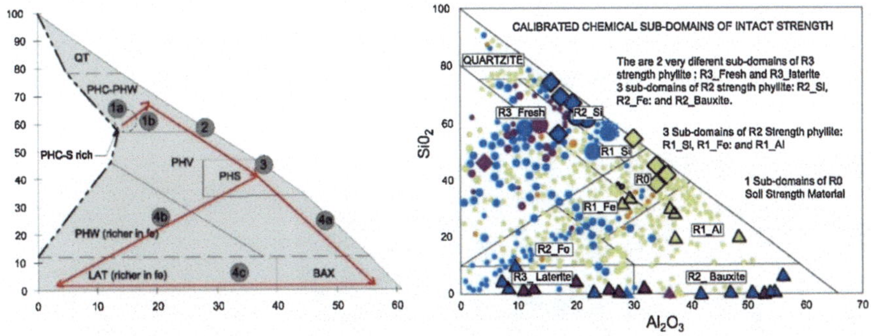

Fig. 13.6 Alteration domains of RC holes using alumina (AL) versus silica (Si) contents (after Jeffcoate et al. 2013)

Another good example, provided by Jeffcoate et al. (2013), is the application of geochemistry to model intact rock strength variations for altered phyllites. This was undertaken for the Simandou project, as shown graphically on Fig. 13.6, with the red arrows showing the alteration process. The authors describe the alteration process of the phyllite, starting with the leaching of the silica, which causes a progressive weakening of the phyllites from PHC, PHW, PHV to PHS (see descriptions within Fig. 13.4).

13.4 Effect of Foliation or Bedding Planes and Failure Mode

The orientation and characteristics of the foliation or bedding planes within a rock mass are important controls on slope behaviour (Fig. 13.7), not only due to strength anisotropy, which is more prominent in the range of R1+ to R2, but also because the weakness planes tend to control propensity of planar failure on footwall (FW) slopes. On the other hand, depending on rock strength and foliation orientation, rotational failure or toppling tends to control hanging wall stability (HW).

For the FW, inter-ramp slopes often tend to follow the ore plunge, such that bench face angles may undercut gently dipping foliation planes (e.g. dip ≤30°–35°). For the materials in the lower half of the R1 range (UCS < 3 MPa), strength anisotropy tends to be minimized and thus the slope configuration needs to rely on the overall rock mass strength in order to maintain stability.

Instabilities of high FW slopes with persistent discontinuities, are typically associated with deep differential weathering or completely weathered layers dipping sub-parallel to the inter-ramp slope. Different failure mechanisms can be recognized, that can be primarily classified depending on whether there is full or partial discontinuity control. Bilinear slab failure, ploughing slab failure and buckling can develop where there is full discontinuity or weakness plane control (Hawley et al. 1986). For partially joint-controlled failure mechanisms, these three basic mechanisms can still occur, with the difference that there is an increase in the shear strength along the governing failure surface due to rock bridges that may exist as discontinuity separators.

For weak rock masses where large deformations can develop for high slopes, waviness of foliation/bedding planes can also be critical for stability. Potential

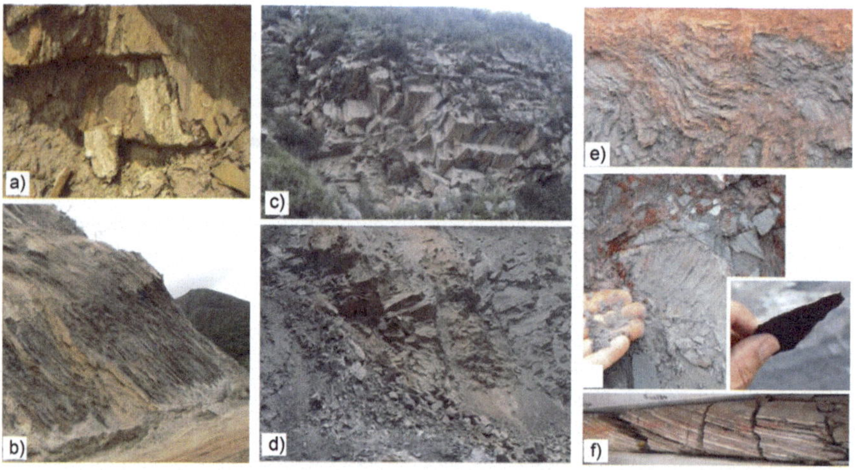

Fig. 13.7 Examples of planar failure in FW slopes (**a**) and (**b**), toppling failure in HW slopes (**c**) and (**d**), and weakness planes in banded iron formation (**e**) to (**g**), with detail of the leaching of silica in itabirite and phyllite (**g**)

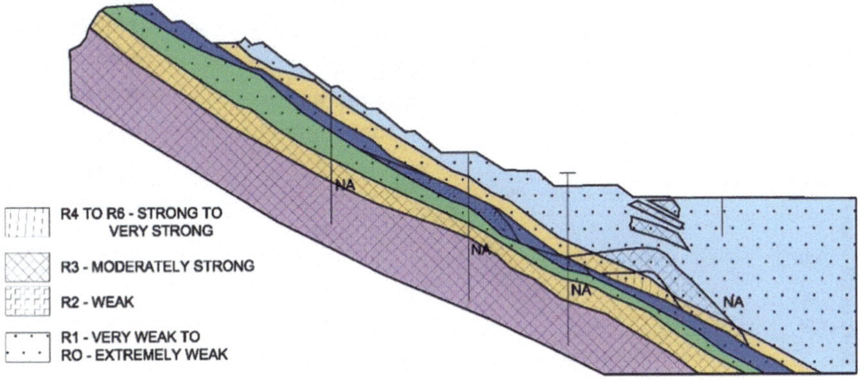

Fig. 13.8 Actual field strength geometry in an open pit in Brazil with potential for bilinear failure (BVP 2017)

failure surfaces in such cases that may eventually govern the global deformation and large instabilities, often develop as concave or convex in shape in relation to the slope face. Figure 13.8 shows a typical geotechnical section where potential shearing is considered along the foliation in completely weathered (R0–R1) units departing from the upper portion of the slope (as in a bilinear slab failure), to become a severe rock mass softening or failure cutting through mostly R0 to R2 units in the lower portion of the slope. Note that in this geometry, anisotropy remains sub-parallel to the overall slope.

In terms of estimation of rock mass strength parameters, as will be discussed in Sect. 13.6, direct shear tests set up to test along the foliation, can provide reasonable determination for the strength of a potential failure surface when sub-parallel to the overall slope face. The Transition Function (Sect. 13.6.1) applied to the Hoek–Brown criterion, can be used for evaluating the strength for any cross-cutting release plane cutting through the lower portion of the slope.

For steeply dipping foliation in HW slopes, there is also the potential for flexural toppling. This tendency is minimized as strength reduces from R2+ to highly/completely weathered R1 materials, when stability again becomes more controlled by overall rock mass strength. Such conditions can be replicated with 2D numerical analyses, such as those carried out to estimate potential modes of failure for the HW slopes planned for the phyllite materials for the Simandou open pit project, considering variation of intact rock strength (UCS) and Young's modulus, E, as illustrated in Fig. 13.9 (Fietze et al. 2013). The vertical axis labelled "Relative Horizontal Displacement", which was obtained by subtracting the horizontal displacement computed for the slope at failure from the maximum horizontal displacement prior to starting the strength reduction factor calculations.

The chart has two regions roughly demarcated by the dashed line. Points in the region above the line show potential for toppling failure, while those below are typical for rotational-type failure, common in soils. This trend line suggests that there

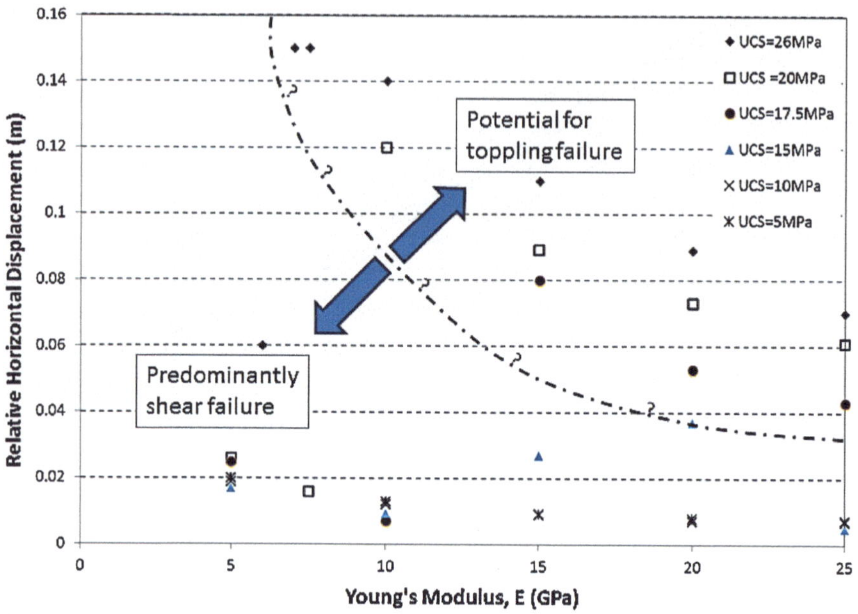

Fig. 13.9 Observed change in the mode of failure as a function of the stiffness and strength of the rock mass (after Fietze et al. 2013)

are combinations of strength and stiffness values where mixed mode failure mechanisms might dominate rather than clear toppling or obvious rotational-type failure.

Figure 13.9 clearly shows that when considering Young's modulus in analytical evaluations, it is possible to clarify anticipated failure mechanisms. The chart indicates a very low likelihood for toppling to occur in the very weak and altered phyllite, due to its soil-like nature (low stiffness, R1 to R2−). On the other hand, for the compact phyllite zone (R2+ to R3), which has Young's modulus values typical of rocks, toppling failure might well become the dominant process.

13.4.1 Intercalation of Moderately Strong and Weak Layers

Differential weathering can develop along a slope profile where different rock formations create layering or where more preserved and competent rock layers and/or boulders intercalated within very weak (R1+) to extremely weak (R0) saprolites (e.g. Figs. 13.2c and 13.10). Where the frequency and thickness of these more competent (R2 to R3) rock layers is substantial (e.g. more than 30% of the slope height), the potential failure surface may develop as a step-path geometry of shallow depth as a way to minimize the strength mobilized when cutting through harder layers at the toe of the slope.

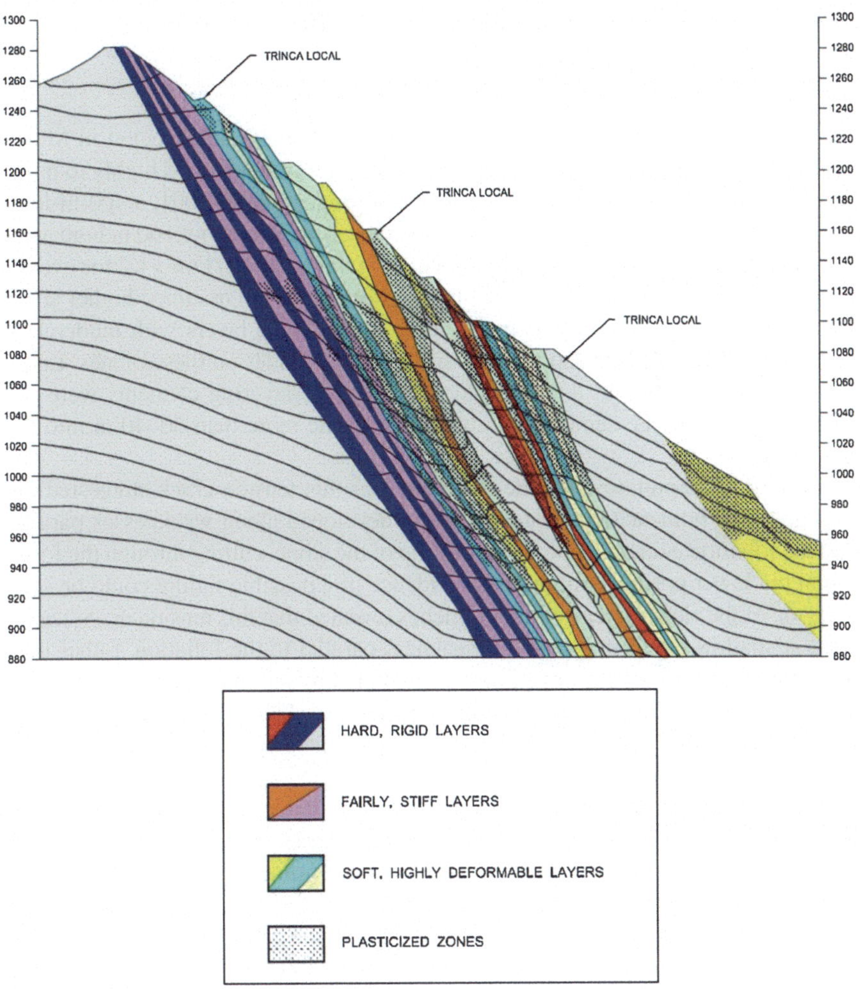

Fig. 13.10 Vertical stress arching due to stiffer layers and developed plastic zones, from computational model. Deformability moduli ratios were 1 to 10 to 1000, respectively from soft, fairly hard and hard layers

In addition, mechanical interaction between stiffer and more deformable layers controls the stress redistribution and hence, the stability condition inside the slope. As the slope deforms, progressive or local failure may develop, as stress concentration increases in the stiffer portions. Episodic slope deformation is thus common as overall slope height increases reflecting the gradual increase in stress concentration in the slope. This will eventually reach locally its peak strength. Conventional limit-equilibrium analysis shows factors of safety (FoS) in such situation that are, in general, 10% lower than those FoS obtained on the same "deformed" sections through numerical modelling.

13.4.2 *Pau Branco Mine: Example of a Bilinear Failure and Flexural Deformation*

The Pau Branco open pit mine, located in the Iron Quadrilateral region in Minas Gerais, Brazil, has been monitored to evaluate the deformation of its highly to moderately weathered phyllite slopes as mining has progressed. In 2001, a 400-m-long tension crack developed behind the crest when the slope was about 160 m high at an overall slope angle of 34°, despite the slope design indicating a FoS >1.2 for overall slope angles in the range of 35°–37°. The higher initial FoS assumed higher shear strength along the foliation planes due to intercalation of layers with moderately decomposed (A3) phyllite and the deposition of iron oxides on these planes. Later, as the pit became deeper, it was noted that such interlayering elements were not continuous, and the Fe cementation did not penetrate more than 30–40 m into the rock mass.

An initial interpretation for the occurrence of this tension crack suggested the potential for a bilinear failure process being developed along weaknesses parallel with the foliation evident in the upper portion of the slope, cutting through the foliation in the lower portion. Although no evidence of a possible failure could be seen at the toe of the slope. Excavation of trenches revealed that this tension crack terminated at shallow depth in a zone of weakness parallel to the foliation, rather than extending deep into the rock mass as was predicted by the limit equilibrium analysis. From the trenching, it was established that this shallow tension crack likely formed in response to the movement at the top of the slope due to shearing along foliation.

Boreholes drilled to depths below the zone of possible weakness occurrence in the lower portion of the slope failed to identify any potential failure surface, mainly because drilling caused a strong disturbance in the dominant highly weathered (A4) phyllite.

In June 2002, a new, major movement of this tension crack over a 13-day period was observed in one of the trenches, with a displacement of 20 mm evident along the foliation. The movement appeared to be accelerating. The crack then extended down through two benches beneath the crest to form a concave shell bending towards the slope face. In connection with design recommendation, the mining team decided to unload the slope height (about 25% of the total height) by excavating a 35-m-high push back, which significantly reduced these accelerating trends.

The overall slope deformation was interpreted to be the result of combined shearing along multiple 40°–50° dipping pre-sheared foliation within less than upper half of the pit wall and the transfer of excess loading to mid height of the slope where convex inflexion of the foliation was undergoing severe softening (Fig. 13.11). An outward movement occurs at the slope toe with reverse shearing on foliation planes as indicated by the arrows.

The phyllite slope was, in general, dry (or drained). As the pit excavation progressed deeper, groundwater seepage was observed on the slope face. Deep horizontal drain holes were, therefore, drilled and the hydrogeological model was revised to reflect the occurrence of permeable layers of metachert intercalated in the phyllite.

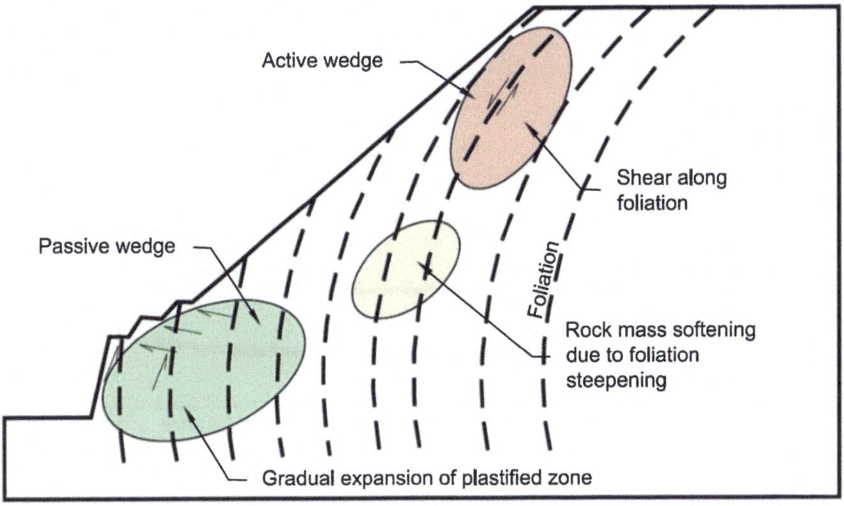

Fig. 13.11 Overall slope deformation model for weak phyllite slope (Brito et al. 2011)

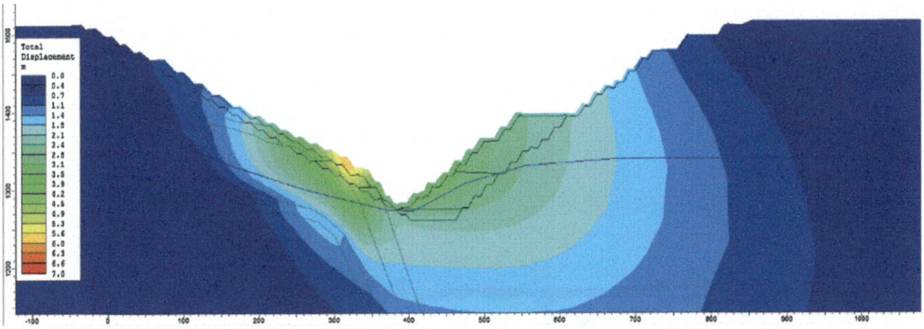

Fig. 13.12 Numerical modelling showing the total displacement pattern in the slope since the mining initiated (shearing plus elastic rebound from the native ground excavation). Maximum value appeared near foliation inflection zone

Since the metachert aquifers were thin and completely confined into the completely weathered phyllite (aquitard), the observed groundwater seepage suggested that deformation cracks developed deeper into the softening zone all the way through the passive wedge to outcrop on the slope face.

A numerical model was prepared, and two multiple extensometers were installed above and below the inflection zone, as shown on Fig. 13.12, where displacement contours are displayed for a mining geometry achieved by 2015–2016.

In the field, each extensometer was installed with three anchors located at depths of 85 m (anchor 1), 60 m (anchor 2) and 30 m (anchor 3). Figure 13.13 shows the variation of the total displacements over time for the extensometer below (Fig. 13.13a) and above (Fig. 13.13b) the inflection zone. The general trend of these total displacements fits well with the complex deformation pattern interpreted for

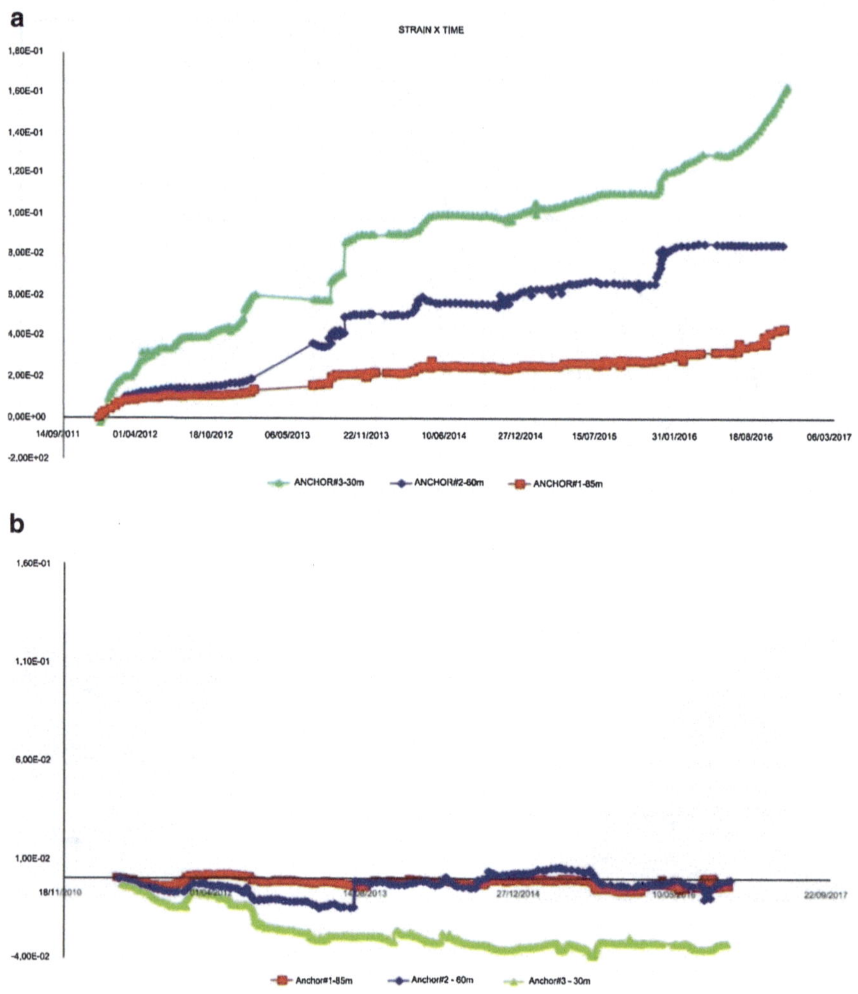

Fig. 13.13 Displacement measurements in the extensometers, respectively, (**a**) below and (**b**) above the foliation inflection zone

Pau Branco open pit mine. The magnitude of the displacements obtained from the numerical model did not show good match with those measured ones, as further calibration of deformation moduli will still be required.

The outward movement of the slope face at the lower half of the wall was a result of the gradual softening in the passive wedge and this has become evident by the displacement pattern of the extensometer installed below the inflection point (Fig. 13.13a). Plastic deformation causes the rock mass to dilate or expand, so that reverse shear along the steeply dipping foliation planes were visible on the face of slope, as idealized in Fig. 13.11. On the other hand, settlement appeared to be dominant above the inflection point as inferred from the very small horizontal displacement detected in the extensometer above the inflection (Fig. 13.13b).

13.5 Discussion on Rock Mass Classifications

Two approaches have applications for utilizing rock mass classification systems for weathered rock masses:

- To define geotechnical domains, each representing relatively similar weak rock mass conditions.
- To allow for rock mass structure pattern delineation in connection to weak rock mass strength estimates.

Several rock mass classification systems that are in common use by rock mechanics practitioners include Rock Mass Rating (RMR, Bieniawski 1976, 1989), *Q* (Barton et al. 1974), MRMR Laubscher (1990), and GSI (Marinos and Hoek 2000). It is assumed that readers are familiar with these systems.

In general, these classification systems were originally intended only for applications to jointed rock masses whose behaviour is controlled by discontinuities, that is, they were derived to consider parameters to estimate block size and inter-block shear strength for the rock mass. It is now becoming generally recognized that for weak rock masses (R1 and R2), block size and inter-block shear strength are not always the main controls on rock mass behaviour. Instead there may be a significant or even greater contribution from matrix strength. To overcome this deficiency, attempts have been made by other systems, such as Robertson (1988), the Blastability Index (BI) (Lilly 1986) used for blasting design, the classification proposed by Chatziangelou and Christaras (2013) for weak or laminated and sheared rock mass or Bar et al. (2016) for highly weathered rocks.

In addition, most current classification systems do not capture the major role played by rock mass structure anisotropy, whether it is caused by differential weathering. It is thus generally found that the behaviour of weak rock masses is not properly addressed in these classifications.

Recognizing that block-controlled behaviour becomes more and more irrelevant as intact strength falls below a given threshold, Table 13.5 presents some general guidelines for the application of classification system for weak rock masses. It is suggested that the routine (normal) rock mass classification systems should not be applied to R0 and R1− (less than 3 MPa) materials, representing the soil-like strength and behaviour, such as the residual and saprolitic soils observed in many mines in Brazil.

From the practical point of view, for large slopes, the threshold between the soil-like and rock-like behaviour can be roughly drawn as a transition starting from about 3 MPa, where the material behaviour is dominantly matrix controlled, to 5 or

Table 13.5 Proposed range of probable uniaxial compressive strengths for weak weathered rocks (adapted from ISRM, 1981)

Field assessment of rock strength or consistency	Degree of weathering and range of UCS (MPa)	Mass strength and behaviour	Use of current rock mass classification systems	Typical weathering profile
R0	W6– 0.25–1	Soil strength and behaviour varying with the grain size distribution and plasticity of the host rock. Develop Mohr–Coulomb parameters derived from laboratory tests.	Not applicable	Residual soil
R1−	W5– 1–3	Hard soil behaviour, controlled by the strength of the intact material and anisotropy. Develop Mohr–Coulomb parameters derived from laboratory tests.	Not applicable	Upper saprolite
R1+	W5/W4– –5	Transition from soil to rock mass behaviour. Strength of intact material dominating at large scale and strongly affected by foliation. Develop Mohr–Coulomb parameters from laboratory tests or from the Transition Function.	Requires adaptation	Saprolite
R2−	W4– 5–10	Transition from soil to rock mass behaviour. Strength of the rock mass transitioning from intact rock strength to rock mass strength, with the foliation and/or bedding planes creating strength anisotropy. Measure RQD. Can apply non-linear rock mass strength criterion with Transition Function modification.	Requires adaptation	Hard saprolite/ saprock
R2+	W4/W3– >10–15	Rock mass behaviour. Strength of rock mass results from a combination of intact rock strength, block size, and inter-block shear strength. Can apply non-linear rock mass strength criterion.	Applied	Saprock

10 MPa, where the material is progressively matrix to block controlled. Above about 10 MPa, it is thought that the rock mass behaves as a block-controlled system at any scale, and accordingly normal (current) classifications fully apply.

13.5.1 Estimating Geotechnical Domains

The RMR_{76} system can be used as basis for correlation to the Hoek–Brown constants m and s and as such maintains some equivalency with GSI. Hence, it is possible to use it in its original form as discussed here. Basically, as originally formulated, this system assigns ratings to five parameters: Uniaxial Compressive Strength (UCS) of the intact rock material; Rock Quality Designation (RQD); Spacing of Discontinuities; Conditions of Discontinuities; and Groundwater Conditions. These ratings are added together to define the RMR value representing the rock mass condition of the zone under consideration, which can then be used with the Hoek–Brown failure criterion as a basis for estimating block size and inter-block shear strength for the rock mass.

The rock mass range described in the RMR_{76} system defines five classes, as follows: Class I—Very Good Rock (RMR > 80); Class II—Good Rock (60 < RMR < 80); Class III—Fair Rock (40 < RMR < 60); Class IV—Poor Rock (20 < RMR < 40); and Class V—Very Poor Rock (RMR < 20).

These classes are used, in general, for reference when establishing geotechnical domains. However, based on real case experience, there is a tendency to subdivide Classes III (IIIA—RMR/GSI 51–60, IIIB—41–50) and IV (IVA—31–40, IVB 21–30) to properly account for weak rock mass behaviours. For example, in underground mines, spiling, shotcrete and longer bolts are usually required for advancing within Class IVB, whereas spiling would not be required for Class IVA, making RMR/GSI of about 30 a good differentiator of rock mass behaviour.

In addition to this subdivision of classes III and IV, it is common practice in Brazil to represent residual and saprolitic soils by a Class VI, an RMR "extended" category to clearly indicate its soil-like behaviour for which conventional rock mass classifications should not be applied.

As is evident from this discussion, there is an obvious need for improvement of current classification to allow weak weathered rock masses to be characterized. In an attempt to contribute to the future development of a suitable classification, it is considered that the following changes should be evaluated.

13.5.1.1 Intact Rock Strength

In the current RMR system, ratings of 1 and 2 are stipulated for R1 and R2 rocks, respectively. However, for R1 materials, laboratory test results show a wide scatter of cohesion values for materials described just in the interval 1–3 MPa, that is, the lower half of the R1 range (ISRM).

It is considered that higher RMR ratings can easily be applied to represent the strength of the intact weak rock. For example, Robertson (1988) proposed modifications to the RMR_{89} system to estimate weak rock mass strength from logging of drillhole core. His approach consisted of disregarding the groundwater condition rating from RMR_{89} and instead adding the maximum rating of 15 from the groundwater condition into the intact rock strength rating, resulting in a rating assignment for the highest intact rock strength class increasing from 15 to 30. In this approach, the water effect was left to be dealt with as phreatic surface in the stability analysis and the strength rating assigned to R1 rocks increased to 15. Robertson also proposed that RQD be measured and rated in the RMR classification for R1+ (or R1/R2) materials in a modified way to reflect the breakage potential when the core was firmly twisted and bent without substantial force or use of any tools or instruments. R1+ rock core, without planes of weakness, would not break under such handling.

A similar approach was used here, where the maximum rating of 10 from the groundwater condition was distributed to the R1+, R2− and R2+ materials, as will be presented in Table 13.6.

13.5.1.2 Rock Mass Structure

For weathered rock masses, the rock mass structure represented by the intercalation of strong and weak layers should be represented by reducing the RMR (or GSI) values for the FW slopes (i.e. parallel to foliation, as described in Sect. 13.4) compared to the estimated values for the HW slopes, similar to the rating adjustments for joint orientation used in the RMR_{76} system.

13.5.1.3 RQD and Fracture Frequency (or Joint Spacing)

In terms of RQD, some practitioners accept it to be measured for materials with strengths ≥R3, others would consider it for strengths in the upper range of R2 (i.e. R2+), and some like Robertson (1988) would recommend measuring RQD for the upper range of R1 (i.e. R1+) or stronger rocks.

As described in Table 13.1, and in line with Robertson's indication, it is proposed that RQD be properly measured for R1+ or stronger rocks, with capped values for weaker rocks. For instance, the maximum rating of the RQD parameter is suggested to be capped at 8 or 10 for the R1+ and R2− materials, respectively, as will be presented in Table 13.6. For R2+ representing UCS in the range of 10–15 MPa to 25 MPa or stronger rocks, the RQD ratings would be capped to 15.

13.5.1.4 Groundwater Effect

The existing geomechanical classification methods consider the influence of groundwater associated with seepage rate only and do not consider any potential impact on shear strength reduction for a weak rock mass. Groundwater is weighted simply in terms of hydrostatic pressure acting on an excavation boundary and as a way to predict the difficulties of a given excavation method. However, in saprolite materials and in soft iron deposits, groundwater can impact shear strength by either altering internal composition to reduce strengths or causing an undrained behaviour due to excess pore-pressure.

Table 13.6 Modified guidelines for RMR_{76} when considering R1/R2 materials

Factor	RMR 1976 Version				Limiting values to the RMR factors and proposed UCS grades for weak rock categories							
	RANGES OF VALUES	RMR 76 Grades	Proposed Modifications to 1976 grades	Proposed Grades	R1 – (<3MPa)	R1+(3 to 5 MPa)	CAP R1+	R2– (5 to 10 MPa)	CAP R2-	R2+ (10 MPa to 25 MPa)	CAP R2 +	OBSERVATIONS
UCS (MPa)	<1	0		0								Grades varying from 3 to 11 for the UCS sub-ranges R1+; R2-; R2+
	1 to 3	1		1	1							
	1 to 5	1	+2	3		3	3					
	5 to 10	2	+5	7				7	7			
	10 to 25	2	+9	11						11	11	
RQD (%)	<25	3		3	3	3	3	3		3	3	Upper RQD range to > 75%
	25 to 50	8		8		8	8	8		8	8	
	50 to 75	13	-3	10				10	10	10	10	
	75 to 90	17	-2	15						15	15	
	90 to 100	20		-								
Joint spacing (m)	<0.05	5		5	5	5	5	5	5	5	5	-
	0.05 to 0.30	10	-5	10		10	10	10	10	10	10	
	0.3 to 1 m	20	-5	15				15	15	15	15	
	1.0 to 3 m	25		20						20	20	
	> 3 m	30		30								
Joint conditions	Soft gouge	0		0	0	0	0	0	0	0	0	Grades adjusted for low strength rocks forming the joint walls when roughness is presented
	Slickensided	5		6		6	6	6	6	6	6	
	Slightly rough, HW	12	-4	8				8	8	8	8	
	Slightly rough, SW	20	-8	12						12	12	
	Very rough	25		25								
Groundwater	> 0.5 (flow)	0	Loose fine grained, argillaceous rock or cement; Contractive. Shearing reduction to *final grain size*	-10		-10	-10					No expected excess pore water pressure nor reduction to final gradation in shearing. Intact rock greater than 10 MPa
	0.2 a 0.5 (dripping)	0	Weakly bonded; Clayey	-7		-7	-7	-7	-7			
	0.1 a 0.2 (saturated)	4	Weakly bonded; Arenaceous	-3		-3	-3	-3	-3			
	< 0.1 (moist)	7	Relatively strong bonded skeleton; Brittle and pronounced dilative in shearing. Shearing non-reduction to final grain size.									
	0 (dry)	10		0						0	0	

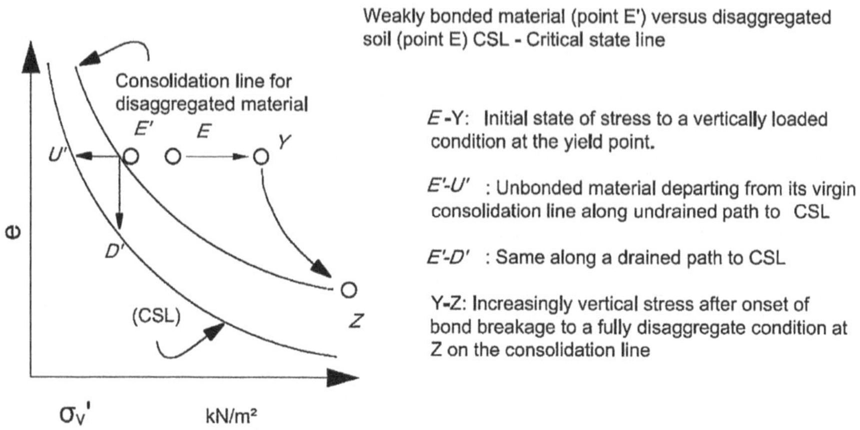

Fig. 13.14 State condition in void ratio vs. vertical effective stress space for a weakly bonded rock (chalk) and soil, after Vaughan (1990) in de Freitas (1993)

The engineering behaviour of a soft rock below the groundwater table highly depends on the mineralogy and the fabric imposed by chemical weathering and micro-fracturing. These are the primary causes of the gradual losses of bonding strength that ultimately lead the rock to a friable or soil-like condition. The role of water becomes more and more relevant for classification purposes as bonding (or "cohesion") losses occur in shear and eventually the rock disaggregates to its grain size texture. As can be seen in Fig. 13.14, the undrained stress path for a weakly bonded rock depends on its grain size and degree of bonding. The figure presented by de Freitas (1993) illustrates the role of interstitial water as weakly bonded rocks are stressed to failure. It highlights that the changes in void ratio associated with the contractive behaviour of the material are important to cause a marked increase in pore-pressure. In addition, conditions must be recognized in which the susceptibility to drying–wetting cycles, and swelling properties can, in general, supersede the role played by bonding disruption in shearing.

To the authors' knowledge, there are no existing engineering classification systems capable of capturing the role of groundwater and its potentially remarkable effect on the shear strength of weak rock masses. However, the concept of yield of weakly bonded materials (Vaughan 1990) is one that could provide the theoretical basis to frame the importance of water in a future classification system for weak rocks.

Some of the key-factors to be considered for a new classification system for soft rocks in a saturated zone are summarized as follows:

- *Type 1*: As rock fits into the category R2, the low void volume and the bonding between grains render the rock to become strong enough to result in brittle and dilative behaviour, thus making the pore-pressure drop to a low value or fully dissipate. Rocks with primary coarse granulation tend to disaggregate to sand size or larger and behave as a drained material. This may differ slightly if the rock is fine-grained and yet no remarkable contraction is expected.

- *Type 2*: Weakly bonded rocks and hard saprolites in R1+ category: the differing categories are:
  - High clay/fines/mica content: May have tendency for contraction if strained to small size particles, to its ultimate gradation, or non-pore-pressure dissipative behaviour. Likely to depend upon confining stresses magnitudes. Lab tests are required to define the condition.
  - Arenaceous (coarse grained) material: Tendency for pore-pressure dissipative behaviour.
- *Type 3*: Cemented Rocks in R1+ category: the behaviour in shear is dependent upon the mineralogy and the cement type. The nature of bonding must be evaluated within the following categories:
  - Clayey matrix
  - Siliceous matrix
  - Lateritic
  - Kaolinite/Illite (filling voids as a result of plagioclase deterioration)

In each case, the grain size reduction resulting from bonding breakage can cause either contractive to potentially high excess pore-pressure or dilative to low pore-pressure magnitudes. One must evaluate based on laboratory testing too.

Following the above discussion on weak material response to the interstitial water in shearing through intact rock—which in turn differs a lot from hydraulic pressures developed along discontinuities in harder rocks—the effect of "Groundwater" is proposed as a reduction factor in the RMR classification, presented in Table 13.6. It is considered that the influence of water in the shear strength of weak rocks would be better captured in the proposed RMR adjustments by assigning negative ratings according to the potential of the lithology to develop excess pore water pressure in shearing through the intact rock, and to what extent the mass failure is intact rock controlled.

As the weak rock mass composition ranges from massive to highly jointed conditions, intact rock pieces become even smaller and the excess pore-pressure developed in shearing through the matrix no longer has influence on the overall strength of the rock mass. In moderately to highly jointed rock masses, the stress–strain behaviour of the intact pieces affects the rock mass behaviour much less as drainage is allowed through the joint network. In this connection, a higher negative rating is proposed to loosen fine grained, argillaceous rock or cement, with contractive behaviour whilst a lower negative rating is suggested to coarse grained rock (i.e. arenaceous).

On the other hand, for rocks with UCS above 10 MPa (R2+) minimal to no effect of groundwater (null rating) is expected in shearing through the intact material, since the high bond strength between grains would cause the material to break into coarse pieces, that is, prevent a reduction to a fine grain size gradation, which could allow the buildup of excess pore-pressure. This means that for the upper range of weak rock spectrum (i.e. R2+), the strength of the intact rock prevails by far the degree of the grain size reduction, as opposed to the behaviour observed in very soft materials.

13.5.2 Proposed Adjustments to the Classification Systems

Until a rock mass classification is fully developed for weak rocks and embraced by the technical community, some trial guidance is given by modifying the application of the RMR classification system.

In terms of the RMR_{76} classification system and considering the discussion in Sect. 13.5.1, Table 13.6 proposes some adjustments for applying this system for weak rocks in order to differentiate what material can be treated as "soil" (R0 to R1−), transition (R1+ to R2−) and rock (gt;R2+, i.e. UCS gt; 10–15 MPa). In this table, all RMR parameters are measured for all rocks, though penalties are applied to weaker rock to limit RMR values. This contrasts with other approaches where parameters such as RQD might not be measured for weak rock and a default lower RMR is assigned.

As described in Table 13.1, for R0 (very weak saprolites, highly altered rock, and very friable hematite) and R1− strength materials, the rock mass classification should not be applied. If necessary, the minimum $RMR_{76} = 9$ can be assigned for the R1− material.

For the upper range of R1 materials (R1+, UCS gt; 3–5 MPa) RMR could range as high as 27 (Class IV), and for R2− materials (UCS gt; 5 to 10–15 MPa) the RMR could reach 40 (Class III) for dry (or a clearly dilatative R2 material) slopes. In both cases, the RQD ratings have been included in the estimation of the RMR values although limiting the RQD ratings to 8 and 10, respectively.

The proposed adjustments for the RMR ratings for weak rock masses (Table 13.6) were applied to a Brazilian open pit mine hosted in strongly weathered phyllite and schist. These weathered rocks exhibited strengths varying from R0 to R2. A comparison between the original and the modified RMR ratings showed that the proposed adjustments provided a significant distinction between the upper limits of RMR ratings in the UCS sub-ranges R1+, R2− and R2+.

13.6 Use of Classification System for Estimation of Weak Rock Mass Strength

For soil to extremely weak (R0) and very weak (R1) material occurring as any form of interlayering in a weak rock mass, their intact strength, derived from laboratory testing and often represented by the Mohr–Coulomb strength criterion, can reliably be used to define the mass strength behaviour. For stronger materials, rock mass strength can be approximated by downgrading the strength of the intact rock material, using values obtained from laboratory testing or using a rock mass classification system and applying, for example, the non-linear Hoek–Brown strength criterion (Hoek et al. 2002).

Weak rock in the R1 to R2− strength range is the subject of differing interpretations of rock mass classification schemes such as Rock Mass Rating (Bieniawski 1976, 1989) and lies between two material strength criteria—the Mohr–Coulomb criterion used for soils, and the Hoek–Brown criterion used for rock. The selected strength criterion to be used in slope design will require different types of material testing and data collection such as core logging and fundamentally a careful interpretation of the "Weak Rock Mass Structure", mainly comprising the geometry of inter-fingering weak and less weak layers, anisotropy and the type of failure that arose from the pattern of those features in the full-scale slope.

Therefore, clarifying the use of a rock mass classification scheme and the approximation of the boundary between soil and rock strength criteria constitutes a valuable tool for aiding design for slopes in weak rock.

At the low end of the rock mass quality scale (UCS <10–15 MPa and RMR or GSI generally <40), rock mass strength tends toward an increasingly matrix controlled (intact) strength. In the transition from inter-block shear failure (which is well-modelled by the Hoek–Brown failure criterion and the RMR system) towards a matrix-controlled rock mass behaviour, a gradual change in the strength curve can be created by taking into account a reduction in the cohesion component.

Currently application of the Rock Mass Rating (RMR)—Bieniawski classification system and its subsequent input into the Hoek–Brown strength criterion yields low strength parameters, underestimating weak rock mass strengths as observed by the high stable slopes excavated in many mine operations and road cuts.

There is a need for better definition in this range (as clearly shown by Zhang and Einstein (1998) for variation in ultimate bearing capacity for weak rocks, showing large variations from intact UCS within 1–12 MPa, Fig. 13.15). These tests were carried out at the bottom of shaft foundations excavated to depths varying from 2 to 20 m.

Figure 13.15 includes the estimated rock mass strength curves obtained using: (a) the Hoek–Brown (Hoek et al. 2002) strength criterion with RMR = 100 (Zhang and Einstein 1998), and (b) the Hoek–Brown (Hoek et al. 2002) strength criterion, RMR = 65 for confining pressure associated to density of 23 kN/m^3 and depths (L) of 2 and 11 m. A vertical dashed line is also plotted to divide the chart in regions above and below a UCS of 3 MPa, used as reference for the lower limit of weak rocks as considered here.

Consideration of an RMR = 50 and the average test depth of L = 11 m would provide a safe lower bound fitting curve for the experimental data since the totality of plate loading results plot above this curve except for the Case Depth#19. For this plate test (plate diameter = 1.2 m), in altered basalt, the RQD was 10% and a RMR of about 45 could be estimated. The experimental data lower bound has been taking into account to search for the maximum ratings allowed to very few jointed rock masses within the intact rock UCS range 5–25 MPa, i.e. encompassing the categories R2− and R2+. It is implicit in this approach that usual plate diameters are prone to test over small volumes of the rock mass. Thus, it is assumed that most of the ultimate strengths determined from those field small scale tests are dominantly matrix-controlled failures, with few exceptions like the above mentioned highly jointed basalt.

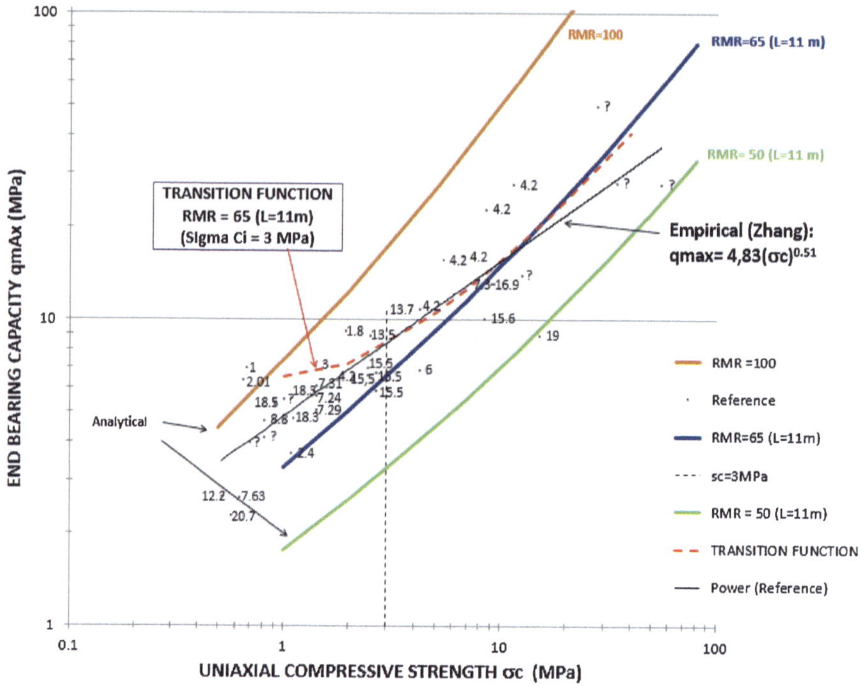

Fig. 13.15 End-bearing capacity from plate loading tests on shafts for different depths and UCS values for several weak rocks. (Adapted from Zhang and Einstein 1998). The strength envelope for an RMR of 65 has also been plotted to draw a low limit of a wide range of experimental data. The envelope for RMR of 50 bounds all data except the test # 19

This interpretation leads the authors to propose upper limits to RMR total ratings for UCS categories R1+ (3–5 MPa), R2− (5–10 MPa) and R2+ (10–25 MPa). As a result, lower and upper limits of the modified RMR ratings for each of the UCS categories R1+, R2− and R2+ can be found in Table 13.6, respectively, from 11 to 27, 15 to 40 and 19 to 58.

It is strongly recommended that the highest R2+ ratings are assigned only when the rock mass is massive (i.e. not fractured) and one is fully confident that rock falls into the upper bound of this range. Laboratory tests are recommended in the detail phase to confirm the RMR rating applied in preliminary design. Strength field estimates on the R2+ lower bound should rather be selected to the R2− category.

These plate test results suggest that application of the Hoek–Brown strength criterion using the RMR classification would tend to be very conservative for UCS ranging up to 10–15 MPa and this becomes more relevant in low confinement conditions. It also helps to explain why in several real cases, the results of triaxial tests in laboratory for intact soft rock specimens classified as R1− have provided strength envelopes significantly higher than those predicted by using corresponding RMR low values as per the current form of the Hoek–Brown criterion (Hoek et al. 2002).

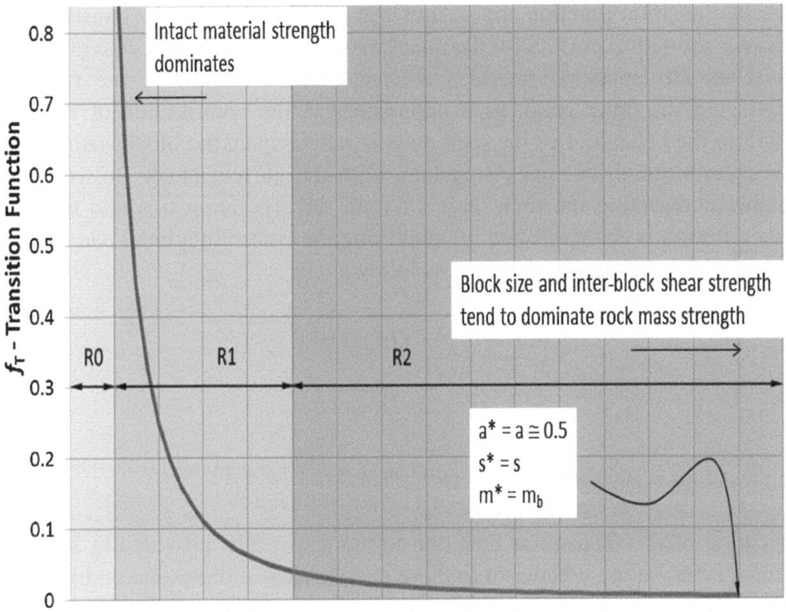

Fig. 13.16 Transition function (after Castro et al. 2013)

13.6.1 Transition Function for Estimating Weak Rock Mass Strength

A transition function was proposed by Carvalho et al. (2007) and modified by Castro et al. (2013) to modify the Hoek–Brown criterion expressions to better replicate strength behaviour of weak rock materials in the range of R1 to R2.

The transition function, f_T, adjusts the proportional contribution from the "intact" material to the strength of the rock mass. It thus ensures a transfer of control from the joint fabric near the rock end of the transition to the "intact" material (as tested in the lab) at the soil-like material other end of the transition. The lower limit suggested for the transition function is UCSi = σ_{ci} = 1.0 MPa. The transition function can be expressed by:

$$f_T = (UCS_i) = \begin{cases} 1, & UCS_i \leq 1.0\ MPa \\ \dfrac{1}{UCS_i^2} & \end{cases}, \ UCS_i \text{ in } MPa \qquad (13.1)$$

A graphical representation of the *transition* function is shown in Fig. 13.16.

As the transition function approaches 1 at the lower end of the transition scale, the matrix strength (i.e. UCS for the intact weak rock) dominates behaviour and the Mohr–Coulomb envelope provides adequate representation of the rock mass strength. On the other hand, as it approaches 0, the contribution of the matrix strength to the behaviour of the rock mass is reduced and the block size and inter-block shear strength becomes prevalent, with the current Hoek–Brown strength envelope representing the rock mass strength. Incorporating this into the Hoek–Brown criterion is accomplished by modifying the controlling input parameters by the transition function relationships, as follows:

$$s^* = s + \left[(1-s) f_{\mathrm{T}}\right] / \left(4a^* - 1\right) \tag{13.2}$$

$$a^* = a + (1-a) f_{\mathrm{T}} \tag{13.3}$$

$$m_{\mathrm{b}}^* = \left[m_{\mathrm{b}} + (m_i - m_{\mathrm{b}}) f_{\mathrm{T}}\right] / \left(4a^* - 1\right) \tag{13.4}$$

Application of this Transition Function to the test results provided in Zhang and Einstein (1998) shows a better fit to those results, as also illustrated on Fig. 13.15.

13.6.2 Mining Example: Goldstrike Betze-Post Mine, Elko, Nevada, USA (Sharon et al. 2005)

The Carlin Formation exposed at Goldstrike contains subunits including the Carlin Silt, Carlin Sand (ash) and Carlin Gravel, with material and strength parameters summarized in Table 13.7. Two cases of RMR_{76} are applied for each rock type considering the range of RMR_{76} values for R1 material, varying from 9 to 27, the latter obtained by using the descriptions and revised ratings proposed in Table 13.6.

Table 13.7 Carlin formation parameters

			Used in the original design			Estimated using the transition function (for Slope Height = 50 m[c])	
Lithology	γ (kN/m^3)	σ_{ci} (MPa)	c (kPa)	Phi (°)	RMR_{76} (assumed)	c (kPa)	Phi (°)
Carlin silt	17.3	2.8	105	30	9[a]	163	26
					27[b]	197	31
Carlin sand (ash)	14.1	1.4	105	35	9[a]	112	35
					27[b]	167	33
Carlin gravel	19.6	2.1	170	35	9[a]	152	28
					27[b]	178	31

[a]RMR_{76} (Bieniawski 1976)
[b]RMR_{76} from Table 13.6
[c]Assumed $m_i = 12$

The estimated values for cohesion and friction angle for the Carlin Sand and Carlin Gravel, applying the transition function shown in Fig. 13.16, produce comparable shear strength results with those used in the original design (Castro et al. 2013).

Table 13.6 also illustrates that varying RMR_{76} from 9 to 27 does not significantly change the shear strength parameters for the R1 materials determined using the transition function, reducing the concern as to whether or not the RQD should be measured and accounted for in the RMR classification of R1 type materials.

References

Bar N, Barton N, Ryan C (2016) Application of the Q-slope method to highly weathered and saprolitic rocks in Far North Queensland. Proceedings Eurock 2016, Cappadocia, Turkey

Barton NR, Lien R, Lunde J (1974) Engineering classification of rock masses for the design of tunnel support. Rock Mech 6(4):189–263

Bieniawski ZT (1976) Rock mass classification in rock engineering. In: Bieniawski ZT (ed) Exploration for rock engineering. A.A. Balkema, Johannesburg, pp 97–106

Bieniawski ZT (1989) Engineering rock mass classifications. Wiley, New York

Brito SNA, Cella PRC, Figueiredo RP (2011) Importância da Geologia de Engenharia e Geomecânica na Mineração. Rev Bras Geol Eng E Ambiental 1(1):123–139

BVP (2009) Consolidação e Análise do Ensaios Geotécnicos de Laboratório Realizados nas Minas N4E, N5E e N4WN – Carajás, PA, VL 707-01-09-E-CA-RT-19-002, Outubro, Vale

BVP (2012) Modelo Geomecânico da Porção Oeste da Cava Norte da Mina de Fpabrica Nova – Complexo Mariana, Mariana-MG. VL119.15-10-E-CA-RT-07-003, Setembro, Vale

BVP (2017) Mina do Sapo. Relatório Técnico Final, Conceição do Mato Dentro_MG, AA 024-15-E-TA-RT-07-163, Fevereiro, AngloAmerican

Carter TG, Marinos V (2014) Use of GSI for rock engineering design. Keynote lecture at the 1st International congress of mine design by empirical methods, Lima, Peru

Carvalho JL, Carter TG, Diederichs MS (2007) An approach for prediction of strength and post yield behaviour for rock masses of low intact strength. In: Proc. 1st Can-US Rock Symp. Meeting Society's Challenges & Demands, Vancouver, pp 249–257

Castro LM, Carvalho J, Sá G (2013) Discussion on how to classify and estimate strength of weak rock masses. In: Dight PM (ed) Proc. 2013 International Symposium on slope stability in open pit mining and civil engineering. ACG, Brisbane, Australia, pp 205–218

Cella PRC (2017) Soft Iron Ores and Other Leached Rocks. Guidelines for Open Pit Slope Design in Weak Rocks, Section 10.5.3 Strength Parameters. CSIRO, A Balkema Book (ed).

Chatziangelou M, Christaras B (2013) Rock mass blastability dependence on rock mass quality. Bull Geol Soc Greece 47:1694–1705

De Freitas MH (1993) Weak arenaceous materials. The engineering geology of weak rock, Eng. Geol. Special Publ, vol 8. Balkema, Rotterdam, pp 115–123

Deere D, Patton F (1971) Slope stability in residual soil. In: Proc. 4th Pan-American conference on soil mechanics and Eng. foundation, June 1971, San Juan, Puerto Rico, vol 1. America Society of Civil Engineers, New York, pp 87–170

Durango JR, Gavinia CM (1987) Geologia de Cerro Matoso. Paper presented in the VII Congreso Nacional de Mineria, Colombia

Fietze C, Creighton A, Castro L, Hammah R (2013) Pit slope design in phyllites for the Simandou large open pit project. In: Dight PM (ed) Proc. 2013 International symposium on slope stability in open pit mining and civil engineering. ACG, Brisbane, Australia, pp 115–125

Hawley PM, Martin DC, Acott CP (1986) Failure mechanics and design considerations for footwall slopes. CIM Bull 79(896):47–53

Hoek E, Carranza-Torres C, Corkum B (2002) Hoek-Brown failure criterion – 2002 Edition. Proc Narms-Tac Conf Toronto 1:267–273

International Society of Rock Mechanics - ISRM (1981) Rock characterization, testing and monitoring - ISRM suggested methods. Pergamon, Oxford

Jeffcoate A, Campbell M, Vannay J-C, Perriaux B (2013) Modelling intact rock strength of clay-rich rocks with geochemistry at the Simandou Iron Ore Deposit, Guinea, West Africa. In: Proc. iron ore conference, Perth, Australia

Klein, S. (2001). An approach to the classification of weak rock for tunnel project, Chapter 26. In: 2001 RETC Proceedings

Laubscher DH (1990) A geomechanics classification system for the rating of rock mass in mine design. J S Afr Inst Min Metall 90(10):257–273

Lilly P (1986) An empirical method of assessing rock mass blastability. In: Large open pit mine conference, Newman, Australia, pp 89–92

Lowe J, Johnson TC (1960) Use of back pressure to increase degree of saturation of triaxial test specimens. In: ASCE research conference on shear strength of cohesive soils, Boulder, Colorado, USA, pp 819–836

Marinos P, Hoek E (2000) GSI: a geologically friendly tool for rock mass strength estimation. In: Proc. GeoEng2000 at the Int. Conf. on geotechnical and geological engineering, Melbourne. Technomic Publishers, Lancaster, PA, pp 1422–1446

Marinos P, Hoek E (2001) Estimating the geotechnical properties of heterogeneous rock masses such as flysch. Bull Eng Geol Environ 60:82–92

Marques EAG, Barroso EV, Menezes Filho AP, Vargas EA Jr (2010) Weathering zones on metamorphic rocks from Rio de Janeiro – physical, mineralogial and geomechanical characterization. Eng Geol 111:1–18

Martin RP, Hencher SR (1986) Principles for description and classification of weathered rock for engineering purposes. Geol Soc 2:299–308

Nickmann M, Spaun G, Thuro K (2006) Engineering geological classification of weak rocks. IAEG 2006 – Paper number 492

Robertson AM (1988) Estimating weak rock strength. AIME Annual General Meeting, Tucson, Arizona, January 1988

Sharon R, Rose N, Rantapaa M (2005) Design and development of the Northeast layback of the Betze-post open pit. In: Proc. of the SME Annual Meeting, Salt Lake City, US. Pre-print 05-09, 10 p

Sheahan TC, Ladd CC, Germaine JT (1996) Rate-dependent undrained shear behavior of saturated clay. ASCE J Geotech Eng 122(2):99–108

Vaughan PR (1990) Mechanical properties: general report. Chalk, Thomas Telford, London, pp 293–294

Zhang L, Einstein HH (1998) End bearing resistance of drilled shafts in rock. J Geotech Geoenviron Eng ASCE 124(7):574–584

Chapter 14
Correlation of Soft Rock Properties

Milton Assis Kanji and Marcio Leão

14.1 Introduction

Soft Rocks are a geologic material with critical problems, since they are in between soils and hard rocks, presenting intermediate and variable strengths. Very frequently they are too hard to be tested in soil mechanics equipment and too soft to be tested in rock mechanics laboratories.

Their nature may be sandy, clayey, carbonatic, or evaporitic, presenting one or more types of characteristics of crumbling, foliated, slaking, expansive, and very much affected by water, originated from sedimentary and metamorphic rocks, and crystalline rocks when weathered.

Usually they present great difficulties in sampling, as the core recovery may be very bad, sometimes requiring special drilling techniques to improve core recovery, and in specimen preparation for testing, requiring special and even improvisations of nonconventional procedures to achieve better results.

In the field, when including discontinuities, due to the smaller difference between the strength of the discontinuities and of the intact rock, their failure departs from pure classical failures in soils or along discontinuities, most of the times involving mixed mechanism of failure.

The characteristics and behavior of soft rocks are not yet well known by lack of more investigation of these materials, leading to adopt conservative parameters for these materials in the engineering practice, thus impairing the economics of the project, as already mentioned by Leão (2011, 2015).

Moreover, the sites with excellent geological conditions have already been taken for the existing engineering works, and more and more the projects have to deal

M. A. Kanji (✉)
Polytechnical School, University of São Paulo, São Paulo, Brazil

M. Leão
Federal University of Viçosa, Viçosa, Brazil

M. Kanji et al. (eds.), *Soft Rock Mechanics and Engineering*,
https://doi.org/10.1007/978-3-030-29477-9_14

with sites dominated by soft or weathered rocks. Many large areas of some countries are entirely dominated by soft rocks, as for instance some sedimentary basins.

It is of utmost importance the effort to investigate soft rocks to understand their characteristics, to improve their sampling and testing procedures, and to identify relatively simple properties that could indicate their behavior and collect experiences, to have a reliable basis to allow adopting realistic design parameters and construction solutions.

It would be also desirable to find some type of index property that in a simple way could indicate the characteristics of the soft rock, in a similar way that grain size distribution and plasticity index work for soils.

One of the ways to improve our knowledge of these materials is considered to develop correlations among their physical and mechanical characteristics, in the attempt to identify simple properties that would indicate their behavior in the first instance. Also, it would be easier to deduct other properties when they show good correlation.

This chapter, therefore, deal with the correlations between different physical and mechanical properties of soft rocks with the objective above mentioned. Having collected a reasonable number of properties available in the bibliography, some hard rocks and soils were also included in the correlations to show correlations of these materials beyond the limits of soft rocks.

This objective was pursued by various authors, and we can mention the works by Dobereiner (1984), Deere (1975), Deere and Vardé (1986), Bosio and Kanji (1998) on soft rocks of the Rio de la Plata Basin in South America (including Argentina, Brazil, Paraguay, and Uruguay) and of Kanji and Galván Lievano (1998), The work by Galván Liévano (1999) is stressed here, as he collected and correlated soft rock properties published in the precedent 20 years, in his investigation of the properties of cemented sands. Later, Kanji (2012, 2014a, b) published further works, broadening the considerations, problems, case histories, and practical problems in soft rocks.

Although the terms "soft rocks" or "weak rocks" may represent either intact rock of low strength or rock masses that are weakened by discontinuities, cavities, weathering or else, here only intact rocks of low strength are considered. The data presented here assembles the data from previous publications, data from more recent papers, including Elhakim (2015), Leão et al. (2017), Romana and Vasarhelyi (2007), data obtained from Wunder and Meirelles (2014) on sandstones from Estreito dam site, and from several papers published in Anagnostopoulos et al. (1993, Editors).

The considerations about the geological types of soft rocks are not presented here as one chapter of this book fully describes this topic. In addition, there are also other specific chapters dealing with sandstones, mudstones, and evaporitic rocks.

14.2 Soft Rocks According to Their Strength

The range of strength to be attributed to intact soft rocks is somewhat variable according to authors, but there are limits generally accepted.

The lower limit for soft rocks coincides with what is considered the upper limit for soils. Most authors consider this value to be between 0.5 and 1 MPa. These values

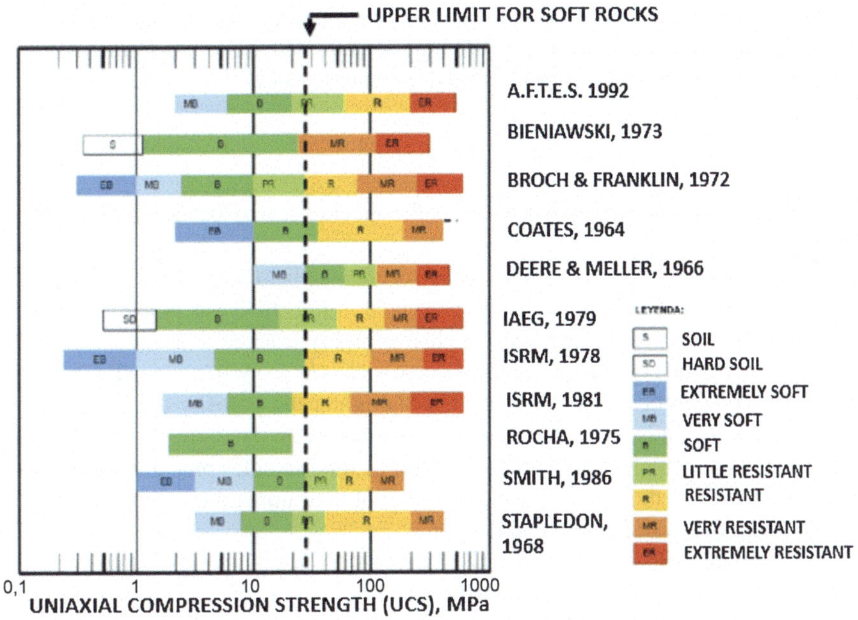

Fig. 14.1 Classification of the degree of strength by diverse authors (Galván Liévano 1999, modified)

coincide with the observation by Terzaghi and Peck (1967) that for soils with N_{SPT} above 50 or for uniaxial compressive strength (UCS) greater than 0.4 MPa the material presents characteristics more like rock than soil. Rocha (1975) suggests that if under submersion the material does not disaggregate it should be considered as rock, indicating a value of about 2 MPa. Similarly, Dobereiner (1984) proposed saturation under vacuum instead of immersion, encountering an approximate value of 0.5 MPa for sandstones, which is practically the same proposed by Terzaghi and Peck.

For the upper limit of the uniaxial strength of soft rocks ISRM (2015) adopted the value of 25 MPa as a reference, although it is clear that this limit must be somewhat flexible depending on the nature and behavior of the rock. In a recollection on how soft rocks are classified according to the UCS, Galván Liévano (1999) found that diverse authors classified the strength degrees of soft rocks somewhat variable, but with some coherence, giving support to the upper limit of 25 MPa adopted by ISRM, as shown in Fig. 14.1.

14.3 Correlation Between Physical Properties

The physical properties of greater interest to evaluate weak rocks are the specific weight (dry, natural and saturated), the porosity and the water absorption capacity (Kanji 2012).

The theoretical relationship between the apparent dry density and porosity is shown in Eq. (14.1), valid for every lithology, usually with the smaller porosities for igneous rocks and the greater values for sedimentary rocks:

$$G_o = G_s(1-\eta) \tag{14.1}$$

where G_o is the apparent dry density, G_s is the specific gravity of the solids, and n is the porosity.

Notwithstanding, there is the interest in seeing whether all lithologies follow this theoretical line, reason why the graph presented in Fig. 14.2 was prepared, including hard crystalline rocks and soils, inclusive highly leached soils. It is seen that there is a very good correlation for the diverse rocks ($R^2 = 0.90$), practically coinciding with the theoretical line, showing that the grater the porosity the lesser is the density. However, it is noted that several data depart from the correlation line, causing a dispersion, which is attributed to several factors: samples preparation, the way the properties were determined, nonuniformity of their mineral composition, densities with variable water content, and so on.

As the graph include a great variety of soils and rocks, the good continuity of the data show that porosity and dry density are basic to characterize the fundamental properties of the rock, with the advantage that the porosity could be very easily estimated from water absorption and the density influences the strength of the material.

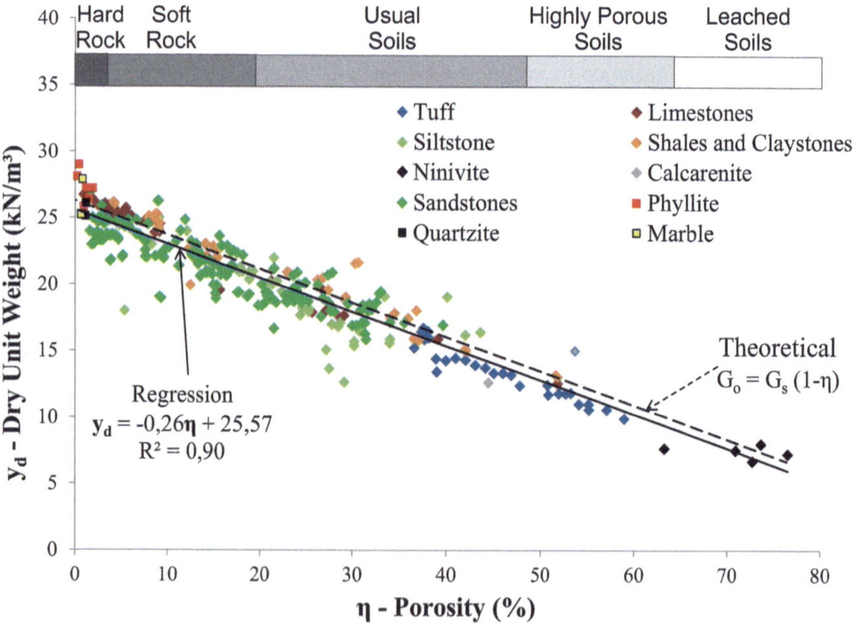

Fig. 14.2 Graph of correlation between dry unit weight and porosity. The bar above indicates roughly the type of rock in each porosity ranges

On the other hand, the porosity is correlated to the water absorption capacity. The absorption was defined by Hamrol (1961) as the Absorption Index (I_v), being the ratio between the weight of absorbed water to the dry weight of the sample. The theoretical correlation between the Absorption Index and porosity can be expressed by the Eq. (14.2) for a saturation degree of 100%.

$$I_v = (G_w / G_s)(\eta / 1 - \eta) \tag{14.2}$$

The relationship of Eq. (14.2) is valid when all the pores are filled with the absorbed water. However, in case of cemented sedimentary rocks and weathering products of crystalline rocks, it is not certain if all pores are interconnected, or same pores might be closed by cementation without allowing entrance of water. This is the main reason why a graph correlating water absorption (I_v) and porosity (η) was prepared, as shown in Fig. 14.3.

Here also the data fit very well along the correlation curve, with some dispersion which could result from some closed pores, sample preparation without achieving full saturation, testing procedures, and so on. The extreme data with very high porosities correspond to "ninivite," an Egyptian special soil highly leached and porous.

This correlation is considered very interesting and important, since water absorption is an easier testing, from which the porosity can be obtained. Moreover, if porosity is related to density and density is related to strength, the absorption may be a property that indicates basically the rock properties.

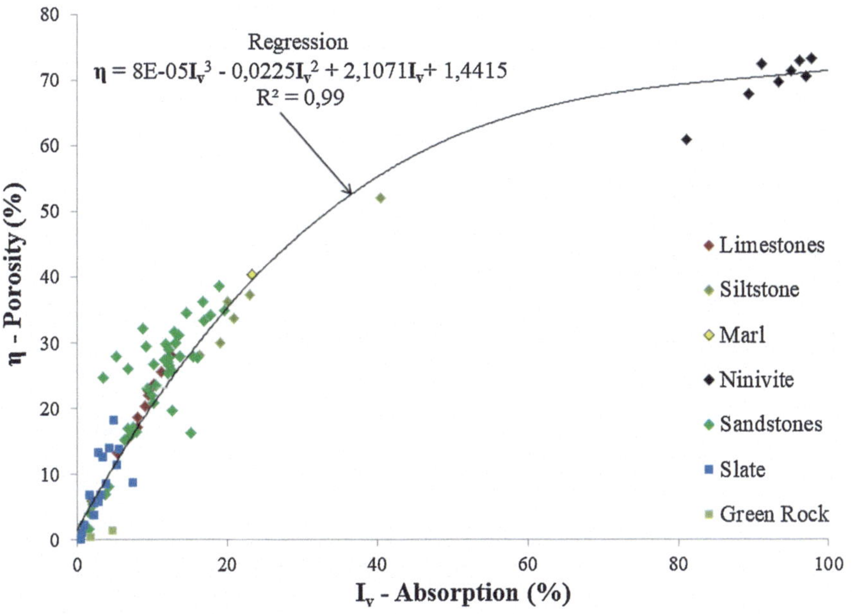

Fig. 14.3 Relationship between Absorption and Porosity

14.4 Correlation Between Physical and Mechanical Properties

The variability of the properties is great, considering their different origins, nature and conditions, as degrees of consolidation, cementation, metamorphism, and so on. Moreover, as already mentioned, the difficulties in sample preparation and laboratory testing must be taken into account. Even the testing procedures of index properties as specific weight, absorption, porosity, sonic velocity, and point test may vary according to labs, even considering the issuing of the Suggested Testing Methods by ISRM, which may not be strictly followed by every lab. All these factors may cause greater or lesser dispersion of results when trying to establish a relationship between them.

The correlation between dry unit weight and uniaxial compressive strength (UCS) is shown in Fig. 14.4, representing the regression lines for dry samples (continuous line) and for saturated samples (discontinuous line). The graph shows what is intuitively known that the denser the material, the harder and stronger it is. It also shows that saturated samples are weaker than dry ones, in different degrees according to the nature of the rock. Another observation is in the sharp inflection of the regression lines, showing that rocks with dry unit weights lower than about 21–24 kN/m^3 and lower than about 25 MPa have very low strength.

The large dispersion observed in the graph must have several causes. First, it depends on the universe of data. Secondly, in the case of phyllites, slates, and shales, for instance, it also depends on the direction the UCS has been determined. In foliated rocks there may be layers with carbonatic material influencing the strength

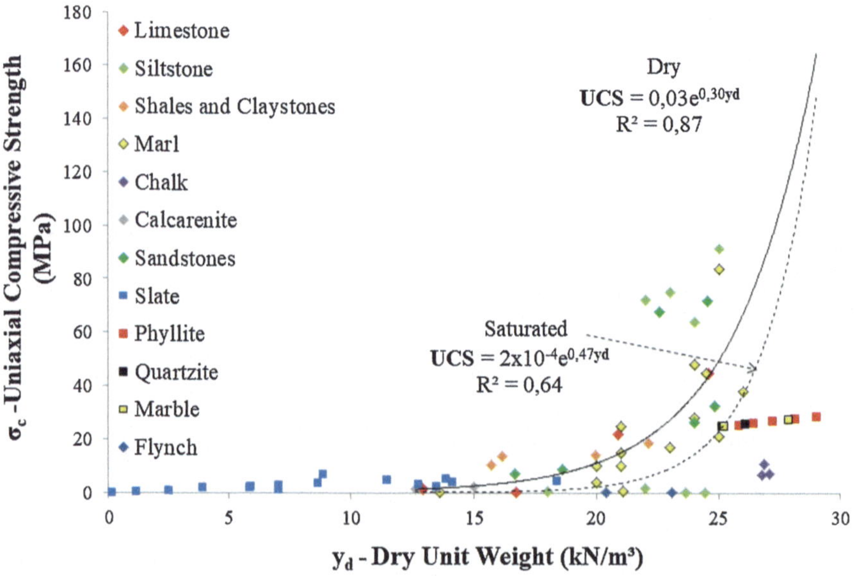

Fig. 14.4 Relationship between dry unit weight and uniaxial compressive strength, for dry and saturated samples

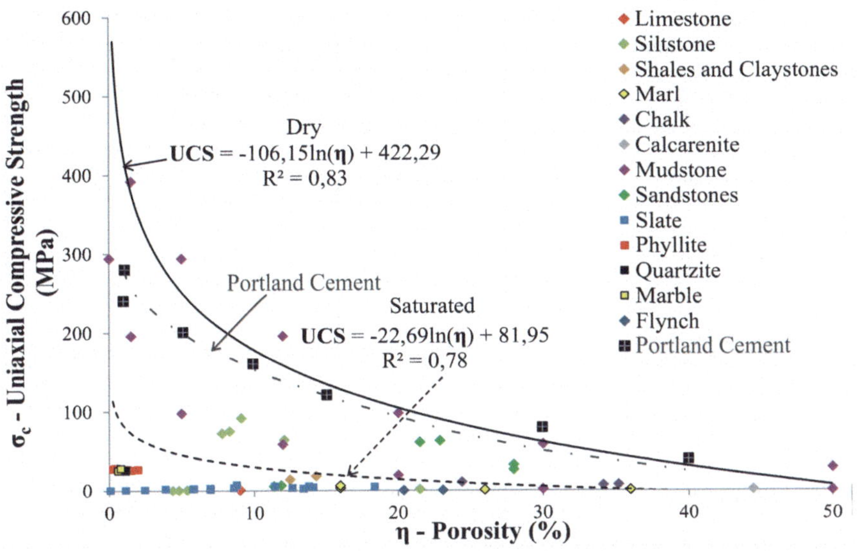

Fig. 14.5 Correlation between UCS and porosity, for dry and saturated samples

(Hsu and Nelson 1993). In sandstones and siltstones, it is basically due to the degree of cementation. In the case of muddy rocks, it may depend on the type and quantity of expansive minerals.

In a similar manner, the correlation between UCS and porosity is also observed, but in an inverse correlation way, since the greater the porosity the lesser is the unit weight of the rock, as shown in Fig. 14.5. The continuous line is a regression line for dry samples and that for saturated samples is the discontinuous line. Here also it is seen that dry samples are stronger that saturated ones, in different degrees. Particularly for slates and some siltstones, they are very weak regardless of the porosity.

A geologic consequence of the influence of porosity on UCS is due to diagenesis with consolidation and cementation, which usually increases with the geologic age. The result is a decreasing porosity with increasing geologic age. A proof of this is demonstrated for sedimentary rocks of the Paraná Basin (involving considerable areas of Argentina, Brazil, Paraguay, and Uruguay, in South America) showing the influence of geological diagenesis and cementation on sedimentary rocks in decreasing porosity and increasing UCS the older are the rocks. Although older, the Jurassic rocks resulted with less UCS than the Cretaceous rocks, due to the different nature of the rocks in these periods. The Cemented Sands from Asunción (Paraguay) although Mesozoic in age present diminished UCS since they are in an initial state of diagenesis. The results are presented in Fig. 14.6.

Most of the data from the properties of sedimentary rocks from the Paraná Basin come from Campos (1988) who organized a report on the properties of those rocks, added with data from Bosio and Kanji (1998) for the Asunción sands.

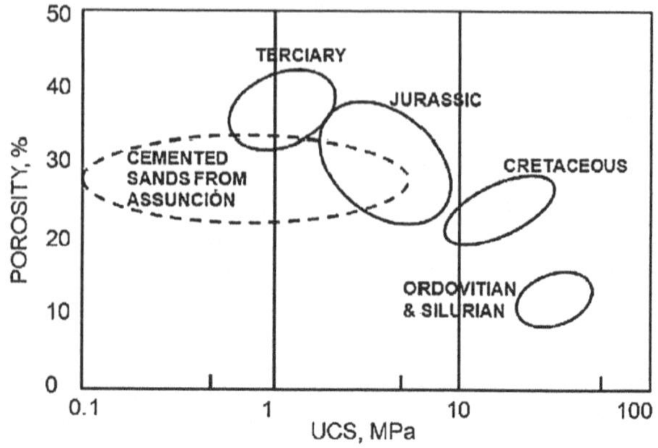

Fig. 14.6 Effect of decreasing porosity with the geologic age, and its influence on UCS, for sedimentary rocks from the Paraná Basin (South America) (from Kanji 2014b)

14.5 Correlation Between Strength and Elastic Properties

In developing an engineering classification of intact rocks, Deere (1968) developed a relationship between uniaxial compression strength and modulus of elasticity or Young Modulus. The elastic modulus is taken as the tangent modulus at 50% of the ultimate strength ($6_{0(ult)}$), which later was simply called E_{50} in the technical literature.

In correlating UCS and E_{50}, Deere established the 'modulus ratio' E_{50}/UCS, which could be high (H) if over 500, medium (M) when between 200 and 500 and low (L) if less than 200. Together to the classification of strength, it results in an engineering classification of intact rocks. The modulus ratio limits resulted from the observation that the data for a single type of rock followed the trend of the modulus ratio lines. Different types of rocks or different testing directions define their position in the graph.

Deere's graph with the plots of the data used in this work to present soft rocks data, as shown in Fig. 14.7, allowing to see that soft rocks plot in the range of Medium and Low modulus ratios, as the case of most sandstones and mudstones. Shales, igneous and metamorphic rocks usually present Medium to High modulus ratios, as well as limestones and well cemented sandstones.

There is a general tendency of the rocks to have higher modulus ratio with the increase in their strength.

A potentially useful correlation using the static modulus of elasticity (E_{stat}) would be with the dynamic modulus of elasticity (E_{dyn}), since knowing one of them should allow the estimation of the other, if such correlation were reliable. Of course this correlation should be here restricted to intact samples, since in the field the wave propagation can be very much affected by geological conditions (type of rock, degrees of weathering or jointing and water content).

The several correlations in the bibliography show some dispersion, but definitely with the same trend. Based on a large collection of data from bibliography, some correlations are presented in Fig. 14.8. For intact rocks it can be said that the E_{dyn} generally is 1.5–2 times the E_{stat}. In this way, the sonic velocity determination at the lab can be a rapid way to roughly estimate the strength of the intact rock.

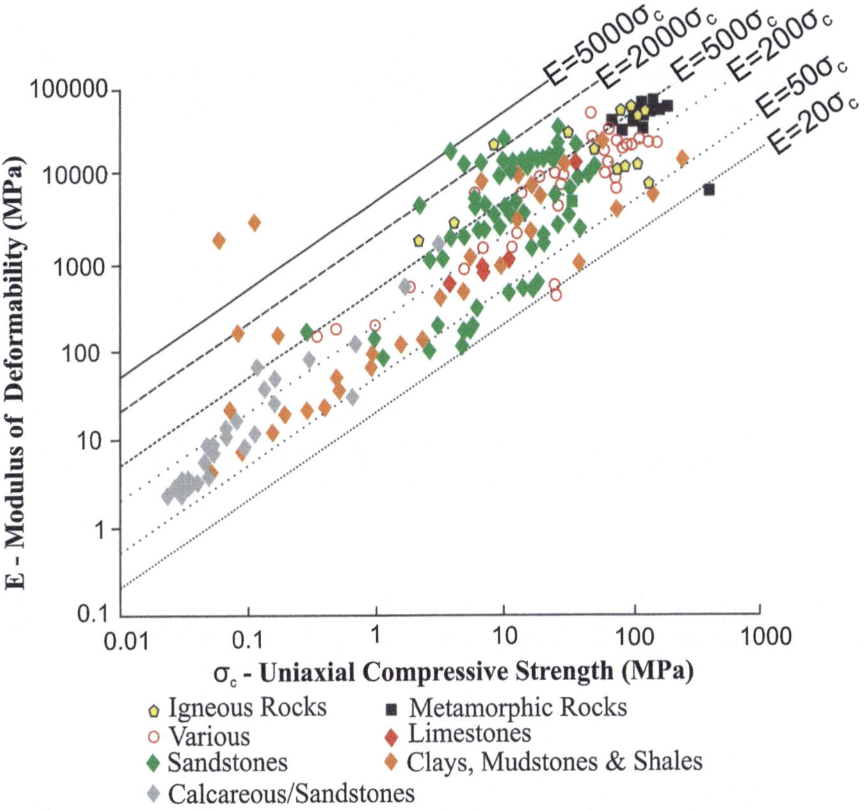

Fig. 14.7 Graph correlating UCS and Young Modulus (E_{50}). Symbols in green are from Wunder and Meirelles (2014)

Since UCS and E_{50} are related between them, they both should also show good correlation with the sonic velocity in intact rock specimens.

The relationship of the modulus E_{50} and the sonic velocity V_p is presented in Fig. 14.9 where, despite the dispersion, the universe of data show a good trend, parallel to the regression lines for dry samples (continuous line) and saturated samples (discontinuous line), showing here again the influence of water in decreasing the properties of the rock. Sandstones and siltstones show smaller sonic velocities, while phyllites and slates present higher velocities.

On its turn, the relationship between the UCS and sonic velocity, presented in Fig. 14.10 show the same trend as $E_{50} \times V_p$, but apparently with a smaller dispersion of results, being a more reliable correlation. Interestingly, it is seen generally that rocks with less than about 25 MPa show sonic velocities less than 3000 m/s, while for higher UCS the sonic velocities are much higher.

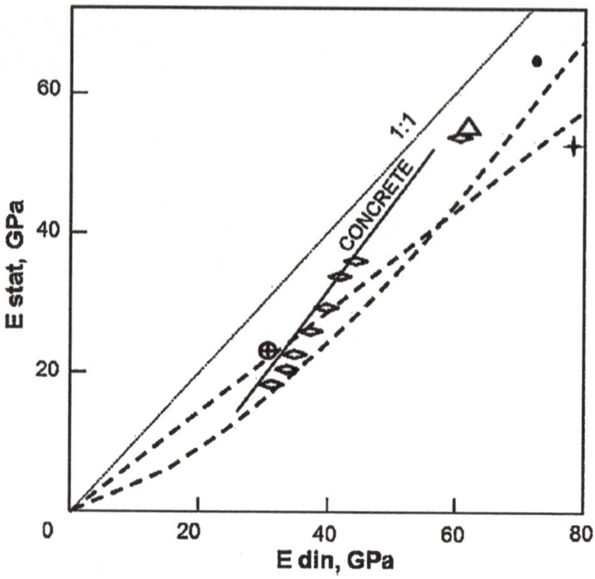

Fig. 14.8 Relationship between static (E_{stat}, E_{50}) and dynamic moduli (E_{dyn}) for intact rocks, showing the data for concrete as a reference (Kanji 2014b, modified from Galván Liévano 1999)

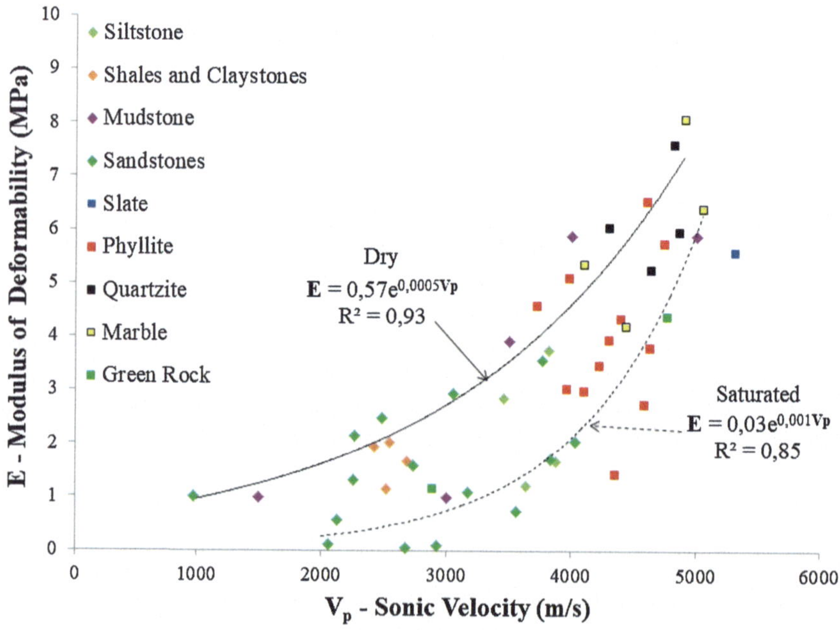

Fig. 14.9 Relationship between modulus E_{50} and sonic velocity for dry and saturated samples

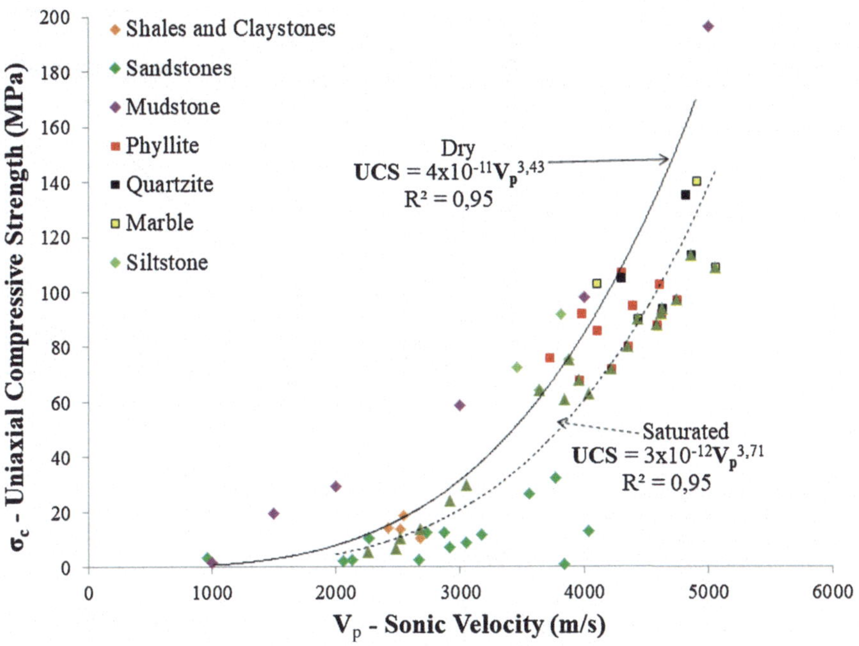

Fig. 14.10 Relationship between UCS and sonic velocity for dry and saturated samples

14.6 Simulation of Arenaceous Soft Rocks Though Artificial Materials

The idea of attempting to simulate rocks by artificial materials consists in the possibility of studying their properties under controlled and well-known conditions, which can be varied at one's wish. It could, for example, define the transition region between brittle and plastic behavior, correlate different properties for a determined material, and so on.

The simulation of clayey soft rocks is more difficult, as it involves many factors perhaps without possibility of control, as consolidation with time, cementation, and mineral composition. Many researchers studied clayey soft rocks, as shales, mudstones, and marls, as for example Jeremias (1997) who did an extensive study of silty clayey materials. However, no one to our knowledge has attempted to simulate them through artificial materials.

Arenaceous or sandy material, on the other hand, is easier to simulate through variable mixtures of sand and Portland cement. Variations can also be made with the grain size and grain size distribution to produce different material with variable properties.

With this idea in mind, Galván Liévano (1999) studied artificial sandy soft rocks through mixtures of medium sand and Portland cement in different proportions. Cylindrical samples were allowed to settle during 1 month immersed in clean water to be prepared as specimens for testing of physical properties (unit weight, porosity,

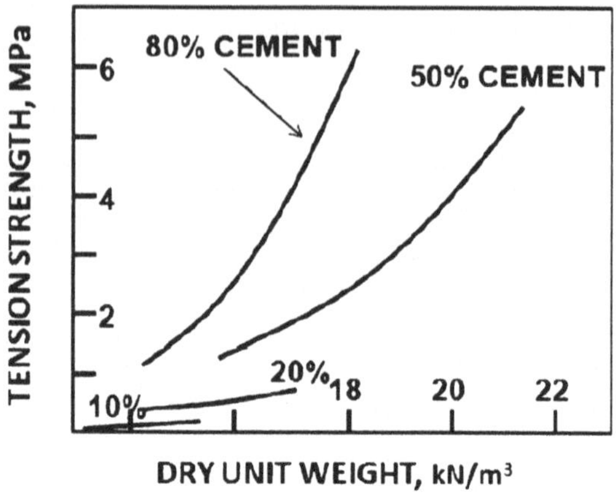

Fig. 14.11 Tension strength of sand and different cement Mixtures determined by Brazilian test (Galván Liévano 1999)

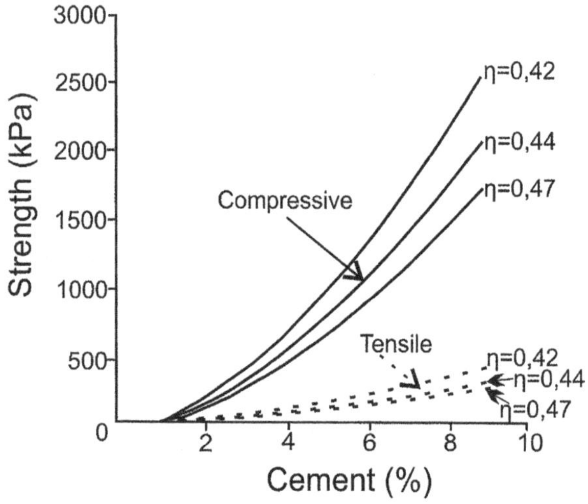

Fig. 14.12 Relationship between UCS and Tensile Strength as a function of cement content. (Floss et al. 2016)

absorption) and mechanical properties (UCS, E_{50}, point load test, Brazilian tension test, sonic velocity, Schmidt hardness, etc.).

Frequently the specimen preparation and testing procedures had to be improvised, as sometimes neither the soil nor the rock mechanics laboratories were adequate to the nature of the material. The results, as summarized in Galván Lievano and Kanji (2012), have shown that these sandy artificial soft rock fit very well among the plots of natural soft rocks, raising the possibility to study them through artificial materials.

The mechanical strength represented by tension strength as a function of cement content is shown in Fig. 14.11.

Compressive and tensile strengths as a function of cement content and porosity are shown in Fig. 14.12.

Only small differences with the collection of data for natural soft rocks were detected with respect to the correlation of Schmidt Hammer rebound testing with UCS, resulting in smaller strengths for the artificial material, as depicted in Fig. 14.13.

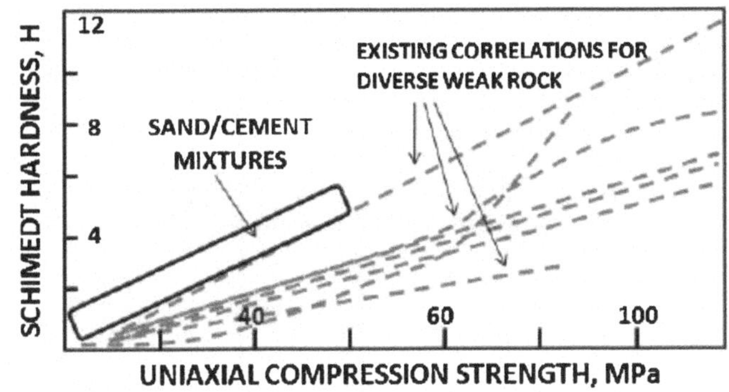

Fig. 14.13 Relationship between UCS and Schmidt Hammer rebound. (Galván Lievano and Kanji 2012)

Fig. 14.14 Results of the testing (q_u—compressive strength RCS and q_t—tensile strength RCD by ratio η—porosity/C_{iv}—cement concentration) with cemented sands by Floss et al. (2016)

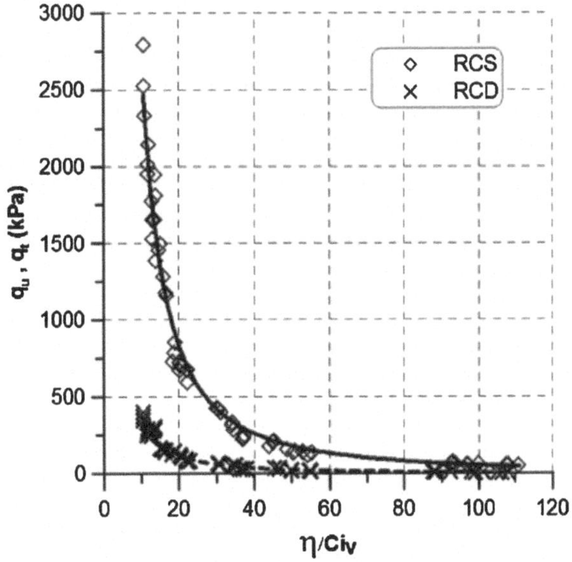

More recently, Floss et al. (2016) have also tested cemented sands, obtaining the results shown in the graph of Fig. 14.14.

14.7 Search for Index Properties

It has been shown above the several correlations between different physical and mechanical and elastic properties for soft rocks, encompassing practically all types of rocks (igneous, sedimentary, metamorphic). Some of them have very little dispersion and can be considered reliable, and others have somewhat greater dispersion, but can be correlated anyhow.

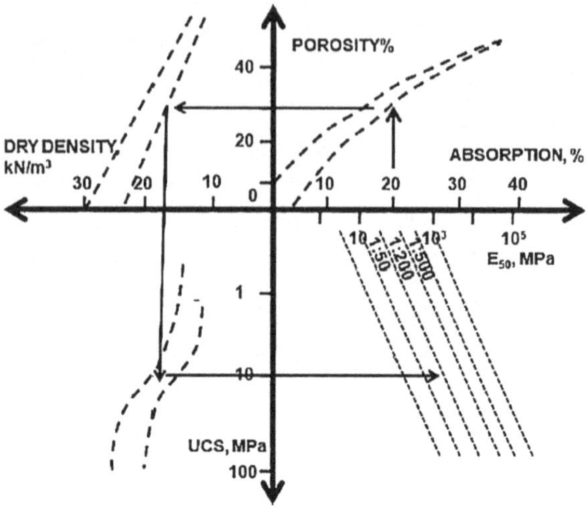

Fig. 14.15 Correlations between Absorption, Porosity, UCS and *E*, starting from Absorption (Kanji 2014a)

Since the porosity can be very well represented by the absorption, and porosity is well related to dry density, which on its turn is related to the strength and elastic properties, the absorption could be an easier and simpler way to indicate generally the rock behavior. The possible series of correlations starting with absorption is shown in Fig. 14.15 (Kanji 2014a). Therefore, it is suggested that water absorption can work as an index property for soft rocks, with the facility that no specimen preparation is required.

It was also seen that since sonic velocity correlated very well with UCS, both for dry and saturated samples, it could also be utilized as an index property. It requires, however, specimen preparation to have cylindrical specimens, which are not available always or not possible to have.

References

Anagnostopoulos A, Frank R, Kalteziotis N, Schlosser F (1993) Editors, Geotechnical Engineering of Hard Soils—Soft Rocks: Proceedings, Intern. Symp, vol 1–3. ISSMFE, Athens

Bosio JJ, Kanji MA (1998) Soft Rocks of the Rio de la Plata—South America, Proc. 2nd. Int. Symp, vol 1. Hard Rocks and Soft Rocks, Naples, pp 55–61

Campos JO (1988) 1° Report, Committee on Geotechnical Studies of Sedimentary Rocks, Coord., Technical Article 15, Assoc. Bras. Geol. Eng., S. Paulo, pp 160

Deere DU (1968) Chapter 1: Geological considerations. In: Stagg KG, Zienkiewicz J (eds) Rock mechanics in engineering practice. Wiley, New York, pp 1–20

Deere DU (1975) Applied Rock Mechanics of Weak Materials—General Report, 5th Panamerican Congr. Soil Mech. And Found. Engineering, Buenos Aires

Deere DU, Vardé O (1986) Engineering Geological Problems Related to Foundations and Excavations in Weak Rock, 5th International Congress IAEG, Buenos Aires

Dobereiner L (1984) Engineering geology of weak sandstones. PhD Thesis. Imperial College of Science and Technology. University of London

Elhakim AF (2015) The use of point load test for Dubai weak calcareous sandstones. J Rock Mech Geotech Eng 7(4):452–457
Floss MF, Rezende IM, Ferreira G, Henrique P, Thomé A, Consoli NC (2016) Evaluation of simple compression strength and artificially cemented agate residue traction. Ciência Engenharia 25:1–6
Galván Liévano VR (1999) Simulation of the geotechnical properties of the weak arenaceous rocks by means of artificial materials, Doctoral Thesis, Dept. Structural & Geotechnical. Engineering, Polytechnical School, University of São Paulo, São Paulo
Galván Lievano V, Kanji MA (2012) Simulation of arenaceous weak rocks by means of cemented sands. In: Proceedings of the 12th International Congress of International Society for Rock Mechanics, vol 2011. CRC Press, Beijing, pp 751–752
Hamrol AA (1961) Quantitative Classification of Weathering and Weatherability of Rocks. In: International Conference on Soil Mechanics and Foundation Engineering, 5p
Hsu SC, Nelson P (1993) Characterization of cretaceous clay-shales in North America. Geot. Eng. of Hard Soils- Soft Rocks. Anagnostopoulos et al (eds. Balkema), pp 139–146
ISRM – International Society of Rock Mechanics (2015) The ISRM suggested methods for rock characterization, testing and monitoring: 2007–2014, 5 ed., Ulusay R (ed)
Jeremias FT (1997) Influence of the geological aspects on the geotechnical properties of silty-clayey rocks, Thesis for the degree of Specialist. Lisboa, Laboratório Nacional de Engenharia Civil (in Portuguese)
Kanji MA (2012) Rocas Blandas—Problemas y Soluciones en obras de Ingeniería (Weak rocks—problems and solutions in engineering works), 2nd Int. Symp. Excavations in Rock, SCG, Costa Rica, 21pp. (in Spanish)
Kanji MA (2014a) Critical Issues in Soft Rock. J Rock Mech Geotech Eng 6(3):186–195
Kanji MA (2014b) Engineering Works Affected by Soft Rocks, Rock Mechanics for Natural Resources and Infrastructure. In: SBMR 2014—ISRM Int. Symp., Goiania, Brazil
Kanji MA, Galván Lievano VR (1998) Correlation of Properties of Soft Rocks, Proc. 2nd. Int. Symp, vol 1. Hard Soils and Soft Rocks, Naples, pp 239–244
Leão MF (2011) Geological-geomechanical modeling of San Juan dam. Dissertation, Federal University of Rio de Janeiro, Depart. of Geology, Rio de Janeiro, RJ, Brazil. (in Portuguese)
Leão MF (2015) Analysis strain stress of a concrete dam in residual soil predominantly anisotropic. Dissertation, University of State of Rio de Janeiro, Department of Civil Engineering, Rio de Janeiro, RJ, Brazil. (in Portuguese)
Leão MF, Polivanov H, Barroso EV, Marques EAG, Vargas EA Jr (2017) Weathering of metapelites from the Quadrilátero Ferrífero mineral province, southeastern Brazil. Bull. Eng. Geology and the Environ. Springer, New York, p 15. https://doi.org/10.1007/s10064-017-1036-1
Rocha M (1975) Some problems related to the rock mechanics of low strength natural materials. In: Proceedings of the 5th Pan-American Conference on Soil Mechanics and Foundation Engineering, Buenos Aires. (in Portuguese)
Romana M, Vasarhelyi B (2007) A discussion on the decrease of unconfined compressive strength between saturated and dry rock samples. In: 11th International Society for Rock Mechanics and Rock Engineering, Lisbon, Portugal
Terzaghi K, Peck R (1967) Soil mechanics in engineering practice. Wiley, New York
Wunder E, Meirelles MC (2014) Soft Sandstone Physical and Geomechanical Characterization. In: In: SBMR 2014—ISRM Int. Symp., Goiania, Brazil

Chapter 15
Deformation Mechanism of Soft Rock

He Manchao and Sun Xiaoming

15.1 Deformation Mechanisms of Soft Rock

The deformation of soft rocks is complicated. According to the theoretical analysis and engineering experiences, the deformation mechanisms of soft rock can be classified into three categories, namely, physical and chemical expansion mechanism, stress dilatancy mechanism, and structural deformation mechanis (He 1992a–d, 1993, 1997, 2002).

15.1.1 Physical and Chemical Expansion Mechanism

The physical and chemical expansion is related to the chemical properties of the molecular structure of the rock. This mechanism can be divided into three subtypes according to the origins of the expansion (Fig. 15.1).

15.1.1.1 Molecular Expansion Mechanism

The expansibility of argillaceous rocks that contain montmorillonite or illite/smectite mixed minerals is quite large. It is often related to the molecular structure of montmorillonite, so this mechanism can also be called as montmorillonite expansion mechanism.

The crystal of montmorillonite is composed of numerous parallel unit cells. The cell thickness is about 14 Å and it is formed by three layers: the upper and lower

H. Manchao (✉) · S. Xiaoming
State Key Laboratory for Geomechanics and Deep Underground Engineering of China, University of Mining and Technology, Beijing, China
e-mail: hemanchao@263.net

M. Kanji et al. (eds.), *Soft Rock Mechanics and Engineering*,
https://doi.org/10.1007/978-3-030-29477-9_15

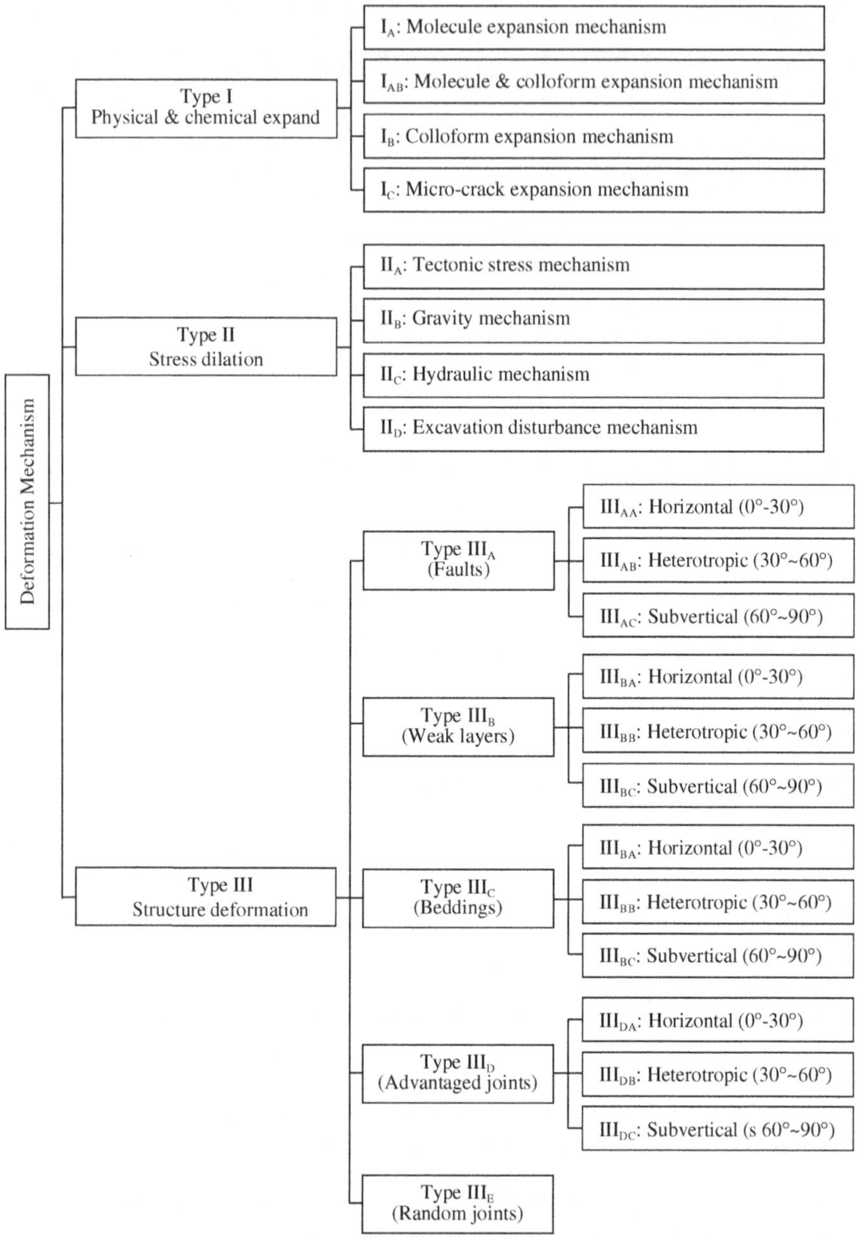

Notes: a°~b° is the angle between tunnel and the strike of fault, weak layer or joints

Fig. 15.1 Deformation mechanisms classification of soft rock

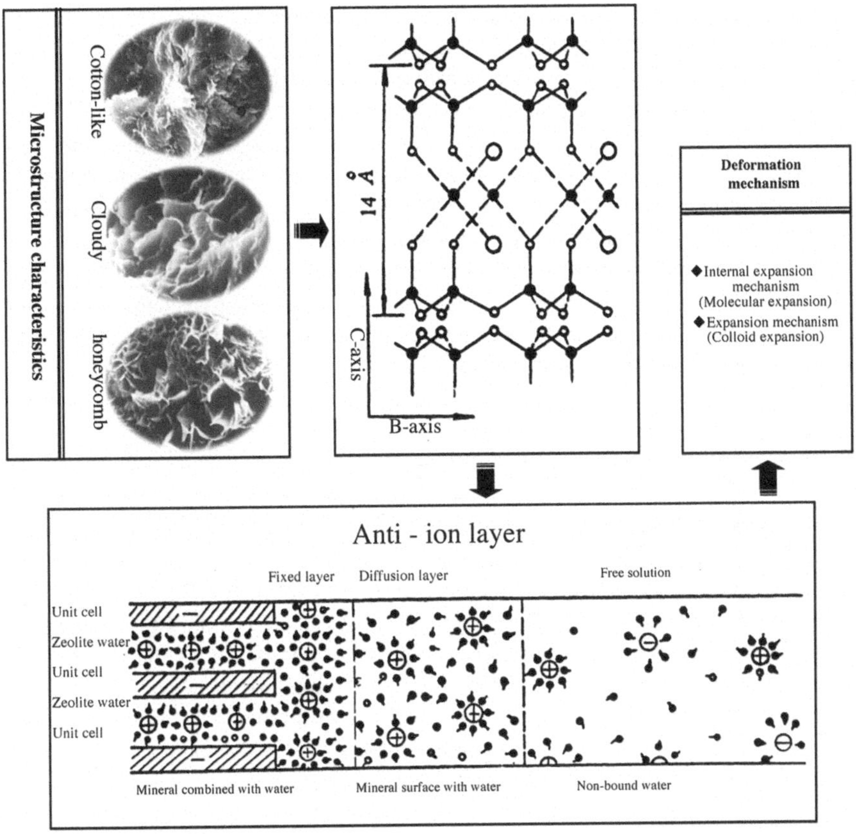

Fig. 15.2 Montmorillonite expansion mechanisms

layers of Si–O tetrahedron and the middle layer of Al–O–OH octahedral, as shown in Fig. 15.2. One characteristic of this molecular structure is that the ratio of tetrahedral to octahedral in each unit cell is 2:1. The unit cells are in contact with O^{2-}, so water molecules can be absorbed as they are not very close to each other. Al^{3+} can be replaced by Fe^{3+}, Fe^{2+}, Ca^{2+}, and Mg^{2+} ions to form a variety of different minerals in montmorillonite group. If the substitution is divalent ions, it generates free atomic valence in the ion frame, which can improve the adsorption capacity and enhance the connection between unit cells. Consequently, the montmorillonite group has strong water adsorption capacity; sometimes the connecting force between adjacent cells may be lost due to large volume expansion.

The adjacent crystal of montmorillonite has the same charge, so they are repulsive to each other (Fig. 15.2). When mudstone contacts with water, parts of anti-ion of zeolite-water can escape from the montmorillonite. It reduces molecular attracting force causing water molecules flowing into space between cells, that is, mineral particles expansion. In addition, the water film between mineral particles also becomes thickening, which leads to colloid expansion. As the montmorillonite can increase intergranular space and thicken the particle structure, its swelling capacity is the largest among clay minerals. It was reported that Ca-montmorillonite could expand to more than seven times of its original volume (Norrish 1954).

Not only does montmorillonite follow the internal expansion mechanism, but also the illite/smectite mineral and illite mineral have the same swelling characteristic. The SiO_2 in the three-layer structure of illite is less than in montmorillonite. Si^{4+} in the upper and lower Si–O tetrahedra can be replaced by Al^{3+} and Fe^{3+}. Therefore, their free atomic valence is different from the montmorillonite, and there are more monovalent captions in the adjacent crystal cells. Sometimes they even give rise to divalent positive ion to compensate the lack of positive charge in the unit cell. As K^+ ion commonly exists in soft rock, illite crystal's lattice is less active than montmorillonite, and its hydrophilic swelling capability is also low.

15.1.1.2 Colloid Expansion Mechanism

For soft rocks in coal mine, some of them do not contain montmorillonite, illite/smectite mixture or illite, but they are also expansible. For example, when the clay is kaolinite, humic substance or insoluble salt, it has certain swelling capability.

Herein, we take kaolinite as an example to illustrate such expansion mechanism. The molecular structure of kaolinite is composed of parallel unit cells. As the unit cell is connected by O^{2-} and OH^- strong bonding, it does not allow water molecules come into the space between unit cells. So its hydrophilicity is low and volume expansion is small. That is, when kaolinite contacts with water, the water molecules cannot flow into the space between unit cells. However, if there are free valence atoms and ions on the clay particle surface, the situation will become different. These atoms or ions can form an electrostatic field in the clay particle surface. Water molecules are dipole, that is, its one end is positive charge and the other end is negative charge, so it can be attracted by electrostatic force. As a result, water molecules near the clay particle surface are attracted and rearranged neatly, as shown in Fig. 15.3.

The water molecules very close to the clay surface lose their activities and are attracted tightly by the particle. At a distance from the clay surface, the electrostatic field weakens, water molecules in this area partially lose their activities and they are weakly bound water in a colloid. In far field, electrostatic attraction has almost no effect, the water molecules move as liquid water. Water molecules that completely lose their activities are in the strongly adsorbed layer of the colloid. The water molecules that partially lose their mobilities are in the weakly adsorbed layer of the colloid. These two portions form a hydrating film that causes clay volume expansion.

The hydrophilic capacity of clay particles is different, the adsorption capacity increases with increasing of particle surface area and decreasing of particle size. When a thick layer of hydrated film is formed at the clay surface, it becomes colloid clay and leads volume expansion. This expansion mechanism is not unique to kaolinite; montmorillonite, illite, or other minerals with particle size less than 0.002 mm have the same expansion mechanism.

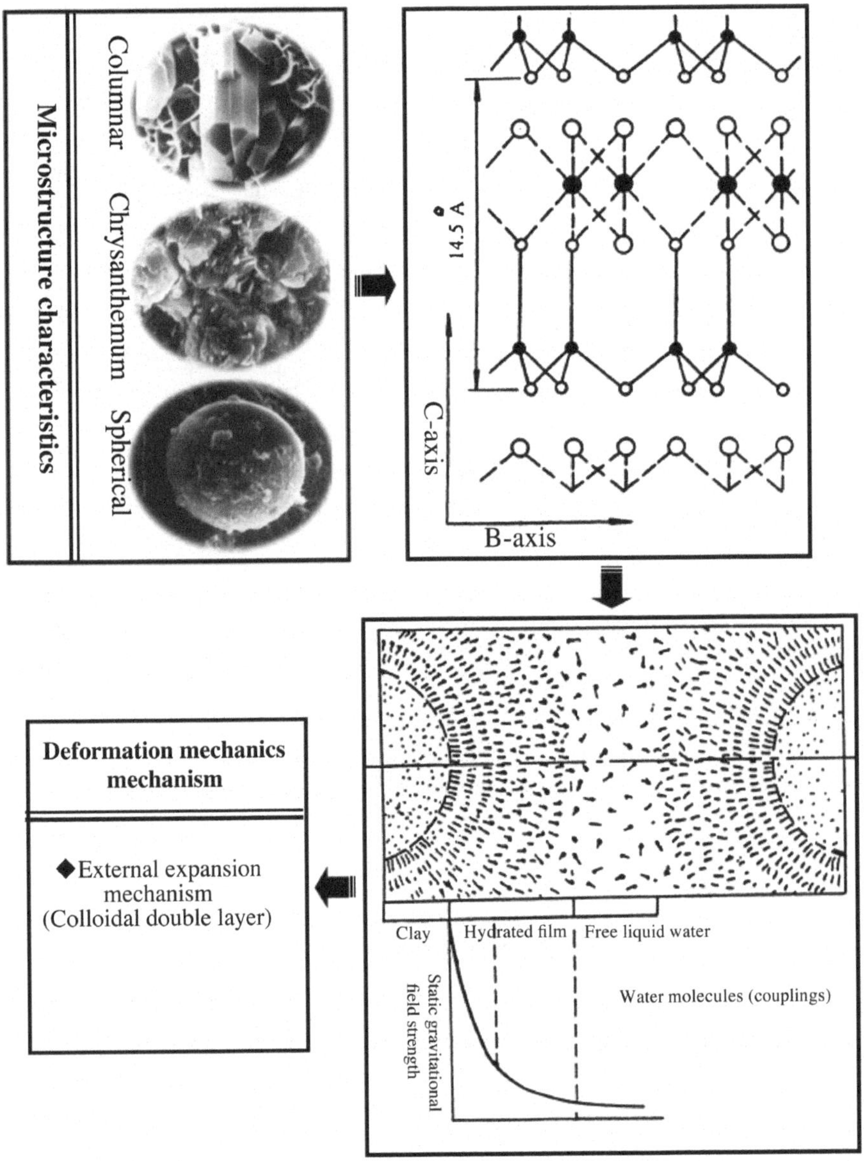

Fig. 15.3 Kaolinite expansion mechanism

In general, soft rock is often a muddy rock with large volume of clay particles. When adjacent particles are close to each other, their hydrating films may be partially overlapped to form a common hydration film (Fig. 15.4). If the particle hydration film becomes thick, their common hydration film will vanish. In such case, rock expansion is plastic deformation as there is no gel effect between clay particles. If

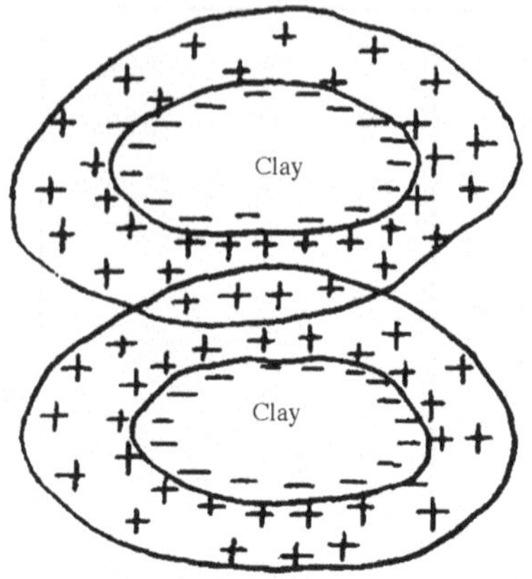

Fig. 15.4 Agglomerate hydration films expanding

the individual hydration film is thin, the common hydration film is formed by the strongly absorbed water molecules, which can generate strong connection between particles, namely cementation.

The process of soft rock colloidal expansion is shown in Fig. 15.5. When the common hydration film around clay particles is formed by strongly absorbed water, the soft rock can be quite hard. If the rock contacts with water, the common hydration film will thicken and be substituted by weakly absorbed water. As a result, the hydration bonding force between clay particles will be diminished. In this situation, the deformation of the soft rock enters into plastic deformation stage. If the water contents of the rock increase continuously, the common hydration film becomes extinct due to thickening of hydration film of each individual particle. Free water can be generated and with flow between particles, the soft rock goes into a viscous state.

15.1.1.3 Capillary Expansion Mechanism

Groundwater seepage may occur through small crack channels within soft rock in cases of large porosity, well developed cracks or high capillary pressure. The capillary pressure is generated by water surface tension, and its magnitude depends on the porosity and effective particle size of the rock. According to testing data, the capillary height is from zero to several centimeters for pebble, tens of centimeters in sand and hundreds of centimeters in clay (equivalent to clay soft rock). In essence, the capillarity of soft rock refers to the interaction between solid material and surface tension generated at water–air interface (Fig. 15.6), that is the reason why water can be removed to all directions around a capillary point.

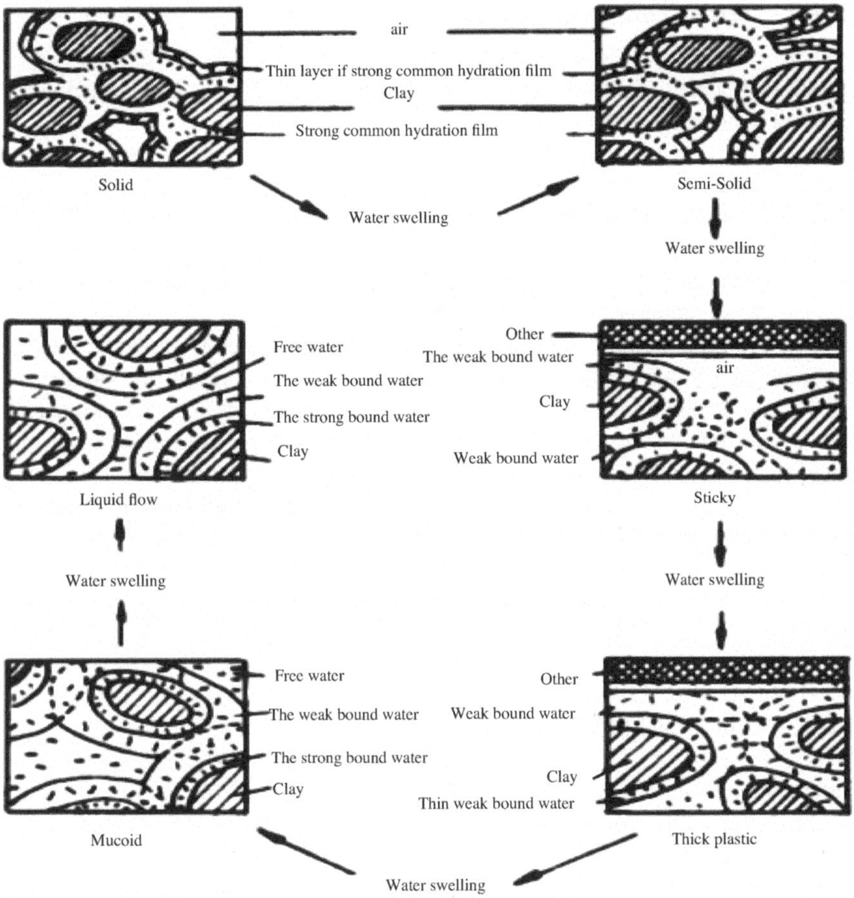

Fig. 15.5 Expansion mechanism induced by water absorption of clay in soft rock

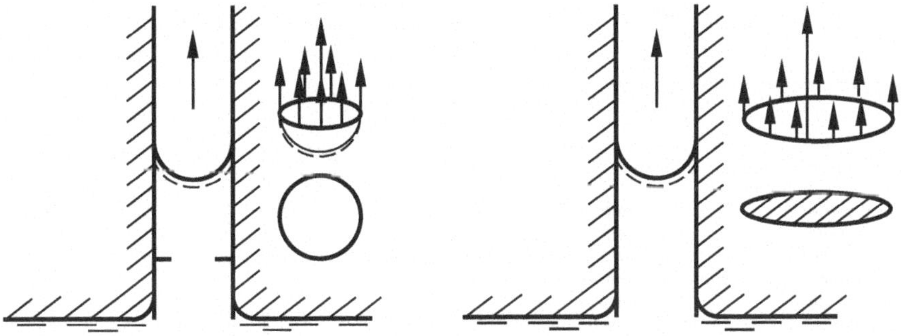

Fig. 15.6 Capillary mechanism in porosity and fissure media

15.1.2 Stress Dilatancy Mechanism

The stress dilatancy mechanism can be divided into four subtypes for soft rock mass based on the origins of the stress (Fig. 15.1).

15.1.2.1 Tectonic Stress Mechanism

In geological period, coal seams experienced multi-stage geological tectonic stresses and stored huge amount of deformational energy. Once a roadway is excavated in a formation, these deformational energies are released towards the excavation causing volume expansion of surrounding rock. As a result, the dominated stresses in surrounding rock changes from three components to two components after excavation. In the procedure, nonlinear elastic-plastic deformation may be generated under high tectonic stress, which is a time-related phenomenon. Tectonic stress is often horizontal. If the tectonic stress is large, the deformation of a roadway developed perpendicular or parallel to the major principle stress is quite different, as shown in Fig. 15.7. So the deformation of surrounding rock caused by tectonic stress is closely related to the major principle stress direction the excavation direction.

15.1.2.2 The Effect of Water

Water can generate physicochemical effect and also mechanical effect in soft rock. The mechanical effect of water can be divided into hydrostatic pressure and hydrodynamic pressure.

When the roadway is excavated in an aquifer, the stability of the surrounding rock is affected by discharge of groundwater. Support difficulties (such as spraying layer) may increase due to hydrodynamic pressure. Moreover, discharge of groundwater increases water contents for shaly soft rock, causing rapid swelling. The related mechanisms involve grain chemical expansion and colloidal expansion.

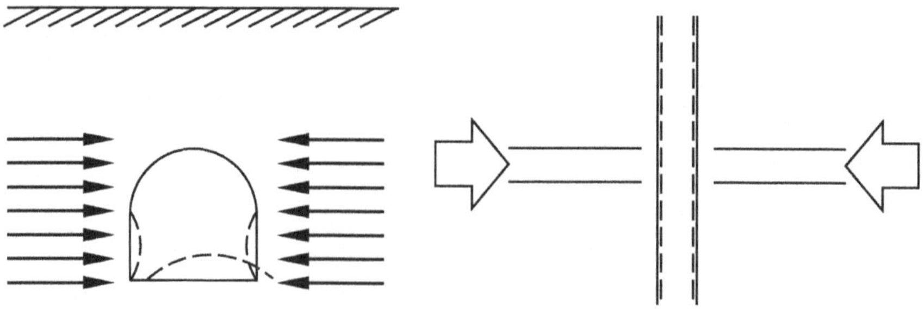

Fig. 15.7 Tectonic stress mechanism, left: deformational profile due to tectonic stress; right: deformation of roadway parallel and perpendicular to the major principle stress

15.1.2.3 Gravity Stress

The stress caused by overburden gravity is vertical. Ground failure of soft rock roadway caused by gravity stress is depth-related but direction-independent. That is, the roadway deformation is small in shallow underground under conventional support, however, if the excavation is in deep underground, the roadway deformation becomes large but without orientation issue. These characteristics are often manifested as a mechanism of gravity expansion.

15.1.2.4 Engineering Induced Stress

After excavation, the stress state of the surrounding rock changes greatly. The tangential stress is localized near the rock wall and the stress state of the rock in far field remains in-situ stress state. It should be noted that excavation of one roadway may influence adjacent roadways. The stress state at a point in surrounding rock can be expressed by a second-order stress tensor, which can be decomposed into two parts: spherical stress tensor and partial stress tensor, that is:

$$\begin{bmatrix} \sigma_{xx} & \tau_{xy} & \tau_{xz} \\ \tau_{yx} & \sigma_{yy} & \tau_{yz} \\ \tau_{zx} & \tau_{zy} & \sigma_{zz} \end{bmatrix} = \begin{bmatrix} \sigma_{xx}-\sigma_c & \tau_{xy} & \tau_{xz} \\ \tau_{yx} & \sigma_{yy}-\sigma_c & \tau_{yz} \\ \tau_{zx} & \tau_{zy} & \sigma_{zz}-\sigma_c \end{bmatrix} + \begin{bmatrix} \sigma_c & 0 & 0 \\ 0 & \sigma_c & 0 \\ 0 & 0 & \sigma_c \end{bmatrix} \quad (15.1)$$

Spherical stress tensor is a static hydraulic pressure that does not produce deformation. The deviator stress of the surrounding rock may induce large deformation and failure of the roadway.

15.1.3 Structural Deformation Mechanism

Structural deformation mechanism is related to the chamber structure and rock mass structure (Fig. 15.8).

When a roadway is developed along a stratum, the deformation of the roadway is noticeably. However, if it is developed passing through the rock stratum, the deformation may be small. The reason is that, the deformation of the roadway is related to the bedding direction of the stratum and anisotropic properties of the rock, especially for rock stratum with a well-developed soft rock interlayer. Accordingly, the deformation mechanisms related to geological structure of surrounding rock can be divided into five types, namely, fault, weak interlayer, bedding, systematic joint, and random joint (Fig. 15.1).

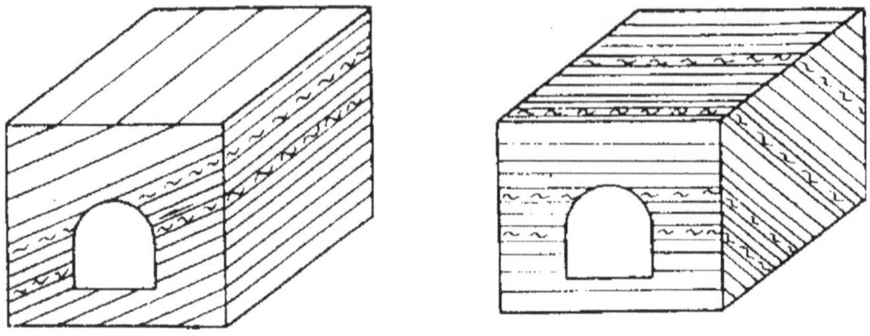

(a) excavation along the stratum (b) excavation across the layers

Fig. 15.8 Structural deformation mechanism

15.2 Determination of Soft Rock Deformation Mechanism

The deformational characteristics of soft rock roadway can be identified through theoretical analysis, laboratory testing and field investigation. Table 15.1 shows the deformation mechanics of surrounding rock and damage characteristics of soft rock roadway. The deformational mechanism of type I is mainly caused by mineral components and microcrack propagation. The mechanism of type II relies on the in-situ stress and mining induced stress. The deformation of type III is dominated by the geological structure of surrounding rock. It is necessarily to evaluate the anisotropical properties of the bedding strata and the orientation of tectonic stress, so as to determine a proper direction of the roadway.

The deformation of soft rock roadway is usually affected by three or more deformational mechanisms. For example, in soft rock excavation at Nalong No. 2 coal mine, Guangxi, China, via field investigation and laboratory tests, it was found that A_3 coal seam and its overlayer and underlayer within 10 m contain montmorillonite and illite/montmorillonite mixed layer. The roadway deformation is large, and it relates to buried depth but without orientation issue. Therefore, the deformation of the rock type is combined I_A and II_B, namely, I_AII_B type. In addition, it was found the shear resistance is nearly zero at the top and bottom interfaces of the variegated shale. Therefore, the mechanism of surrounding rock deformation is $I_AII_BIII_{BA}$ and $I_AII_BIII_{BC}$ according to the relationship between the formation of the weak layer and the chamber directions.

Table 15.1 Deformation mechanisms and failure characteristics of soft rock roadway

Type	Subtype	Mechanism	Controlling factor	Damage characteristics of soft rock roadway
I type	I_A type	Molecular water absorption mechanism, unit cells can absorb water molecules, strong water absorption capacity	Montmorillonite	After exposure, the rock is easy to weathering, softening and cracking; it is vulnerable to wind, water, and shockwave. Floor heave occurs in I type and is difficult to support. The softening order is I_A, I_{AB}, I_B; I_C is related to the microcrack development
	I_{AB} type	I_A & I_B is determined by mixing layer	Illite/smectite	
	I_B type	Colloidal water absorption mechanism, unit cells are not allowed the water molecules enter, an adsorption layer will be formed on the surface of clay particles	Kaolinite	
	I_C type	Capillary water absorption along microcrack	Microcrack	
II type	II_A type	Residual tectonic stress	Tectonic stress	The deformation is related to the direction but not related to the depth
	II_B type	Gravity stress	Gravity	It is related to depth but not related to direction
	II_C type	Groundwater	Hydraulic	Only related to groundwater
	II_D type	Engineering disturbance	Engineering force	Design-related
III type	III_A type	Fault zone	Fault	Collapse, caving
	III_B type	Weak interlayer	Interlayer	Over-excavation, flat top
	III_C type	Bedding	Bedding	Rules jagged
	III_D type	Advantaged joint	Joint	Irregular jagged
	III_E type	Random joint	Random joint	Block off

15.3 Key Support Technology of Soft Rock Roadways

Soft rock roadway support is often difficult due to large deformation and great ground stress of surrounding rock. Single support measure can often hardly provide sufficient ground controlling for the roadway due to complicated deformational

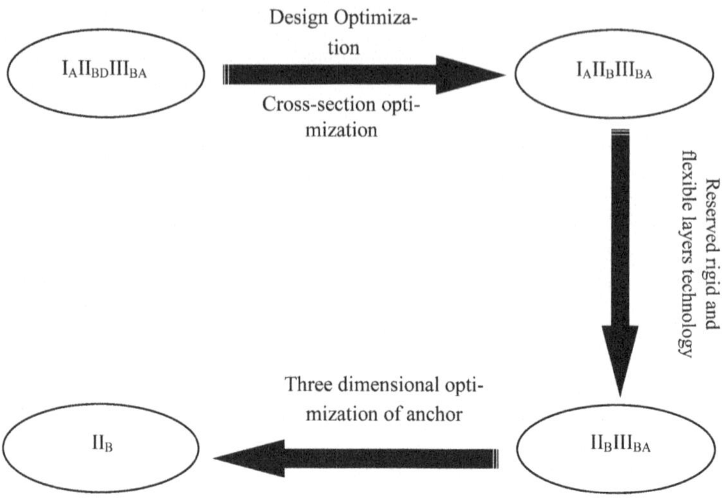

Fig. 15.9 Conversion of compound deformation type in Nalong No. 2 mine shaft-station

behavior of the surrounding rock. Therefore, multi-support method should be implemented for soft rock roadway reinforcement. The successfulness of soft rock roadway support is related to following technical approaches:

1. Determination of the soft rock deformation types.
2. Transform of the composite deformation into a single type.
3. Proper selection of support technics for each single type deformation.

Each single deformation type has corresponding effective supporting methods. Thus, it is the key to convert composite deformation into one single deformation type. For example, the related deformation mechanism at the shaft bottom in Nalong mine is $I_A II_B III_{BA}$ type. By the means of intensive pumping stations at water intake wells, reserved rigid-flexible layer technology and anchorage three-dimensional optimization, its deformational behavior is changed gradually into one single type—II_B, the specific conversion process shown in Fig. 15.9.

References

He M (1992a) Soft rock expansion mechanism. Guangxi Coal:20–24

He M (1992b) Discussion on mechanics of deformation mechanics of soft rock roadway. Soft Rock Eng:46–51

He M (1992c) Mechanism classification of soft rock deformation and support countermeasure. Guangxi Coal:70–74

He M (1992d) Mechanism of Soft Rock Expansion and Its Supporting. Proceedings of the Young Workers' Seminar of China Coal First Institute, pp 118–125

He M (1997) Mechanism and support countermeasure of soft rock deformation in coal mine. Hydrogeol Eng Geol 2(2):12–16

He M, Zou Z, Zou Y (1993) Introduction to soft rock tunnel engineering. China University of Mining and Technology Press, Beijing. (in Chinese)

He M, Jing H, Sun X (2002) Soft rock engineering mechanics. Science Press, Beijing. (in Chinese)

Norrish K (1954) Manner of swelling of montmorillonite. Nature 173(4397):256–257

Chapter 16
Soft Rock Roadway Reinforcement

He Manchao, Han Jun, and Cao Chen

16.1 Introduction

Advances in rock bolting technology over the past four decades have firmly established the use of rock bolts as the primary rock reinforcement measure in underground roadway support. It is estimated that more than 130 million rebar bolts are installed in underground coal mine roadways in China every year. Coal mine roadways built in the deep underground or soft rock mass are subjected to high ground stress, which is often associated with roof fall, floor heave, large rib convergence or even various underground disasters such as coal bumps, coal and gas outbursts, and rock bursts. In this case, all engineering designs have to rely on effective rock support as the final line of safeguard for workers, equipment and the whole mining operation.

New support techniques and materials have been developed for high ground stress, strong mining induced stress, large roadway deformation and soft or fractured surrounding rock conditions. According to field observations, Li (2007, 2010) suggested that under high ground stress conditions, rock bolts should not only have a high load-bearing capacity but also be able to accommodate large rock dilations, that is, they should be able to absorb a large amount of energy prior to failure. Kang et al. (2009) reported using strong cable bolts, large arch shaped plates and high strength steel meshes for high stress ground and large deformation roadway support.

H. Manchao (✉)
State Key Laboratory for Geomechanics and Deep Underground Engineering, Beijing, China
e-mail: hemanchao@263.net

H. Jun
College of Mining, Liaoning Technical University, Fuxin, Liaoning, China

C. Chen
College of Mining, Liaoning Technical University, Fuxin, Liaoning, China

CME, EIS, University of Wollongong, Wollongong, NSW, Australia

M. Kanji et al. (eds.), *Soft Rock Mechanics and Engineering*,
https://doi.org/10.1007/978-3-030-29477-9_16

Based on in situ stress analysis, Shen (2014) proposed a support design for large deformation roadways, which included an optimal cable/bolt arrangement, full length grouting, and high pretensioning of bolts and cables. He et al. (2014) developed large elongation constant axial resistance rock bolts for soft rock roadways or complex ground support. Guo et al. (2014) used high energy absorption rock bolts to couple the support element with large deformed surrounding rock. Gao et al. (2014) used fully cross-sectionally yieldable support in large deformation soft rock roadways. Liu et al. (2014) used lengthened rebar bolts in large deformation roadways to control roof falls. Via in situ monitoring and numerical modelling of axial forces in the roof and the rib, Zhang et al. (2015) reported that bolt support designs for the roof and the rib must differ. To control floor heave in deep soft rock roadways, Zheng et al. (2015) adopted the ground control techniques of foam concrete backfilling, U-steel retractable canopy and prestressed cable bolts. Ma et al. (2015) pointed out that the long extensible bolt could be used in deep roadway support. Wang et al. (2016) used lengthened bolt + rigid rebar bolt + steel mesh + W-steel strip + shotcrete for large deformation roadway support; and proposed that certain initial deformations of the surrounding rock should be allowed in high stress ground. Li (2016) used an inflatable bolt for soft and weak rock support. Recently, Wang et al. (2018) studied the phenomenon of roof fall in large deformation roadways; it was found that high horizontal tectonic stress and fault slip induced by mining activities were key contributing factors.

These studies provide insights and ground control techniques for soft rock or large deformation roadway support. However, these anchorages often require high engineering techniques and expensive support materials. Therefore, identification of the best supporting time and development of traditional rebar bolt for soft rock mass or large deformation roadway support should be taken into account. Based on the roadway deformation procedure and rock bolting failure mechanisms, this chapter provides roadway deformation-support coupling theory and develops an optimal rebar bolt for soft rock and large deformation roadway reinforcement.

16.2 Support Coupling Theory

16.2.1 Concept and Classification

According to soft rock roadway support practice, the failure of the supporting system is a procedure. That is, one or more sections of the surrounding rock deform firstly, which leads subsequent large deformation. With increasing of the area of large deformation region, rock crazing, exfoliation and breaking may occur, causing instability of the surrounding rock. Therefore, minimising the initial damage of rock mass is one point for the stability of the roadway.

Ground support failure can be due to uncoupling of reinforcing element with rock mass strength or with ground stress. As the failure procedure is beginning from

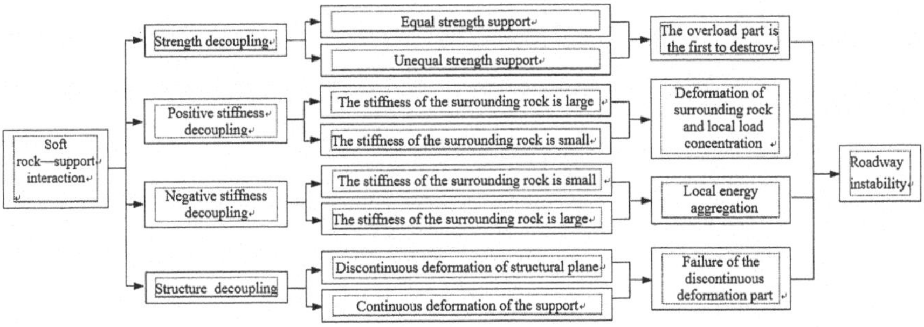

Fig. 16.1 Support coupling theory

a small area and then expending, its deformational behaviour can be divided into four types, shown as Fig. 16.1.

Type I is strength uncoupling between the supporting structure and the surrounding rock, that is, the supporting structure bears the load unevenly. As the support element often has uniform strength in each section, local stress concentration may cause overloading, resulting subsequent whole reinforcement system failure.

Type II is positive stiffness uncoupling between the supporting structure and the surrounding rock. That is, the surrounding rock is damaged firstly, causing strength decrease. The load is then transferred to the supporting element, which leads local overloading and structure failure as the stiffness of the supporting structure becomes smaller than that of the surrounding rock.

Type III is negative stiffness uncoupling between the supporting structure and the surrounding rock. If the stiffness of the supporting structure is greater than that of surrounding rock, the swelling energy of surrounding rock cannot be transformed into deformational energy and be released. Local energy concentration occurs, causing local overloading and failure.

Type IV is structure deformation uncoupling between the supporting structure and the surrounding rock. In this type, the deformation of supporting structure is even, but the deformational behaviour of the surrounding rock may be different due to weak interlayer, stratified plane and joint, and so on, which may cause local failure.

16.2.2 Deformational Characteristics

Based on the analysis of deformation behaviour, the deformation curves of roadways can be classified into four types, as shown in Fig. 16.2.

The first type of the roadway deformation is deceleration–acceleration deformation, which reflects that the surrounding rock strength is lower than load bearing. After a period of decelerating deformation, the deformation increases rapidly causing ground failure. The second type of the deformation curve is slow-deceleration

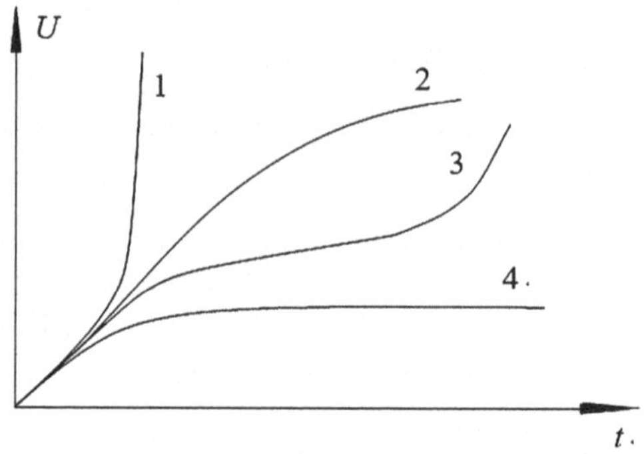

Fig. 16.2 Deformation characteristics of the roadway; (1) No-support: Deceleration–acceleration deformation; (2) Uncoupling: Slow-deceleration deformation; (3) Uncoupling: Deceleration–constant acceleration deformation; (4) Coupling: Even-deceleration deformation

deformation, reflecting the strength of the reinforcement is still below the bearing load of the surrounding rock. The third type of the deformation curve is deceleration–constant accelerated deformation, indicating that the strength of the reinforcing system is still slightly less than its bearing load. The fourth type of the curves is an even-deceleration deformation, which reflects that the support element and the surrounding rock are coupled.

In practice, rock cracks often accompany with high stress corrosion phenomenon, that is, support body scaly or flaking. High stress corrosion can be divided into four stages: the squamous peeling stage, the flake stage, the massive caving (plastic hinge) stage and the structural instability (collapse) stage. High stress corrosion can be used as a sign for identifying support coupling.

16.2.3 Coupling Support Techniques

The supporting methodology for soft rock roadway differs much from that of hard rock. Plastic deformation is often not allowed for a hard rock roadway as hard rock can lose its carrying capability after entering into plastic state. However, for soft rock, plastic deformation is allowed. Another feature of soft rock support is that large amount of plastic energy (such as swelling energy) should be released using various means. The supporting force of soft rock roadway can be presented as follows:

$$P_T = P_D + P_R + P_S \tag{16.1}$$

Where P_T is resultant force of surrounding rock after excavation, including gravity, water pressure, swelling force and mining induced stress; P_D is the force transformed to the form of deformation, including elastoplasticity conversion (irrelative with time), viscoelastic plasticity conversion (relative with time), and swelling force conversion (relative with time). For soft rock, it is mainly the release of plastic energy in the form of deformation. P_R stands for self-bearing ability of surrounding rock, that is, the strength of surrounding rock for supporting partial or total load; P_S is supporting force.

In this equation, the composite force, P_T, is borne by three components. One part of P_T is released through elastoplastic deformation firstly. Another part of P_T is carried by the surrounding rock itself. If the strength of rock is high and $P_R > (P_T - P_D)$, the roadway can be supported by itself. For soft rock, P_R is small and often less than $(P_T - P_D)$. So support (P_S) is necessary; that is, the value of $(P_T + P_R)$ must be great than that of $(P_T - P_D)$.

An optimum design of the roadway support must satisfy the following three conditions simultaneously: (1) $P_D \rightarrow$ max; (2) $P_R \rightarrow$ max; (3) $P_S \rightarrow$ min. However, P_D and P_R cannot reach maximum at the same time. To choose a proper supporting strength and time is the key to satisfy $P_D \rightarrow$ max and $P_R \rightarrow$ max.

The deformation of the roadway can be divided into three stages: decelerating deformation stage (DDS), approximate linear deformation stage (LDS) and accelerating deformation stage (ADS). When the deformation enters into ADS, rock may crack causing strength decreasing. In the ADS, $P_D \rightarrow$ max can be reached, but the value of P_R decreases greatly; thus, this is not an optimum design. The point to this problem is the determination of the optimal supporting time.

The supporting time optimum means that $P_R + P_D$ reaches maximum, as shown in Fig. 16.3, in which, the optimal supporting time, T_S, is the time corresponding to the peak point of $(P_R + P_D)$–t curve.

In practice, it is difficult to obtain the intersection point between curves P_R–t and P_D–t. There is a manner to obtain the optimum support time, as shown in Fig. 16.4. The roadway is often exfoliated when its deformation enters into region B. After the roadway deforms into region C, some rock blocks begin to collapse and ground becomes unstable. So the optimal support time is the time when the roadway begins to exfoliate and high stress erosion just appears.

Field observation shows that in the deformation state of *A* area in Fig. 16.4, multi squamous peeling begins to generate; flaking appears in the zone B; and massive structural instability and caving occur in region C. Therefore, high stress corrosion

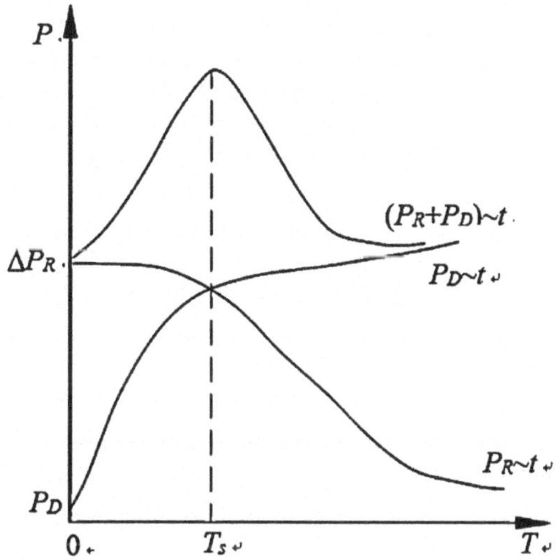

Fig. 16.3 The concept of optimum supporting time

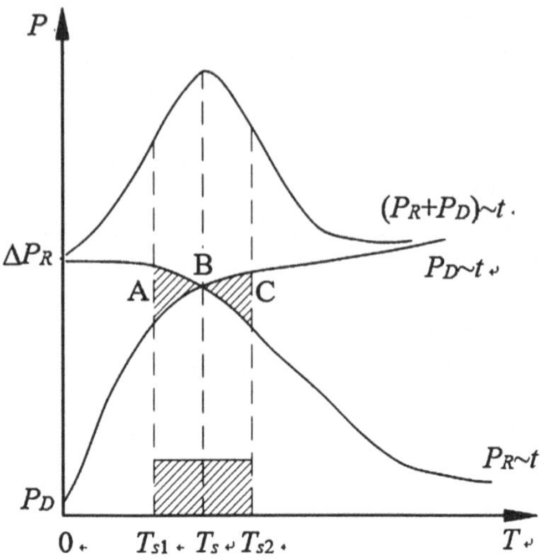

Fig. 16.4 The concept of a period of optimum supporting time

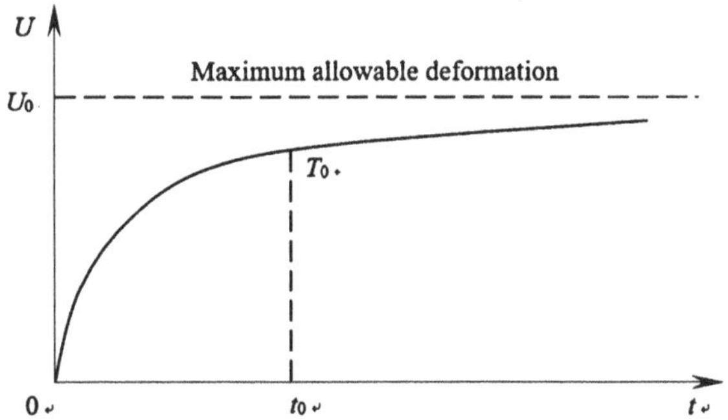

Fig. 16.5 Determination of the optimal supporting time

phenomenon of scale and flake can be used to determine the optimal supporting time. According to the field investigation, while tensile crack width reaches 1–3 mm, the roadway is in the A and B regions, reflecting that the deformation reaches its 60% design margin. In addition, according to the time displacement (U–t) curve shown in Fig. 16.5, the changing rate of the roadway displacement becomes gentle from T_0, which is the best supporting time for the roadway.

16.3 Rebar Bolt Designed for Soft Rock Roadway Support

16.3.1 State of the Art

In contemporary civil and mining engineering projects, rebar bolting forms effective system in ground reinforcement. Rock reinforcement belongs to the category of improvement methods to increase the strength or decrease the deformability of a

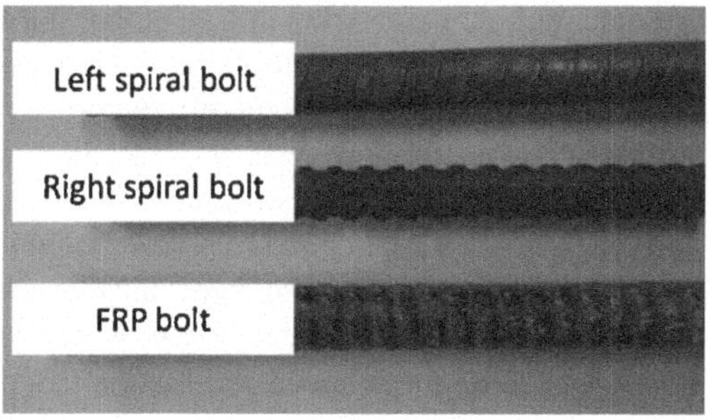

Fig. 16.6 Bolts commonly used by underground coal mining industry in China

rock mass. Various kinds of rock bolt are available to provide such effect as primary support members.

In China, three kinds of rebar bolt are often used for roadway support by underground coal mining industry, as shown in Fig. 16.6. Left spiral bolt of diameter 22 mm is often used as roof bolt; right spiral bolt diameter 17–20 mm is often used for pillar side rib support; and glass fibre reinforced polymer bolt is used in the panel side rib for easy cutting.

The load transfer capacity of a bolting system is closely related to the bolt profile configuration. Current rebar bolt profile geometry is introduced from steel tendons used in concrete industry, which has been studied for over one century. Abrams (1913) reported tests of bond strength between steel and concrete for different rib geometry, establishing tendon profile researching area. Via pull-out testing of 17 different deformed reinforcing bars in concrete, Clark (1946, 1949) concluded that the rib height and rib face angle appeared as important factors in determining bond resistance. These studies were accepted by ASTM to establish the standards for deformation requirements of reinforcing bars (ASTM 1996). Through pull-out testing of bars with only one rib, Lutz and Gergely (1967) introduced apparent rib bearing angle concept for slip of a deformed bar causing concrete beam splitting. Based on the radial components of bond forces being balanced by tensile ring stresses in surrounding concrete, Tepfers (1979) proposed a concrete cover splitting mechanical model, in which, the principal stress was closely related to the rib geometry. Goto and Otsuka (1980) conducted experimental studies on the deformational behaviour of concrete around deformed tension bars. Results showed that the elastic stiffness of the bond and the ring strain of the concrete would be different for different rib geometries. Murata and Kawai (1984) performed pull-out tests using different deformed bars to evaluate the initial bond strength and splitting bond strength, which gave information about adhesion between the deformed bar and the concrete. To study the effect of rib height and spacing on the bond strength of reinforcing bars, Darwin and Graham (1993) observed that concrete powder formed against the loaded face of the ribs. Hamad (1995) conducted a series of seven experiments to evaluate the effect of rib geometry on bond stress slip characteristics of the deformed bars in reinforced concrete structure. Test results indicated that bond strength varied with the bar rib face angle, spacing and height. Cairns and Jones (1995) considered that bond failure was bearing failure of the ribs to surrounding concrete and categorised that a ribbed bar could have bond failure by one of three mechanisms. Idun and Darwin (1999) studied the effect of rib geometry on bond strength by measuring the frictional coefficients between epoxy-coated and uncoated reinforcing steel and mortar. Results from 68 test specimens indicated that the friction coefficient varied from 0.503 to 0.627. Choi and Lee (2002) derived analytical expressions to predict bond strength using interfacial mechanical properties. The frictional coefficient and stress bearing angle were recognised as key variables in the proposed equation. Wu et al. (2012) investigated the bond strength between vitreous enamel coated rebar and concrete through testing of 96 pull-outs of cylinder specimens. Results showed that enamel coating increased the bond strength due to small load bearing angles. Dybeł and Furtak (2017) discussed that formation of slip planes around the ribs due to crushing of the concrete was "poor" bond conditions as an increase of the splitting force component and a decrease in the component parallel to the reinforcement axis. To sum up, the tendon rib geometry was studied and optimised continuously in the past century, up to now, a set of nearly standardised rib geometric parameters was commonly accepted by the industry as rib spacing around 12 mm, rib height of 1–2 mm and rib face angle between 45° and 70°.

Rock bolting extensively used for underground coal mine roadway support was from 1990s. The bolt profile was initially nearly same or very similar to tendons used in concrete industry. Several modifications have been made for encapsulated resin mixing, such as right spiral or left spiral, but key profile parameters remained. However, the application conditions, failure mechanisms and loading status of the rock bolts are quite different from concrete tendons. The major failure mode of the reinforced concrete is cover splitting; however, the dominated failure mode of rock bolting is pull-out failure due to infinite confinement of surrounding rock. Moreover, for large deformation roadway, the residual strength and energy absorbing ability should be emphasised for bolting design; for reinforced concrete beam, its deformations are often required to be controlled within elastic stage. In addition, the in situ stress plays an important role in surrounding rock deformations and anchorage performances; but there is no such ground stress for reinforced concrete structures. Therefore, it can be concluded that the optimised rib geometric parameters for concrete tendon may not be the most favourable for rock bolt used in underground roadway support.

Rock bolt profile researches were also carried out from 1990s. Fabjanczyk and Tarrant (1992) investigated the load transfer mechanism in push-out tests. They found that bolts with a lower profile height had small stiffness and concluded that the load transfer was a function of hole geometry, resin properties and bar surface configuration. Blumel (1996) carried out pull-out testing of equal diameter bolts with different profile spacing of 13.7, 27.4 and 54.8 mm. It was found that widening of the rib spacing enhanced the load transfer capacity of the bolting. Moosavi et al. (2005) studied the profile configurations in cementitious grout, leading to similar conclusions. Aziz and Webb (2003) and Aziz et al. (2006) extended the study to include both push and pull-out tests using 75 and 115 mm steel sleeves. Result showed that profile spacing affected bolt anchorage stiffness, and the anchorage performance reached the optimum with the bolt profile spacing of 37.5 mm. Lin et al. (2014) and Lin and Ren (2015) conducted pull-out test of rib height 1.4 mm bolts with different profile spacing of 11, 22, 33 and 44 mm. It was found that the anchorage performance reached the optimum with the bolt profile spacing of 33 mm. Wu et al. (2017) study the right spiral bolt with different profile spacing of 12, 24, 36 and 48 mm. The test results showed that the pulling force and energy dissipation increased with increase of rib spacings. Zhao et al. (2018) studied the left spiral bolt at profile spacing of 12, 24, 36 and 48 mm. Results show that the pulling force reached maximum with the bolt profile spacing of 24 mm. These studies showed that bolt profile has great influence on the anchorage performance, and suggested that increasing of bolt rib spacing could improve energy dissipation of bolting specimens. It should be noted that all these works were laboratory tests, there has been no further analytical work or field testing being undertaken to advance the load transfer capacity of the bolts with respect to different profile configuration. Accordingly, this aspect of the topic is currently being further evaluated both analytically and experimentally in this section.

16.3.2 Rock Bolting Failure Analysis

Failure mode analysis is the crux of load transfer mechanisms of the rock bolting system. Rock bolting tensile failure could be classified in one or more of these manners: the bolt, the grout, the rock, the bolt–grout interface or grout–rock interface (Littlejohn 1993; Guan and Ma 1997; Fan et al. 1997). The steel bar governed the axial behaviour of the bolt, which was much stiffer and stronger than the grout and rock. If the bolt had sufficient length to transfer the entire load to the rock it would fail. The shear stress at the bolt–grout interface was greater than that at the grout–rock interface because of smaller effective area. If the grout and rock were of similar strengths, failure could occur at the bolt–grout interface. If the surrounding rock was softer then failure could occur at the grout–rock interface.

The type of axial failure depended on the properties of individual element. For pull-out failure, Kang et al. (2014) pointed out that factors affecting the anchoring force were diameter and profile of the bolt, mechanical properties of the anchorage agent, and the strength and integrity of surrounding rock. Hu (2015) studied the influence of the strength of surrounding rock on anchorage performance, and found that the stress of surrounding rock decreased with the decrease of surrounding rock strength, but the plastic zone was expanded. Aziz et al. (2016) reported that factors affecting anchorage performance included borehole diameter, resin annulus thickness, installation time, the effect of gloving and hole over drill, and proposed that the borehole diameter had a detrimental effect on the bonding strength.

Yazici and Kaiser (1992) developed a conceptual model for fully grouted cable bolts, and considered that the bolting failure was radial cracks developed in anchoring agent extending to grout–rock interface, eventually leading the anchorage agent splitting. Hyett et al. (1995) further studied fracturing and splitting of the anchorage

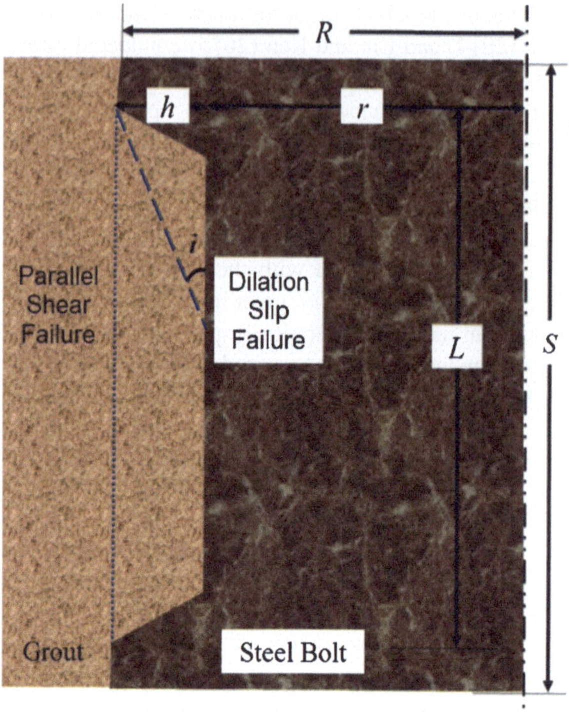

Fig. 16.7 Failure modes of bolt subjected to axial loading (Cao et al. 2013b)

agent and established a mechanical model for fully grouted cable bolts under consideration of the torsional effect of the bolt.

Cao et al. (2013a, b) studied the effect of rebar bolt profile configuration via analysis of the resin-bolt interface failure. Two major failure modes were identified, namely parallel shear failure (dotted line in Fig. 16.7) and dilational slip failure (dashed line in Fig. 16.7). For Mohr–Coulomb materials, following bolt optimum profile geometric parameters equation can be derived:

$$L = \frac{\cos\phi \cdot \cos(i+\phi)}{(1-\cos i)\left[2\sin i + \sin(2i+2\phi)\right]}\left(h + \frac{rh}{r+h}\right) \tag{16.2}$$

Where:

$$\frac{c}{p} = \frac{\sin i - \cos i \sin(i+\phi)\cos(i+\phi)}{\cos\phi \cdot \cos(2i+\phi)} \tag{16.3}$$

$h = R - r$ is the rib height, p is ground pressure, and c and ϕ are Mohr–Coulomb parameters of the grout.

It was found that the optimal profile geometry for failure resistance was related to the mechanical properties of the grout and confining pressure of the bolt when

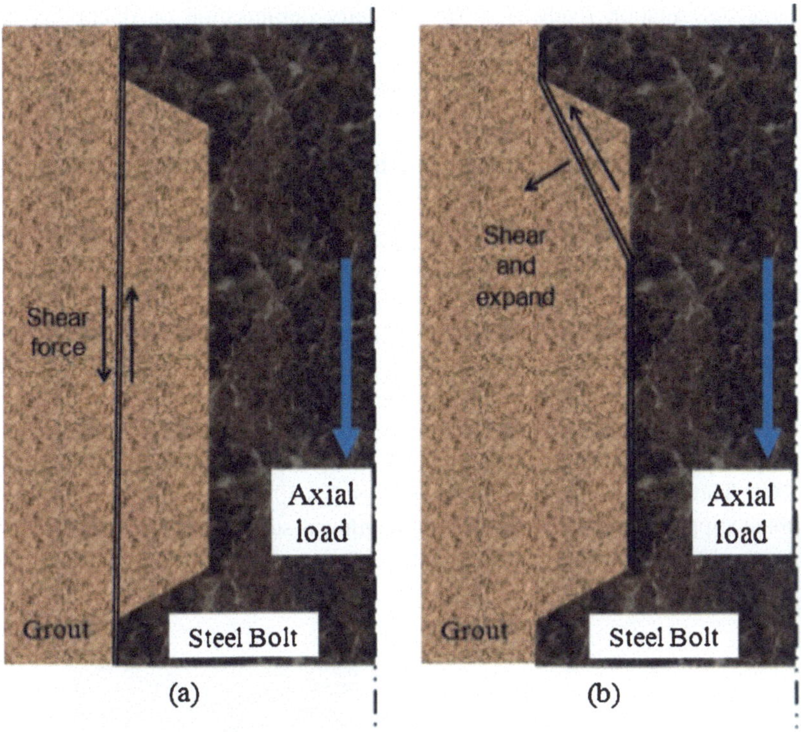

Fig. 16.8 Residual strength for different bolting failure modes. (**a**) Residual force after parallel shear failure. (**b**) Residual force after dilational slip failure

failure occurred. Therefore, bolt profile studies must take into account the mechanical properties of the grout and geo-conditions of the field.

Residual strength of an anchorage is a key parameter for energy absorption of the bolting system. For parallel shear failure, its residual strength is the friction of a cylindrical complex composed by the bolt and carried grout passing through the failure surface, as shown in Fig. 16.8a. For dilational slip failure, bolting residual strength is resistance of a cone shaped complex shearing and expanding the surrounding medium, as shown in Fig. 16.8b.

According to rock bolting failure analysis, the residual strength of the anchor is mainly frictional and hence depends on the pressure build-up at the interface which in turn depends on the dilational movement against the confining grout or rock. Previous experimental results (Aziz and Webb 2003; Aziz et al. 2006; Lin et al. 2014; Lin and Ren 2015; Wu et al. 2017; Zhao et al. 2018) could be explained as that increasing of rib spacing may elongate the dilational movement of the bolt. The confining pressure generated in parallel shear failure would be most likely smaller than that in dilational slip failure due to small radial dilation. The magnitude of residual strength of the anchor is related to ground stress, redial stiffness and mechanical properties of grouting material. Consequently, rebar bolt profile design for large deformation roadway is rib geometry optimisation on ductibility and energy absorption capability of the anchor system under site geo-conditions.

16.3.3 Rebar Bolt for Soft Rock Support

The optimum bolt profile for bolting failure resistance purpose was obtained (Eqs. 16.2 and 16.3). For large deformation roadways, however, residual strength of the bolting system is important. Therefore, under consideration of local geo-conditions (ground stresses and radial stiffness of surrounding rock), increasing residual strength and elongating dilational slipping length of the bolt via failure mode controlling are design principles of the bolt profile for soft rock support.

A rebar bolt has been designed and tested for soft rock and large deformation roadway support. The geometric parameters of the bolt profile are shown in Fig. 16.9. The effective rib spacing was designed as 50 mm to elongate post-failure dilational slipping distance. Its residual strength was supposed to be high in case of high ground stress. The ribs were partially overlapped to further enhance the bolt

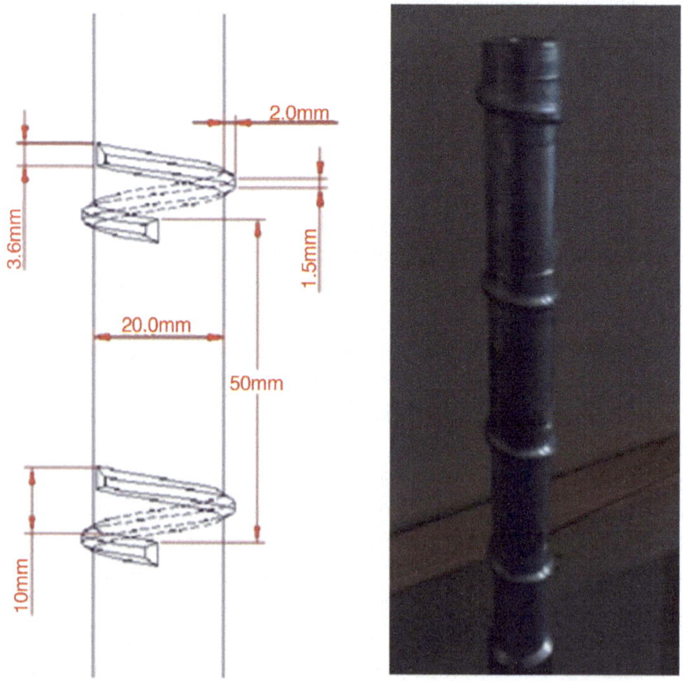

Fig. 16.9 Rebar bolt for soft rock roadway support

Table 16.1 Geometric and mechanical parameters of the bolt for soft rock and large deformation roadway support

Parameter	Value	Parameter	Value
Core diameter (mm)	20.0	Top rib width (mm)	1.5
Rib spacing (mm)	50.0	Bottom rib width (mm)	3.6
Rib height (mm)	2.0	Rib face angle (°)	60
Yield strength (MPa)	376	Tensile strength (MPa)	647

built pressure via increasing of radial displacement of the resin annulus and surrounding rock. According to the viscosity and gel time of the grouting material used in the coal mine, China, left spiral direction of the rib was selected and the size of axial gap of each rib was determined as 10 mm. The geometric and mechanical parameters of the bolt are shown in Table 16.1.

16.3.4 Laboratory Tests

Left spiral roof bolt commonly used in coal mines, China, was also tested as a control of the large deformation roadway bolt. The rib profile parameters and mechanical properties of the left spiral bolt are shown in Table 16.2.

The bolt samples were encapsulated into testing cells using resin grout (MSCKa23-35). The average uniaxial compressive strength and shear strength of the resin were measured as 43.1 MPa and 19.5 MPa, respectively. The measured Young's modulus was of 12.6 GPa and Poisson's ratio was of 0.26. And 5.8 mm wall thickness 6061 aluminium tubes were used as confining cell, the inner diameter of the tube was selected as 32 mm according to the driller size commonly used in coal mines in China.

While bolting samples were prepared, the bolts were centrally located with resin annulus, and efforts were made to ensure the bolts were also axially parallel to the tube axis. The testing specimens were inserted into a steel framework, which was gripped by WAW-600C testing machine. The loading rate of the test was 1 mm/min, and the load/displacement information was monitored on a PC. The post-tested samples were shown in Fig. 16.10.

Figure 16.11 shows the testing results. In the diagram, LX-12 indicates left spiral bolt presented in blue colour, and N-50 is the anchorage performance of the large displacement roadway support bolt presented by red colour.

Results show that the average peak pulling force of the designed bolt was 113 kN, 13% higher than that of left spiral bolt. From a view point of energy absorption ability, the bolt designed for large deformation roadway support was 3169 J, 15% higher than that of left spiral bolt of 2768 J. It can be concluded that the anchorage performance of the designed bolt is better than that of commonly used left spiral bolt under the experimental conditions.

Table 16.2 Geometric and mechanical parameters of left spiral bolt in the test

Parameter	Value	Parameter	Value
Core diameter (mm)	22.1	Top rib width (mm)	1.2
Rib spacing (mm)	11.6	Bottom rib width (mm)	2.8
Rib height (mm)	0.90	Rib face angle (°)	48
Yield strength (MPa)	556	Tensile strength (MPa)	728

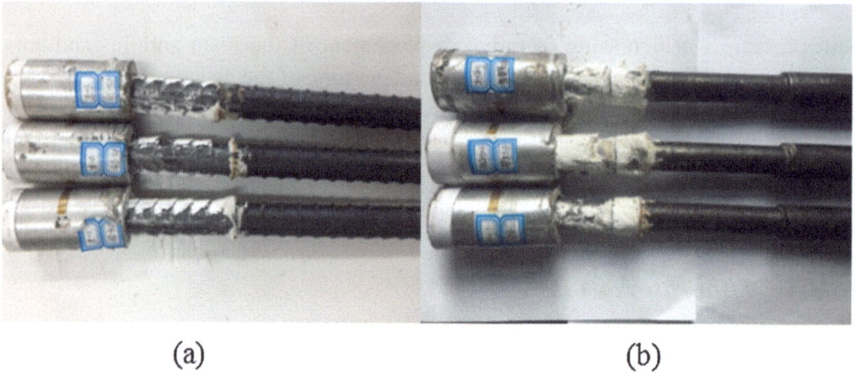

Fig. 16.10 Post-tested specimens, (**a**) left spiral bolts; (**b**) soft rock bolts

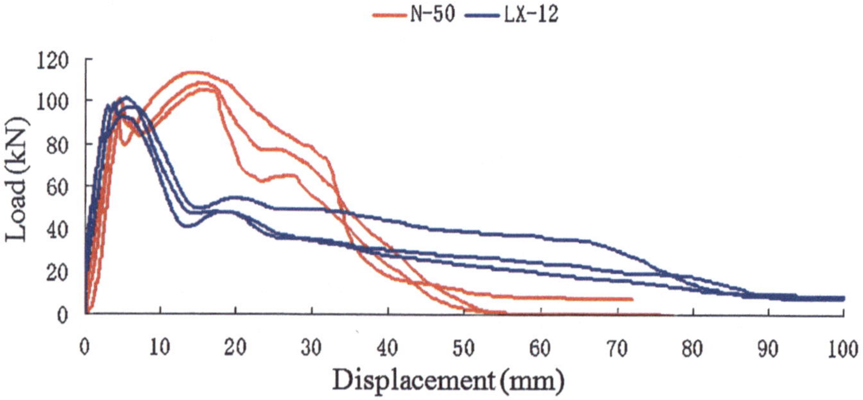

Fig. 16.11 Pull-out curves of the designed bolt (N-50) and left spiral bolt (LX-12)

16.3.5 Field Tests

Field tests were also carried out at 2822# tailgate in Qianjiaying coal mine, China. Qianjiaying coal mine is located 20 km south of Tanshan, China, with annual production of 6.4 Mt in 2017. The mining area is approximately 42.6 km^2, and current mining depth is 600–850 m. The bury depth of 2822 longwall panel is 650–710 m

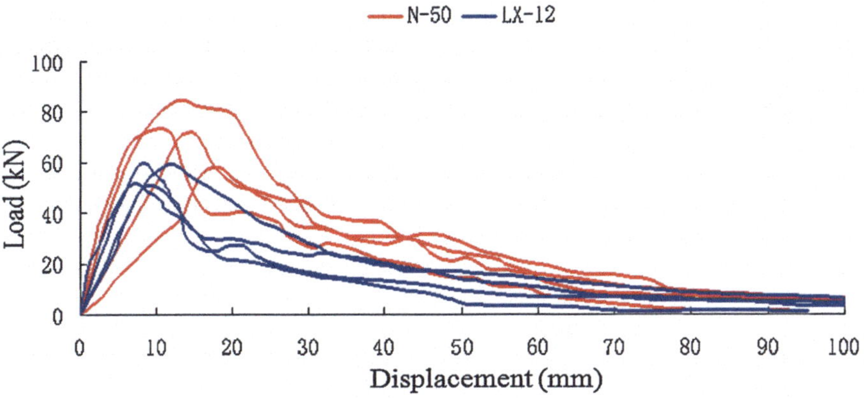

Fig. 16.12 Field tests of new bolt (N-50) and left spiral bolt (LX-12)

and currently in roadway development. The measured in situ stresses σ_1, σ_2 and σ_3 were 32.2 MPa, 19.3 MPa and 16.7 MPa, respectively. The primary support of the roof was left spiral bolt, steel net and W-steel strip; however, due to large convergence of the roof, GU_{29}-12m^2 U-steel sheds were used as secondary support. The roof convergence was observed more than 300 mm due to high major principle stress.

Pull-out test was conducted in 2822# tailgate roof. Left spiral rebar bolt used for roof support in the coal mine and new bolt were tested. The average diameter of the drilling holes was measured as 31.8 mm, and the anchorage length of the testing bolts was 150 mm. Due to high fracture of the roof, some experimental data were eliminated. The testing result is shown in Fig. 16.12.

It shows that the average peak pulling force of the new bolt was 72 kN, 30% higher than that of left spiral bolt. For bolting energy absorption, the new bolt was 2315 J, 47% higher than that of left spiral bolt of 1574 J. It can be concluded that the anchorage performance of the newly designed bolt is suitable under the site conditions.

16.4 Conclusions

This chapter focuses on soft rock roadway support theory and techniques. The support coupling technique emphasises the combination of support structure with surrounding rock forming optimum composite structure to improve its self-support capacity. Compared with the uniformly installed primary support, flexible support can be considered as secondary support after strata become stable.

This chapter also provides recent rock bolting mechanism advances and rebar bolt design for soft rock roadway support. According to rock bolting post-failure analysis, increasing residual strength and elongating dilational slipping length of the bolt via failure mode controlling are identified as design principles of the rebar bolt for soft rock roadway support. Under consideration of the local geo-conditions and key factors of residual strength of a bolting system, a soft rock rebar bolt is designed for soft rock and large deformation roadway support. Laboratory SEPT was conducted and results showed that the peak load and energy absorption capability of the soft rock rebar bolt are 13% and 15% higher than those of commonly used left spiral bolt in China. Field tests were also confirmed this result. It can be concluded that the designed rebar bolt has better anchorage performance than left spiral bolt under the site conditions.

References

Abrams DA (1913) Tests of bond between concrete and steel. Bulletin No. 71. University of Illinois Engineering Experiment Station, Urbana

ASTM A615/A615M–09a (1996) Standard specification for deformed and plain billet-steel for concrete reinforcement. ASTM, West Conshohocken

Aziz N, Webb B (2003) Study of load transfer capacity of bolts using short encapsulation push test. Proc 4th Underground Coal Operators Conference, Wollongong, pp 72–80

Aziz N, Jalalifar H, Concalves J (2006) Bolt surface configurations and load transfer mechanism. Proc 7th Underground Coal Operators Conference, Wollongong, pp 236–244

Aziz N, Craig P, Mirzaghorbanali A, Nemcik J (2016) Factors influencing the quality of encapsulation in rock bolting. Rock Mech Rock Eng 49(8):1–15

Blumel M (1996) Performance of grouted rock bolts in squeezing rock. Proc EUROCK'96, Predictions and performance in rock mechanics and rock engineering. Balkema, Rotterdam, pp 885–891

Cairns J, Jones K (1995) Influence of rib geometry on strength of lapped joints: an experimental and analytical study. Mag Concrete Res 47(172):253–262

Cao C, Nemcik J, Aziz N, Ren T (2013a) Analytical study of steel bolt profile and its influence on bolt load transfer. Int J Rock Mech Min Sci 60(60):188–195

Cao C, Nemcik J, Aziz N, Ren T (2013b) A study of rock bolting failure modes. Int J Min Sci Technol 23(1):79–88

Choi OC, Lee WS (2002) Interfacial bond analysis of deformed bars to concrete. ACI Struct J 99(6):750

Clark AP (1946) Comparative bond efficiency of deformed concrete reinforcing bars. J ACI Proc 43:4

Clark AP (1949) Bond of concrete reinforcing bars. J ACI Proc 46(11):161

Darwin D, Graham EK (1993) Effect of deformation height and spacing on bond strength of reinforcing bars. ACI Struct J 90(6):646–657

Dybeł P, Furtak K (2017) The influence of high-strength concrete–rebars bond conditions on the mechanism of its failure. Mag Concrete Res 69(4):163–174

Fabjanczyk MW, Tarrant GC (1992) Load transfer mechanisms in reinforcing tendons. Proc 11th Int Conf on Ground Control in Min, The University of West Virginia, Morgantown, pp 1–8

Fan S, Chen Y, Cui D et al (1997) Study on the feasibility of high temperature resistant resin anchoring agent. Coal Sci Technol 25(9):17–20

Gao M, Zhao Y, Li M et al (2014) Roof and support and bottom yielding support with whole section and O-shape control principle for soft rock roadway and engineering practice. Rock Soil Mech 35(8):2307–2313

Goto Y, Otsuka K (1980) Experimental studies on cracks formed in concrete around deformed tension bars. Proc JSCE 294:854100

Guan S, Ma N (1997) Mechanical analysis of anchorage failure of resin anchorage rod. J Min Saf Eng (z1):201–203
Guo Z, Wang J, Zhang Y et al (2014) Failure mechanism and constant resistance large deformation control measures of deep soft rock in Qingshui Coal Mine. J Min Saf Eng 31(6):945–949
Hamad BS (1995) Comparative bond strength of coated and uncoated bars with different rib geometries. ACI Mat J 92(6):579
He M, Gong W, Wang J, Qi P, Tao Z, Du S (2014) Development of a novel energy-absorbing bolt with extraordinarily large elongation and constant resistance. Int J Rock Mech Min Sci 67:29–42
Hu B (2015) Numerical study on the influence of strength of surrounding rock on anchorage performance of resin anchorage. Coal Min Support 962–965:968–972
Hyett AJ, Bawden WF, Macsporran GR, Moosavi M (1995) A constitutive law for bond failure of fully-grouted cable bolts using a modified Hoek cell. Int J Rock Mech Min Sci Geomech Abst 32(1):11–36
Idun EK, Darwin D (1999) Bond of epoxy-coated reinforcement: coefficient of friction and rib face angle. ACI Struct J 90(4):773–782
Kang H, Lin J, Wu Y et al (2009) High pretensioned stress and intensive cable bolting technology set in full section and application in entry affected by dynamic pressure. J China Coal Soc 34(9):1153–1159
Kang H, Cui Q, Hu B (2014) Analysis on anchorage performances and affecting factors of resin bolts. J China Coal Soc 39:1):1–1)10
Li CC (2007) A practical problem with threaded rebar bolts in reinforcing largely deformed rock masses. Rock Mech Rock Eng 40(5):519–524
Li CC (2010) Field observations of rock bolts in high stress rock masses. Rock Mech Rock Eng 43(4):491–496
Li CC (2016) Analysis of inflatable rock bolts. Rock Mech Rock Eng 49(1):273–289
Lin J, Ren S (2015) Numerical simulation optimization research of bolt profile configuration with resin full-length anchoring. J Min Saf Eng 32(2):273–278
Lin J, Ren S, Yang J (2014) Laboratory research of resin full-length anchoring bolts dimension optimization. J China Coal Soc 39(6):1009–1015
Littlejohn S (1993) Rock reinforcement-technology, testing, design and evaluation. In: Hudson JA (ed) Comprehensive rock engineering principals, practice and projects, vol 4, pp 413–451
Liu H, Wang F, Jiang L et al (2014) On the performance of lengthened bolt coupling support system in roadway roof. J Min Saf Eng 31(3):366–372
Lutz LA, Gergely P (1967) The mechanics of bond and slip of deformed bars in concrete. ACI J Proc 64(11):711–721
Ma N, Zhao X, Zhao Z et al (2015) Stability analysis and control technology of mine roadway roof in deep mining. J China Coal Soc 40(10):2287–2295
Moosavi M, Jafari A, Khosravi A (2005) Bond of cement reinforcing bars under constant radial pressure. Cem Concr Comp 27(1):103–109
Murata J, Kawai A (1984) Studies on bond strength of deformed bar by pullout test. J Jpn Soc Civil Eng 1(348):113–122
Shen B (2014) Coal mine roadway stability in soft rock: a case study. Rock Mech Rock Eng 47(6):2225–2238
Tepfers R (1979) Cracking of concrete cover along anchored deformed reinforcing bars. Mag Concrete Res 31(106):3–12
Wang W, Yuan C, Yu W et al (2016) Stability control method of surrounding rock in deep roadway with large deformation. J China Coal Soc 41(12):2921–2931
Wang H, Xue S, Jiang Y, Deng D, Shi S, Zhang D (2018) Field investigation of a roof fall accident and large roadway deformation under geologically complex conditions in an underground coal mine. Rock Mech Rock Eng 51(7):1–21
Wu T, Cao C, Zhao X et al (2017) Laboratorial study of anchorage performance in different rib spacings of bolt. J China Coal Soc 42(10):2545–2553
Wu C, Chen G, Jeffery S, Volz RK, Brow M, Koenigstein L (2012) Local Bond Strength of Vitreous Enamel Coated Rebar To Concrete. Construction and Building Materials 35:428–439
Yazici S, Kaiser PK (1992) Bond strength of grouted cable bolts. Int J Rock Mech Min Sci Geomech Abst 29(3):279–292

Zhang K, Zhang G, Hou R, Wu Y, Zhou H (2015) Stress evolution in roadway rock bolts during mining in a fully mechanized longwall face, and an evaluation of rock bolt support design. Rock Mech Rock Eng 48(1):333–344

Zhao X, Zhang H, Cao C et al (2018) Study on optimization of bolts rib spacing and anchoring force in different conditions of surrounding rock. Rock Soil Mech 39(4):1–9

Zheng P, Chen W, Tan X et al (2015) Study of failure mechanism of floor heave and supporting technology in soft rock of large deformation roadway. Chin J Rock Mech Eng 34(s1):3143–3150

Chapter 17
Large Deformation Support for Engineering Soft Rocks

He Manchao, Gao Yubing, and Gong Weili

17.1 Introduction

In recent years, China's coal production has been around 3.5 billion tons annually, causing shallow resources exhausted and increasing of mining depth. According to statistics, mining depth in 47 coal mines reaches underground 1000 m in China. Among them, Suncun coal mine, Xinwen, China, is 1503 m in underground. High ground stress in deep underground coal mine causes large deformation of engineering rock mass.

Soft rock falls into two categories, that is, geological soft rock and engineering soft rock (He 2014). Geological soft rock refers to the rock in low strength, large porosity, poor cementation, fracturing and sensitive to weathering. It often contains expansible clay minerals and loose, soft, weak interlayers. Engineering soft rock refers to the rock that can produce significant plastic deformation under engineering forces. The concept of engineering soft rock emphasizes the relationship of rock strength and loading conditions. In deep resources mining, in addition to geological soft rock (mud, shale, etc.), the geological hard rock (e.g. Shale, cemented, sandstone) may also deform like soft rock under high engineering force (He and Sun 2004; He and Qian 2010).

With increase of mining depth, large deformation of soft surrounding rock causes roadway collapse, roof fall, rib spalling or even rockburst disaster (Zhang et al. 2012, 2015, 2017; He et al. 2005, 2012, 2015a; Konicek et al. 2013; Bai et al. 2016; Chen et al. 2015). The main reason is that traditional steel canopy and rock bolt support materials cannot accommodate large deformation of surrounding rock. The maximum elongation for rebar bolt is usually less than 15%, so the reinforcement based on rebar bolt cannot control large deformation of

H. Manchao (✉) · G. Yubing · G. Weili
State Key Laboratory for Geomechanics and Deep Underground Engineering, China University of Mining and Technology, Beijing, China
e-mail: hemanchao@263.net

M. Kanji et al. (eds.), *Soft Rock Mechanics and Engineering*,
https://doi.org/10.1007/978-3-030-29477-9_17

surrounding rock (Gao et al. 2017; He et al. 2014). Moreover, the traditional bolts belong to Poisson's ratio materials (or plastic harden material), which have lower plasticity and toughness after cold hardening.

New techniques and products have also been developed for roadway support in recent years (Basarir et al. 2015; Deng et al. 2010; Gong et al. 2015; He et al. 2010; Islam and Shinjo 2009; Xie et al. 2017; Yan et al. 2012; Kang et al. 2018; Zhang et al. 2016). Kang et al. (2015) reported an enhanced reinforcement system for longwall entry developed in soft rock mass. Field monitor showed that the proposed support can reduce rock fracture and roadway convergence. For permanent roadway reinforcement in deep underground, Huang et al. (2018) developed a concrete-filled steel tubular support (CSTS) structure. It is composed of a seamless steel tube that is bent to conform to the cross section of the roadway and filled with concrete. Field application in a 1200 m underground roadway suggests that the CSTS combining with rebar bolt and cable bolt can effectively control large ground deformation. Ma et al. (2018a, b) investigated fracture development and associate controlling technology for thick soft rock roof. The support principle is to intensify the roof strength. The support system is designed as high strength bolt-beam-net with dense fully grouted cable bolt. Lu et al. (2011) developed a yieldable support scheme for deep underground roadway developed in fracturing soft rock under high ground stress. The proposed support system could control surrounding rock damage and ensures the long-term stability of the roadway. However, all these support products are made of Poisson's ratio materials.

According to the deformational behaviour of the support element, soft rock roadway support can be classified into strength support and yieldable support. The strength support is essentially elastic and provides supporting load approximately equal to the strength of the supporting material, such as rebar bolt. The yieldable support is essentially plastic that is able to accommodate large deformation, but its stiffness is often lower than that of rebar bolt.

In 2007, Professor Manchao He developed a kind of large displacement and constant axial resistance rock bolt, namely, NPR bolt (He et al. 2014, 2016). The Negative Poisson's Ratio (NPR) is defined as the positive ratio of the lateral strain to vertical strain, the formula is: ($\nu_{npr} = \varepsilon_x/\varepsilon_y$) ($\nu_{npr}$ is NPR, ε_x is lateral strain, ε_y is vertical strain) (Hu et al. 2011).

Comparing with rebar bolt, the NPR bolt has better mechanical performance in axial resistance and energy absorption capacity. In this chapter, the structure, working principle and mechanical properties of the NPR bolt will be discussed in detail in later section.

In longwall coal mining, large deformation of surrounding rock often occurs at the gateways. Gateway accidents account for 80–90% of total accidents in longwall panel. Therefore, traditional longwall and pillar mining method (namely, 121 mining method), as shown in Fig. 17.1, should be considered to be further improved.

In 2008, Cutting Cantilever Beam Theory (CCBT) was developed. Based on CCBT, a novel nonpillar mining method (namely 110 mining method) is proposed in following years. In the method, only one gateway needs to be developed for one longwall panel after the first mining cycle. Another gateway is retained along the

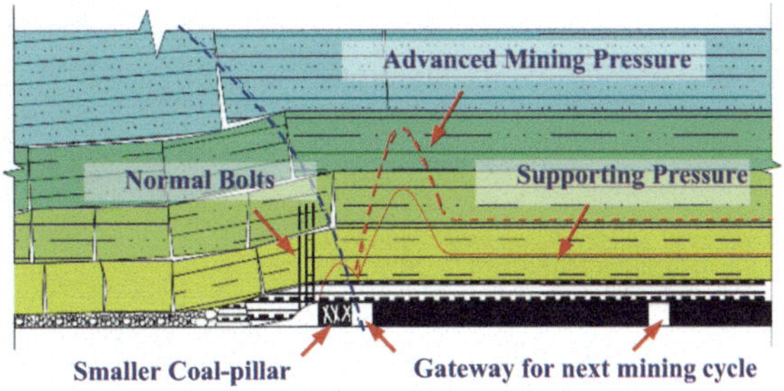

Fig. 17.1 Roof strata movement and force diagram of promoted 121 method

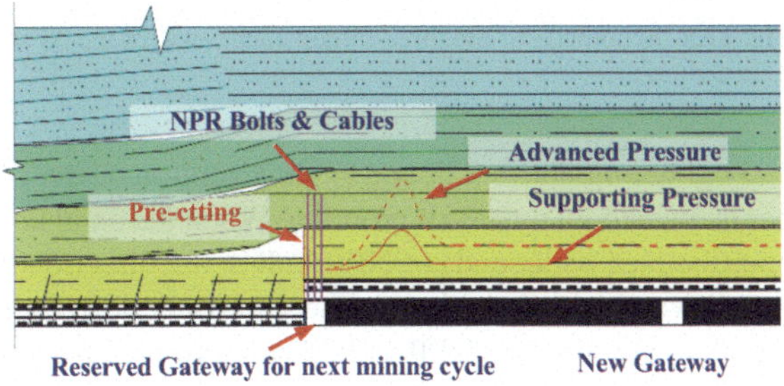

Fig. 17.2 Roof strata movement and force diagram of 110 method

gob side of adjacent panel. So there is no coal pillar left in mining area, as shown in Fig. 17.2 (He et al. 2015b, 2017; Tao et al. 2017; Wang et al. 2018).

The 110 mining method is regarded as a technical basis for the third coal mining evolution in China. In this new method, the gob gangues are utilized to form the retained gateway. When the longwall roof caves, strong disturbances are generated to the retained gateway. In such condition, the surrounding rock of the gateway can be regarded as typical engineering soft rock that is difficult to support using rebar bolt. Field application shows that the NPR bolt can accommodate large deformation and high dynamic energy released during roof caving. In the last section of this chapter, the application cases of NPR bolt in 110 mining method are presented.

17.2 NPR Support Bolt/Cable for Engineering Soft Rock

17.2.1 Development of the NPR Bolt/Cable

To meet the demand for high capacity bolt devices in soft rock engineering, a novel energy-absorbing bolt having ultra-high energy-absorbing capacity and extraordinarily large elongation at high constant axial resistance has been developed, as shown in Fig. 17.3 (He et al. 2014).

The NPR bolt is a compound device consisting of following components: a piston-like cone body installed on a bolt shank (a rebar), a sleeve pipe with its inner diameter slightly smaller than the large-end diameter of the cone body, a face pallet

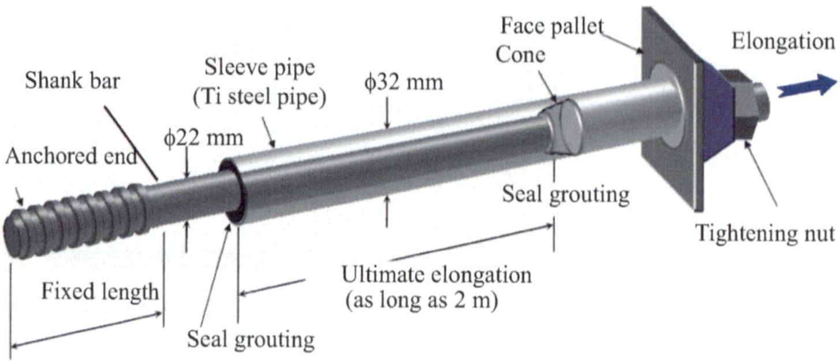

Fig. 17.3 Schematic of the three-dimensional view of the NPR bolt (He et al. 2014)

and tightening nut functioned as the retention device. The fixed length of the shank bar is bonded by means of grouting. When the axial tensile load is applied on the far end (the face pallet end) of NPR bolt in the direction opposite to the anchored end, the sleeve pipe will displace in the same direction relative the cone which is actually the elongation of the NPR bolt. The motion of the sleeve is equivalent to the displacement of the cone relative to the internal surface of the sleeve pipe. The large-end diameter of the cone is slightly larger than the inner diameter of the sleeve pipe in order to generate the frictional resistance (i.e. the working resistance of the NPR bolt) during the relative sliding between the cone and sleeve pipe. The elastic limit of the shank bar is larger than the frictional resistance. Thus, the shank only deforms elastically when NPR bolt is subjected to the tensile load. NPR cable has similar structure with the NPR bolt (He and Guo 2014).

17.2.2 Principle and Theoretical Fundamentals

Figure 17.4 shows the working principle of the NPR bolt in stabilizing the surrounding rock mass. Figure 17.4a shows the installation of the NPR bolt when the full free length of the shank bar is contained in the sleeve pipe and the fixed length of the shank is encapsulated in the borehole. The anchored end of the shank is grouted in the stable interior region beyond the area undergoing deformation. The sleeve pipe is also encapsulated in the borehole and the pallet and nut provide necessary surface retaining support. Figure 17.4b shows the working state of the NPR bolt at which the bolt is loaded by deformation of the rock surrounding a borehole and displacements associated with the extension of the virginal discontinuities. During deformation of the surrounding rock, the bolt system will restrain the dilation so that a tensile force (pull force) is induced at the face pallet exerting on the sleeve pipe. When the pull force exceeds the threshold resistance force, P_0 (the frictional force when the cone slides inside the sleeve), the sleeve pipe will displace in the direction opposite to the anchored end of the bar, exporting elongations.

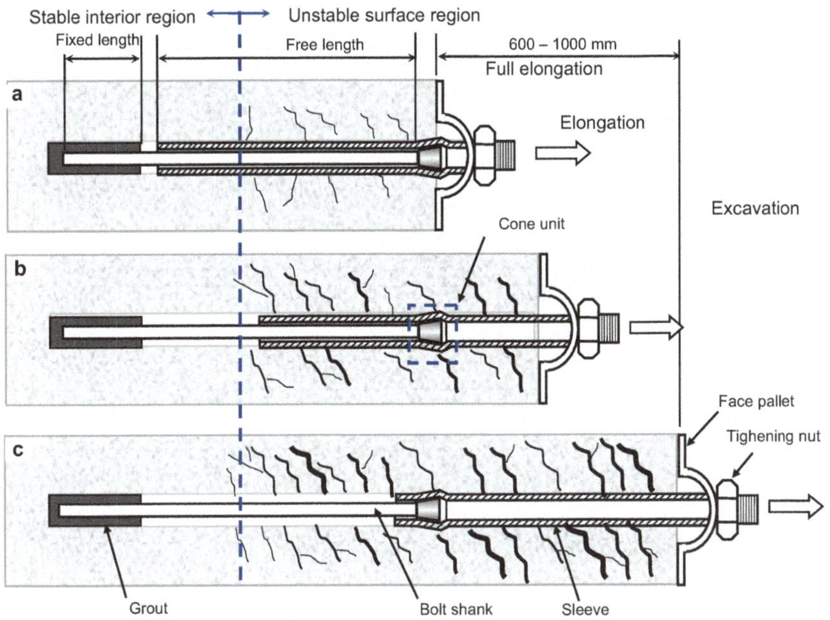

Fig. 17.4 Working principle of the NPR bolt in stabilizing the surrounding rock mass (He et al. 2014): (**a**) installation of the NPR bolt; (**b**) working state when rock mass undergoes deformation, and (**c**) retaining rock mass deformation at full elongation

The constant resistance force, P_0, is determined by the elastic property of the sleeve pipe, structure of the cone body and frictional properties of the interface between the cone and internal surface of the sleeve, but is independent of the external load under the static loading conditions. The element sliding (or structural deformation) mechanism at a constant force, P_0 (known as constant resistance), employed by the NPR bolt is the so-called "apparent yield" (Thompson 2012), that is, the bolt will be at rest unless the external load reaches its threshold magnitude P_0. The external load generated by the rock deformation is consumed by the relative frictional displacement across the interface between cone the sleeve. Unlike other currently used energy-absorbing bolts such as the Cone bolt whose load transfer is realized by the cone ploughing through the high strength grouts in the boreholes, the load transfer of the NPR bolt is realized within the designed elements, that is, the cone-sleeve relative slippage. Figure 17.4c shows the working state for the NPR bolt at full elongation in retaining the rock mass deformation.

A rock bolt/cable with an NPR structure can exhibit extraordinarily large elongation at very high working resistance to the external load, and also have an ideal elastic–plastic behaviour. The physical model for the NPR structure is shown in Fig. 17.5. The resistance force of the NPR bolt is given by (He et al. 2014):

$$P_0 = 2\pi f I_s I_c \tag{17.1}$$

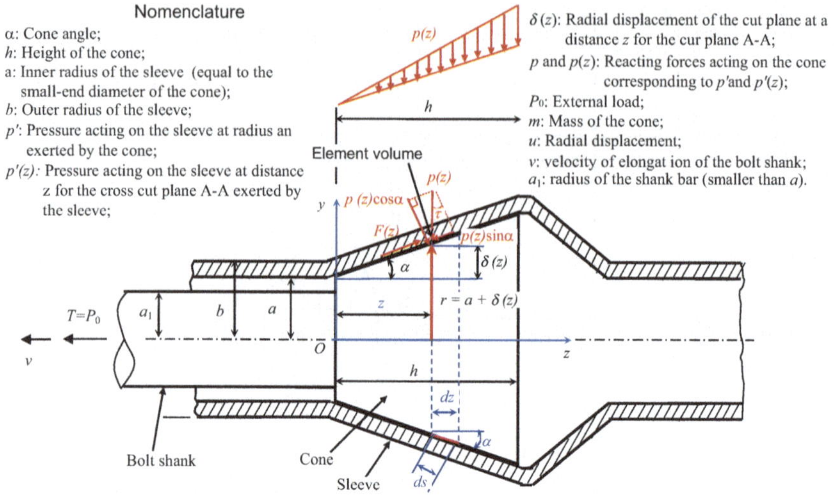

Fig. 17.5 Physical model for the NPR structure (He et al. 2014)

where P_0 is the overall resistance (also constant resistance); f is the static frictional coefficient for the interface between the cone body and sleeve; I_s is the elastic constant of the sleeve and a function of the elastic modulus E, Poisson's ratio μ, and geometrical factors a, b, the half angle of the cone α; and I_c is the geometrical constant of the cone body and a function of the geometrical factors of the cone only. I_s and I_c are given by (He et al. 2014):

$$I_s = \frac{E\left(b^2 - a^2\right)\tan\alpha}{a\left[a^2 + b^2 - \mu\left(b^2 - a^2\right)\right]} \tag{17.2}$$

$$I_c = \frac{ah^2}{2}\cos\alpha + \frac{h^3}{3}\sin\alpha \tag{17.3}$$

where α is the half angle of the cone and h the height of the frustum of the cone body.

17.2.3 Laboratory Test of the NPR Bolt Properties

17.2.3.1 Static Pull-Out Test

Static pull-out tests of the NPR bolt were conducted using the LEW-500 NPR bolt test system, as shown in Fig. 17.6. The technical specifications of the test machine are:

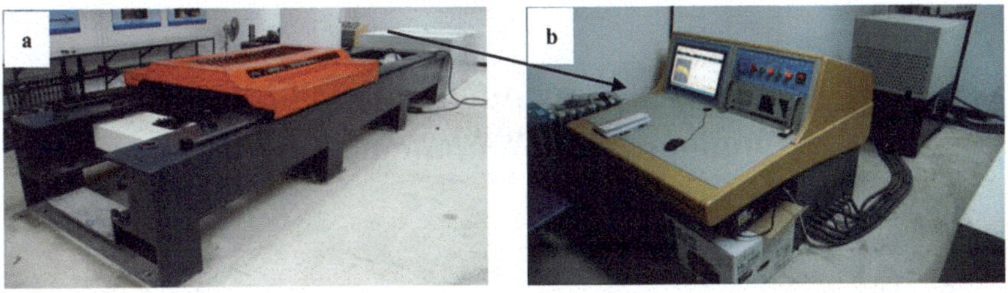

Fig. 17.6 LEW-500 NPR bolt test system: (a) test machine, and (b) control panel

Table 17.1 Structural parameters and test results for the CRLD bolt

NPR bolt No.	Length (mm)	Outer diameter (mm)	Rod diameter (mm)	Maximum tensile force (kN)	Ultimate elongation (mm)	Resistance (kN)	Loading rate (mm/min)
MG-12-7	1002	32	22	174.90	719.49	130–150	10
MG-12-8	1010	32	22	160.00	704.44	130–150	20
MG-12-9	965	32	22	164.60	664.77	120–140	100

1. Maximum tensile load: 500 kN.
2. Maximum range of measurement: 1100 mm.
3. Force loading rate: 0.1–20 kN/min.
4. Displacement loading rate: 0.5–100 mm/min.
5. Length of the specimen: 2000–3000 mm.

Three bolts MG-12-7, MG-12-8, and MG-12-9 were tested using displacement controlling at loading rates of 10 mm/min, 20 mm/min, and 100 mm/min, respectively (Table 17.1).

Figure 17.7 shows the force-displacement curves of the three CRLD bolts obtained from the static pull-out tests. The test results are also summarized in Table 17.1. It is seen that the loading rate has an influence on the static performance of the bolts; the peak axial resistance and the maximum displacement decrease as the loading rate increases.

17.2.3.2 SHPB Impact Experiment

SHPB (Split Hopkinson Pressure Bar) impact test device is widely used to study the dynamic behaviors of concrete and rocks. The device can measure the stress-strain relationship of materials under dynamic impact loading. As the device has many advantages such as simple measurement and a range of strain loading methods, it can be used for dynamic loading experiments such as tensile, compression, and bending tests. To study the tensile performance of the NPR bolt under dynamic

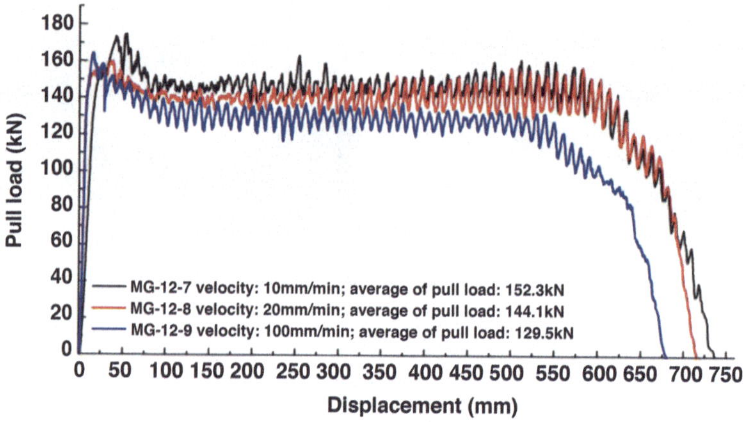

Fig. 17.7 Static tension curves for the tested three NPR bolts

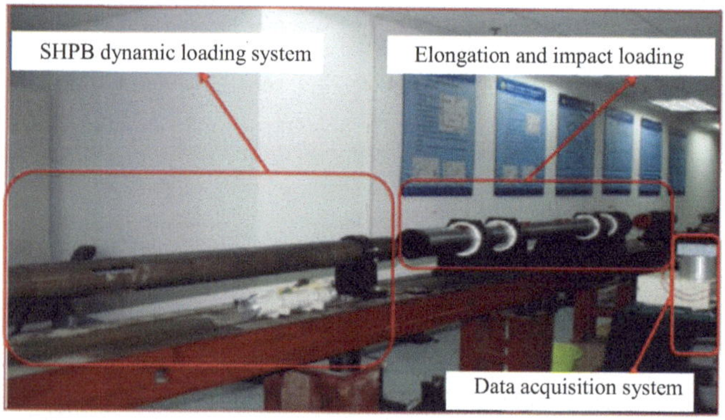

Fig. 17.8 Elongation and impact loading experiment system for NPR bolts

loading, an Elongation and Impacting Experiment System for the NPR bolt was developed. Figure 17.8 shows the basic composition of the system.

Figure 17.9 shows the physical parameters and the measurement points of specimen MG-15-1 (300), obtained before and after the dynamic experiment. Before the experiment, measurement and calibration of the axial (per 5 rings) and the radial (measuring a-a′ and b-b′ by the crossing method) alignments were conducted. The length of the device is 450 mm and the elongation rate of sleeve pipe is about 35.57%.

Figure 17.10 shows the expansion curves of the bolt. The average concave diameter of the sleeve pipe are about 29.6 mm before experiment and 31.1 mm after experiment; thus, the average radial expansion is about 1.5 mm. The average dimensions of the convex of the sleeve pipe are about 32.4 mm before experiment and 33.15 mm after experiment; thus, the average expansion is about 1.6 mm.

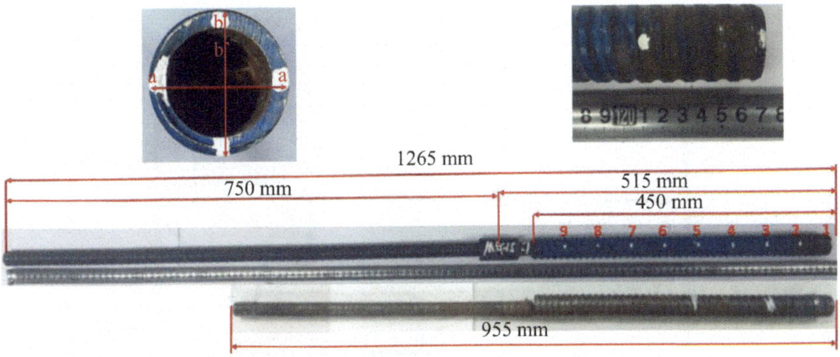

Fig. 17.9 The basic physical parameters and measurement points of NPR bolt

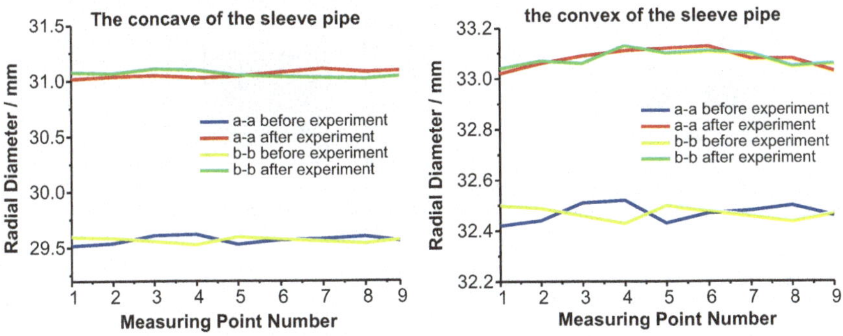

Fig. 17.10 Schematic of change of expansion of NPR bolts

It is seen that the diameter of the sleeve pipe becomes larger after the experiment, which shows a Negative Poisson's Ratio effect. In this experiment, the lateral strain ($\varepsilon_x = \Delta d/d_0$) is defined as the ratio of the amount of diameter deformation (Δd) of the sleeve pipe to its initial diameter (d_0), the vertical strain ($\varepsilon_y = \Delta l/l_0$) is defined as the ratio of the amount of length change (Δl) of the sleeve pipe to its initial length (l_0). The ratio ($\nu_{npr} = \varepsilon_x/\varepsilon_y$) of the lateral strain to the vertical strain is negative, which means that the CRLD bolt is a Negative Poisson's Ratio material. The equivalent negative Poisson's ratios of the concave and the convex of the sleeve pipe are −0.143 and −0.052, respectively.

17.2.3.3 Drop Impact Experiment

Drop impact tests on the NPR bolts were carried out using the Dynamic Testing System for NPR bolts (Fig. 17.11). This test machine can be used to evaluate the resistance and energy absorbing capacity of the NPR bolt by measuring the extension length and the radial deformation of the bolt.

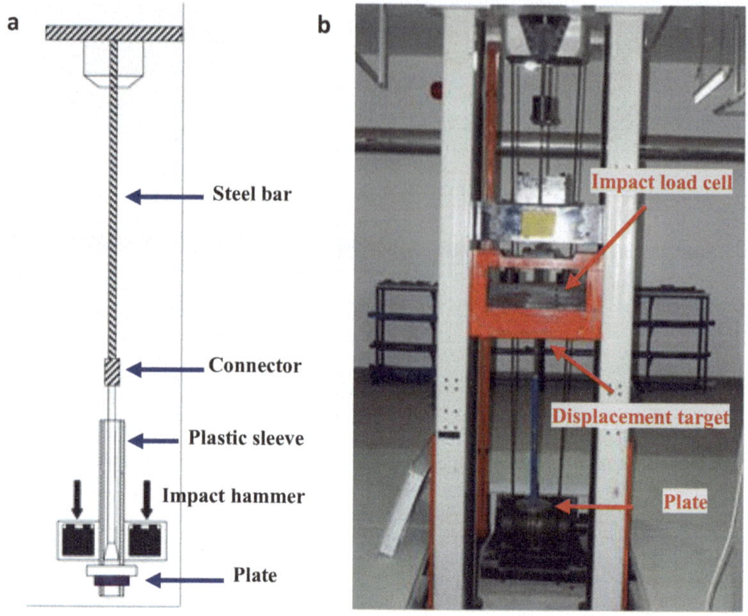

Fig. 17.11 Dynamic drop impact experiment: (**a**) alignment of the test and (**b**) test setup

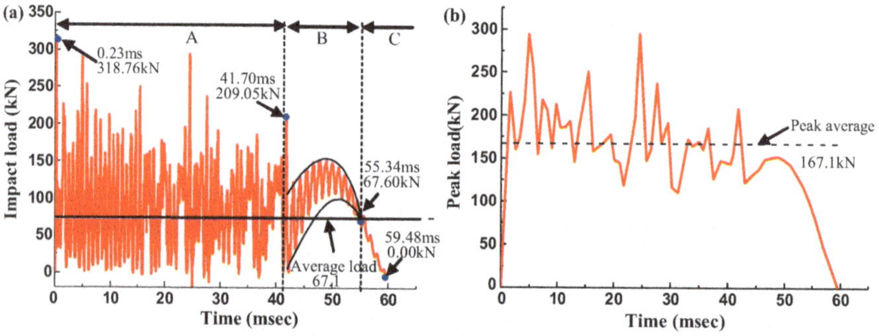

Fig. 17.12 Relation between the impact force and time for MG-12-02 NPR bolt at drop height of 700 mm. (**a**) Plot for the impact force vs. time, and (**b**) major peaks and valleys of the plot in (**a**)

The NPR bolt was cramped tightly in the test machine firstly, and then the plate was installed and the initial position was set. After setting the maximum drop height, the test was conducted with the designed cyclic-loading scheme. The elongation (the structural deformation) of the NPR bolt can be measured by an auto-controlled displacement meter, and the impact force was measured by a YFF-force transducer with an accuracy of 0.025 N and a maximum measuring capacity of 250 kN.

Figure 17.12a shows the variation of the impact force with time at a drop height of 700 mm for MG-12-02 NPR bolt, and Fig. 17.12b shows the major peaks and valleys of the plot Fig. 17.12a. The test results are summarized as:

1. Yielding deformation or elongation phase: the impact force fluctuates around a mean of 67.1 kN (the averaged tensile force for NPR bolt), which is lower than the yield force (142.0 kN) of the MG-12-2 NPR bolt. Phase A is the working phase with effective deformation of the NPR bolt; the maximum impact force of 318.76 kN occurred at the initial stage of this phase. Some local peaks afterwards also have comparable magnitudes.
2. Elastic deformation phase: the duration of this phase is small due to the rod's elasticity. Note that under a higher drop height, the NPR bolt has a larger elastic deformation range.
3. Residual deformation phase: the rod body was deteriorated gradually and lost its load-bearing capacity. This phase cannot be counted into the effective deformation range.

17.3 Field Application in Engineering Soft Rock

Roadway surroundings can be regarded as engineering soft rock in many situations, such as at great depth, in high ground stress environment or in strong disturbed area. In 110 mining method, the roadway surrounding is regarded as a typical engineering soft rock owing to its complicated stress condition. In this section, application of NPR cable in 110 mining method under different geological conditions are illustrated. The design principle of the NPR bolt support are also provided.

17.3.1 Roadway Condition in 110 Mining Method

In the new nonpillar mining method, the headgate of last panel is retained as tailgate for current mining panel. There is no coal pillar left after mining. The key point of the method is that gob gangue is utilized to form one wall of the retained entry, which can be achieved by effective roof cutting.

Before roof cutting, the working face roof and the entry roof is a whole beam structure, and their movements are synchronized. Therefore, the roof caving inevitably cause significant deformation to the retained entry roof. After roof cutting, the connection of the gob roof and the entry roof is changed. Roof caving imposes less impact on the stability of the retained entry. As the working face advances, the caved roof rock is turned into a gangue rib that takes place of the original coal rib of the entry (He et al. 2017).

The method of the entry retaining includes three stages, as shown in Fig. 17.13 (Ma et al. 2018a, b).

In the first stage, the stress concentration in the surrounding rock is mainly caused by entry excavation. Installation of the NPR cable is imperative. Generally, the length of the NPR cable should be great than the roof fracturing zone, and the bolt must be anchored in the stable rock. In this stage, the roof strata are primarily

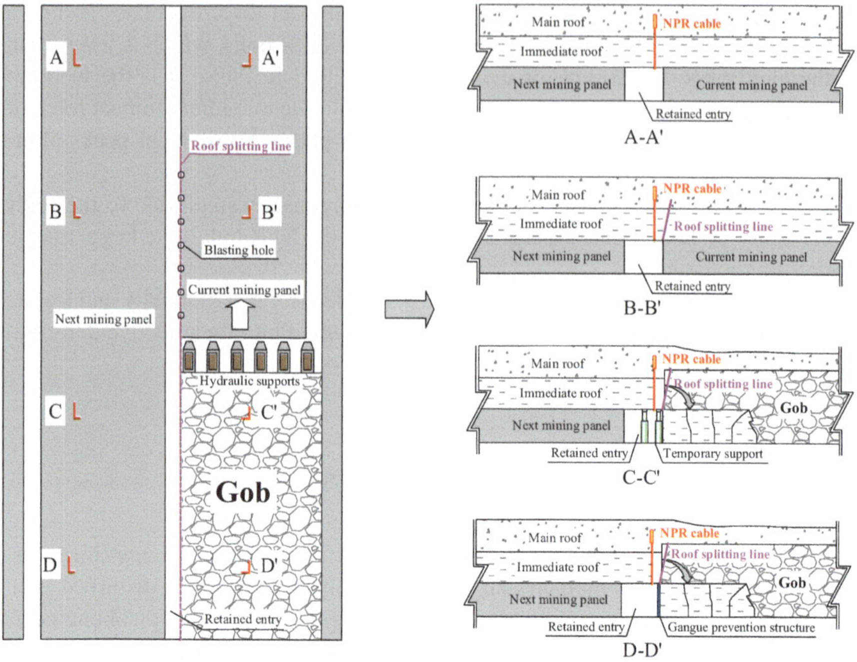

Fig. 17.13 Stress condition of the roadway in the 110 mining method under different stages

supported by the intact coal ribs. The entry roof structure can be regarded as a beam structure with two fixed ends, so the entry is in the most stable state of all stages.

In the second stage, boreholes are drilled along the corner of the roof and panel rib for roof fracturing. Then, the roof is fractured in front of the retreating face by directional blasting using an energy-accumulated device. The retained entry in this stage is considered to be in a relatively stable state.

As the retreating face reaches the roof fracturing position, The roof caves along the pre-fracturing line and falls into gob at irregular shapes and sizes, those gangues is used to form a side wall of the retained entry. The rock volume in initial broken state is considerably larger than that of the original state, but the gangues rock is compacted with subsequent roof caving. The roof strata can be viewed as a cantilever beam structure in unstable state. Therefore, temporary entry-in support and gangue wall construction support should be implemented during this stage.

In the final stage, the gangue is gradually compacted by overburden and periodic weighting and the entry become stable. The temporary support can be removed. Then the entry is retained and can be used for the next panel.

During mining of the next mining panel, the retained entry will experience mining induced stress again. From the entire process, it can see that the roadway surroundings is a typical engineering soft rock, owing to its special stress environment and strong engineering disturbance.

17.3.2 Design and Installation of the NPR Support Structure

The directional cumulative blasting is the core technology to cut off the stress transfer of the roof. The height of the roof cutting must be more than the height of the caving zone. The formulas for blasting depth are as follows (Ma et al. 2018a, b):

$$H \geq H_C / (k_0 - 1)\cos\theta \tag{17.4}$$

where H_C is the working face height; k_0 is the initial gangue expansion factor; and $k_0 = 1.4–1.8$; θ is the roof splitting angle.

To ensure the NPR cable support effect in the process of entry retaining, the length of the NPR cable is generally designed to be 2 m longer than the roof cutting height, and the anchorage end is located in stable rock. The designed length of the NPR cable H_l can be expressed as:

$$H_l = H_C / (k_0 - 1)\cos\theta + 2 \tag{17.5}$$

The construction process of the NPR cable can be expressed as follows (see Fig. 17.14).

1. Drilling and chambering. Before drilling, the cable locations should be marked based on design. The drilling hole diameters were 28 mm, and the drilling direction was perpendicular to the entry roof. After drilling, a chambering drill bit with diameter of 75 mm was used to enlarge the boreholes.
2. Anchoring. Resin is firstly transported to the borehole end through a PVC air duct hose. Then the cable bolt is pushed to the hole end and the resin is stirred for approximately 20–30 s, as shown in Fig. 17.14b. After the resin is solidified, the accessories of the NPR cable are installed.
3. Installing accessories (see Fig. 17.14c). In this step, a W-type steel band, a constant resistance device, a cable plate, and a cable lock are installed.
4. Pre-tightening (see Fig. 17.14d). The pre-tightening force should be at least 300 kN. The finished NPR cables in the entry are shown in Fig. 17.14e.

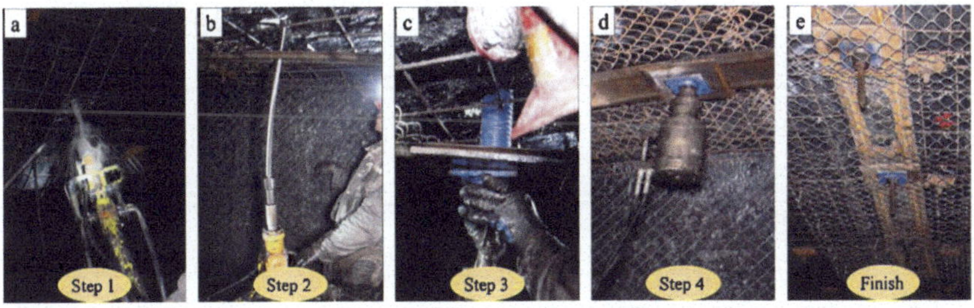

Fig. 17.14 Construction process of the NPR cable: (**a**) Drilling cable holes; (**b**) Installing stranded steel rope and anchoring resin; (**c**) Installing cable accessories; (**d**) Imposing pre-tightening force; (**e**) Installing complete

17.3.3 Field Application Cases

17.3.3.1 Application in Thick Coal Seam

Site Specifications

S1201 longwall panel of the Ningtiaota coal mine is located near the Yulin, Shanxi, China. The panel is approximately 295 m wide along dip and 3010 m long along strike, with a dip angle of less than 2°. Full-seam comprehensive mechanized retreating mining methods are used at S1201 mining panel. The mean thickness of the coal seam is 4.0 m, with an overburden ranging from 80 to 160 m. Before the test began, the S1203 panel had been mined out and the S1201 panel had retreated by approximately 2210 m. The adjacent S1201-I panel was under development for subsequent mining, as shown in Fig. 17.15.

Design of the NPR Support

Apart from the primary support, the entry roof needs reinforcing during the entry retaining process. The reinforced entry-in support includes temporary and permanent supports. The temporary support stabilize the entry and can be removed and reused later. The permanent support is mainly formed by NPR cable. Taken into considerations of the roof fracturing height and direction, the NPR cables with 21.8 mm in diameter and 10,500 mm in length were used as reinforcing supports. The interval of NPR cable rows was 1000 mm along the entry. The adjacent cables were connected by a W-shape steel strip. The constant resistance and constant resistance distance were 350 kN and 500 mm, respectively. The detailed support design is illustrated in Fig. 17.16.

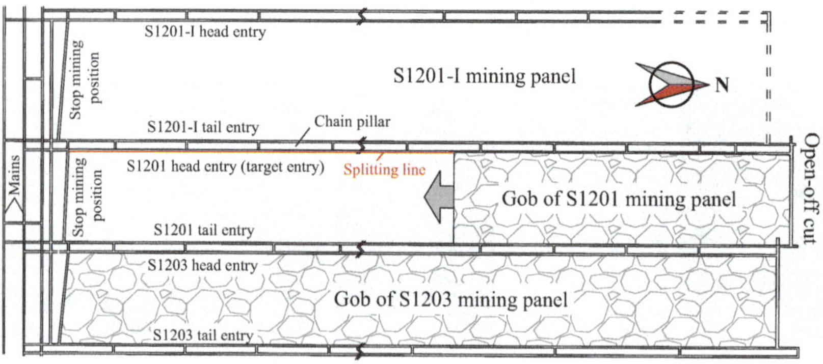

Fig. 17.15 Layouts of the S1201 mining panel

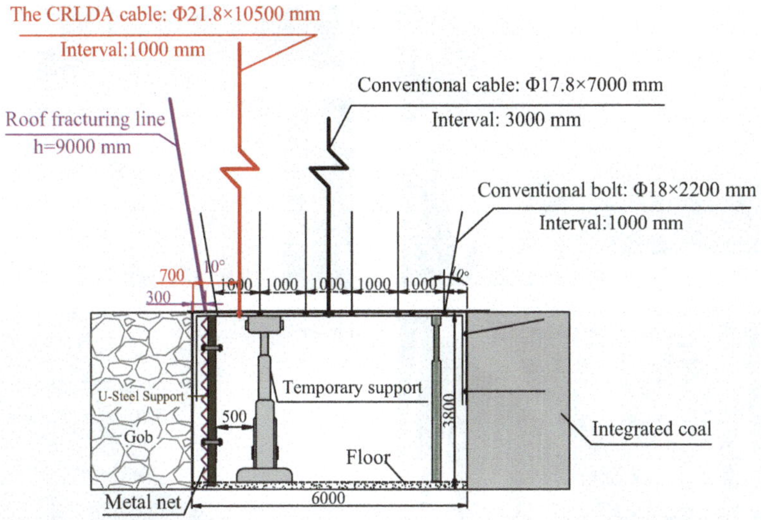

Fig. 17.16 NPR support design for the roadway in a thick coal seam

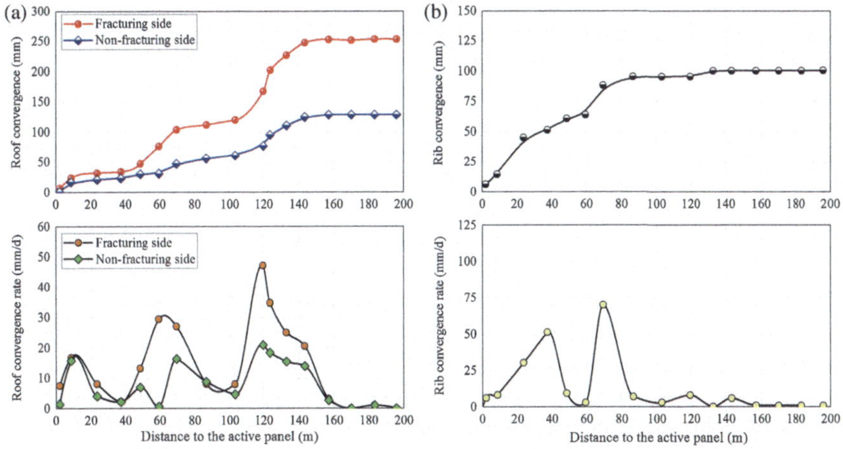

Fig. 17.17 Monitoring results of the entry deformation (He et al. 2017): (**a**) Convergences and convergence rates of the roof to floor at the fracturing and non-fracturing sides; (**b**) Convergence and convergence rate of the gangue rib to coal rib

Roadway Reinforcement Effects

Figure 17.17 shows the measured convergences and convergence rates in the S1201 headentry as the S1201 panel retreated. With the working advance, the roof-floor and ribs convergences gradually became stable, and the convergence rates ultimately decreased to zero. When S1201 panel passed the monitoring site at a distance of 155 m, the roof convergence reached a plateau. The final deformations of the fracturing and non-fracturing ribs were 253 mm and 129 mm, respectively.

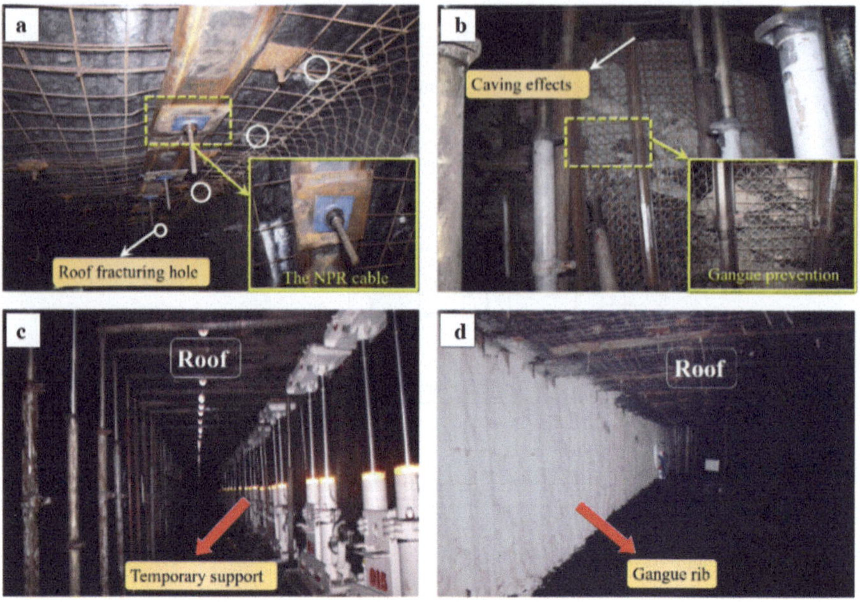

Fig. 17.18 (**a**–**d**) Roadway reinforcement effects using NPR support in a thick coal seam 110 mining method

The convergence was not in a continuous growth state, but in a leap growth state. This is mainly because of periodical fracture of the main roof. When roof deformation is induced by dynamic pressure from roof caving, the tensile load of NPR bolt increases. When the axial load reaches a certain value, the constant resistance device slips along the cable until the bend main roof contacts with the gangue. When the NPR cable and gangues take full load of the overburden, the entry is stable.

Figure 17.18 shows site pictures of retained entry. As the mining face advanced, the gob roof gradually caved, causing dynamic disturbance in the roadway. Observations show that the roadway in thick coal seam could be maintained by using NPR cable after the temporary support was removed. Under considering of the cross section requirements for ventilation and transportation, the retained entry was able to meet the requirement of the next panel.

17.3.3.2 Application in High Dip Angle Coal Seam

Mining Specifications

The 011810 working face of Jinfeng Coal Mine is located in the southeast of Wuzhong, Ningxia, China. The longwall pillar mining method is used in this mining area, which causes huge amount of resource loss. Nonpillar mining method can improve coal recovery ratio and reduce development length.

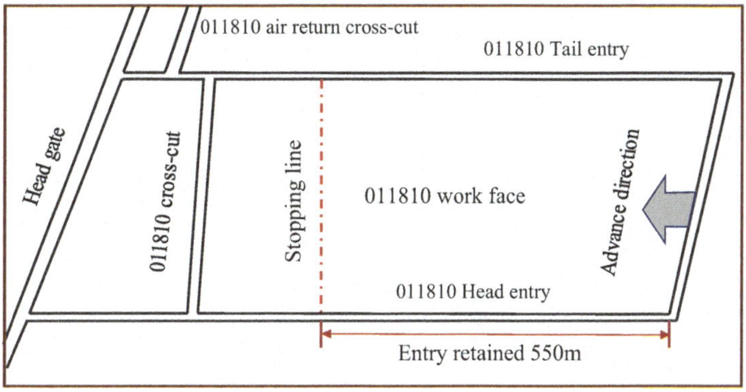

Fig. 17.19 The 011810 working face layout and the entry retaining position

The buried depth of the 011810 working face ranges from 214.06 to 328.9 m. The working face is 260 m long and the mining length is 550 m. The dipping angle of the coal seam is 12–32°, with an average of 22°. The layout of the 011810 working face and the entry retaining position are shown in Fig. 17.19.

Design of the NPR Support

To ensure the stability of gob-side entry during the process of roof cutting and periodic weighting, the NPR cable is used as entry support before roof cutting. According to the original roadway deformation and mine pressure situations, three series of NPR cables were used. Series one at the cutting side is 1000 mm array pitch and located 500 mm from the entry side vertically. Series two in the middle of the entry is 1500 mm array pitch and vertically. Series three at the coal side is 3000 mm array pitch and located 500 mm from the coal seam at 10° angle with plumb line. The diameter of the NPR cable is 21.8 mm and the length is 12.3 m. The cross section diagram is presented in Fig. 17.20.

Roadway Reinforcement Effects

The roof convergence and surrounding rock deformation at the monitoring points of 1#, 3#, 5# and 7# (20 m, 50 m, 80 m and 120 m away from open-off cut, respectively) were obtained as shown in Fig. 17.21a. The maximum surrounding rock displacement of Point 5# is selected for further analysis as shown in Fig. 17.21b.

According to the monitoring data, the entry surrounding rock deformation is that: the roof convergence speed is quite large at 25–50 m behind the working face. The maximum value is 24 mm/day. The roof convergence tends to be stable at 200 m behind the working face. The high deformation speed of the surrounding

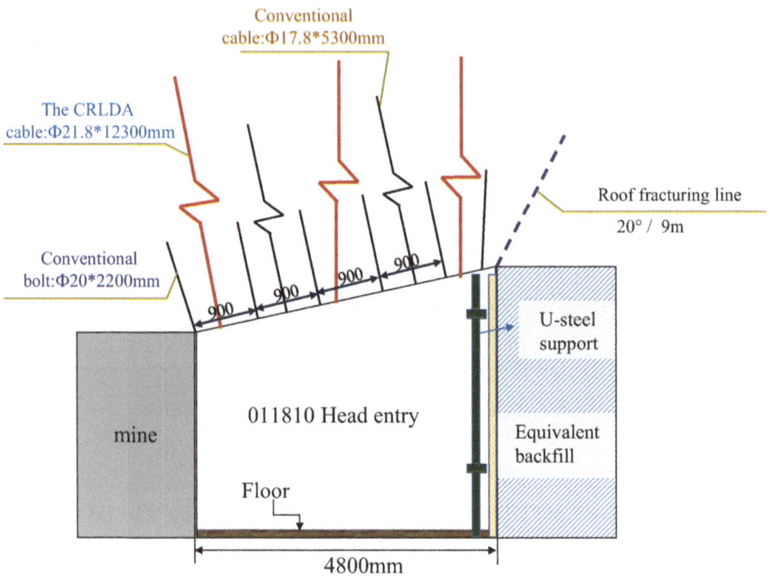

Fig. 17.20 NPR cable design for the roadway in a high dip angle coal seam

rock is 25–60 m behind the working face. The maximum value is 14 mm/day. Deformation of the surrounding rock tends to be stable 160 m behind the working face. The maximum roof deformation is 275 mm, and maximum deformation of the surrounding rock is 204 mm after final stability of the entry. The gob-side entry can satisfy the ventilation demand of next working face. The entry retaining effect is presented in Fig. 17.22.

17.3.3.3 Application in Deeply Buried Coal Seam

Mining Specifications

The 21,304 mining panel of the Chengjiao coal mine is located in Henan province, China. The buried depth of the mined coal seam ranges from 835 to 915 m. The average thickness of the coal seam is 3.1 m and the average dip angle is 4°. The immediate roof of the coal seam is mudstone, with a mean thickness of 2.85 m. The main roof of the coal seam is fine sandstone with thickness of 3.76 m, followed by siltstone with thickness of 5.23 m. In traditional longwall pillar mining method, the roadway roof experienced large deformation and the traditional support structure failed under high stress concentration, as shown in Fig. 17.23.

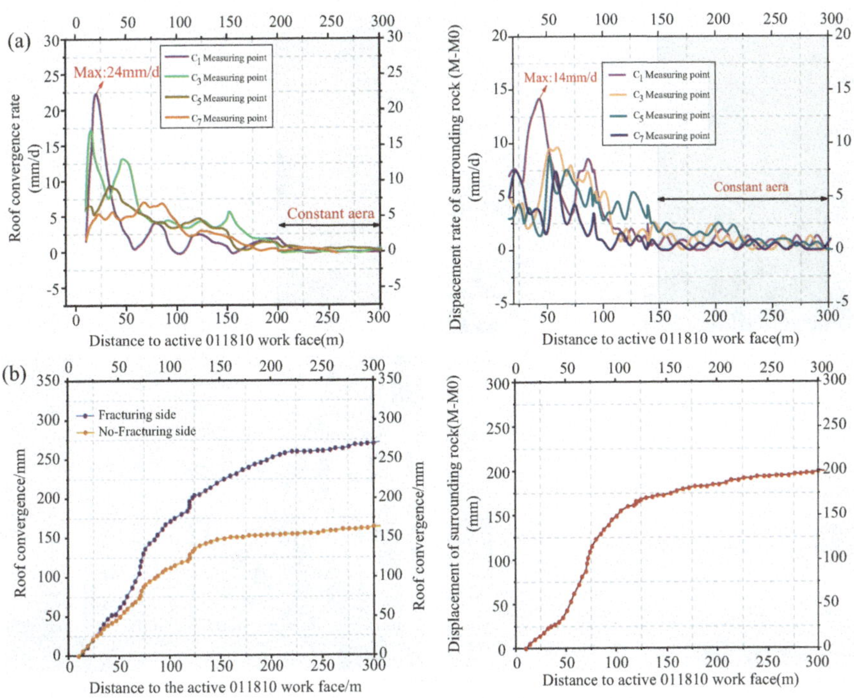

Fig. 17.21 (**a**, **b**) Monitoring curves of roadway surface deformation

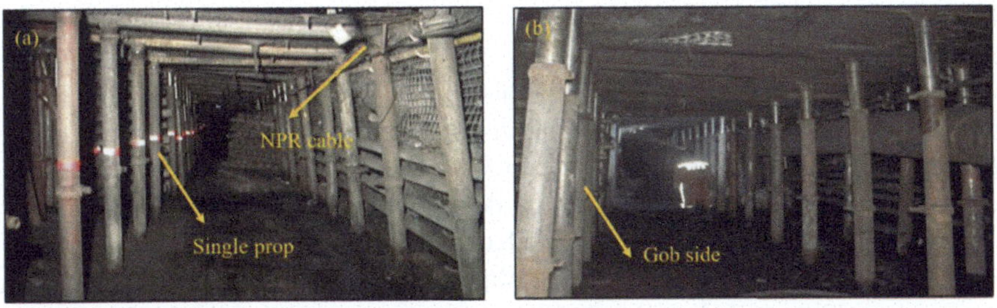

Fig. 17.22 (**a**, **b**) Roadway reinforcement effects using NPR support in a high dip angle coal seam

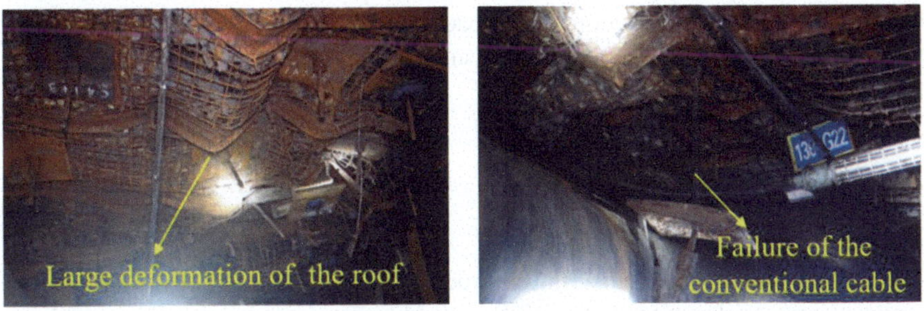

Fig. 17.23 Large deformation of the roadway surroundings in a deep coal mine

Fig. 17.24 (**a–d**) Roadway reinforcement effects using NPR support in a large buried depth coal seam

Roadway Reinforcement Effects

As shown in Fig. 17.24a, b, the initial and final states of the constant resistance device were monitored in the field. The shrinkage of the constant resistance device can reflect the energy absorbing effects of the NPR cable. At the initial state, the exposed length of the constant resistance device was approximately 23 mm. As the mining face advanced, the exposed length decreased and the lock of the NPR cable gradually shrunk into the hole. The final slippage distance of the constant resistance device was 42 mm, indicating that the impact resistance of the NPR cable is superior and one NPR cable could absorb approximately 14,700 J energy.

The reinforcing effects of the NPR cable in deeply buried roadway are shown in Fig. 17.24c, d. The final section of the roadway met the safety production requirements, verifying that the proposed NPR support structure is feasible in controlling engineering soft rock roadways.

17.4 Conclusions

This chapter focuses on engineering soft rock roadway reinforcement using a novel-designed support structure, that is, NPR bolt/cable. The working principle and theoretical fundamental of the NPR bolt were analyzed. It was found that the NPR bolt has a compound structure consisting of a cone-like piston sliding inside an sleeve pipe. The bolt has extraordinarily large elongation and high constant resistance characteristics. Static pull-out experiment, SHPB impact experiment and drop impact experiment were conducted in the laboratory to explore the extraordinarily characteristics of the new bolt. The experiments demonstrate that the NPR bolt is robust in static and dynamic loads. Finally, the NPR bolt/cable was used to reinforce typical engineering soft rock roadways under different geological conditions in 110 mining method. The field test verified that the designed NPR bolt/cable could remain during the entire entry retaining process.

References

Bai Q-S, Tu S-H, Chen M, Zhang C (2016) Numerical modeling of coal wall spall in a longwall face. Int J Rock Mech Min Sci 88:242–253

Basarir H, Ferid Oge I, Aydin O (2015) Prediction of the stresses around main and tail gates during top coal caving by 3D numerical analysis. Int J Rock Mech Min Sci 76:88–97

Chen T, Wang X, Mukerji T (2015) In situ identification of high vertical stress areas in an underground coal mine panel using seismic refraction tomography. Int J Coal Geol 149:55–66

Deng Y, Tang J, Zhu X, Fu Y, Dai Z (2010) Analysis and application in controlling surrounding rock of support reinforced roadway in gob-side entry with fully mechanized mining. Min Sci Technol 20(6):839–845

Gao Y, Liu D, Zhang X, He M (2017) Analysis and optimization of entry stability in underground longwall mining. Sustainability 9(11):2079

Gong W, Peng Y, He M, Wang J (2015) Thermal image and spectral characterization of roadway failure process in geologically 45° inclined rocks. Tunnel Underg Space Technol 49:156–173

He M (2014) Latest progress of soft rock mechanics and engineering in China. J Rock Mech Geotech Eng 6(3):165–179

He M, Guo Z (2014) Mechanical property and engineering application of anchor bolt with constant resistance and large deformation. Chin J Rock Mech Eng 33(7):1297–1308

He M, Qian Q (2010) The basis of deep rock mechanics. Science Press, Beijing

He M, Sun X (2004) Support design and construction guide for roadway within soft rock in China. Science Press, Beijing

He M, Xie H, Peng S, Jiang Y (2005) Study on rock mechanics in deep mining engineering. Chin J Rock Mech Eng 24(16):2803–2813

He M, Jia X, Gong W, Faramarzi L (2010) Physical modeling of an underground roadway excavation in vertically stratified rock using infrared thermography. Int J Rock Mech Min Sci 47(7):1212–1221

He M, Jia X, Coli M, Livi E, Sousa L (2012) Experimental study of rockbursts in underground quarrying of Carrara marble. Int J Rock Mech Min Sci 52:1–8

He M, Gong W, Wang J, Qi P, Tao Z, Du S et al (2014) Development of a novel energy-absorbing bolt with extraordinarily large elongation and constant resistance. Int J Rock Mech Min Sci 67:29–42

He M, Peng Y, Zhao S, Shi H, Wang N, Gong W (2015a) Fracture mechanism of inversed trapezoidal shaped tunnel excavated in 45° inclined rock strata. Int J Min Sci Technol 25:531

He M, Zhu G, Guo Z (2015b) Longwall mining “cutting cantilever beam theory” and 110 mining method in China—the third mining science innovation. J Rock Mech Geotech Eng 7(5):483–492

He M, Li C, Gong W, Wang j, Tao Z (2016) Support principles of NPR bolts/cables and control techniques of large deformation. Chin J Rock Mech Eng 35(8):1513–1529

He M, Gao Y, Yang J, Gong W (2017) An innovative approach for gob-side entry retaining in thick coal seam longwall mining. Energies 10(11):1785

Hu H, Wang Z, Liu S (2011) Development of auxetic fabrics using flat knitting technology. Text Res J 81(14):1493–1502

Huang WP, Yuan Q, Tan YL, Wang J, Liu GL, Qu GL et al (2018) An innovative support technology employing a concrete-filled steel tubular structure for a 1000-m-deep roadway in a high in situ stress field. Tunnel Underg Space Technol 73:26–36

Islam MR, Shinjo R (2009) Numerical simulation of stress distributions and displacements around an entry roadway with igneous intrusion and potential sources of seam gas emission of the Barapukuria coal mine, NW Bangladesh. Int J Coal Geol 78(4):249–262

Kang HP, Lin J, Fan MJ (2015) Investigation on support pattern of a coal mine roadway within soft rocks—a case study. Int J Coal Geol 140:31–40

Kang H, Lv H, Gao F, Meng X, Feng Y (2018) Understanding mechanisms of destressing mining-induced stresses using hydraulic fracturing. Int J Coal Geol 196:19–28

Konicek P, Soucek K, Stas L, Singh R (2013) Long-hole destress blasting for rockburst control during deep underground coal mining. Int J Rock Mech Min Sci 61:141–153

Lu Y, Wang L, Zhang B (2011) An experimental study of a yielding support for roadways constructed in deep broken soft rock under high stress. Min Sci Technol 21(6):839–844

Ma Z, Jiang Y, Du W, Zuo Y, Kong D (2018a) Fracture evolution law and control technology of roadways with extra thick soft roof. Eng Fail Anal 84:331–345

Ma Z, Wang J, He M, Gao Y, Hu J, Wang Q (2018b) Key technologies and application test of an innovative noncoal pillar mining approach: a case study. Energies 11(10):2853

Tao Z, Song Z, He M, Meng Z, Pang S (2017) Principles of the roof cut short-arm beam mining method (110 method) and its mining-induced stress distribution. Int J Min Sci Technol 28:391–396

Thompson AG (2012) Ground support terminology and classification: an update. Geotech Geol Eng 30(3):553–580

Wang Q, He M, Yang J, Gao H, Jiang B, Yu H (2018) Study of a no-pillar mining technique with automatically formed gob-side entry retaining for longwall mining in coal mines. Int J Rock Mech Min Sci 110:1–8

Xie LX, Lu WB, Zhang QB, Jiang QH, Chen M, Zhao J (2017) Analysis of damage mechanisms and optimization of cut blasting design under high in-situ stresses. Tunnel Underg Space Technol 66:19–33

Yan S, Bai J, Li W, Chen J, Li L (2012) Deformation mechanism and stability control of roadway along a fault subjected to mining. Int J Min Sci Technol 22(4):559–565

Zhang N, Yuan L, Han C, Xue J, Kan J (2012) Stability and deformation of surrounding rock in pillarless gob-side entry retaining. Saf Sci 50(4):593–599

Zhang Z, Bai J, Chen Y, Yan S (2015) An innovative approach for gob-side entry retaining in highly gassy fully-mechanized longwall top-coal caving. Int J Rock Mech Min Sci 80:1–11

Zhang J, Jiang F, Zhu S, Zhang L (2016) Width design for gobs and isolated coal pillars based on overall burst-instability prevention in coal mines. J Rock Mech Geotech Eng 8(4):551–558

Zhang J, Jiang F, Yang J, Bai W, Zhang L (2017) Rockburst mechanism in soft coal seam within deep coal mines. Int J Min Sci Technol 27(3):551–556

Chapter 18
Rock Mass Classification of Chalk Marl in the UK Channel Tunnels Using *Q*

Nick Barton and Colin Warren

18.1 Introduction

The Channel Tunnel was driven in chalk marl with the prior expectation by the designers of quite ideal tunnelling conditions on the UK side. This expectation was partly the result of little emphasis on the implications of joint structure. As a result of the difficulties and initial delays caused by overbreak in some of the UK sub-sea TBM drives, the first author was requested to assess the rock quality in existing tunnels in chalk marl. The work was performed during 1990 and 1991 under contract to GeoEngineering who were conducting a major review for Eurotunnel. The assessment was made using the *Q*-system of rock mass classification (Barton et al. 1974) which was also being used by TransManche Link (TML) in the Marine Service and Running Tunnels. The first author's classification of the grey chalk at Shakespeare Cliffs and of the chalk marl in the Beaumont and Terlingham Tunnels was performed prior to any data being provided on conditions in the Marine Service Tunnel (MST) or in the Marine Running Tunnel (MRT). The PB series of core logs and photographs for marine drill core PB1 to PB8 was also classified without prior knowledge of MST or MRT conditions. The extensive MST and MTR *Q*-logging by TML was subsequently made available by the second author of this paper, who was Eurotunnel's chief geologist. The comparison of multiple parties' *Q*-logging was satisfactorily close.

N. Barton (✉)
Nick Barton & Associates – Formerly Norwegian Geotechnical Institute, Oslo, Norway

C. Warren
Warren Geotechnical Associates, Surrey, UK – formerly Sir William Halcrow and Partners, London, UK

M. Kanji et al. (eds.), *Soft Rock Mechanics and Engineering*,
https://doi.org/10.1007/978-3-030-29477-9_18

TransManche Link TML's own rock mass *Q*-characterisation in the Marine Service Tunnel for km 20–30 was based on 250 face logs and 1120 side wall logs. Average *Q*-values were 9.9 for km 20–24 where most ring-building difficulties with overbreak were experienced above the cutter-head and shield, and *Q* averaged 33.4 for km 24–30. Lower values were obtained when only face logs were analysed, due to the absence of swarf. In the low cover zone from km 20.5 to 21.3, TML's mean *Q*-value was only 5.6. The above range of mean *Q*-values is very similar to that obtained independently from the pre-construction sources by the first author. According to *Q*-system case records, tunnels of 8.4 m span (two Marine Running Tunnels, MRT) and 5.3 m span (Marine Service Tunnel, MST) need *Q*-values of 40 and 10 respectively for no support to be required. The 17–18 m of unsupported tunnel lengths behind the MST and MRT tunnel boring machine tunnel faces made overbreak a very likely phenomenon when *Q*-values were in the range 1–10. (Sharp et al. 1996). Since rock mass classification is very much based on visual assessment and experience, it is judged to be helpful if the following chalk and chalk marl classifications are illustrated by some representative photographs. The starting point is logically the overlying grey chalk at Shakespeare Cliffs which is illustrated in Fig. 18.1.

Fig. 18.1 Left: Representative conjugate jointing and sub-horizontal bedding in the grey chalk at Shakespeare Cliffs. Right: Joints in the underlying chalk marl where the tunnels were to be constructed. Note the extreme planarity of the joints, which was responsible for extensive overbreak in permeable, lower-cover sections

18.2 Q-classification of Grey Chalk at Shakespeare Cliffs

The strongly developed bedding and steeply dipping conjugate jointing are easily recognised at many locations along the cliffs. Figure 18.1 is a typical illustration of joints, with the bedding less clearly seen. Superficial (non-systematic) *Q*-system classification of the grey chalk exposed in the lower cliffs gave the following preliminary indications of potential rock mass quality, where *Q* is defined as:

$$RQD / J_n \times J_r / J_a \times J_w / SRF \quad (18.1)$$

1. Typical $Q = (80\text{–}90)/9 \times 2/1 \times 1/1 = 18\text{–}22$ (good)
2. Range $Q = (70\text{–}100)/(6\text{–}12) \times (1.5\text{–}2)/(1\text{–}2) \times 1/1 = 4.4\text{–}33$ (fair to good)

These parameters describe RQD (rather high); J_n, number of joint sets (often three); J_r, roughness (smooth undulating); J_a, alteration (not visible); J_w, water inflow (optimistically low) and SRF, stress/strength (assumed favourable). The degree to which the observed jointing was representative of jointing in the chalk marl was examined in much more detail in relation to marine drill core and in underground exposures logged in the Beaumont and Terlingham Tunnels.

The steeply dipping and very persistent and planar WNW-ESE trending dominant joints seen both in the cliffs and in the chalk marl in the foreshore at low tide below the cliffs had virtually no undulation or small scale roughness. Measurements of amplitude/length (*a*/*L*) indicated very low values of joint roughness coefficient (JRC) (Barton and Choubey 1977) for these dominant steeply dipping joints. Shear strength would be correspondingly low. The significance of JRC values as low as 1–2 (more or less non-dilatant surfaces) for the stability of blocks in the periphery of a tunnel can be readily demonstrated in distinct element models such as UDEC and UDEC-BB (Cundall 1980; Makurat et al. 1990; Barton 1993). Today (Barton 2007 and more recently) we know that overbreak is inevitable when the ratio $J_n/J_r \geq 6$. In the present case, J_n/J_r was frequently 9/1.0 (three sets, smooth planar).

18.3 Q-classification of PB Series Drill Core

Potential tunnelling conditions in the chalk marl were assessed from Channel Tunnel marine drill core (PB series), and from direct classification of the chalk marl in the Beaumont and Terlingham Tunnels. In each case, the *Q*-parameters were logged in histogram format, to give a fair indication of the range of parameter values. An example of jointing in one of the PB1 to PB8 drill core is shown in Fig. 18.2.

Several hundred core box photographs and corresponding core logs and "fracture logs" were studied, resulting in the extensive set of histograms shown in Fig. 18.3. The six *Q*-parameters are shown on the left-hand side, with complementary estimates or measurements of joint frequency, spacing, joint roughness and joint wall strength on the right-hand side (see Barton et al. 1992a, b for a fuller description of

Fig. 18.2 Photograph of PB7 drill core, depth 14.9–16.95 m. Note three joint orientations

this geotechnical logging format). The jointing seen in Fig. 18.2 was described as follows at the time:

PB7: well jointed zone at 15–18 m in Chalk Marl. Joint surfaces are reportedly slickensided (i.e. J_r = 0.5–1.5 depending on planarity, J_n = 9 (or more), RQD = 85% (logged)).

The weighted mean sample for the six *Q*-parameters plotted in histogram form in Fig. 18.3 was as follows:

$$Q \text{ (mean)} = 89/6 \times 1.4/1.1 \times 0.8/1.2 = 12.6$$

Note in particular the low estimates of joint roughness JRC at core scale: JRC = 1–2 for bedding joints, and JRC = 2–3 for steeply dipping joints, and the correspondingly small roughness amplitudes (*a*/*L* at 10 cm scale; $a \approx 0.2$–1.0 mm).

18.4 *Q*-classification in the Terlingham and Beaumont Tunnels

Many hours were spent in *Q*-mapping and photography of the partly flooded Terlingham and Beaumont Tunnels. Areas of major overbreak and consistent breakage to sub-horizontal bedding planes *via* steeply dipping joints were a common feature in both tunnels. Examples of joint-related overbreak are shown in Fig. 18.4a–c. In some locations, many cubic metres of collapsed roof debris had to be climbed over, and it could be reasonably claimed that the tunnel cross-section had moved upwards (and outwards) a metre or so. The sub-horizontal bedding planes tended to

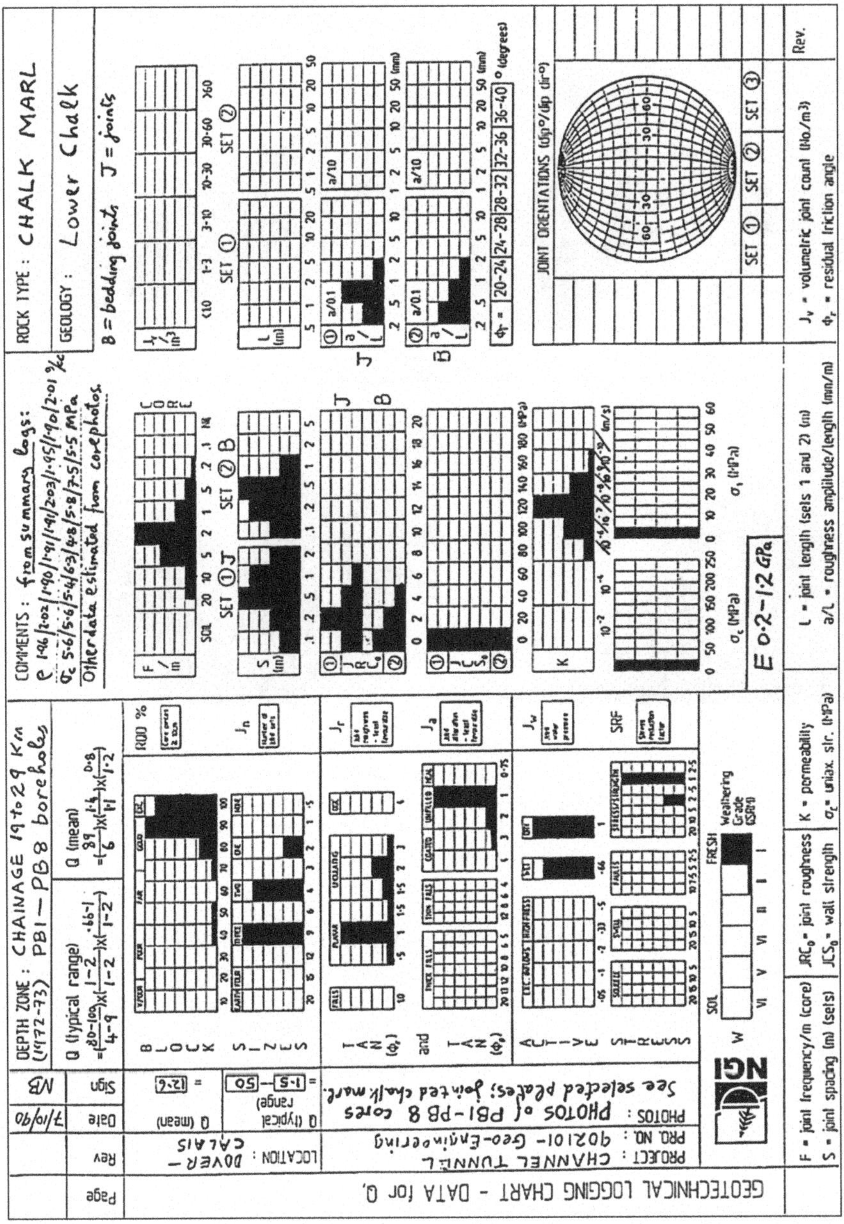

Fig. 18.3 Geotechnical and Q-characterisation of PB series drill core (PB1 through PB8)

Fig. 18.4 (**a**, **b**) Joint and bedding plane controlled overbreak in Terlingham Tunnel and in Beaumont Tunnel. Some of the 2.2 m diameter tunnel can be seen in the background in (**b**). (**c**) The wedge fall-out in Beaumont Tunnel is joint controlled, as experienced at various scales in the MST and MRT tunnels, especially along ch. 20–24 km. (**d**) The effect of a local increase in depth of cover when passing under the 70 m high Shakespeare cliffs

form the new roof. They could be continuous for many tens of metres in places. The thousands of tidal cycles during the 110 years since 1880, may have been the cause of "arch migration" at Beaumont, where it ran beneath the tidal range.

There was also an "early chainage" stress-or-strain related fractured area, where the tunnel had passed under the cliffs, causing increased tangential stress and Poisson ratio related strain. (This is known from recent studies by Dr Baotang Shen using FRACOD: Barton and Shen 2017.) The stress-or-strain related fracturing is shown in Fig. 18.4d.

The Q-parameter histograms for the two tunnels are reproduced in Fig. 18.5.

Terlingham Tunnel: Q (mean) = 90/7.4 × 1.6/1.1 × 0.9/1.5 = 10.6

Q (typical range) = 1.3–50

Beaumont Tunnel: Q (mean) = 93.6/4.8 × 1.4/2.7 × 0.7/2.1 = 3.4

Q (typical range) = 0.2–100

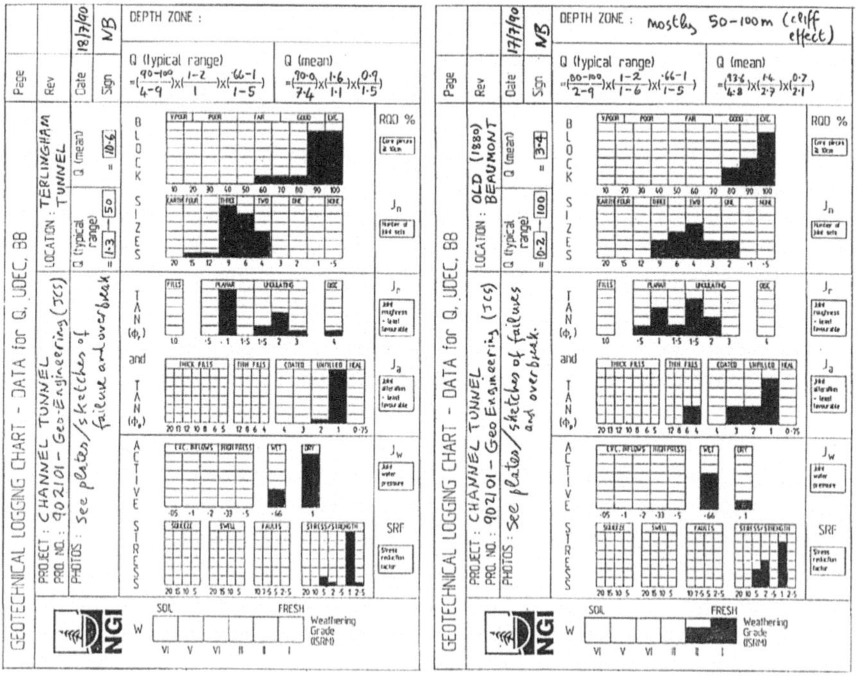

Fig. 18.5 Q-classification of the Terlingham and Beaumont (Abbots Cliff) Tunnels using histogram logging

The Q-parameter histogram method of logging used in Fig. 18.3 (see left side), was developed during air-travel to 50 km of headrace tunnels in Turkey, where much of the Q-logging had to be done from a jeep in motion, looking at countless road cuttings. This method of Q-parameter logging can clearly be used rapidly when needed, with sufficient practice. It was subsequently expanded, as in Fig. 18.3, to include other parameters needed for numerical DEM modelling during the UK Nirex LLW and ILW nuclear waste disposal project, 1990–1996, when 11 km of deep boreholes were Q (and sometimes RMR) logged in preparation for extensive UDEC-BB distinct element modelling of potential disposal cavern behaviour (Barton et al. 1992b).

The mapped section of the Beaumont tunnel had an overburden increasing under the cliffs from about 50–120 m. This resulted in some stress, or Poisson ratio extension-strain related failure in the haunches, which is reflected in the lower Q-values. In the case of the MST and MRT channel tunnels, the effective stress level caused by 20–60 m of rock cover and 30–50 m of sea depth is lower than some of Beaumont. Considering the uniaxial strength of the Chalk Marl (3.5–11 MPa, mean 6 MPa) the ratio of maximum principal stress to uniaxial strength (σ_θ/σ_c) lies well within the Q-system data base. Figure 18.4d illustrates part of this long (40–50 m) fractured zone.

18.5 Synthesis of *Q*-parameters from Precedent Data

Figure 18.3 shows the *Q*-system classification that was performed on PB1 to PB8 cores (using photos and logs). For comparison between the three sets of *Q*-parameter logging, Fig. 18.5 shows the *Q*-logging histograms for the Terlingham and Beaumont tunnels. The individual histograms have been combined in terms of relative frequency in Fig. 18.6. Numbers of observations are given from each mapping site so that a weighted mean can be obtained from the whole sample. The letters SH, BT, TT and PB refer respectively to Shakespeare Cliffs, Beaumont and Terlingham tunnels, and PB drill core.

Combining data from the four sites is considered important since the core logging may provide an overly optimistic picture of the joint frequency, as vertical holes were used. It is also valuable that the logging data could be tempered by the actual tunnelling experiences gleaned from inspection of the Beaumont and Terlingham tunnels. The addition of a very small number of observations in the grey chalk at Shakespeare Cliffs adds little to the data base. However, the experience of the jointing at the cliffs and along the foreshore is considered an important calibration process. With only moderate reduction in mechanical strength, it is inconceivable that jointing observed in the grey chalk at the Shakespeare Cliffs does not also penetrate the chalk marl in general, as indeed observed in these precedent tunnels, and in the foreshore below the cliffs (Fig. 18.1, Left). A synthesis of all the first author's *Q*-system observations from these sources resulted in the range of properties which were expected to apply to the MST and MRT drives under the Channel, shown in Fig. 18.6.

The rounded values (one decimal place accuracy) shown in Table 18.1 give a mean *Q* of 8.3. Rigorous multiplication and division of the whole (unrounded) sample gives a mean *Q* of 7.8. For practical purposes a round figure of $Q = 8$ can be adopted. The typical *range* of *Q* was 2–50 (poor to very good).

The suggested weighted mean of $Q = 8$ may not be the most typical or frequently occurring rock mass character. A glance at the histograms in Fig. 18.6 indicates that the following are the most frequently occurring characteristics according to the classifications performed:

1. Most frequent $Q = 100/9 \times 1/1 \times 1/1 = 11.1$
2. Next most frequent $Q = 90/4 \times 2/2 \times 0.66/2 = 7.4$
3. *Possible problem ground* $Q = 90/9 \times ½ \times 0.66/1 = 3.3$

A probable frequently occurring combination of the first two "classes", due to the likelihood of higher water inflows and slight joint alteration when three (as opposed to two) joint sets are present is given as *problem ground* #3. The 1993 updated version of the *Q*-system tunnel support diagram (Grimstad and Barton 1993), which is reproduced in Fig. 18.7, indicates that *Q*-values of approximately 40, 10 and 1 are required for permanent unsupported spans (diameters) of 8.4 m, 5.3 m and 2.1 m (i.e. the diameters for MRT, MST and Beaumont) respectively. The fact that some 40–50% of the Beaumont tunnel is still standing with its

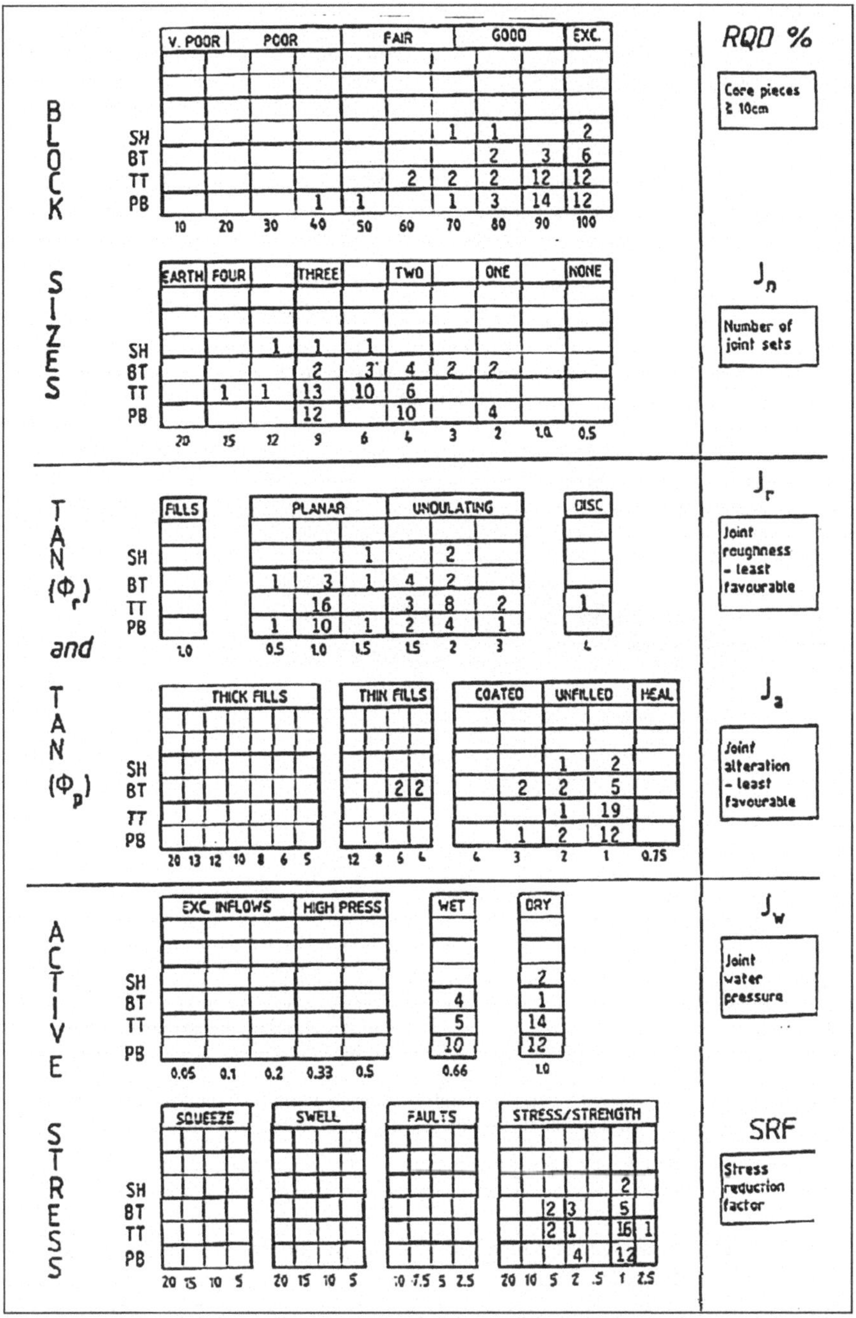

Fig. 18.6 Integrated rock mass classification data base from cliff mapping, precedent tunnel mapping and core logging, which could have been used for predicting expected conditions in the Channel Tunnel

Table 18.1 Synthesis of the pre-construction sources of rock quality for the chalk marl

Parameter	(mean)	Description	Sum/number of observations	Typical range
RQD	90.1	Excellent	6940/77	90–100
J_n	6.5	Two joint sets plus random	473/73	4–9
J_r	1.5	Rough, planar joints	93/63	1–2
J_a	1.5	Slight alteration	79/53	1–2
J_w	0.9	Slight water inflow	42/48	0.66–1.0
SRF	1.5	Slight stress problems	74/48	1–2

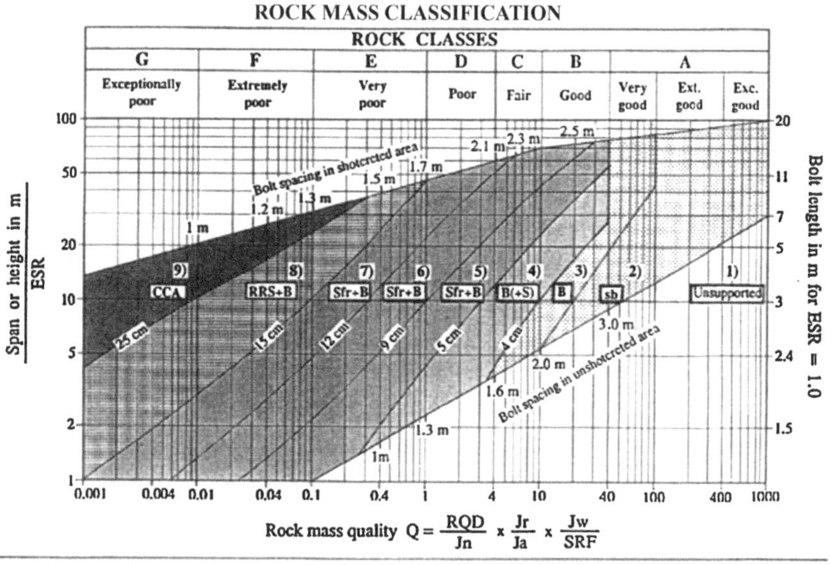

REINFORCEMENT CATEGORIES:

1) Unsupported
2) Spot bolting, sb
3) Systematic bolting, B
4) Systematic bolting, (and unreinforced shotcrete, 4-10 cm), B(+S)
5) Fibre reinforced shotcrete and bolting, 5-9 cm, Sfr+B
6) Fibre reinforced shotcrete and bolting, 9-12 cm, Sfr+B
7) Fibre reinforced shotcrete and bolting, 12-15 cm, Sfr+B
8) Fibre reinforced shotcrete >15 cm, reinforced ribs of shotcrete and bolting, Sfr,RRS+B
9) Cast concrete lining, CCA

Fig. 18.7 *Q*-support diagram showing the required spans for unsupported openings (lower diagonal line) and support measures when this span is exceeded (Grimstad and Barton 1993)

original circular cross-section after 100 years is reasonable. The worst regions of overbreak and failure presumably have *Q*-values of between about 0.2 (Beaumont minimum) and 1.0. The "no support required" lower diagonal line in Fig. 18.7 is slightly conservative (compared to mining practice) since it reflects civil engineering practice. (This diagonal line is unchanged since 1974, when the *Q*-system was first published.)

18.6 The Possibility of Overbreak Problems in the Channel Tunnels

The mean value of $Q = 8$ for the whole sample of Chalk Marl (as characterised by the first author prior to MST and MRT assessment) and the foregoing analysis of frequency of occurrence, leads to the following possible scenario for evaluating the potential for overbreak in poorer ground. We will assume the occurrence of the following (frequently observed) local conditions and their cumulative effect on the mean Q-value:

1. Three joint sets ($J_n = 9$)
2. Smooth, planar joints ($J_r = 1.0$)
3. Slightly altered joint walls ($J_a = 2.0$)
4. Medium water inflow (>5 L/min locally) ($J_w = 0.66$)

The successive, cumulative effects that these frequently observed conditions will have, compared to the weighted mean Q-value of 8.0, are as follows:

$$Q = 8.0 \rightarrow 5.8 \rightarrow 3.9 \rightarrow 2.9 \rightarrow 1.9$$

This progressive worsening moves both MST and MRT size tunnels well into the regions of the Q-support diagram that require support close to the tunnel face. The majority of the original Q-system case records were, however, related to drill-and-blast tunnels and caverns. During the instant of excavation, blocks of rock that are inherently unstable due to unfavourable J_n, J_r and J_a values (perhaps combined with "external" factors like stress and water pressure) will tend to fall out in the excavation process and contribute to overbreak.

When equally unstable blocks are freed by a TBM cutter head, many of them will try to fall out on the shield or trailing fingers. However, a good percentage of them will probably remain in place since they were not disturbed as much by the TBM as by blasting which involves blast gas penetration and higher levels of vibration. The lower frequency of vibration in a TBM excavation scenario may well lead to less overbreak, but overbreak that does occur is of course more problematic when trying to build pre-cast linings. The fact that the MST tunnel is "left unsupported" for 3.5 diameters (17 m to first ring) and that the larger MRT tunnel is "left unsupported" for 2.0 diameters (18 m to first ring) means that overbreak problems are likely when local conditions follow the worsening and likely scenarios outlined above.

The MST and MRT have diameters approximately two to four times that of the Beaumont Tunnel. They are not subjected to such high rock stresses as at the under-cliff part of Beaumont, but they are in many locations subjected to several more serious factors such as more adverse joint orientations, weathering effects and high pore pressures in the sub-vertical joints connected to the sea bed. The increased tunnel sizes induce a fundamental scale effect, where unchanged joint frequency causes greater problems with overbreak, the larger the tunnel diameter. In the Q-system, this scale effect is reflected in a greater need for immediate temporary support and final support, the larger the tunnel.

The need for a certain level of support in the case of large tunnels even with "good" rock quality ($Q > 10$) is demonstrated in Fig. 18.7.

18.7 Preliminary Review of TML *Q*-logging in the MST AND MRT

TML made extensive use of the *Q*-system in describing (under difficult mapping conditions) the rock mass conditions encountered. In the Marine Service Tunnel (MST), approximately 250 face logs and approximately 1120 sidewall logs between chainage km 20 and 30, provide a wealth of information on TML's *Q*-system estimates, which, taken together with their careful descriptions of tunnelling conditions and joint characteristics, give a very useful data base from which to draw conclusions on the encountered conditions.

The conditions under which TML's observations were obviously made were not ideal, due to limited access to the rock, as is typical with TBM. However, numerous and more extensive mapping results were available from cross-passages to supplement the necessarily sparser data sets from the face and sidewall logs. These were carefully reviewed, and an assessment made of the validity of TML's estimates.

Output from TML computer file print-out provided by Eurotunnel were analysed in four different ways as follows:

1. distributions of *Q* for km 20–30
2. distributions of *Q* for km 20–24
3. distributions of *Q* for km 24–30
4. distributions of *Q* for km 20.5–21.3

In the first stage of analysis that follows, both face logs and side wall logs were analysed. Due to the presence of swarf, the latter may be a less reliable source of data, and side wall logs are subsequently excluded from our analysis in Sect. 18.8.

TML's average estimate of *Q* for these 10 km in the MST was 22.9 (good), though most frequent observations were in the "very good" class ($Q = 40–100$). Comparison of TML estimates for km 20–24 and km 24–30 in the MST reflect both the poorer quality of rock in the early sub-sea kilometres, and the more accurate description of conditions that was possible when overbreak due to jointing was frequent. Average *Q*-values were 9.9 (fair) and 33.4 (good) respectively. The most frequent rock class observed in km 20–24 was "fair" ($Q = 4–10$, mean 6.3, 270 observations). The most frequent rock class observed in km 24–30 was "very good" ($Q = 40–100$, mean = 50, from 380 observations).

Table 18.2 shows the range of TML's observations in the MST for the poorer ground between km 20 and 24, and Table 18.3 shows the range of TML's observations in the MST for the low-cover and *poorest ground* between km 20.5 and 21.3.

TML's *Q*-system classification of the low cover zone (km 20.5–21.3) in the MST (Table 18.3) gave a mean $Q = 5.6$ (fair) with most frequent observations in the "fair" ($Q = 4–10$) and "poor" ($Q = 1–4$) classes.

Maximum ranges and mean values of *Q* estimated by TML for the various zones mapped in the MST between 20 and 30 km fairly closely resemble the first author's independently derived predicted conditions obtained from core logging (PB series)

Table 18.2 TML's Q estimates for km 20–24, from face and side-wall logging in the MST (marine service tunnel)

Range of Q	Description	No. of observations	Sum of actual Q-values	Mean Q-values
0.1–1	Very poor	5	3.2	0.6
1–4	Poor	191	470.4	2.5
4–10	Fair	272	1708.7	6.3
10–40	Good	96	1577.4	16.0
40–100	Very good	47	2280.0	49.0
	Totals	611	6039.7	9.9 (fair)

Table 18.3 TML's Q estimates for the worst-quality chainage km 20.5–21.3, from face and side-wall logging in the MST (marine service tunnel)

Range of Q	Description	No. of observations	Sum of actual Q-values	Mean Q-values
0.1–1	Very poor	2	1.5	0.7
1–4	Poor	42	90.6	2.2
4–10	Fair	45	275.9	6.1
10–40	Good	14	213.1	15.0
40–100	Very good	0	0.0	0.0
	Totals	103	581.1	5.6 (fair)

and precedent experience (Beaumont and Terlingham Tunnels). The following list compares the two sets of data:

TML mapping during construction of the MST:		
km 20–30	Q (range) = 0.3–100	Q (mean) = 22.9
km 20–24	Q (range) = 0.3–40	Q (mean) = 9.9
km 20.5–21.3	Q (range) = 0.7–20	Q (mean) = 5.6
Author's estimates from pre-construction sources:		
PB1 to PB8	Q (range) = 1.5–50	Q (mean) = 12.6
Terlingham Tunnel	Q (range) = 1.3–50	Q (mean) = 10.6
Beaumont Tunnel	Q (range) = 0.2–100	Q (mean) = 3.4

Comprehensive data packages were analysed by the first author at ten well documented chainages within km 20–30, in order to independently check TML's methods of Q-system application in the various qualities of rock. Some rather small but consistent errors in their application of the Q-system included a non-conservative use of $J_w = 1.0$ in many cases where significant water flow was observed and where $J_w = 0.66$ should have been used. In contrast, TML consistently used a conservative value of SRF = 2.5 in all cases, while only a limited number of the poorer, low cover tunnel sections perhaps qualify for this "low stress, near surface" characterisation.

In the poorest qualities of rock where TML's structural descriptions were quite comprehensive and accurate due to joint delineated overbreak, TML's estimates of Q were very similar to those of the author. At the ten well-documented sections

Table 18.4 Comparison of TML and the first author's *Q*-estimates at well documented MST and MRT chainage between km 19.8 and 27.2

	Author's estimates of *Q*		TML estimates of *Q*	
Chainage (± 50 m)	*Q* (range)	*Q* (mean)	*Q* (range)	*Q* (mean)
1. 19,824 km	7.5–50	17.6	2.4–80	28.9
2. 19,925 km	0.9–25	7.5	1.7–40	11.4
3. 20,651 km	1.0–100	7.8	1.6–40	11.7
4. 21,026 km	1.7–50	9.4	4.4–13.2	7.4
5. 22,151 km	1.2–17	5.7	2.4–19	7.0
6. 22,526 km	1.2–17	5.6	3.0–6.7	4.9
7. 22,901 km	0.5–20	2.6	1.8–10.7	4.9
8. 23,276 km	0.5–25	3.7	1.1–8.2	4.0
9. 23,651 km	1.5–25	5.7	2.5–13.3	7.9
10. 27,025/167 km	9.9–133	12.4	3.0–80	16.2

between km 20 and 30, careful interpretation was made of TML's face and tailskin logs for the MST, and of their logs of cross-passages and vertical and sideways probes. The following results were obtained:

TML: range of mean $Q = 4.0$–28.9, overall mean $Q = 10.4$
Author: range of mean $Q = 2.6$–17.6, overall mean $Q = 7.8$

Details of this comparison are given in Table 18.4.

In these well documented cases, an example of which is shown in Fig. 18.8, there is no question about the poor quality of the ground, and the first author's estimates are in fact slightly more conservative than TML's estimates. Conditions encountered were well within the range predictable from pre-construction information if the necessary classifications (of pre-construction sources of information) had been performed.

18.8 Detailed Review of TML Face Logging Results

In the foregoing section TML's *Q*-logging was analysed, using both the TBM face logs and the TBM side wall logs. As indicated earlier, the latter might be expected to be affected by the swarf. In this section we have therefore included only the results of TML's face logging. This is a significantly smaller data base as can be seen when comparing with the numbers in parentheses (from Tables 18.2 and 18.3) (Table 18.5).

A similar comparison between face logging results and the full data set (from Table 18.3) is given for the low cover section (km 20.5–21.3) in Table 18.6.

If we again analyse just the face logging results, but include both the MST and both the MRT (marine running tunnels, of 8 m diameter), we obtain the results given in Tables 18.7 and 18.8 for the chainage km 20–24, and for the poorest quality, low-cover section km 20.5–21.3.

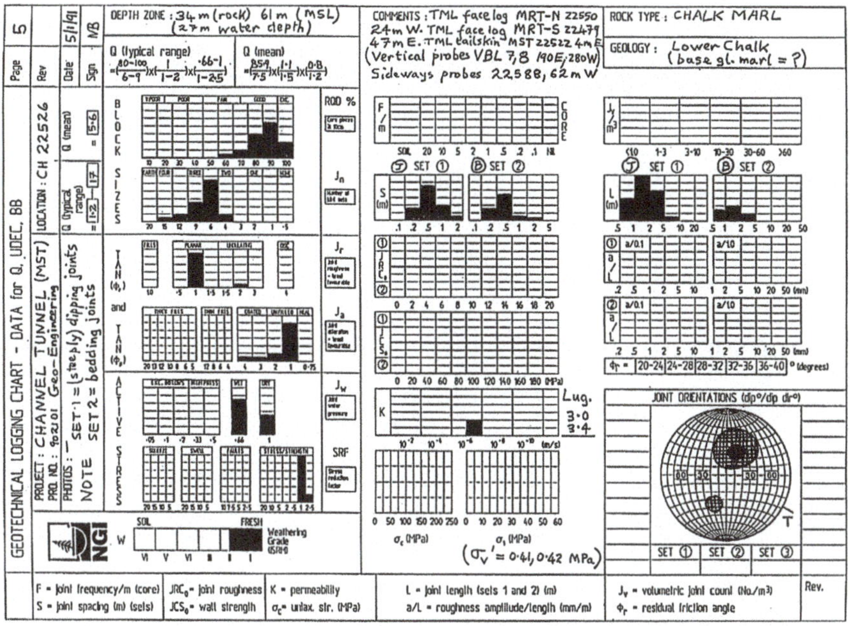

Fig. 18.8 Example of Q and Geotechnical log at well documented section for MRT/MST at ch. 22.526 m. Note that the most frequent ratio of J_n/J_r is 6/1, but 9/1 is also frequent. Refer to Fig. 18.9 example linking overbreak to J_n/J_r

Fig. 18.9 An example of the overbreak experienced in the chalk marl, as observed here in one of the piston relief ducts (PD39) between the running tunnels. Two smooth near-vertical joint sets are seen. This is a classic example of $J_n/J_r = 9/1.0$, symbolising the likelihood of overbreak, since $J_n/J_r \geq 6$, according to subsequent conclusions in Barton (2007)

Table 18.5 TML's MST face logging results, km 20–24

Range of Q	Description	No. of observations	Mean Q-values
0.1–1	Very poor	0 (5)	– (0.6)
1–4	Poor	18 (191)	2.2 (2.5)
4–10	Fair	17 (272)	6.0 (6.3)
10–40	Good	9 (96)	14.6 (16.0)
40–100	Very good	5 (47)	40.0 (49.0)
	Totals	49 (611)	9.7 (9.9)

The numbers in parentheses are from Table 18.2

Table 18.6 TML's MST face logging results, km 20.5–21.3. (The numbers in parenthesis are from Table 18.3)

Range of Q	Description	No. of observations	Mean Q-values
0.1–1	Very poor	0 (2)	– (0.7)
1–4	Poor	8 (42)	2.0 (2.2)
4–10	Fair	6 (45)	6.2 (6.1)
10–40	Good	3 (14)	11.7 (15.0)
40–100	Very good	0 (0)	– (–)
	Totals	17 (103)	5.2 (5.6)

Table 18.7 TML's face logging of all marine tunnels, km. 20–24

Range of Q	Description	No. of observations	Mean Q-values
0.1–1	Very	0	–
1–4	Poor	51	2.8
4–10	Fair	52	6.2
10–40	Good	24	15.5
40–100	Very good	10	40.0
	Totals	137	9.0

Table 18.8 TML's face logging of all marine tunnels, low-cover section, km 20.5–21.3

Range of Q	Description	No. of observations	Mean Q-values
0.1–1	Very poor	0	–
1–4	Poor	13	2.3
4–10	Fair	22	6.0
10–40	Good	12	13.4
40–100	Very good	5	40.0
	Totals	50	10.5

Summarising the above analyses of potential differences between face logs and side wall logs, we can observe that TML's average estimate of Q for km 20–30 was 22.9 for all logs, but only 10.3 considering solely the face logs within the Chalk Marl. For MST (marine service tunnel) km *20–24*, TML's average Q-values were 9.9 (for all logs) and 9.7 (for face logs). In this case the difference is rather small, as also seen in Table 18.5. For MST km *24–30*, TML's average *Q*-values were 33.4 (for all logs) but only 17.3 (for face logs). In fact apart from the low-cover section in particular, and ch. 20–24 in general, the good quality chalk marl experienced in *most* of the tunnel did indeed prove ideal for TBM tunnelling, and the world records for this tunnel, in the 8–9 m diameter class, attributed to Robbins TBM, remain to this day (best day 75 m, best week 428 m, best month 1719 m, best monthly average 873 m). For further details of the TBM performances see Warren et al. (1992), Table 14.2.

The most frequent rock class observed for MST km 20–24 was "fair" (all 272 logs, mean $Q = 6.3$), and "poor" to "fair" (35 face logs, mean $Q = 4.1$). The most frequent rock class observed for MST km 24–30 was "very good" (all 380 logs, mean $Q = 50$), and "fair" to "good" (30 face logs, mean $Q = 11.7$), in this case a marked reduction. *It is therefore seen that the face log interpretation generally gave a somewhat lower value of Q than the combined face and side wall logs.* This is to be expected due to the problem of swarf smearing over joint traces, making observations very difficult or impossible.

An important point to be noted in the above analyses is that the MST entered Glauconitic Marl and then Gault/6A material between chainage km 26.2 and 29.1 as it deviated to the north and below the level of the adjacent running tunnels past the UK Crossover cavern. Ideally this section of MST values should be ignored if considering solely Chalk Marl (Fig. 18.9).

The poor conditions encountered between km 20 and 24, and in particular from the low cover section between km 20.5 and 21.3 indicate mean *Q*-values logged by TML that are virtually identical to the weighted mean value $Q = 8$ obtained from the precedent experience described earlier (PB1 to PB8 drill core, Terlingham and Beaumont tunnels).

18.9 Use of Precedent Data in Predicting Tunnelling Problems

In Fig. 18.6 the *Q*-system histograms for the combined observations of ***Terlingham and Beaumont Tunnels, of the PB series core and of Shakespeare Cliffs and foreshore***, gave the following "most frequent" and "next most frequent" occurrences:

1. Most frequent $Q = 100/9 \times 1/1 \times 1/1 = 11.1$
2. Next most frequent $Q = 90/4 \times 2/2 \times 0.66/2 = 7.4$

From these two theoretical cases, it was reasonable to surmise that higher water inflows and slight joint alteration were more likely to be present with three joint sets ($J_n = 9$) than with two joint sets. A third class was therefore predicted as follows:

3. **Possible problem ground $Q = 90/9 \times 1/2 \times 0.66/1 = 3.3$**

Each of these six parameter values were frequently observed (i.e. most frequently or next most frequently) and they combine to form a logical physical reality. If an

SRF value of 2.5 had been used (the shallow siting assumption made by TML), an even poorer quality (Q = 1.7) could have been reasonably predicted.

The same procedure of histogram analysis will now be followed for ***seven detailed structural data packages within the chainage km 20–24.*** The most frequently and next most frequently estimated Q parameters for the seven relevant data packages within this chainage were as follows:

1. Most frequent $Q = 90/5.8 \times 1/1 \times 0.9/1 = 13.2$
2. Next most frequent $Q = 88/6.6 \times 1.4/2.5 \times 0.8/2.3 = 2.6$

If we proceed as before and combine the most frequently and next most frequently observed parameters in the generally *least favourable* manner, we arrive at the third category:

3. **Possible problem ground $Q = 88/6.6 \times 1.0/2.5 \times 0.8/1.0 = 4.3$**

If TML's SRF value of 2.5 had been used as before, an even poorer quality (Q = 1.8) is obtained. This range of problem ground (Q = 1.8–4.3) is remarkably similar to that deduced earlier from pre-construction data (Q = 1.7–3.3) and suggests that poor ground conditions were predictable.

It is worth noting that users of a rock mass classification systems such as the Q-system (and RMR) will be on the lookout for joints and unfavourable features (in outcrops, tunnels and drill core) and may arrive at an overly pessimistic classification of the ground even when less jointed conditions are represented in their logging. The above analysis and comparison of pre-construction *predictable* conditions and post-construction *observable* conditions may therefore be a little unfair, if it turns out that one is comparing the generally poorer zones observed in the Terlingham and Beaumont Tunnels and in the PB core, with more average conditions in the chainage interval km 20–24.

The five worst sections logged by TML received the lowest Q-values according to TML logging (refer to Table 18.4). The first author's mean Q-estimates at the same chainages, based on the extensive documentation given by TML are given on the right hand side of TML's estimates in Table 18.9.

A set of Q-system histograms for one of these five chainages was given in Fig. 18.8. An analysis of the frequency of Q-parameter observations for these five sections is given in Fig. 18.10. From this we can derive data for the three categories "Most frequent", "next most frequent" and "possible problem ground" as before. The following results are obtained:

1. Most frequent $Q = 90/6 \times 1/1 \times 0.66/1 = 10.0$
2. Next most frequent $Q = 100/9 \times 2/3 \times 1/2.5 = 3.0$

Table 18.9 Comparison of TML's and the first author's estimates of mean Q-values at five well-documented sections in poorer ground

Chainage	TML	Author
km 21.026 =	7.4	9.4
km 22.151 =	7.0	5.7
km 22.526 =	4.9	5.6
km 22.901 =	4.9	2.6
km 23.276 =	4.0	3.7

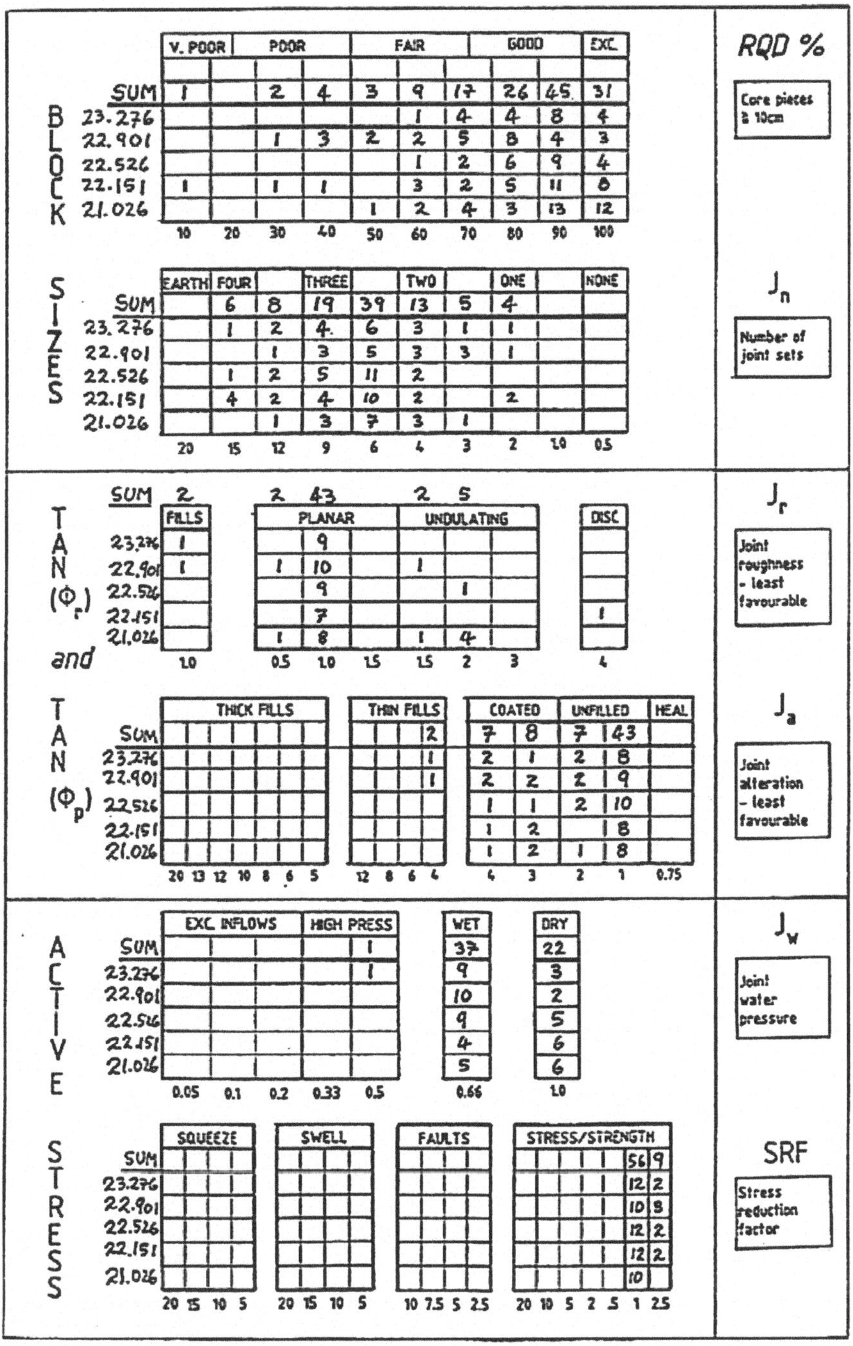

Fig. 18.10 Analysis of five of the poorest Channel Tunnel chainages in sub-sea, permeable, and partially weathered sections

If we proceed as before and combine the most frequently and next most frequently observed parameters in the generally least favourable manner, we arrive at the third category:

3. **Possible problem ground $Q = 90/9 \times 1/3 \times 0.66/1 = 2.2$**

If the above minimum value of SRF = 2.5 had also been used as before, an even poorer quality ($Q = 0.9$) is obtained. Table 18.10 finally compares the predictable conditions with those in the 20–24 km sections, using the "most frequently" observed, the "next most frequently" observed and the "possible problem ground" categories described above.

In view of the fact that the *Q*-value is based on a rating and calculation method ($a/b \times c/d \times e/f$) that results in a "logarithmic-like" scale (minimum 0.001, maximum 1000), the closeness of the predicted and observed conditions is emphasised. With rock masses of relatively poor quality, the "possible problem ground" ($Q = 0.9$–4.3) will inevitably have caused overbreak when excavated by TBM, and especially when 17–18 m of tunnel length is without support until the PC-element assembly is reached (and the liner is suitably stiffened with pea-gravel and grouting). The marine running tunnels are of sufficient diameter (8.4 m) that the "next most frequent" *Q* range of 2.6–7.4 will also undoubtedly have led to overbreak, and specific problems for the contractor when (unfortunately) wedge-lock and not bolted PC-elements were designed for the UK side of the project.

A glance at Fig. 18.7 indicates the level of NMT style permanent support (Barton and Grimstad 1994) actually required in support classes 4 and 5 when *Q*-values are less than 10. For the case of a drill-and- lasted or road-header excavated tunnel, the necessary support would be systematic bolting and shotcrete, with steel fibre reinforcement in the poorest classes of rock. If stresses were higher, and the SRF factor became

Table 18.10 Comparison of predictable and encountered conditions based on *Q*-values calculated from the most frequent observed conditions, followed by next most frequent, and finally possible problem ground

Data from	Most frequent	Next most frequent	Possible problem ground
Predicted conditions (Pre-construction data)	$Q = 11.1$ $\left(\frac{100}{9}\right)\times\left(\frac{1}{1}\right)\times\left(\frac{1}{1}\right)$	$Q = 7.4$ $\left(\frac{90}{4}\right)\times\left(\frac{2}{2}\right)\times\left(\frac{0.66}{2}\right)$	$Q = 1.7$–3.3 $\left(\frac{90}{9}\right)\times\left(\frac{1}{2}\right)\times\left(\frac{0.66}{1-2}\right)$
Encountered conditions (typical: km 20–24)	$Q = 13.2$ $\left(\frac{90}{5.8}\right)\times\left(\frac{1}{1}\right)\times\left(\frac{0.9}{1}\right)$	$Q = 2.6$ $\left(\frac{88}{6.6}\right)\times\left(\frac{1.4}{2.5}\right)\times\left(\frac{0.8}{2.3}\right)$	$Q = 1.8$–4.3 $\left(\frac{88}{6.6}\right)\times\left(\frac{1}{2.5}\right)\times\left(\frac{0.8}{1-2.3}\right)$
Encountered conditions (Poorest: km 20–24)	$Q = 10.0$ $\left(\frac{90}{6}\right)\times\left(\frac{1}{2}\right)\times\left(\frac{.66}{1}\right)$	$Q = 3.0$ $\left(\frac{100}{9}\right)\times\left(\frac{2}{3}\right)\times\left(\frac{1}{2.5}\right)$	$Q = 0.9$–2.2 $\left(\frac{90}{9}\right)\times\left(\frac{1}{3}\right)\times\left(\frac{0.66}{1-2.5}\right)$

"mobilised" by unfavourable ratios of principal stress to uniaxial compression strength, then heavier support would of course be needed. A drained lining and satisfactory drainage measures are of course a prerequisite for this NMT style of support.

18.10 Utilisation of Seismic Measurements

Offshore geophysics carried out during several campaigns in preparation for the Channel Tunnels indicated P-wave velocities generally in the region of 2.0–2.6 km/s for the UK Chalk Marl. These low values reflect the low compressive strength and the relatively high porosity of the Chalk Marl. Extensive laboratory testing of the Unit 2 Chalk Marl through which most of the UK tunnels were driven showed the average values assembled in Table 18.11.

During the years since the Channel tunnel was completed, developments have been made in linking seismic velocity measurements with *Q*-value descriptions of rock mass quality (Barton 1995, 2002, 2006). The objective has been to improve tunnel support prognoses based on refraction seismic measurements. This work was accelerated by direct calibration of core logging results with adjacent cross-hole seismic tomography. An initial calibration between Q and V_p was obtained for shallow, jointed, hard rock sites for which the following proved to be a quite accurate method (Barton et al. 1992a, b).

$$V_p = 3.5 + \log_{10} Q \tag{18.2}$$

Subsequently, seismic measurements and rock quality assessments from many sites around the world, including chalks, sandstones and other weak rocks, were added to the data base, providing the opportunity to extend the correlation to weak porous rocks at variable depth.

Figure 18.11 shows the most recent version of these correlations. The central (thick) diagonal line gives the relationship between V_p and Q shown in Eq. (18.2), for which $Q_c = Q$ (when uniaxial strength σ_c approximates 100 MPa) The "normalising" of the *Q*-value by direct application of the uniaxial strength is a necessary

Table 18.11 Laboratory index test values for Unit 2 Chalk Marl

		Average	Min/Max
Uniaxial compressive strength	MPa	5.9 (252)	0.6/17.8
Young's modulus (vertical)	GPa	0.64 (37)	0.15/4.2
V_p (axial)	km/s	2.44 (152)	1.26/3.27
V_p (transverse)	km/s	2.62 (144)	1.37/3.58
Specific gravity	g/cm^3	2.71 (72)	2.67/2.73
Dry unit weight	g/cm^3	1.96 (289)	1.63/2.30
Moisture content	%	13.3 (288)	5.8/23.8
Porosity (calculated)	%	27.7 –	15.7/39.0

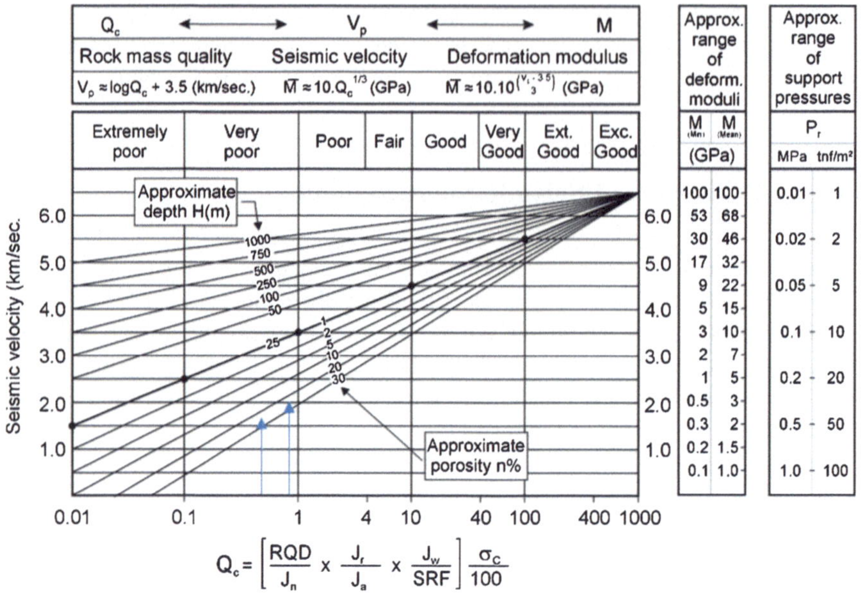

Fig. 18.11 Seismic correlations chart for interrelating Q, V_P, M and depth (with negative correction for porosity which is approximate) (Barton 1995). The higher velocities at depth were empirical results resulting from deep cross-hole seismic velocity measurements (at the UK Nirex site), and Q-logging of 11 km of boreholes by NGI/Atkins colleagues (Barton et al. 1992b)

modification for very weak (and very strong) rocks, and further correction is provided by the porosity and depth.

In order to illustrate the use of this seismic correlation chart, we can take the weighted mean value of $Q = 8$ from our precedent study (PB1 to PB8 core, Terlingham and Beaumont tunnels). This value is very close to the TML mean of $Q = 9$ for km 20–24 obtained from all the face logs in the MST, MRTN and MRTS. (Refer back to Table 18.7.)

Since Q (mean) = 8, and σ_c (mean) = 6 MPa, therefore Q_c (mean) = 8 × (6/100) = 0.5

This Q_c value intersects the reference diagonal line (Eq. 18.2) at $V_p = 3.2$ km/s. Correction for average porosity ($n = 27.7\%$, Table 18.11) results in a reduction of 1.6 km/s giving 1.6 km/s. However, tunnel depths of, for example, 40 m (see Fig. 18.12) brings this value up to about 2.0 or 2.1 km/s. Even lower values appear likely in the shallow cover (20 m) zone between km 20.5 and 21.3.

In those sections of the tunnel with markedly higher Q-values, that is, $Q = 15$ (approximately) for the MST between km 24 and 30 (excluding 26.2–29.1) we have the following:

Q (mean) = 15, σ_c (mean) = 6 MPa, therefore Q_c (mean) = 15 × (6/100) = 0.9

This Q_c-value intersects the reference diagonal line (Eq. 18.2) at $V_p = 3.4$ km/s. Correction for average porosity ($n = 27.7\%$) results in a reduction of 1.4 km/s, giv-

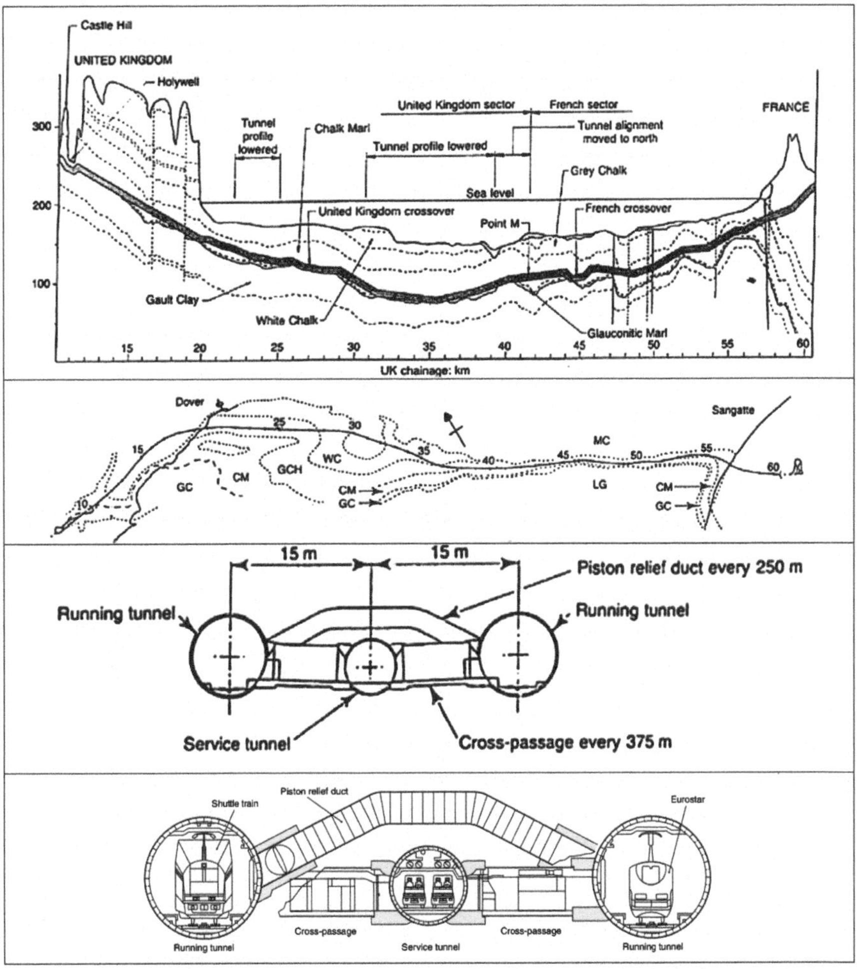

Fig. 18.12 Top: Channel tunnel stratigraphy, indicating that the tunnels mostly followed the chalk marl on the English side. The low-cover section from ch. 20–24 starts from close to the Shakespeare Cliffs. Note also the location of the UK cross-over cavern (Fugeman et al. 1992; Varley et al. 1992). Water depths are shown in the second diagram. The third diagram shows relative dimensions of the MST (5.3 m dia.) and MRT (8.4 m dia.) tunnels, and the cross-passages and smaller piston relief ducts. The final diagram is from Japan Railway and Transport Review 26, 2001, and shows the Eurostar London-to-Paris trains

ing 2.0 km/s. Tunnel depths of up to 40 m (approx.) bring this value up to about 2.5 km/s. An uncertainty in the above correlations which potentially show excellent agreement with the offshore geophysics (typically 2.0–2.6 km/s) is the effect of water depth and effective stress. Where permeability is very low due to less interconnected structure, the water could perhaps be considered as an additional load thereby potentially increasing the velocity and modulus of deformation. In permeable sections with a lot of structure, the water pressure giving significantly reduced

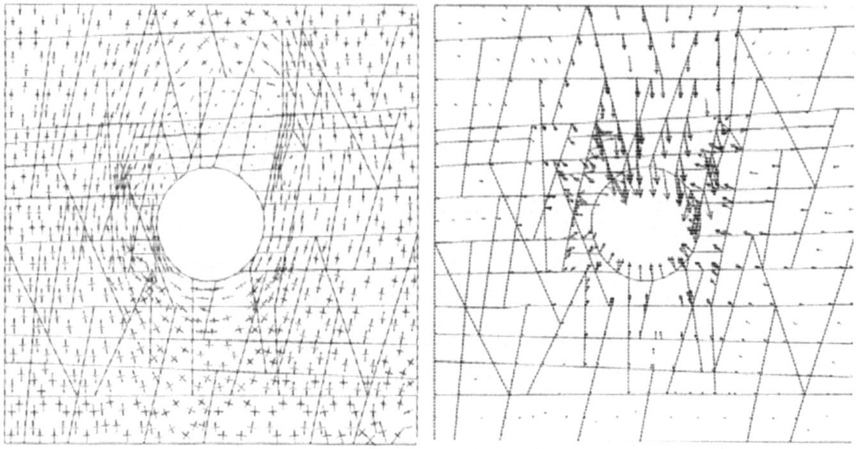

Fig. 18.13 Low effective stresses (left), particularly in the horizontal direction, combined with the low JRC estimates (see logging data in Fig. 18.3), meant that predicted vertical deformation (right), and maximum joint shearing could be several millimetres. A 50 m deep simulation of the 5.3 m diameter MST. UDEC-BB model, pers. comm. Makurat 1990)

effective stress would presumably have resulted in velocities of between 1.5 and 2.0 km/s according to the trends exhibited in Fig. 18.11. Due to the changed tunnel location in relation to earlier offshore boreholes and seismic lines, it is not clear whether these potentially lower and adverse velocity values were registered.

The third parameter illustrated in Fig. 18.11—deformation modulus *M*—is subject to considerable uncertainty due to effective stress effects, and disturbance effects when measured in situ, and due to the EDZ or excavation disturbed zone effect of reduced modulus immediately surrounding the tunnel in a rock mass with *apparently unchanged* characterisation. An undisturbed, fully *confined* modulus prediction of 3–5 GPa (*M* mean) and a fully *disturbed* modulus prediction of 0.5–1 GPa (Fig. 18.11) may well be in line with the assumed near-tunnel values of about 0.8–1.4 GPa that were derived by back-analysis of TBM tunnel deformation measurements by Eves and Curtis (1992).

For reference purposes the location of the tunnel in the geology of the English Channel is illustrated in Fig. 18.12, together with water depths and the relative dimensions and separation (15 m c/c) of the TBM tunnels. The combination of shallow tunnels, a permeable rock mass (in the low-cover section) and a significant water depth, and also the relatively low density rock (Table 18.11), meant that effective stresses could be rather low, and represent a challenge for not only overbreak, but also stability, as indeed shown in the contemporary (1990) distinct element modelling (UDEC-BB) performed by Makurat (Priv. Comm.) which is reproduced in Fig. 18.13.

18.11 Conclusions

In a drill and blasted tunnel, overbreak occurs as part of the excavation cycle and support can be applied right to the face, something which was not feasible with the chosen TBM method. Unsupported tunnel lengths of 17 and 18 m for the MST and MRT service and running tunnels represent approximately 3.5 and 2 diameters of unsupported rock in a rock mass with an average predicted rock mass quality Q of about 8, but with a quality range of at least 1–50. Tunnels of 8.4 m and 5.3 m span require rock mass qualities of 40 and 10 (respectively) for no support to be required according to case record analysis in the Q-system. Since rock mass quality between 20 and 24 km in the low-cover section was generally below 10 and much below 40, problems with stability (overbreak) were predictable and inevitable. Sensitivity studies using the most frequently observed Q-system parameters indicate that combinations of three joint sets, smooth joints, high water pressure, locally weathered joints, would inevitably lead to overbreak and the need for immediate support if it should be avoided. Overbreak with blocks falling onto the trailing fingers was therefore inevitable in many locations in this sub-sea chainage. In general, remarkable consistency was shown between the "blind logging" of precedent sources of data for the chalk marl (Terlingham and Beaumont tunnels, PB-series sub-sea cores), and the extensive logging in the three TBM drives by the contractor TML. The latter data sets were subsequently provided for comparison in this paper. It is interesting to also note the closeness of the TML Q-logging in the UK cross-over cavern ($Q = 2.5–13$, mean $Q = 7$) to precedent sources of Q (Birch and Rankin 1992). These Q-values are in the poor to fair category and proved to be a challenge for this important NATM cavern at one particular location, where vertical deformations increased suddenly from 30 to 90 mm in March 1990 (Varley and Warren 1995). Recently developed correlations between seismic velocity and Q-value using porosity, depth and uniaxial compressive strength appear to be promising ways of improving prognoses of rock quality and tunnelling problems in future projects.

Acknowledgements The work described in this chapter was performed for Geo-Engineering who were under contract for Eurotunnel. The permission of Eurotunnel and of Dr John Sharp of Geo-Engineering to publish this comprehensive analysis is gratefully acknowledged.

References

Barton N (1993) Physical and discrete element models of excavation and failure in jointed rock. Keynote lecture. ISRM Int. Symp. on Assessment and Prevention of Failure Phenomena in Rock Engineering, Istanbul, Turkey

Barton N (1995) The influence of joint properties in modelling jointed rock masses. Keynote lecture. 8th ISRM Congress, Tokyo, vol 3. Balkema, Rotterdam, pp 1023–1032

Barton N (2002) Some new Q-value correlations to assist in site characterization and tunnel design. Int J Rock Mech Min Sci 39(2):185–216

Barton N (2006) Rock quality, seismic velocity, attenuation and anisotropy. Taylor & Francis, Abingdon, p 729

Barton N (2007) Future directions for rock mass classification and characterization – towards a cross-disciplinary approach. Invited lecture. Proc. of 1st US-Canada Rock Mech. Symp., Vancouver, pp 179–188

Barton N, Choubey V (1977) The shear strength of rock joints in theory and practice. In: Rock mechanics, vol 1/2. Springer, Vienna, pp 1–54

Barton N, Grimstad E (1994) Rock mass conditions dictate choice between NMT and NATM. Tunnels & Tunnelling, October 1994, pp 39–42

Barton N, Shen B (2017) Extension failure mechanisms explain failure initiation in deep tunnels *and* critical heights of cliff faces and near-vertical mountain walls. US Rock Mech. Symp. San Fransisco, ARMA17-686, 20p

Barton N, Lien R, Lunde J (1974) Engineering classification of rock masses for the design of tunnel support. Rock Mech 6(4):189–236

Barton N, Grimstad E, Aas G, Opsahl OA, Bakken A, Pedersen L, Johansen ED (1992a) Norwegian method of tunnelling. WT focus on Norway. World tunnelling. June/August 1992

Barton N, Loset F, Smallwood A, Vik G, Rawlings C, Chryssanthakis P, Hansteen H, Ireland T (1992b) Geotechnical core characterization for the UK radioactive waste repository design. Proc. of ISRM Symp. EUROCK, Chester, UK

Birch GP, Rankin WJ (1992) Geotechnical aspects of the UK undersee Crossover. Ch. 22. Proceedings of the Institution of Civil Engineers. The Channel Tunnel

Cundall P (1980) A generalized distinct element program for modelling jointed rock. Report PCAR-1-80. Contract DAJA37-79-C-0548, European Research Office, US Army. Peter Cundall Associates

Eves RCW, Curtis DJ (1992) Tunnel lining design and procurement. Proceedings of the Institution of Civil Engineers. The Channel Tunnel. Part 1: Tunnels, pp 127–143

Fugeman ICD, Hawley J, Meyers AG (1992) Major underground structures. Proceedings of the Institution of Civil Engineers. The Channel Tunnel, Part 1: Tunnels, pp 87–102

Grimstad E, Barton N (1993) Updating of the Q-system for NMT. In: Kompen R, Opsahl OA, Berg K (eds) Proceedings of the International Symposium on Sprayed Concrete – Modern Use of Wet Mix Sprayed Concrete for Underground Support, Fagernes. Norwegian Concrete Association, Oslo

Makurat A, Barton N, Vik G, Chryssanthakis P, Monsen K (1990) Jointed rock mass modelling. International Symposium on Rock Joints. Loen 1990. Proceedings, pp 647–656

Sharp JC, Warren CD, Barton NR, Mairwood R (1996) In: Harris CS, Hart MB, Varley PM, Warren CD (eds) Fundamental evaluations of the chalk marl for the prediction of UK marine tunnel stability and water inflows. Ch. 32 in Engineering Geology of the Channel Tunnel, pp 472–507

Varley PM, Warren CD (1995) The channel tunnel project. Keynote lecture. GDF bulletin 11. Proc. XIth European Conf. of Soil Mech. and Found. Eng., Copenhagen

Varley PM, Darby A, Radcliffe E (1992) Geology, alignment and survey. Proceedings of the Institution of Civil Engineers. The Channel Tunnel. Part 1: Tunnels, pp 43–54

Warren CD, Varley PM, Parkin R (1992) UK tunnels: geotechnical monitoring and encountered conditions. Ch. 14. Proceedings of the Institution of Civil Engineers. The Channel Tunnel

Chapter 19
Applications of the GSI System to the Classification of Soft Rocks

Marinos Vassilis

19.1 Nature of Weak Rock Masses

Soft or weak rock masses represent a significant percentage of rock material on the earth's crust. They could be cases with very low intact rock properties (soft/weak per se rocks), or rocks, soft or strong, but highly tectonised-jointed or weathered or altered; heterogeneous rock masses with members of low strength can may also be included. This chapter will assist the practitioner in properly identifying weak rock units, in adequately describing their characteristics in the field, and geotechnically classify, in feasible cases, in order to quantify their properties and behaviour. In this chapter, the term "weak rock mass" is used more than "soft rock", since the latter is normally used for intact rock properties. Here the rock mass as a whole (intact rock and fractures) is classified. The weak rock masses that are examined in this study are jointed rocks, having generated from tectonical stresses and/or weathering and/or alteration effects, on either brittle or soft rock materials. Cases, where the decrease of the quality is expressed on the rock mass scale and not necessarily on the primary low intact rock strength is thus presented. The initial intact rock strength before any disturbance can be either low or high. A complex rock mass is referred here as the one that displays evident lithological, structural and geotechnical heterogeneities and nonuniformities in macroscopic scale (scale of meters). An example of such of a weak rock mass in illustrated in Fig. 19.1.

Hathaway (1990) define as "Weak Rock" one case of a consolidated earth material possessing an unusual degree of bedding or foliation separation, fracturing, weathering, and/or alteration products, and a significant content of clay minerals, altogether having the appearance of a rock, yet possessing properties in the borderline of a soil and a rock.

M. Vassilis (✉)
School of Geology, Faculty of Sciences, Aristotle University of Thessaloniki, Thessaloniki, Greece
e-mail: marinosv@geo.auth.gr

M. Kanji et al. (eds.), *Soft Rock Mechanics and Engineering*,
https://doi.org/10.1007/978-3-030-29477-9_19

Fig. 19.1 Weak Rock mass conditions resulted by severe tectonism in a heterogeneous flysch formation of siltstone and sandstone alternations (Photo from Northern Greece)

Marinos (1993) described that the weak rock masses could be distinguished to the rock masses that were "born" weak (e.g. mudrocks, siltstones), the ones that became weaker through a retrogression from an original stronger material (weathering, alteration or tectonic deformation) and the rock masses that became weaker, although still strong, since they are associated by genesis with weak rocks at a scale affecting the engineering behaviour of the formation they belong (alternation of competent and incompetent rocks—heterogeneous rock masses such as flysch). In a more general context it is difficult to draw limits in order to stick the label "weak rock" since other intrinsic or secondary features may impose a weak behaviour (stress dependent and time dependent behaviour, as creep or swelling or dissolution, change of water content, slaking, inherent anisotropy).

Santi and Doyle (1997) focus to the following weak rock units: (1) Materials with high clay content, (2) Materials with other undesirable minerals, (3) Young materials, (4) Highly weathered materials, (5) Metamorphosed materials, (6) Hardened soils. Santi and Doyle then define weak materials as "intact, unweathered to slightly weathered materials that have low compressive strength or are highly fractured". They defined weathered materials as "materials that show significant deterioration, particularly near the ground surface or along fractures". They note that, "although these types of materials show markedly different genetic and post-depositional histories, they both represent a range of properties intermediate to soil and rock".

Another characterisation of "weak rock masses" could be made, based on the behaviour, according to the engineering project type. For instance, in tunnelling, the combinations of rock mass structure, intact rock strength and overburden height that may lead to the development of stress-induced behaviour are highlighted.

Consequently, a rock mass could be characterised as weak when its potential failure is driven by the inadequacy of its strength (σ_{cm}) to bear the stresses that are imposed upon it; inadequacy that refers to the rock mass as a whole and not to its individual constituents (e.g. by local failure of its components-intact rock and discontinuities). Moreover, the characterisation of the rock mass in an engineering problem does not only depend on the rock mass properties but also to the potential failure mode. A general borderline could be that a weak rock mass: (1) in tunnelling can develop shear failures and deformations even under relatively low overburden and (2) in dam foundation raise serious concerns for the foundation using concrete dam type (e.g. arch dam, concrete gravity dam) apart from an earth fill dam.

19.2 Boundaries of Soft Rocks and Classifications

Regarding the intact rock as such, many different boundaries of the weakness can be found in literature, usually associated with the Uniaxial Compressive Strength (UCS, σ_{ci}). Rocha (in Oliveira 1993) proposed $\sigma_{ci} = 2$ MPa as the limit between the soils and rocks. According to the definition of ISRM (1981) $\sigma_{ci} = 2$–6 MPa corresponds to very weak rocks whereas 6–20 MPa to weak rocks. Similar definitions could be set for the rock mass strength (e.g. σ_{cm}). Further borderlines for some engineering properties of some cases of weak rock, are summarised by Santi and Doyle (1997).

- Compressive strength (1–20 MPa) (Afrouz 1992)
- RQD < 25–75% (Santi and Doyle 1997). Although the 75% value should be considered high
- Hammer rebound > category 4 (Santi 1995),
- Seismic wave velocity <2100 m/s (Caterpillar 1996; White and Richardson 1987)
- Ratio of weathered matrix to unweathered blocks >75% matrix (Geological Society 1995)
- Jar slake <4 (Santi 1995)
- Slake durability, Id(2), ASTM D4644-87 <90% (Santi and Doyle 1997; Lee and De Freitas 1989)
- Dearman weathering classification > category 4 (Santi 1995)
- RMR (Bieniawski 1976) <35–60 (Santi 1995). Although the 60 value should be considered high
- *Q* (Barton et al. 1974) <2 (Santi 1995)

Durability test methods have been described by Nickman et al. (2006) for sandstone, mudstone/clay-siltstone and marl, while Chaney et al. (1998) provide predictive equations in relation to jar slake (rate of change of strength (*R*)-requires time) and *Q* value (Barton et al. 1974). Moreover, factors affecting the durability of selected weak and clay-bearing rocks are presented by Gokceoglu et al. (2000).

Kanji (2014) discusses several efforts made to study and investigate soft rocks, that is, with very low intact rock properties (soft/weak per se rocks), as well as their physico-mechanical characteristics. Current and widespread classification systems to some types of weak rocks are also discussed in the same paper in an effort to highlight shortcomings and gaps remaining to be filled.

Santi (2006) proposed two modified classification methods to indicate for the relevant rocks, proportion and nature of corestones and matrix, strength, influence of discontinuities, and reactivity to water. Finally, field-testing methods that estimate strength, permeability, durability, and reaction to water are identified. These include point load index, *Q* Rock Mass Classification, jar slake, and hammer rebound classification. Researchers and practicing professionals who deal with weak rock of shales have suggested different schemes (Santi and Shakoor 1997). In an overview of classification systems, Chapman and others (1976) underline that shale classification systems fall into three categories: systems developed by geologists emphasising genesis, systems developed by engineers emphasising quantitative laboratory tests, and systems developed by agencies, which prove to be limited to specific regions or applications.

Santi (1997) suggests a modified Dearman's "Weathering Grade" field classification system that includes shales and other non-granitic rocks and also a system, incorporating this modified Dearman scheme (1976) and of the Geological Society of London (1995), assimilating several existing systems (modified from Geological Society 1995; Palicki 1997), to indicate proportion of corestones and matrix, strength, influence of discontinuities, and reactivity to water.

Castro et al. (2013) present some modifications to the RMR76 system (Bieniawski 1976) and propose a transition function for estimating the strength for weak rock masses.

19.3 Weak Rock Mass Properties

Definition of rock mass properties for a particular design problem usually involves one or either more of the following data acquisition methods (Marinos and Carter 2018) (a) laboratory testing, (b) in situ testing, (c) use of rock mass classifications, and/or (d) back analysis. However, there can be significant differences in scale between results that each of these approaches may yield. Additional complications in representativeness may also arise because, for instance, laboratory scale samples may not be truly representative of the rock mass due to natural heterogeneity present for most formations or the containing micro-fractures. Cost and time issues additionally, complicate data acquisition, particularly when it comes to carrying out in situ tests. To estimate reasonable geotechnical parameters for the design of many engineering projects, and in particular for design of tunnel support for long deep tunnels beneath mountain ranges where drilling is difficult before construction starts, and where a back analysis approach would not be possible, there is no option but to rely upon the use of some form of rock mass classification scheme that is correlated with the basic parameters needed for design. This is particular true for the case of weak rock masses. Back-analysis is indisputably the best way to estimate appropriate geotechnical parameters, but can be performed when construction has started, provided the analysis approach being used for the parameter definition is valid. For almost all rock engineering problems, back-analysis evaluation of deformation measurements generally provides the most insight, and oftentimes yields the most credible parameter understanding, such that it can be used to validate or modify any parameters previously used for design.

Gneisses, granites, ophiolites, limestones, schists, siltstones/mudstones/shales, as well as molassic and flysch formations are each represented in this chapter, highlighting how geological differentiation amongst different rock types affects overall geotechnical properties on a comparative basis. The complexity of these geological materials imposed a more specialised research for their geological characterisation due to the special features of their rock masses regarding both their structure and their lithological characteristics. However, as it is emphasised in the paper, every rock mass has its own features and the presented specific GSI ranges should be used with caution. While "structure" appearance (from intact to disintegrated) can be similar for a wide range of rock masses, major changes in strength and deformability can occur between similarly fractured rock masses, solely because of differences in parent rock material competence.

19.3.1 Challenges on Putting Numbers to Weak Rock Masses Through Classification Systems

The need for design input parameters is satisfied nowadays in many cases through the use of geotechnical classification systems. When the earliest of the more well-known and more commonly applied rock mass classification systems were initially developed (Terzaghi 1946; Lauffer 1958; Wickham et al. 1979; Barton et al. 1974; Bieniawski 1973), they were intended principally for tunnelling and mainly as an empirical design method for the purpose of estimating underground support, although not all including necessarily weak rock masses.

With the rapid growth of improved numerical design tools, which now allow progressive failure processes and also sequentially installed support to be analysed, and synthetic rock masses to be built, the need for acquisition of more reliable rock mass parameters has also grown (Marinos and Carter 2018).

Introduction of the Hoek–Brown failure criterion in 1980 has over the last three decades provided a sound basis for the evaluation of rock mass strength parameters when predicated on reliable engineering geological input, but only when parameters are carefully selected with good geological judgment. While the Hoek–Brown criterion (Hoek and Brown, 1980, 1997, 2018; Hoek et al. 2002) has aided the revolution in rock engineering, it has over the years also suffered a lack of precision in definition of input constants. It was recognised that this was only partially initially addressed through use of the then available rock mass classification systems (Carter and Marinos 2014). As these were considered at the time somewhat restrictive and not sufficiently linked with real geological observation, and thus were perceived to be over-rigid, almost mechanical in their application, the "Geological Strength Index", GSI, was developed, initially by Hoek (1994) but then improved in subsequent papers published by Hoek et al. (1998) and Marinos and Hoek (2000, 2001) and Marinos (2007), resulting in the now familiar basic chart (Marinos and Hoek 2000) illustrated in Fig. 19.2.

The Geological Strength Index (GSI) more than the other available classification systems was formulated to attempt to characterise rock masses from a more geological rather than a typical engineering approach so as to better meet the need for delivering reliable input data, particularly related to those rock mass properties required as inputs into numerical analysis or into closed form solutions for designing tunnels, slopes or foundations in rocks.

The classification covers the weak rock masses from either soft or brittle rocks, resulted from jointing, weathering or alteration. The strength of the rock initially intact has always to be combined. The soft rock, per se, when it remains intact, can be just classified by its unconfined compressive strength while properties can be derived by tests. In this case GSI meaningless.

The GSI has considerable potential for use in rock engineering because it permits the manifold aspects of rock to be quantified, enhancing geological logic and reducing engineering uncertainty. Its use allows the influence of the variables which make up a rock mass to be assessed and hence the behaviour of rock masses has to be explained more clearly. One of the key advantages of the index is that the geological reasoning it embodies allows characterisation to be made of a very wide range of rock-masses and conditions, including both weak and complex situations, but also always allowing care to be taken to keep within valid applicability limits (Hoek and Marinos 2007).

The initial GSI classification was set up to match the earliest tables for m and s (as per Table 19.1) and to address the two principal factors considered important influences on the mechanical properties of a rock mass—the structure (or blockiness), and the condition of the joints. It specifically was set up to avoid including other factors, such as intact strength, in situ stresses, and/or groundwater pressures, because these factors normally would be allowed for in any analyses. Fair to Very Poor-quality rock masses can be identified in Table 19.1 for different combinations of m and s, that is, for various geological situations (modified from Hoek et al. 1995; Hoek personal communication).

It was proposed that a new variation of the previously published Hoek–Brown transfer equations should be used for disturbed and undisturbed rock conditions with the limit GSI for the equation set equal to 25.

The most major revision of the Hoek–Brown criterion was published by Hoek et al. in 2002, which resulted in the following suite of equations:

$$\sigma_1 = \sigma_3 + \sigma_{ci}\left(m_b * \sigma_3 / \sigma_{ci} + s\right)^a \tag{19.1}$$

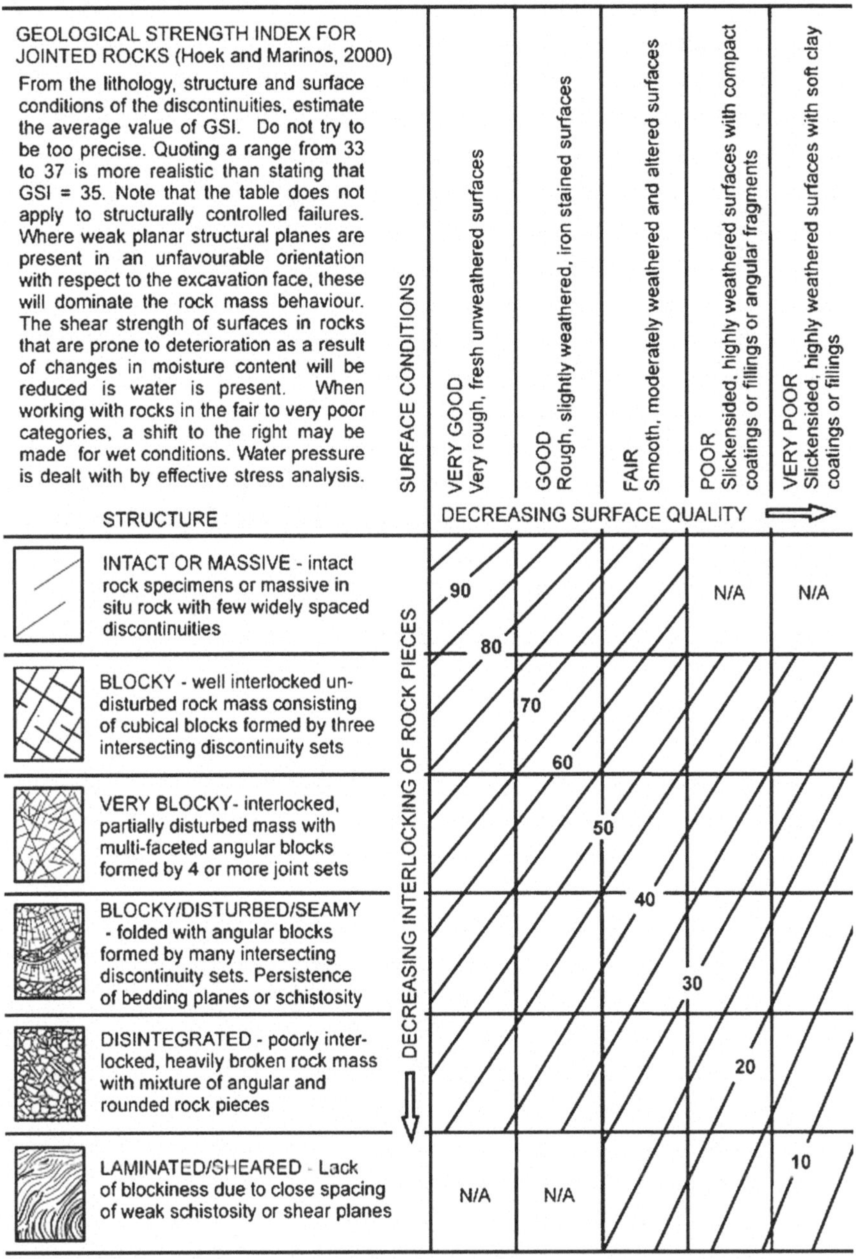

Fig. 19.2 Basic GSI chart for characterisation of rock masses (Hoek and Marinos 2000)

Table 19.1 Original GSI tabulations for m and s for various geological conditions indicating the cases of weak rock mass (modified from Hoek et al. 1995; Hoek Personal communication)

Empirical failure criterion $\sigma_1' = \sigma_3' + (m\sigma_c \sigma_3' + s\sigma_c^2)^{1/2}$ σ_1' = major principal stress σ_3' = minor principal stress σ_c = uniaxial compressive strength of intact rock m, s = empirical constants	Carbonate rocks with well-developed crystal cleavage, e.g. dolomite, limestone, and marble	Lithified argillaceous rocks, e.g. mudstone, siltstone, shale and slate (tested normal to cleavage)	Arenaceous rocks with strong crystals and poorly developed crystal cleavage, e.g. sandstone and quartzite	Fine grained polymineralic igneous crystalline rocks, e.g. andesite, dolerite, diabase, and rhyolite	Coarse grained polymineralic igneous and metamorphic crystalline rocks, e.g. amphibolite, gabbro, gneiss, granite, norite, and quartzdiorite
Intact rock samples Laboratory size samples free from pre-existing fractures Bieniawski, (1974) (CSIR)[a] rating 100 Barton et al. (1974) (NGI)[b] rating 500	$m = 1$ $s = 1$	$m = 10$ $s = 1$	$m = 15$ $s = 1$	$m = 17$ $s = 1$	$m = 25$ $s = 1$
Very good quality rock mass Tightly interlocking undisturbed rock with rough unweathered joints spaced at 1–3 m Bieniawski, (1974) (CSIR) rating 85 Barton et al. (1974) (NG1) rating 100	$m = 3.5$ $s = 01$	$m = 5$ $s = 0.1$	$m = 7.5$ $s = 0.1$	$m = 8.5$ $s = 0.1$	$m = 12.5$ $s = 0.1$
Good quality rock mass Fresh to slightly weathered rock, slightly disturbed with joints spaced at 1–3 m Bieniawski, (1974) (CSIR) rating 65 Barton et al. (1974) (NGI) rating 10	$m = 0.7$ $s = 0.004$	$m = 1$ $s = 0.004$	$m = 1.5$ $s = 0.004$	$m = 1.7$ $s = 0004$	$m = 2.5$ $s = 0004$
Fair quality rock mass Several sets of moderately weathered joints spaced at 0.3–1 m, disturbed Bieniawski, (1974) (CSIR) rating 44 Barton et al. (1974) (NGI) rating 1	$m = 0.14$ $s = 00001$	$m = 0.20$ $s = 00001$	$m = 0.30$ $s = 00001$	$m = 0.34$ $s = 00001$	$m = 0.50$ $s = 00001$
Poor quality rock mass Numerous weathered joints at 30–500 mm with some gouge. Clean, compacted rockfill Bieniawski, (1974) (CSIR) rating 23 Barton et al. (1974) (NGI) rating 0.1	$m = 0.04$ $s = 0.00001$	$m = 0.05$ $s = 0.00001$	$m = 0.08$ $s = 0.00001$	$m = 0.09$ $s = 0.00001$	$m = 0.13$ $s = 0.00001$
Very poor quality rock mass Numerous heavily weathered joints spaced at 50 mm with gouge. Waste rock Bieniawski, (1974) (CSIR) rating 3 Barton et al. (1974) (NGI) rating 0.01	$m = 0.007$ $s = 0$	$m = 0.010$ $s = 0$	$m = 0.015$ $s = 0$	$m = 0.017$ $s = 0$	$m = 0.025$ $s = 0$

[a]*CSIR* Commonwealth Scientific and Industrial Research Organization
[b]*NGI* Norway Geotechnical Institute

where m_b is a reduced value of the material constant m_i and is given by:

$$m_b = m_i \exp^{((GSI-100)/(28-14D))} \tag{19.2}$$

and s and a are constants for the rock mass given by the following relationships:

$$s = \exp^{((GSI-100)/(9-3D))} \tag{19.3}$$

and

$$a = 1/2 + 1/6\left(\exp^{-GSI/15} - \exp^{-20/3}\right) \tag{19.4}$$

D is a factor which depends upon the degree of disturbance to which the rock mass has been subjected by blast damage and stress relaxation. It varies from 0 for undisturbed in situ rock masses to 1 for very disturbed rock masses. Guidelines for the selection of D are discussed in Hoek et al. in 2002. The excavation disturbance influences the properties of the rock mass and since they are initially defined as inputs in the model, this should be also considered in the analysis.

One of the most important goals of this revision had been to remove the "switch" at GSI = 25 which had been required in the version published by Hoek et al. (1995). In place of rock, defined by GSI < 25 and rock for GSI > 25, a more general Blast Damage Factor D was introduced to handle the transition. While also not perfect this new D factor has proved workable, although some clarification and refinements of definition have over the years also been necessary (Carter and Marinos 2014).

An extension of the original GSI application charts for heterogeneous and structurally complex rock masses, such as flysch, in most cases forming weak rock masses, was initially introduced by Marinos and Hoek (2001), updated and extended by Marinos (2007), Marinos et al. (2013b) and Marinos (2014). Specific GSI charts for molassic formations (Hoek et al. 2005), ophiolites, including particularly weak rock masses (Marinos et al. 2005a, b), gneiss (in its disturbed or weathered form) have been developed from experience gained during design and excavation of 62 tunnels as part of the Egnatia Highway project in Northern Greece.

19.3.1.1 The Effect of Tectonism in Weak Rock Mass Fabrication

When tectonism is low, GSI values are generally high to very high (Intact to Blocky structure). In this case lower GSI values can however be acquired when rock masses are crossed by discontinuities formed by genesis such as frequent thin bedding or well-expressed schistosity planes, often with conditions of these discontinuities to be Fair to Very Good.

In tectonic areas, particularly if compressional, GSI values may be considerably reduced since the structure is more fractured (more joints) or even sheared; such rock behaves in a more ductile manner (e.g. mudstones, shales, siltstones). In the

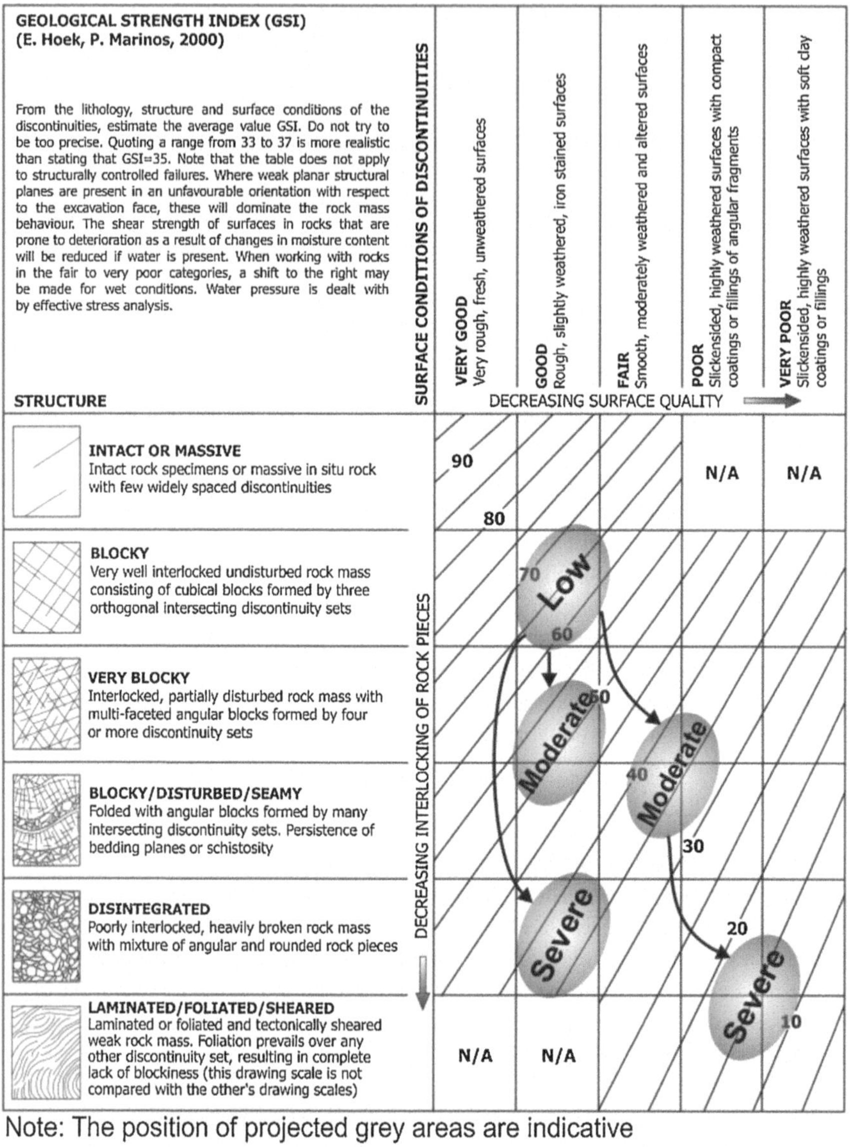

Fig. 19.3 Indicative example of how tectonism (from low to severe) transforms compact rocks (either competent or weak in parent nature) to weak or very weak rock masses (Marinos and Carter 2018)

last case, the conditions of joints are most probably Poor to Very Poor (in cases with slickensided or soft clay coatings due to shearing) (Fig. 19.3). The intact rock strength σ_{ci} and the m_i value may also be reduced in the case of shearing; the ground is weak from both the intact rock and the rock mass.

19.3.1.2 The Effect of Weathering and Alteration in Weak Rock Fabrication

Particular consideration needs to be given to the impact of the natural processes of weathering and alteration that can have in degrading intact rock and rock mass material quality, strength and deformability from initial intact state. There is much confusion in the engineering rock mechanics (non-geological) literature regarding these two processes, as the terms are so often used interchangeably, largely because there is overlap in characteristics, but the origin is different.

Alteration is an effect brought about by generally deep geological processes—hydrothermal alteration, metamorphic alteration, and so on. Many rocks that are heavily altered are common in mining situations as they tend to be located in immediate proximity to the main ore body, so oftentimes form the hanging wall or footwall of the stopes for which rock engineering design is needed.

Weathering by contrast, is almost universally a shallow depth decomposition process brought about by two different, but linked processes—mechanical and chemical degradation. Again, in many engineering situations one has to design engineering projects in weathered rocks. The various stages of weathering of intact rock, rock mass and discontinuities have been described by ISRM (1981) in certain grades (from fresh rock W-I to clayey-sandy soil W-VI). Other descriptions for weathering have been made by the Engineering Group of the British Geological Society (Anon 1995).

In the context of GSI, the influence that either process exerts on the ascribed value of GSI is that both degrade not just the parent intact rock material but also, they change the character and competence of the rock mass fabric. According to the weathering degree the discontinuity surface condition becomes poorer and the interlocking of rock blocks becomes loosened. The structure on the other hand may not be in principle affected, at least if weathering is not very advanced. An indicative example of how this increase in weathering degree affects the GSI value is illustrated in Fig. 19.4. In weathering degrees W-II and W-III, discontinuity condition is shifted to the middle or right columns of the GSI chart. In W-IV to W-V, joint condition is Poor to Very Poor due to the weathering products along the joints.

The degradation of the parent material within the intact pieces within the fabric though needs special consideration. Figure 19.5 provides some guidance for assessing the influence of chemical/mineralogical changes that occur due to pervasive weathering. Reduction factors from intact rock strength (Stacey and Page 1986) according to the weathering grade are presented in Fig. 19.5. Note that at W-III state the rock material is not yet friable like in W-IV. The change from W-III to W-IV is however a critical boundary since there is generally considerable reduction to intact properties at this change, often way in excess of 50% of fresh intact material, while GSI may also by affected to some degree by reduction of the structure also (Marinos and Carter 2018).

However, the actual changes that will occur in a given situation are very much rock-type dependent. Some rocks are resistant to intact material fabric change. These are the rock types that the aggregate industry chooses for high durability,

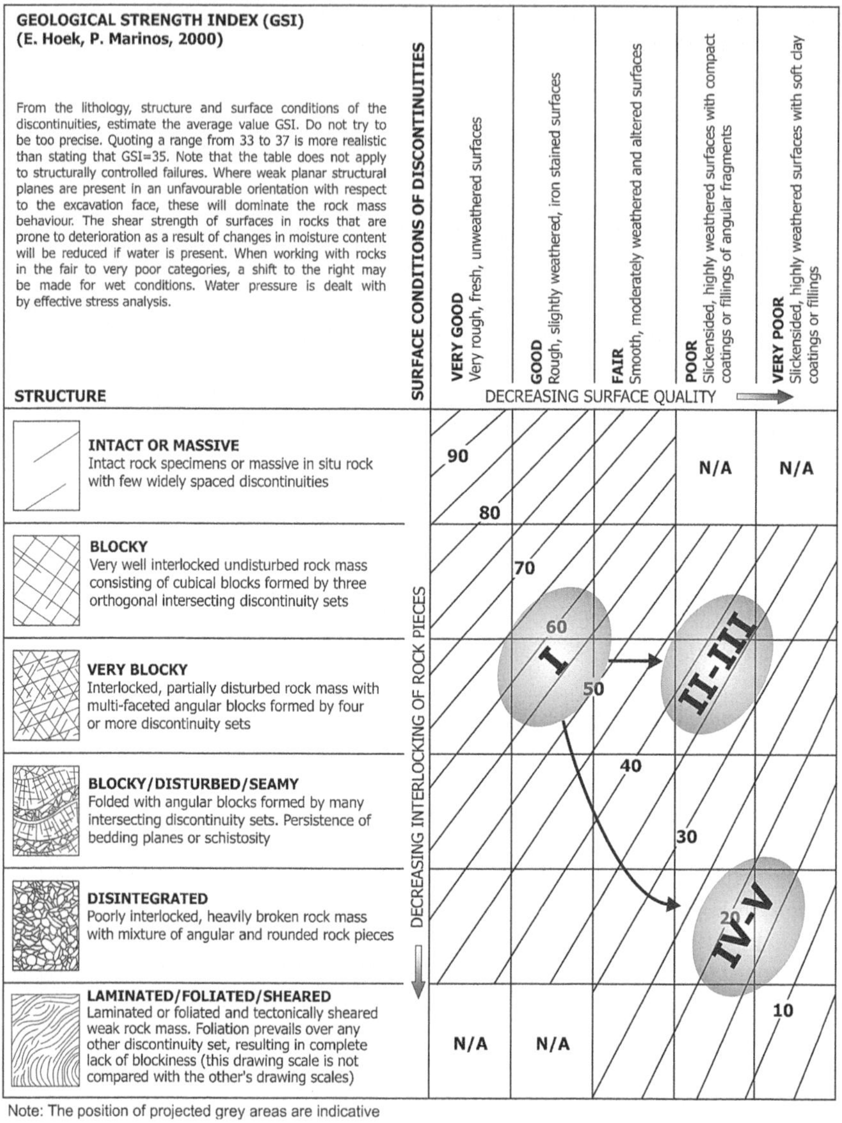

Fig. 19.4 Indicative example of how the weathering degree (W-I to W-V) affects the fabrication towards a Weak Rock (Marinos and Carter 2018)

some diabase, some hornsfels, some diorites are typical of very high durability rock materials. Other rocks tend to decompose completely with penetrative weathering or alteration (e.g. feldspars in granites turning to clay minerals). Some rock materials absorb water and disintegrate as a consequence, some by swell processes, others by drying processes, depending on the internal mineralogy. Processes of laterisation—which are typical in deep, tropically weathered areas also reduce rock mate-

GRADE SCALE (ISRM)	TERM	Description	σ_{ci} reduction factor	GSI notes
VI	Residual soil	Soil derived from in situ weathering (100% soil) (from grades IV,V)	N/A (advise soil mechanic testing)	N/A (advise soil mechanic testing)
V	Completely weathered	All rock material is decomposed and/or disintegrated to soil (less than 30% rock of grades I,II,III). The original mass structure is still visable Shearing can be affected through matrix.	0.001-0.004	Area where GSI is marginally applicable. The structure has been severely disturbed and the interlocking between the fragments has been lost. Clayey-sandy zones follow the original structure and rock fragments are not interlocked. Joint condition is Very Poor.
IV	Highly weathered	More than a half of the rock material is decomposed and/or disintegrated to a soil (30% to 50% rock of grades I,II,III). Severe weathering along the surfaces. Fresh or discoloured rock is present either as a discontinuous framework or as corestones. The rock material is friable. Corestones still affect shear behaviour of the rock mass.	0.04	The structure has been highly disturbed and the interlocking between the fragments has been highly loosened. Clayey and sandy products are filling all the discontinuities. Joint condition is Very Poor. The GSI shifts down and right in the chart.
III	Moderately weathered	Less than half of the rock material is decomposed and/or disintegrated to a soil (50% to 90% rock of grades I,II,III). High to severe weathering along the surfaces. Fresh or discoloured rock is present either as a discontinuous framework or as corestones. The rock material is not friable. The structure is locked.	0.1	The interlocking between the fragments has been considerably loosened. Weathering coatings and fragments are filling principle discontinuities (e.g. gneissic bands) and other joints.Joint condition is Poor. The GSI shifts to the poorer structure(e.g. from Very Blocky to Blocky/Disturbed and to the right in the chart.
II	Slightly weathered	Discolouration indicates weathering of rock material and discontinuity surfaces (>90% rock of grades I,II,II). All the rock material may be discoloured by weathering and may be somewhat weaker than its fresh condition.	0.4	The structure is not changed but the quality of the discontinuity surfaces is (shift to the right). The GSI is reduced to Fair conditions.
I	Fresh	No visible sign of rock material weathering (100% rock); perhaps slight discolouration on major discontinuity surfaces	1.0	Fresh rocks are generally massive (Intact to Very blocky). Joint condition is Very Good(very rouch) to Good(rouch). Blocks and surfaces are strongly interlocking. Rock mass may be even more fractured but only in depth (along a fault zone) where wethering has not been favored. In surface, a fractured rock mass is rarely fresh.

Fig. 19.5 GSI and intact strength change for weathering grade W-I to grade W-VI (grades according to ISRM 1981) (Marinos and Carter 2018)

rial strength, particularly at the saprolite boundary. In such rocks, the intact structure may become pervasively degraded right within the rock material itself.

Alteration in principle also affects both the intact rock properties of the material and the joint surface condition. Slightly to moderately altered rock masses often exhibit smoother or slickensided joint surfaces (e.g. though serpentinisation). The structure, σ_{ci} and m_i in this case are not or only slightly affected. With severe alteration, rock mass structure and surface conditions can be reduced considerably. The structure becomes disturbed (e.g. from Blocky to Very Blocky or perhaps to Sheared) according to the alteration degree (e.g. formation of schistose or laminated planes). Joint condition in this case is Poor to Very poor. The intact properties σ_{ci} and m_i are considerably reduced. An indication of how alteration affects GSI is illustrated in Fig. 19.6. However, there are cases (e.g. contact metamorphism) where alteration may result in stronger rocks (such as keratites).

19.3.1.3 GSI Limitations

GSI classification, although can be used if limited for classification purposes, when used for assessing properties is based upon the assumption that the rock mass contains a sufficiently large number of intersecting discontinuities that it can be consid-

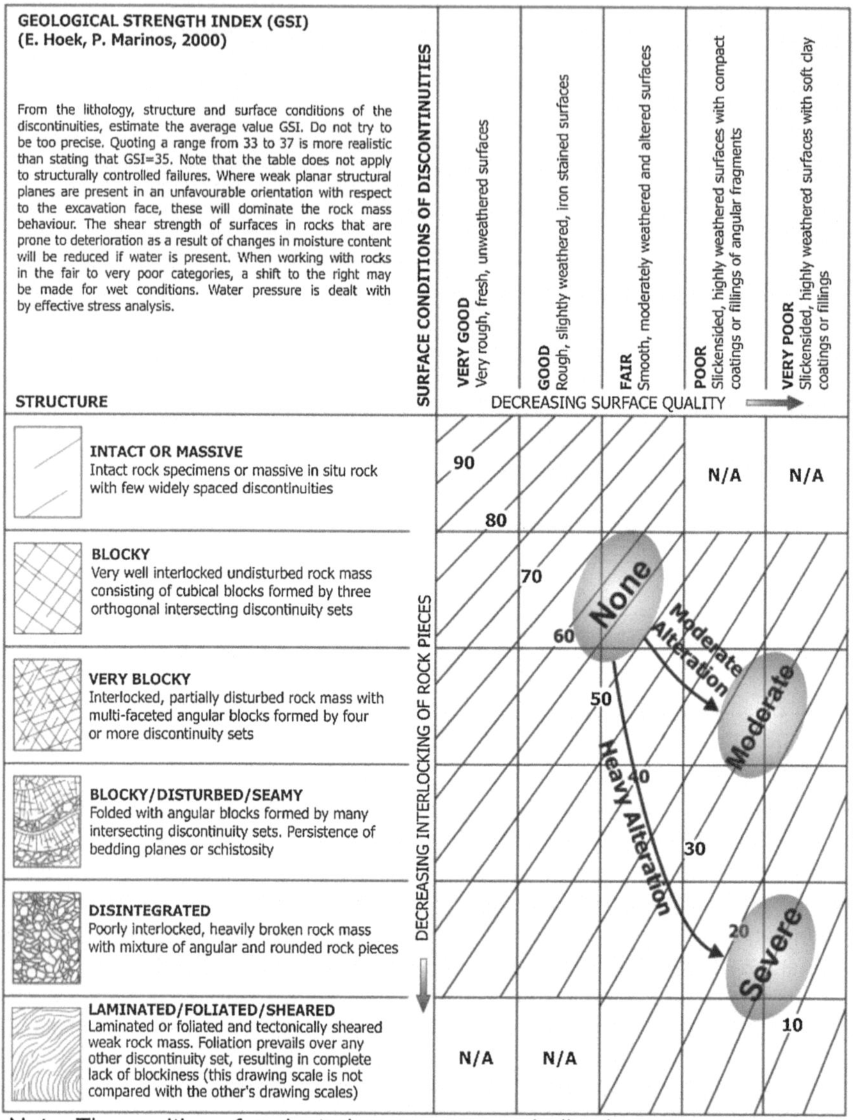

Fig. 19.6 Indicative example of how alteration (none to severe) can transform a rock into a Weak Rock Mass based on the rock mass structure and surface quality (Marinos and Carter 2018)

ered to behave as an isotropic mass. The GSI system therefore should not be applied as such for rock masses with clearly defined dominant in orientation structural fabrics. Undisturbed slate is an example of a highly anisotropic rock mass, where applying a GSI value should only approximately be done if the mode of potential failure won't be governed by the shear strength of the incipient discontinuities within the slate. In the case of anisotropy governed analysis a modified GSI could be applied by subtracting the persisting discontinuity and use for it its properties. A discussion is presented by Fortsakis et al. (2012). For rock masses with completely sheared structure, such as that shown in the sixth (very last) row of the GSI chart in Fig. 19.1, anisotropy is not an issue as the difference in strength between the rock and the discontinuities within it will likely be small.

19.3.2 Assessment of the Strength of Intact Rock: Tests

Weak rocks, once suspected in terms of geologic associations, should be described, mapped, logged, sampled and tested. However, not all of the previously described weak rocks can be directly tested.

A test that can measure the Unconfined Compressive Strength (UCS) of an intact rock at the field is the Point Load Index (I_s). There are relationships (Smith 1997) connecting the two parameters. UCS is generally 20–25 times the I_s (ISRM 1985) but for weaker rocks the uniaxial compressive strength of the intact rock samples can be estimated, with a reasonable level of accuracy, by multiplying the point load index $I_{s\,(50)}$ by 13 (Tsiambaos and Sabatakakis 2004), for $I_{s\,(50)}$ values up to 2 MPa, where $I_{s\,(50)}$ is determined according to ISRM (1985) suggestions. The triaxial cell, known as the Hoek cell, was developed specifically for testing under triaxial compression rocks such as siltstone, shale strata that are sensitive to desiccation. Nevertheless, when laboratory testing is not possible, point load tests, should be carried out as soon after core recovery as possible in order to ensure that the sample is close to the in situ conditions.

Indeed, there are certain problems in measuring the uniaxial compressive strength of argillaceous rocks and heterogeneous and tectonically disturbed rock mass (e.g. Flysch formations), obtained from laboratory testing. Firstly, it is extremely difficult to recover core of "intact" and representative rock elements and to prepare specimens for laboratory testing (Fig. 19.7). Secondly, even if specimens are prepared and tested successfully, the nature of the materials means that there will be a large scatter of the uniaxial strength values determined from these tests. The first problem is quite difficult to solve since an examination of the core from any drill hole in a disturbed flysch will reveal a significant heterogeneity and the presence of closely spaced bedding planes and possibly other discontinuity surfaces. Hence, almost any uniaxial test on a typical laboratory specimen (50–100 mm diameter core) will contain elements of the "rock mass" and will not be representative of the uniaxial compressive strength of the intact rock components. Consequently, any laboratory tests carried out on core samples will be more representative of the rock mass than for the intact rock components. Using the results of such tests in the Hoek–Brown criterion will impose a double penalty on the strength (in addition to that imposed by GSI) and will give unrealistically low values for the rock mass strength (Hoek and Marinos 2000). On the other hand, when samples can be pre-

Fig. 19.7 Characteristic view of a sheared folliated intact rock, where it is extremely difficult to recover core of "intact" and representative rock elements and to prepare specimens for laboratory testing of the uniaxial compressive strength or point load test

pared from a rock mass sequence, these are generally of higher strength rock and may not be representative of the weak heavily sheared product that embraces them.

Experience has shown that there is a common tendency to underestimate the value of the intact strength in many cases where actual laboratory data is not available. These underestimates can have serious implications for any engineering design and care has to be taken to ensure that realistic estimates of intact strength are made as early as possible in the project (Marinos et al. 2006). In some situations, early estimates can be refined through detailed back-analysis, for example of tunnel deformation, and, while this may require considerable effort and even the involvement of numerical analysis, the attempt will generally be repaid many times over in time and cost savings achieved by more realistic designs.

19.4 Weak Rock and Engineering Project

The term weak rock or weak rock mass attempts to describe and quantify the quality of the rock mass via an absolute geotechnical approach, irrespective of the kind or characteristics of the project. However, the type of engineering project is also important in order to predict the behaviour of the rock mass during construction.

A classification connecting properties of some weak rocks based on weathering grade (modified from Krank and Watters 1983; Lee and De Freitas 1989; Santi 1995) with the engineering behaviour (foundation conditions, excavatability, building material, slope stability and tunnel support) is presented by Santi (2006).

Characterisation of "weak rock masses" according to the engineering project type has been proposed by Marinos et al. (2015) based on different approaches. In tunnelling, the combinations of rock mass structure, intact rock strength and overburden height that may lead to the development of stress-induced behaviour can be assessed. In dam foundation the engineering geological elements enabling, if present, to characterise different kinds of rock as "weak rock" were presented.

Conventional excavation of tunnels where the behaviour is as for a weak rock masses can be better approached by identifying the failure modes that could potentially emerge. These failure modes are usually associated with stress-induced phenomena such as face instability and squeezing, manifested by tunnel face extrusion, tunnel closure or shell overstress.

A factor that could be adopted to identify a weak rock mass behaviour in tunnelling is the ratio of the rock mass strength divided by the geostatic stress at the tunnel level (σ_{cm}/p_o, Hoek and Marinos 2000). The low values of this ratio could be a result of high excavation depth or/and low σ_{cm} that is further analysed in low σ_{ci} and/or low GSI. The Tunnel Behaviour Chart (TBC, Marinos 2012) in Fig. 19.8 is a comprehensive tool to indicate combinations of rock mass structure, intact rock strength and overburden height that may develop stress-induced behaviour. Typical examples of such behaviour include tectonised heterogeneous sedimentary sequences such as flysch, highly tectonised, altered and/or weathered ophiolites such as foliated serpentinites, heavily sheared homogeneous sedimentary rocks of low intact rock strength such as claystones, shales, and molassic formations.

Dam foundation in weak rock masses often includes challenges like compressible rock masses, presence of incompetent members and fractured and sheared zones of low to very low strength, poorly to medium cemented rocks, diverse heterogeneity and presence of cavities. These conditions often lead to a single choice selection of the construction of an embankment, earth/rock fill dam and/or, seldom, the strengthening of the foundation zone or the selection of an alternate dam location. Cases where various rocks can be characterised as "weak" in dam foundation engineering are presented in Table 19.2.

19.5 Classification of Specific Cases of Weak Rock Masses

Figure 19.9 demonstrates how a rock, of different parent competence can present weak rock mass characters (Marinos and Carter 2018). In this overall diagram, it should also be appreciated that differences in assigned GSI commonly occur due to different tectonism, weathering and alteration effects on either brittle or soft rock materials. Some of these effects, which occur due to influence of different geological processes, create differences not just in material fabric but also in rock material mineralogical composition affecting their engineering behaviour.

Most of the common GSI characterisation for typical gneisses, granites, ophiolites, limestones, schists, siltstones/mudstones/shales, molassic and flysch formations are illustrated here, highlighting how the differences in origin exert on the

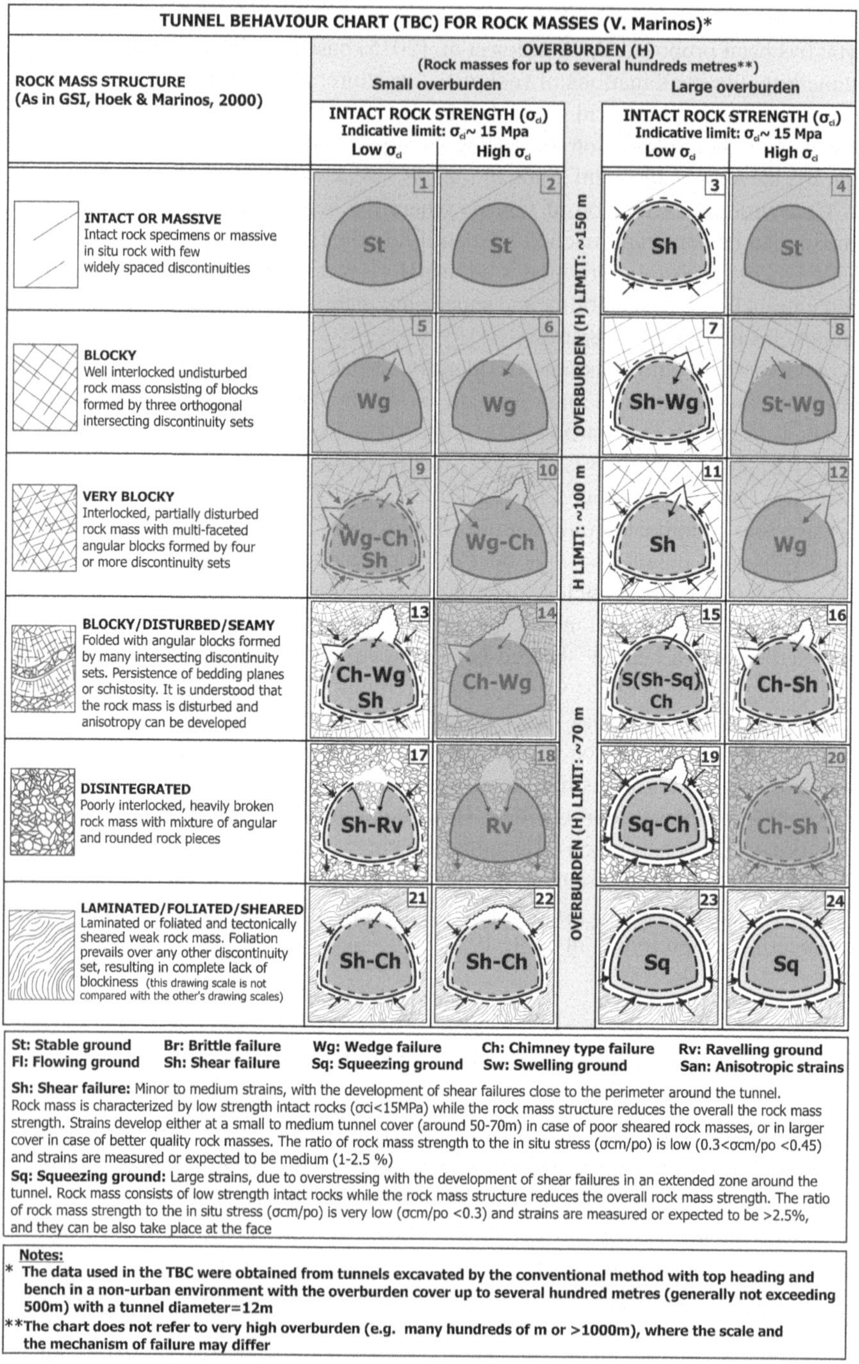

Fig. 19.8 Stress-controlled problematic behaviour that corresponds to weak rock masses are shown in the highlighted boxes in the Tunnel Behaviour Chart (TBC) presented by Marinos (2012). Numbers indicate the possible combination of Structure, Overburden and Intact Rock Strength

Table 19.2 Rock mass elements that can characterise a rock as "weak" in dam foundation (Marinos et al. 2015)

Rock material	Engineering geological elements enabling, if present, to characterise the rock as "Weak rock mass" in dam foundation
Conglomerate, Sandstone	Poor cementation material (e.g. clayey), low diagenesis. Low strength and deformation modulus. Cases with frequent intercalations with pelitic interlayers
Marl, Siltstone Mudstone, Slate	Presence of clayey minerals, poor cementation. Possible laminated with diagenetic planes. When tectonically disturbed they form very weak rocks masses.
Clayshale	Low rock mass strength (foliated structure, low intact rock strength) and deformation modulus—Sheared surfaces, slickensided surfaces. Swelling minerals. Slaking potential
Evaporites	Dissolution phenomena, presence of voids (gypsum, halite). Swelling (anhydrite)
Limestone, Marble	Internal karstic structure, voids empty or soil filled. Cases of intercalations of weak pelitic layers (phyllites in the case of marble)
Molasse	Alternations of sandstones with siltstones (most common). Diverse heterogeneity. Presence of members with low strength. Slaking potential when exposed
Flysch	Alternations of sandstones with siltstone the most common. Diverse heterogeneity. Presence of members with low strength. Unlike molasse tectonically disturbed structures (structural complexity in space due to folding and presence of sheared tectonic zones and layers). Persistence of discontinuities with low strength in depth
Volcanic rocks	Generally strong, but may alternate with pyroclastic compressible, swelling or erodible material. Heterogeneity in strength and deformation modulus. Weathering or alteration with presence of unstable minerals/high plasticity soils. Joints from cooling/lava tunnels
Pyroclastic rocks	Highly erodible, collapsible, or swelling. Extreme variability within the formation
Granite	Generally strong but possibility of extended weathering presence. Irregular weathering profile and bedrock interface (less weathered boulders within completely weathered material). Concealed sheet joints close to surface
Basic-Ultra basic/ Ophiolites	Generally strong but peridotites may present serpentinised zones of low strength in irregular geometry within the mass. Shear zones with altered compressible material
Graphitic-Chlorite schists. Phyllites	Dense schistosity with weak planes. Possible low strength and modulus. Weathered to other clay minerals
Gneiss	Generally strong. Weak in extended weathered/sheared brecciated zones only
Quartzite	Strong but often quartzitic layers alternate with incompetent phyllites
Metamorphic schists	Generally strong. Possible presence of weak zones due to tectonic shears and to slickensided schistosity planes

engineering geological specific characteristic "keys" each geomaterial embodies. The opportunity created by the construction of Egnatia Highway across Northern Greece involving a significant number of geotechnical investigations plus extensive experience gained from the excavation of 62 tunnels (Marinos 2007) offered

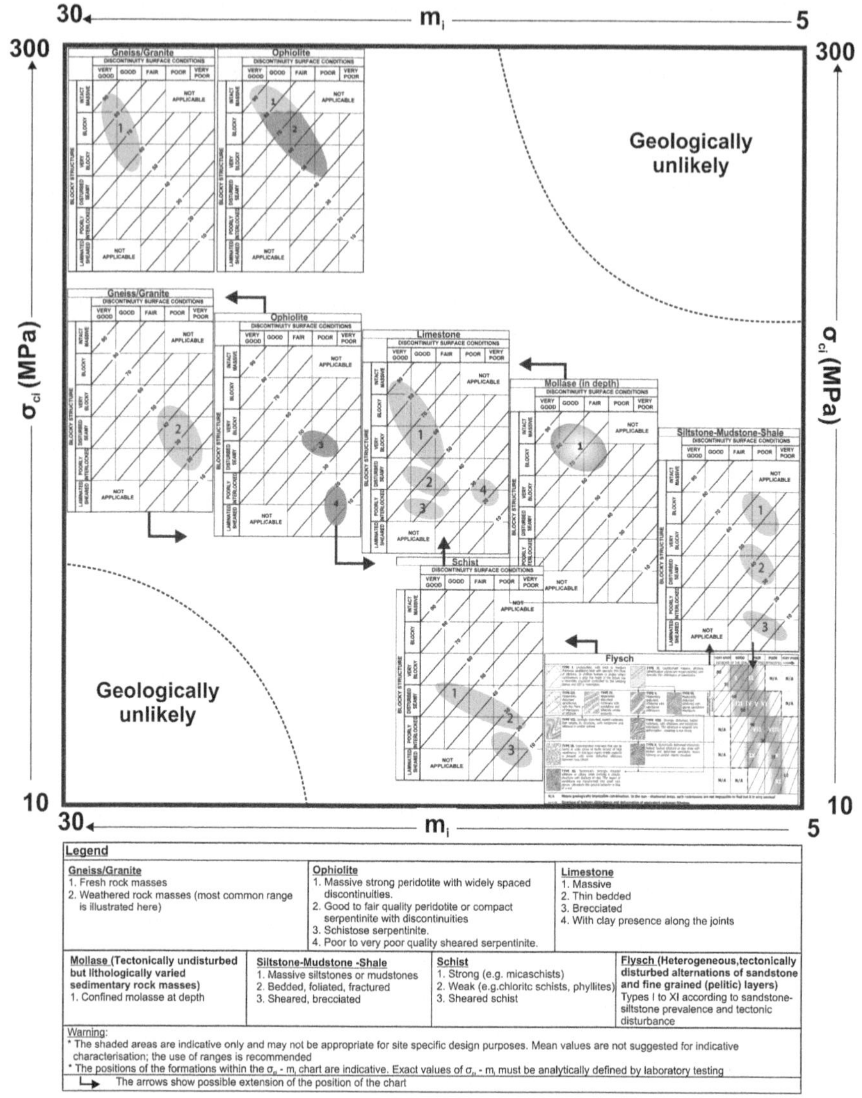

Fig. 19.9 A demonstration of how a rock, of different parent competence (examples of gneisses, granites, ophiolites, limestones, schists, siltstones/mudstones/shales, molassic and flysch formations) in conjunction with a range of m_i and σ_{ci}, can present weak rock mass characters through common GSI ranges (Refer to text and reference list for original papers for more details on charts) (Marinos and Carter 2018)

tremendous insight and plenty of data concerning the engineering geological conditions of several rocks and formations. Data could be collected on both tectonically disturbed and undisturbed rock units, enabling research to be made a distinction of the variability of rock mass types and their quantification. Similar conditions exist in the major mountain chains worldwide. For more details, the reader is referred to the original publications presenting these charts for each specific formation (Marinos and Hoek 2000; Hoek et al. 2005; Marinos et al. 2005a, b, 2011, 2013a, b; Marinos 2019).

The flexibility that this concept (Fig. 19.9) affords is that it allows new material-specific or new site-specific individual rock mass characterisation GSI charts to be developed for any particularly weak and complex rock masses or for those not already within the scope of published charts. In the following paragraphs, each of the different groups in this table are explored, starting in the middle high end m_i rock types. GSI characterisation issues for each of the rocks and formations are then to illustrate the differences in origin exert on the engineering geological specific characteristic "keys" each geomaterial embodies.

19.5.1 Weak Rocks Masses of Sedimentary Rocks and Formations

Weak rock masses of sedimentary rocks can be identified in clastic isolated rocks and in clastic series (like flysch or molasse) either under tectonic depression or in undisturbed systems. Non-clastic series like limestones and dolomites can also present weak characters when they are strongly tectonised or when including in their formations rocks like marls or weak siltstones.

19.5.1.1 Arenaceous and Argillaceous Rocks

As the origin of all waterborne sedimentary rocks (conglomerates, sandstones, siltstones, claystones and mudstones) varies due to the water velocity controlling their original deposition, different original fabrics characterise each of these rock types over and above their basic differences in grain size. Similarly, airborne derived rock units—aeolian sandstones, siltstones and loess deposits which are widespread across the globe also vary appreciably dependent on their parent composition and original depositional environment. GSI's characteristically will be similar for similar rock types, irrespective of origin, but because mineralogical differences and distinctions in depositional fabrics will govern engineering behaviour with weathering and alteration, extreme care needs to be taken in characterising GSI to reflect the different origin of the parent materials composing the rock mass.

Conglomerates present variable quality depending on the cementation material and the fracturing degree. The quality of the conglomerates depends mainly on the matrix material and the fracture degree after the tectonic disturbance, if any, of the rock mass. When the matrix is composed of calcite or silicate the rock mass presents high intact rock strength and the structures generally range in the upper rows of the GSI chart (from compact to blocky-very blocky), depending on the number of discontinuities that are consistent throughout the mass. In surface, these rock masses do not present considerable weathered forms but some dissolution phenomena along the joints. On the other hand, when the matrix is silty, clayey, or marly, there are no continuous joints and the rock mass does not present systematic blocks and in some depth, is generally more homogenous. The rock, in this case though, has lower intact rock properties. In surface, though, the rock mass is often comprised of loosened to completely disintegrated forms, with very small blocks and poor to very poor surface properties, due to weathering.

Sandstones are amongst the commonest rocks worldwide, and in consequence they may have widely different characteristics depending on the parent origin. Sandstones in general are strong and present brittle behaviour. Sandstones with calcitic or silicate cementation material are even stronger and depending on the tectonic disturbance can be compact to very blocky (higher GSI values) or even disintegrated (lower GSI values) along fault zones. They can present more ductile deformation, with folded structures, when they are thinly bedded and projected to the "seamy-disturbed" row in the GSI chart. Surface conditions for these rock masses are generally rough to smooth but can also be poor with angular grain fragments when they have been tectonically disturbed or slided in the past. The sandstones however can be silty or marly and these exhibit low strength values. Their unconfined compression strength may be about 10 MPa if they are marly or silty and well over 50 MPa in their typical granular form. In this case, silty sandstones are more vulnerable to weathering and the rock mass can be loosened, disintegrated or schistosed due to fissility parallel to the bedding when these rocks are exposed or are close to the surface. In these cases, sandstones present low GSI values along the two last rows in the chart.

Marls are typically a fine grained siltstone, almost mudstone, with a percentage of calcite though. They are characteristically red in oxidised conditions, but often with blebs and layers of green or grey colouration, where reducing environments have controlled rock mass induration. Calcitic marls are competent and demonstrate very good behaviour since the intact rock strength is moderate to high and GSI values remain high with blocky to very blocky rock mass structure, according to the fracturing degree, including the bedding. A characteristic example is the behaviour of blocky calcitic marls along the Isthmus of Corinth in central Greece, where 6 km long and steep (70°) cuts of 80 m height have been remaining stable over than 100 years. Clayey marls, on the other hand, have low intact rock properties and show a more ductile behaviour with no persistent blocky structures since there are no intersecting joints. Clayey marls generally host weak structures and present low GSI values when the rock mass is schistosed close to the surface due to weathering or in-depth due to shearing. When this rock mass is not highly tectonised it is projected in the upper rows in the GSI chart but always in the right columns.

Shales show a fissility not present in typical mudstones. As this fissility (which most often derives from their original consolidation and induration) controls the behaviour of these types of rocks it is thus of critical engineering significance. GSI values are confined in the last rows of the chart (Fig. 19.2) due to the thin stratification and fabric in flakes. When the rock mass is sheared the structure becomes laminated and present very low GSI values. In tectonically undisturbed areas and only in depth, their structure could be better and projected in the Blocky-Disturbed-Seamy row, since the bedding planes are not expressed and released within the mass. Shales, close to the surface, present a disintegrated-laminated structure.

For claystone or siltstone intercalations in a folded-highly disturbed rock mass, the structure is less tight due to the poor contact of the blocks with different deformational characteristics (for instance the plastic members are sheared and the limestone beds are broken and cannot follow the same deformation pattern) and the parallelism between the bedding planes is reduced. In this case, the competent members do not contribute much to the overall, "weighted" intact rock properties.

As far as cherts are concerned, they are rarely found alone. If they are intercalated with limestones or the rock mass has brittle behaviour they can be projected on the basic chart usually along the Very Blocky-Blocky/Disturbed structure and between second and third column. If cherts are alternating with clayey shales, they can be classified with the GSI chart for heterogeneous rock masses such flysch (Fig. 19.10). In the case of thin undisturbed, non-folded bedded rock mass, the anisotropy has to be applied in analyses.

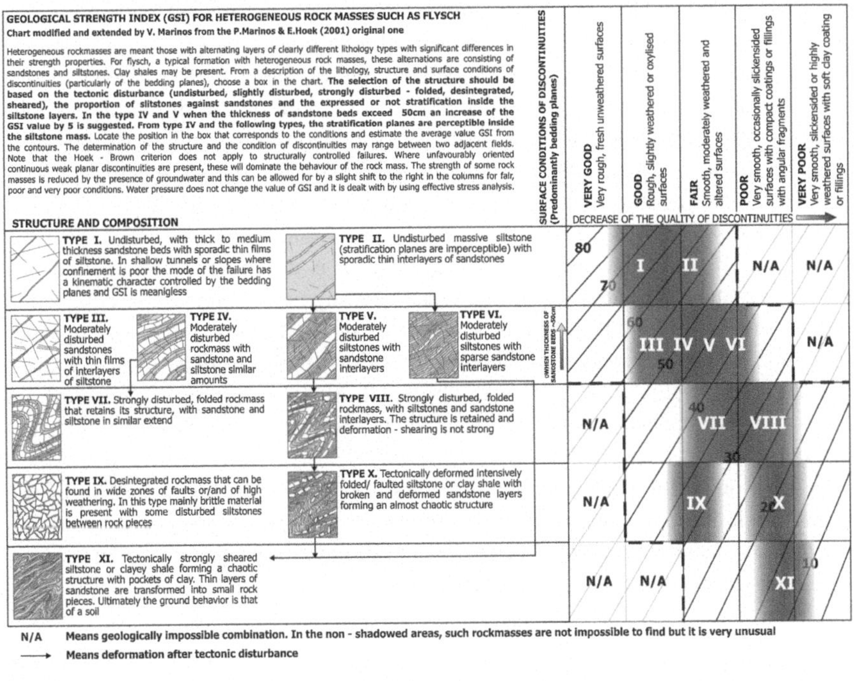

Fig. 19.10 A revised, geotechnical classification GSI system for tectonically disturbed heterogeneous rock masses, such as flysch, with the fields of most probable occurrence (Marinos 2019). The chart modifies and extends the original one (Marinos and Hoek 2001)

19.5.1.2 Geotechnical Classification of Heterogeneous Tectonically Disturbed Rock Masses Such as Flysch

Characterising and establishing a representative range of GSI values for complex, heterogeneous rock units comprising of a strong member alternating with a weaker one such as flysch (tectonically disturbed series), can be challenging.

Flysch formations are generally characterised by diverse heterogeneity, presence of members with low strength geomaterial and tectonically disturbed structures. Marinos (2007, 2014) has classified flysch formations into 11 distinct rock mass types (I–XI), according to the siltstone-sandstone contribution and the tectonic disturbance.

The new GSI diagram for heterogeneous rock masses such as flysch is presented (Marinos 2019), where a certain range of GSI values for every rock mass type is proposed (Fig. 19.10). This flysch chart updates the 2001 chart, proposed by Marinos and Hoek (2001). In the new diagram, GSI values are increased from 10 units to 35 units for the "Blocky" to "Undisturbed" structures, respectively, particularly for the siltstone type. The high presence of siltstone beds does not decrease the GSI value, but only in the highly disturbed forms. When rock mass is undisturbed or slightly disturbed, independently of siltstone or sandstone predominance, GSI ratings have to be considered much higher. This was confirmed in tunnel construction, where lighter temporary support categories (correlated with high GSI values) were implemented and experienced marginal measured deformations. Hence, the selection of the structure should be initially based on the tectonic disturbance (from undisturbed to sheared rock masses), then on the proportion of siltstones against sandstones and finally on the expressed or not bedding stratification inside the siltstone layers. One more addition in the flysch GSI chart is the bedding thickness consideration of the competent sandstone beds. In the type IV and V (slightly disturbed structures) when the thickness of sandstone beds exceeds 50 cm, an increase of the GSI value by 5 is suggested. It is noticed that for the non-disturbed types anisotropy is present due to the bedding planes and in consequence in analysis this fact should be taken into consideration. In the case, where strong sandstone blocks are numerous, continuous and with defined geometry, the rock mass properties are evaluated by different approach. Such an approach, the block in matrix approach (beamrocks), is effectively described by Wakabayashi and Medley (2004).

In addition to the GSI values, it is necessary to consider the selection of the "intact" rock properties σ_{ci}, m_i and E_i for the heterogeneous rock masses considered as a unit. A 'weighted average' of the intact strength properties of the strong and weak layers is proposed in Marinos (2019).

19.5.1.3 Geotechnical Classification of Tectonically Undisturbed Rock Masses Such as in Molasses

Molasse is quite different from flysch, although both are consisting of same lithological types, since molasse is formed after the orogenesis, and did not suffer from compressional tectonics. The deterioration of the quality of their rock mass is limited. Only in few cases the molassic formations may be deformed and present thrusts due to the final advance of tectonic napes but such a decrease of their quality is localised.

Of particular concern when categorising molassic rock masses is ensuring that appropriate consideration is given to assessing variable heterogeneity, the presence of low strength geomaterials and the improvement of structure with depth. Molassic rocks differ significantly between surface and depth. Bedding is the essential joint set in a molassic rock mass, but may only be expressed as a significant weakness plane near the surface, due to fissility. At depth, bedding is mainly concealed and may be virtually healed. Moreover, due to the limited deformation to which they have been subjected after their deposition, joints in molassic rocks are generally free from the effects of shear movement (slickensides).

Weathering though, alters the rock mass strength significantly and starts almost immediately to degrade bedding plane strength. The siltstone (or marly) members are very vulnerable to weathering and a development of fissility parallel to the bedding when these rocks are exposed or are close to the surface may be developed. Thus, in outcrops they appear thinly layered and when they alternate with sandstones, their appearance resemblances to flysch. This appearance in outcrops can be misleading when considering the behaviour of these molassic rocks in a confined underground environment in which the process of air slaking is restricted and the rock mass is continuous and massive without any sign of stratification or schistosity inside the siltstone beds. Siltstone (or marly) members are very vulnerable to weathering and a development of fissility parallel to the bedding when these rocks are exposed or are close to the surface may be developed. Siltstone members in outcrops generally appear thinly layered and when they alternate with sandstones, the appearance of the rock mass in exposures may resemble flysch. This appearance in outcrops can be misleading when considering the behaviour of these molassic rocks in a confined underground environment, where the process of slaking will be much restricted and where bedding in the rock mass remains the only discontinuity such that beds are continuous and massive. Siltstone and mudstone beds, constructing such rock masses due to their initial rapid compaction and induration generally extend to form an almost continuous medium of incredibly consistent composition, attributing a high GSI value to the whole mass. The presence of low UCS materials may however reduce overall rock mass strength resulting in only a fair to weak rock mass.

An initial GSI classification for molassic rocks based on the observations and evaluations from the design of the 12 tunnels along the Egnatia Highway was introduced by Hoek, Marinos, and Marinos (2005). The GSI chart for molasses (Hoek et al. 2005) can be of general application to all formations consisting from alternations of sedimentary rocks not associated with significant tectonic disturbance, though this chart was based on the observations from the excavation of 12 tunnels along the Egnatia highway in molassic formation. Two charts were developed one for close to surface rock masses, where fissility is present and one for use for the molassic rock at depth. Since all rock masses at depth are confined, discontinuities such as bedding planes do not appear as clearly defined surfaces and thus the rock mass appears and geomechanically behaves as massive. For such rock masses the use of the descriptive GSI chart (Marinos and Hoek 2000) is recommended as a tool for categorisation as they are typically difficult to characterise.

As a result, molassic rock masses have dramatically different structure when they outcrop or are close to the surface as compared to those confined in depth, where bedding planes, especially the siltstone ones, do not appear as clearly defined discontinuity surfaces. A GSI value of 50–60 or more is to be applied in this case (If no discontinuities are present, GSI values are very high and the rock mass can be treated as intact with engineering parameters given by direct laboratory testing). When fault zones are encountered in depth, the rock mass may be highly broken but it will not have been subjected to air slaking and a GSI value will of 25–40 can be assigned.

In surface, the heterogeneity of the formation is discernible and similarities exist with the structure of some types of flysch. Hence the GSI chart for heterogeneous rock masses such as flysch can be used with the exclusion of sheared and disturbed types and with a slight shifting to the left of the flysch chart categories, as the molasse is always less disturbed. This version of the chart, for fissile molassic rocks is proposed by Hoek, Marinos, and Marinos (2005).

The values defined using this special GSI chart were validated during portal construction of the Egnatia Highway tunnels (Marinos et al. 2013a). However, it was found more appropriate to use the basic GSI chart for construction at depth in molassic rocks as this chart resulted in much more appropriate (and lighter) support measures than was initially foreseen, based on a flysch-like GSI classification approach.

19.5.2 GSI and Igneous Rocks

19.5.2.1 Intrusive Rocks

The intrusive are subdivided between felsic and mafic composition as mineralogy directly affects engineering characteristics and behaviour.

Geotechnical Classification of Felsic Intrusives

Granites granodiorites and diorites are typical light coloured igneous intrusives. When they are fresh, granitic rock masses can be projected in three first rows and two columns of the standard GSI chart and they certainly do not form weak rock masses. According to the degree of weathering, they may shift to the right and down to the lower rows. If weathering is more intense (weathering grades III, IV, V, ISRM 1981), the rock mass had started to degrade en masse and the index shifts downwards to a poorer overall structure. For tropically weathered rock masses, such as exist in many parts of the world, competence ratings will change several orders of magnitude in the vertical plane. For the granites of Hong Kong, for example, one might move from intact, high-strength, fine-grained grey granites with high m_i and high σ_{ci} which plot towards the top left of the diagram, progressively downwards and to the right through the weathering grades to completely degraded granite—essentially a soil. If they are completely weathered, granitic rock masses are not readily described in terms of GSI. Given similarity in mineralogy, certain facets of the GSI classification chart (Fig. 19.16) for gneiss presented in Sect. 19.5.3.1, give parallels for description and categorisation of granites and similar intrusive rocks.

Geotechnical Classification of Mafic Intrusive Rock Masses: General

The darker, more basic suite of plutonic igneous rocks such has peridotites and gabbros contain significantly less siliceous content, than the felsic suite, and as a consequence, once exposed, are somewhat more susceptible to deterioration by weathering than their similar grain-sized siliceous (felsic) counterparts. Again, like the granites, where the coarser grained and feldspathic degrade most, the coarser grained, more olivine rich, basic intrusive rocks tend also to degrade the most.

Mostly the intermediated and basic igneous rocks, syenites, gabbroes, and so on, when they are fresh, mostly are classified in the three first rows and three first columns on the standard GSI chart. When they are heavily fractured and weathered, they tend to disintegrate on structure first and thus more to the right section of the chart.

Geotechnical Classification of Ophiolitic Rock Masses

Peridotites and other basic and ultrabasic plutonic rocks often seen in ophiolite assemblages, frequently associated with overthrusts, can vary remarkably in geotechnical qualities, right from excellent, through to very poor when serpentinisation is extensive and/or shearing present. Ophiolites often associated with subsequent overthrusts, contain a variety of rock types with geotechnical qualities varying from excellent to fair, becoming poor to very poor when serpentinisation is extensive and/or shearing present. The main included types are peridotites, gabbros, peridotites more or less serpentinised, serpentinites, schisto-serpentinites, sheared serpentinites, pillow lavas and chaotic masses in ophiolitic melanges. When very high degrees of serpentinisation occur, together with intense shearing, the resulting rock mass may be difficult to identify for any initial texture or fabric (Fig. 19.11). For such rock masses site specific GSI charts, such as that published by Marinos et al. (2005a, b), are recommended. This study is based on field data from outcrops, cuts

Fig. 19.11 Characteristic view of schistosed serpentinites consisted of slippery laminar pieces and small flakes of millimetres in size (Type IV in Fig. 19.14)

in slopes, borehole cores and tunnel excavations and from various significant ophiolitic complexes and melanges in northern and central Greece.

When unweathered or not significantly serpentinised, peridotites are strong and behave as typical brittle materials. Their tectonic disturbance is expressed in terms of intersecting joint sets. The range of GSI for peridotitic types of rock masses in an ophiolitic complex is shown in Fig. 19.13 (areas I and II). Serpentinisation can be present on the surface of discontinuities and the conditions of the joints are dramatically reduced to poor or very poor with coatings of "slippery" minerals such as serpentine or even talc. As the rock mass becomes more jointed or fractured GSI values can easily drop as low as 35 due to pervasive degradation emanating outwards into the mass from the poor discontinuity conditions, which become smooth and slippery due to serpentinisation (Fig. 19.12). In a disturbed peridotitic mass, the serpentinisation process often loosens and disintegrates parts of the rock itself, but becomes penetrative leading to lower GSI values but also reducing the intact strength values. Such disturbed peridotites fall in the lower bound of area II of the GSI diagram of Fig. 19.13.

If the process of serpentinisation is due to autometamorphism and/or associated with tectonic thrust, the rock mass is poor, the rock mass may already be of poor quality due to the schistose disturbed structure, which may reduce the GSI to values to 30 or less (area III in the GSI diagram of Fig. 19.13). In the sheared zones of serpentinites there is a lack of blockiness, which allows the rock to disintegrate into slippery laminar pieces and small flakes of millimetres in size. GSI values can drop to less than 20 (Fig. 19.13, area IV). When the serpentinisation, due to weathering, has affected all the mass, in addition to the reduction of intact strength and character, the original structure is all but obliterated. In ophiolitic melanges, in tectonic

Fig. 19.12 Slightly to moderately weathered, serpentinised, moderately to highly fractured peridotites

thrusts, for example, the rocks often are mixed in complete disorder with other rocks of various origins (flysch, chert, shales, and other), the rock mass is disturbed-folded or disintegrated (when peridotites have mainly blocky nature and not laminar—foliated form) or sheared-foliated (Fig. 19.11) and, thus, low to very low GSI values are assigned (area V in Fig. 19.13).

19.5.2.2 Extrusive and Pyroclastic Rocks

Like the intrusive rocks, the extrusive and pyroclastic volcanic assemblages (i.e. lavas and ejectamenta) can be of widely ranging mineralogy and character. Competence of original deposition can also vary remarkably from extremely competent hard brittle rock types (dolerites, rhyolites, etc.) of aggregate quality, through to weak friable and often swell sensitive materials (tuffs, pumices, etc.). Composition ranges from siliceous through to basic with cementation also varying dramatically. In consequence, some of these rocks can be the most challenging engineering materials anywhere. Characterisation for GSI, and hence for Hoek–Brown m_i requires careful calibration, as intact-looking volcanoclastic rocks can have quite significant degradation characteristics. Many of such rocks are slake-sensitive and GSI values and intact strengths can drop significantly from the intact undisturbed state.

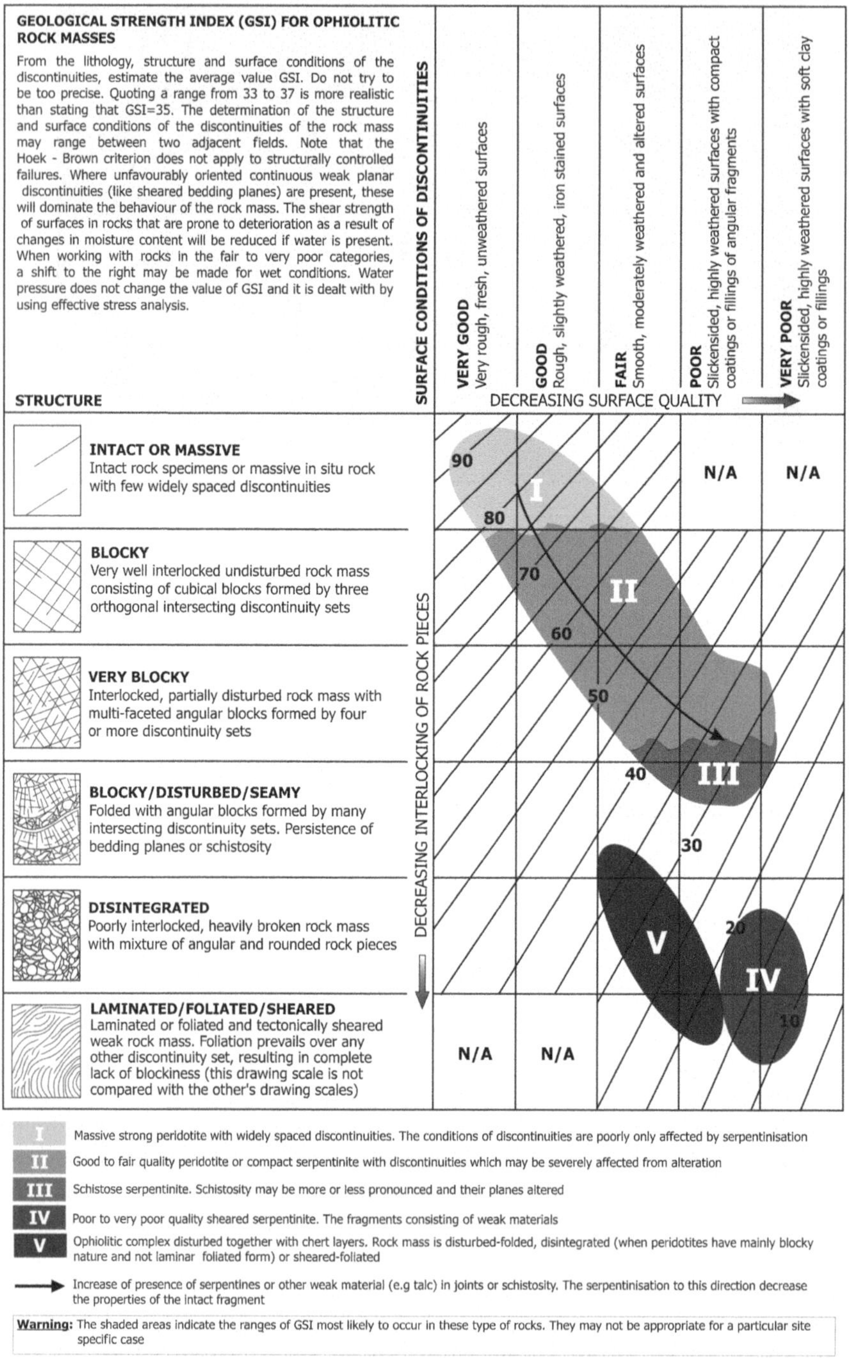

Fig. 19.13 GSI classification chart for ophiolitic rock masses (modified from Marinos et al. 2005b)

19.5.3 GSI and Metamorphic Rocks

Two basic metamorphic rock units can be recognised from the view point of different character from a GSI perspective. Competent crystalline foliated rocks (including most gneisses) and schistosed anisotropically textured rocks (including schists, slates, and phyllites).

19.5.3.1 Geotechnical Classification of Tectonically Disturbed and Weathered Gneiss Rock Masses

Fresh gneiss forms evidently very competent rock masses with minor problems in geotechnical works. However, in certain geological conditions fair to very poor quality gneissic rock masses can exist controlled largely by weathering degree. Such rocks are commonly the product of intense tectonism and disturbance and thus weathering (transformation of feldspars to clayey minerals) may vary markedly with depth. In such conditions, intact rock and rock mass strength present a very wide range of values with the behaviour in underground works can ranging from simple to extremely problematic. In a disturbed tectonically environment, the complexity in gneiss rock mass, due to intense weathering and fracturing, is characterised by erratic geometry to all directions and the expected increase in quality with depth may not be apparent.

Weathering loosens, “opens” and progressively disintegrates the rock mass along the gneissic band, enabling the rock mass to express its inherent weakness characteristics. In depth under normal geological conditions, gneissic rock masses are strongly tight and “sewed”. In weathered forms the foliation planes can be “coated” with clay minerals, while the distance between discontinuities, normal to the gneissic band, can be of few centimetres (Fig. 19.14). In the completely weathered form, rock mass has lost its structure with clayey presence and friable pieces of rock remnants (Fig. 19.15). These masses have very low shear strength in both their mass and any retained discontinuity surface. In less weathered forms, the segregation is less intensive and the rock mass is composed by well-interlocked blocks which exhibit higher friction. When the rock is more schistosed and foliated, weathering is enhanced and the rock mass is separated in thinner bands, creating smaller blocks.

A decrease in the GSI value is proposed for the gneissic rock masses in order to consider more appropriately the weathering effect. By the comparison of the classifications and the temporary support categories, from six tunnels driven in Northern Greece, between the design and the construction records, it was shown that this decrease lies to around 10 units. However, the use of this numerical difference must be carefully done due to construction issues, like for example the procedure to connect a classification value with the support category selection. Of course, this difference could also be due to the application of heavier support demand due to frequent over-breaks, like chimney type failures and not due to stress controlled problems. Nevertheless, the wide application of heavier support systems between similar geo-

Fig. 19.14 Intense weathering along the gneissic bands but also along intersecting joints

Fig. 19.15 Characteristic view of completely weathered gneiss (Type V)

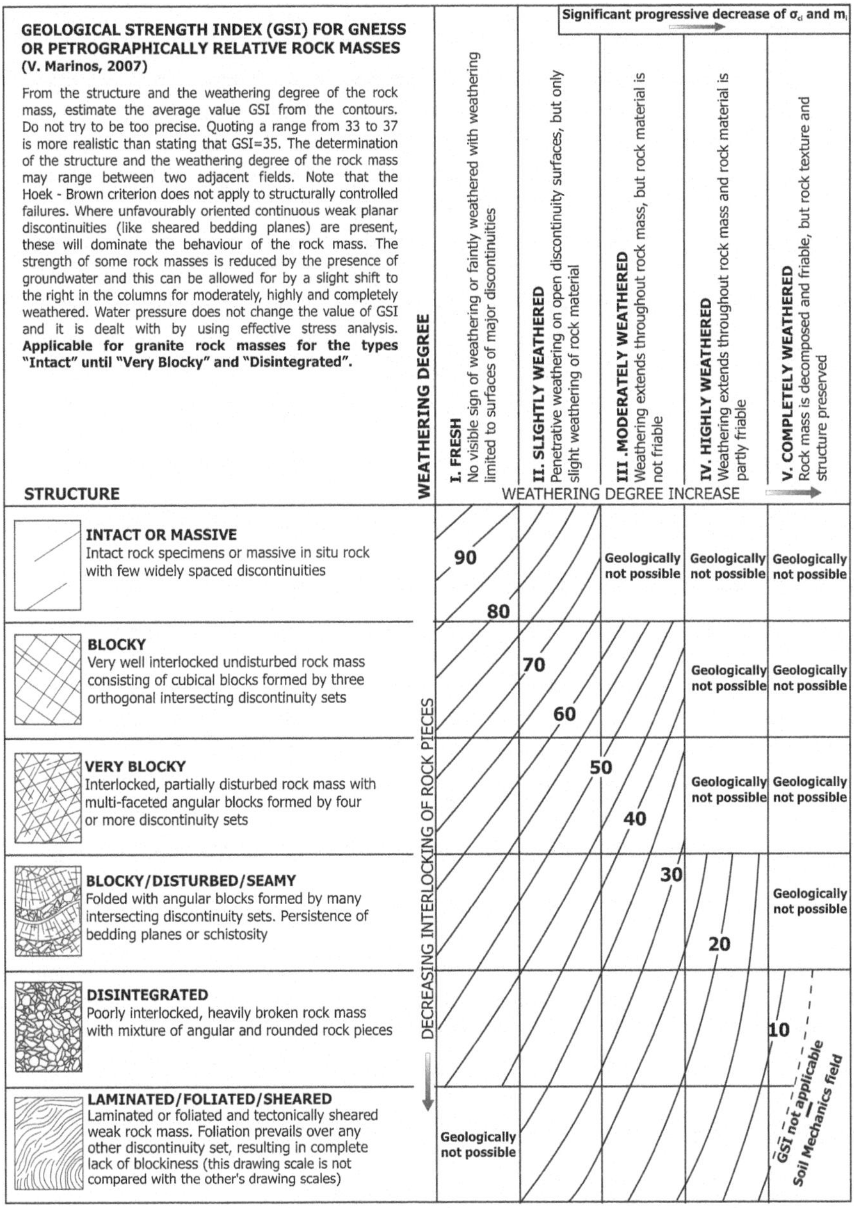

Fig. 19.16 GSI classification chart for gneiss or petrographically similar rock masses (Marinos and Drosos 2010)

technical conditions, corresponding to a difference of 10 units, in all the tunnels and the geological concept described before, agreed to this diversity.

A GSI chart for weathered gneissic rock masses or rock masses with similarities in weathering, as the granite can be used (Fig. 19.16). This chart maintains the basic structures but the surface conditions of joints are replaced by the weathering grades (from Fresh-grade I to Completely Weathered—grade V, Brown 1981). This is associated with a new calibration and the substitution of the straight lines of the fundamental chart with curved lines, bended to the left side of the chart. As the weathering degree increases to the right, bending is increased as well. In the first column, where the rock mass is not weathered (category I), the calibration lines remains the same and GSI values do not change. The Decrease in GSI values starts from "Slightly weathered" (category II) rock masses and becomes higher (around 10 units) in "Completely weathered" rock masses (category V) with a "Blocky-Disturbed" or "Disintegrated" structure. However, a number of unfeasible geologically conditions (e.g. "Very Blocky" and "Highly weathered" rock mass) must be excluded from the GSI chart.

19.5.3.2 Schistosed Anisotropically Textured Rocks

Schists: Schistose rock masses rarely plot in the top left corner of the GSI chart. They usually project in the Very Blocky-Blocky/Disturbed rows. Strong schists (e.g. quartz-mica schists) would be expected in the left columns (Very rough to fair), while weak schists would report typically in the right ones. Extreme foliation weakness occurs in schists and phyllites. By contrast, slates can be remarkably competent, even though possessing a characteristic slaty cleavage. When weak schists are sheared (also sheared phyllites) GSI's can shift to the last row and to the last two columns. Anisotropy has to be taken into account in any analysis for all of these rock types.

Acknowledgements Thanks are due to Evert Hoek and Paul Marinos for their insight in initially developing the GSI approach. The author would also like to thank Egnatia Odos S.A. for its support and the data provided.

References

Afrouz AA (1992) Practical handbook of rock mass classification systems and modes of ground failure. CRC Press, Boca Raton, p 195

Anon (1995) The description and classification of weathered rocks for engineering purposes. Geological society engineering. Group working party report. Quart J Eng Geol 28:207–242

Barton NR, Lien R, Lunde J (1974) Engineering classification of rock masses for the design of tunnel support. Rock Mech 6(4):189–239

Bieniawski ZT (1973) Engineering classification of jointed rock masses. Trans South Afr Inst Civ Eng 15:335–344

Bieniawski ZT (1974) Geomechanics classification of rock masses and its application in tunnelling. Proc.3rd Int. Congr. On Rock Mechanics, Vol. IIA, Int. Soc. Rock Mech., Denver, pp. 27–32

Bieniawski ZT (1976) Rock mass classification in rock engineering. In: Bieniawski ZT (ed) Exploration for rock engineering. A.A. Balkema, Johannesburg, pp 97–106

Brown ET (1981) Rock characterization, testing and monitoring—ISRM suggested methods. Pergamon, Oxford, pp 171–183

Carter T, Marinos V (2014) Use of GSI for rock engineering design. Proceedings of the 1st International Conference on Applied Empirical Design Methods in Mining, Lima

Castro LAM, Carvalho J, Sá G (2013) Discussion on how to classify and estimate strength of weak rock masses. In: Dight PM (ed) Slope stability 2013. Australian Centre for Geomechanics, Perth

Caterpillar (1996) Handbook of ripping, 10th edn. Caterpillar, Inc., Peoria, p 9

Chaney RC, Demars KR, Santi PM, Higgins JD (1998) Methods for predicting shale durability in the field. Geotech Test J 21(3):195

Chapman DR, Wood LE, Lovell CW, Sisiliano WJ (1976) A comparative study of shale classification tests and systems: Technical paper. Purdue University, West Lafayette, IN

Dearman WR (1976) Weathering classification in the characterization of rock: a revision. Bull Int Assoc Eng Geol 13:373–381

Fortsakis P, Nikas K, Marinos V, Marinos P (2012) Anisotropic behaviour of stratified rockmasses in tunnelling. Eng Geol 141–142(19):74–83

Geological Society Engineering Group Working Party (1995) The description and classification of weathered rock for engineering purposes. Quart J Eng Geol 28(3):207–242

Gokceoglu C, Ulusay R, Sonmez H (2000) Factors affecting the durability of selected weak and clay-bearing rocks from Turkey, with particular emphasis on the influence of the number of drying and wetting cycles. Eng Geol 57(2000):215–237

Hathaway A (1990) Weak Rock poorly lithified cockroaches and snakes. AEG News 33(3)

Hoek E (1994) Strength of rock and rock masses. News J Int Soc Rock Mech 2(2):4–16

Hoek E, Brown ET (1980) Empirical strength criterion for rock masses. J Geotech Eng Div ASCE v. 106, n. GT9, p. 1013–1035

Hoek E, Brown ET (1997) Practical estimates of rock mass strength. Int Jour Rock Mech Min Sci Geomech Abst 34:1165–1186

Hoek E, Brown ET (2018) The Hoek-Brown failure criterion and GSI – 2018 edition, J Rock Mech Geotech Eng, pp 22. https://doi.org/10.1016/j.jrmge.2018.08.001

Hoek E, Marinos P (2000) Predicting tunnel squeezing in weak heterogeneous masses. Tunn Tunn Int 1. November Issue, pp 45–51; Part 2, December Issue, pp 34–36

Hoek E, Marinos P (2007) A brief history of the development of the Hoek-Brown failure criterion. Soil Rock 30(2):85–92

Hoek E, Kaiser PK, Bawden WF (1995) Support of under-ground excavations in hard rock. Balkema, Rotterdam

Hoek E, Marinos P, Benissi M (1998) Applicability of the Geological Strength Index (GSI) classification for weak and sheared rock masses – the case of the Athens Schist formation. Bull Eng Geol Env 57(2):151–160

Hoek E, Caranza-Torres CT, Corcum B (2002) Hoek-Brown failure criterion - 2002 edition. In: Bawden HRW, Curran J, Telsenicki M (eds) Proc. North American Rock Mechanics Society (NARMS-TAC 2002). Mining Innovation and Technology, Toronto, pp 267–273

Hoek E, Marinos P, Marinos V (2005) Characterization and engineering properties of tectonically undisturbed but lithologically varied sedimentary rock masses. Int J Rock Mech Min Sci 42(2):277–285

International Society for Rock Mechanics (ISRM) (1981) Commission on classification of rocks and rock masses. Int J Rock Mech Min Abst 18:85–110

International Society for Rock Mechanics ISRM (1985) Suggested method for determining point load strength. Int J Rock Mech Min Sci Geomech Abst 22:51–62

Kanji MA (2014) Critical issues in soft rocks. J Rock Mech Geotech Eng 6(3):186–195

Krank KD, Watters RJ (1983) Geotechnical properties of weathered sierra nevada granodiorite. Environ Eng Geosci XX(2):173–184

Lauffer H (1958) Gebirgsklassifizierung für den Stollenbau [Mountain classification for the gallery construction]. Geology Bauwesen (in German). 74(1):46–51.

Lee SG, De Freitas MH (1989) A revision of the description and classification of weathered granite and its application to granites in Korea. Quart J Eng Geol 22(1):31–48

Marinos P (1993) General report session 1: hard soils-soft rocks: geological features with emphasis to soft rocks. In: Anagnostopoulos A et al (eds) Geotechnical engineering of hard soils-soft rocks. Balkema, Rotterdam, pp 1807–1818. ISBN 9054103442

Marinos V (2007) Geotechnical classification and engineering geological behaviour of weak and complex rock masses in tunneling. Doctoral thesis, School of Civil Engineering, Geotechnical Engineering Department, National Technical University of Athens (NTUA), Athens (In greek)

Marinos V (2012) Assessing rock mass behaviour for tunnelling. Environ Eng Geosci 18(4): 327–341

Marinos V (2014) Tunnel behaviour and support associated with the weak rock masses of flysch. J Rock Mech Geotech Eng 6:227–239

Marinos PV (2019) A revised, geotechnical classification GSI system for tectonically disturbed heterogeneous rock masses, such as flysch. Bull Eng Geol Environ 78:899. https://doi.org/10.1007/s10064-017-1151-zBOEG-D-17-00268.1

Marinos V, Carter T (2018) Maintaining geological reality in application of GSI for design of engineering structures in rock. Eng Geol 239:282–297

Marinos V, Drosos G (2010) Tunnelling through gneiss. A competent or a problematic rock mass? Proccedings of the 11th International congress of IAEG (in CD), Auckland

Marinos P, Hoek E (2000) GSI: a geologically friendly tool for rock mass strength estimation. Proc. GeoEng2000 at the Int. Conf. on Geotechnical and Geological Engineering, Melbourne, Technomic publishers, Lancaster, Pennsylvania, pp 1422–1446

Marinos P, Hoek E (2001) Estimating the geotechnical properties of heterogeneous rock masses such as flysch. Bull Eng Geol Env 60:82–92

Marinos V, Marinos P, Hoek E (2005a) The geological strength index – applications and limitations. Bull Eng Geol Environ 64(1):55–65

Marinos P, Hoek E, Marinos V (2005b) Variability of the engineering properties of rock masses quantified by the geological strength index: the case of ophiolites with special emphasis on tunnelling. Bull Eng Geol Environ 65(2):129–142

Marinos V, Fortsakis P, Prountzopoulos G (2006) Estimation of rock mass properties of heavily sheared flysch using data from tunnelling construction. Proceedings of the 10th International congress of IAEG in Nottingham, paper number 314, in CD

Marinos V, Fortsakis P, Prountzopoulos G (2011) Estimation of geotechnical properties and classification of geotechnical behaviour in tunnelling for flysch rock masses. In: Anagnostopoulos A et al (eds) Proceedings of the 15th European Conference on Soil Mechanics and Geotechnical Engineering, Part 1, Athens, pp 435–440

Marinos V, Fortsakis P, Prountzopoulos G (2013a) Tunnel behaviour and support in molassic rocks. The experiences from 12 tunnels in Greece. In: Kwasniewski M, Lydzba D (eds) Rock mechanics for resources, energy and environment (EUROCK2013). CRC Press, Boca Raton, pp 909–914

Marinos V, Prountzopoulos G, Fortsakis P, Korkaris K, Koumoutsakos D, Papouli D (2013b) Tunnel information and analysis system: a geotechnical database for tunnels. J Geotech Geoenviron Eng 31:891. https://doi.org/10.1007/s10706-012-9570-x

Marinos V, Fortsakis P, Stoumpos G (2015) Classification of weak and complex rock masses according to the engineering project type. In: Lollino G et al (eds) Engineering geology for society and territory, vol 6. Springer, Cham, pp 859–863

Nickman M, Spaun G, Thuro K (2006) Engineering geological classification of weak rocks. Proceedings of the 10th International congress of IAEG in Nottingham, paper number 492, in CD

Oliveira R (1993) Weak rock materials. Proceedings of the 26th Annual Conference of the Engineering Group of the Geological Society, Leeds, UK, September 1990
Palicki KS (1997) A graphical method for the classification of rock and weak rock masses based on field observations. Eng Environ Geosci 3(1):7–12
Santi PM (1995) Classification and testing of weak and weathered rock materials: a model based on Colorado Shales. Unpublished PhD dissertation, Colorado School of Mines, Golden, CO, 286 p
Santi PM (1997) Comparison of weak and weathered rock classification systems. In: Santi PM, Shakoor A (eds) Characterization of weak and weathered rock masses. Association of Engineering Geologists Special Publication #9. Association of Engineering Geologists, Denver, pp 139–160
Santi PM (2006) Field methods for characterizing weak rock for engineering. Environ Eng Geosci XII(1):1–11
Santi PM, Doyle BC (1997) The locations and engineering characteristics of weak rock in the U.S. In: Santi PM, Shakoor A (eds) Characterization of weak and weathered rock masses. Association of Engineering Geologists Special Publication #9. Association of Engineering Geologists, Sudbury, MA, pp 1–22
Santi PM, Shakoor A (eds) (1997) Characterization of weak and weathered rock masses, association of engineering geologists special publication #9: Association of engineering geologists, Sudbury, MA, 233 p
Smith HJ (1997) The point load test for weak rock in dredging applications. International Journal of Rock Mechanics and Mining Sciences. 34(295):3–4
Stacey TR, Page CH (1986) Practical handbook for underground rock mechanics. Trans Tech. Publications, Clausthal-Zellerfeld
Terzaghi K (1946) Rock defects and load on tunnel supports. In: Proctor RV, White TL (eds) Introduction to rock tunnelling with steel supports. Commercial Shearing & Stamping Co., Youngtown
Tsiambaos G, Sabatakakis N (2004) Considerations on strength of intact sedimentary rocks. Eng Geol 72:261–273
Wakabayashi J, Medley EW (2004) Geological characterization of mélanges for practitioners. Felsbau 22(5):10–43
White RM, Richardson TL (1987) Predicting the difficulty and cost of excavation in the Piedmont. In: Smith RE (ed) Foundations and excavations in decomposed rock of the piedmont province: ASCE geotechnical special publication No. 9: ASCE, New York, pp. 15–36
Wickham GE, Tiedeman HR, Skinner EH (1979) Ground support prediction model-RSR Concept. Proceedings, Rapid Excavation and Tunneling Conference, AIME, New York, pp 691–707

Chapter 20
Soft Rocks in Underground Hydroelectric Schemes

Luís Ribeiro e Sousa, Hui Zhou, Rita Leal e Sousa, and Chuanqing Zhang

20.1 Initial Considerations

Hydroelectric power schemes are important undertakings that make use of the underground space. They are composed, in general, by one or two dams, the hydraulic circuit with the eventual inclusion of surge chambers and the powerhouse complex. The use of underground space has been widely implemented because of safety and environmental advantages that it brings when compared with other surface solutions. The costs of excavations and supports are usually balanced by the costs of the foundation and superstructure of surface infrastructures. In good rock mass conditions the supports for tunnels, caverns and shafts can be considerably reduced and economical and environmental impacts are always mitigated.

Underground works associated with the powerhouses form a fundamental part of a hydroelectric scheme whose performance is dependent on their suitable location, conception and on the adequate design of supports in order to ensure the safety of the overall underground complex (Sousa et al. 1994). Surge chambers are also important underground structures that may be located in the hydraulic circuit depending on its arrangement and particularly its dimensions. In some cases they can be also associated to the powerhouse underground complex. Surge chambers are in general concentrated structures in which tridimensional equilibriums develop with similar problems caused by the caverns of the powerhouse complex. Pressure

L. Ribeiro e Sousa (✉)
Tongji University, Shanghai, China

SKL-GDUE of China University of Mining and Technology, Beijing, China

H. Zhou · C. Zhang
SKL-GGE, Institute of Rock and Soil Mechanics, Chinese Academy of Sciences, Wuhan, China

R. Leal e Sousa
Masdar Institute of Science and Technology, Abu Dhabi, UAE

M. Kanji et al. (eds.), *Soft Rock Mechanics and Engineering*,
https://doi.org/10.1007/978-3-030-29477-9_20

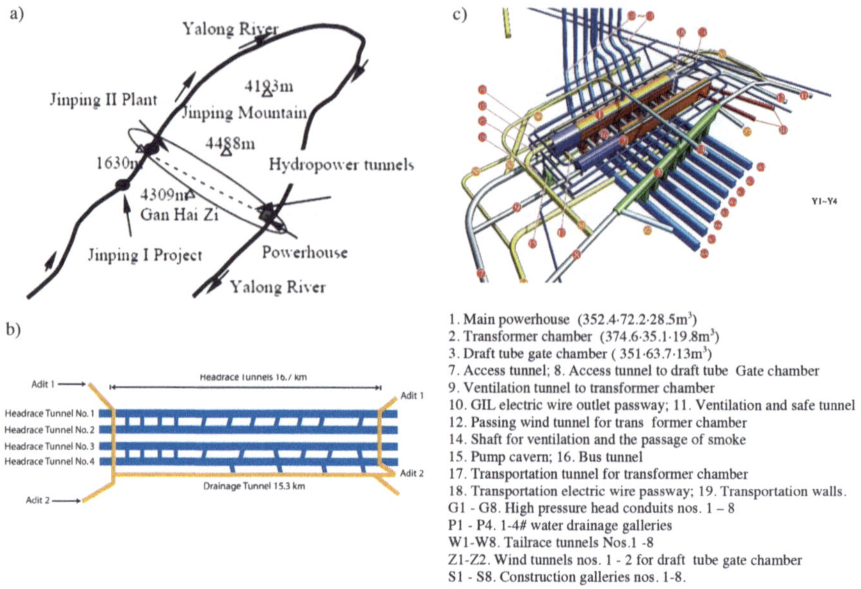

Fig. 20.1 Jinping II underground hydroelectric scheme: (**a**) location; (**b**) high pressure tunnels; (**c**) underground powerhouse complex (Zhang et al. 2011)

tunnels and shafts that form part of the hydraulic circuit of hydroelectric schemes can be of considerable importance due to the internal and external high water pressures that may be present, due to the length they can have and due to the variety of geotechnical conditions that may occur. Therefore, in these undertakings, we can distinguish different types of works, including caverns, being generally formed by the powerhouse underground complex, tunnels and shafts under high and lower pressures, and surge chambers.

An example of large underground schemes is illustrated in Fig. 20.1 regarding the Jinping II underground hydroelectric scheme where different types of underground structures occurred (Wu et al. 2010; Zhang et al. 2011; Feng et al. 2012). Jinping II is an important project and it is located in Yalong River. Jinping site is unique in that it utilizes a natural 180° bend in the Yalong River. Jinping II has a large underground complex for the powerhouse with 8 units and a total installed capacity of 4800 MW, to produce a multiyear average annual output of 24.23 TWh.

For the hydraulic circuits, there are several alternative schemes that have been evolving over time. The choice among them has to take into account the local conditions, in order to select the position of shafts, the high and low pressure tunnels, as well the location of the powerhouse complex and surge chambers (Sousa et al. 1994). The hydraulic circuits were originally constructed partly in underground, namely the high pressure tunnels, comprising a structure located on the hillside, which worked as surge chamber, after the high pressure tunnel, followed by a penstock at the ground surface, which led the water to the powerhouse, near the foot of the hill. After 1950s, the entire hydraulic circuit was moved to an underground location and the surge chamber was located upstream of the powerhouse. The hydraulic circuit was then constituted by a concrete shaft with or without a steel lining until the powerhouse, with the water to be restored through a low pressure tunnel. In areas of low resistance or very permeable rock masses, it was necessary to carry out a treatment for strengthening the rock mass.

Many of the hydroelectric schemes are reversible, they accumulate water in an upper reservoir by pumping, using the energy during lower consumption periods in order to take advantage of the accumulated water to produce energy during periods of highest demand. This energy storage system has been assuming a prominent role in the electric power load diagram, helping to support the peak demand of the electricity consumption (Schleiss 2000; Feng et al. 2012).

The design of the elements associated with the hydraulic circuit and the powerhouse of a reversible hydroelectric scheme does not differ too much from a conventional scheme, unless the necessity of the existence of two reservoirs and of electromechanical equipment. These projects use a lower and an upper reservoir. The upper reservoir can be created by a dam or by an artificial basin on top of a hill as a result of excavations and construction of circular dikes, or can be in the underground space. There have also been proposed reversible hydroelectric projects with deep underground reservoirs designated by UPHS systems (Underground Pumped Storage Hydroelectric). The first UPHS scheme with both reservoirs in underground was implemented at Socorridos plant in Madeira Island that is integrated in a multiple purpose scheme with the same name (Cafofo et al. 2007).

The choice of the adequate location and the alignment of the hydraulic circuit depends on several economical and technical factors, namely, water heads and internal pressures, surface topography, in situ stresses and prevention of hydraulic jacking, the needs of supports and final linings, and of course requirements for access, ventilation, and drainage (Lamas 1993; Schleiss 2000). The design of supports for pressure tunnels and shafts should be done on the basis of a detailed geotechnical study. The underground works for the powerhouse are, in general, an essential part of a hydroelectric scheme which depends of a suitable location and an adequate conception and design of the shapes and supports taking into consideration the location of the electromechanical equipment and the nature of the surrounding rock mass. In addition to the powerhouse cavern other important cavities can exist such as transformer room, spherical and butterfly valve chambers and sometimes surge chambers.

The project needs a good geological and geotechnical study of the rock mass. The design of these works involves the definition of the main orientation axes of the caverns, the definition of the shapes and associated shafts and tunnels, taking into account factors related to the mechanical properties of the rock mass, the discontinuity sets and other lower strength surfaces and the in situ state of stress installed in the rock mass. Risk evaluation of the stability of large cavities is an important issue to be considered. The adequate design of the different parts of the underground powerhouse complex should be a process that provides an optimum solution from the safety, functionality, and economical point of view. The use of numerical models is fundamental and in general the development of huge 3D numerical models is required.

Most accidents and other associated problems occur during the construction works of the underground structures and are very often related to uncertainties concerning the ground conditions. To help eliminate or at least reduce these accidents, it is necessary to systematically assess and manage the risks associated with underground construction (Einstein 2002; Sousa 2006, 2010). Risk assessment is developed with the goal of avoiding major problems that can occur in the underground structures of hydroelectric schemes.

For risk evaluation it is necessary to identify the models to be used to represent the existing knowledge and perform risk and decision analysis. Risk assessment and management requires an evaluation of the hazard and the assessment of the likelihood of the harmful effects. Risk assessment starts with the hazard identification, focusing on the likelihood of damage extend. After hazard identification, risk characterization is followed, which involves a detailed assessment of each hazard in order to evaluate the risk associated to each one of them (Sousa 2010).

Based on studies presented in several publications (Lamas 1993; Sousa 2006, 2010), several hazard situations are characterized for the underground works during the different stages as identified in Table 20.1. A detailed analyzed of different scenarios is presented in the publication of Feng et al. (2012).

Risk analysis can follow the guidelines established by ITA (Eskesen et al. 2004). A process established by Popielak and Weining (2010) consists in the activities: ranking the risk factors; defining the system to be modeled; establishment of conceptual models of the system; numerical analysis to study potential impacts of different risk factors; and establishment of a risk management plan. The first activity is related to the rank the risk factors according to their impact in the matrix of risks, and should include project cost, construction sequence, safety, operations, and the environment. The second activity regards the system to be modeled combining the constructing facilities and the existing rock mass where the underground structures will be performed. The rock mass includes the rock formations, their arrangements, the discontinuity sets and the existing faults and other low strength surfaces,

Table 20.1 Hazard identification scenarios for the underground structures

Stage	Hazard	Description
Construction	H_1	Rock fall
	H_2	Rockburst
	H_3	Water inflow and leakage
	H_4	Collapse
	H_5	Large deformations
First filling and normal operation	H_3	Water inflow and leakage
	H_5	Large deformations
	H_6	Inadequate confinement
	H_7	Deterioration of rock mass
	H_8	Buckling of steel linings
	H_9	Dynamic fluctuations of water pressure
	H_{10}	Landslides

the groundwater factors, and the hydromechanical and thermomechanical properties. Other factors with impact in the schedule and project costs should also be included. The underground facilities includes: the works for the powerhouse complex, that in general form an essential part of all the scheme from the risk analysis point of view; the hydraulic circuit sometimes with considerable extension involving a large variety of geotechnical conditions and high pressures in the upper circuit, and considers the intakes with eventual instabilities at the surface; the surge chambers that can involve large concentrated excavations with high risk situations; and finally the access works that comprise access tunnels and shafts that can reach a considerable development and also connecting galleries and shafts for the major underground works. In the system other infrastructures should be included like the electromechanical equipment, transportation facilities, ventilation during construction, power supply, lightning, and other equipment. In the context of risk analysis, the system is entirely dynamic involving the knowledge of all components that will be updated as works progresses, which happens with the ongoing geotechnical studies and investigations and with excavations. The conception and design of the underground structures will change and this will affect the system and of course future risk analyses (Popielak and Weining 2010).

The conceptual models of the system should consider a description of the process to a successful construction and operation of all the underground structures with particular emphasis to the large cavities like caverns and a list of uncertainties that could influence the cost; and the sequence of the excavations. The system model should consider the most relevant activities and also include the relationship between them. The diagrams of the system may be subdivided in subprocesses according to the different types of underground structures, like the powerhouse complex, hydraulic circuit, surge chambers included or not in the hydraulic circuit and for the access works.

The numerical analysis regarding the conception and design of the underground structures is a very important issue in order to ensure the safety of the works and it is an integrated process including the study of the potential impact of the risk factors in the safety of the structures. For the conception of the works, the process consists of (Geoguide 4 1992; Sousa et al. 1994): choice of the site with optimum conditions from the safety point of view; definition of the alignment of the hydraulic circuit and of the cavern main axes of orientation minimizing the stability problems; definition of the shapes for the different types of cavities taking into consideration mechanical and geometrical properties of the rock mass and particularly of the discontinuity sets and faults and the in situ state of stress; dimensioning of the different parts of the underground structures in order to achieve an optimum solution from economic point of view; choice of the construction process for the different cavities, equipment, and topographic and geological conditions, and monitoring of the works.

The design stage of the underground structures is based on numerical analysis of the different hazard scenarios already defined on Table 20.1. The analysis started with the hazard ranking highlighting their relative importance. The ranking hazard for the large cavities is H_1 (rock fall) combining wedge failure with stress induced failure, isolated or with mutual influence. It is therefore important to evaluate the existing discontinuity sets. To analyze the risk for wedge failure and to determine the corresponding support requirements various types of numerical calculations should be carried out (Popielak and Weining 2010). Other important hazard analysis is for instant H_4 (collapse) due to the interception of the cavities by low strength surfaces. It is necessary to develop large numerical models using different techniques, like FEM, FDM or DEM in order to predict stability of the large cavities and their deformations. Nowadays the use of complex 3D numerical models is possible considering different types of underground structures and can be combined with 2D simplified models. For the high pressure tunnels and shafts of the hydraulic circuit, hydromechanical models need to be applied in order to simulate the real behavior of the rock mass (Lamas 1993; Schleiss 2000).

Also and for better evaluation of the risk involved it is relevant to combine the results of these numerical models with probabilistic simulations using Monte Carlo method or Bayesian networks (BN) approaches. The results of the probabilistic numerical simulations determine the probability distribution of schedule impact and the sensitivity of the project to model input parameters risk factors and to take decisions about the appropriate measures to be taken during the several stages of the works (Sousa 2010; He et al. 2011).

The final activity regarding risk management is to review the probabilistic analysis and identify a list of the potential problems. The prioritized list of significant risks represents an important step in the risk mitigation plan. A recommended risk management plan should be presented and their strategies (Popielak and Weining 2010).

Due to the variability of rock formations, there is a large degree of uncertainty associated with the evaluation of geotechnical properties in these underground hydroelectric schemes. As a result, the evaluation of geomechanical parameters is often carried out through in situ and laboratory tests along with the application of empirical methodologies (Barton 2000; Hoek 2007a, b; Miranda et al. 2018).

Soft rocks can occur also be present in these undertaking. Soft rocks exhibit unfavorable behavior related to low strength, high deformability, fast weathering, as well as others (Rocha 1975; Kanji 2014; He 2014). Soft rocks can be sedimentary rocks and weathered igneous and metamorphic rocks, or the so-called residual rocks. The deformability moduli of soft rock masses, even those for residual formations can be considerably higher than those for soil formations. For design purposes shear strength parameters are often selected rather than determined. The selection of deformability and strength parameters requires mainly sound engineering judgment, experience and on the use of empirical systems (ASCE 1996; Wyllie 1999; Sousa et al. 2010, 2018).

When dealing with soft rock masses, an important issue is the occurrence of surfaces with low strength. This was the case of Morrow Point underground power station where two low strength shear zones where assigned, one intersecting the ceiling of the powerhouse cavern and the other intersected two vertical walls of the cavern in the lower part (Sousa 1976). This situation of Morrow Point was one of the first applications to an underground structure by finite element models. Another situation was the case of Cahora Bassa Hydroelectric Scheme in Mozambique that led to significant consequences such as the accident that occurred during the construction of one of the surge chambers of the (Rocha 1978; Sousa et al. 2010, 2018).

In the next sections, examples of application where soft rock problems occurred on underground hydroelectric schemes are illustrated. The first case is related to the Socorridos hydroelectric scheme in Madeira Island, Portugal, in volcanic formations (Sect. 20.2). In Sect. 20.3, the Mingtan underground pumped station in Taiwan presented complex geological formations with soft rocks is presented. Important application of in situ tests was performed to characterize some weak formations. The wedge failure occurred at Cahora Bassa, Mozambique, was studied due to the existence of a low strength low surface as illustrated in Sect. 20.4. Finally the last case is related to the mega-project Jinping II, in China (Sect. 20.5). Deep tunnels intersected by chlorite schists are studied in detail. A conclusion is drawn in the last section.

20.2 UPHS Socorridos Hydroelectric Scheme

Madeira Electricity Company from Madeira Island, Portugal, decided to repower the Socorridos hydroelectric project that is integrated in a multiple purpose project (Fig. 20.2). The hydroelectric complex is equipped with reversible units with a differential elevation of about 450 m between the Covão upper tunnel and the lower storage reservoir. The rock mass involved basalts, breccias, and tuffs (Cafofo and Sousa 2007). The repowering included the following sequence of underground works:

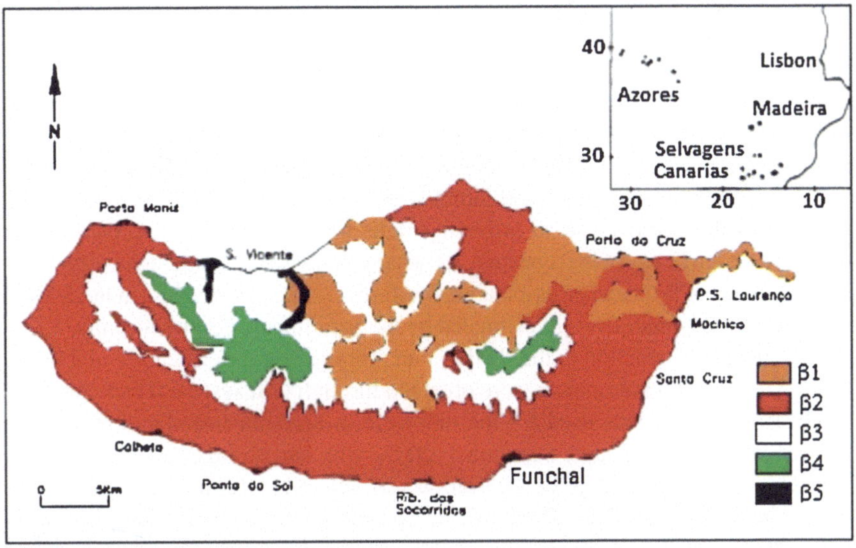

Fig. 20.2 Madeira Island. Location of Socorridos hydroelectric scheme

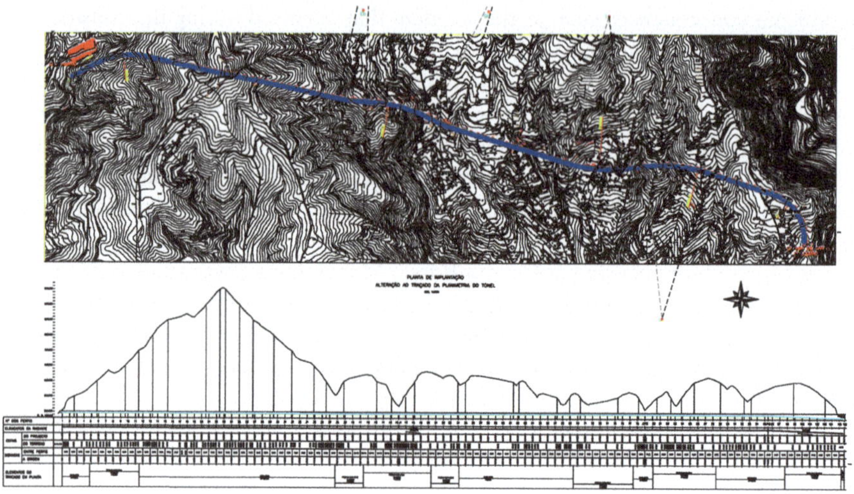

Fig. 20.3 Plan and longitudinal cross section of Covão tunnel (Cafofo and Sousa 2007)

a 5.2 km tunnel located at the upper level (Covão tunnel); galleries for storage of water with a total capacity of 40,000 m^3; and a cavern pumpage station, where a pumpage equipment was installed, and an auxiliary tunnel.

The Covão tunnel allows besides the purpose of water supply and irrigation an upper pressure gallery for the Socorridos hydroelectric powerhouse (Fig. 20.3). Initially it had a length of 100 m excavated by the drill and blasting method with a section of 4.20×3.60 m^2. The remaining 5144 m were excavated by a TBM, making the tunnel circular with a 3 m diameter (Fig. 20.4).

A detailed survey of the area was done and empiric approaches were followed, RMR and MR systems (Rocha 1976; Bieniawski 1989) in order to get geomechanical parameters of the volcanic formations (deformability and strength) (Cafofo and Sousa 2007). Covão tunnel crosses a complex region in the volcanic complex β2 identified in Fig. 20.2. β2 is predominantly composed by basalts, breccias, and tuffs. Each type zone was divided into two lithologies as indicated in Table 20.2. Table 20.3 summarizes the geomechanical properties determined for each intercepted by the tunnel using RMR and GSI systems. As what can be seen disaggregated breccias and both types of tuffs can be associated to soft rocks since UCS is less than 25 MPa. The value of deformability was estimated by the Serafim and Pereira formula (1983) and also using a correction to the RMR value proposed by Alber (Cafofo and Sousa 2007). The strength parameters were obtained through GSI system (Hoek 2007a).

TBM excavation was used for the first time in Madeira Island and took place between January 2005 and September 2005, with an average advance ratio of 21.3 m/day. The advance of the excavation was defined of 1.2 m, corresponding to the maximum length of the TBM able to excavate without change the position of grippers or the backup. It can be observed from the Table different advances rates in the zones and as expected the less resistance geotechnical materials were excavated

Fig. 20.4 TBM used at Covão tunnel

Table 20.2 Geotechnical zones for the Covão tunnel

Zone	Lithologies	Designation
ZG1	Compact basalts Fractured basalts	CB FB
ZG2	Consolidated breccias Disaggregated breccias	CBr DBr
ZG3	Compact tuffs Disaggregated tuffs	CT DT

Table 20.3 Geomechanical properties of Covão tunnel volcanic formations (Cafofo et al. 2007)

		E (GPa)		Strength parameters	
Volcanic formation	UCS (MPa)	RMR	TBM	*c* (MPa)	ϕ (°)
CB	50	17	30	3.7	41
FB	25	6.5	13.5	1.3	38
CBr	50	24	4.8	2.9	39
DBr	10	5.2	0.9	0.45	35
CT	1	0.38	1.1	0.05	36
DT	0.5	0.11	0.4	0.015	11

Fig. 20.5 Rock fall at Covão tunnel (Sousa 2010)

in less time than the more resistant. For basalts an average increment of 44% was contested; for breccias an increment of 120% and for tuffs an average increment of 67%. However, and in general the excavation ratio was bigger when compared with breccias and tuffs.

This type of volcanic formations is heterogeneous and prone to block falls during the excavation process. A rock fall that occurred during excavation is presented in Fig. 20.5. These accidents affected naturally the advance race of the TBM during the excavation.

The hydroelectric project of Socorridos is reversible, thus the lower reservoir, with the extension of 1200 m, and pumping stations are located underground. The underground reservoir is shown in Fig. 20.6. Figures 20.7 and 20.8 present the location of the pumping station and its cross section.

The underground reservoir and the auxiliary tunnel crossed formation with similar conditions. Table 20.4 shows different geological conditions for both underground structures. For ZG3 the rock mass was of low resistance consisting of low consolidated formations such tuffs, breccias, and recent deposits. In the excavation of the pumping station two zones were defined. The upper part was formed by alluvium and the lower part by intercalations of basalts and breccias. The percentage of alluviums found was about one third of the percentage of basaltic formations.

A geologic section of the pumping station zone is indicated in Fig. 20.9. The geotechnical survey consisted of six boreholes drilled in the area of the pumping station. The depths of the boreholes vary between 30 and 65 m. An image of the samples in a borehole is illustrated in Fig. 20.10. In general, the rock mass is highly fractured and with low consistent. Lugeon and LNEC dilatometer tests were executed inside the boreholes. Rock samples were extracted for UCS tests, porosity, and specific density tests (Cafofo 2006).

The dilatometer tests were performed through boreholes S_2–S_6, at different depths and the deformability modulus was obtained. At low depths dilatometer tests were not able to be executed, reason why a new borehole, 30 m deep, was done.

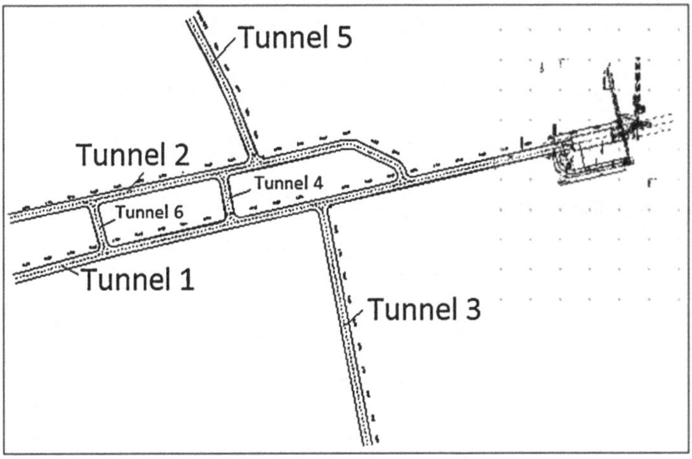

Fig. 20.6 Underground reservoir of Socorridos plant

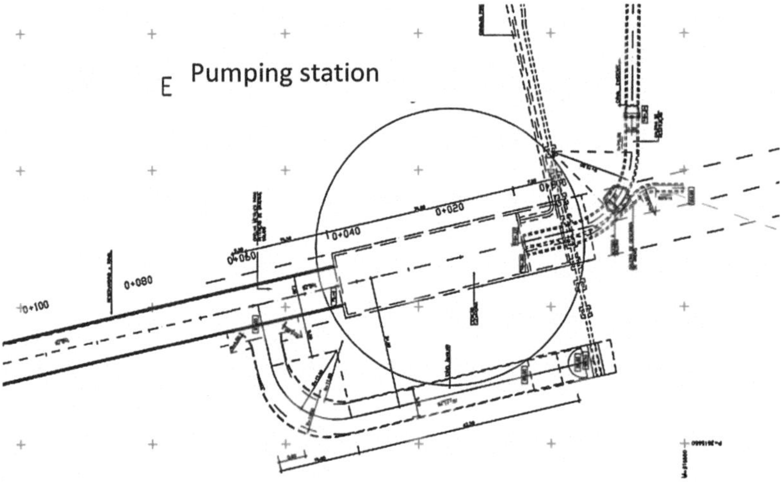

Fig. 20.7 Pumping station of Socorridos scheme

Results are illustrated in Table 20.5, showing the average values in each borehole for the loading cycles and unloading cycles. As expected the deformability modulus during unload cycles is always higher than during loading cycles, with an average increase of about 45%. The values are relatively low and with small dispersion, between 1.30 and 3.87 GPa for loading cycles and between 1.95 and 5.17 for unloading cycles.

Finally, a special reference to an empiric system that was developed specifically for volcanic rock formations, derived from the RMR system and from a classification developed at IPT, Brazil, including geomechanical information from several

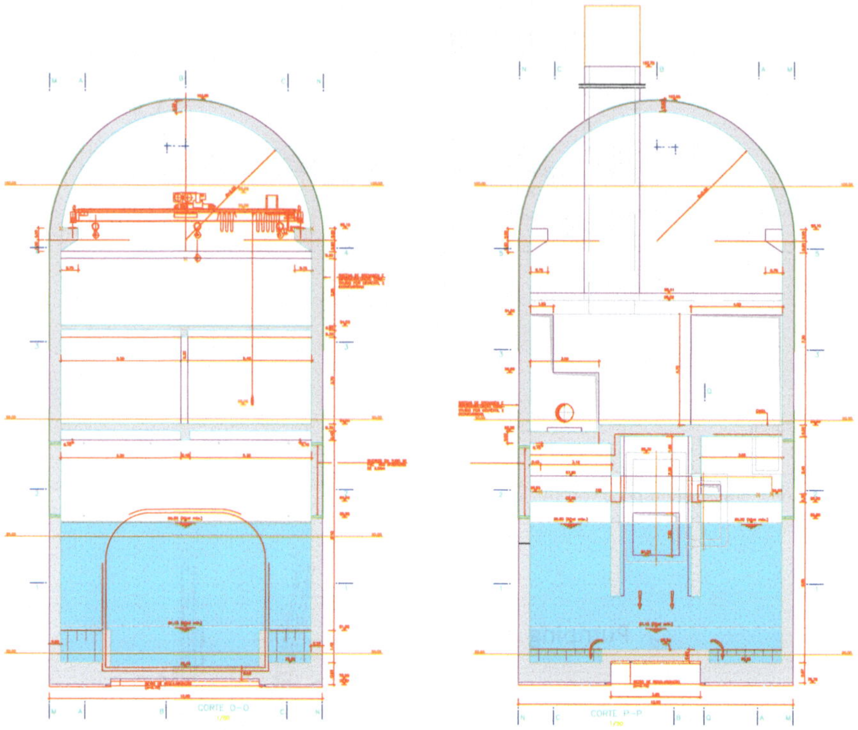

Fig. 20.8 Transversal cross section of the pumping station

Table 20.4 Geomechanical zoning for the lower reservoir and auxiliary tunnel

Zone	Lithology	Characteristics
ZG1	CB and CBr	Good geomechanical behavior
ZG2	FB or fine intercalations of DBr and DT	Reasonable geomechanical behavior
ZG3	Weak formation with lower cohesion	Poor geomechanical behavior

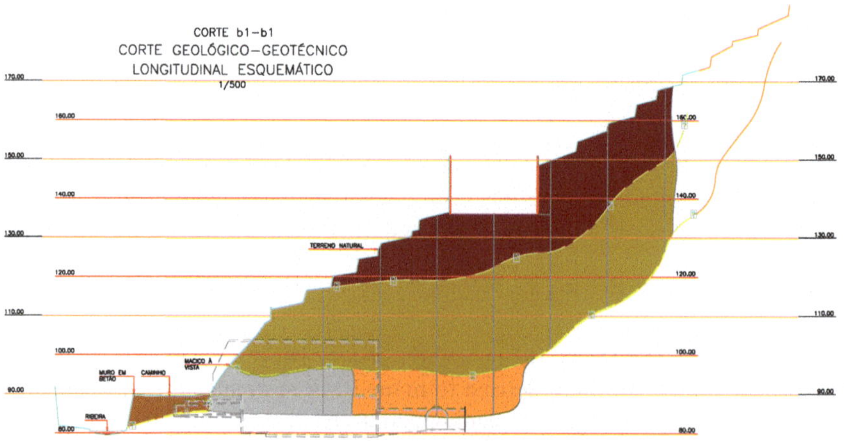

Fig. 20.9 Geologic section of the pumping station zone

Fig. 20.10 Samples obtained in a borehole of the pumping station zone (Cafofo 2006)

Table 20.5 Dilatometer test results

	E (GPa)	
Borehole	Load	Unload
S_2	1.62	3.22
S_3	3.87	5.17
S_4	2.51	3.24
S_5	1.30	1.95
S_6	1.66	2.32
Average	2.16	3.13

tunnels in basaltic formations (Menezes et al. 2007; Moura and Sousa 2007; Miranda et al. 2018). The geomechanical information of this hydroelectric scheme was taken into consideration, as well from other tunnels in Madeira Island, and also from Azores and Canarias Islands, and also from Mexico. The parameters considered were: uniaxial compressive strength; rock characteristics; intensity of jointing; discontinuity conditions; presence of water; and disposition of blocks (Miranda et al. 2018). The addition of all weights gives an index called VRS. Volcanic rock mass classification VRS and weights are indicated in Fig. 20.11.

The rock mass in the new empirical system is classified in six classes: class I—100 ≥ VRS ≥ 91 (Excellent); II—90 ≥ VRS ≥ 76 (Good); III—75 ≥ VRS ≥ 61 (Reasonable); IV—60 ≥ VRS ≥ 41 (Regular); V—40 ≥ VRS ≥ 21 (Poor); VI—20 ≥ VRS ≥ 0 (Very Poor). Volcanic weak rocks fall in classes V and VI.

20.3 Mingtan Underground Hydroelectric Project. Rock Mass Characterization

The Mingtan Pumped Storage Hydro Power Plant is the largest hydroelectric power plant in Taiwan (Fig. 20.12). It uses Sun Moon Lake as the upper reservoir and Shuili River reservoir as lower reservoir. During high power demand, water used for power generation from Sun Moon Lake, as well as Shuili River, fills Mingtan Reservoir. During low demand, the water is pumped back upstream to the lake. This system can generate up to 1602 MW at peak production.

Mingtan gravity dam was completed in 1990 and is 82 m high. It was affected by the Chi Chi Earthquake, M7.6, at September 21, 1999. The earthquake had a peak

Volcanic Rock Mass classification and weights

P_1	Uniaxial strength Weight	R_1 15	R_2 9	R_3 6	R_4 3	R_5 1
P_2	Rock weathering Weight	A_1 20	A_2 12		A_3 4	
P_3	Joint frequency Weight	F_1 25	F_2 20	F_3 15	F_4 10	F_5 5
P_4	Joint surface conditions Weight	B_1 30	B_2 25	B_3 17	B_4 10	B_5 0
P_5	Presence of water Weight	C_1 10	C_2 7	C_3 4	C_4 0	
P_6	Block position Weight	D_1 0	D_2 -2	D_3 -5	D_4 -10	

Uniaxial Strength (P_1)

R_1 >120 MPa	R_2 60–120 MPa	R_3 30-60 MPa	R_4 15-30 Pa	R_4 <15 MPa

Rock Weathering (P_2)

A_1 – Sound or practically sound Macroscopically do not exist signals of weathering in the minerals; difficult to break with hammer.	A_2 (Moderately weathered) Minerals with moderate weathering; weathering materials changing color to original rock; relatively easy to break with hammer.	A_3 (Highly or extremely weathered) Minerals with powdery aspect and totally without luster; easy to break with hammer; steel blade leaves clear mark on rock surface.

Joint Frequency (P_3)

F_1 One or less per m	F_2 2-4 per m	F_3 5-10 per m	F_4 11-15 per m.	F_5 16 or more per m.

Joint Surface Conditions (P_4)

B_1 Very rough discontinuities; closed, hard walls.	B_2 Slight rough discontinuities; separation <1mm, hard walls.	B_3 Slight rough discontinuities; separation >1mm, soft walls.	B_4 Separation >5mm; discontinuities with slickensided walls or 1-5mm thick gouge.	B_5 Discontinuities with soft gouge; separation >5mm; widecontinuous.

Presence of Water (P_5)

C_1 Dry or damp.	C_2 Dripping.	C_3 Flowing.	C_4 Inflow >0.1 l/mim.

Block Position (P_6)

D_1 Very favorable to stability.	D_2 Favorable to stability.	D_3 Acceptable to stability.	D_4 Not acceptable to stability.

Classification of the Volcanic Rock Mass

I (Excellent) VRS=100-91	II (Good) VRS=90-76	III (Reasonable) VRS=75-61	IV (Regular) VRS=60-41	V (Poor) VRS=40-21	VI (Very Poor) VRS=20-0

Fig. 20.11 Volcanic rock mass classification VRS and weights

acceleration PHGA of 0.4–0.5 g. No reported damage was registered in the dam body (Nuss et al. 2017). The power cavern complex is located in sandstone, sandstone with siltstone interbeds and several siltstone beds (Hoek 2007a). The sandstones are fine grained to conglomeratic and sometimes quartzitic. In general they are strong to very strong although they are slightly to moderately weathered. Locally, softer zones of highly weathered or altered material are encountered. The siltstones are moderately strong and almost always sheared. Occasionally, massive sandstone beds occur with a thickness of up to 7 m. Shear zones were encountered in the cavern area. All of these features are parallel to the bedding planes, occurring in the relatively soft siltstone or interbedded sandstone and siltstone layers. These shear zones are composed of multiple clay seams and shattered, softened or decomposed rock. A geological plan of the powerhouse area is shown in the publication of

Fig. 20.12 Taiwan map and localization of Mingtan hydroelectric project

Hoek (2007a) and an isometric drawing of the underground complex is illustrated in Fig. 20.13 with bedding plane orientation.

Laboratory and in situ tests were carried out in the 1970s for the Mingtan project. The rock mass in the powerhouse area was divided into three types: jointed sandstone, bedded sandstone and faults, and properties were presented in a chapter of a book on soft rocks (Sousa et al. 2018). Several in situ tests were performed by LNEC (Rodrigues et al. 1978). A typical in situ tests for evaluating the deformability was Large Flat Jack (LFJ) field plate loading test as illustrated in Fig. 20.14.

Table 20.6 summarizes results LFJ tests at Mingtan, as well the geology in the site. For adit E two test chambers C_2 and C_3 were used (Rodrigues et al. 1978). The tests for chamber C_2 were planned with three flat jacks and three slots were executed A, B, and C. A total of 6 loading cycles were carried out, the maximum pressure of the third, fifth, and sixth cycles was held for at least 24 h. The maximum pressure was 8 MPa. The average values of E were 5.0 GPa and 3.2 GPa for slots parallel and perpendicular to bedding respectively. For adit E, test chamber C_3, the rock mass was not always the same, as indicated in the Table 20.6. The deformability tests were carried out with only one LFJ due to difficulties in the execution of the tests. A total of 7 loading cycles were

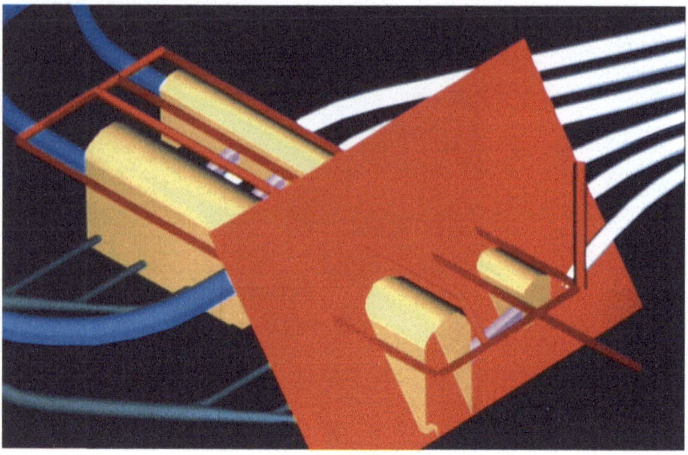

Fig. 20.13 Mingtan underground complex in relation to the bedding plane orientation (Hoek 2007a)

Fig. 20.14 Characteristics of the LFJ used

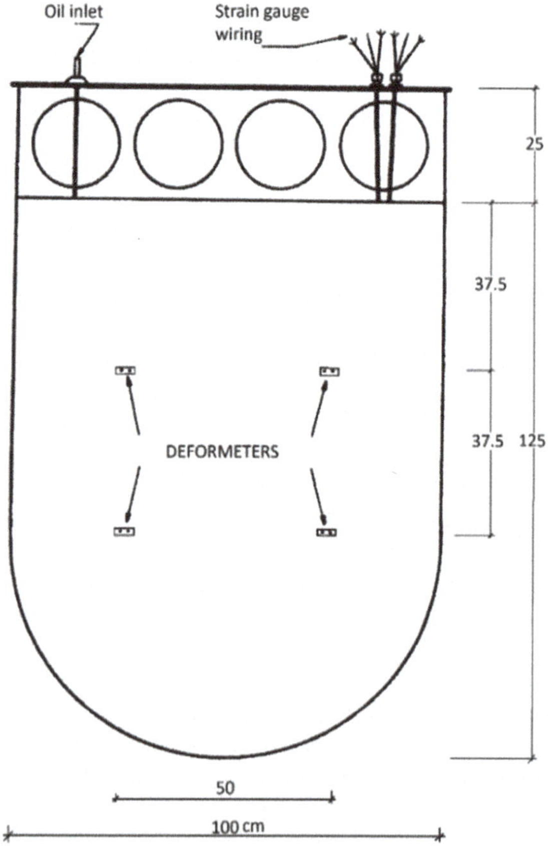

Table 20.6 LFJ tests at Mingtan underground powerhouse (Adapted from Rodrigues et al. 1978)

Adit-chamber	Position of slots	No. of slots	Geology	Values of *E* (GPa)
E-C2	Parallel to bedding Perpendicular to bedding	3	Massive sandstone, light grey, medium to coarse	4.8–5.2 (5.0) 3.0–3.5 (3.2)
E-C3	Parallel to bedding Perpendicular to bedding	1	Interbedded sandstone and siltstone.	2.1–2.3 (2.2) 10.0–11.7 (11.0)
AMP1	Parallel to bedding Perpendicular to bedding	2	Sandy siltstone, dark grey, massive.	2.9–3.7 (3.3) 5.3–6.0 (5.7)
AMS1	Parallel to bedding Perpendicular to bedding	2	Siltstone, dark grey.	10.8–14.3 (12.4) 12.2–17.7 (14.8)
AMH1	Parallel to bedding Perpendicular to bedding	2	Sandstone, brown to grey, zone very changeable.	1.8–2.8 (2.3) 4.3–5.9 (5.1)

carried out. The maximum pressure of the third, fifth, and sixth cycles was held for at least 24 h. The maximum pressure reached was 6 MPa. The average values of *E* were 2.2 GPa and 11.0 GPa for slots parallel and perpendicular to bedding, respectively. For adit AMP1, 2 slots were executed. A total of 4 loading cycles were carried out. The maximum pressure of the second, and fourth cycles was held at least 24 h. The maximum reached pressure was 7 MPa. The average values of *E* were 3.3 GPa and 5.7 GPa for slots parallel and perpendicular to bedding, respectively. For adit AMS1, 2 slots were also executed. A total of 5 loading cycles were carried out. The maximum pressure was 8 MPa. The average values of *E* were 12.4 GPa and 14.8 GPa for slots parallel and perpendicular to bedding, respectively. Finally, for adit AMH1, 2 slots were also executed. A total of 5 loading cycles were carried out. The maximum reached pressure was 4 MPa. The average values of *E* were 2.3 GPa and 5.1 GPa for slots parallel and perpendicular to bedding, respectively.

Creep tests were also performed, and the results showed a decrease of about 11% in the value of *E* after 40 h at 4 MPa. In addition, cancelling pressures were obtained for the different cycles. These tests showed in situ stresses of more than 1 MPa, from 1.2 to 3.0 MPa for E-C2; 1.0–6.0 MPa for AMP1; 0.4–4.8 MPa for AMS1; and 0.5–2.9 MPa for AMH1 (Rodrigues et al. 1978).

Due to the complexity of the rock masses with the inclusion of soft rock layers, the procedure followed in the design was complex and described in detail in the publication of Hoek (2007a, b). Numerical modelling details of a section are illustrated in Fig. 20.15. This model was excavated in six stages and the cable reinforcement was installed in each stage to simulate the construction sequence.

20.4 Cahora Bassa Surge Chambers

The Cahora-Bassa dam is located on the river Zambeze, in Mozambique, in a stretch of the river, forming part of a multipurpose scheme. It comprises a 170 m maximum height arch dam and an important underground powerhouse complex with an underground hydraulic circuit. The rock mass consists of granitic gneisses with several

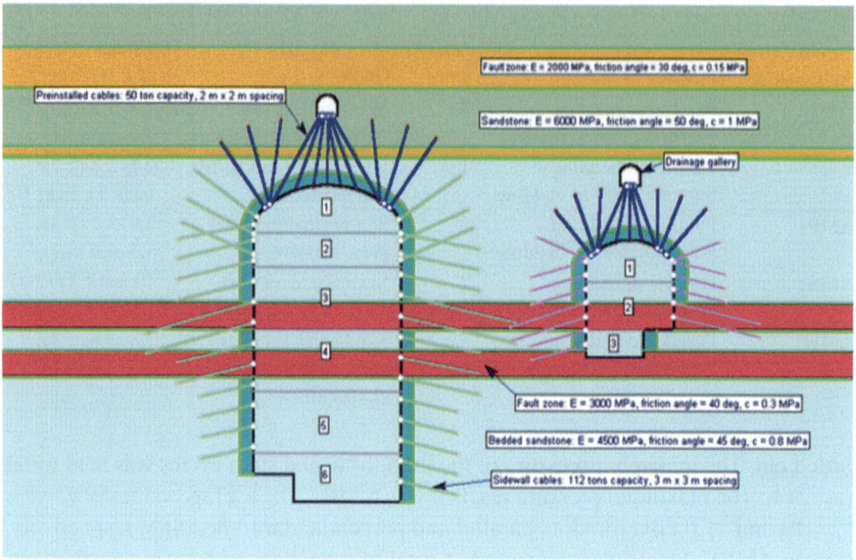

Fig. 20.15 Numerical model showing distribution of material and layout of reinforcement in the rock mass surrounding the caverns (Hoek 2007a)

Table 20.7 Synthesis of the geomechanical parameters determined in situ

Parameter	Value
Deformability modulus of rock mass	
– Average – Maximum – Minimum	65 GPa 116 GPa 23 GPa
Shear strength	
– Discontinuities – Lamprophyre	$c = 0.3$ MPa; $\phi = 33–41°$ $c = 0.2$ MPa; $\phi = 20°$
In situ stresses	
– Maximum principal stress – Minimum principal stress	13–18 MPa 9–12 MPa

lamprophyre and gabbroic veins. A synthesis of the geomechanical parameters determined in situ is presented in Table 20.7. The geotechnical characterization of the rock mass was thoroughly analyzed by LNEC, and an extensive test program had been carried out (Sousa 1983; Sousa et al. 1995). The existence of lamprophyre veins affected considerably the global behavior of the complex of the caverns and also the foundation of the arch dam.

The storage reservoir has a 63 Nm^3 capacity covering a 2.7 × 10^3 km^2 area with a length of about 270 km and a maximum width of 38 km. The studies for the project started in 1956, the construction started in 1971 and was completed in 1974. The first filling of the reservoir was carried out from December, 1974 to June, 1976, and the operation of the power scheme started in March, 1977.

Fig. 20.16 Cahora Bassa dam (HCB 2018)

The dam is set up in a narrow gorge, the cross section of this gorge presents in its lower part almost vertical slopes up to the proximity of the crest of the dam. It is an arch dam, with the following geometric characteristics (Fig. 20.16): 170 m maximum height, 300 m crest length, 1.54 chord-height ratio, a 4 m and 23 m thick central cantilever, respectively at the crest and at the base. There are eight spillways in the body of the dam, with a 14×10^3 m^3/s total runoff capacity for the maximum flood level and a surface spillway of the "volet" type at the central blocks of the dam.

The underground structures are located in the right bank of the river at a depth, which ranges, in the case of the powerhouse, from 130 m at one end to 230 m at the other. The Cahora Bassa system is the largest hydroelectric scheme in southern Africa with the powerhouse containing five 415 MGW turbines. In Fig. 20.17 the scheme of the underground works is presented. A physical 3D model was developed using a cube with 1.75 m in which the cavities were reproduced as illustrated in Fig. 20.18 (Silveira et al. 1974). Numerical simulations were done for this underground complex either using 2D finite element models (Sousa 1983) and more recently a 3D complex model using FLAC 3D (Fig. 20.19).

When dealing with soft rocks, one of the causes that can happen in rock masses is related with the occurrence of continuous surfaces with low strength formations (Rocha 1975). The existence of several lamprophyre veins influenced notoriously the behavior of the foundation of the dam and mainly was in the genesis of an accident during the construction of one of the surge chambers (Rocha 1975; Sousa 2006, 2010; Feng et al. 2012).

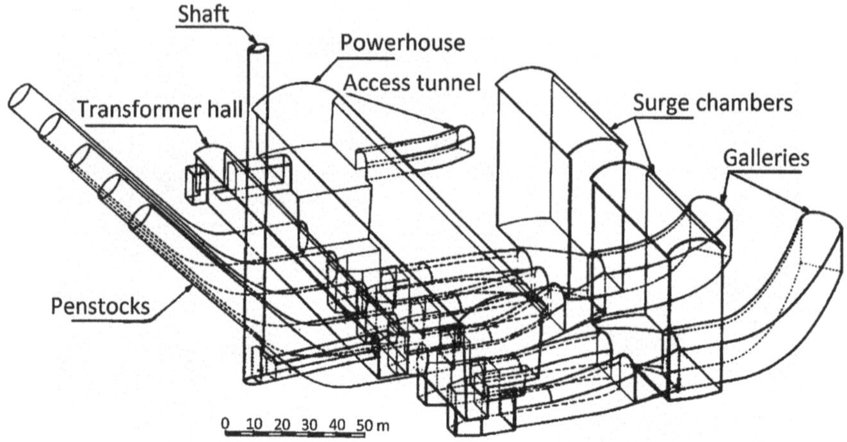

Fig. 20.17 Underground hydroelectric scheme of Cahora Bassa (Sousa 1983)

Fig. 20.18 3D physical model of Cahora Bassa underground cavern complex (Silveira et al. 1974)

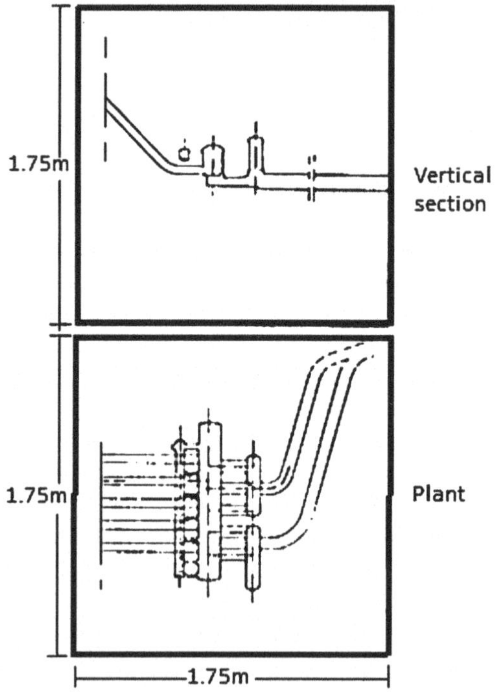

Figure 20.20 illustrates of a photograph of the accident. It consisted of a wedge type failure, with a volume of about 2000 m^3. Three major discontinuities, one sub-horizontal that consisted of a lamprophyre vein, not detected timely, and two inclined sets, as well as faults of lesser importance existed in the rock mass. The sub-horizontal lamprophyre dyke intersected a surge chamber in the ceiling and walls by gneiss formations. For the lamprophyre, based in in situ and laboratory tests, the average

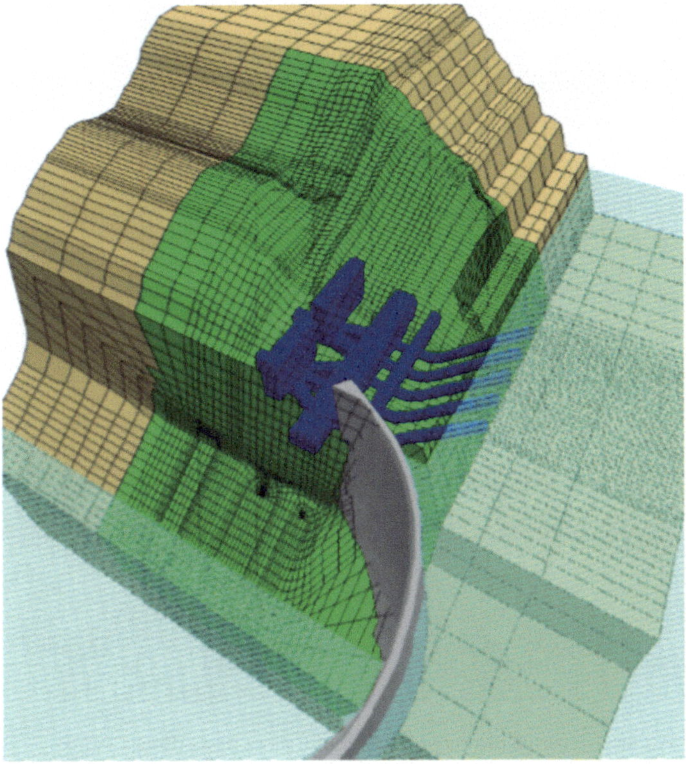

Fig. 20.19 3D numerical model from underground complex of Cahora Bassa (Lemos 2010)

strength parameters were $\phi = 20.3°$ and $c = 0.22$ MPa (see Table 20.7). The accident was due to a wedge failure that took place along the intersection line of two inclined discontinuities planes due to occurrence of the low strength surface with a very low friction angle. A sketch describing the accident is illustrated in Fig. 20.21.

20.5 Jinping II Hydroelectric Station. Deep Tunnels in Chlorite Schist

20.5.1 Project Overview

The Jinping II hydropower station is a large-scale diversion-type underground power station on the Yalong River that uses the river's 150 km river bend to help generate power (Fig. 20.1). The diversion tunnel's elevation was arranged as a gentle 3.65‰ slope. The depth of the overlying strata along the tunnel was 1500–2000 m, and the maximum buried depth was about 2525 m. Because of its significant depth and the long, large hole diameters of its tunnels, the Jinping station is considered an ultra-deep, long-tunnel, large underground hydropower project.

Fig. 20.20 Accident occurred in a surge chamber of Cahora Bassa (Rocha 1975)

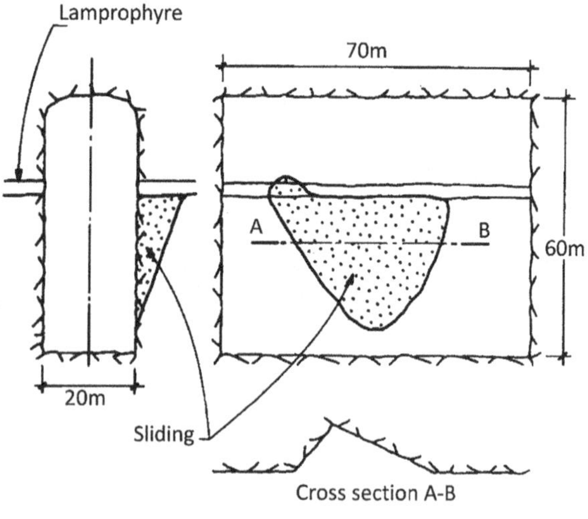

Fig. 20.21 Accident schematic (Sousa 2010)

During the excavation, diversion tunnels #1 and #2 to the west of the main diversion tunnel of the Jinping II hydropower station entered into the chlorite schist zone (T1 formation) from (1) 1 + 537 and (2) 1 + 613. The burial depth of the chlorite schist was between 1550 and 1850 m, and the gravity stress was about 41–50 MPa.

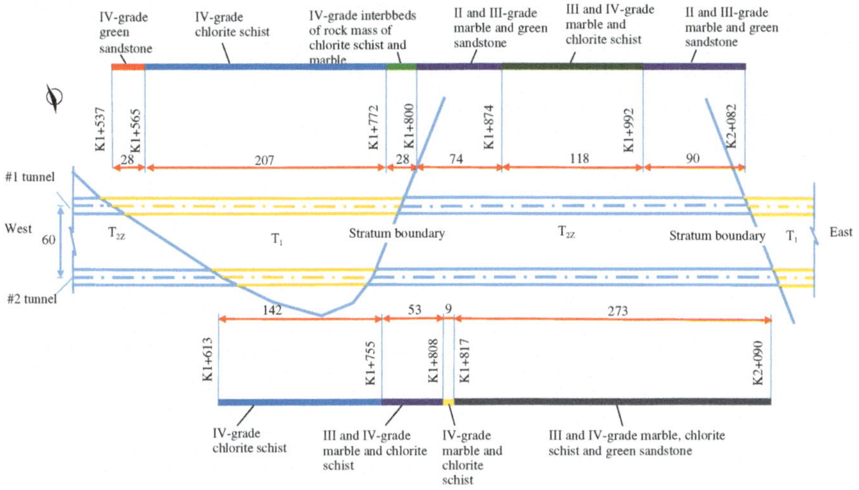

Fig. 20.22 Distribution of the chlorite schist exposed along the boreholes in tunnels #1 and #2

The geological conditions of this section were complex and the lithology changed frequently. Large-scale landslides and the large-scale deformation of the surrounding rock impacted the lining clearances in diversion tunnels #1 and #2 that were excavated to the chlorite schist. The deformation of the surrounding rock was still prevalent in the secondary expansion tunnel, and the local deformation was used in the design of the linear clearance.

As shown in Fig. 20.22, the T1 strata were the core of the Ponor anticline, and the strata on both sides consisted of the T2Z Zagunao group marble, which were integrated on contact. In addition, the T1 strata were non-soluble rock, the karstification degree was very weak, and the groundwater consisted mainly of a small amount of crack water. And the 2-rock contact zone lithology was often complex; the content of the green sandstone and chlorite schist in the T2Z strata increased significantly (usually occurring in striped or brecciated form) and large local lenses were formed. There was a series of NW folds that were strongly squeezed, distorted, and wrinkled.

20.5.2 Tests of the Physical and Mechanical Properties of the Chlorite Schist

20.5.2.1 Mineral Composition and Microstructure

The mineral compositions of the five chlorite schist blocks rock samples were obtained using the X-ray diffraction method, as shown in Table 20.8.

Figure 20.23 shows the results of the SEM scans of the five blocks. Except for IRSM-4#, the other samples had a similar mineral arrangement and pore structure

Table 20.8 Mineral composition and fractures characteristics in the chlorite schist of the Jinping II Station

Sample No.	Chlorite content (%)	Talc content (%)	Amphibole content (%)	Calcite content (%)	Dolomite content (%)	Fractures characteristics
IRSM-1#	46	3	51	0	0	Smooth, waxy luster, bright black particles entrainment
IRSM-2#	39	14	39	8	0	Granophyric with white marble granules or strip-like bodies
IRSM-3#	47	23	30	0	0	Powdery, unglaze, yellow chalcopyrite particle entrainment
IRSM-4#	39	36	25	0	0	Oily, not shiny, grey-green body and green body were mixed together
IRSM-5#	25	22	25	12	16	Satiny, grey-green, uniform

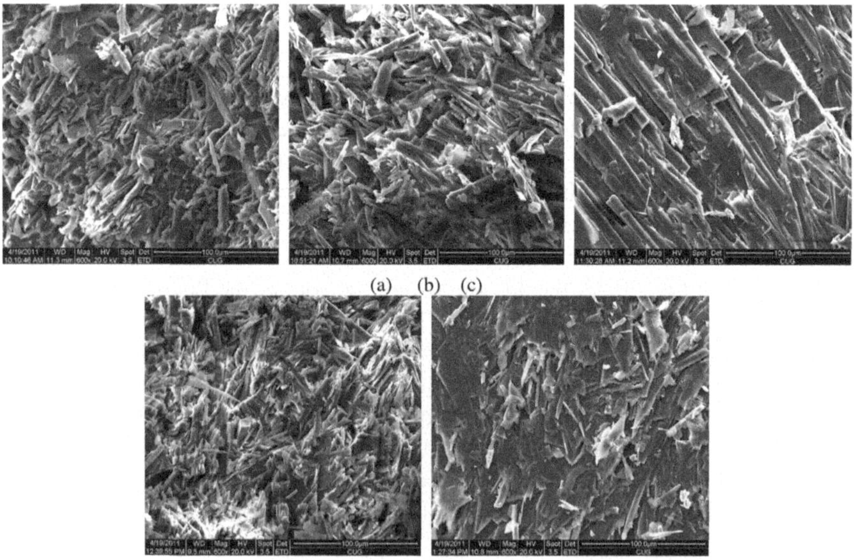

(a) (b) (c)

(d) (e)

Fig. 20.23 SEM results of the pore structure of the chlorite schist: (**a**) IRSM-1# specimen; (**b**) IRSM-2# specimen; (**c**) IRSM-3# specimen; (**d**) IRSM-4# specimen; (**e**) IRSM-5# specimen

characteristics, mainly in the lamellar-oriented layered arrangement. Chlorite and talc were generally flaky and acicular or fibrous amphibole. The calcite and dolomite mineral particles were massive. The orientation of the lamellar structure resulted in a very dense pore structure with a small porosity, which revealed the reason for the very low permeability of these rocks. The mineral composition of the

IRSM-4# sample only included chlorite, talc, and amphibole. The arrangement of the mineral particles was more chaotic than that in the other samples and the bridges between the particles formed pores, but the particle arrangement was still dense. As a result, the overall SEM scan results showed a greater porosity for IRSM-4# than for the other samples.

20.5.2.2 Physical Properties

A comprehensive physical test and expansion test of the chlorite schist was carried out on the chlorite schist, the test results are shown in Tables 20.9 and 20.10.

20.5.2.3 Mechanical Tests Under Dry and Saturation Conditions

Mechanical parameters of the chlorite schist under saturated and dry conditions were shown in Table 20.11.

Figure 20.24a shows that the peak intensity of the chlorite schist was sensitive to the confining pressure and conformed well to the linear characteristics of the Mohr–Coulomb strength criterion. Figure 20.24b shows that the elastic modulus of the chlorite schist increased with the confining pressure and showed a linear correlation under both conditions. The increase of the confining pressure compressed the rock sample structure and the compressive hardening effect of the elastic property was significant. However, the confining effect of the elastic modulus was relatively weak compared to the confining effect of the strength.

Table 20.9 Physical properties of the chlorite schist

Specimen number	Bulk density (g/cm^3)			Particle density (g/cm^3)	Natural moisture content (%)	Water absorption rate (%)	Water saturation (%)	Porosity (%)
	Natural	Dry	Saturation					
JPWL-1	2.73	2.71	2.78	2.93	0.97	2.34	2.78	7.52
JPWL-2	2.75	2.73	2.80	2.93	0.88	2.16	2.54	6.91
JPWL-3	2.74	2.72	2.79	2.93	0.92	2.24	2.68	7.28
Average	2.74	2.72	2.79	2.93	0.92	2.25	2.66	7.24

Table 20.10 Expansibility of the chlorite schist

Rock name	Degree of weathering	Free expansion rate (%)		Axial expansion of lateral confinement (%)	Expansion force (kPa)
		Axial	Radial		
Chlorite schist	Weak weathering	0.034	0.193	0.062	11.078
		0.154	0.101	0.120	19.436
		0.043	0.057	0.123	15.538
	Average value	0.077	0.117	0.102	15.351

Table 20.11 Mechanical parameters of the chlorite schist under saturated and dry conditions

Chlorite schist conditions	Average elastic modulus (GPa)	Uniaxial compressive strength (MPa)	Cohesive force (GPa)	Internal friction angle (°)	The maximum axial stress of schist in the creep test under different confining pressure (MPa)			
					0 MPa	5 MPa	10 MPa	20 MPa
Dry	13.18	38.8	10.39	28.02	75.0	70.0	140.7	84.9
Saturated	3.54	19.47	2.85	25.73	12.0	41.0	55.0	40.0
Softening coefficient	0.27	0.5	0.27	0.92	0.16	0.59	0.39	0.47
	Softening effect was prominent	A significant reduction	Softening effect was prominent	Not a significant reduction	The creep curves of the saturated and dry samples under different confining pressures were different, and the creep deformation was significant.			

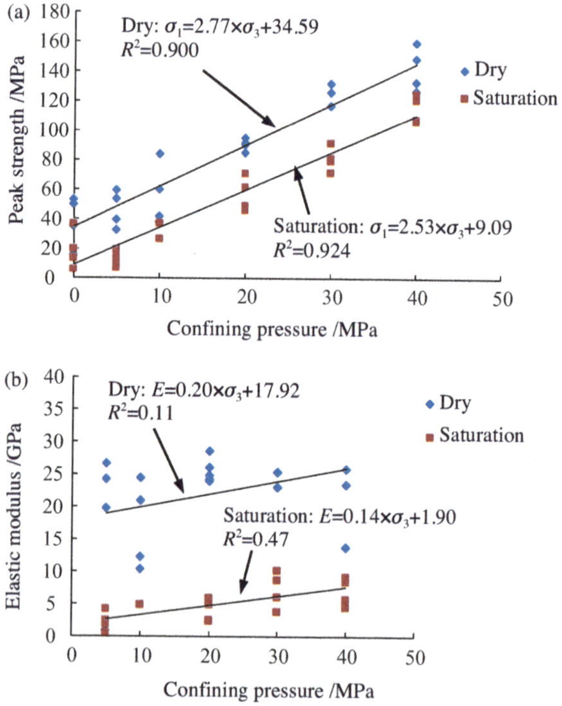

Fig. 20.24 Relationship between the peak intensity or elastic modulus and confining pressure of the chlorite schist under dry and saturated conditions: (**a**) The peak intensity—confining pressure; (**b**) The elastic modulus—confining pressure

20.5.2.4 Summary of the Mechanical Properties of the Chlorite Schist

Summary of the mechanical properties of the chlorite schist were shown in Table 20.12.

Table 20.12 Summary of the mechanical properties of the chlorite schist

Mechanical properties	Influence on the chlorite schist	Project problems	controlling means
Small elastic modulus	Strong deformability	Large deformation problems	A strong surface support
Low strength	Lower strength under a low confining pressure at the end of the unloading path	The unloading led to the destruction of the surrounding rock	Strength the broken rock
Water on the deformation properties	The softening coefficient of the elastic modulus increased with the increase of the confining pressure	Lower permeability coefficient	Control the water entering the rock
Water on the strength properties	The strength softening coefficient of the strength was smaller	The anchoring force of the anchor and rock mass decreased rapidly	Limit the development of the deformation and providing a certain confining pressure

20.5.3 Elastic-Plastic Constitutive Model of the Chlorite Schist

20.5.3.1 Definition of the Internal Variable

The mechanical properties of the chlorite schist varied with the plastic deformation and were affected by the confining pressure. In order to reflect this characteristic, it was necessary to revise the definition of the internal variable (usually the plastic strain) in the traditional plastic mechanics formula to include the influencing factors of the confining pressure.

Under general conditions, the internal variable κ is defined as

$$\begin{cases} \kappa = \int \mathrm{d}\kappa \\ \mathrm{d}\kappa = \dfrac{\sqrt{\dfrac{2}{3}\mathrm{d}e_{ij}^{\mathrm{p}}\mathrm{d}e_{ij}^{\mathrm{p}}}}{f\left(\dfrac{I_1}{\sigma_{\mathrm{c}}}\right)} \end{cases} \tag{20.1}$$

where $\mathrm{d}e_{ij}^{\mathrm{p}}$ is the plastic strain tensor, defined as $\mathrm{d}e_{ij}^{\mathrm{p}} = \mathrm{d}\varepsilon_{ij}^{\mathrm{p}} - \frac{1}{3}\mathrm{tr}\left(\mathrm{d}\varepsilon_{ij}^{\mathrm{p}}\right)I$; $\sqrt{\frac{2}{3}\mathrm{d}e_{ij}^{\mathrm{p}}\mathrm{d}e_{ij}^{\mathrm{p}}}$ is the equivalent plastic shear strain increment, recorded as $\mathrm{d}\gamma^{\mathrm{p}}$; $f(I_1/\sigma_{\mathrm{c}})$ is a function of the first invariant I_1 of the stress tensor, and the uniaxial compressive strength σ_{c} is introduced in order to make it dimensionless.

According to the conventional triaxial test results of the chlorite schist under different confining pressures, the expression of $f(I_1/\sigma_{\mathrm{c}})$ is

$$f\left(\frac{I_1}{\sigma_{\mathrm{c}}}\right) = 0.0015\left(\frac{I_1}{\sigma_{\mathrm{c}}}\right) + 0.0003 \tag{20.2}$$

We used Eq. (20.1) to calculate the value of the corresponding internal stress κ under different confining pressures. We found that κ was about 0.1, so $\kappa = 0.1$ was used as the criterion for the peak stress.

20.5.3.2 Mathematical Description of the Evolution Rule of the Strength Parameters

Using the above-defined internal variables, we incorporated the triaxial test results into Eq. (20.1), and after the dκ superposition, each κ corresponded to a group (σ_1, σ_2, σ_3). Since the pressure was assumed to be positive, the formula of the Mohr–Coulomb yield criterion with the first principal stress σ_1 and third principal stress σ_3 is

$$\sigma_1 = \sigma_3 \frac{1+\sin\phi}{1-\sin\phi} + 2c\frac{\cos\phi}{1-\sin\phi} \tag{20.3}$$

where c and ϕ are the cohesion and internal friction angle of the rock, respectively.

Given the values of the two groups (σ_1, σ_3), the value of one group (c, ϕ) can be solved, and so the variation curve of the intensity parameter with the internal variable κ can be plotted. And he following formula was based on the intensity parameter evolution feature:

$$\begin{cases} c(\kappa) = \dfrac{c_t - c_0}{\kappa_t}\kappa + c_0 & (0 \le \kappa \le \kappa_t) \\ c(\kappa) = \dfrac{c_t - c_r}{(\kappa_t - 1)^2}(\kappa - 1)^2 + c_r & (\kappa_t < \kappa \le 1) \end{cases} \tag{20.4}$$

$$\phi(\kappa) = (\phi_0 - \phi_r)(\kappa - 1)^2 + \phi_r \quad (0 \le \kappa \le 1) \tag{20.5}$$

20.5.3.3 Elastic-Plastic Constitutive Model

The test results for the chlorite schist showed that the shear deformation was not prominent during the shear failure process. If the associated flow rule was used in the constitutive model, it would lead to an unrealistically large dilatancy deformation. For this reason, the non-associated flow law was used here, and the dilatancy angle ψ was substituted for the internal friction angle ϕ in the yield function for the plastic potential function. We also studied the evolution rule of the dilatancy angle with the equivalent plastic shear strain. The plastic potential function is defined as follows:

$$\begin{cases} g_1 = \sigma_1 - \sigma_2 N_\psi - 2c\sqrt{N_\psi} \\ g_2 = \sigma_1 - \sigma_3 N_\psi - 2c\sqrt{N_\psi} \end{cases} \tag{20.6}$$

where N_ψ is a function of the dilatancy angle ψ, and $N_\psi = (1 + \sin\psi)/(1 - \sin\psi)$.

Assuming that it follows the non-associated flow rule, then

$$\mathrm{d}\varepsilon^{\mathrm{p}} = \mathrm{d}\lambda_1 \frac{\partial g_1}{\partial \sigma} + \mathrm{d}\lambda_2 \frac{\partial g_2}{\partial \sigma} \tag{20.7}$$

Then, the dilatancy angle ψ is

$$\psi = \arcsin \frac{\mathrm{d}\varepsilon_{\mathrm{v}}^{\mathrm{p}}}{-2\mathrm{d}\varepsilon_1^{\mathrm{p}} + \mathrm{d}\varepsilon_{\mathrm{v}}^{\mathrm{p}}} \tag{20.8}$$

So the function relationship of the plastic volume strain $\varepsilon_{\mathrm{v}}^{\mathrm{p}}$, the first plastic principal strain $\varepsilon_1^{\mathrm{p}}$, and the equivalent plastic shear strain γ^{p} were obtained through collation. The dilatancy angle was calculated under different equivalent plastic shear strains. The corresponding relationship between the equivalent plastic shear strain and the variables was obtained through Eq. (20.1). The dilatancy angle corresponding to different internal variables was obtained through interpolation. The function of the dilatation angle that changes with the variable was obtained using the above analysis, which yielded

$$\psi(\kappa) = (\phi_{\mathrm{r}} - \phi_0)(\kappa - 1)^2 + \phi_0 \quad (0 \le \kappa \le 1) \tag{20.9}$$

20.5.3.4 Verification of the Model Through Numerical Simulation Method

According to the numerical calculation of the chlorite schist triaxial test, the simulated stress-strain curve reflects the transformation process of the chlorite schist from a brittle fracture at a low confining pressure to a ductile failure at a high confining pressure and the mechanical characteristics of the rock like the dilatancy property, the hardening before the peak value, and the softening after the peak value.

20.5.4 Deformation Characteristics and Support Principles

20.5.4.1 Deformation Characteristics of the Surrounding Rock Mass in the Tunnel Section Buried in the Chlorite Schist

After receiving some excavation support, laser scanning was carried out once again on the cross section. The deformation of the section was assessed by comparing it to the design excavation section.

The results show that the large deformation problem in the (1) 1 + 635 to 1 + 800 section of tunnel #1 occurred mainly on the north side arch. The distribution laws of the maximum deformation location of tunnels #1 and #2 are approximate. However, the maximum deformation of tunnel #1 was more inclined to the north side wall, and there were more sections of the largest deformation occurring in the south side of the skewback, while the maximum deformations of tunnel #2 mainly occurred on the north side of the arch. There were greater deformations from the side wall to the top arch, and most of the maximum deformations occurred on the north side of the arch section, the values of which are between 0.2 and 0.6 m.

20.5.4.2 Deformation Level Evaluation

The deformation of the excavated section indicated that the maximum deformation of multiple sections of tunnels #1 and #2 was about 0.5–0.7 m after the initial support, which was 7.6–10.6% of the tunnels' radius. A significant extrusion deformation was observed from the site. The deformation range of the Jinping diversion tunnel corresponding to different deformation degrees was calculated according to the Hoek extrusion deformation evaluation method. Then, the degree of extrusion deformation was evaluated in Fig. 20.25.

20.5.4.3 Support Principles and Measures Based on the Tests and Field Responses

The support principles and specific measures of the tunnels were developed using the mechanical properties of the chlorite schist revealed by the tests and the evaluation results of the deformation degree of the surrounding rocks. The relationship between the support principles, the supporting measures, and the engineering mechanical properties of the chlorite schist is shown in Fig. 20.26.

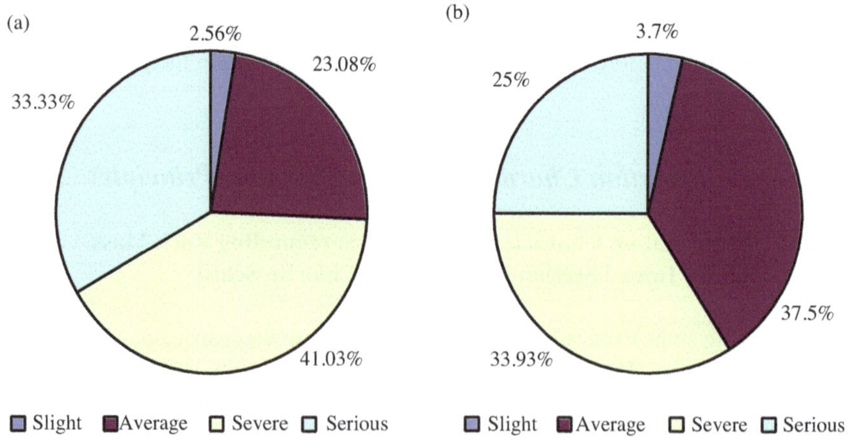

Fig. 20.25 Proportion of the different degrees of the extrusion deformation at the cross section: (**a**) tunnel #1; (**b**) tunnel #2

Basic engineering mechanical properties →	Support principles →	Supporting measures
□ The elastic modulus is small and very sensitive to the confining pressure; □ The elastic modulus of the unloading confining pressure path is about 50% lower than that of the conventional loading path; □ The strength is very low; □ The strength of the unloading confining pressure path reduces with the confining pressure increasing and decrease percentage is related to the confining pressure; □ Due to the rock softening with water, the strength and elastic modulus both reduce significantly so that the softening coefficient is very sensitive to the confining pressure; □ The long-term strength is low, about 15~29% of instantaneous strength.	□ Control the loss of the confining pressure in the surrounding rock or restore a specific confining pressure to control the reduction of the elastic modulus and strength; □ Reinforce the broken rock and control its deformation to increase the bearing capacity of the rock mass; □ Prevent the entrance of water and provide a specific confining pressure to control the reduction of the strength and elastic modulus of the surrounding rock after it encounters water; □ The support strength must take into consideration the attenuation of the long-term strength and growth of the long-term deformation.	□ Advanced and small pipe grouting is enforced and firmly weld with the grid arch; □ Advance bolt plays a role in strengthening rock mass; □ Reinforce the tunnel face rock body; □ Timely closure of the surface and the support principle of soft first and rigid after and suitable release-pressure are required; □ Prevent water by grouting; □ Pay attention to long-term creep effect; □ Monitoring feedback and reinforcement treatment should be carried out.

Fig. 20.26 The supporting principle of the chlorite schist and determination of the support measures

20.5.5 Stability Control Technologies and Support Optimization

20.5.5.1 Stability Control Principles of Squeezing the Tunnel Section

There was a degree of extrusion deformation or "shrinkage" problems in the upper layer of the excavation and support section of the deep-buried chlorite schist rock section at the Jinping II Hydropower Station. After the "shrinkage," the section "invaded" the secondary lining section. And the stability of the surrounding rock was poor. Figure 20.27 shows that when the follow-up rock cutting excavation were being taken place, the surrounding rock of the excavated upper layer was deformed about 0.1 m, which led to the further deterioration of the upper-stage bolt state and the surrounding rock properties. The initial support partially lost the vertical support of the footing after the further cracking of the support system and the excavation of the lower area. In addition, after the end of the rock cutting excavation, the bulging deformation of the cutting-down arch was significant

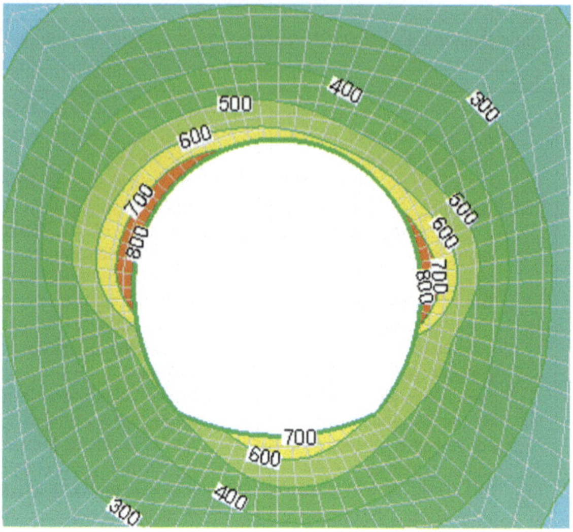

Fig. 20.27 The final deformation of the segment with a severe deformation without further reinforcement

(floor heave might occur). Therefore, before the end of the excavation, the design principles of reinforcement were to ensure the stability of the cross section and control the possible floor heave.

20.5.5.2 Support Parameters Optimization

Since the largest deformation of the "shrinkage" section mainly occurred in the spandrel and skewback, the reinforcement was focused on these two parts, especially as the support of the upper half of the section. The stability of the skewback was of great importance to the stability of the upper surrounding rock.

Based on the analysis of the reinforcement scheme before the rock cutting excavation (as well as other factors like the reinforcement effect, feasibility of construction, construction period, construction interference, and project investment), the chosen Scheme was as follows:

> the waist-lock foot-lock bolt of the replacement arch (Φ32, @ = 0.5 m, $L = 9$ m) doubles as a pre-support bolt of the arch shoulder and arch foot. Lead anchor piles (3Φ32, @ = 1.0 m, $L = 9$ m) were arranged at the arch foot. The pre-stressed anchor bolt (Φ32, $T = 150$ kN, @ = 1.5 m, $L = 9$ m) was arranged in the middle part of the horizontal wall. And the pre-stressed anchors with horizontal pre-stressed anchor bars in the central part of the side wall were adjusted to the pre-stressed anchor cables with $T = 1000$ kN, @ = 3 m, and $L = 15$ m to provide a better constraint expansion of the displacement field caused by the rock cutting excavation process and to enhance the pre-support effect.

20.5.5.3 Optimization of the Excavation Schemes of the Lower Bench

The deformation of the surrounding rock in the following three schemes (Fig. 20.27) was analyzed and compared:

Scheme 1: A full-section rock cutting excavation. Set up the grille arch after cutting the rock. The system bolt lags behind the tunnel face for a certain distance, and the mortar requires a certain amount of time to gain strength. Therefore, the system bolt support effect did not need to be considered in this scheme.

Scheme 2: The left and right frames are cut down separately, that is, the left frame should be excavated first and the right frame should be excavated afterwards. The grille arch should be set up after the left and right frames are cut downed. The effect of bolting was not considered either.

Scheme 3: The middle slot is cut down, that is, the left frame, right frame, and middle slot are excavated at the same time while excavating the lower half. Only the support effect of the grille arch is considered.

The comparison of the deformation distribution obtained by the mechanical simulation showed that the order of the schemes from the best to the worst is: full section, middle slot, and half frame. The deformation stability of the full-face excavation support obtained through the mechanical simulations was the best, which was mainly because the numerical simulation of the actual excavation and support process had been simplified. From the perspective of the stability control of the construction process and the results of the mechanical analysis, there was no significant difference between the mid-span scheme and the half-plan scheme. The results of the mechanical analysis show that the deformation and stability of the surrounding rock on both sides of the tunnel were the focus of the monitoring and construction control during the rock cutting excavation, no matter which scheme was adopted.

20.5.6 Design Method of Line in Consideration of Operation Circumstance

The chlorite schist had a low strength and a large deformation; it became soft in water and its rheological effects were remarkable. The problem of the extrusion deformation after the excavation and during the reinforcement process was serious. The extrusion deformation and control factors revealed during construction were undoubtedly a great threat to the long-term safety of the lining during operation. Therefore, the long-term stability of the deep-buried chlorite schist tunnel section lining is a key controlling factor for the long-term safety of the Jinping II Hydropower Station diversion tunnel project.

So the visco-elasto-plastic rheological constitutive model was used, and the elasto-plastic and rheological problems as well as their coupling effects were considered in the analysis of the fluid–solid coupling of the rheology.

20.5.6.1 Rheological Mechanical Model

In consideration of the plastic and long-term mechanical behavior, the rheological characteristics of the surrounding rock and the results of the triaxial rheological test of the chlorite schist and on-site monitoring date. We utilized the H-K rheological model, the Mohr–Coulomb yield criterion, and the visco-elastic-plastic rheological model (CVISC) in our evaluation.

The CVISC model is composed of the Burgers rheological model and Mohr–Coulomb yield criterion. The model diagram is shown in Fig. 20.28.

20.5.6.2 Rheological Inversion Analysis of the Mechanical Parameters

Given the mechanical response of the rock mass after the secondary excavation, the mechanical parameters of the surrounding rock support complex had to be inverted according to the monitoring data after the secondary excavation and reinforcement. The parameters obtained by the inversion are shown in Table 20.13.

The displacement curve of the monitoring point was obtained by putting the parameters from the inversion into the model and comparing it with the measured curve, which is shown is Fig. 20.29. The dashed line represents the monitored displacement curve and the solid line represents the displacement curve of the corresponding monitoring point. The agreement between the two curves is quite good, indicating that the mechanical model used in the calculation and the mechanical parameters obtained by the inversion had a good reliability.

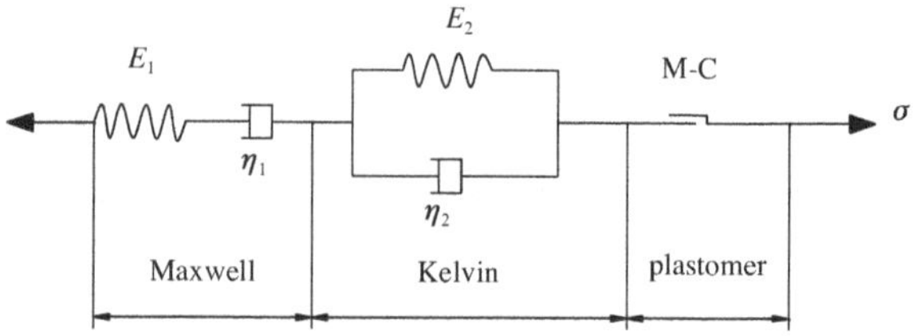

Fig. 20.28 Schematic graph of the CVISC model

Table 20.13 Values of the equivalent rheological parameters

	Kelvin shear modulus G^K (MPa)	Kelvin viscosity η^K (MPa day)	Maxwell shear modulus G^M (MPa)	Maxwell viscosity η^M (MPa day)	Bulk modulus K (MPa)	Cohesion c (MPa)	Internal friction angle ϕ (°)
Original rock	2.0e4	1.25e6	2.72e7	1.12e11	3.63e3	2.60	36.02
Reinforcement zone	2.0e4	1.25e6	1.70e7	1.00e11	2.27e3	2.00	28.97

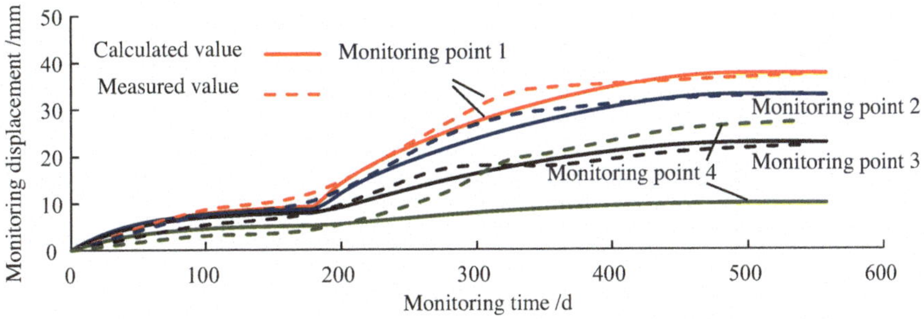

Fig. 20.29 Comparison between the calculated and measured curves of the monitoring position of the right haunch (Yang et al. 2017)

20.5.6.3 Stability Analysis and Evaluation of the Reinforced Concrete Line in Complex Operating Circumstances

A numerical simulation of the possible environmental conditions during the different construction stages and operation periods was carried out. The stresses of the lining concrete and steel bars were analyzed, and the safety of the lining concrete and steel bars during the operation period was evaluated. Such as Initial mechanical situation after pouring the lining, lining safety under long-term water infiltration conditions, safety of the lining and surrounding rock considering the softening of the intrusive rock mass and the stress distribution of the surrounding rock calculated before and after the long-term operation of the lining. It was concluded that the current lining design scheme was reasonable considering the environmental conditions.

20.6 Conclusion

This chapter is developed by the authors taking into consideration the experience in underground hydroelectric schemes where the existence of soft rocks occurred, particularly in the Socorridos scheme in Madeira Island, Mingtan in Taiwan, Cahora-Bassa in Mozambique, and Jinping II in China. The purpose was to analyze the problems that raised from the existence of soft rocks formations with emphasis to the in situ tests performed for predicting the behavior and the risk associated. Special emphasis is given to the initial considerations devoted to the different types of underground structures and their risk analysis and management.

Acknowledgements A special acknowledgement is made to Chunsheng Zhang, Xiangrong Chen, Jing Hou, and Yang Zhang for the contribution to the collection of the field information about the deep headrace tunnels at the Jinping II hydropower station.

A special acknowledgement is made to Paulo Cafofo for his contribution to the study of Socorridos hydroelectric scheme.

References

ASCE (1996) Rock foundations (Technical Engineering and Design Guides as adapted from US Army Corps of Engineers, no. 16). American Society of Civil Engineers, New York, p 129

Barton N (2000) TBM tunneling in jointed and faulted rock. Balkema, Rotterdam, p 172

Bieniawski Z (1989) Engineering rock mass classification. US Corps of Engineering, Technical Report GL-799-19. WES, Vicksburg

Cafofo P (2006) Underground works of the Socorridos scheme. Report for Geotechnical Design discipline. University of Minho, Guimaraes, p 131. (in Portuguese)

Cafofo P, Sousa LR (2007) Innovative underground works at Socorridos, Madeira Island, Portugal. Workshop on Volcanic Rocks. Taylor and Francis Group, Ponta Delgada, pp 73–80

Cafofo P, Sousa LR, Moura F (2007) In: Kaliampakos D, Bernardos A (eds) Innovative underground works at Madeira Island, Portugal. 11th ACUUS Conference on Underground Space: Expanding the Frontiers. NTUA Press, Athens, pp 137–143

Einstein H (2002) Risk assessment and management in geotechnical engineering. 8th Portuguese Geotechnical Congress, Lisbon, pp 2237–2262

Eskesen S, Tengborg P, Kampmann J, Veicherts T (2004) Guidelines for tunnelling risk management. International Tunnelling Association, Working Group No. 2. Tunn Undergr Space Technol 19:217–237

Feng XT, Jiang Q, Sousa LR, Miranda T (2012) Underground hydroelectric power-schemes. In: Sousa LM, Vargas EA, Fernandes MM, Azevedo RF (eds) Innovative numerical modeling in geomechanics. Taylor & Francis, London, pp 13–50

Geoguide 4 (1992) Guide to cavern engineering. Geotechnical Engineering Office, Hong Kong, p 147

HCB (2018) Hydroelectric of Cahora Bassa. www.hcb.co.mz

He M (2014) Latest progress of soft rock mechanics and engineering in China. J Rock Mech Geotech Eng 6:165–179

He M, Sousa LR, Sousa RL, Gomes A, Vargas E, Zhang N (2011) Risk assessment of CO_2 injection processes and storage in carboniferous formations: a review. J Rock Mech Geotech Eng 3(1):39–56

Hoek E (2007a) Practical rock engineering. www.rocscience.com

Hoek E (2007b) Big trends in bad rock. Therzaghi lecture. ASCE J Geotech Geoenviron Eng 127(9):726–740

Kanji MA (2014) Critical issues in soft rocks. J Rock Mech Geotech Eng 6:186–195

Lamas LN (1993) Contributions to understanding the hydromechanical behavior of pressure tunnels. PhD Thesis. Imperial College, London, p 419

Lemos JV (2010) Modeling rock masses in large underground works. In: Cruz, Cunha, Maia, Pinto (eds) Conference on Hydroelectric Schemes in Portugal. A New Cycle. University of Porto, Porto, pp 143–155. (in Portuguese)

Menezes AT, Varela FM, Sousa LR, Moura F (2007) Geomechanical studies for a road tunnel in volcanic formations. ITA Congress, Prague

Miranda T, Sousa LR, Gomes AT, Tinoco J, Ferreira C (2018) Volcanic rocks geomechanical characterization by using empiric systems. J Rock Mech Geotech Eng 10(2018):138–150

Moura F, Sousa LR (2007) Road tunnels at Madeira Island, Portugal. In: Malheiro AM, Nunes JC (eds) Workshop on volcanic rocks. Taylor and Francis Group, Ponta Delgada, pp 201–206

Nuss L, Hansen K, Marsumoto N (2017) Earthquake performance of concrete dams. Seismic Analysis of Concrete Dams Workshop, USSD Conference, New Orleans, PPT presentation

Popielak R, Weining W (2010) Engineering and design services for excavation – DUSEL. Preliminary Design, Preliminary Report #2, Contract D10-04, Lakewood, Colorado, p 104

Rocha M (1975) Some problems regarding rock mechanics of low strength materials. V Pan-American Congress on Soil Mechanics and Foundation Engineering, Buenos Aires, pp 489–514. (in Portuguese)

Rocha M (1976) Underground structures. LNEC, Lisbon. (in Portuguese)

Rocha M (1978) Analysis and design of the foundation of concrete dams. ISRM Symp. on Rock Mechanics Applied to Dam Foundations, Rio de Janeiro, vol 3, pp 3.11–3.70

Rodrigues FP, Grossman NF, Rodrigues LF (1978) Rock mechanics tests of the Mingtan pumped storage project. LNEC report. LNEC, Lisbon. 2 Vol

Schleiss A (2000) Hydraulic schemes. EFPL, Lausanne, p 482. (in French)

Serafim L, Pereira P (1983) Considerations on the geomechanical classification of Bieniawski. Int. Symp. Engineering Geology and Underground Construction, Lisbon, vol 1, pp 33–42

Silveira A, Azevedo M, Costa P (1974) Contribution to the study of the underground powerhouse of Cahora Bassa. LNEC, Memory no. 430. LNEC, Lisbon. (in Portuguese)

Sousa LR (1976) Modern methods for design of tunnels – mathematical models. J Geotecnia (16):51–81. (in Portuguese)

Sousa LR (1983) Three-dimensional analysis of large underground power stations. ISRM Congress, Melbourne

Sousa LR (2006) Learning with accidents and damage associated to underground works. In: Matos AC, Sousa LR, Kleberger J, Pinto PL (eds) Geotechnical risks in rock tunnels. Taylor & Francis, London, pp 7–39

Sousa RL (2010) Risk analysis for tunneling projects. MIT, PhD Thesis, Cambridge, p 589

Sousa LR, Lamas LN, Martins CS (1994) Applications of computational mechanics to underground structures in hydraulic projects. In: Vargas E, Azevedo R, Sousa LR, Fernandes MM (eds) 1st Int. Workshop on Applications of Computational Mechanics in Geotechnical Engineering. Balkema, Rio de Janeiro, pp 15–88

Sousa LR, Ramos JM, Silva HS (1995) Cahora-Bassa hydroelectric scheme. A new monitoring plan. 8th ISRM Congress, Tokyo

Sousa LR, Chapman D, Miranda T (2010) Deep rock foundations of skyscrapers. J Soil Rock 33(1):3–22

Sousa LR, Sousa RL, Zhou C, Karam K (2018) Methodology for evaluating geomechanical properties of soft rock masses by laboratory and in situ tests. In: Soft rock masses. Springer, New York, p 33

Wu SY, Feng X, Sousa LR (2010) Jinping II mega hydropower project, China. Int. Conf. on Hydroelectric Schemes in Portugal. A new cycle, Porto, pp 223–231

Wyllie D (1999) Foundations on rock. E & FN Spon, London, p 401

Yang F, Zhang C, Zhou H, Liu N, Zhang Y, Azhar MU, Dai F (2017) The long-term safety of a deeply buried soft rock tunnel lining under inside-to-outside seepage conditions. Tunn Undergr Space Technol 67:132–146

Zhang CQ, Zhou H, Feng XT (2011) An index for estimating the stability of brittle surrounding rock mass—FAI and its engineering application. Rock Mech Rock Eng 44:401–414

Chapter 21
Tunnelling in Weak Rock

Nicholas Vlachopoulos, Bradley Forbes, and Ioannis Vazaios

21.1 Introduction

Tunnel design and construction has gone through major advancements in excavation techniques with the use of enhanced numerical analysis methods (hardware and software), more accurate rock mass classification systems, hi-tech machineries, more tunnel construction and improved ground reinforcement and improvement techniques. In this way, all of these factors have aided in a better understanding of the tunnel mechanics and better stabilization of the tunnel face, limiting the possibilities of complete tunnel collapse.

Classical tunnel designs have been based on the Rock Mass Ratio (RMR) (designing with respect to deformations) and Terzaghi based (designing primarily to support all loads including overburden pressure by the final lining). A newer tunnelling method such as the New Austrian Tunnelling Method (NATM) incorporates an observational approach that is deformation based. This method integrates the surrounding rock into the overall support structure (i.e., the supporting formations will themselves be a part of the supporting structure as the rock is able to support itself to a certain degree) (Romeo 2002). Using NATM, a controlled deformation of rock mass is permitted (a limited closure rate of approximately 1%) and this gives the stresses an opportunity to be partly released and less stiff. Therefore, a less-expensive support system can be used (Kontogianni and Stiros 2003).

N. Vlachopoulos (✉)
Royal Military College of Canada, Kingston, ON, Canada
e-mail: vlachopoulos-n@rmc.ca

B. Forbes
Queen's University, Kingston, ON, Canada

I. Vazaios
Arup, London, UK

M. Kanji et al. (eds.), *Soft Rock Mechanics and Engineering*,
https://doi.org/10.1007/978-3-030-29477-9_21

This chapter will investigate the specific design and tunnelling techniques that are used with respect to tunnelling within weak rock masses. The empirical determination of stresses using the convergence-confinement method (Carranza-Torres and Fairhurst 2000) will be elaborated upon as an initial idealization of the displacement and stress determinations within a rock mass. Tunnel design and support considerations will then be highlighted with emphasis on NATM. Finally, the support classes and design of Driskos tunnel is included with relevant instrumentation.

21.2 Method of Stress Analysis

In order to understand the issues that arise when attempting to design tunnel support, it is necessary to examine some rudimentary concepts of how a rock mass surrounding an excavation behaves and how (and when) the tunnel support can be introduced into this system.

A simplistic closed-form solution model for the analysis of tunnel behavior is a circular tunnel which has been excavated. The mechanics from such a model will allow one to develop an elementary understanding of basis tunnel behavior. In order to set up this model, we assume that we have a tunnel of radius r_i which is subjected to a hydrostatic stress p_o as well as an internal (uniform) support pressure p_i (as seen in Fig. 21.1 below).

The radius to the outer boundary of the plastic zone is denoted by r_o while the internal radius of the excavated portion is r_i. If one assumes that the rock will fail with no plastic volume change, the critical stress level at failure is denoted by a critical pressure, p_{cr}:

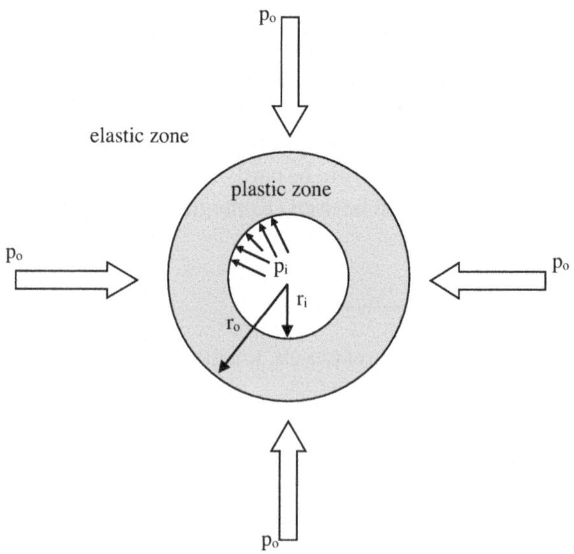

Fig. 21.1 Simplistic mechanics associated with circular tunnel excavation

$$p_{cr} = \frac{2p_o - \sigma_{cm}}{1+k} \tag{21.1}$$

where

p_{cr}: critical pressure

p_o: external, hydrostatic stress (pressure)

σ_{cm}: the uniaxial compressive strength of the rock mass

k: lateral earth pressure

(i.e., $k = (1 + \sin \phi)/(1 - \sin \phi)$; where ϕ is the friction angle of the rock mass)

Hoek undertook a study whereby he tried to determine the strength and deformation characteristics of a rock mass associated with behavior of the rock mass surrounding the tunnel (Hoek 2003). He created dimensionless plots from the results of parametric studies; a Monte Carlo analysis in which the input parameters for rock mass strength and tunnel deformation were varied at random in 2000 iterations. These plots (Fig. 21.2) are used not only to provide insight into potential tunnel behavior, but as the basis of indicating what type of tunnel support is required.

Note that once the rock mass strength falls below 20% of the rock mass strength to in situ stress level ratio, deformations increase substantially. Unless control measures are taken, it is predicted that the tunnel will most likely collapse (Fig. 21.3).

The ultimate goal of the determination of rock mass behavior is to establish what type of support is required for a particular behaviour (strain). Where a particular rock mass plots with respect to the strain versus rock mass strength to in situ stress (Fig. 21.4a) determines potential support requirements (Fig. 21.4b).

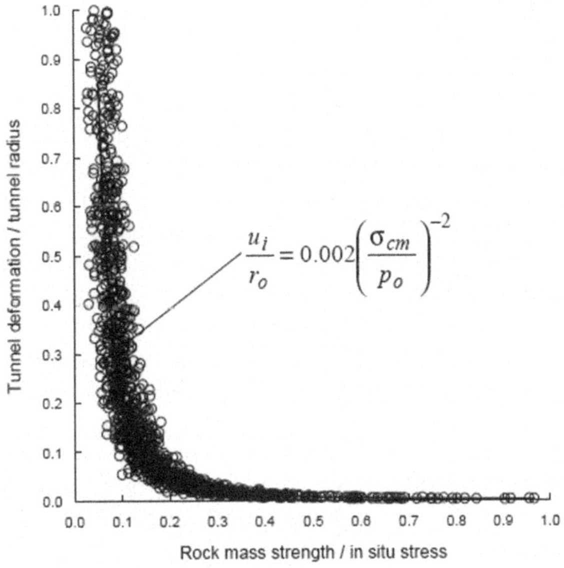

Fig. 21.2 Normalized tunnel deformation versus ratio of rock mass strength to in situ stress (Hoek 2003)

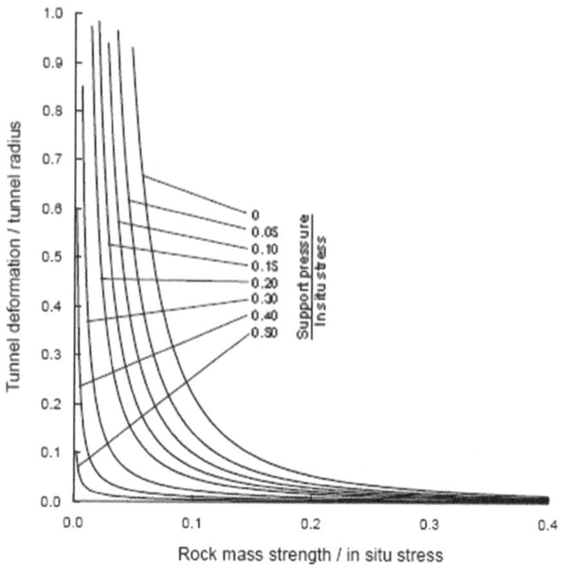

Fig. 21.3 Ratio of tunnel deformation to tunnel radius versus ratio of rock mass strength to in situ stress for various support pressures (Hoek 2003)

21.2.1 Convergence-Confinement Method

The convergence-confinement method yields an estimate of the load imposed on the support immediately behind the newly excavated face of a tunnel (i.e., upon excavation, the area immediately behind the face is partially supported by the face itself and does not carry the full load of an open cavity at that moment). As the tunnel and face advance with further excavation, the support must carry a greater portion of the load and eventually the entire load when the face has moved well away from the support (Carranza-Torres and Fairhurst 2000). Idealistically, this problem can be illustrated as shown in Fig. 21.5.

Two important assumptions to this problem (and of the convergence confinement method) is that it assumes a uniform or hydrostatic stress field, p_o or σ_o (i.e., deformation occurs in a plane perpendicular to the axis of the tunnel—2D plane strain conditions) and the tunnel cross section is circular of radius R. A support is placed at section A-A′ (Fig. 21.5a), located a distance L from the tunnel face. The problem has been idealized in such a way as to determine the loads that will be transmitted to the support once the excavation has progressed (i.e., L increasing) and the support provided by the face does not affect the supported region (A-A′) any longer. As seen in Fig. 21.5b, excavation of the rock mass (assume elastic medium) produces a failed rock zone around the cavity. The farfield hydrostatic stress is denoted by σ_o radial displacement u_r and the pressure p_i (i.e., reaction of support on the walls of the tunnel). Figure 21.5c shows a cross section at section A-A′. The support is of thickness t_c and the external radius is R or r_o, the uniform pressure, p_s is the load transmitted by the rock mass to the support. The radial displacement u_r is the displacement due to p_s.

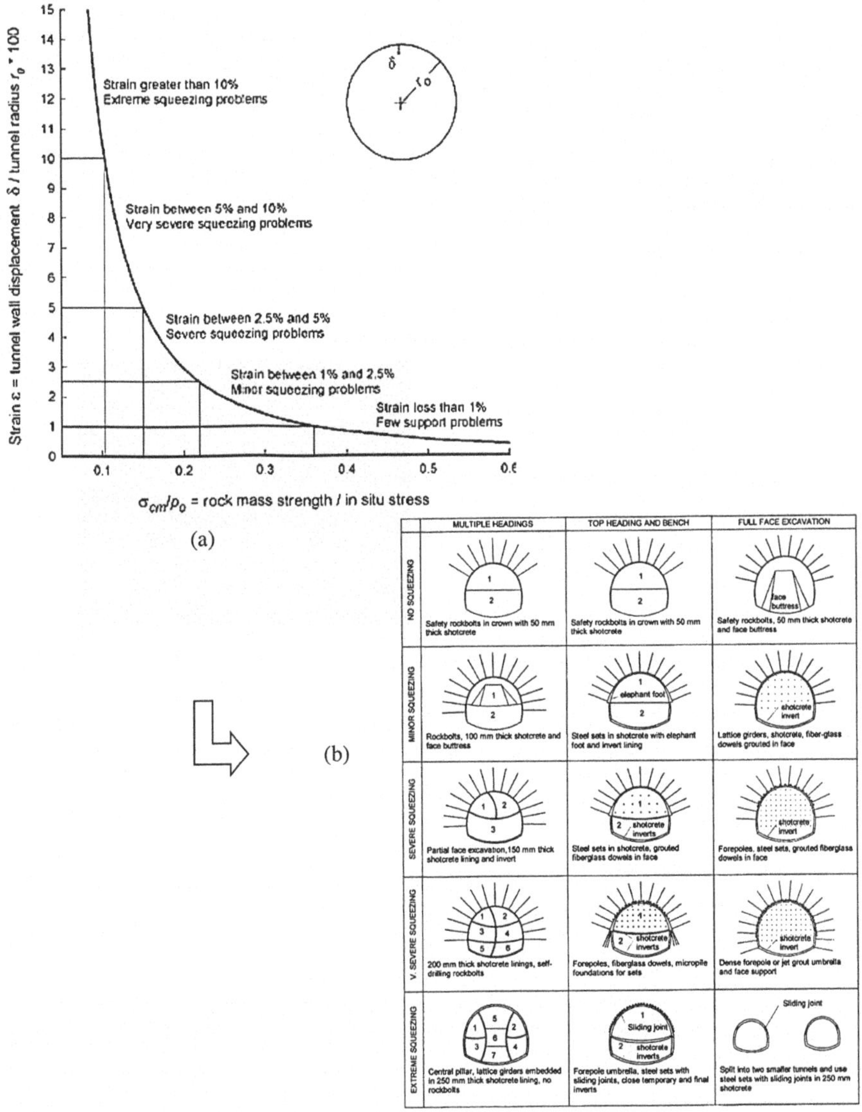

Fig. 21.4 (**a**) Potential tunnelling problems associated with the level of strain and (**b**) face excavation and support options for large tunnels (Hoek 2001)

The convergence confinement method can be described as a series of excavation stages varying L with time. At time = 0, L is as depicted in Fig. 21.5a. The ground has converged radially by a certain amount (Δv_r). Again, it is assumed that at this time the rock mass does not transmit load to the support as long at the face does not advance. As the tunnel advances (i.e., L increases), the support (at A-A′) begins to carry some of the load that has been previously carried by the face and the ground

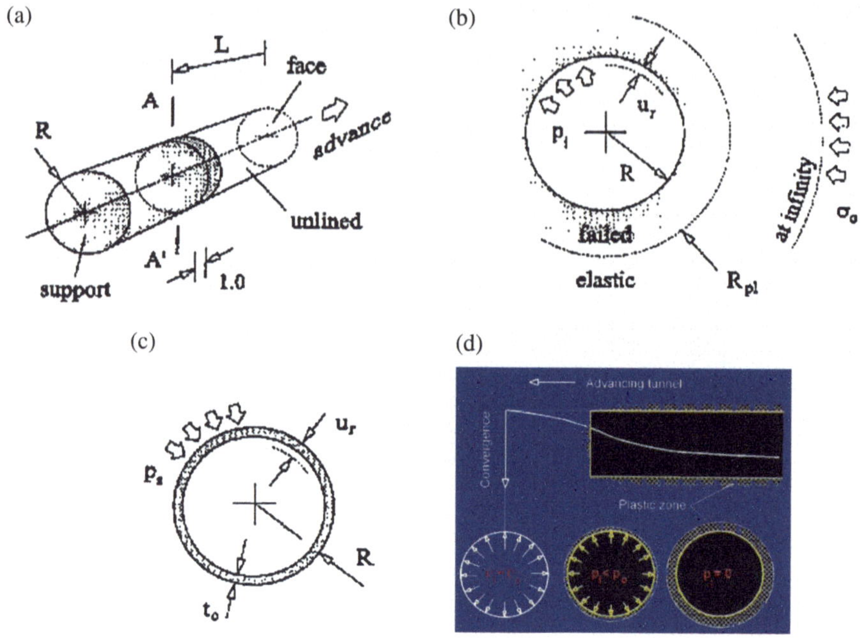

Fig. 21.5 Idealized tunnel excavation; (**a**) cylindrical tunnel of radius *R* driven in rock mass, (**b**) Cross section of the rock mass at section A-A′, (**c**) Cross section of the circular support installed at section A-A′, and (**d**) longitudinal tunnel showing convergence with the loss of confinement during tunnel advance (after Carranza-Torres and Fairhurst 2000)

convergence has also increased. At a time when the face has moved far enough ahead and is not influencing the support at A-A′, the support system is at equilibrium and the support carries the entire, final or design load. At this time, the support and the ground have converged together by the same amount.

In this way, three deformation curves can be plotted with respect to time; The Longitudinal Deformation profile (LDP)—the tunnel as it moves forward, the Ground Reaction Curve (GRC)—the section of excavation perpendicular to the tunnel axis and the Support Characteristic Curve (SCC)—the support installed at that section. Interpretation between these curves allows one to define the pressure that the ground transmits to the support as the face advances. This is consistent with observed convergence within tunnel excavations as seen in the field (Fig. 21.6).

Also seen in Fig. 21.6 is a simple circular tunnel deformation analysis within an elastic medium and hydrostatic stress conditions conducted by Hoek (1999). This idealized model also corresponds to observations made in the field. The rock mass ahead of the tunnel excavation begins to yield as a plastic zone forms ahead of the excavation.

Therefore, timing of the installation of the support also becomes an issue. If the support is installed prematurely, the stresses are large and the support will fail. If the support is installed too late, the strain will have induced radial closure and the opening will not be sufficient as per the design (Fig. 21.7).

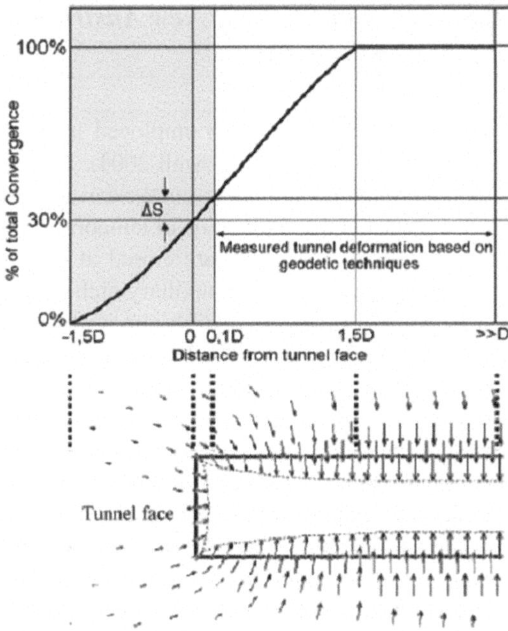

Fig. 21.6 Top: Percentage of the total convergence (GRC) in a tunnel with diameter D as a function of distance from the excavated face, assuming a nearly constant excavation rate. The excavation front is at 0, but deformation extends to the unexcavated area, along a distance of approximately $-1.5D$. The corresponding closure equals to about 1/3 of total deformation. Elastoplastic deformation extends up to a distance of $1.5D$. Beyond this, tunnel deformation is related to creep. Shaded area corresponds to deformation not recorded by instrumentation measurements. Bottom: Vectors of mass displacement in the plastic zone around the excavated area (Kontogianni and Stiros 2002)

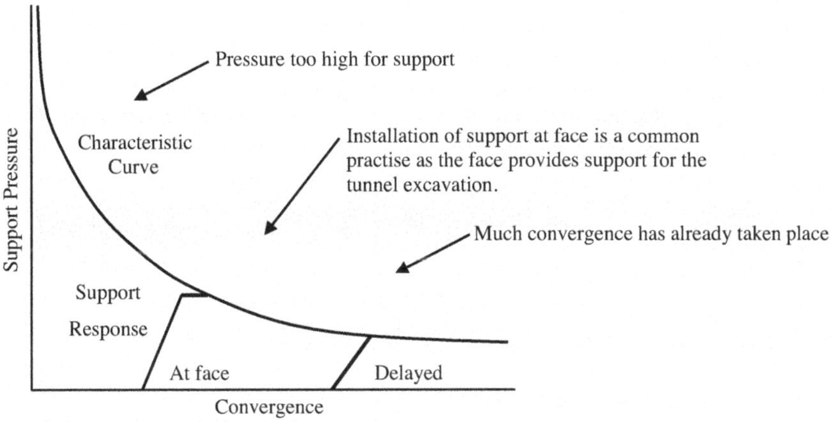

Fig. 21.7 Installation of support and its effect on the convergence

21.2.2 *Observational Design Method: New Austrian Tunnelling Method (NATM)*

The New Austrian Tunnelling Method was first employed in the UK within the within the mining field in 1964 (Karakuş and Fowell 2004). NATM is based on a concept whereby the rock surrounding an underground excavation becomes a load bearing structure itself. It consists of the application of temporary support measures in the form of shotcrete and steel arches that are closed at the earliest possible moment by an invert to complete a ring (or an auxiliary arch), the deformation of which is measured as a function of time until equilibrium is obtained (Karakuş and Fowell 2004). As such, NATM is a specific method of excavation and support techniques (Brown 1981) as can be seen below in Fig. 21.8.

The material is excavated in stages using conventional techniques. The top heading is the top portion of the tunnel portal and is dug out first. This is followed by the bench excavation and if required, further followed by a supported invert. As excavation is being conducted, support in the form of rockbolts, forepoles (to be discussed within the next section), steel sets and shotcrete are installed directly behind the advancing face. Throughout all stages of tunnelling, the tunnel is being monitored in order to determine the rock masses behavior and change (if required) the rock support class of a particular tunnel section. Successful application of this technique depends upon a high degree of understanding of support mechanics and intelligent and experienced interpretation of geological conditions and monitoring observations (Brown 1981). This technique has been used extensively on tunnelling projects around the world and is deep routed in sound geomechanical principles. NATM was chosen as the technique for use with the excavation of Drikos tunnel.

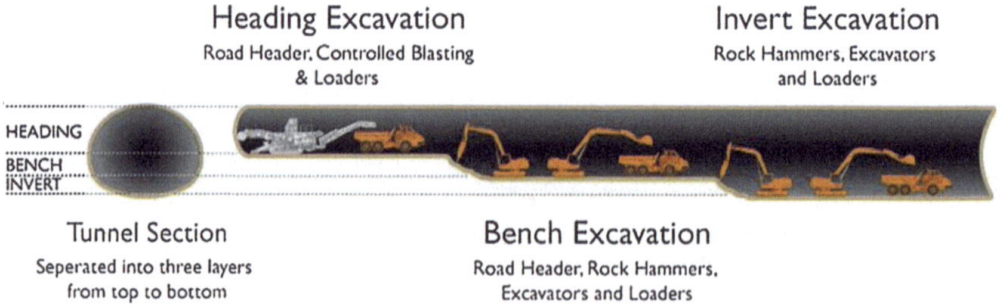

Fig. 21.8 Depiction of the excavation stages involved with NATM

21.2.3 Support Design and Considerations

There are major design considerations that must be addressed and properly interpreted. The material through which one tunnels cannot be fully defined in terms of well know strength and deformation properties (i.e., materials are often discontinuous, inhomogeneous and anisotropic in nature). A design must take into consideration the effects of the disturbance caused by tunnel excavation including stages of excavation not completely confined by the long-term support and final lining. It is during this stage that the preexisting stresses in the rock mass (deviated by the opening of the tunnel) are channeled around the cavity in an arch effect, creating zones of increased stress on the walls of the excavation. The most important task of a tunnel design engineer is to determine how and if an arch effect can be triggered when a tunnel is excavated. The engineer must then ensure that the arch effect is formed by calibrating excavation and stabilization operations (Lunardi 2000). Understanding this rock mass and support interaction becomes a critical issue.

Lunardi (2000) conducted field investigations of Italian tunnels excavated in rocks and categorized support methods into three different groups. Each group exerts a different effect on the tunnel cavity. He found that the rigidity of the core determines the stability of a tunnel since the deformation of the advanced core causes (a) the extrusion of the face, (b) pre-convergence behind the face, and (c) the convergence of the cavity. This has also been described in Sect. 21.2.1 and is shown in Fig. 21.6.

The purpose of tunnel support is to maintain confinement for the rock mass in order to help the rock mass support itself. Under these confined conditions, the interlocking components of the rock pieces produce a strong and stable rock mass. Care must be taken when excavating the face in order to ensure that confined conditions can be maintained. This is achieved through the immediate installation of support technologies such as (not the same in all cases) fiberglass dowels, spiles, shotcrete, rock bolts and grouting. Again, the initial support systems installed at or in advance of the tunnel face serve to retain the rockmass integrity and provide all of the short term support and permit the ultimate installation of the final lining. Excavation in most tunnels within a weak rock mass is carried out in this staged fashion (Examples of support systems employed and various stages of construction are shown in Figs. 21.9 and 21.10 below). As with the NATM technique mentioned previously, a top heading can be excavated and then a bench (or invert sections) may be left in place for further support. The primary support comes from the initial installation of rock bolts and steel arched rib sections supplemented with shotcrete.

A typical approach would involve the development of a number of typical cross sections for support design. Each section would be related to an anticipated magnitude of strain (or radial displacement). When advancing through difficult ground the use of the forepoling umbrella arch method is oftentimes employed (Figs. 21.9 and 21.10). As an example, for a 10 m span of tunnel in such difficult ground, this method would typically involve the installation of 12 m long, 75 mm diameter

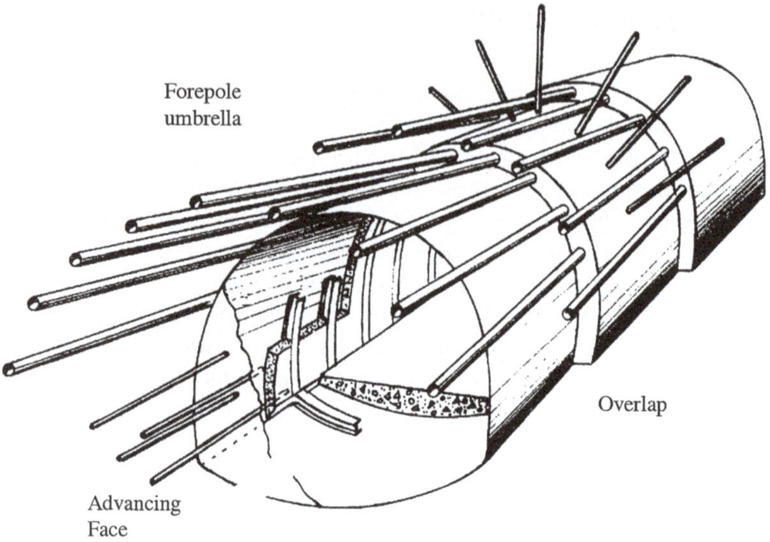

Fig. 21.9 A comprehensive primary support system for weak rock tunneling (courtesy E. Hoek), including presupport beyond the face

Fig. 21.10 Sequencing of tunnel construction at selected stages; (**a**) forepoles, (**b**) Installation of forepoles at tunnel face using forepoling machine, (**c**) view of tunnel face with forepole umbrella inserted, fiberglass dowels, and drainage pipes, (**d**) shotcrete being applied, (**e**) controlled excavation of face, (**f**) excavated material, (**g**) removal of excavated material, (**h**) installation of structural support—steel sets, (**i**) installation of further radial support—grouted rock bolts, (**j**) successive forepole umbrellas seen as well as bench, (**k**) installation of geomembrane, (**l**) installation of steel reinforcement for final lining, (**m**) detail of reinforced final lining, and (**n**) completed tunnel

grouted pipe forepoles at a spacing of 300–600 mm. The forepoles are installed every 8 m to provide a minimum overlap of 4 m between successive umbrellas. This method is usually used in combination with other support systems such as steel sets embedded in shotcrete, face stabilization by grouted fiberglass dowels and the use of a temporary invert (bench) to control floor heave (Hoek 2001).

Seen above, Fig. 21.10 depicts various stages of the construction sequence associated with the Egnatia Odos tunneling works currently underway in Northern Greece. Another major consideration when dealing with face stability is water pressure and drainage. Water may accumulate behind material within fault zones and these barriers may act as dams. Therefore, high pressure water may potentially be trapped behind the face. Care must be taken in order to ensure that a probe hole be advanced well ahead of the face at all times. This allows for the detection of the presence of high pressure water and to allow for sufficient drainage prior to excavation. Also incorporated in design is the use of a geosynthetic impermeable membrane and drainage layer placed between the primary support and the final lining. Reduction of water pressures is dealt with through the use of drainage pipes located on either side of the tunnel footing.

21.3 Tunnels of the Egnatia Odos: Design and Construction Considerations

21.3.1 Design of Tunnel and Temporary Support

As introduced in the preceding section, tunnelling in weak rock presents many challenges since misjudgment in the design of support systems can lead to under design and costly failures or over design and high tunnelling costs. In order to achieve technically and economically optimized constructions, the general principles for the tunnel design for Egnatia Odos were (Egnatia Odos AE 2001):

(a) a good general tunnel design layout that does not force tunnel constructions to be sited into the worst conditions within the project area,
(b) investigations and exploration that clarify all details of soil and geomechanical behavior before the design phase, and
(c) a design that should adapt well to all steps of execution, while the lifetime construction should be guaranteed by taking into consideration all the in situ conditions and/or influences.

The design and construction methods employed for the tunnels of Egnatia Odos took on the form of a modified version of the NATM, an observational approach utilizing much instrumentation and an extensive monitoring program. Egnatia Odos S.A. created new guidelines for tunnel design firms based on international best practice. For the most part, the tunnels were/are designed in accordance with German and Greek standards as stipulated by the following documents: OSMEO—Design

Guidelines for conducting Road Works Design, TSY—Tunnels Materials and Workmanship Specification, OSAT—Guidelines for the Environmental terms and landscape design and EAK 2000—Greek Seismic Code (Rawlings et al. 2001).

The standard cross section of the excavated tunnel portals on the Egnatia Odos are horseshoe shaped and provides for a 5 m structure clearance with two traffic lanes 3.75 m wide each (after final lining). All Egnatia tunnels have a design life of 120 years. A typical tunnel cross section can be seen in Fig. 21.11 below. The tunnels are designed as twin tunnels and pedestrian cross passages are provided every 350 m and emergency vehicle cross passages and parking areas every 1000 m (Fig. 21.11). The drainage system of the tunnel pavement includes a continuous fissured (slotted) gully which discharges every 50 m into a 400 mm diameter main collector pipe (Lambropoulos 2005).

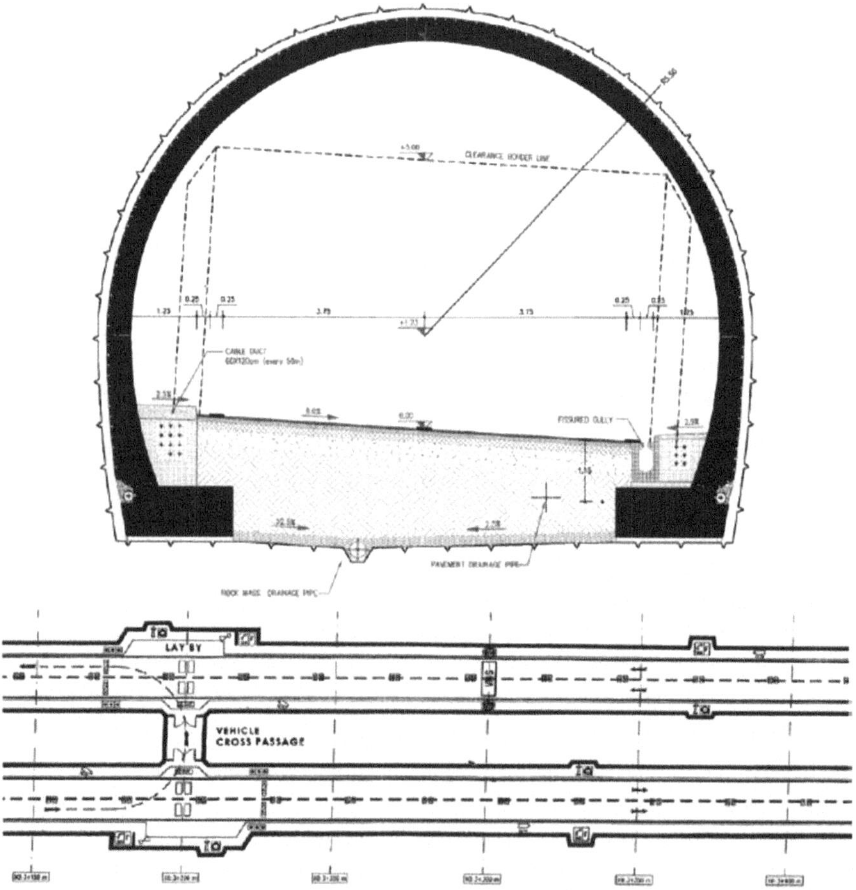

Fig. 21.11 Typical cross section and plan view of tunnel as part of Egnatia Odos Motorway (Lambropoulos 2005)

The first step in the tunnel design process was development of the engineering geological model/map, which was preceded by the geotechnical site investigation. This formed the basis for the tunnel design and provided the essential geotechnical and ground parameters for tunnel primary support and final lining design. In this investigation, the discussion concentrates on the primary or temporary support design considerations.

In designing the temporary support measures, the anticipated response of the ground to the stresses induced by tunnel excavation at all geological locations were taken into account. The design was largely influenced by the rock and groundwater conditions, rock mass properties, deformation behavior in relation to the in situ stresses (e.g., squeezing conditions), overburden and ground discontinuities. Rock mass primary support categories were then developed for ranges of the rock mass classification values for individual tunnels (and site-specific subsections of rock masses located at each tunnel location). The rock mass mechanical properties derived were used as input into 2D numerical analyses of the ground/primary support behavior and input into the final design.

In keeping with the observational method, in progressively worsening rock mass conditions that were encountered in selected complex geological structures common to the tunnels of Egnatia, the initial category of primary support was upgraded to higher support classes (Lambropoulos 2005). One of the most common additional support measures was the use of an *elephants foot* (Fig. 21.12) located either side of the top heading. This was inserted in order to improve the load distribution (and bearing capacity and stability) of the top heading support arches.

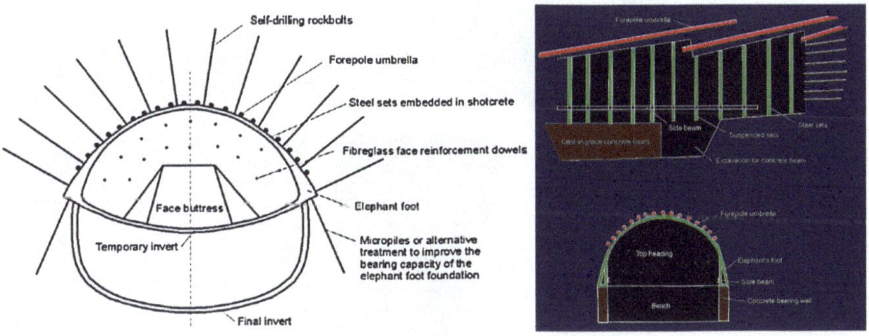

Fig. 21.12 Schematic representation of all possible support measures for Egnatia Odos Tunnels (Egnatia Odos AE 2001)

21.3.2 Tunnel Instrumentation and Influence on Tunnel Design

Instrumentation and monitoring play a vital role in verifying design assumptions and calibrating numerical models. As well, monitoring serves as an alert if the initial support or lining is not performing as intended or if the tunnel is in danger of collapse. Deformation is a main factor in controlling the failure and cost-effectiveness of underground excavation. As such, in the last two decades deformation monitoring has become a fundamental requirement for assessing the stability of underground openings and for quantifying the acceptable risk of rock response (Kontogianni and Stiros 2003). Monitoring data also provide a wealth of data as to the 3D behavior of the rock mass, support and the time history associated with excavation. This information can be used to improve geotechnical models and optimize the excavation process.

The monitoring program within the tunnels of Egnatia Odos incorporates the use of inclinometers, extensometers, strain gauges, load cells, instrumented rock bolts, and standard convergence and deformation measurements (Hindley et al. 2004). Within the concept of the observational method of tunnel construction, monitoring has also played an important role in making design changes to primary support systems. Most of this data is in digital form and selected data is available to the author through the National Technical University of Athens and Egnatia Odos, S.A.. Egnatia Odos S.A. controls the dissemination of certain data due to sensitivity involved at certain construction sites (i.e., the Anthochori tunnelling site has some outstanding legal issues that must be settled). Figure 21.13 depicts some of the mon-

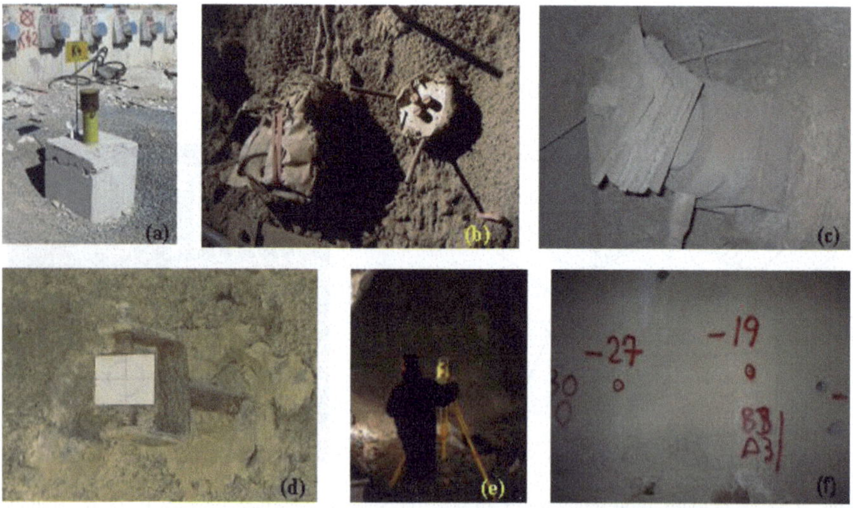

Fig. 21.13 Instrumentation and targets associated with monitoring program of Egnatia Odos; (**a**) monitoring well and survey target on benchmark, (**b**) pressure cell (left) and extensometer (right), (**c**) tunnel wall pressure cell, (**d**) tunnel wall survey target, (**e**) surveying the tunnel face, and (**f**) measurement of targets on tunnel wall (−19 denotes 19 mm of inward displacement at that target location) within Driskos Tunnel

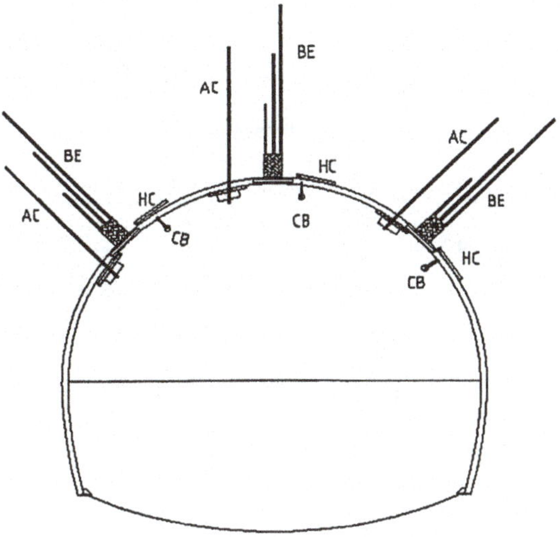

Fig. 21.14 Generic, Typical arrangement of tunnel monitoring instrumentation for Egnatia Odos tunnels

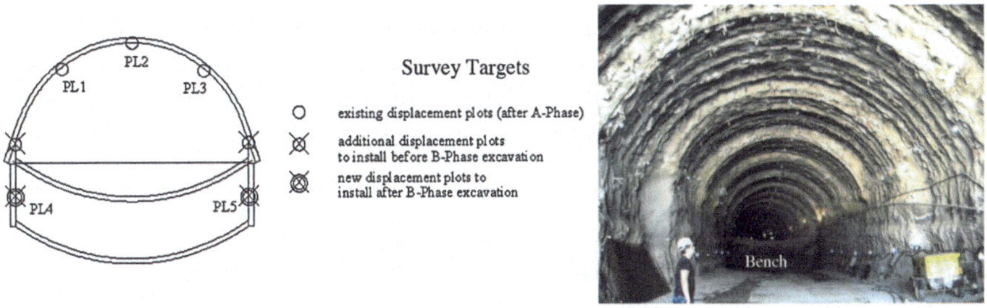

Fig. 21.15 Optical survey markers and their locations within the tunnel excavation (Grasso et al. 2003). Photo taken by author. Note that A-Phase and B-Phase are analogous to top heading and bench excavation phases respectively

itoring instrumentation being employed during tunnel construction for Egnatia Odos. Configurations of other instrumentation that was used within the tunnels of Egnatia Odos are shown in Fig. 21.14. This arrangement of tunnel monitoring instrumentation includes extensometers (BE), survey target points (or convergence markers) (CB), hydraulic pressure cells (HC), and anchor strain gages (AC).

Another data gathering intensive portion of the observation method includes gathering geodetic data in the form of survey monitoring. This monitoring scheme was used in order to verify the adequacy of the adopted geomechanical model and support classification/support system. This data is paramount in making decisions on modifications to design and construction optimizations during the construction phase. A monitoring program for the tunnels of Egnatia Odos included three-dimensional tunnel wall displacements using optical survey (Fig. 21.15), rock mass deformation using sliding deformeters and inclinometer columns (Fig. 21.16), and groundwater pressure variations using piezometric cells.

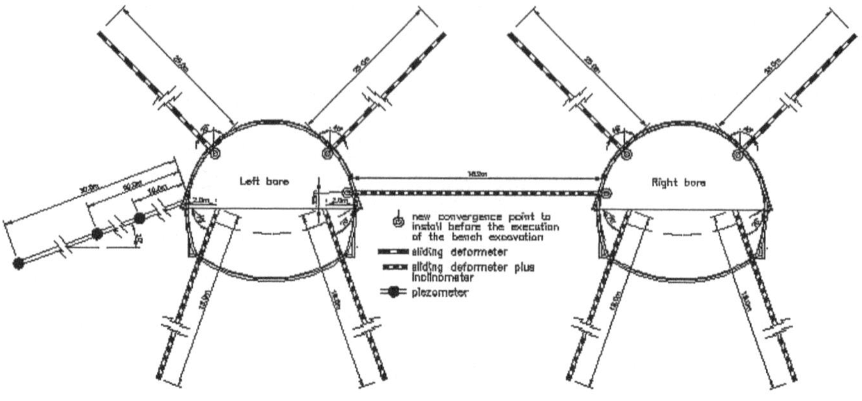

Fig. 21.16 Cross section of twin tunnels showing locations of sliding deformeters and piezometers (Grasso et al. 2003)

Depicted in Fig. 21.16 are the locations and arrangement of sliding deformeters with lengths of 25 m inclined at 45°. These deformeters were located on the inner and outer side of the bores. Three piezometers are located at 30, 20, and 10 m radial distances from the bores. The array of instrumentation allows for constant monitoring of the tunnel and for the designers, contractors and engineers to gain insight as to the behavior of the rock mass subjected to tunnel excavation.

21.4 Design and Construction of Driskos Twin Tunnels

21.4.1 Introduction

The tunnel construction at the Driskos site consists of twin, parallel tunnels that are approximately 4570 m in length with a maximum overburden of 225 m. In a generic sense, the geological profile shows a gently folded synclinal structure at the centre of the alignment, which elevates to an anticline at the southern end. The Driskos tunnels are 850 m above mean sea level and excavation began in 1999 with excavation being driven from both ends. The faces were split into a 60 m^2 top heading followed by a 40 m^2 bench. The tunnels have been advanced using a Tamrock Para 206 T two-boom and basket drillrig. The north portals were constructed using an umbrella canopy for 50 m followed by 40–50 m cut and cover with another 20–30 m of forepole umbrella (Smith 2000), Fig. 21.17. More details are included in Vlachopoulos et al. (2013).

Cross-passages connecting the two tunnels occur every 350–400 m. A 180 m ventilation shaft is also being constructed (2008) near the centre of the tunnel alignment. Tunnel separation is 18.2 m between each bore. In terms of tunnel bore dimmensions, a single bore cross section has dimmensions of 9.47 m by 11.0 m with respect to rock mass excavation tolerances. The cross section of the bores are

Fig. 21.17 Driskos Northern Portals during construction in 2003

arranged in a horseshoe configuration; An idealized cross section of this nature can be seen in Fig. 21.18 below. In terms of twin tunnel dimmensions, Fig. 21.19 outlines the spacing and separation distances for the parallel bores.

21.4.2 Support Categories

As described in the previous sections, a function of stress, strain and rock mass (i.e., tunnel deformation/tunnel diameter versus rock mass strength/in situ stress) defines the support requirements for tunnel excavations through rock at a specific site. This exercise was no different for the design of the twin tunnels as part of the Driskos tunnels. The main factors that influenced the design were: lithology, rock mass quality, and the height of overburden (i.e., in situ stress). Based on these results, five rock mass categories (Category II–V, with II being the better rock mass in terms of strength) and corresponding support categories (Fig. 21.20) were defined (Egnatia Odos AE 1999; Structural Design SA 1999). The percentage of rock masses and support systems corresponding to the overall carriage length for Driskos is:

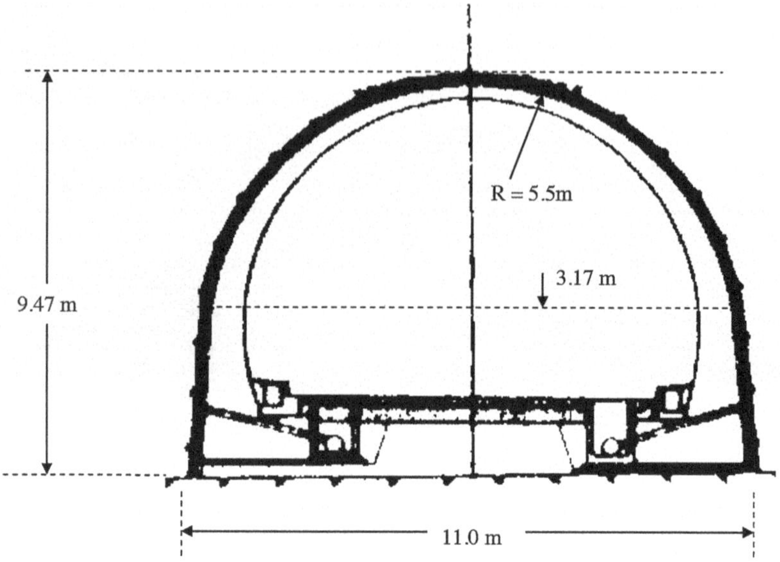

Fig. 21.18 Idealized horseshoe cross section of Driskos tunnel

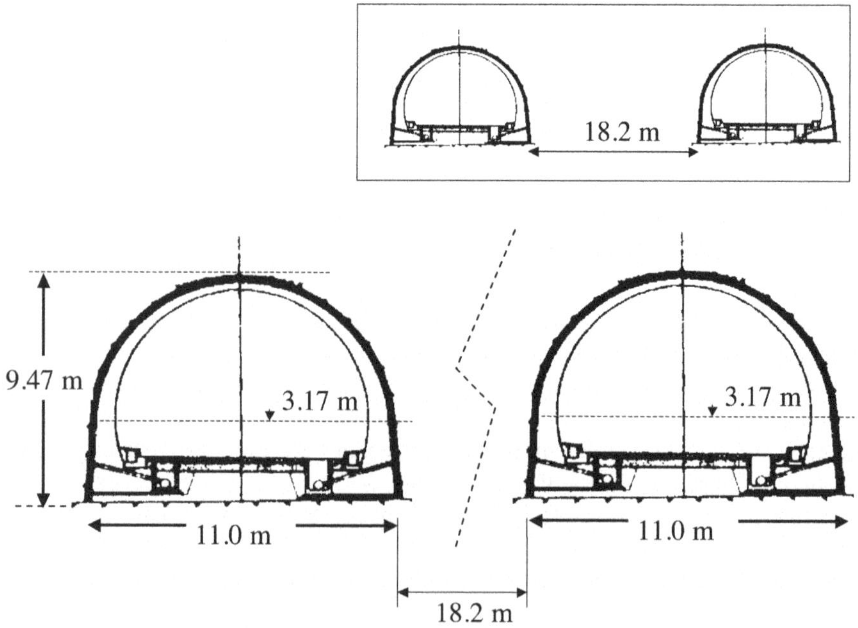

Fig. 21.19 Idealized cross section of Driskos tunnel for twin bores

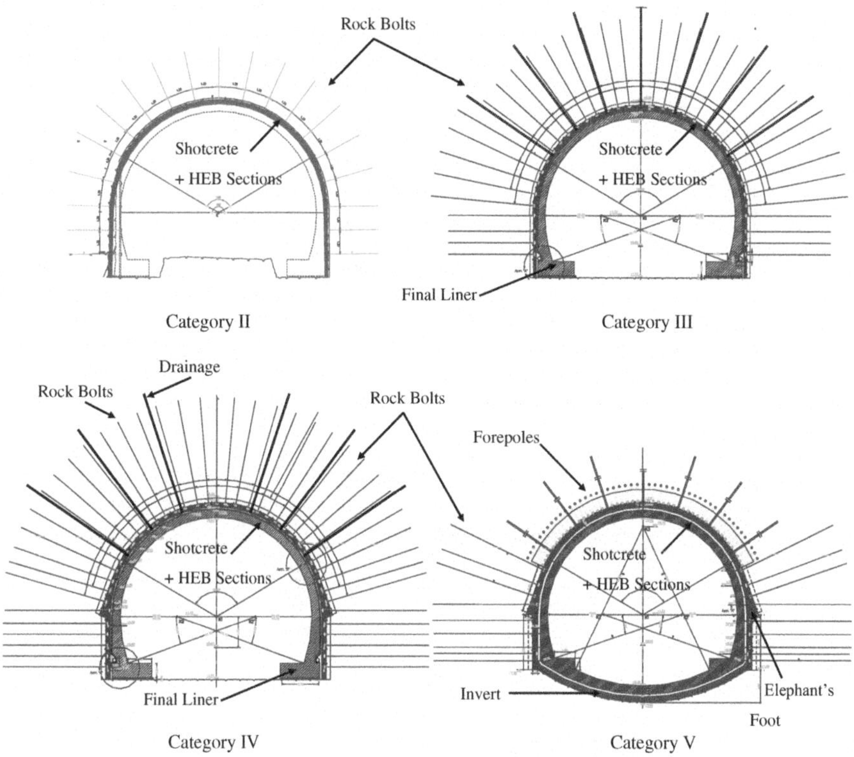

Fig. 21.20 Original Support Categories II–V for Driskos Tunnels (Egnatia Odos AE 1999)

(a) Category II—22% (b) Category III—42% (c) Category IV—27% (d) Category V—8%	*Trends*: Deteriorating Rock Mass Strength Increased Support Requirments

Shown in Fig. 21.20 are the extent of support measures that were incorporated in the preliminary design for the Driskos tunnels. One can see the increased mechanical support requirements as one goes from Category II to Category V. Included within the support requirements are rockbolts, shotcrete, steel-sets, and forepoles.

The support measures that correspond to the rock mass categories for Driskos tunnel are summarized in Table 21.1. Category V is further subdivided into Va and Vb based on the requirement (observational method) of forepoling. A typical longitudinal cross section of category V support is illustrated in Fig. 21.21.

Table 21.1 Support measures for each rock mass category for Driskos Tunnel (after Structural Design SA 1999)

Support category	Construction Phases	Distance of unsupported part	Support measures				Additional measures
			Application time	Rockbolts	Shortcrete	Steel sets	
II	2 (Top Heading-Benching)	6 m	18 h from the blasting	3 m nockbolts in distribution 1.5 × 2.0 m in crown and sides	10 cm in the crown and 5 cm in the sides		
III	2 (Top Heading-Benching)	3 m	12 h from the blasting	3 m rockbolts in distribution 1.3 × 1.3 m in crown and sides	15 cm in the crown and 10 cm in the sides		
IV	4 (2 Top Heading-2 Benching)	2 m	During excavation	4 m rockbolts in distribution 1.2 × 1.0 m in crown (50° of the crown) and sides. 6 m rockbolts in the rest of the crown.	20 cm in the crown and the sides and 10 cm at the face of top heading	Lattice girder in 1 m distances	
Va	4 (2 Top Heading-2 Benching)	2 m	During excavation	6 m rockbolts in distribution 1.5 × 1.0 m in crown, sides and bottom	25 cm in the crown and the sides and 10 cm at the face of top heading	Steel sets (HEB) in 1 m distances	Forepoling umbella if necessary
Vb	2 (Top Heading-2 Benching)	2 m	During excavation	6 m rockbolts in the basis of the steel sets	25 cm in the crown and the sides and 10 cm at the face of top heading. 25 cm for the invert.		Forepoling umbella

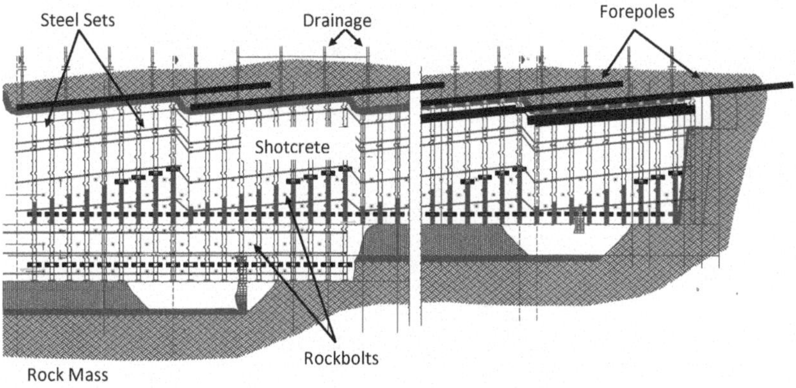

Fig. 21.21 Longitudinal cross section of Driskos showing tunnel support detail (Structural Design SA 1999)

21.4.3 Tunnelling Issues Associated with Driskos Tunnel Construction

Prior to construction, it was predicted that Section 4 of Driskos tunnel would be the most problematic due to the weak rockmasses within that section. As such, the rockmass was designated as a Category V and the initial design complimented this designation (Structural Design SA 1999). However, the deformations that ensued due to tunnel excavation were greater than anticipated. As such, the tunnel temporary design was reevaluated and modified after an extensive monitoring program by GeoData S.A. (Grasso et al. 2003) and design recommendations by Panel of Expert reports (Hoek and Marinos 2000).

The primary author visited the construction site and photos of the problematic section are shown in Fig. 21.22. The identified problem was overstressing of the temporary/primary support that occurred at several locations during excavation and extended over distances of some of metres (approximately 10 m). This is typical of the response to be expected and is associated with large deformations due to a combination of high stress and low rock mass strength. As well, the strength of the siltstone-sandstone flysch is reduced locally by the presence of a high concentration of horizontal bedding crossed by frequent faults. Also contributing to the excessive deformation is the presence of water that contributes to a reduction in rock mass strength.

The deformations that were observed did not exhibit a tendency to stabilize with time. The design, therefore underestimated the rock mass strength for this particular section and as a result, the design capacity of the primary support was determined to be too low (Hoek and Marinos 2000). A rigorous monitoring program was implemented. As a summary, the major observations associated with the monitoring pro-

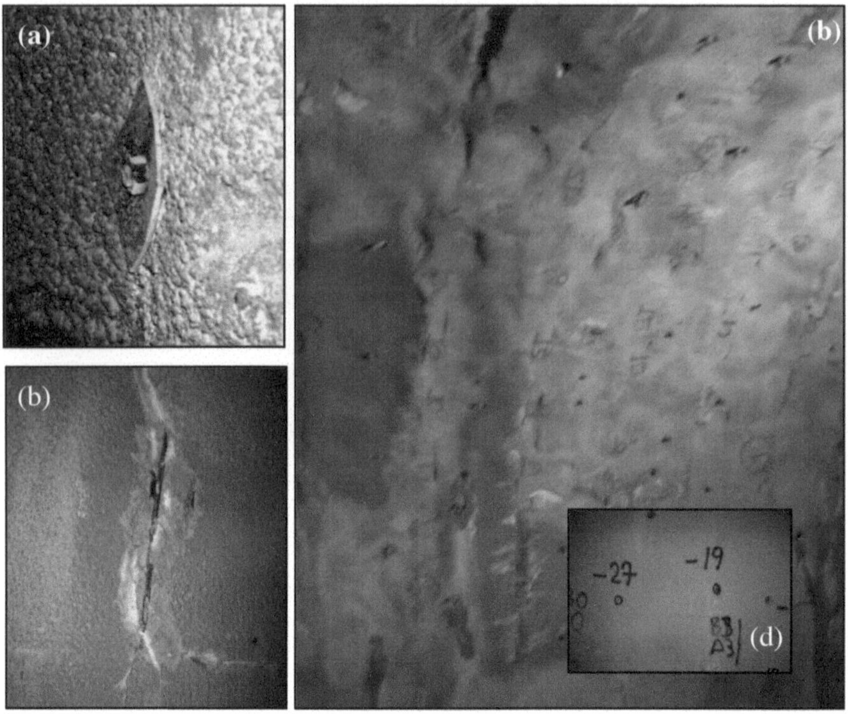

Fig. 21.22 Excessive deformations experienced within Section 4 of Driskos tunnel. (**a**) Deformed rockbolt face plate as a result of overstressing of the primary support, (**b**) Spalling of the shotcrete adjacent to a stressed steel set, (**c**) Sidwall of Driskos Tunnel denoting (in Red) the total amount of inward displacements at selected, monitored target points, and (**d**) Clearer designation of inward displacement in millimeters

gram as well as its implications to the redesign of Section 4 of the Driskos twin tunnels are (Egnatia Odos AE 2001):

(a) During the top heading phase of excavation, 210 mm of tunnel closure was exhibited by the primary/temporary support;
(b) Upon excavation of the parallel bore, 310 mm of displacement was captured;
(c) Additional primary support measures included: closer spacing of steel set support sections, more and longer fully grouted rockbolts, thicker shotcrete shell, introduction of prestressed cable anchors and the introduction of an invert;
(d) Micropiles were also included into the redesign of the temporary support. They were incorporated as a steel-reinforced continuous concreted beam element that was laid at the base of the steel ribs along each sidewall;
(e) Deformations were on average larger on the outer sidewalls indicating asymmetric tunnel closure behavior. This was not predicted at the initial design phase for the primary support. More symmetric conditions were achieved after the modification of the sidewall rockbolt patter was implemented. Another factor that influences this behavior is a realistic evaluation of the stressfield conditions (i.e., K).

21.5 The Numerical Simulation of Tunnelling Projects: Continuum vs. Discontinuum Analysis and Their Applications

Over the last decades a number of different numerical methods and software packages have been developed in order to address a variety of problems within the engineering field. More specifically, the numerical simulation of underground excavations has become an invaluable and indispensable part of the employed design process, especially when complex geological conditions are to be encountered and conventional practices are not adequate to capture the ground behavior.

For underground openings excavated within rock materials, the applied numerical analysis usually involves continuum-based approaches with the finite element method (FEM) and the finite difference method (FDM) being common practice in tunnelling projects. Furthermore, other numerical techniques based on discontinuum assumptions, such as the discrete element method (DEM) or the hybrid finite-discrete element method (FDEM), have also become a tool in the tunnel design process due to the recent technological and computational advances that increased their efficiency.

The development and application of different numerical techniques aim for the simulation of rockmasses under varying conditions depending on the material properties and characteristics, including the strength and deformability of the intact rock material, the degree of fracturing, the strength of joints and other discontinuous features, and so on, with the modelling results controlled by the adopted numerical method, the material assumptions, and the imposed boundary and initial conditions. Especially for rock materials in situ, the aforementioned parameters may have a high variability which is governed by the nature of the underground project (single tunnel, twin tunnels, multiple room excavation, etc.), the material heterogeneity and anisotropy, the medium blockiness, the joint density and persistence, and so on, hence increasing the challenges of the simulation of these geomaterials.

21.5.1 The Numerical Simulation of Rockmasses

The response of weak, soft rockmasses (rockmasses consisting of intact materials of low unconfined compressive strength (UCS), moderately to heavily fractured with persistent joints, weathered and altered rock materials, etc.) during an underground excavation is influenced by the intact rock properties and the rockmass structural features which are usually combined to derive an "equivalent" continuum medium, under the assumption that the material is homogeneous and isotropic.

For very blocky rockmasses, this assumption is valid since the behavior of the fractured material is not controlled by specific joints. On the contrary, the rockmass (intact rock and rock joints) does not appear to have specific planes of weakness that can be identified, and the material response is controlled by the overall rockmass shear strength. Under such conditions, it is a good approximation to consider the rockmass as a material with an elastoplastic behavior with continuum properties, and continuum numerical methods such as the FEM and FDM are appropriate to use.

A characteristic example is the extensive work conducted by Vlachopoulos (2009) at the Driskos Tunnel, which is part of the Egnatia Odos Project, in Northern Greece (Fig. 21.23).

The twin tunnel was excavated in high heterogeneous flysch, material mainly consisting of thin to medium bedded alternations of siltstones and sandstones (Fig. 21.24).

The relatively low strength of the equivalent medium combined with the field stresses of a high overburden (~220 m) resulted in highly squeezing conditions that could be successfully captured by elastoplastic continuum models (Fig. 21.25).

Research conducted by Vlachopoulos and Diederichs (2009) utilized continuum models based on the principles of elastoplasticity in two and three dimensions (2D and 3D) in order to associate the development of the stress induced plastic zone around the excavation contour, as a result of shear failure, and the anticipated convergence of the tunnel as the excavation advances, better known as the Longitudinal Displacement Profile (LDP) approach (Fig. 21.26).

The behavior of weak, soft rockmasses within a tunnelling environment can be adequately captured by continuum models based on shear failure criteria. However, as the material of interest becomes harder with nonpersistent joints, shear failure constitutive models integrated into continuum numerical techniques cannot adequately capture the fracturing mechanisms occurring. Fracturing in harder rock excavations due to loss of confinement within high magnitude in situ stresses, and the brittle response that the material demonstrates are better captured by using discontinuum approaches.

Research conducted by Vazaios et al. (2019) by applying the FDEM method demonstrated how such a numerical approach could be used to simulate brittle failure based on field observations from the Canadian Underground Research Laboratory (URL) Test Tunnel, located in Manitoba, Canada (Diederichs 2007). The developed tunnel model was calibrated to capture the fracturing mechanisms and the replication of the "v-shaped" notch failure which was observed at the URL Test Tunnel (Fig. 21.27).

The tunnel was excavated within a massive, virtually fracture-free granite and the manifested failure was governed by extensile fracturing due to the high compressive, tangential stresses at the crown and invert of the excavation. The numerical model was able to capture these fractures in extension and the excavation response.

21.5.2 Discontinuity Simulation Within the Numerical Model

For the successful numerical modelling of an underground excavation within a rockmass it is critical to incorporate the discontinuities present in a fashion that is both mechanistically accurate and computationally efficient.

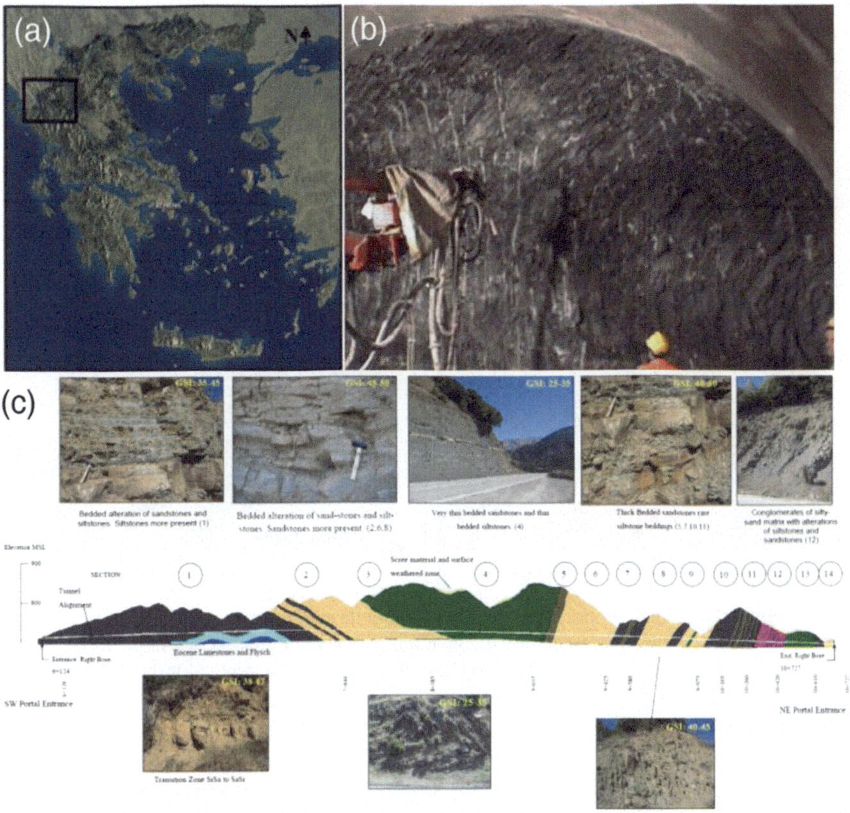

Fig. 21.23 (**a**) General location of the Driskos Tunnel in Northern Greece. (**b**) Photo of Tunnel Face at Chainage 9 + 400 to 9 + 600 m. (**c**) Representative rock masses and cross section of the Driskos Tunnel, Egnatia Odos, Greece (Vlachopoulos 2009)

Fig. 21.24 Flysch consisting of alternating sandstone and siltstone layers (Vlachopoulos 2009)

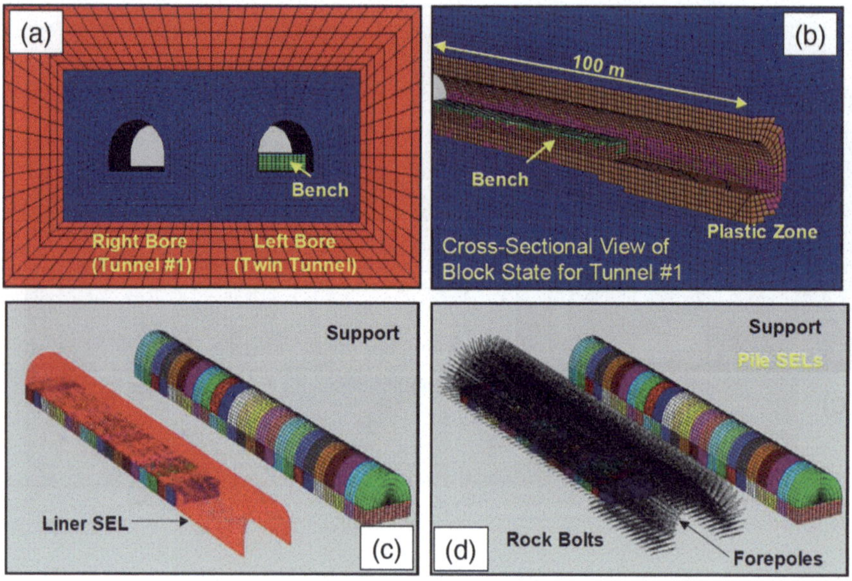

Fig. 21.25 3D Numerical model of the Driskos Tunnel developed in FLAC3D. (**a**) Twin tunnels with the first branch fully excavated and top heading completed in the second branch. (**b**) Plastic yield zone of single tunnel. (**c**) Model detail highlighting with a different color the different excavation stages and support detail of the tunnel liner, and (**d**) forepoles and rockbolts completing the support system (Vlachopoulos 2009)

Weak, heavily fractured rockmasses simulated by employing continuum based modelling are usually not explicitly incorporating discontinuous features. The influence of discontinuities and their impact on the rockmass mechanical properties is taken into consideration in the model indirectly by appropriately reducing the rockmass strength and deformability depending on the degree of jointing (rockmass blockiness) and joint condition (joint strength). The work conducted by Vlachopoulos (2009) utilized the Geological Strength Index (GSI) (Marinos and Hoek 2000) as determined at the Driskos Tunnel to achieve the required strength reduction and deformability of the material when the Hoek-Brown criterion (Hoek et al. 2002) and Hoek-Diederichs modulus equation (2006) are used respectively both in 2D and 3D models.

However, for harder rockmasses with nonpersistent joints such an approach is not able to capture the true rock mechanics behavior and does not yield the correct results. In such cases the simulation of the joints needs to be performed explicitly and the joints need to be integrated within the numerical model. Furthermore, the joint geometry pattern, dictated by the joint size, joint number and joint spatial distribution, is characterized by a wide variability depending on the in situ conditions. To capture this variability, discrete fracture network (DFN) modelling may be employed, as demonstrated by Vazaios et al. (2018) (Fig. 21.28).

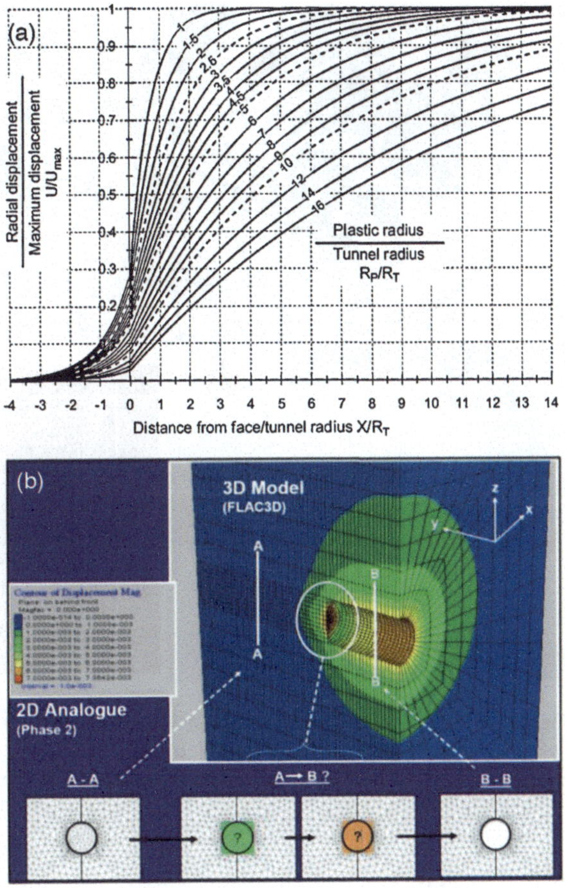

Fig. 21.26 (**a**) Longitudinal Displacement Profile (LDP) templates for estimating the tunnel convergence as a function of the tunnel plastic radius at different excavation stages (Vlachopoulos and Diederichs 2009). (**b**) 3D tunnel model capturing the 3D effects at the face (FLAC3D) and 2D plain strain tunnel model at different stages as an analogue of the tunnel advancement (Vlachopoulos 2009)

21.5.3 Assessing the Excavation Induced Damage by Using Numerical Modelling

By adopting the appropriate numerical method, tunnel scale excavations can be simulated successfully and provide a better understanding of the rockmass behavior during excavation, hence assisting and optimizing the engineering design. Failure mechanisms captured accurately by the numerical simulations can provide reasonable estimates of the extent, depth, and shape of the excavation induced damage as a result of the redistributed around the underground opening. That is of great significance as the excavation induced damage inflicted on the surrounding ground alters the material properties within the vicinity of the excavation, with the extent of the that altered material zone depending on the original rockmass properties and in situ stress regime.

Especially for underground excavations, and given specific site conditions and project requirements, estimating relatively accurately that damage extent is critical.

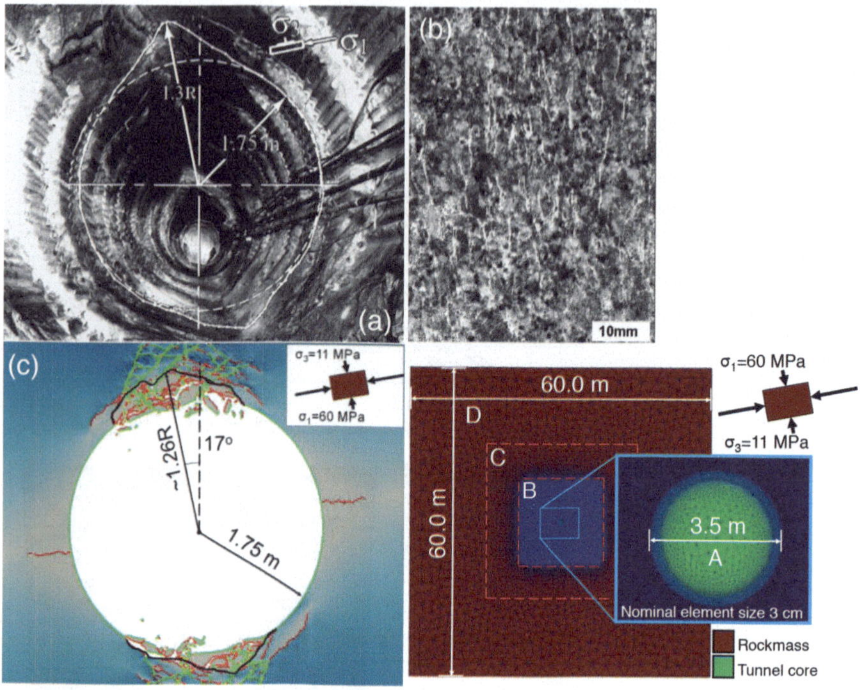

Fig. 21.27 (**a**) Photograph of the URL Test Tunnel and the "v-shaped notch" failure due to brittle failure (spalling) (Diederichs 2007), and (**b**) detail of cracks (dyed white) in compressive test sample of Lac du Bonnet granite (Diederichs 2000). (**c**) Calibrated tunnel model using the FDEM method developed in Irazu replicating the brittle failure observed at the URL Test Tunnel

Within weak rockmasses, the disturbance of the surrounding ground in the case of twin tunnels is important to be evaluated since the excavation of the first branch and the associated stress induced damage alters the rockmass properties and affects the second boring if the damage extent exceeds that of the pillar between the two tunnels (Fig. 21.29).

Regarding underground excavations in rock materials, the interaction of stress induced cracks and preexisting, nonpersistent joints is required to be taken into account especially for underground works meant for storage of substances that need to remain in isolation from the ground surface (e.g., nuclear waste). Under such conditions, the numerical model needs to able to predict the zone affected by stress induced fractures, capture the fracture mechanics of the intact rock bridges between the preexisting joints and simulate their interaction (Fig. 21.30).

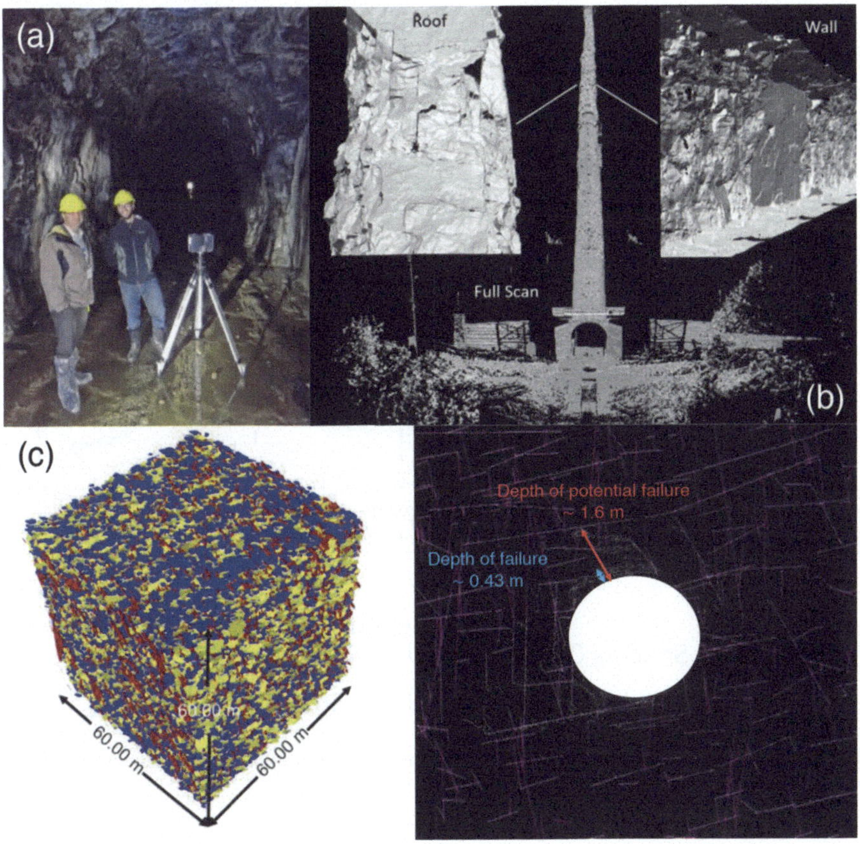

Fig. 21.28 (**a**) LiDAR scanning conducted at the Brockville Tunnel, Ontario, Canada. (**b**) 3D surface model of the tunnel from collected LiDAR data (Diederichs et al. 2013). (**c**) Discrete fracture network (DFN) model created by LiDAR obtained data and incorporation of the simulated joints into UDEC (Farahmand et al. 2018)

21.5.4 Summary

The numerical modelling of tunnelling excavations within rock materials has been and is a great asset in the design toolbox of engineers. Given that the appropriate method for specific site conditions and project specifications is adopted, numerical simulations can provide a great insight of the rockmass response during an excavation, assist in the design process, and be an integral part of the observational tunnelling approach in which numerical models and field observations are coupled to provide solutions and optimize the design.

Numerical models developed with caution and by making reasonable assumptions can provide reasonable estimates of the excavation induced damage and help with determining the support system required for a given project. However, they are

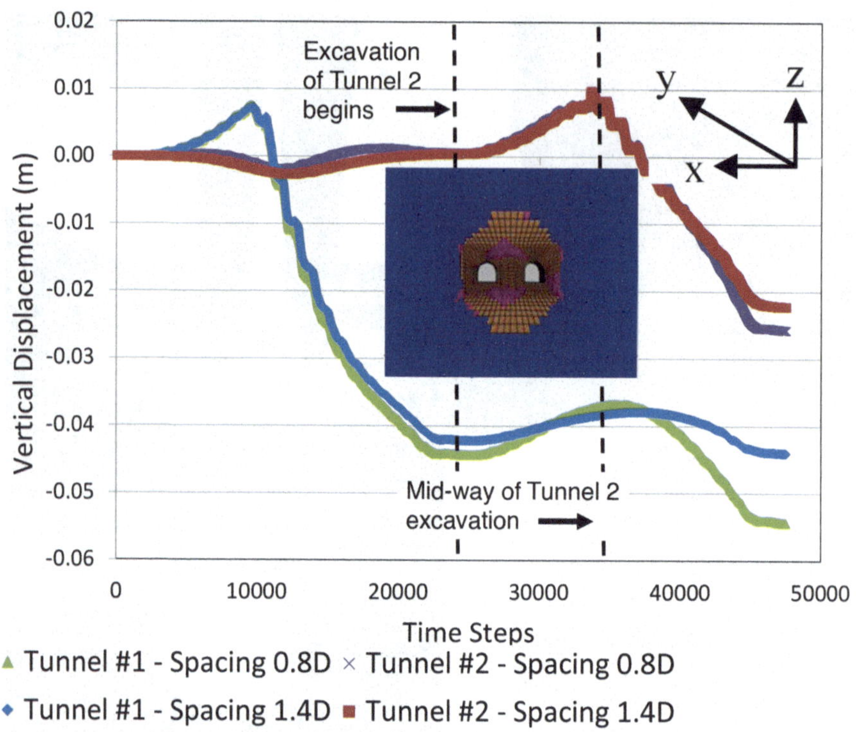

Fig. 21.29 Twin tunnelling interaction due to the plastic zone surrounding the underground openings (Vlachopoulos et al. 2018)

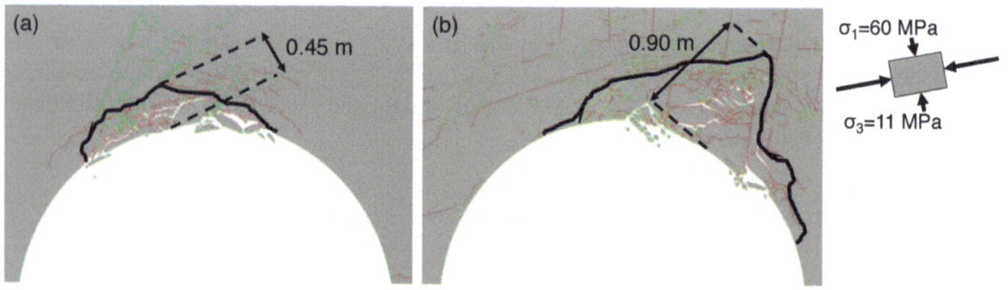

Fig. 21.30 Comparison of damage shape and extent between (**a**) "fracture-free" and (**b**) "fractured" model using FDEM (Vazaios et al. 2018)

subjected to limitations. Employed methods, determination of the required input parameters, constitutive assumptions, required computational time, and level of realism achieved from the model are only some of the factors that need to be taken into account for the development of a numerical model that is going to be representative and correspond to the in situ conditions (Fig. 21.31). A tunnel model that is not obeying in the field conditions it is meant for can result in significant misinterpretations, hence misleading the engineer and resulting in a design that is not appropriate for the specific ground conditions.

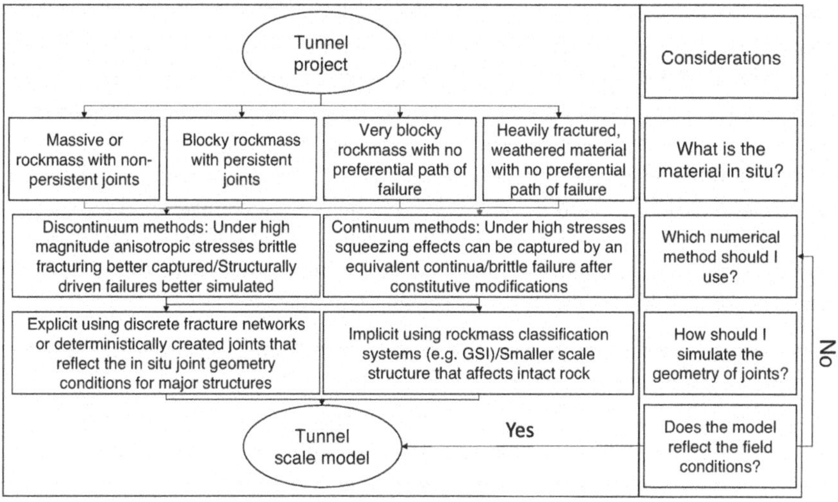

Fig. 21.31 Considerations and indicative considerations of how to build a tunnel scale numerical model

21.6 Emerging Techniques for the Validation and Optimization of Support Design

The case study of the Egnatia Odos and the Driskos Tunnel present a candid example for the necessity of an intensive monitoring program throughout tunnel construction. However, even under much less adverse project conditions, monitoring will play a crucial component in tunnel design and construction. One aspect explored in more detail herein, is the role of monitoring in the validation and optimization of ground support design. Specifically, an optical fiber based strain sensing technique is demonstrated to have the potential to monitor spatial complex support member response.

21.6.1 Sensing Approach to Ground Support

During the construction phase of a tunnelling project, the support system acts as the first line of defense for workers and equipment within the excavation. It is also pivotal in controlling or maintaining excavation-induced displacements to meet project-related limitations, which may be dictated by the presence of existing infrastructure, final dimensional requirements, and other stipulations. There are severe consequences to life and property from underestimating support demands, but an overly conservative design can also result in excessive projects costs, sending a project well over initial budgets and proposed timelines (Marr 2001). As discussed by Hutchinson and Diederichs (1996), an ideal support design procedure would follow a cyclical approach involving the design, implementation, and verification of a support system. Following this rationale, which parallels with concepts of both the NATM or an observational approach previously discussed, the support system is systematically and continuously updated according to current excavation conditions rather than designing for the worst-case scenario (Austrian Society for Geomechanics 2010). This in its simplest form involves a comparison of the measured support response at the excavation with that predicted in the design stage by empirical, analytical, or numerical methods (or a combination of the three). In general, this accomplished by two types of sensing techniques:

(a) An extrinsic sensing technique to the support system (e.g., geodetic monitoring, extensometer arrays, closure meters) or
(b) An intrinsic sensing technique to the support system (e.g., load, strain, or orientation sensors coupled to constituent support members).

21.6.2 Extrinsic Sensing Techniques

Extrinsic sensing solution are unquestionably favored in many projects due to their relative ease of use, easily understood measurements (e.g., closure of the excavation profile), and potential for reuse (e.g., one lidar unit can be used along an entire tunnel development). In addition, sensing techniques that measure from within the excavation (i.e., surveying and lidar scans) are very suitable for support systems that are heavily based on surface support (e.g., Support Category II of the Driskos Tunnel, Table 21.1). Yet the support system mechanistic behavior becomes less trivial with increased components (e.g., Support Category V of the Driskos Tunnel, Table 21.1). In multicomponent support system design, a sensing technique that is not inherent to the constituent support members will require assumptions regarding the transfer of ground mass displacements or loads to support member/system response. As such, it is necessary to have realistic understanding of the relationship between the loading response of a support system (or constituent member) and the displacement and/or load measurement provided by an extrinsic sensing technique. This will undoubtedly depend on the type of support elements that compose the tunnel support system, as well as the geological and in situ stress state conditions at the given project.

Various support typologies may not necessarily have a load response that is uniform across the entirety of their length, depending on their respective anchor mechanism, and their load-displacement response may vary significantly. This is visually depicted in Fig. 21.32, which shows the load-displacement profile comparison between 20 mm cement grouted rebar, 46 mm friction set, and a 22 mm resin grouted D-Bolt (Li 2010). Referring to the plot, for each support element typology, a same level of applied coaxial load can result in vastly different displacement magnitudes between the three support elements. This has been shown to highlight that

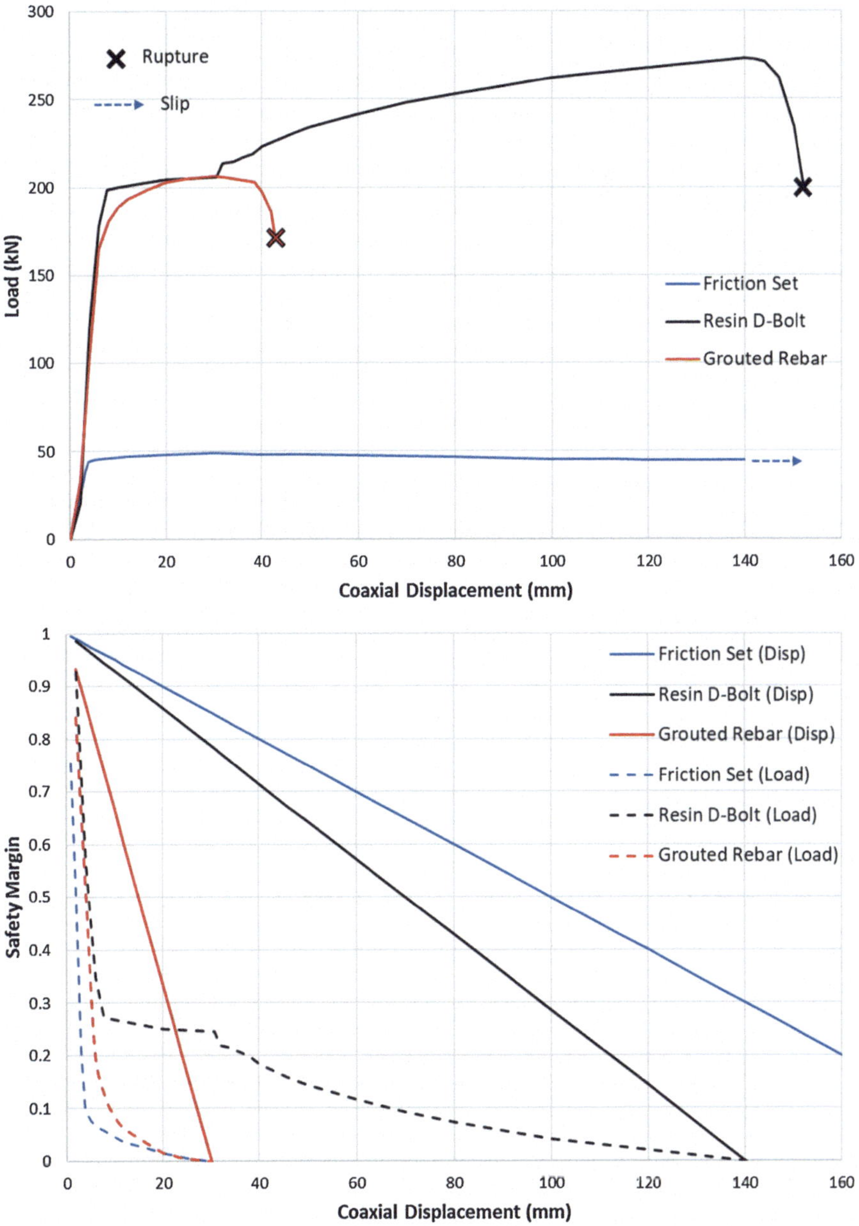

Fig. 21.32 (Upper) Load-displacement profile comparison between a 20 mm cement grouted rebar (Blanco Martín et al. 2011), 46 mm friction set (Li 2017), and 22 mm resin grouted D-Bolt (Li 2012). (Bottom) Corresponding load and displacement safety margin (SM) per applied coaxial displacement—where safety margin is determined according to SM = 1 − (1/Factor of Safety). Accordingly, a SM of 0 is equivalent to a Factor of Safety of 1. Under a relative uniform displacement of the tunnel, coaxial displacement could be taken as convergence of the tunnel profile

various support members can be placed at different positions along their respective safety factor curves at the same tunnel displacement (Fig. 21.32). Support member loading behavior will also vary significantly depending on the relative discontinuity of the ground mass in addition to anisotropic in situ stresses. In consideration of complex support system behavior, extrinsic sensing techniques can be used to provide insight into the current condition or remaining capacity of a support system to allow resupport/rehabilitation procedures, or additional support measures to be recommended (as was recommended for the Driskos Tunnel), but may not provide a sufficiently accurate measurement as to allow reductions to be made to the support design without unwarranted risk to the stability of an excavation.

21.6.3 Intrinsic Sensing Techniques

There a wide variety of intrinsic sensing techniques that have been employed to monitor the performance of support members. Perhaps the most common approach has involved the use of discrete technologies (i.e., one sensor equals one measurement point) directly coupled with the support member, Fig. 21.33. Examples include: electrical-resistive strain gauges (e.g., Farmer 1975; Serbousek and Signer 1987), load cells (e.g., Rodger et al. 1996; Mitri 2011), long base length

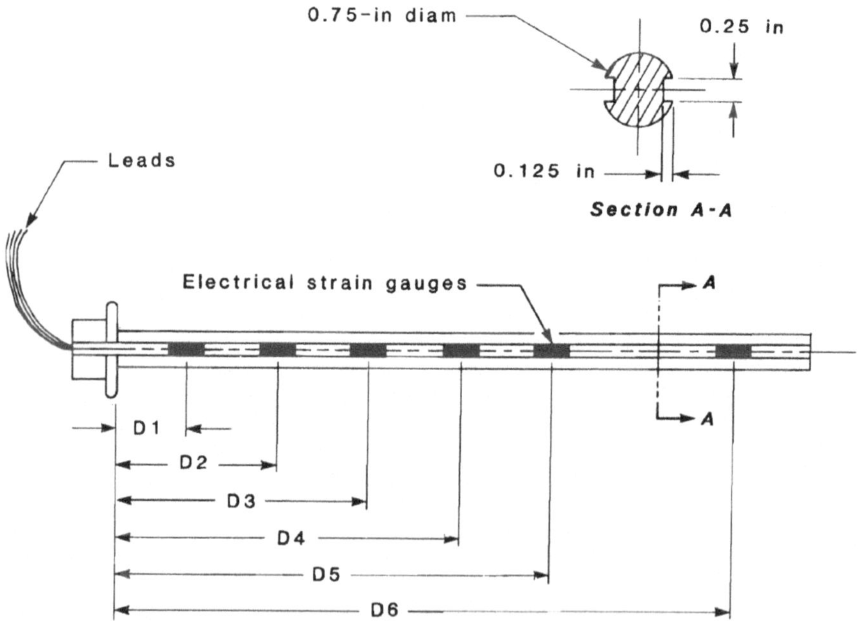

Fig. 21.33 Schematic representation of strain gauge positioning along a bolt specimen within a pair of diametrical opposed machined out grooves (after Serbousek and Signer 1987)

induction sensors (Spearing et al. 2013). Depending on the sensor type and base-length, the load distribution along a support member is effectively discretized into one to many individual measurement points or zones. The density of measurements along the support member, or the spatial resolution, will be decided by the number of discrete sensors that are installed along the given support member. As intrinsic sensing techniques monitor the support system/member itself, the deflection, load/strain state, or even corrosion (e.g., Wei et al. 2015; Craig et al. 2016) is directly measured (i.e., not inferred through the interaction with the ground mass). This allows a convenient comparison between the designed/predicted support behavior and the measured support response during tunnel advance and over time. Although, a separate sensing technique to measure the excavation displacement should also be employed to compare with the support response.

It is important to acknowledge that the measurements provided from a given discrete sensing technique are restricted to the span of the sensor's base-length. As such, the location(s) designated to install discrete sensors should take into consideration the support design or anticipated behavior. This will assist in identifying the most salient position(s) to gather measurements within a given support system or along an individual support member. The anticipated tunnel behavior can also provide insight into the number of sensors and sensor types that should be consider. For example, a circular tunnel that is expected to deform in a relatively uniform manner and is supported by a coarse pattern of rock bolts would require significantly less measurements than the conditions previously described for the Driskos Tunnel (Section 4).

The spatial resolution of measurements using discrete sensing technologies is inherently limited as there both cost and practical limitation to the number of discrete sensors that can be employed within a given support system or along an individual support member. This has led various researchers to consider innovative sensing techniques that are capable of providing a distributed measurement along the length of a support member. Two techniques of noteworthy consideration are the use of guided ultrasonic waves (e.g., Steblay 1987) and fiber optic sensing (FOS) (e.g., Schroeck et al. 2000).

21.6.4 Guided Ultrasonic Waves

Guided ultrasonic wave techniques have primarily found success as support member length or encapsulation (i.e., resin or cement grout) length inspection tools. These techniques operate by emitting and monitoring a low frequency signal (<100 kHz) from one end of the support member (predominately the exposed head at the excavation periphery). The signal will attenuate according to encapsulation inconsistencies and reflect at the opposing end of the support member (Zou and Cui 2011; Stepinski et al. 2014), Fig. 21.34. Higher order frequencies (i.e., MHz range) have also been identified to be less susceptible to surface conditions (i.e., encapsulation) along a support member and that attenuation is largely contributed

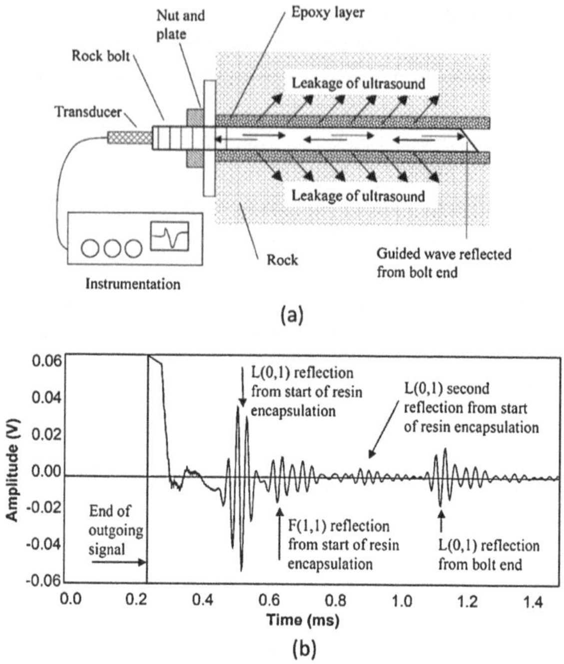

Fig. 21.34 (**a**) Schematic diagram of an example pulse-echo based rock bolt sensing configuration. (**b**) Example time trace recorded from a 2.4 m long, partially grouted rock bolt (after Beard and Lowe 2003)

to curvature of the member (e.g., Beard and Lowe 2003). Accordingly, there is vast potential to use guided ultrasonic waves as an economic nondestructive testing technique to monitor support member load, deflection, and installation quality (e.g., Cui and Zou 2012). Although the technique can be considered a distributed sensing approach (the support member itself is used as a waveguide for the signal), it presently lacks load or strain measurement accuracy and spatial resolution to make the technique a feasible solution for inspecting complex support response.

21.6.5 High Spatial Resolution Fiber Optic Strain Sensing

One of the most intriguing aspects of FOS is the potential to use a single, micrometer diameter, optical fiber as both the lead and the transducer for an array of measurement points. There currently exists a variety of demodulation techniques that resolve either spontaneous or induced back reflections of an incident light signal at specific locations or continuously along the length of an optical fiber. As a change in the local strain or temperature along an optical fiber will induce a modulation of the back-reflected signal (i.e., amplitude, phase, and frequency), a continuous or semicontinuous sensor can be realized by comparing an initial, unperturbed signal with that at a later date. A comparison of the operational features of several commercially available FOS technologies is provided in Table 21.2. Of the listed

Table 21.2 Summary of operational features/capabilities and pricing for wavelength division multiplexing fiber Bragg gratings (FBG), quasi-distributed FBG (DFBG), Brillouin based distributed sensing (BOTDR/BOTDA), and Rayleigh based distributed sensing (ROFDR) techniques (after Forbes et al. 2018)

Technique	FBG (Micron Optics Inc. 2012; FBGS 2015)	DFBG (Luna Innovation Inc. 2017; Sensuron 2017)	BOTDR/ BOTDA (Omnisens 2014)	ROFDR (Luna Innovation Inc. 2017)
Max. sensing length	>1000 m	<52 m	>1000 m	<40 m
Measurement repeatability[a]	± 0.1–10 με	± 1 με	± 1 με	± 5 με
Spacing of measurements (i.e., spatial resolution)	0.10 m (practically)	6.35 mm	0.10–1 m	0.65 mm
Max. number of measurement points	10–20 (practically)	>1000	>1000	>1000
Sensing range	± 17,500 με	± 30,000 με	± 30,000 με	± 30,000 με
Acquisition time	<1000 Hz	<250 Hz	<1 Hz	<60 Hz
Unit price (approximate USD)	$15,000–$125,000	>$70,000–$125,000	$100,000–$250,000	$60,000–$150,000
Sensor price (approximate USD)	~$300–$1000 per sensor	~$300–$5000 per sensor	$0.10 per meter of fiber	$0.10 per meter of fiber
Max. number of connected sensors[b]	>10	8	2	8

[a]Repeatability will ultimately be related to the level of strain experienced by the optical sensor. At higher strain levels (>10,000 με) repeatability will decrease
[b]Maximum number of connected sensors without the purchase of an additional switch unit

technologies, the Rayleigh-based optical frequency domain reflectometry (ROFDR) (Froggatt and Moore 1998; Soller et al. 2005) unit is further discussed in the context of ground support monitoring due to its superior capability regarding spatial resolution of measurements (0.65 mm). This technology measures the spontaneous Rayleigh back-scatter (amplitude and phase) that arises from random fluctuations of the refractive index continuously along the length of a standard (i.e., unmodified) single-mode optical fiber. In comparison with the other listed technologies, ROFDR provides the best potential to both identify and capture local and microscale ground support response mechanisms that may have been misinterpreted or even omitted by the limited spatial resolution of the previously discussed discrete sensing techniques.

Several research efforts have discussed various manners by which an optical fiber can be coupled with a ground support member: surface mounting, encapsulated within a central conduit or within machined grooves (e.g., Iten and Puzrin 2010; Mohamad et al. 2011; Hyett et al. 2013). These procedures were employed in order to best protect the optical fiber sensor at various stages of the support member's serviceability life; installation to end of design life. For example, positioning the sensor within lengthwise machined grooves along a support member situates the optical fiber below the exterior surface. An adhesive can then be used to both bond

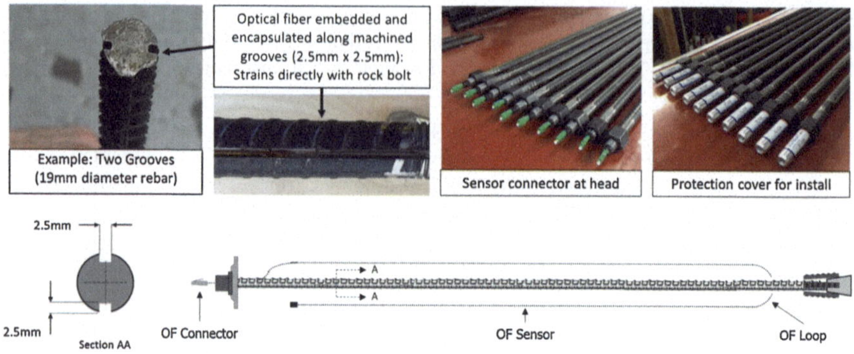

Fig. 21.35 Example measures taken to protect the optical fiber sensor and sensor connector. In this case the optical fiber sensor is positioned within machines grooves cut along the length of a rebar member and encapsulated with an adhesive. A protective cover is also used to prevent the sensor connector from being damaged during installation with a bolting jumbo

the sensor and act as a protective barrier. Other considerations specific to the support element type and installation protocol can also be considered to protect the connection at one end of the support member (Fig. 21.35).

Referring to Table 21.2, FOS techniques do not necessarily provide a superior sensing accuracy, range, acquisition time, or monitoring capability. For example, many conventional (electrical) discrete sensing techniques can now be equipped with an individual radio or wifi unit, removing the need for lead wires in a tunnel project. FOS techniques, on the other hand, presently require each sensor to be directly connected to the selected interrogator/analyzer/sensing unit. However, where FOS techniques are limited in data monitoring, perhaps limiting temporal resolution of measurements, they can compensate for by providing unparalleled information on support response, especially under complex geological conditions, through fine spatial resolution. Figure 21.36 displays a comparison of strain profiles measured during an example double shear loading experiment of an instrumented rebar. In this experiment, the rebar was instrumented with both an array of electrical-resistive strain gauges and the ROFDR technology. Comparing both the FOS and strain gauge measurements, it is apparent that the coarse spatial resolution provided by the strain gauges does not sufficiently capture the shear mechanism, as drastically different strain profiles are obtained. This suggests that in many cases it may be completely fortuitous for such discrete sensing techniques to measure localized and complex support response in situ without an a priori knowledge of the loading condition, in this case two vertical shear planes.

Forbes et al. (2017) have discussed how a single optical fiber sensor can be used to monitor three lengths of a single support member to allow a three-dimensional strain rosette analysis to be conducted. This implies that both the location and orientation of ground mass displacements along a support member do not need to know beforehand to be accurately captured, which is off significant benefit for both verification and optimization of ground support design. Using this technique, the

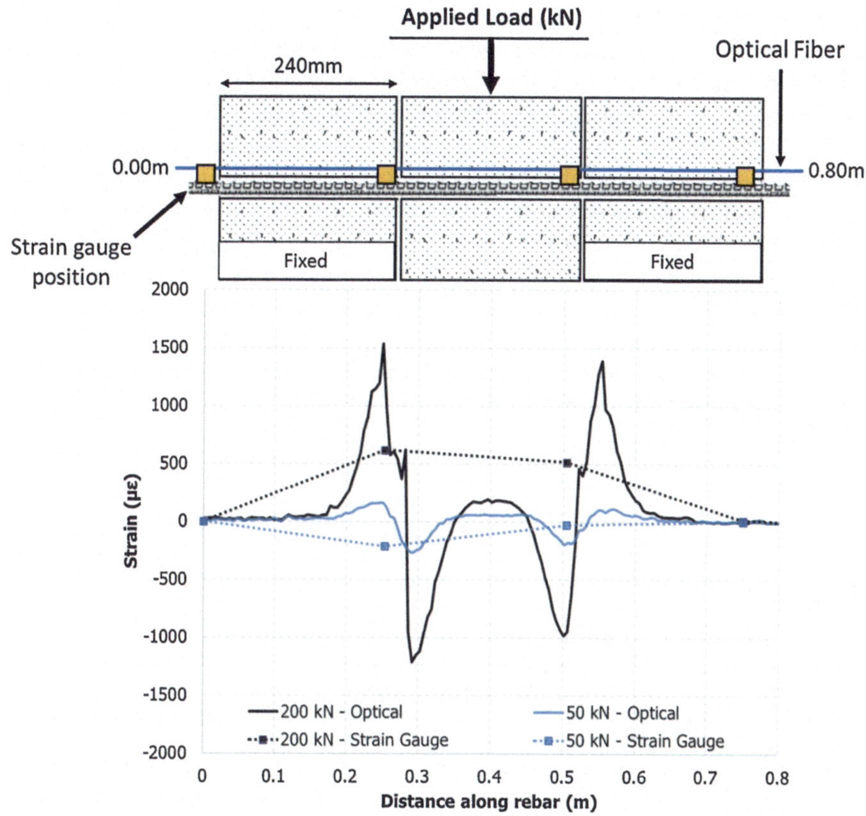

Fig. 21.36 Strain profile comparison measured along the top axis of an instrumented rebar during a double shear test. (Upper) Schematic of the experimental arrangement—a 19 mm diameter rebar instrumented with an optical fiber sensor and an array of four strain gauges was center and grouted within a borehole running through three concrete blocks. The outer two blocks were fixed and load was applied to the central block (to cause displacement). (Bottom) Interpolated strain profile from the array of strain gauges (Strain Gauge) and the strain profile measured using the ROFDR technique (Optical) at 50 and 200 kN of load applied to the central block

coaxial and bending moment induced strain components can be distinguished along the entire length of a support member, allowing the support response to be assessed throughout an entire tunnel construction process.

21.7 Summary

The methodology and design procedures for tunnelling within weak rock is non-trivial in nature. Prior to determining the proper tunnelling technique and ground/rock support method, a tunnel design engineer must utilize the data obtained from the site investigation, rock classification and characterization. This Chapter summarized the key components associated with weak rock tunnelling and also included practical design considerations associated with a Case Study of the Driskos Twin Tunnel as part of the Egnatia Odos Motorway that was constructed in Northern Greece. Modelling considerations at the design stage for such support elements was also included in order to highlight the factors that must be considered when modelling such support elements within various rockmass types. In terms of emerging technologies, a state-of-the-art fiber optic strain sensing technique was included in order to highlight its potential use in future tunnel design, operation, and maintenance considerations as warranted by the Observational Method.

Acknowledgements The authors wish to acknowledge the support of the following industrial as well as governmental sponsors: Natural Sciences and Engineering Council of Canada (NSERC), the Canadian Department of National Defense, The members of the GeoMechanics Group at Queen's University, MITACS, Yield Point Inc. and the Royal Military College (RMC) Green Team.

References

Austrian Society for Geomechanics (2010) Guideline for the geotechnical design of underground structures with conventional excavation. ÖGG, Salzburg

Beard MD, Lowe MJS (2003) Non-destructive testing of rock bolts using guided ultrasonic waves. Int J Rock Mech Min Sci 40(4):527–536

Blanco Martín L, Tijani M, Hadj-Hassen F (2011) A new analytical solution to the mechanical behaviour of fully grouted rockbolts subjected to pull-out tests. Construct Build Mater 25(2):749–755. https://doi.org/10.1016/J.CONBUILDMAT.2010.07.011

Brown ET (1981) Rock characterization testing and monitoring. Pergamon Press, Oxford

Carranza-Torres C, Fairhurst C (2000) Application of the convergence-confinement method of tunnel design to rock masses that satisfy the Hoek-Brown failure criterion. Tunnel Undergr Space Technol 15(2):187–213

Craig P, Serkan S, Hagan P, Hebblewhite B, Vandermaat D, Crosky A, Elias E (2016) Investigations into the corrosive environments contributing to premature failure of Australian coal mine rock bolts. Int J Min Sci Technol 26:59–64

Cui Y, Zou D (2012) Assessing the effects of insufficient rebar and missing grout in grouted rock bolts using guided ultrasonic waves. J Appl Geophys 79:64–70

Diederichs MS (2000) Instability of hard rockmasses: the role of tensile damage and relaxation. Ph.D. thesis, University of Waterloo, Waterloo, Ontario, Canada

Diederichs MS (2007) The 2003 Canadian Geotechnical Colloquium: mechanistic interpretation and practical application of damage and spalling prediction criteria for deep tunnelling. Canad Geotech J 44(9):1082–1116

Diederichs MS, Carter MA, Lato MJ, Bennett JB, Hutchinson DJ (2013) Lidar surveying for liner condition, rock stability and reconditioning assessment of Canada's oldest railway tunnel in Brockville, Ontario. Proceedings of the Annual Conference of the Canadian Geotechnical Society GeoMontreal, Montreal, 2013

Egnatia Odos AE (1999) Selection or rock mass properties for tunnel design. Proceedings of the Conference on Tunnels of Egnatia Highway, Ioannina, Greece

Egnatia Odos AE (2001) The appropriate use of geological information in the design and construction of the Egnatia Motorway Tunnels, Tunnel Construction in Greece, Technical notes (MASc. program) of the National Technical University of Athens, Section 5

Farahmand K, Vazaios I, Diederichs MS, Vlachopoulos N (2018) Investigating the scale-dependency of the geometrical and mechanical properties of a moderately jointed rock using a synthetic rock mass (SRM) approach. Comput Geotech 95:162–179. https://doi.org/10.1016/j.compgeo.2017.10.002

Farmer IW (1975) Stress distribution along a resin grouted rock anchor. Int J Rock Mech Min Sci Geomech Abst 12:347–351

FBGS (2015) Draw Tower Grating (DTG's) in low bend loss fiber (1550/830) [Datasheet] FBGS International NV

Forbes B, Vlachopoulos N, Hyett AJ, Diederichs MS (2017) A new optical sensing technique for monitoring shear of rock bolts. Tunnel Undergr Space Technol 66:34–36. https://doi.org/10.1016/j.tust.2017.03.007

Forbes B, Vlachopoulos N, Hyett AJ (2018) The application of distributed optical strain sensing to measure the strain distribution of ground support members. FACETS 3(1):195–226. https://doi.org/10.1139/facets-2017-0093

Froggatt M, Moore J (1998) High-spatial-resolution distributed strain measurement in optical fiber with Rayleigh scatter. Appl Optics 37(10):1735–1740

Grasso P, Scotti G, Blasini G, Pescara M, Floria V, Kazilis N (2003) Successful application of the observational design method to difficult tunnel conditions – Driskos Tunnel, as provided by Geodata SA

Hindley G, Gibbons P, Agius M, Carr B, Game R, Kashani K (2004) Linking past and future. Civil Eng ASCE 74(5):50–59

Hoek E (1999) Support for very weak rock associated with faults and shear zones. International Symposium on Rock Support and Reinforcement Practice in Mining, Kalgoorlie, Australia

Hoek E (2001) Big tunnels in bad rock. J Geotech Geoenviron Eng 127(9):726–740

Hoek E (2003) Rock-Support Interaction analysis for tunnels in weak rock masses. Rocscience. Available from https://www.rocscience.com/documents/pdfs/rocnews/winter2012/Rock-Support-Interaction-Analysis-for-Tunnels-Hoek.pdf. Accessed on 21 Oct 2018

Hoek E, Diederichs MS (2006) Empirical estimation of rock mass modulus. Int J Rock Mech Min Sci 43(2):203–215

Hoek E, Marinos P (2000) Seventh Report by Panel of Experts (Geotechnical/Tunnelling) on EGNATIA ODOS HIGHWAY PROJECT Sections 1.1.4, 1.1.6, 2.3, 3.2, and 11.23

Hoek E, Carranza-Torres C, Corkum B (2002) Hoek-Brown failure criterion-2002 edition. 5th North American Rock Mechanics Symposium and 17th Tunneling Association of Canada Conference: NARMS-TAC, 2002, pp 267–271

Hutchinson DJ, Diederichs MS (1996) Cablebolting in underground mines. BiTech Publishers Ltd, Richmond

Hyett AJ, Forbes BJ, Spearing AJ (2013) Enlightening bolts: using distributed optical sensing to measure the strain profile along fully grouted rock bolts. Proc. 32nd International Congress on Ground Control in Mining, Morgantown, WV, pp 107–112

Iten M, Puzrin AM (2010) Monitoring of stress distribution along a ground anchor using BOTDA. Proc. SPIE Sensors and Smart Structures Technologies for Civil, Mechanical, and Aerospace Systems, San Diego, California, USA

Karakuş M, Fowell RJ (2004) An insight into the new Austrian tunnelling method (NATM). Proc. ROCKMEC

Kontogianni VA, Stiros SC (2002) Predictions and observations of convergence in shallow tunnels: case histories in Greece. Eng Geol 63(3-4):333–345

Kontogianni V, Stiros S (2003) Tunnel monitoring during the excavation phase: 3-D kinematic analysis based on geodetic data. 11th FIG Symposium on Deformation Measurements, Santorini, Greece

Lambropoulos S (2005) Tunnelling on the Egnatia Odos Motorway: procedures adopted, managerial choices made and lessons learned. Symposium on Design, Construction and Operation of Long Tunnels, Taipei

Li CC (2010) A new energy-absorbing bolt for rock support in high stress rock masses. Int J Rock Mech Min Sci 47:396. https://doi.org/10.1016/j.ijrmms.2010.01.005

Li CC (2012) Performance of D-bolts under static loading. Rock Mech Rock Eng 45:183–192. https://doi.org/10.1007/s00603-011-0198-6

Li CC (2017) Principles of rockbolting design. J Rock Mech Geotech Eng 9:396. https://doi.org/10.1016/j.jrmge.2017.04.002

Luna Innovation Inc (2017) ODiSI-B optical distributed sensor interrogator. Available from http://lunainc.com/wp-content/uploads/2016/07/ODB5_DataSheet_Rev13_020217.pdf

Lunardi P (2000) The ADECO-RS approach in the design and construction of the underground works of Rome to Naples High Speed Railway Line: a comparison between final design specifications, construction design and "as built". AITES-ITA World Tunnel Congress "Progress in tunnelling after 2000", pp 329–340

Marinos PG, Hoek E (2000) GSI: a geologically friendly tool for rock mass strength estimation. Proc. GeoEng2000 Conference, pp 1422–1442

Marr WA (2001) Why monitor geotechnical performance? Proc. 49th Annual Geotechnical Conference, Minneapolis, MN

Micron Optics Inc (2012) Sensing instrumentation and software: Enlight, User guide (Revision 1.138) Atlanta, GA. Available from http://www.micronoptics.com/support_downloads/Manual.pdf

Mitri H (2011) Evaluation of rock support performance through instrumentation and monitoring of bolt axial load. Proc. 11th Underground Coal Operators' Conference, Wollongong, NSW, Australia, pp 136–140

Mohamad H, Soga K, Pellew A, Bennett PJ (2011) Performance monitoring of a secant-piled wall using distributed fiber optic strain sensing. J Geotech Geoenviron Eng 137:1236. https://doi.org/10.1061/(ASCE)GT.1943-5606.0000543

Omnisens (2014) Omnisens DITEST: fiber optic distributed temperature & strain sensing technique [Technical Note] Omnisens SA, Switzerland

Rawlings CG, Silva RD, Kazilis N, Aggistalis G, Rachianiotis N, Mantziara P, Hoek E, Marinos P (2001) The tunnels of the Egnatia Motorway project, Northern Greece, Underground Construction 2001 IMM, BTS, London as cited in Technical Notes (MASc. program) of the National Technical University of Athens, Section 4

Rodger AA, Littlejohn GS, Xu H, Holland DC (1996) Instrumentation for monitoring the dynamic and static behaviour of rock bolts in tunnels. Proc Inst Civil Eng Geotech Eng 119(3):146–155

Romeo V (2002) NATM in soft-ground: a contradiction of terms. Final Liner. Jacobs Assoc Newslett Spring 2002:338

Schroeck M, Ecke W, Graupner A (2000) Strain monitoring in steel rock bolts using FBG sensor arrays. Proc. Applications of Optical Fiber Sensors. Glasgow, UK, pp 298–304

Sensuron (2017) RTS125+ Data Sheet. Available from http://www.sensuron.com/rts125/

Serbousek MO, Signer SP (1987) Linear load-transfer mechanics of fully grouted roof bolts. U.S. Bureau of Mines, Report of Investigations 9135

Smith M (2000) WT ON SITE-Egnatia highway across Northern Greece-most ambitious infrastructural project to link east with west approaches the halfway stage. World Tunnel 13(4):185–190

Soller BJ, Wolfe M, Froggatt ME (2005) Polarization resolved measurement of Rayleigh backscatter in fiber-optic components. Proc. Optical Fiber Communication Conference and Exposition and The National Fiber Optic Engineers Conference. Anaheim, CA

Spearing AJS, Hyett AJ, Kostecki T, Gadde M (2013) New technology for measuring the in situ performance of rock bolts. Int J Rock Mech Min Sci 57:153–166. https://doi.org/10.1016/j.ijrmms.2012.07.027

Steblay BJ (1987) New instrumentation for roof bolt load measurement. IEEE Trans Indust Appl IA-23(4):731–735

Stepinski T, Matsson KJ, Ekenbro B (2014) New instrument for rock bolt inspection using guided waves. Proc., 11th European Conference on Non-Destructive Testing, Prague, CZ
Structural Design SA (1999) Egnatia highway - evaluation of the data of the geotechnical observation of Driskos Tunnel by O.K. Consultants (in Greek)
Vazaios I, Vlachopoulos N, Diederichs MS (2018) The effect of jointing in massive highly interlocked rockmasses under high stresses by using a FDEM approach. Proceedings of the IAEG 2018 Conference, San Francisco, USA
Vazaios I, Vlachopoulos N, Diederichs MS (2019) The mechanical analysis and interpretation of the EDZ formation around deep tunnels within massive rockmasses using a hybrid finite-discrete element approach: the case of the AECL URL test tunnel. Canad Geotech J 56:35
Vlachopoulos N (2009) Back analysis of a tunnelling case study in weak rock of the Alpine System in Northern Greece: validation and optimization of design analysis based on ground characterization and numerical simulation. Ph.D. thesis, Queen's University, Kingston, Ontario, Canada
Vlachopoulos N, Diederichs MS (2009) Improved longitudinal displacement profiles for convergence confinement analysis of deep tunnels. Rock Mech Rock Eng 42:131–146. https://doi.org/10.1007/s00603-009-0176-4
Vlachopoulos N, Diederichs MS, Marinos V, Marinos P (2013) Tunnel behaviour associated with the weak Alpine rock masses of the Driskos Twin Tunnel system, Egnatia Odos Highway. Canad Geotech J 50(1):91–120. https://doi.org/10.1139/cgj-2012-0025
Vlachopoulos N, Vazaios I, Madjdabadi B (2018) Investigation into the influence of excavation of twin bored tunnels within rock masses adjacent to slopes. Canad Geotech J 55:1533. https://doi.org/10.1139/cgj-2017-0392
Wei H, Zhao X, Li D, Zhang P, Sun C (2015) Corrosion monitoring of rock bolt by using a low coherent fiber-optic interferometry. Opt Laser Technol 67:137–142
Zou DH, Cui Y (2011) A new approach for field instrumentation in grouted rock bolt monitoring using guided ultrasonic waves. J Appl Geophys 75(3):506–512

Chapter 22
Face Stability of Tunnels in Soft Rocks

Qiujing Pan and Daniel Dias

22.1 Literature Review to Tunnel Face Stability

These deterministic models give good predictions with high computational efficiency compared to results given by numerical modelings. Soft rocks, a common geomaterial in nature, show some basic features, such as low strength, strong plasticity, and highly weathering. Engineering constructions in soft rocks are often difficult due to their mechanical characteristics. When excavating in soft rocks, the tunnel face, a high-risk region, can collapse because of the stress release and the low strength of the surrounding rock masses. For closed-face tunnels using tunneling boring machines, shield machines can provide support pressures to tunnel face. It permits to compensate earth pressures as well as underground water pressures. For openface conventional tunnels, no face support is conducted, but some auxiliary techniques, such as face reinforcement with dowels and umbrella pipe, can be employed to maintain a safe excavation. Thus, it is very interesting to investigate tunnel face stability in soft rock masses. The issue of tunnel face stability has been investigated by numerous researchers by means of experimental tests, numerical simulations, and theoretical models.

22.1.1 Experimental Studies

Chambon and Corte (1994) appear to be the first to investigate face failure mechanism by conducting a series of centrifuge experiments to monitor the face extrusion in sands. The tunnel model is a rigid metallic tube of 100 mm diameter. The failure zone

Q. Pan · D. Dias (✉)
3SR Laboratory, Grenoble Alpes University, CNRS UMR 5521, Grenoble, France
e-mail: Daniel.dias@3sr-Grenoble.fr

M. Kanji et al. (eds.), *Soft Rock Mechanics and Engineering*,
https://doi.org/10.1007/978-3-030-29477-9_22

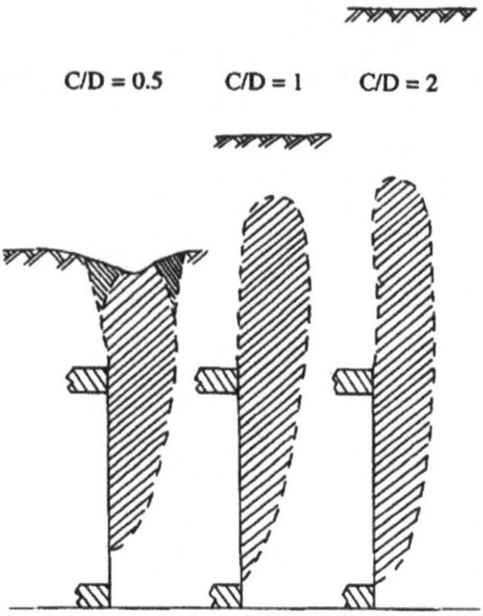

Fig. 22.1 Failure shapes observed experimentally (Chambon and Corte (1994))

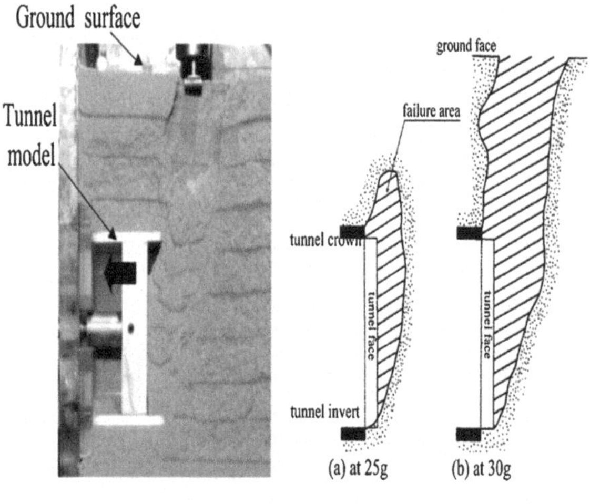

Fig. 22.2 Failure patterns observed in centrifuge tests (Kamata and Mashimo (2003))

geometries ahead of the tunnel face are plotted in Fig. 22.1, showing bulb-shaped (or chimney-shaped) and being dependent of the tunnel depths. The failure zone reaches the ground surface for a shallow tunnel (the cover-to-diameter ratios $C/D = 0.5$), while for a deep tunnel ($C/D = 1.0, 2.0$) it only extends vertically up to a height of one diameter.

Kamata and Mashimo (2003) studied the tunnel stability in a frictional material using centrifuge model tests. The tunnel model has a diameter of 80 mm. For faces with no reinforcement, the failure zone resembles to a dome-shape zone at 25 g and reaches the ground surface at 30 g, as shown in Fig. 22.2.

Takano et al. (2006) presented three-dimensional images of tunnel face failure patterns by means of a X-ray computed tomography scanner (see Fig. 22.3), in

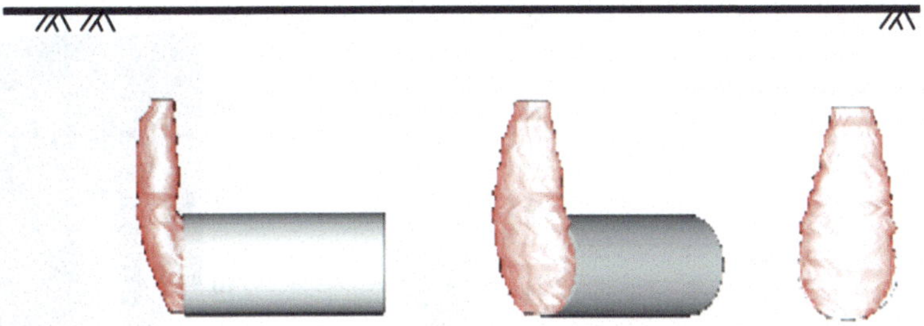

Fig. 22.3 CT images of a failed tunnel face (Takano et al. (2006))

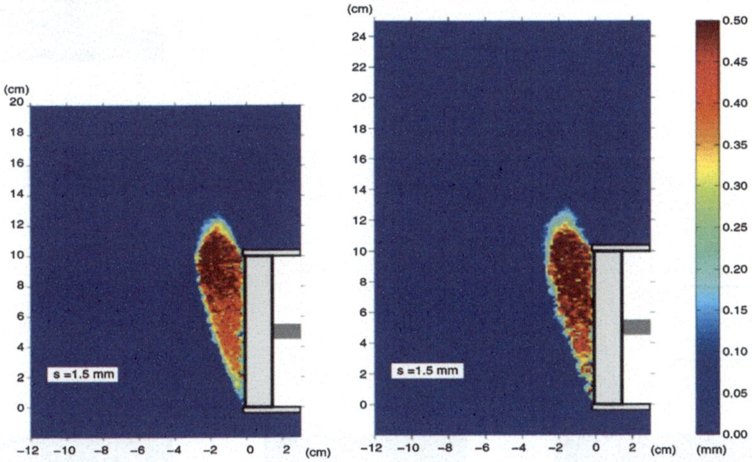

Fig. 22.4 Incremental displacements of a tunnel face in dense sands (Kirsch (2010))

which the failure is modeled by pulling out a tunnel model (a pipe with an outer diameter of 150 mm and an inner diameter of 120 mm) in sandy materials. It shows that the real failure pattern is a three-dimensional one. The authors suggested that the failure mechanism can be simulated using logarithmic spirals in vertical planes and elliptical shapes in the horizontal cross sections.

By application of the particle image velocimetry (PIV) technique, Kirsch (2010) performed experiment tests to observe the movements around a tunnel with a diameter of 100 mm in dry sands. The results of displacement patterns show that the overburden has a negligible impact on the shape and extent of the failure regions; a clear chimney-wedge-type collapse mechanism was observed in dense sands (see Fig. 22.4).

Idinger et al. (2011) implemented centrifuge model tests, respectively at 1 g and 50 g, on face stability of shallow tunnels for different cover-to-diameter ratios. Figure 22.5 presents the obtained displacement vectors and contours of shear strain. A failure mechanism with two well-defined shear bands (or slip surfaces), one starting from the tunnel invert and the other from the tunnel crown, is clearly observed.

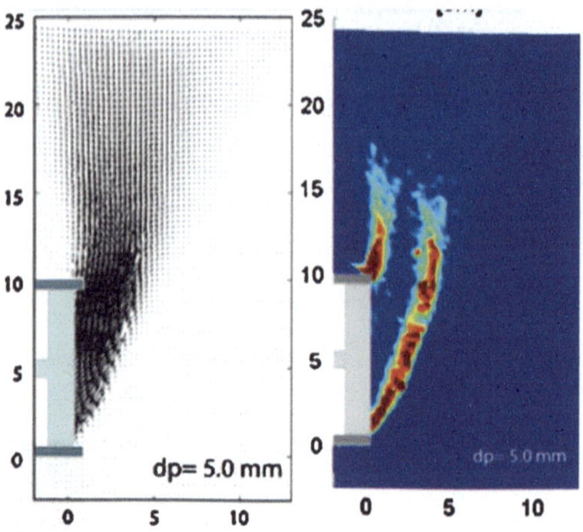

Fig. 22.5 Displacement vectors and contours of shear strains (Idinger et al. (2011))

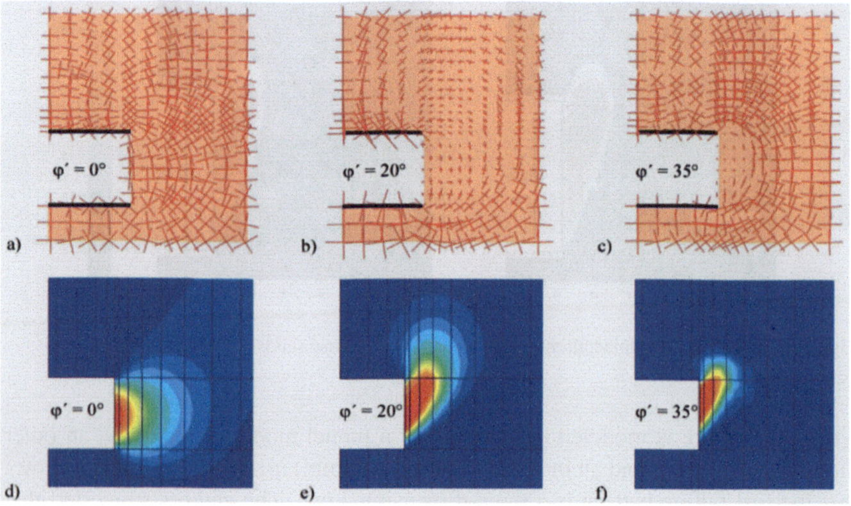

Fig. 22.6 Principal stresses and incremental displacements ahead of a tunnel face at failure (Vermeer et al. (2002))

22.1.2 Numerical Modelings

Vermeer et al. (2002) conducted three-dimensional finite element calculations to study the heading stability of a tunnel in drained grounds. Figure 22.6 presents the principal stress distributions and incremental displacements ahead of a tunnel face at the failure for different friction angles. A chimney-like collapse mechanism can be observed in this figure; for friction angles higher than 20°, the arching effect (a strong stress arch) is extremely clear.

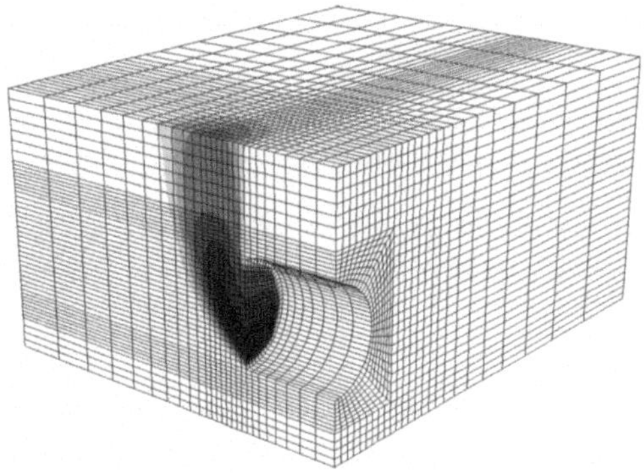

Fig. 22.7 Contour of velocity of a tunnel face at failure (Mollon et al. (2009a))

Mollon et al. (2009a) investigated the face stability numerically by means of a finite difference numerical model in cohesive-frictional materials. A velocity field of a tunnel face at the ultimate limit state is shown in Fig. 22.7.

22.1.3 Theoretical Models

In order to perform efficient and reliable calculations to evaluate stability of the core ahead of the face, for example, at the preliminary design stage, many attempts have been made to develop simplified theoretical models or analytical models. The upper-bound limit analysis (LA) method and the limit equilibrium (LE) method are two main theoretical approaches applied for this problem. Both LA and LE require an assumption of the failure shape of a tunnel face. The face failure features observed in experimental and numerical studies provide guidance to assume a tunnel face failure mechanism.

22.1.3.1 Limit Analysis Methods

The limit analysis is a powerful tool to perform a stability analysis and has been widely used in the realm of geotechnical engineering (e.g., slopes, retaining walls, foundations, and tunnels). The upper and lower bound theorems of limit analysis could offer a rigorous and powerful technique for estimating collapse loads. There have been many achievements about applying the limit analysis method to determine required face pressures.

Translational Failure Mechanisms

Leca and Dormieux (1990) proposed a three-dimensional rigid cone mechanism for a shallow tunnel face composed of two conical blocks in the case of the collapse and of a single conical block outcropping at the ground surface for the blow-out case

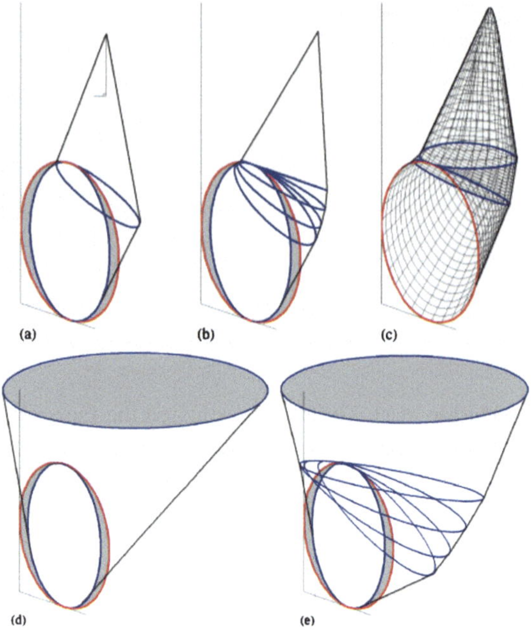

Fig. 22.8 A series of cone failure mechanisms: (**a**) Two conical blocks, (**b**) Multi-blocks and, (**c**) Discretized multi-blocks and 2 passive failure mechanisms (**d**) Single block and, (**e**) multi-blocks (Mollon et al. (2011))

(see Fig. 22.8a, d). Mollon et al. (2009b) presented a three-dimensional multi-block failure mechanisms, by adding several truncated rigid cones between the two conical blocks (see Fig. 22.8b, e). Such a modification allows the slip surface to move more freely and improves the existing solutions. However, these failure mechanisms (see Fig. 22.8a, b, d, e) only cover an inscribed ellipse of the tunnel face, other sections of the tunnel face remaining to be immobile. In order to cover the entire circular tunnel face, Mollon et al. (2009c) proposed a spatial discretization technique to generate a new three dimensional multi-block failure mechanism (see Fig. 22.8c). In addition these cone failure mechanisms have a translational velocity field, for which the velocity field in each block is the same. This is different from the velocity fields observed in experimental studies and numerical modelings.

Rotational Failure Mechanisms

The rotational failure mechanisms assume a cylindrical rotational velocity field. The velocities in failure mechanisms vary at different positions. They are equal to the product of the angular velocity and the distance between the rotating center and the point under consideration. Such a rotational velocity field is inspired by experimental observations and numerical simulations.

Chambon and Corté (1990) seem to be the first to propose a 2D rotational failure mechanism based on two logarithmic spirals (see Fig. 22.9a). A new two dimensional rotational mechanism was proposed by Mollon et al. (2010), which is constructed by a spatial discretization technique. The failure shape is not described by standard curves such as log-spirals (see Fig. 22.9).

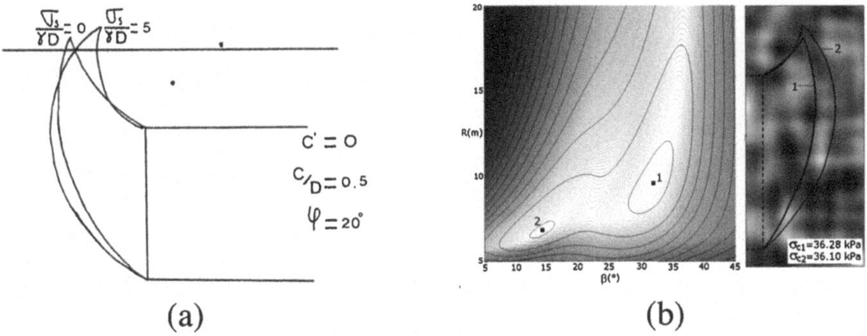

Fig. 22.9 Two-dimensional horn mechanisms of (**a**) Chambon and Corté (1990) and (**b**) Mollon et al. (2010)

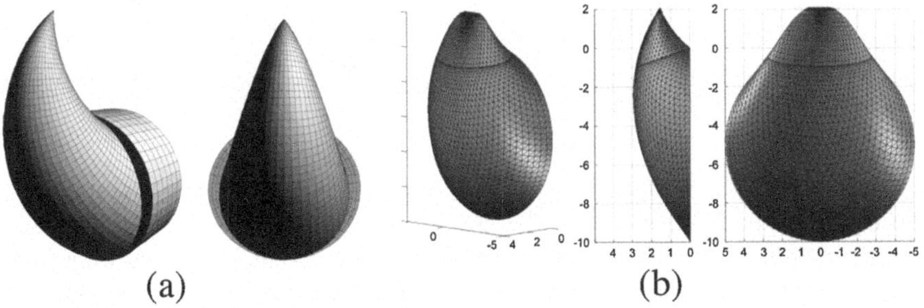

Fig. 22.10 Three-dimensional horn mechanisms of (**a**) Subrin and Wong (2002) and (**b**) Mollon et al. (2011)

The 2D failure mechanism tends to overestimate the critical collapse pressures because it underestimates the arching effect which is three-dimensional in reality. Subrin and Wong (2002) seem to be the first to apply a three dimensional horn mechanism to investigate the tunnel face stability (see Fig. 22.10a). Mollon et al. (2011) proposed an advanced failure mechanism by using a spatial discretization technique, for which the failure mechanism is generated "point by point" instead of using existing standard geometric shapes such as cones or cylinders. The advanced 3D rotational failure mechanism poses two significant characteristics: it includes the entire tunnel face and is consistent with the rotational movement observed in experiments. It has been proven that this advanced 3D rotational failure mechanism yields the best upper-bound solutions.

22.1.3.2 Limit Equilibrium Methods

The most classical failure mechanism in the context of the limit equilibrium method is the wedge-prism sliding model firstly proposed by Horn (1961), which consists of a wedge ahead of the tunnel face and an overlaying prism up to the ground surface, as shown in Fig. 22.11. The limit state can be obtained by equilibrating all

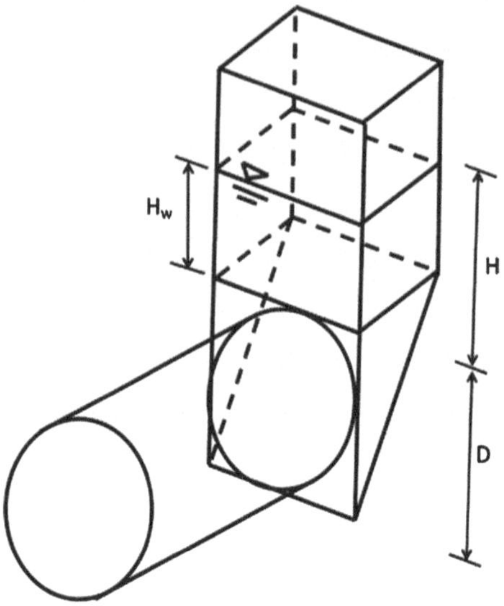

Fig. 22.11 The wedge-prism sliding failure model (Horn (1961))

forces acting on the wedge, including the vertical force obtained by applying the silo theory to the prismatic body, the gravity of the wedge, the normal and shear forces acting on the inclined sliding surface, and the horizontal force resulting from face support systems. The wedge-prism model has been widely extended to investigate face stability subjected to seepage forces and reinforced by bolts.

The wedge-prism sliding model requires several user-defined parameters, such as the lateral stress coefficients, the distribution of vertical stresses inside the wedge. Anagnostou and Perazzelli (2013) improved the wedge-prism model by applying a slice method to the wedge block to solve the distributions of vertical stresses inside the wedge. However, this model has some drawbacks. The cross section of the tunnel face is approximated by a square with an equivalent area, and the assumed failure surface does not coincide well with the typical patterns of failure observed in experiments.

In the following part, the kinematical approach of limit analysis will be applied to the tunnel face stability.

22.2 Kinematical Approach of Limit Analysis

The application of limit analysis to geotechnical fields started by Drucker and Prager (1952) who employed the kinematical approach of limit analysis to investigate slope stability. Chen (1975) systematically introduced the limit analysis to three types of classic geotechnical problems: slope stability, earth pressure and foundation bearing capacity. Since then, the application of limit analysis in geotechnical engineering has experienced a great development and become an effective manner to examine the stability of geotechnical structures.

The limit analysis method, based on classical plasticity theory, which is a powerful vehicle for stability analysis of geotechnical engineering, consists of the kinematical approach (also known as the upper-bound theorem) and the static approach (or the lower-bound theorem). With the two theorems used together for a stability problem, a range between which the true limit load exists can be found regardless of the complex behavior of the material. The application of the kinematical approach of limit analysis requires the construction of a kinematically admissible velocity field which complies with the kinematical boundary conditions and the flow rule. While the lower-bound theorem relies on establishing statically admissible stress fields in which both the equilibrium and the material yield criterion are satisfied at any point. Most previously published researches about the application of limit analysis on stability analysis concentrated on the kinematical approach, but the static approach is rarely used due to the difficulties in constructing a statically admissible stress field. The method employed in this chapter also refers to the kinematical approach of limit analysis.

The kinematical approach of limit analysis is able to provide an upper bound estimate of an active limit load, for instance the foundation capacity, or a lower bound to a reaction, such as on a retaining wall and a tunnel. Besides, it also could give an upper bound to a slope height, or a lower bound to, for example, the necessary cohesion to maintain the stability for an earth structure or the density of reinforcement required to stabilize a slope or a tunnel face.

With the failure mechanism defined by a kinematically admissible velocity field, the kinematical approach of limit analysis can provide an upper-bound estimate of the ultimate load which causes failure or a lower-bound estimate to the limit load which offers a resistance against failure for a structural system by means of a work balance equation, in which the work rate of external forces applied to the system is equal to or less than the internal energy dissipation (Chen 1975),

$$\int_S T_i \dot{u}_i^* dS + \int_V F_i \dot{u}_i^* dV \le \int_V \sigma_{ij} \dot{\varepsilon}_{ij}^* dV \tag{22.1}$$

where $\dot{\varepsilon}_{ij}^*$ and $\dot{u}_i^*$ are, respectively, the strain rate tensor and the velocity vector in the kinematically admissible velocity field; σ_{ij} is the stress associated with $\dot{\varepsilon}_{ij}^*$; F_i refers to the body force distributed over the volume V bounding by the failure surface S and T_i represents the surface force at the boundary S.

The sum of the two terms on the left-hand side represents the work rate of external forces, while the term at the right-hand term refers to the internal energy dissipation. The computation of internal energy dissipation is dependent of the yield criterion adopted to characterize material failure. There are two common yield criteria that are often used for soft rocks: the Mohr–Coulomb yield criterion and the Hoek–Brown yield criterion.

22.3 Failure Criteria Used for Soft Rocks

22.3.1 Mohr–Coulomb Yield Criterion

The Mohr–Coulomb failure criterion presumes that the plastic flow occurs at any point inside a material when the following relation is reached,

$$\tau = c + \sigma \tan \varphi \tag{22.2}$$

where σ is the normal stress and τ the shear stress on the failure plane; c and φ are corresponding strength parameters, and c represents the cohesion and φ the angle of internal friction. The Mohr–Coulomb failure criterion is widely used for soil and rock engineering practice due to its simplicity and good performance.

When applying the kinematical approach of limit analysis to Mohr–Coulomb materials, the internal energy dissipation can be computed by (Chen 1975),

$$W_D = \int_S cv \cos \varphi ds \tag{22.3}$$

in which S represents the failure surface, and v is the velocity on the failure surface.

22.3.2 Generalized Hoek–Brown Criterion and Tangential Line Technique

The nonlinear Hoek–Brown yield criterion is extensively used for rock masses with varying degrees of fracture. The generalized Hoek–Brown criterion reported by Hoek et al. (2002) can be written as the following equation:

$$\sigma_1 = \sigma_3 + \sigma_{ci} \left(m \frac{\sigma_3}{\sigma_1} + s \right)^a \tag{22.4}$$

where σ_1 and σ_3 represent respectively the maximum and minimum effective principal stresses; σ_{ci} denotes the uniaxial compressive strength of the intact rock. The Hoek–Brown parameters, s, a, and m, are determined by the constant m_i, the geological strength index GSI and the disturbance factor D_i,

$$\frac{m}{m_i} = \exp\left(\frac{\text{GSI} - 100}{28 - 14D_i} \right) \tag{22.5}$$

$$s = \exp\left(\frac{\text{GSI} - 100}{9 - 3D_i} \right) \tag{22.6}$$

$$a = \frac{1}{2} + \frac{1}{6}\left[\exp\left(-\frac{\mathrm{GSI}}{15}\right) - \exp\left(-\frac{20}{3}\right)\right] \tag{22.7}$$

The magnitude of constant m_i can be determined by compression tests; the geological strength index GSI characterizes the rock mass quality. The disturbance factor D_i reflects the disturbance degree to rock masses in situ, varying from 0.0 for undisturbed rock masses to 1.0 for highly disturbed rock masses.

Because of the nonlinear failure envelope of Hoek–Brown criterion, the tangential line technique is used to compute the internal energy dissipation in the kinematic approach of limit analysis. The main idea is to replace the nonlinear failure criterion by a tangential line, in which the slope and the intercept of a tangential line to the nonlinear envelope correspond respectively to the equivalent friction angle φ_t as well as the equivalent cohesion c_t.

The tangential line technique yields an upper-bound solution to the original problem since the tangential line exceeds the original nonlinear envelope. The equivalent friction angle φ_t as well as the equivalent cohesion c_t in the context of generalized Hoek–Brown criterion follows the following relationship (Yang et al. 2004):

$$\frac{c_t}{\sigma_{ci}} = \frac{\cos\varphi_t}{2}\left[\frac{ma(1-\sin\varphi_t)}{2\sin\varphi_t}\right]^{\frac{a}{1-a}} - \frac{\tan\varphi_t}{m}\left(1+\frac{\sin\varphi_t}{a}\right)\left[\frac{ma(1-\sin\varphi_t)}{2\sin\varphi_t}\right]^{\frac{1}{1-a}} + \frac{s}{m}\tan\varphi_t \tag{22.8}$$

When computing the internal energy dissipation, the cohesion c in Eq. (22.3) can be replaced by the equivalent cohesion c_t in Eq. (22.8). This leads to an additional parameter, the equivalent friction angle φ_t, which is determined by optimization together with other variables that determine the geometry of the failure mechanism.

22.4 Tunnel Excavated by a TBM (Circular Tunnel Face)

22.4.1 Introduction of a 3D Rotational Failure Mechanism

The main principle of the spatial discretization technique for generation of the 3D rotational mechanism proposed by Mollon et al. (2011) is described in this section. The whole failure mechanism rotates around an horizontal axis passing through the point O with an angular velocity ω, assuming a cylindrical rotational velocity field. For any point of the failure mechanism, the velocity is equal to the product of the angular velocity and the distance between the rotating center and that point under consideration.

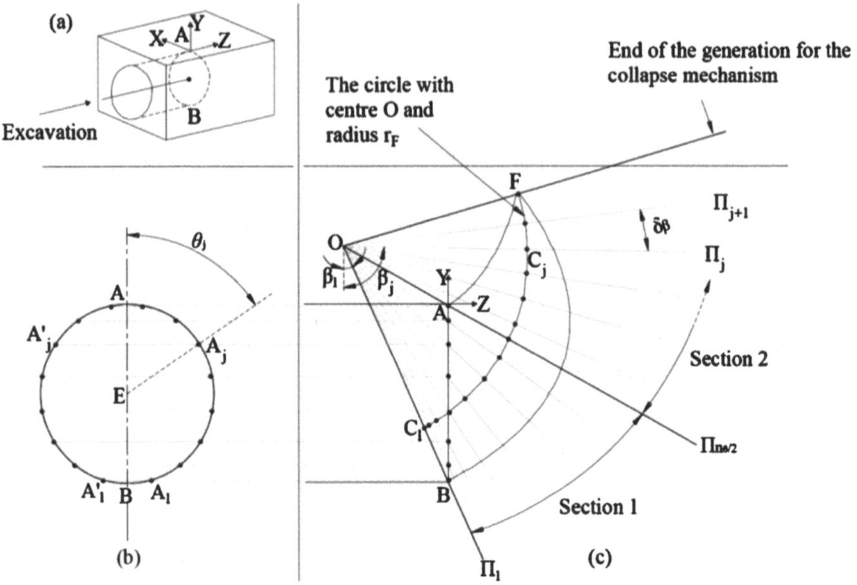

Fig. 22.12 Generation of the 3D rotational collapse mechanism (Mollon et al. 2011)

Fig. 22.13 Schematic of three neighboring points (Mollon et al. 2011)

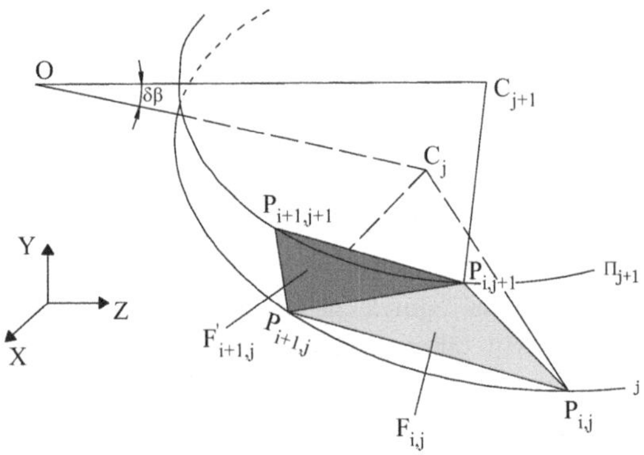

The whole failure mechanism is divided into two sections as shown in Fig. 22.12c. The tunnel face is discretized by n_θ points (the first discretization parameter), denoted by A_j and $A_j'(1 \leq j \leq n)$, as seen in Fig. 22.12b. A series of radial planes passing through the rotational center O are defined. Two symmetrical points A_j and A_j' at the tunnel face together with the rotational center O determine a radial plane, and two adjacent radial planes are separated by a fixed angle of δ_β, referred to as the second discretization parameter.

The 3D failure mechanism is generated by a series of points (e.g., $P_{i,j}$ and $P_{i+1,j}$), and every three neighboring points, $P_{i,j}$, $P_{i+1,j}$, $P_{i,j+1}$, can define a triangular facet $F_{i,j}$, which constitutes the boundary of the failure mechanism, as shown in Fig. 22.13.

The key work lies in generating a new point in a given radial plane ($\prod_{j+1}$) using another two points in the previous radial plane ($\prod_j$).

Both $P_{i,j}$ and $P_{i+1,j}$ located in the plane $\prod_j$ are assumed to be known, so it is possible to calculate the coordinates of the third point $P_{i,j+1}$ on the plane $\prod_{j+1}$ by using (1) the normality condition of the upper-bound theorem and (2) $P_{i,j+1}$ quasi-uniformly distributed along the plane $\prod_{j+1}$. Therein, the normality condition requires that the normal vector $\mathbf{N}_{i,j}$ of the triangular facets $F_{i,j}$ make an angle of $\pi/2 + \varphi$ with the velocity vector, where φ is the friction angle.

With the help of the spatial discretization technique, the 3D failure surface is generated by a series of points (e.g., $P_{i,j}$, $P_{i+1,j}$, $P_{i,j+1}$, and $P_{i+1,j+1}$), which represent the boundary (triangular facets) of the failure mechanism. Every three neighboring points (triangular vertexes), $P_{i,j}$, $P_{i+1,j}$, $P_{i,j+1}$ (respectively $P_{i,j}$, $P_{i+1,j}$, $P_{i,j+1}$), can define a triangular facet $F_{i,j}$ (respectively $F'_{i,j}$), as illustrated in Fig. 22.14. The first kind of triangular facets $F_{i,j}$ satisfies the normality condition. However, the second kind of triangular facets $F'_{i,j}$ (the inverted one), defined by three points $P_{i,j+1}$, $P_{i+1,j+1}$, and $P_{i,j+1}$, may not meet the normality condition strictly. But if the discretization parameters vary towards an infinite fineness, the two connected triangular facets (i.e., $F_{i,j}$ and $F'_{i,j}$) will be coplanar, and the discrepancy of the normality condition will vanish. Therefore, the two discretization parameters, namely, n_θ and δ_β, largely influence the accuracy of the failure mechanism.

The generation process starts at two symmetric points A_1 and A'_1 belonging to plane $\prod_1$, which is the initial condition for the failure mechanism. The generation of the mechanism is terminated if generated points reach the position of F before outcropping or if generated points outcrop before arriving at F. The position of F is the intersection of the two logarithmic spirals, which are the boundaries of the mechanism projected in the (Y, Z) vertical plane along the tunnel axis.

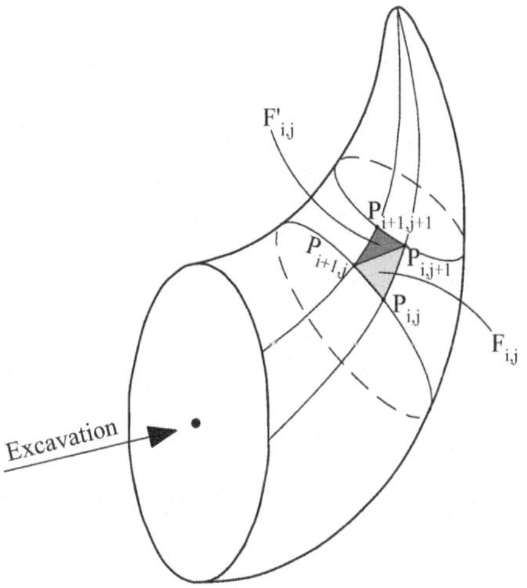

Fig. 22.14 Schematic of surface of the 3D rotational failure mechanism

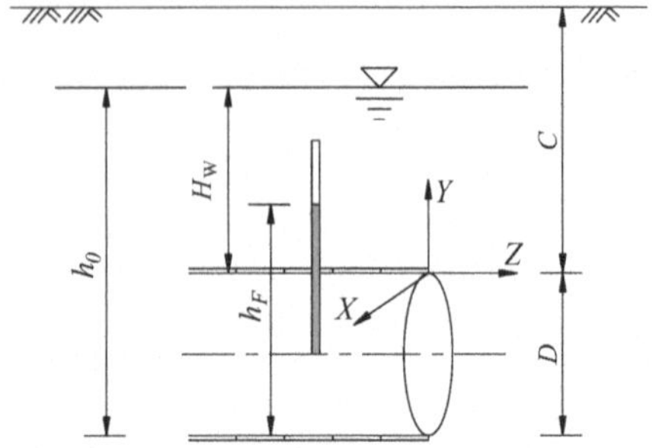

Fig. 22.15 Schematic diagram of a tunnel driven under the water table

22.4.2 *Face Stability in Hoek–Brown Rock Masses Subjected to Seepage Forces*

In practical engineering, tunnels are often constructed below water table such as subsea tunnels and cross-river tunnels. A schematic diagram for the problem under consideration is illustrated in Fig. 22.15. A circular tunnel with a diameter of D is excavated under the water table, C being the tunnel depth; h_0 denotes the water table elevation measured from the invert of the tunnel, and h_F refers to the piezometric head in the work chamber at the tunnel face, H_w being the water table elevation measured from the tunnel crown. If $h_F < h_0$, the groundwater flows into the excavated face with rapidly varying hydraulic head gradients ahead of the tunnel face, which simultaneously induces seepage forces acting in the direction of groundwater flow. This process results in face collapse.

22.4.2.1 Work Rate Calculations

In the process of tunneling, the external forces involve the gravity, the surface loads if the failure mechanism outcrops, the seepage forces when tunneling under water table, and the face supporting forces.

The work rate of external forces can be obtained by a simple summation on each element (the elementary volumes or elementary surface). The internal energy dissipation only takes place on the failure surface because of an assumption of rigid-block rotation, which is the sum of that on each triangular facet. As seen in Fig. 22.16, the triangular facet $P_{i,j}P_{i+1,j}P_{i,j+1}$ constitutes the boundary of the failure mechanism; points $P'_{i,j}$, $P'_{i+1,j}$, $P'_{i,j+1}$ are the projections of points $P_{i,j}$, $P_{i+1,j}$, $P_{i,j+1}$ on the symmetry plane. The element $V_{i,j}$ is adopted to compute the work rate of gravity.

The work rate of the weight of the block is computed by:

$$W_\gamma = \iiint_V \vec{\gamma}.\vec{\upsilon}.dV = \omega\gamma \sum_i \sum_j \left(R_{i,j} V_{i,j} \sin\beta_{i,j} \right) \tag{22.9}$$

where γ is the unit weight of rock masses; $R_{i,j}$ and $\beta_{i,j}$ are the polar coordinates of the barycenter of the element $V_{i,j}$; $V_{i,j}$ is volumes of the element; ω is the angle velocity of the failure mechanism.

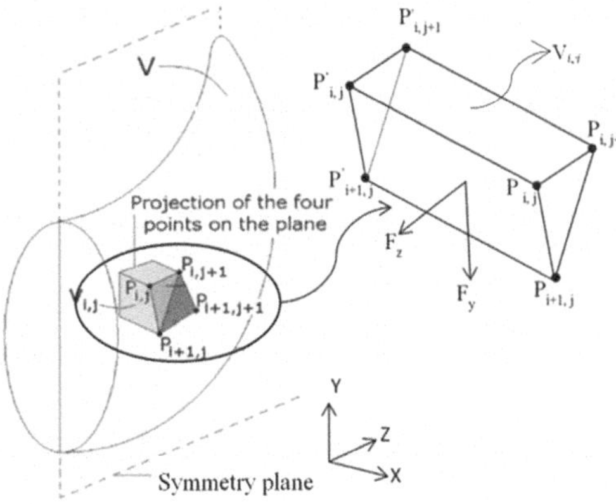

Fig. 22.16 Element discretization for work rate calculation

When the failure mechanism outcrops, it is truncated by the ground surface. The outcropping surface is constituted by the intersections between the failure mechanism and the ground surface. Introducing σ_s the ground surface surcharge, rate of work of a possible surcharge is formulated as

$$W_S = \iint_S \vec{\sigma}_s . \vec{\upsilon} dS = \omega \sigma_s \sum_i \left(S_i R_i \sin \beta_i \right) \tag{22.10}$$

where R_i and β_i are the polar coordinates of the point on the possible outcropping surface, and S_i represents the area of the element at the outcropping surface.

Denote σ_c the face pressures, the work rate of the face pressures is expressed as

$$W_T = \iint_S \vec{\sigma}_c . \vec{\upsilon} dS = -\omega \sigma_c \sum_i \left(S_{i,0} R_{i,0} \cos \beta_{i,0} \right) \tag{22.11}$$

where $R_{i,0}$and $\beta_{i,0}$ are the polar coordinates of a point on the tunnel face, and $S_{i,0}$ represents the area of the discretized element at the face.

The computation of internal energy dissipation is written as

$$W_D = \iint_S c.v.\cos\varphi.dS = \omega c_t \sum_i \sum_j \left(R_{i,j} S_{i,j} \cos\varphi_t \right) \tag{22.12}$$

where $S_{i,j}$ is the area of the triangular facets $F_{i,j}$; c_t and φ_t are respectively the equivalent cohesion and the equivalent friction angle.

The seepage force is a body force, so its work rate can be obtained by integrals over the whole failure mechanism:

$$W_e = \gamma_w \iiint_V \left(\frac{\partial h}{\partial x} v_x + \frac{\partial h}{\partial y} v_y + \frac{\partial h}{\partial z} v_z \right) dV \tag{22.13}$$

where h is the hydraulic head and $\frac{\partial h}{\partial x}$, $\frac{\partial h}{\partial y}$, and $\frac{\partial h}{\partial z}$ are hydraulic gradient components; v_x, v_y, and v_z are velocity components. Please note that $v_x = 0$ in the rotational velocity field, only y and z directions are involved.

The entity $V_{i,j}$ is adopted to calculate the work rate of seepage forces; therefore, the work rate of seepage forces can be formulated as

$$W_{\text{seepage}} = \omega \sum_i \sum_j \left[R_{i,j} \left(F_y \sin \beta_{i,j} + F_z \cos \beta_{i,j} \right) \right] \tag{22.14}$$

where F_y and F_z denote the components of resultant seepage forces acting on this element, which can be calculated by

$$F_y = -\gamma_w \int_V \frac{\partial h}{\partial x} dV \tag{22.15}$$

$$F_z = -\gamma_w \int_V \frac{\partial h}{\partial z} dV \tag{22.16}$$

By applying the divergence theorem to the integrals over the volume of the element $P_{i,j}P_{i+1,j}P_{i,j+1} - P'_{i,j}P'_{i+1,j}P'_{i,j+1}$, the components of resultant seepage forces can be further expressed as

$$F_y = -\gamma_w \int_S h n_y ds = \gamma_w \sum_k \overline{h}_k n_{y,k} s_k \tag{22.17}$$

$$F_z = -\gamma_w \int_S h n_z ds = \gamma_w \sum_k \overline{h}_k n_{z,k} s_k \tag{22.18}$$

where the summation index k indicates the five boundary surfaces of the element $P_{i,j}P_{i+1,j}P_{i,j+1} - P'_{i,j}P'_{i+1,j}P'_{i,j+1}$, and $n_{y,k}$ and $n_{z,k}$ denotes the direction cosines of the unit normal vector of each surface; s_k is the area; $\overline{h}_k$ is the average hydraulic head at each surface.

Based on the upper-bound theorem, the necessary effective face pressures σ_c that is smaller than the true limit loads can be acquired by equating the total external work rate to the total internal energy dissipation (Chen 1975):

$$\sigma_c = \gamma' D N_\gamma + \sigma_s N_s - c_t N_c + \gamma_w H_w N_w \tag{22.19}$$

where D is the tunnel diameter; c_t is the equivalent cohesion which is related to the equivalent friction angle φ_t given in Eq. (22.13). N_γ, N_s, N_c, and N_w are non-dimensional coefficients, representing, respectively, the contribution of rock weight, ground surface surcharge, the equivalent cohesion and seepage forces, which are given as follows:

$$N_c = \frac{\sum_i \sum_j \left(R_{i,j} S_{i,j} \cos \varphi_t \right)}{\sum_i \left(S_{i,0} R_{i,0} \cos \beta_{i,0} \right)} \tag{22.20}$$

$$N_s = \frac{\sum_i \left(S_i R_i \sin \beta_i \right)}{\sum_i \left(S_{i,0} R_{i,0} \cos \beta_{i,0} \right)} \tag{22.21}$$

$$N_\gamma = \frac{\sum_i \sum_j \left(R_{i,j} V_{i,j} \sin \beta_{i,j} \gamma' \right)}{D\gamma \sum_i \left(S_{i,0} R_{i,0} \cos \beta_{i,0} \right)} \tag{22.22}$$

$$N_w = \frac{\sum_i \sum_j \left[R_{i,j} \left(\sin \beta_{i,j} \sum_k \overline{h}_k n_{y,k} s_k + \cos \beta_{i,j} \sum_k \overline{h}_k n_{z,k} s_k \right) \right]}{H_w \sum_i \left(S_{i,0} R_{i,0} \cos \beta_{i,0} \right)} \tag{22.23}$$

Equation (22.19) is a function of three variables of β_E, r_E/D and φ_t, where r_E and θ_E are the polar coordinates of the tunnel face center. In the collapse case, the face pressures provide a resistance against face collapse, so the critical face pressure can be obtained by maximization of the σ_c in Eq. (22.19).

22.4.3 Hydraulic Head Distributions Ahead of Tunnel Face

The calculation of hydraulic heads relies on solving the Laplace equation which is hard to solve for a three dimensional seepage flow problem. The hydraulic head distribution under a three-dimensional steady state hydraulic field was computed numerically using the finite difference software *FLAC 3D*. A constant pore-water pressure is applied to the tunnel face. It is also assumed that there is a continuous and sufficient groundwater recharge from the adjacent river, well or reservoir, which means the water table elevation keeps constant.

The seepage field obtained by numerical modelings is extracted and then used to compute the hydraulic head of each point of the 3D failure mechanism. Figure 22.17 presents the contour lines of the piezometric head distributions in two planes., in which the elevation head is measured from the foot of the tunnel face. The increasing density of contour lines in the vicinity of the tunnel face indicates a rapid variation in hydraulic gradients, resulting in large seepage forces.

22.4.4 Comparisons with Numerical Analysis

In order to verify the presented method, the results obtained by the presented method are compared with those provided by numerical analysis using the finite difference software *FLAC 3D*. Since the problem is symmetrical at the vertical plane, only half of the domain is considered. Figure 22.18 shows a sketch of the numerical model of a circular tunnel with diameter of 10 m under buried depth of 20 m. The model contains approximately 90,000 zones and 95,000 nodes. The dimensions of the numerical model are taken as 40 m in the *X* direction, 55 m in the *Y* direction and 40 m in the *Z* direction. These dimensions are large enough to avoid boundary effects.

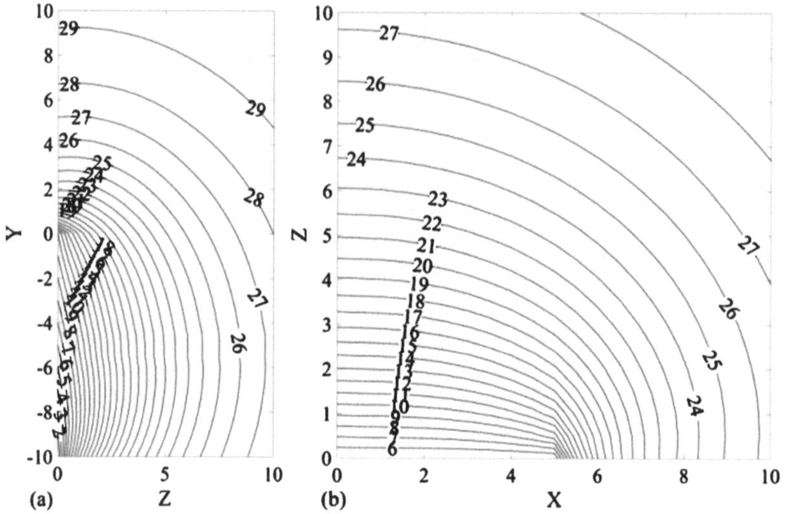

Fig. 22.17 Piezometric head contours in (**a**) vertical symmetry plane and (**b**) horizontal symmetry plane at $y = -5$ (unit: m)

Fig. 22.18 A sketch of the numerical model

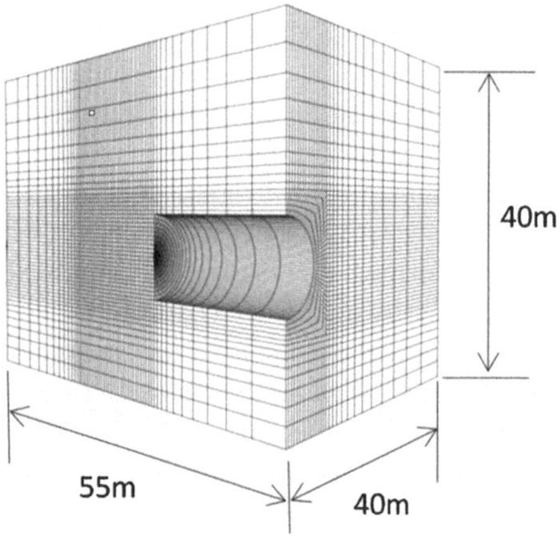

A fluid-mechanical analysis is employed to model this problem. The simulation proceeds in two steps: (1) fluid flow-only calculation during which stage the model is cycled to equilibrium to establish a steady-flow state for the changed hydraulic boundary conditions (this steady-state seepage field is also used to interpolate the hydraulic head of the 3D rotational failure mechanism for the kinematical approach); (2) mechanical-only calculation with the supporting pressure acting on the tunnel face.

In the fluid-mechanical simulation, both mechanical boundary conditions and hydraulic boundary conditions must be specified. For the mechanical boundary conditions, the displacements at the bottom of the model are fixed, and only the normal directions fixed on the vertical boundaries. For the hydraulic boundary

Table 22.1 Mechanical parameters for the hydro-mechanical analysis

Parameters	Case 1	Case 2
Rock submerged unit weight	20 kN/m^3	
Rock dry unit weight	25 kN/m^3	
Permeability coefficient	1.0×10^{-1} cm/s	1.0×10^{-4} cm/s
Young's modulus	300 MPa	700 MPa
Poisson's ratio	0.3	0.2

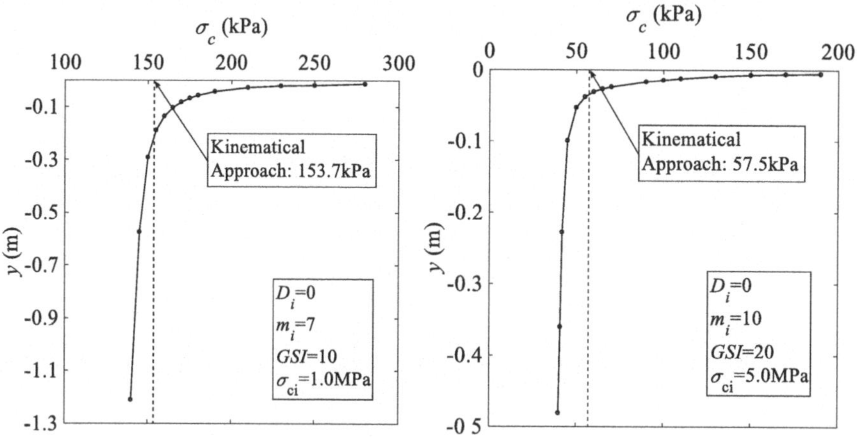

Fig. 22.19 Horizontal displacements at the tunnel face center versus face pressures

conditions, the pore-water pressure in the work chamber is presumed to be zero, and the groundwater level is assumed to be constant for the steady-state fluid flow analysis. The tunnel lining modeled by the structure element of liner is set to be impermeable so that the groundwater only seeps into the tunnel face.

In the numerical model, a linear perfectly elastic-plastic constitutive model based on the generalized Hoek–Brown yield criterion is assigned to the rock masses. The tunnel concrete lining is simulated with shell structural elements with a Young's modulus of 20 GPa, a Poisson's ratio of 0.2 and a thickness of 0.22 m. The other mechanical parameters used in the numerical analysis are listed in Table 22.1.

In order to determine the critical face pressures by the numerical analysis, the horizontal displacements of the center at the tunnel face versus different face pressures are plotted in Fig. 22.19 for the case of $m_i = 7$, GSI = 10, $\sigma_{ci} = 1.0$ MPa (a very poor rock mass) and the case of $m_i = 10$, GSI = 20, $\sigma_{ci} = 5.0$ MPa (a better soft rock mass) under $C/D = 2.0$ and $H_w/D = 2.0$. The minimum required face pressure may be defined as the value at which the horizontal displacement suddenly begins to go up significantly. The critical effective face pressures calculated by the presented kinematical approach are also given in the figure. It is seen that the critical effective face pressures obtained by the presented approach are close to those provided by the hydro-mechanical analysis, showing that the presented method is valid to deal with the considered problem.

22.4.5 Design Charts

Thanks to the efficient computational speed of the presented approach, it is possible to perform parametric analysis for a practical use. Two design charts are provided in Fig. 22.20 for $H_w/D = 1.0$ and in Fig. 22.21 for $H_w/D = 3.0$. In each chart, the normalized critical face pressure $\sigma_c/\gamma'D$ is plotted as a function of the normalized uniaxial compressive strength $\sigma_{ci}/\gamma'D$. In each set of figure, GSI changes from 10 to 40, and m_i from 5 to 20; the overburden ratio C/D is taken as 2.0 and the disturbance

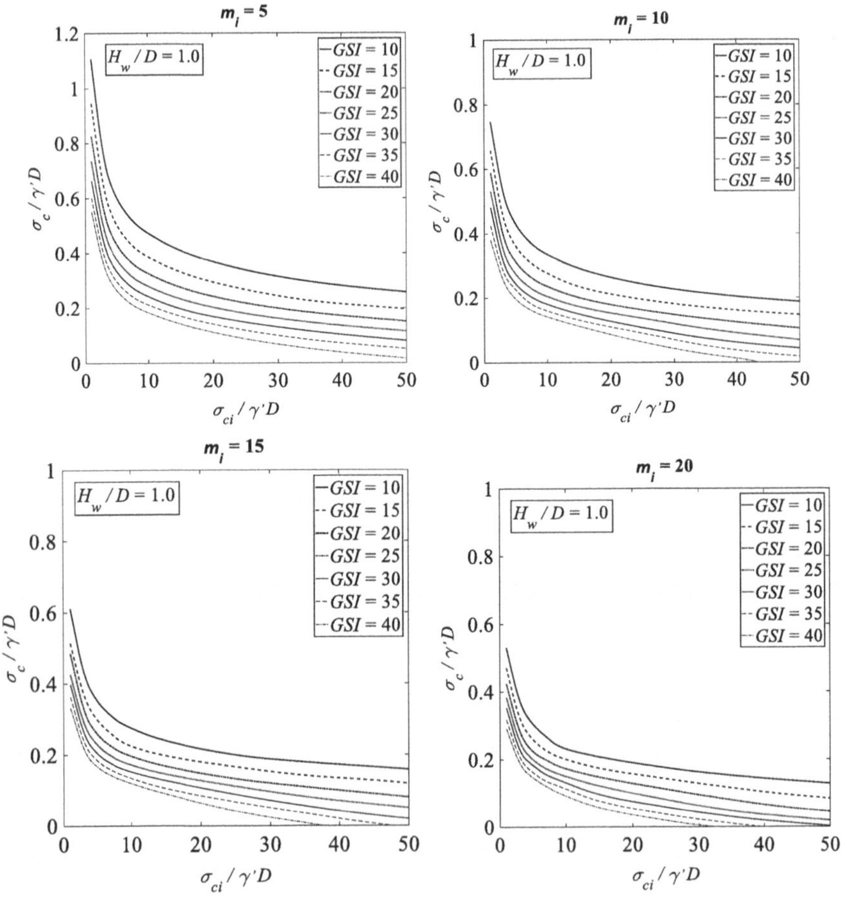

Fig. 22.20 Normalized critical face pressures versus normalized uniaxial compressive strength at $Hw/D = 1.0$

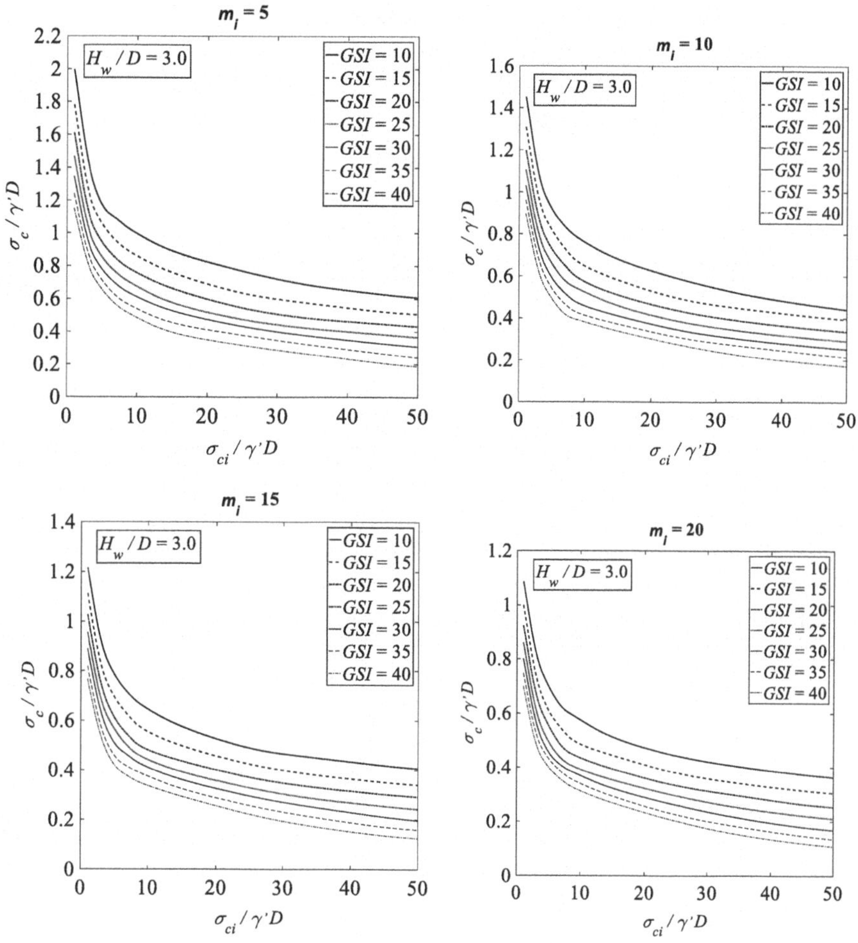

Fig. 22.21 Normalized critical face pressures versus normalized uniaxial compressive strength at $Hw/D = 3.0$

factor D_i is set to be zero. The diameter of the tunnel is taken to be 10 m; the ratio of submerged unit weight γ' to water unit weight γ_w is set to 2.0. The ground surface surcharge σ_s is set to zero.

The design charts presented are useful for the direct computation of the critical face pressure at the preliminary stage. Not surprisingly, the normalized critical face pressure rises with the increase of H_w/D ratio, but decreases with the rock mass quality. It is observed that $\sigma_c/\gamma'D$ shows an obvious nonlinear relationship with $\sigma_{ci}/\gamma'D$, especially for $\sigma_{ci}/\gamma'D$ smaller than 20.0.

22.5 Tunnel Excavated by a Traditional Process (Non-circular Tunnel Face)

22.5.1 *Problem Statement*

The open-face conventional tunneling method, such as New Austrian Tunneling Method (NATM), has also been extensively used for tunneling. In NATM, no supporting pressures are applied to the tunnel faces, therefore the more appropriate factor for the design in the open-face tunneling is the safety factor. Besides, the cross section of a NATM tunnel is often non-circular (horse-shoe, elliptical or more complex shape), which should be taken into account in the analysis. Most previously reported investigations failed to consider this point, in which the realistic section is replaced by a rectangular shape or a circular shape with equivalent area (Oreste and Dias 2012; Lee et al. 2004).

Since no supporting pressures are exerted at the face in conventional tunneling, some auxiliary techniques, like ground improvement with grouting, face reinforcement with dowels and umbrella pipes, have been developed to ensure a safe excavation. The use of dowels as a reinforcement technique has been successfully applied to improve tunnel face stability during excavation, for the sake of safety and construction speed.

In this section, the discretization technique is extended to generate a rotational failure mechanism for a non-circular tunnel face reinforced by bolts. A non-circular tunnel with height H, width W and buried depth C is schematically sketched in Fig. 22.22, points A, B, and E representing, respectively, the crown, invert and center of the tunnel face.

In Fig. 22.22, a horizontal bolt is inserted into the failure mechanism ahead of the tunnel face. Denote by T_m the limit tensile strength of the reinforcement, v_k the velocity at the failure surface and L_k the bolt length; R_k and β_k are the polar coordinates of

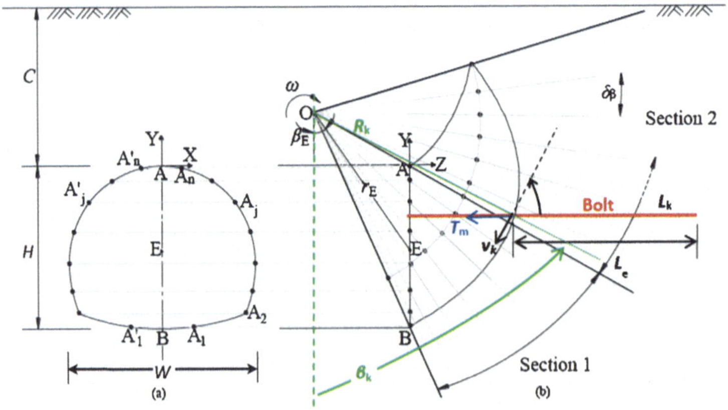

Fig. 22.22 Discretization of the collapse mechanism of a conventional tunnel face reinforced by bolts

the intersection between the bolt and the failure surface, and L_e is the bolt length outside the failure mechanism.

Based on the kinematic upper-bound approach, the required face pressure for a reinforced tunnel face can be formulated by

$$\sigma_c = \gamma H N_\gamma - cN_c - \frac{T_m}{\varnothing^2} N_b + \sigma_s N_s \tag{22.24}$$

where N_γ, N_c, and N_s are non-dimensional coefficients, representing given before. The non-dimensional coefficient N_b denotes the contribution of the face reinforcement, $\varnothing$ being the bolt diameter.

22.5.2 Proposed Failure Model of Dowels

22.5.2.1 Model Description

A new failure mode for the reinforcing element ahead of the tunnel face is proposed to calculate the resisting work rate. The proposed model is based on the following assumptions: (1) the reinforcing member is modeled as a one-dimensional structural element, for which only tensile resistance is taken into account; (2) a interaction zone is employed to simulate interaction between rock masses and bolts, which is located on the right of the failure surface; (3) at the failure surface, the axial velocity of the reinforcing element is compatible with that of the failure mechanism; (4) in the interaction zone, the rock masses are assumed to be motionless and a perfect adherence exists between rock masses and reinforcing elements. The assumption of the tensile failure mode can be justified by the observation that the bending effect ahead of the tunnel face during excavation is small and can be negligible (Kamata and Mashimo 2003).

Figure 22.23 shows a sketch of a reinforcement member passing through the interaction zone. The left region represents the rotational failure mechanism (a rigid block),

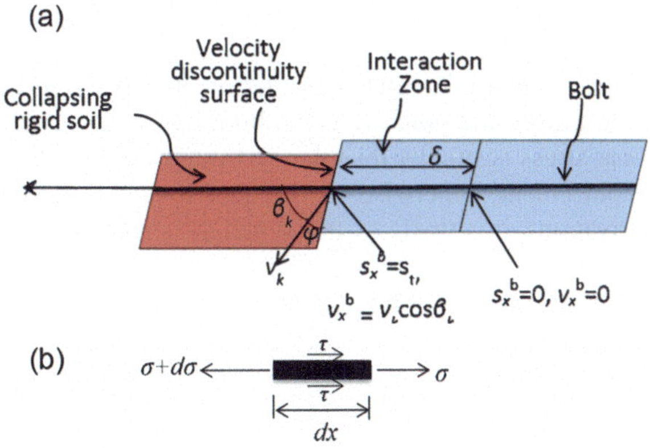

Fig. 22.23 The proposed tensile failure mode (**a**) the interaction zone and (**b**) the equilibrium element

and the right region is the motionless rigid rock masses. These two regions are separated by the failure surface, on which the velocity of the failure mechanism is denoted by v_k. The interaction zone is located on the right part of the failure surface, s_x^b and v_x^b, respectively, being the bolt axial displacement and velocity. The thickness of the interaction zone is denoted by δ. At the failure surface, the axial bolt velocity is compatible with the failure mechanism velocity with $v_x^b = v_k \cos\beta_k$. Because, the reinforcing member is modeled as one-dimensional structural element, only the tensile failure is taken into account. The tensile stress mobilized along the bolt over the interaction zone is trigged by the axial displacement s_t at its left boundary.

The interactions between the rock masses and the bolt are considered analytically for which the tensile failure and the pullout failure are considered separately under corresponding boundary conditions. The analytical solutions ensure that both the axial forces and the axial strain rates change with the position along the reinforcing member in the interaction zone. For example in the case of the tensile failure, the reinforcing member reaches its limit strength at the left boundary of the interaction zone and then reduces over the whole zone (also the axial strain rate). This feature is consistent with the real failure feature of bolts (Kamata and Mashimo 2003; Shin et al. 2008).

22.5.2.2 Resisting Work Rate of Reinforcements

Figure 22.23b plots the stress state for a differentially small element of a bolt; σ denotes the axial stress developed in the reinforcement (tensile force positive), and τ the shear stress at the interface along the reinforcement. An equilibrium equation between the axial stress σ and the shear stress τ can be derived as

$$A d\sigma = \pi \varnothing \tau dx \tag{22.25}$$

where A refers to the bolt cross-sectional area.

The shear stress τ at the interface along the bolt reads

$$\tau = \begin{cases} ks & s < \tau_m / k \\ \tau_m & s \geq \tau_m / k \end{cases} \tag{22.26}$$

where s denotes the relative displacement; k is the shear stiffness of the interface of rock masses and bolts; τ_m is the limit bonding strength at the interface. When the shear stress is smaller than the maximum value, it depends linearly on the relative displacement; otherwise, it is equal to the maximum value. In the following derivations, it is firstly assumed that the shear stress is smaller than the maximum value ($s < \tau_m/k$).

On substitution of Eq. (22.26) into Eq. (22.25) gives

$$A \frac{d\sigma}{dx} = k\pi \varnothing s \tag{22.27}$$

The tensile stress in the bolts can be written as

$$\sigma = E\varepsilon_x \tag{22.28}$$

where E is the elastic modulus of the reinforcement material; ε_x is the axial strain of bolt. Because the rock masses are idealized as motionless in the interaction zone, the axial strain ε_x can be calculated by the following expression:

$$\varepsilon_x = \frac{ds(x)}{dx} \tag{22.29}$$

By introducing Eq. (22.29) into Eq. (22.27) results in

$$\frac{d\sigma}{dx} = E\frac{d^2 s}{dx^2} \tag{22.30}$$

Then, one can obtain

$$\frac{d^2 s}{dx^2} - \alpha^2 s = 0 \tag{22.31}$$

where $\alpha = \sqrt{\frac{k\pi\varnothing}{EA}}$, and boundary conditions for Eq. (22.31) are

$$\begin{aligned} x &= 0, s = 0 \\ x &= \delta, s = s_t \end{aligned} \tag{22.32}$$

Then the relative displacement can be written as

$$s(x) = \frac{\mathrm{e}^{\alpha x} - \mathrm{e}^{-\alpha x}}{\mathrm{e}^{\alpha\delta} - \mathrm{e}^{-\alpha\delta}} \cdot s_t \tag{22.33}$$

The axial strain along the bolt is

$$\varepsilon_x(x) = \frac{\mathrm{e}^{\alpha x} + \mathrm{e}^{-\alpha x}}{\mathrm{e}^{\alpha\delta} - \mathrm{e}^{-\alpha\delta}} \cdot \alpha s_t \tag{22.34}$$

The axial force mobilized in the bolt is

$$T(x) = EA\alpha s_t \frac{\mathrm{e}^{\alpha x} + \mathrm{e}^{-\alpha x}}{\mathrm{e}^{\alpha\delta} - \mathrm{e}^{-\alpha\delta}} \tag{22.35}$$

Under the tensile failure, it is assumed that the bolt axial force at $x = \delta$ reach the tensile yielding strength T_m, the axial displacement s_t at $x = \delta$ (also the maximum bolt displacement) reads

$$s_t = \frac{\left(\mathrm{e}^{\alpha\delta} - \mathrm{e}^{-\alpha\delta}\right)T_m}{\left(\mathrm{e}^{\alpha\delta} + \mathrm{e}^{-\alpha\delta}\right)EA\alpha} \tag{22.36}$$

If $s_t < \tau_m/k$: Tensile Failure

The axial displacement s_t at $x = \delta$ is the maximum bolt displacement. If s_t computed by Eq. (22.36) is smaller than τ_m/k, it means that relative bolt displacements are smaller than τ_m/k. Therefore, the axial force at the tensile failure developing along the bolt is determined by

$$T^t(x) = T_m \frac{e^{\alpha x} + e^{-\alpha x}}{e^{\alpha\delta} + e^{-\alpha\delta}} \quad x\epsilon\left[0\,\delta\right] \tag{22.37}$$

The resisting work rate of a single bolt in terms of the tensile failure may be formulated by

$$dW_{D,\text{nail}}^{N} = \int_0^{\delta} T^t(x)\cdot\varepsilon_x\,dx = \int_0^{\delta} T_m \frac{e^{\alpha x} + e^{-\alpha x}}{e^{\alpha\delta} + e^{-\alpha\delta}}\cdot\frac{e^{\alpha x} + e^{-\alpha x}}{e^{\alpha\delta} - e^{-\alpha\delta}}$$
$$\cdot\alpha v_k \cos\beta_k dx = T_m v_k \cos\beta_k \cdot\left[\frac{1}{2} + \frac{2\alpha\delta}{\left(e^{2\alpha\delta} - e^{-2\alpha\delta}\right)}\right] \tag{22.38}$$

If $s_t \geq \tau_m/k$: Pullout Failure

If s_t calculated by Eq. (22.36) is greater than τ_m/k, Eq. (22.25) may not be valid at a certain point in the interaction zone. It is assumed that the interface shear stress is equal to the bonding strength τ_m in the interaction zone, then the equilibrium equation between the axial stress and the shear stress becomes

$$A\frac{d\sigma}{dx} = \pi\varnothing\tau_m \tag{22.39}$$

Substituting Eq. (22.28) into Eq. (22.39) leads to

$$\frac{d^2 s}{dx^2} - \frac{\pi\varnothing}{EA}\tau_m = 0 \tag{22.40}$$

The corresponding boundary conditions are

$$\begin{cases} x = 0, s = \dfrac{\tau_m}{k} \\ x = \delta, s = s_t \end{cases} \tag{22.41}$$

According to the boundary conditions, the solution to Eq. (22.40) is

$$s(x) = \frac{\pi\varnothing}{2EA}\tau_m x^2 + \frac{s_t - \dfrac{\tau_m}{k} - \dfrac{\pi\varnothing}{2EA}\tau_m\delta^2}{\delta}x + \frac{\tau_m}{k} \tag{22.42}$$

The axial strain of the bolt is calculated by

$$\varepsilon_x(x) = \frac{\pi\varnothing}{EA}\tau_m x + \frac{s_t}{\delta} - \frac{\tau_m}{k\delta} - \frac{\pi\varnothing}{2EA}\delta\tau_m \tag{22.43}$$

The axial strain rate of the bolt is obtained as

$$\dot{\varepsilon}_x(x) = \frac{d\varepsilon_x}{dt} = \frac{1}{\delta}\frac{ds_t}{dt} = \frac{v_k \cos\beta_k}{\delta} \tag{22.44}$$

The axial force mobilized in the bolt is given by the following equation:

$$T(x) = EA\left(\frac{\pi\varnothing\tau_m}{EA}x + \frac{s_t}{\delta} - \frac{\tau_m}{k\delta} - \frac{\pi\varnothing\delta\tau_m}{2EA}\right) \tag{22.45}$$

For pullout failure, it is assumed that the bolt axial force at $x = \delta$ is equal to the pullout force T_p, then one can compute the axial displacement s_t as follows:

$$s_t = \frac{T_p\delta}{EA} + \frac{\tau_m}{k} + \frac{\pi\varnothing\delta^2\tau_m}{2EA} > \tau_m / k \tag{22.46}$$

Therefore the corresponding axial force at the tensile failure mobilized along the bolt is determined by

$$T^p(x) = \pi\varnothing\tau_m x + T_p \;\; x\epsilon\left[0\,\delta\right] \tag{22.47}$$

The resisting work rate of a single bolt for the pullout failure is expressed as

$$dW_{D,\text{nail}}^N = \int_0^\delta \left(\pi\varnothing\tau_m x + T_p\right)\frac{v_k\cos\beta_k}{\delta}$$

$$= T_p v_k \cos\beta_k\left(1 + \frac{\pi\varnothing\tau_m\delta}{2T_p}\right) \tag{22.48}$$

where T_p represents the pullout force. In order to take into account the shear failure at the interface inside the failure mechanism that is expected for bolts with high tensile strength and big overlapping length (no anchor plate of bolts on the tunnel face), the pullout force is assumed to take the form below:

$$T_p = \pi\varnothing\tau_m \min\left(L_e, L - L_e\right) \tag{22.49}$$

where L_e is the length outside the failure mechanism. Thus the resisting work rate of a single bolt for the pullout failure is rewritten as

$$dW_{D,\text{nail}}^N = T_p v_k \cos\beta_k\left[1 + \frac{\delta}{2\min\left(L_e, L - L_e\right)}\right] \tag{22.50}$$

Consequently, the total resisting work rate is easily derived as,

$$W_{D,\text{nail}} = \begin{cases} \omega \sum_{1}^{m} T_m R_k \cos\beta_k \cdot \left[\frac{1}{2} + \frac{2\alpha\delta}{\left(e^{2\alpha\delta} - e^{-2\alpha\delta}\right)} \right] & \text{tensile failure} \\ \omega \sum_{1}^{m} T_p R_k \cos\beta_k \cdot \left[1 + \frac{\delta}{2\min\left(L_e, L - L_e\right)} \right] & \text{pullout failure} \end{cases} \tag{22.51}$$

where m is the number of bolts. Then the term N_b can be obtained as

$$N_b = \begin{cases} \dfrac{\varnothing^2 \sum_{1}^{m} T_m R_k \cos\beta_k \cdot \left[\frac{1}{2} + \frac{2\alpha\delta}{\left(e^{2\alpha\delta} - e^{-2\alpha\delta}\right)} \right]}{T_m \sum_{i} \left(S_{i,0} . R_{i,0} . \cos\beta_{i,0}\right)} & \text{tensile failure} \\ \dfrac{\varnothing^2 \sum_{1}^{m} T_p R_k \cos\beta_k \cdot \left[1 + \frac{\delta}{2\min\left(L_{e,k}, L_k - L_{e,k}\right)} \right]}{T_m \sum_{i} \left(S_{i,0} . R_{i,0} . \cos\beta_{i,0}\right)} & \text{pullout failure} \end{cases} \tag{22.52}$$

Please note that the interaction zone introduces a new parameter, the width δ, in the optimization process of the upper-bound analysis.

22.5.2.3 Safety Factor Calculations

For an open-face tunnel, as no supporting pressure is applied at the tunnel face, an effective way to estimate its stability is to calculate safety factors. The strength reduction method (SRM) has been widely applied to compute safety factors in the context of Mohr–Coulomb materials. The safety factor is the ratio of the real shear strength of the material to the critical shear strength corresponding to the limit equilibrium state.

The strength reduction procedure is adopted to calculate the safety factors of a tunnel face in combination with the kinematical approach of limit analysis. In this context, the cohesion c and friction angle φ are continuously reduced or increased until the internal energy dissipation is equal to the external work rate. The reduced cohesion c_F and the reduced friction angle φ_F read as follows,

$$\begin{cases} c_F = c / \text{FS} \\ \varphi_F = \arctan\left(\tan\varphi / \text{FS}\right) \end{cases} . \tag{22.53}$$

22.5.3 *Non-reinforced Case*

22.5.3.1 Comparison with Numerical Results

This section compares safety factors obtained by the proposed approach and by 3D numerical calculations with *FLAC 3D*. In the calculations, geometrical and geo-mechanical parameters are taken from a real tunnel, the Toulon tunnel. The Toulon tunnel (a horseshoe shaped tunnel) connecting two motorways from Marseille to Nice was excavated using NATM. The detailed information about geotechnical and geometrical parameters of Toulon tunnel is summarized in Table 22.2, and its geometrical shape is plotted in Fig. 22.24.

Table 22.2 Parameters of Toulon tunnel

Geometrical parameters	Width, W (m)	11.6
	Height, H (m)	10.3
	Buried depth, C (m)	40.0
	Full face area (m^2)	100.0
	Top heading area (m^2)	53.0
Geotechnical parameters	Cohesion, c (kPa)	50
	Friction angle, φ (degree)	20
	Dilatancy angle, ψ (degree)	0
	Unit weight, γ (kN/m^3)	22
	Young's modulus, E(MPa)	300
	Poisson's ratio, ν	0.3

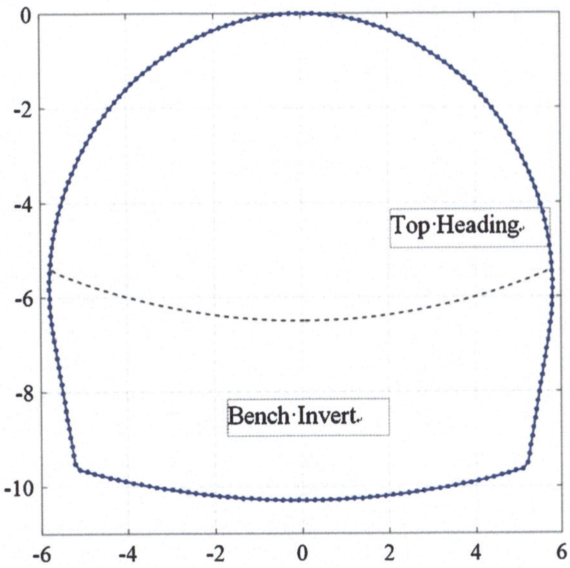

Fig. 22.24 Geometrical shape of Toulon tunnel

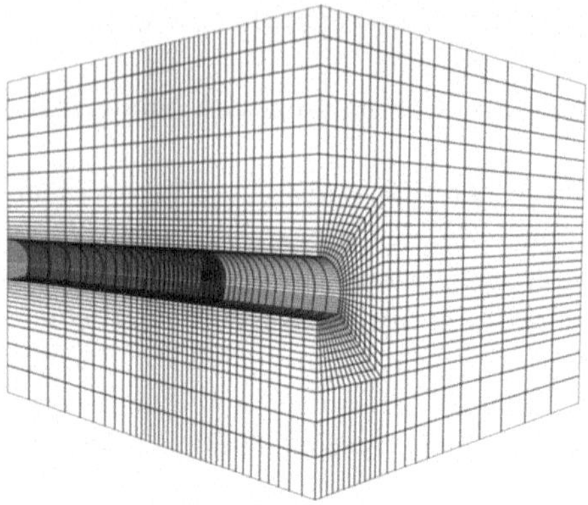

Fig. 22.25 Numerical model of Toulon tunnel

Table 22.3 Computed safety factors for Toulon tunnel for full face excavation

Methods	Circular section	Non-circular section	Difference (Circular section/Non circular section)
FLAC 3D	1.74	1.81	4.0%
Kinematical approach	1.78	1.84	2.7%
Difference (*FLAC* 3D/KA)	2.2%	1.6%	–

Table 22.4 Computed safety factors for Toulon tunnel for top heading excavation

Methods	Circular section	Non-circular section	Difference (Circular section/Non circular section)
FLAC 3D	2.15	2.49	13.7%
Kinematical approach	2.38	2.60	8.5%
Difference (*FLAC 3D*/KA)	9.7%	4.2%	–

Figure 22.25 presents a full face excavation numerical model of the Toulon tunnel with a discretization of 56,000 zones and of 61,000 nodes. The size of the numerical model is taken as: 100 m × 130 m × 70 m, which are large enough to avoid the influence of the boundary effects. The displacements at the bottom of the model boundary are fixed, and on the vertical boundaries, only the normal directions are fixed.

In the numerical model, a linear elastic-perfectly plastic constitutive model based on the Mohr–Coulomb strength criterion is assigned to the soft rock masses. The tunnel concrete lining is modeled with shell structural element with a Young modulus of 15GPa, a Poisson's ratio of 0.2 and a thickness of 0.22 m. In order to analyze the influence of the cross section shape on the face stability, two cases of excavations, the full face excavation as well as the partial excavation at the top heading, are considered in the calculations. Besides, in each case, a circular section with an equivalent area is calculated for comparisons.

Tables 22.3 and 22.4, respectively, list the obtained safety factors with respect to the full face excavation and the top heading excavation, including the non-circular section and the corresponding circular section with equivalent area.

The safety factors obtained from the proposed kinematical approach (KA) agree reasonably well with those provided by the numerical simulations both in terms of the full face excavation and the top heading excavation. Only a maximum difference of 9.7% occurs in the case of top heading excavation. This shows that the proposed method is an effective approach to evaluate the safety factors for a tunnel face both with circular as well as non-circular cross sections.

Besides, it is of interest to see that both the results obtained by the upper-bound method and by the finite difference method show that the safety factors for the non-circular sections are bigger than those for the circular counterparts. In the case of the full face excavation ($W/H = 1.13$), the maximum difference is equal to 4.0%; in the case of the top heading excavation for which the ratio of width to height reaches 1.78, the difference between the two shapes increases to 13.7%, showing a higher influence of the cross section. This interesting phenomenon suggests that when the ratio W/H is close to 1.0, the application of an equivalent circular shape tends to underestimates the safety factor. This underestimation is slight and acceptable for the practical engineering purposes.

22.5.3.2 Design Table

Figure 22.26 plots safety factors as a function of the normalized cohesion $c/\gamma D$ for a circular cross section tunnel and a non-circular cross section tunnel at $C/D = 2.0$. The horseshoe cross section of Toulon tunnel is taken as the non-circular case, for which D is its equivalent diameter.

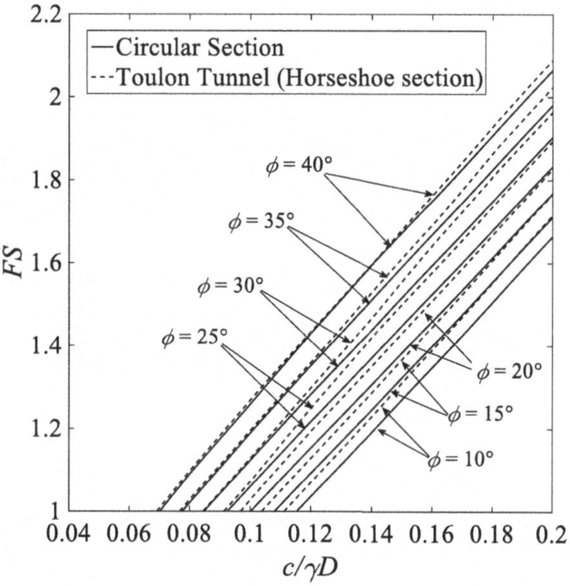

Fig. 22.26 Safety factor versus cohesions with different friction angles

As expected, in Fig. 22.26, the computed safety factors increase with the increase of the cohesion and of the friction angle with respect to the circular as well as the non-circular cross sections. For all cases, the safety factors show a liner relationship to the normalized cohesion. Besides, a comparison between the two cross-sectional shapes indicates that the horseshoe cross section provides higher safety factors than the circular section, especially for the friction angles smaller than 20° and the cohesions larger than $0.14\gamma D$. This shows that the application of an equivalent circular shape to evaluate the safety factor of a non-circular tunnel is also dependent on the friction angle and on the cohesion.

22.5.4 Reinforced Case

22.5.4.1 Influence of Strength Parameters

Several charts provided in Fig. 22.27 show safety factors as a function of the reinforcement density n_b for different values of the friction angle φ and the cohesion c. The reinforcement density n_b varies from 0.0 (non-reinforced case) to 1.5 bolts/m^2. The horseshoe-shaped Toulon tunnel is used in all calculations. The unit weight γ is set to 22 kN/m^3, and the bolt length L is taken to be $1.2D$.

According to the results plotted in Fig. 22.27, not surprising, the safety factor increases with the reinforcement density, the cohesion and the friction angle. When the cohesion is higher than 20 kPa, the tunnel face is able to maintain stable ($FS > 1$) even without reinforcement.

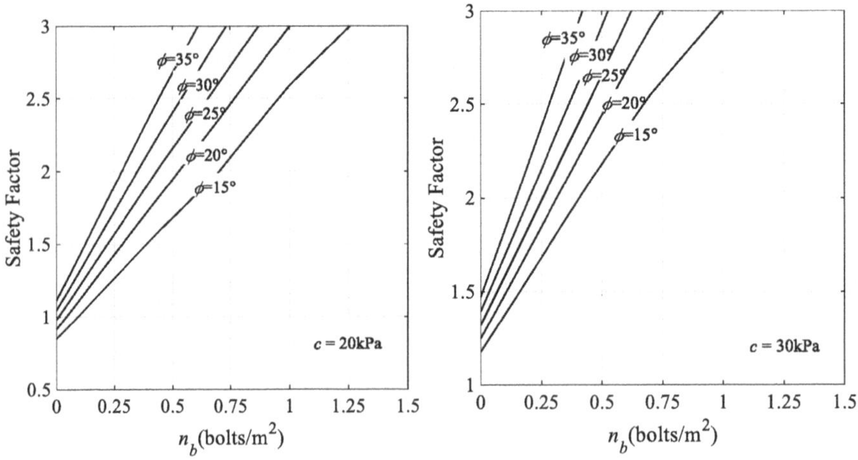

Fig. 22.27 Safety factor versus reinforcing densities with different friction angles and normalized cohesion

22.6 Tunnel Face Stability Considering Material Spatial Variability

It is well accepted that geomaterials exhibit spatially variable properties. For example, soil properties change vertically and horizontally in situ. The random field theory is widely used to simulate the spatial variability. This section aims to conduct a probabilistic analysis on tunnel face stability considering material spatially variable. The cohesion and the friction angle are modeled as two anisotropic independent lognormal random fields. The influences of the spatial variability on probability density functions (PDFs) of required face pressures and on computed failure probabilities are investigated.

22.6.1 *Lognormal Random Field*

A lognormally distributed random field G_i ($i = c, \varphi$) is used to model the spatial variability of shear strength parameters of weak rocks (e.g., the cohesion c, the friction angle φ, with the mean value μ_i, the standard deviation σ_i, and the autocorrelation distance θ_i). It is assumed that the cohesion and the friction angle have identical autocorrelation distances θ in one ground stratum (Fenton et al. 2005). The lognormally distributed random field can be obtained by a transformation from its corresponding standard normal random field with zero mean and unit variance. For illustration, the following equations are for the random field of cohesion c. The lognormally distributed random field of cohesion c is given by,

$$G_c(\boldsymbol{x}) = \exp\left[\mu_c^{\ln} + \sigma_c^{\ln} G_c^{\ln}(\boldsymbol{x})\right] \tag{22.54}$$

where $\boldsymbol{x}$ represents the spatial position; $G_c^{\ln}$ is the standard normal random field; $\propto_c^{\ln}$ and $\sigma_c^{\ln}$ are, respectively, given by,

$$\sigma_c^{\ln} = \sqrt{\ln\left(1 + \sigma_c^2 / \mu_c^2\right)} \tag{22.55}$$

$$\mu_c^{\ln} = \ln \mu_c - \left(\sigma_c^{\ln}\right)^2 / 2 \tag{22.56}$$

The stochastic spatial variability of a random field is characterized by its standard deviation and autocorrelation distances. In natural deposits, the correlation of a certain material property between two positions decreases as their separation distance increases (Baecher and Christian 2005). In the random field theory, a spatial autocorrelation function is used to define this correlation between two arbitrary points. A squared exponential autocorrelation function in a 2D random field reads (Baecher and Christian 2005),

$$\rho_c = \exp\left\{-\left[\left(\frac{|x_1 - x_2|}{\theta_h}\right)^2 + \left(\frac{|y_1 - y_2|}{\theta_v}\right)^2\right]\right\} \tag{22.57}$$

where (x_1, y_1) and (x_2, y_2) defines the spatial coordinates of two points; θ_h and θ_v respectively denote the horizontal autocorrelation distance and the vertical autocorrelation distance.

A standard normally distributed random field $G_c^{\ln}$ can be approximately represented by an M-term K-L expansion:

$$G_c^{\ln}(\boldsymbol{x}) \approx \sum_{j=1}^{M} \sqrt{\lambda_j}\, \xi_j \psi_j(\boldsymbol{x}) \tag{22.58}$$

in which ξ_j is a set of independent standard normal distribution variables; M is the number of truncation terms of the expansion; λ_j and $\psi_j(\boldsymbol{x})$ respectively denote the eigenvalues and the eigenfunctions of the autocorrelation function in Eq. (22.57). The convergence and the accuracy of a K-L expansion depend on the number of terms M. The determination of M relates to the domain of the random field and the autocorrelation distances.

22.6.2 Probabilistic Analysis Using Monte Carlo Simulations

In order to assess failure probabilities against tunnel face collapse for a given face pressure σ_T, a performance function is defined as:

$$G = \sigma_T - \sigma_c \tag{22.59}$$

where σ_T is the applied face pressure and σ_c is the required face pressure determined by the 3D rotational failure mechanism, $G_1 < 0$ corresponding to the failure domain. In the calculations, the applied face pressure σ_T is modeled as a lognormally distributed random variable with a coefficient of variation of 15%. The failure probability can be evaluated by means of Monte Carlo simulations,

$$\widehat{P_f} = \frac{1}{N_{\text{mc}}} \sum_{i=1}^{N_{\text{mc}}} I(G) \tag{22.60}$$

where N_{mc} is the number of MCS samples; $I(G)$ is equal to 1 for $G < 0$, otherwise $I(G) = 0$. Eq. (22.60) is an unbiased estimate of failure probability and its coefficient of variation is calculated by

$$\text{COV}\left(\widehat{P_f}\right) = \sqrt{\frac{1 - \widehat{P_f}}{N_{\text{mc}}\widehat{P_f}}} \tag{22.61}$$

Latin hypercube sampling technique is employed to generate Monte Carlo samples due to its uniform sampling feature, and the size of samples is set to be large enough to get a coefficient of variation of failure probability smaller than 10%.

22.6.3 Discretization of Random Fields

Figure 22.28 presents a longitudinal section of a tunnel face with a diameter D and a buried depth C. Two cross-correlated 2D lognormal random fields are used to model the spatial variability in the horizontal and vertical directions in terms of cohesion c and friction angle φ, bounded by a domain Ω around the tunnel face whose horizontal length and vertical length are, respectively, denoted by l_{h} and l_{v}. For the realization of random fields, the domain Ω is discretized into rectangular elements, separated by a distance of δ in both the horizontal and vertical directions. The magnitude of the cohesion (or the friction angle) at each grid point on the element can be determined by a K-L expansion. Then, the corresponding value in the center of the element is taken as the average value of those of its four nodes.

The random fields in this work are assumed to be anisotropic, since in reality the horizontal correlation distance θ_h is greater than the vertical one θ_v for typical ground stratum (Fenton and Griffiths 2003). Table 22.5 lists the K-L terms M for different set of autocorrelation distances considered in this work for a prescribed error of 9~10%. The number of K-L terms increases with the decrease of autocorrelation distances.

Table 22.6 presents input parameters used in this work. Due to low uncertainties, the tunnel diameter D, the buried depth C, and the weak rock unit weight γ are taken to be deterministic. The element size δ is set to 0.5 m, which is fine enough for good results; the horizontal length l_{h} and vertical length l_{v} are respectively taken as 1.5D and $C + D$.

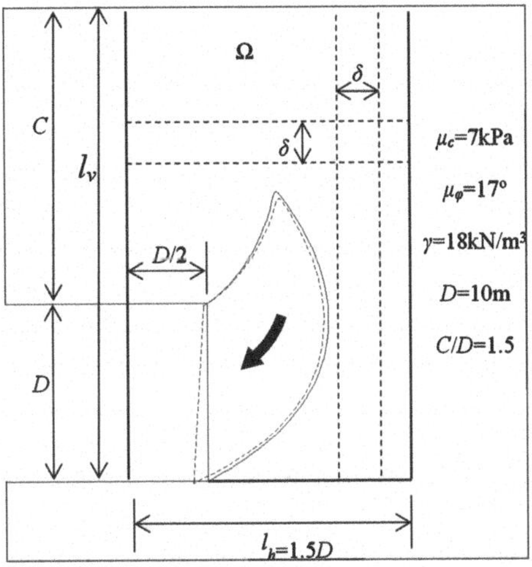

Fig. 22.28 Domain of the random field around a tunnel face

Table 22.5 K-L terms for different autocorrelation distances

θ_v (m)	θ_h (m)	M
1.5	10	260
	20	150
	30	120
	40	100
2.0		75
2.5		60
3.0		50

Table 22.6 Statistics properties of input parameters

Input variables	Mean value (μ)	COV (%)
c (kPa)	7	20
φ(°)	17	10
γ (kN/m^3)	18	–
C(m)	15	–
D(m)	10	–
σ_T (kPa)	/	15

22.6.4 *Influence of Spatial Variability*

In order to study the effect of spatial variability on probability density functions of required face pressures and on failure probabilities, the horizontal autocorrelation distance θ_h is set to vary from 10 m to 40 m, θ_v from 1.5 m to 3.0 m.

22.6.4.1 Influence on Probability Density Functions

Figure 22.29 shows PDFs of normalized face pressures corresponding to different autocorrelation distances together with the case of random variables. With the application of MCS, the PDF curves are obtained by using a Kernel smoothing technique. The case of random variable means that the cohesion and the friction angle are considered as random variables. The results are obtained for θ_h ranging between 10 and 40 m with θ_v being 1.5 m in Fig. 22.29a, and θ_v changing between 1.5–3.0 m with $\theta_h = 40$ m for Fig. 22.29b.

It is seen that a decrease of autocorrelation distances gives a taller and narrower PDF curve, suggesting that the variability of required face pressures decreases with the spatial heterogeneity. A comparison between these two plots shows that the PDF of the required face pressure is more sensitive to the vertical autocorrelation distances.

22.6.4.2 Influence on Failure Probabilities

The computed failure probabilities as a function of the mean value of applied face pressures for different autocorrelation distances are given in Fig. 22.30.

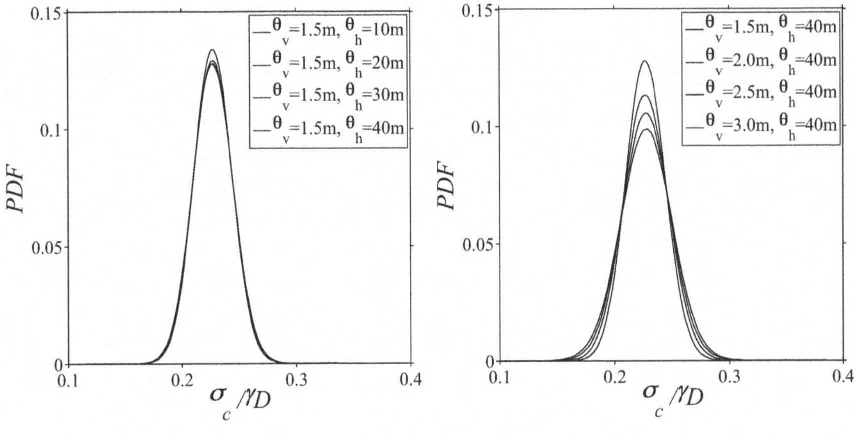

Fig. 22.29 Influence of spatial variability on probability density functions

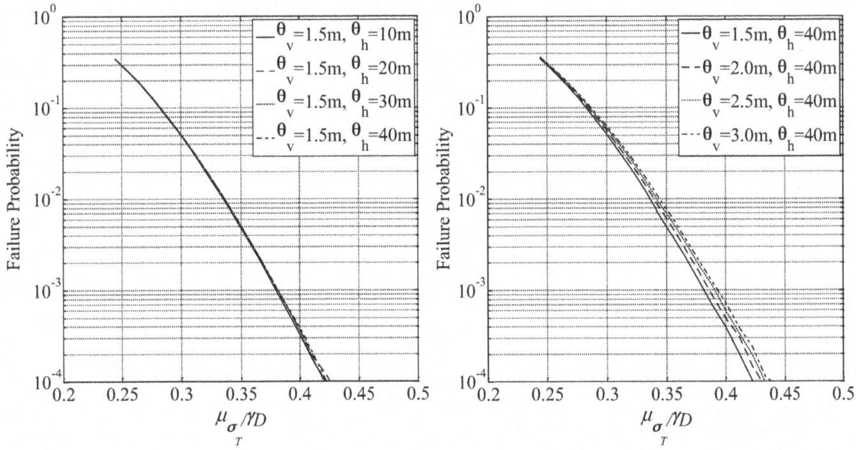

Fig. 22.30 Failure probabilities versus the applied face pressure for different spatial variabilities

It is not surprising to find that failure probabilities are greatly influenced by the applied face pressures, small applied face pressure resulting in high failure probability. Besides, it can be seen that the increase of autocorrelation distances leads to an increase of failure probabilities, both for the horizontal and vertical autocorrelation distances. For example, the failure probability increases from 3.94×10^{-4} to 7.29×10^{-4} as θ_v increases from 1.5 m to 3 m when θ_h =40 m and $\mu_{\sigma T}/\gamma D = 0.4$. In other words, an overestimation of autocorrelation distances (or an underestimation of spatial variability) leads to conservative results in the engineering design. It is also noticed that the vertical autocorrelation distances impacts failure probabilities more than the horizontal autocorrelation distances.

22.7 Conclusions

This chapter studies stability of a tunnel face in the context of the kinematical approach of limit analysis. The main contribution is to present several predicting models of assessing necessary face pressures/safety factors for a tunnel face subjected to seepage forces, for a non-circular face reinforced by horizontal bolts and for a tunnel face in spatially random soft rock masses. These proposed theoretical models are proved to give good predictions on assessing face stability with high computational efficiency comparing to results from numerical simulations. Besides, these proposed deterministic models are good tools for probabilistic analysis due to their high computational efficiency.

References

Anagnostou G, Perazzelli P (2013) The stability of a tunnel face with a free span and a non-uniform support. Geotechnik 36(1):40–50

Baecher GB, Christian JT (2005) Reliability and statistics in geotechnical engineering. Wiley, Hoboken, NJ

Chambon P, Corté JF (1990) Stabilité du front de taille d'un tunnel dans un milieu frottant. Approche cinématique en calcul à la rupture. Rev Fr Géotech (51):51–59

Chambon P, Corte JF (1994) Shallow tunnels in cohesionless soil: stability of tunnel face. J Geotech Eng 120(7):1148–1165

Chen WF (1975) Limit analysis and soil plasticity. Elsevier, Amsterdam

Drucker DC, Prager W (1952) Soil mechanics and plastic analysis or limit design. Q Appl Math 10(2):157–165

Fenton GA, Griffiths DV (2003) Bearing-capacity prediction of spatially random c-φ soils. Can Geotech J 40(1):54–65

Fenton GA, Griffiths DV, Williams MB (2005) Reliability of traditional retaining wall design. Geotechnique 55(1):55–62

Hoek E, Carranza-Torres C, Corkum B (2002) Hoek-Brown failure criterion-2002 edition. Proceedings of NARMS-Tac, 1, pp 267–273

Horn N (1961) Horizontaler Erddruck Auf Senkrechte Abschlussflaechen von Tunnelroehren. Landeskonferenz Der Ung-Arischen Tiefbauindustrie, pp 7–16

Idinger G, Aklik P, Wu W, Borja RI (2011) Centrifuge model test on the face stability of shallow tunnel. Acta Geotech 6(2):105–117

Kamata H, Mashimo H (2003) Centrifuge model test of tunnel face reinforcement by bolting. Tunn Undergr Space Technol 18(2):205–212

Kirsch A (2010) Experimental investigation of the face stability of shallow tunnels in sand. Acta Geotech 5(1):43–62

Leca E, Dormieux L (1990) Upper and lower bound solutions for the face stability of shallow circular tunnels in frictional material. Geotechnique 40(4):581–606

Lee IM, Lee JS, Nam SW (2004) Effect of seepage force on tunnel face stability reinforced with multi-step pipe grouting. Tunn Undergr Space Technol 19(6):551–565

Mollon G, Dias D, Soubra AH (2009a) Probabilistic analysis of circular tunnels in homogeneous soil using response surface methodology. J Geotech Geoenviron 135(9):1314–1325

Mollon G, Dias D, Soubra AH (2009b) Probabilistic analysis and design of circular tunnels against face stability. Int J Geomech 9(6):237–249

Mollon G, Dias D, Soubra AH (2009c) Face stability analysis of circular tunnels driven by a pressurized shield. J Geotech Geoenviron 136(1):215–229
Mollon G, Phoon KK, Dias D, Soubra AH (2010) Validation of a new 2D failure mechanism for the stability analysis of a pressurized tunnel face in a spatially varying sand. J Eng Mech 137(1):8–21
Mollon G, Dias D, Soubra AH (2011) Rotational failure mechanisms for the face stability analysis of tunnels driven by a pressurized shield. Int J Numer Anal Methods Geomech 35(12):1363–1388
Oreste PP, Dias D (2012) Stabilisation of the excavation face in shallow tunnels using fibreglass dowels. Rock Mech Rock Eng 45(4):499–517
Shin JH, Choi YK, Kwon OY, Lee SD (2008) Model testing for pipe-reinforced tunnel heading in a granular soil. Tunn Undergr Space Technol 23(3):241–250
Subrin D, Wong H (2002) Stabilité du front d'un tunnel en milieu frottant: un nouveau mécanisme de rupture 3D. C R Mécanique 330(7):513–519
Takano D, Otani J, Nagatani H, Mukunoki T (2006) Application of x-ray CT on boundary value problems in geotechnical engineering: research on tunnel face failure. In Geo Congress 2006: Geotechnical Engineering in the Information Technology Age, pp 1–6
Vermeer PA, Ruse N, Marcher T (2002) Tunnel heading stability in drained ground. Felsbau 20(6):8–18
Yang XL, Li L, Yin JH (2004) Seismic and static stability analysis for rock slopes by a kinematical approach. Geotechnique 54(8):543–549

Chapter 23
Characterization of Soft Rocks in Brazilian Coal Beds

Clovis Gonzatti, João Alberto Fiorentini, Luiz Zorzi, and Ivone Maria Agostini

23.1 Introduction

In the mid-1980s, an extensive work began on the geomechanical characterization of exploited or potentially exploitable coal seams through underground mining in the states of Rio Grande do Sul and Santa Catarina, southern Brazil (Fig. 23.1), conducted by CIENTEC—Science and Technology Foundation of Rio Grande do Sul State. These studies were pioneers with this scope in the country and aimed to establish a methodology for the sizing of pillars for underground coal mines in southern Brazil. In this context, an extensive activity program was developed that included, in summary: studies of the geological constraints of the mineral deposits; case studies of pillar failure; sampling and laboratory tests of the materials present in the main exploited coal seams, as well as on their host rocks; in situ tests on large blocks in two economically important coal seams in operation.

In the South Catarinense Coalfield, located in the state of Santa Catarina, the studies on the coal seams consisted of the collection of large blocks of coal, preparation of cubic specimens, with side dimensions ranging between 5 and 30 cm, and the execution of uniaxial compression tests. Sampling included almost all the underground coal mines in operation at the time of the study. The studies focused

The original version of this chapter was revised. The correction to this chapter is available at https://doi.org/10.1007/978-3-030-29477-9_26

C. Gonzatti (✉)
Department of Mineralogy and Petrology, Federal University of Rio Grande do Sul (UFRGS), Porto Alegre, Brazil
e-mail: gonzatti@ufrgs.br

J. A. Fiorentini · L. Zorzi · I. M. Agostini
Foundation for Science and Technology of Rio Grande do Sul State (CIENTEC), Porto Alegre, Brazil
e-mail: fiorentini@cientec.rs.gov.brl

M. Kanji et al. (eds.), *Soft Rock Mechanics and Engineering*,
https://doi.org/10.1007/978-3-030-29477-9_23

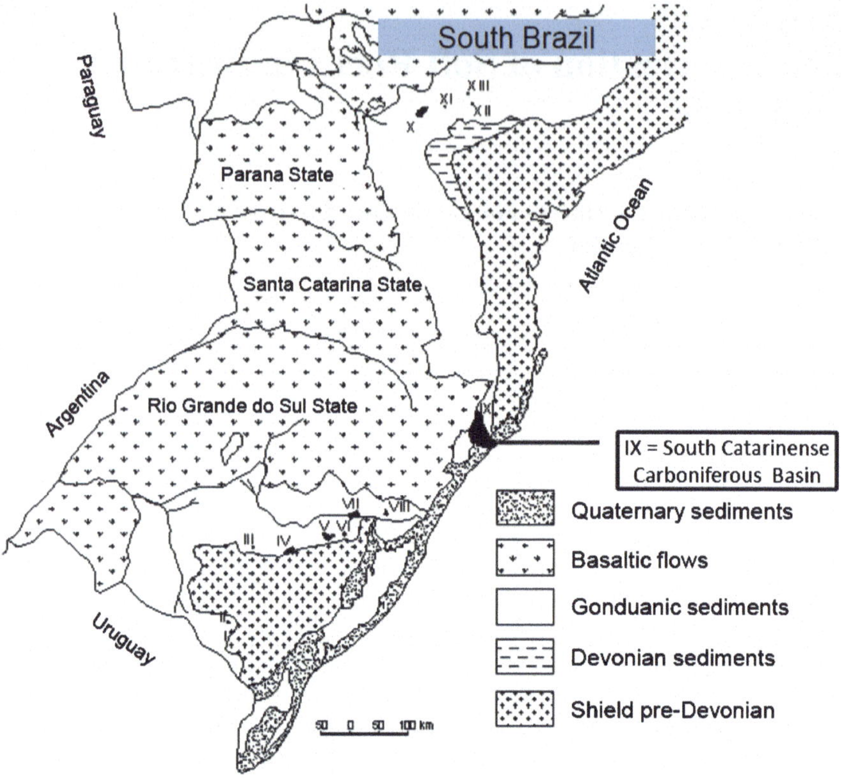

Fig. 23.1 Geological map of South Brazil. Numbers I to XIII indicate known coalfields in Brazil (modified from Costa 2000)

on the most economically important layers, such as the Barro Branco coal seam, the most heterogeneous one, in the 1980s (766 tests—Zorzi 1990), the Irapuá coal seam in the 1990s (56 tests—Zorzi 1996) and the Bonito coal seam in the 2000s (78 tests—Zorzi et al. 2006; Gonzatti 2007). In the Barro Branco coal seam, in situ tests were carried out on large samples. A set of three large blocks, of approximately 1 m^3, were tested. The results obtained at different work scales in the coal seams allowed the proposal by CIENTEC of a pillar design methodology for the underground coal mines of southern Brazil (Zorzi et al. 1991). This methodology started being recommended, as of 2005, by DNPM—National Department of Mineral Production, the official Brazilian agency responsible for mining inspections.

The methodology for the design of coal pillars proposed by CIENTEC contemplates the analysis of the load capacity of the host rocks. Therefore, knowing this attribute of the host rocks is a basic condition for the pillar design of the Brazilian coal mines. There are few geomechanical studies about the sedimentary rocks of southern Brazil, especially on the sedimentary host rocks of coal seams. In general, the existing studies are associated with materials present in civil engineering projects (Dobereiner 1987; Duarte et al. 2010). Regarding the host rocks of the coal seams, the studies carried out by IPT—Instituto de Pesquisa Tecnológica (1987) on samples of sandstones, laminated sandstones, siltstones and laminated siltstones in

Santa Catarina state should be emphasized, and also by Zorzi et al. (1989), with samples of the Palermo and Rio Bonito Formation from the Santa Terezinha Coalfield in the state of Rio Grande do Sul.

In the early 1990s, CIENTEC developed detailed laboratory studies for the geomechanical characterization of the materials present in the host layers of the main coal seams in southern Brazil. In the South Catarinense Coalfield, this study began with the host rocks of the Barro Branco seam, the main coal layer mined underground, followed by the host rocks of the Bonito coal seam, in the same basin. The objective of the laboratory tests was to define the mechanical behavior of the host rocks of the immediate roof and floor of the coal seam. In underground coal mining in Brazil, the geometry of the galleries consists of openings with 5.5–6 m in width, and height corresponding to the mined thickness of the coal seam, of about 1.8–2.3 m in the Barro Branco seam and 3–3.5 m in the Bonito seam. Within this context, the first 5 m above and below the layer of coal constitute the portion of the rock massif that undergoes the greatest redistribution of stresses due to the excavation. The main support system of the galleries is the roof bolting. In this system, the most used mechanisms are the suspension effect and the friction or beam-building effect. Knowing the mechanical characteristics and the spatial distribution of the different lithotypes present in the host rock of the roof is crucial for the design of the gallery support systems.

This chapter presents the procedures adopted for the geomechanical characterization of the Barro Branco, Irapuá, and Bonito coal seams, which are the most economically important ones mined in the South Catarinense Coalfield, southern Brazil. Studies with carbonaceous rocks were carried out at the laboratory and in situ, while the materials of the host rocks were limited to the laboratory scale on drill cores. The adopted in situ sampling procedures, the sample preparation techniques and the procedures used to carry out the petrographic, physical, and mechanical characterization tests in the laboratory, as well as the in situ tests, are detailed ahead. The obtained results were statistically processed and allowed to estimate the in situ strength of the coal layers, as well as the typification of the main host rock lithologies present in the South Catarinense Coalfield.

23.2 Geology of the South Catarinense Coalfield

Geographically, all the studied sites are located within the South Catarinense coalfield, which is part of the eastern edge of the Paraná Basin, as shown in Fig. 23.1. This basin, which in Brazil spans over several other states, is of Devonian age at its oldest sedimentary deposits and has the evolutionary characteristics of an intracratonic basin.

Within the stratigraphic sequence, as seen in Fig. 23.2, the carbonaceous lithologies are linked to the units called Itararé Group (more basal) and Guatá, with ages between the Upper Carboniferous and Upper Permian. But it is in the Guatá unit that the thicker coal seams are located, of better quality and under exploitation. Among the seven individual coal seams in the state of Santa Catarina, three are more economically important: Barro Branco, Irapuá, and Bonito seams, as highlighted in Fig. 23.3.

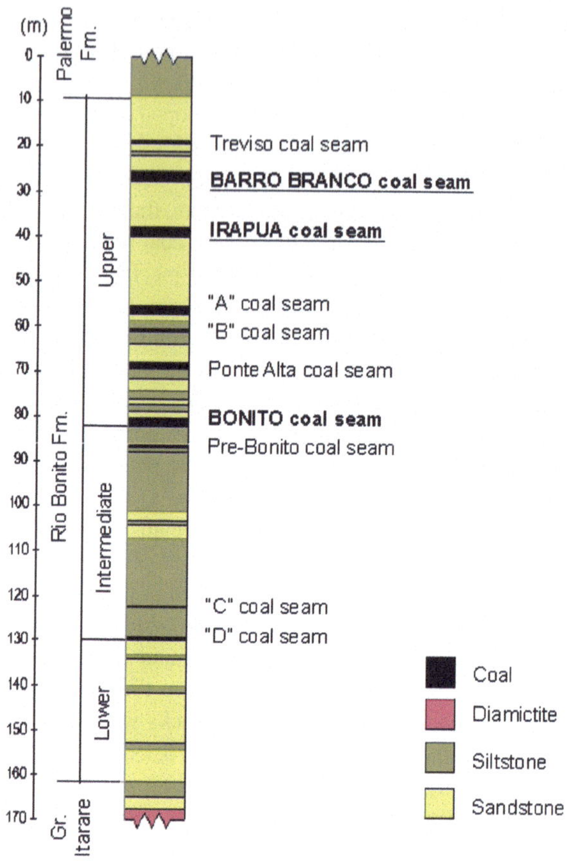

Fig. 23.2 Typical stratigraphy profile of the Rio Bonito Formation with the main coal seams in the South Catarinense Coalfield (modified from Costa 2000)

The structural conditioning found throughout the Paraná basin is represented by normal faults, with subvertical planes. Thrust faults rarely occur, with little tailing. The presence of "slickensides" caused by differential sediment compaction is frequent. Fracture systems can follow faults, forming conjugated systems, filled or not by diabase dikes. The folds seen in some coal seams are associated with "drag fold" faults. However, folds caused by compressive stress fields are rarely found.

23.3 Main Coal Seams Under Exploitation in the South Catarinense Coalfield

23.3.1 Barro Branco Coal Seam

This layer is characterized by a great thickness and lateral extension uniformity and is exploited in all the mines of the South Catarinense deposit. It features a regional dip of approximately 1° West. However, the layer shows small dips near the faults.

Fig. 23.3 Typical profile of the main coal seams mined in the South Catarinense Coalfield

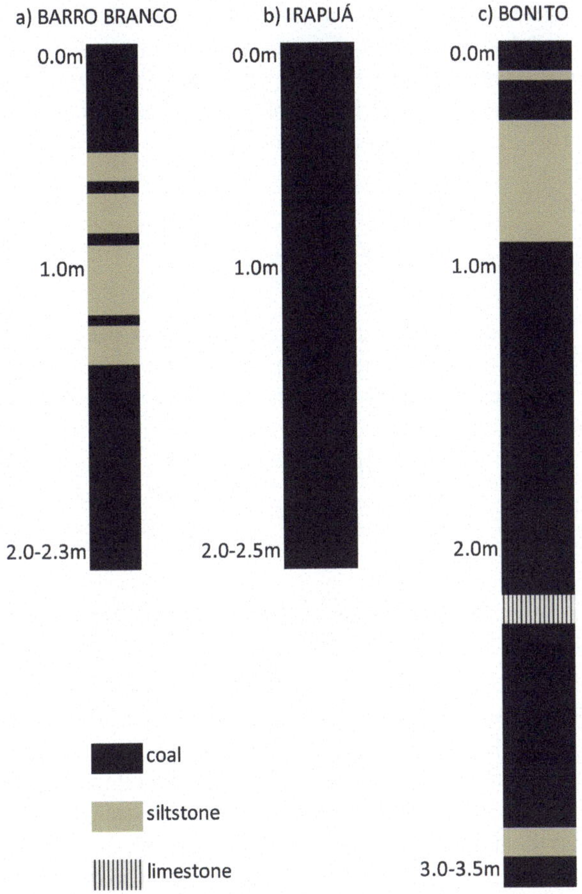

It has an average thickness of 2 m, however, where it was intersected by fault or there was a diabase dike intrusion, its thickness may be reduced, or it does not exist. These diabase bodies often damage the coal in their vicinity.

Its floor ("Lapão" siltstone) and roof contacts with sandstones and siltstones are clear. Figure 23.3a shows a typical profile, where there are alternating beds and layers of coal and siltstone (sterile) that characterize the seam.

23.3.2 *Irapuá Coal Seam*

Stratigraphically, this coal layer lies below the Barro Branco seam. The range of sterile interlayers is between 12 and 16 m in most of the places within the deposit where it is found, with an average thickness of about 13 m.

The Irapuá seam is of restricted occurrence and has a string shape, with isolated segments, and no apparent continuity. In profile, it has a flat-bottomed channel shape, where the layer thins out towards one or both flanks. Inside the mines, this shape is well distinguished in some places by the changes in the dip of the floor or roof of the coal seam, reaching 10°, and variations of more than 1 m in thickness. The thickness of the mineral layer is variable, approximately between 1.6 and 2.7 m (Fig. 23.3b). This channel shape is associated with a possible deposition of peat in paleochannels.

23.3.3 Bonito Coal Seam

The Bonito coal seam (Lower), present in the Fontanella Mine—one of the most important mines currently exploited in this seam—is basically composed of layers of frosted coal and carbonaceous shale, with interbedding of thin layers of silt and clay. In addition, limestone concretions are observed in the middle of the layer, which are inclusions with strength and deformability characteristics entirely different from the rest of the coal seam. Figure 23.3c shows a typical profile of the Bonito coal seam.

23.4 Geotechnical Characterization of Coal Seams

The execution of geomechanical tests with the materials of the different Brazilian coal seams aimed to investigate the strength of these seams at the size scale of the pillars. During decades of coal mining in the South Catarinense Coalfield, a significant number of collapses occurred in several mines, and many of them were due to the lack of knowledge about the coal seam strength, which is the material that makes up the mine pillars. Until then, the strength of the coal seam used for pillar design was the same as that of South Africa, because of geological similarities at the edges of both continents.

Pillar strength, defined by the maximum load supported by a unit of area, depends on at least three basic elements: size effect; shape effect; the properties of the rock that forms the pillar and the host rocks.

In summary, the size effect implies higher strength values for samples tested in a laboratory, when compared to that of real pillars. This increase in strength is attributed to the lower presence of surfaces of weakness in the samples used in laboratory tests since the small dimensions used do not entirely represent the rock mass that originated them. In the case of carbonaceous rocks, the definition of the characteristic strength of the coal seam is of paramount importance, since this is the rock that makes up the support structures (pillars) of the excavations. In this type of rock, this is a complex task, since the size effect seems to be even more pronounced than in other materials, due to the formation aspects of the coal, as reflected by the presence of small to large geological structures, such as cleats, bedding planes, fractures, faults, and inclusions of materials other than coal.

Different techniques for the estimation of the in situ strength of the coal seam can be used. The technique of laboratory tests with small samples is one of the least costly ones from the economical viewpoint; however, it is a limited technique, due to the low representation of the lithological variations and structural characteristics of the rock mass. When there is access to the coal seam, using larger samples than those obtained from drill cores is more desirable and tends to significantly help in the understanding of the mechanical behavior of the rock, in order to estimate the in situ strength with reasonable accuracy.

In Brazil, the first studies on coal using scientific techniques are from the early 1980s, with emphasis on the laboratory and in situ test developed by the IPT on the coal from the I_1F seam (Midea et al. 1985), and by CIENTEC (Zorzi 1990). CIENTEC studies covered the most important coal seams under underground exploitation at the time or with future potential in Brazil in about 1300 laboratory tests specimens and five in situ tests. This work was a pioneer on the definition of in situ strength of the I_1F (Charqueadas deposits, in the state of Rio Grande do Sul) and Barro Branco seams (South Catarinense Coalfield, in the state of Santa Catarina), in addition to serving as a base for the proposal of a methodology for pillar sizing applied to southern Brazil coal seam conditions (Zorzi et al. 1991).

23.4.1 Sampling of Coal Seams for Laboratory Testing

The samples of the Barro Branco, Irapuá, and Bonito coal seams were collected during mining at the main mines in operation, as indicated in Fig. 23.4. With the aid of a universal cutter, blocks of up to 1.5 m^3 in volume were separated from the rock mass by vertical and horizontal cuts. Immediately after extraction, the samples were covered with a thin layer of wax to protect the rock against the change in humidity and the action of weathering agents. The blocks were transported on sawdust beds in order to minimize shock damage during the journey from the mine to the laboratory.

The preparation of the cubic specimens with different sizes was done with the use of large diamond saws, as shown in Photo 23.1, and smaller saws were used for the final cut, and the finishing of the faces for the application of the loads was done with the use of a flat surface grinder, as shown in Photo 23.2. Photos 23.3 and 23.4 show test specimens in different size ranges, before failure, for two of the three different coal seams studied.

23.4.2 Barro Branco Coal Seam

The Barro Branco coal seam has very particular characteristics that significantly distinguish it from the other Brazilian coal seams. It has much interbedding of sterile materials with the carbonaceous matter, and in many cases, gradations are lithologically placed between them. However, this layer is marked by more differentiated inorganic (sterile) matter (see Fig. 23.3a), giving the siltstones of this layer

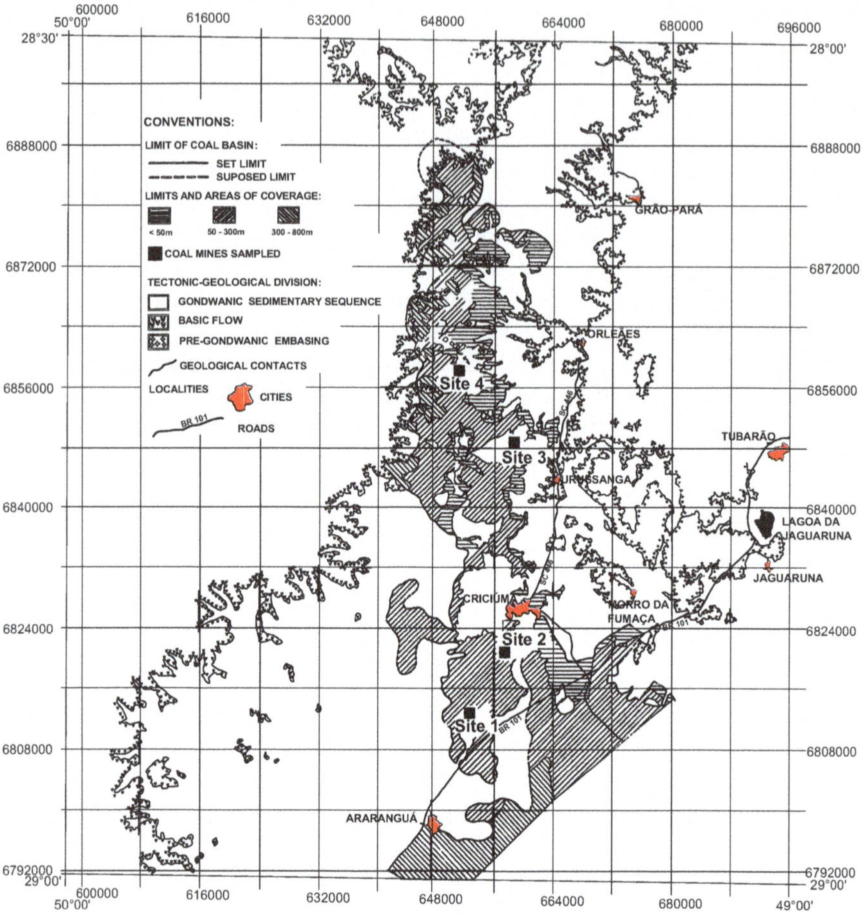

Fig. 23.4 Sampling locations of the main coal seams, Barro Branco, Irapuá and Bonito, and the host rocks in the South Catarinense Coalfield (Coordinates UTM J22) (modified from Zorzi 1990)

greater strength than the siltstones of other layers with associated organic matter. Such singularities in the lithologic composition led to different studies for the mechanical characterization, including many laboratory tests and three in situ tests.

23.4.2.1 Laboratory Testing

Execution of the Tests

The tested samples represent the coal layer present in six important mines distributed over the four sampling sites in the South Catarinense Coalfield, as indicated in Fig. 23.4.

Photo 23.1 Saw with 1-m-diameter blade, used to reduce the size of the coal samples in the laboratory

Photo 23.2 Flat-surface grinding used for finishing the cubic faces for load application

The laboratory studies for the determination of the uniaxial compressive strength have considered these heterogeneity aspects of the coal seam. For this purpose, test specimens were prepared with different sizes (Table 23.1) and classified according to five lithological types:

- Coal (C): an essentially carbonaceous material with bulk density lower than 1.8 g/cm^3;
- Coal and Pyrite (CP): Carbonaceous material containing pyrite nodules. The density is variable, depending on the proportion of each material present;

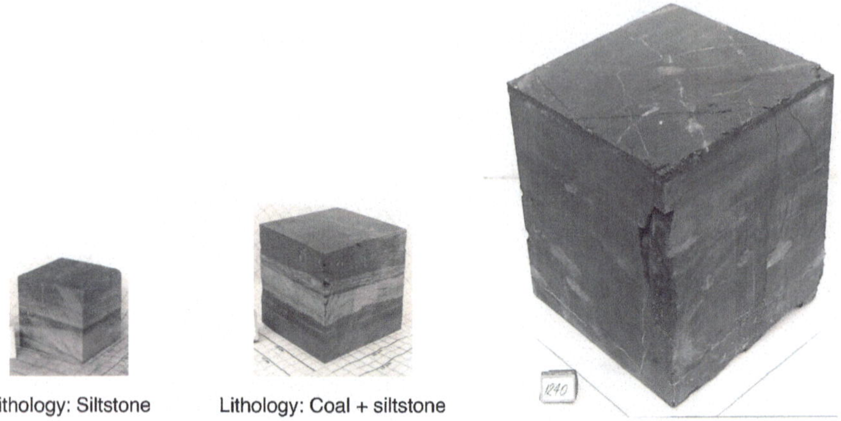

Photo 23.3 Cubic test specimens of different lithologies and sizes used in the strength effect study on the Barro Branco coal seam. Lithology: Siltstone (L = 6 cm). Lithology: Coal + siltstone (L = 9 cm). Lithology: Coal (L = 26 cm)

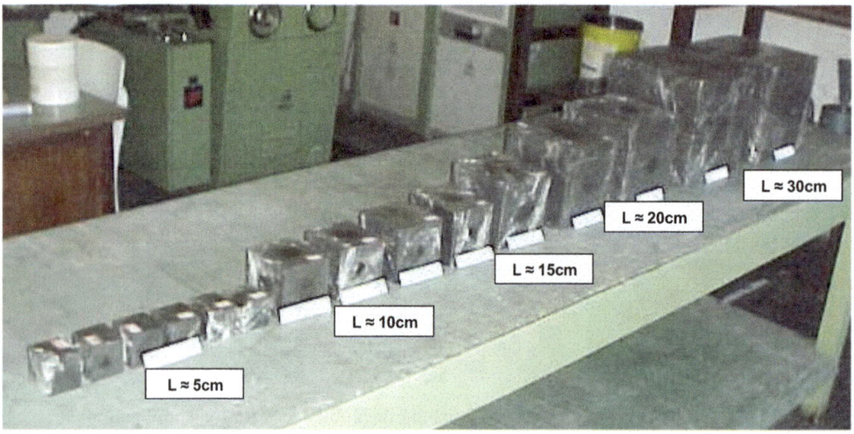

Photo 23.4 Cubic test specimens in larger size ranges (5–30 cm) used in the size effect study on the Bonito coal seam

- Coal and Siltstone (CS): it is a stratified sample with interbedding layers of coal and siltstone, with bulk density between 1.8 and 2.3 g/cm^3;
- Siltstones: sterile material, containing little or much carbonaceous matter, with bulk density equal to or greater than 2.3 g/cm^3. This lithology was classified into two strength levels, medium strength (MS) and high strength (HS).

The test specimens were taken to failure point in a hydraulic press with a load capacity of 100 kN, for the smaller test specimens, and a reaction gantry equipped

Table 23.1 Sizes and amounts of cubic test specimens prepared for testing the Barro Branco coal seam

Cube side dimension range (cm)	Total specimens
4.17–5.83	154
5.83–7.50	114
7.50–9.17	72
9.17–10.83	156
10.83–12.50	27
12.50–14.17	37
14.17–15.83	83
15.83–17.50	11
17.50–19.17	17
19.17–20.83	60
20.83–22.50	5
22.50–24.17	5
24.17–25.83	15
25.83–29.17	2
29.17–30.83	8
Total	766

with a hydraulic jack/pump assembly, with a load capacity of 2 MN for the larger test specimens. The uniaxial compressive strength was determined by applying the load on the test specimen in the perpendicular direction to the bedding planes of the coal seam, according to ISRM (1979a) recommendations for this type of assay. Typically, the loading speed applied resulted in a stress rate in the order of 0.7 MPa/s. The results of 766 tests are shown graphically in Fig. 23.5.

Analysis of Results

Regarding the carbonaceous rocks of the Barro Branco coal seam, the results in Fig. 23.5 show that the definition of the characteristic strength of the coal seam is a complex task, due to the presence of small to large geological structures, such as cleats, bedding planes, fractures and inclusions of materials other than coal (siltstones, pyrite). As a rule, siltstones are more resistant than coal.

The strength results were statistically analyzed by cube-side size class interval. The distribution of values at each interval showed both the adherence to the Normal distribution model (Fig. 23.6) and Weibull distribution. This finding was observed in all analyzed size range intervals.

Based on preliminary analyses of all 766 laboratory tests, regression studies were conducted to determine the scale effect using two mathematical models:

- the exponential model: $\sigma_c = A + B * e^{C * L}$,
- the algebraic model: $\sigma_c = k * L^{-\alpha}$.

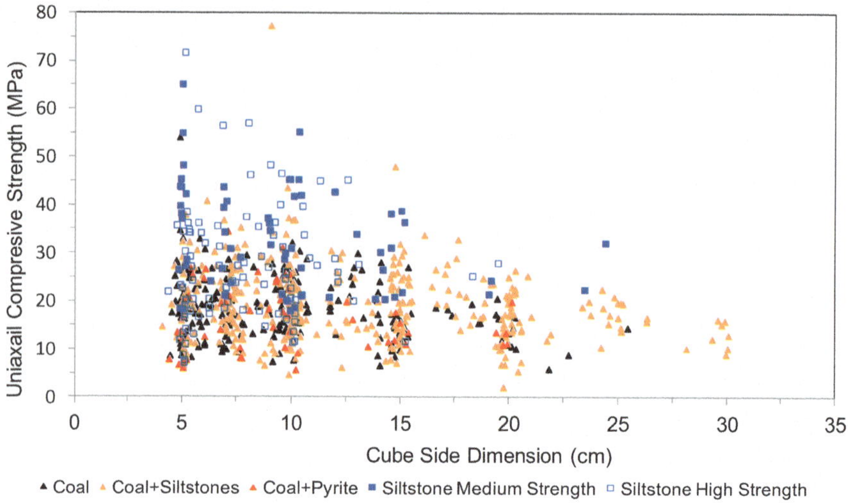

Fig. 23.5 Results of uniaxial compression tests carried out with cubic samples of different sizes (all lithologies) of the Barro Branco coal seam

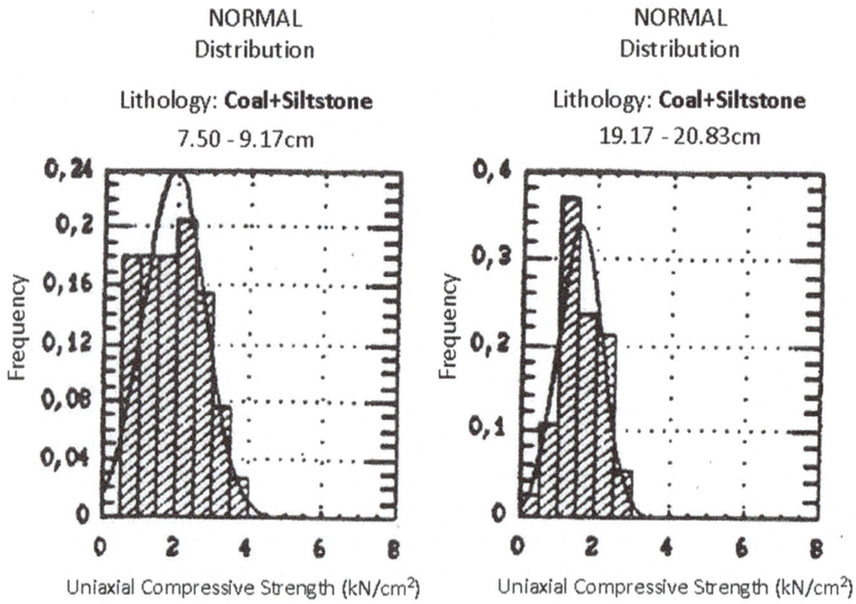

Fig. 23.6 Distribution model applied to the uniaxial compressive test results on cubic test specimens in different size ranges

where: σ_c = uniaxial compressive strength;

L = size of the test specimen;

A, B and C = constants.

Additionally, these regression analyses were conducted using strength correction factors, *with* and *without weight*. The weight used for the strength adjustment in the test specimen size range was defined as the inverse of the strength variance. In summary, the exponential model provides, in both conditions, *with weight* and *without weight*, calculated values close to those obtained in situ for the Barro Branco coal seam. Thus, the results of the analyses showed that, among the different tested correlation models to represent the scale effect, the "square root" algebraic model (constant $\alpha = 0.5$), as suggested by Steart (1954) and Gaddy (1956), is the most conservative one, followed by the exponential model, which provides results between 40% and 70% higher than the "square root" algebraic model, and lastly, the algebraic model, which produces results 1.8–2.5 times higher than the first. These models are shown in Fig. 23.7, along with the mean strength value observed in each test specimen size range, as defined in Table 23.2. In the same figure, the results of in situ tests on cubic test bodies with side dimensions up to 0.8 m can be seen, as described below.

For the Barro Branco seam, the application of the exponential model, with weights, produced the most significant coefficients, allowing the prediction of strength values very similar to those obtained in the in situ tests. The laboratory results associated with in situ tests led to the recommendation of the following equation for the calculation of the scale effect:

$$\sigma_c = 7.12 + 16.01 * e^{-0.023*L}$$

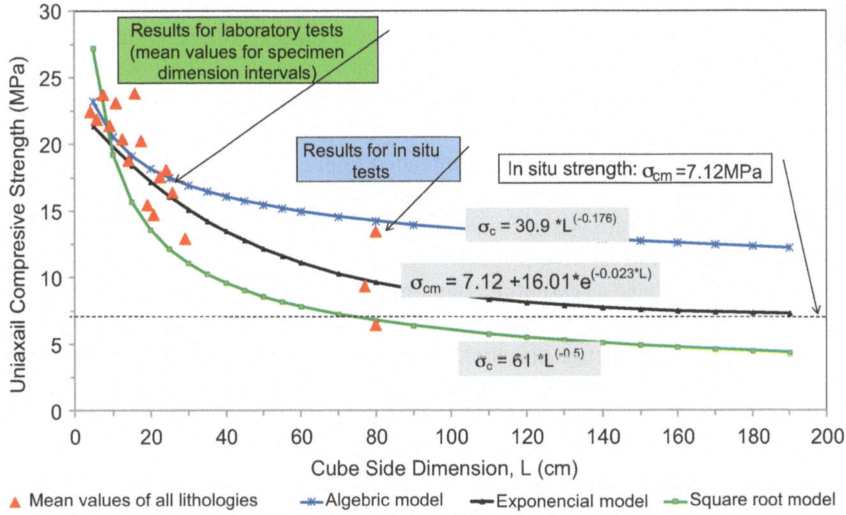

Fig. 23.7 Mean uniaxial compressive strength by size range of test specimens tested in the laboratory; values obtained by in situ tests and analytical models to predict the scale effect for the Barro Branco coal seam

Table 23.2 Summary of the results of tests carried out in situ in the Barro Branco coal seam, south of the Santa Catarina Coalfield

Specimen identification	Geometry of test specimen (m)	σ_c (MPa)	E_m (GPa)	ν_m
CP 3	0.80 (*H*) × 0.81 (*L*) × 0.80 (*D*)	13.5	5.32	0.12
CP 4	0.76 (*H*) × 0.78 (*L*) × 0.77 (*D*)	9.4	6.77	0.21
CP 5	0.79 (*H*) × 0.81 (*L*) × 0.80 (*D*)	6.5	1.83	0.18

H height, *L* length, *D* depth, σ_c uniaxial compressive strength, perpendicular to the bedding planes of the coalbed, E_m average Young's modulus, ν_m average Poisson's coefficient

where: σ_c = predicted strength (MPa);
L = size of the specimen (cm).

Thus, for the Barro Branco coal seam, the recommended in situ strength for the sizing of the pillars corresponds to the function value when it becomes an asymptote, that is, $\sigma_{cm} = 7.12$ MPa, as seen in Fig. 23.7.

23.4.2.2 In Situ Testing

In the second half of the 1980s, CIENTEC conducted extensive geomechanical characterization work on the materials found in the coal deposits of southern Brazil. The execution of uniaxial compression tests in large samples in the Barro Branco coal seam in the South Catarinense deposit, state of Santa Catarina, was given special attention due to its economic importance and the need to define the in situ strength of this coal seam.

A set of three in situ tests was performed at the A-Sangão Mine, located in the South Catarinense Coalfield (Site 2 in Fig. 23.4). Aiming to characterize the heterogeneity of the Barro Branco coal seam and the contact with the host rocks, test specimens were taken from the lower half and the upper half of the layer. Figure 23.8 shows the lithological profile of the test specimens (CP). The location of the specimens near the mining fronts was intended to preserve as much as possible the mechanical integrity of their constituent materials.

The load application system consisted of a set of 6 or 7 hydraulic jacks, with capacity for 9 MN and 12 MN, respectively, driven by a hydraulic pump. The basic test configurations have as essential components the test specimen, a concrete reaction block for transferring the loads, a set of hydraulic jacks and metal plates of 4.1 cm in thickness to better distribute the stresses on the concrete block, and the reaction rock layers (host rocks). Photo 23.5 shows the configuration of two tested specimens. The strain measurements were both in the direction of the load (axial strain) and in the perpendicular direction (transversal strain). A set of four Linear Voltage Displacement Transducers (LVDTs), with a measuring range of 50 mm and resolution of 0.1 mm, and four dial indicators, with a measuring range of 20 and 0.01 mm resolution, were installed on each side face of the test specimen. For the test specimen failure, the staggered load technique was adopted, with stress increments of 0.6 MPa every 10 min.

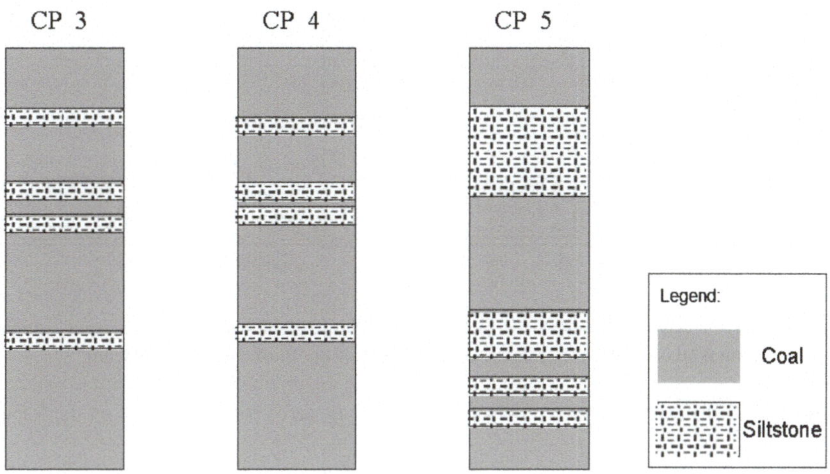

Fig. 23.8 Lithological profile of test specimens from the A-Sangão Mine. Two of them are from the lower half of the Barro Branco coal seam (CP 3 and 4) and one from the upper half (CP 5)

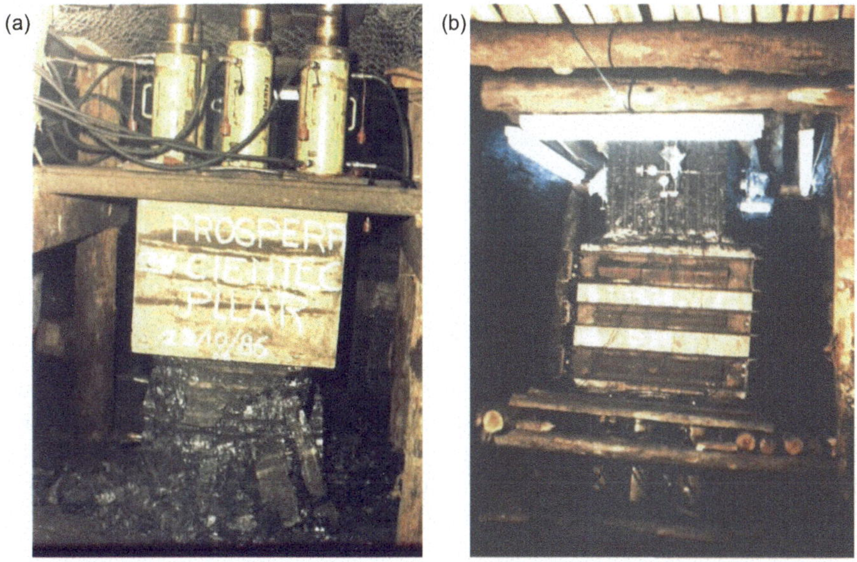

Photo 23.5 Detail of test specimens 4 and 5 tested on the lower (**a**) and upper (**b**) half of the Barro Branco coal seam

The summary of the results of the three tests is presented in Table 23.2, which shows the geometry of the specimens, the average maximum uniaxial compressive strength, Young's modulus and Poisson's coefficient values, determined according to the procedure recommended by ISRM. Figure 23.9 shows the "stress x strain" curves of the tested specimens.

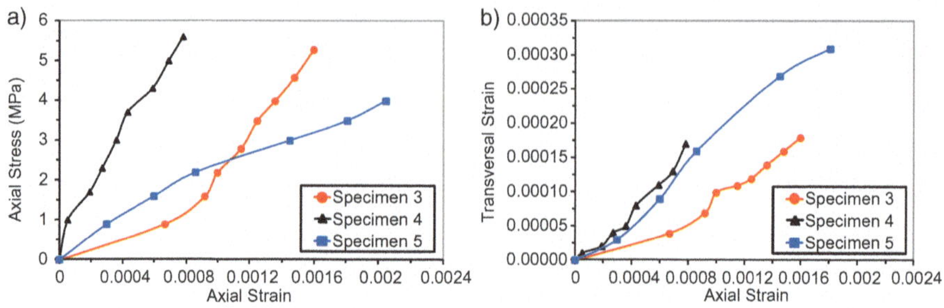

Fig. 23.9 (**a**) "Axial stress × Axial strain" and (**b**) "Transversal strain × Axial strain" graphs obtained in uniaxial compression tests on large samples of the Barro Branco coal seam

Table 23.3 Sizes and amounts of cubic test specimens prepared for experiments on the Irapuá coal seam

Cube side dimension range (cm)	Total specimens
4.6–7.3	17
8.1–10.5	16
14.7–15.6	13
19.6–20.6	6
24.6–25.6	2
31.1–31.4	2
Total	56

23.4.3 *Irapuá Coal Seam*

23.4.3.1 Execution of Tests

In total, 56 specimens with a cubic shape were tested in the laboratory, with side dimensions ranging from 4.6 to 31.4 cm, in the quantities shown in Table 23.3. The samples represent an area of about 2 km^2 at A-Sangão Mine (Site 2 in Fig. 23.4). The failure of the test specimens was conducted according to ISRM (1979a) recommendations and using the same equipment used for the Barro Branco coal seam.

23.4.3.2 Analysis of Results

Figure 23.10 shows the plotted strength data, the regression model curve for the strength scale effect law, and the confidence interval of the mean and of an individual value for the confidence limit of 95%. The main parameters involved in the statistical analysis of a regression model are shown in the graphic representation of the models. Although the coefficients of determination, r^2, show that the variability

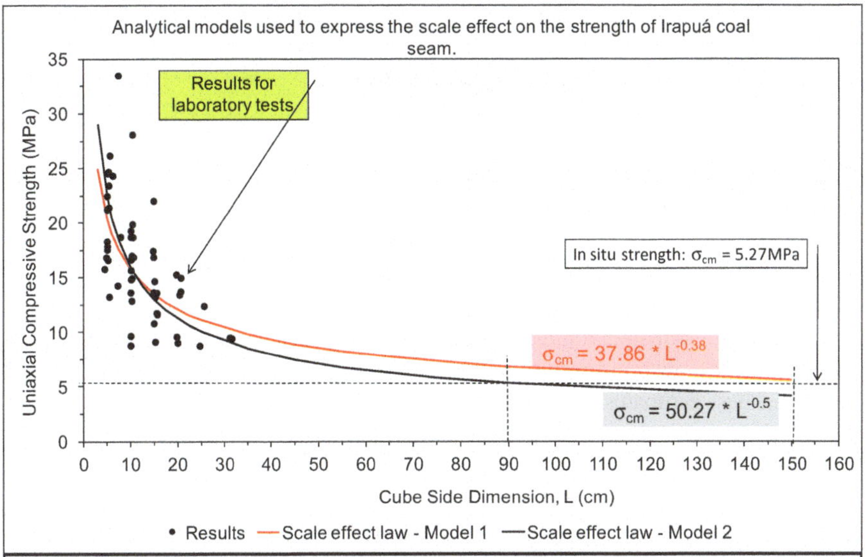

Model 1: $\sigma_C = k * L^{\alpha}$

Parameter	Coefficient	Standard deviation	T value	Significance level
k	37.8606	0.1475	24.6289	0
α	-0.3777	0.0610	-6.1886	8.4E-8
Correlation coefficient = -0.64 Coefficient of determination = 41.5%				
Analysis of Variance				
Origin	Sum of squares	Degrees of freedom	F	Significance level
Model	2.3161	1	38.30	0
Error	3.2655	54	-	-

Fig. 23.10 Models and parameters of the scale effect law of the Irapuá coal seam uniaxial compressive strength

shown by the model is lower than 50%, the F-test, for analysis of variance, and the T-tests, to verify the coefficient of correlation and null hypothesis of the coefficients at significance level of 0.05%, indicate that the model is helpful to explain the data.

A regression study with experimental data from the Irapuá seam showed values of $k = 37.9$ and $\alpha = -0.38$ (Model 1 in Fig. 23.10). When setting the α value at -0.5, $k = 50.27$ is obtained. This model (Model 2 in Fig. 23.10) agrees with the law proposed by Steart (1954) and Gaddy (1956) and is in accordance with the results obtained from many studies conducted by different authors on carbonaceous materials. The vertical dotted lines, traced from the x-axis, indicate two of the most recommended critical sizes to define in situ strength of coal seams. Starting from the first size (90 cm) to the second (150 cm), the strength decreases by 23%, when $\alpha = -0.5$, and 17%, when $\alpha = -0.38$.

Table 23.4 In situ compressive strength (σ_{cm}) estimated for the Irapuá coal seam considering different models and critical sizes

Model	σ_{cm} at Critical Size, L_{crit}	
	L_{crit} = 90 cm (MPa)	L_{crit} = 150 cm (MPa)
1) $\sigma_{cm} = 37.86 * L^{-0.38}$	6.85	5.64
2) $\sigma_{cm} = 50.27 * L^{-0.5}$	5.27	4.10
3) According to Wilson (1983)	4.15	

The values estimated for the Irapuá coal seam in situ strength are shown in Table 23.4. The obtained results allowed to estimate σ_{cm} values between 5.0 and 5.5 MPa. These estimates are consistent with the 5.27 MPa value suggested by Zorzi et al. (1991), obtained from the analysis of pillar failure cases at the A-Sangão Mine.

23.4.4 Bonito Coal Seam

23.4.4.1 Execution of Tests

The Bonito coal sample collection was carried out at the Fontanella Mine, in the municipality of Treviso/SC (Site 3 in Fig. 23.4). Two different geological conditions were represented within the mine: a heavily geologically disturbed one, near faulted areas or diabase dikes, and another one in an area of the mine free of geological disturbances. The samples represent an area of about 1 km^2 at the Fontanella Mine. In total, 78 specimens with cubic shape and side dimensions ranging from 5 to 30 cm (Photo 23.6) were tested, as shown in Table 23.5.

The uniaxial compression failure of the specimens was conducted using an MTS 816 servo-controlled system with a load capacity of 2 MN. The failure followed the recommendations of the ISRM (1999) method for uniaxial compression tests with the determination of the post-failure curve. In the cubic test specimens with side dimensions above 20 cm, analog and/or electrical sensors with a measurement capacity of up to 10 mm and a resolution of 0.001 mm (Photo 23.6a) were used, while in the other test specimens, strain gage-based sensors (MTS models) and/or strain-gage load cells of 20 mm in length (Photo 23.6b) were used. The uniaxial compressive strength results obtained on the 78 cubic test specimens in different sizes can be seen in Fig. 23.11.

23.4.4.2 Analysis of Results

The preliminary analysis of the results in Fig. 23.11 clearly shows two strength levels, and the geological determinants were probably the main cause. The materials from undisturbed areas of the mine are those with the highest strength levels.

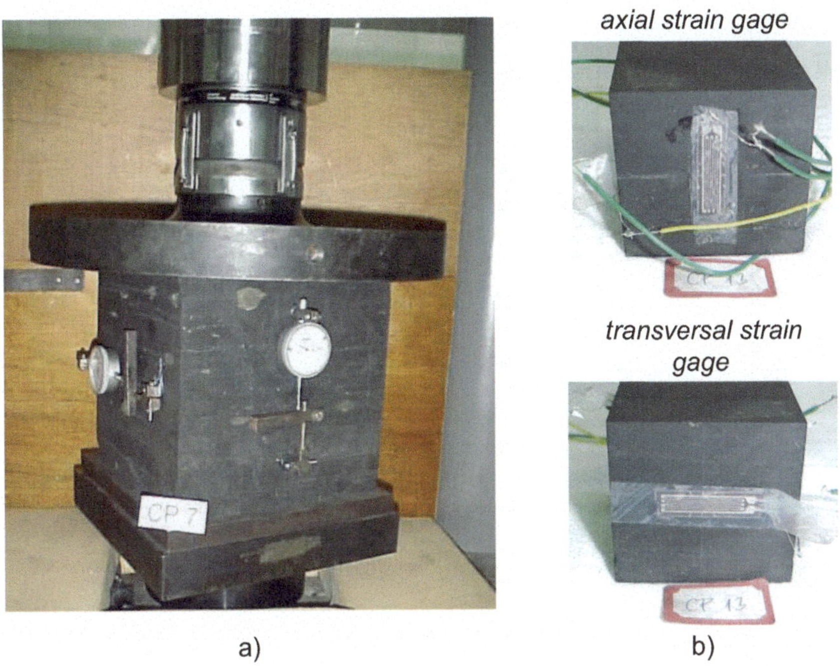

Photo 23.6 Detail of cubic test specimens with different sizes instrumented with deformation measurement systems. (**a**) cube side dimension, $L = 26$ cm; (**b**) $L = 5$ cm

Table 23.5 Sizes and amounts of cubic test specimens prepared for the assays of the Bonito coal seam

Geological condition	Cube side dimension range (cm)	Total of specimens
Disturbed geological condition	5–10	30
	10–20	10
	20–30	1
Non-disturbed geological condition	5–10	18
	10–20	11
	20–30	8
Total		78

Table 23.6 shows the synthesis of the statistical data obtained by the application of the power function, $\sigma_c = k * L^{\alpha}$ which, in addition to being simple, is one of the most used, internationally, to express the scale effect.

The constants (k) and (α) of the adjusted models, the statistical indices (p), the coefficients of determination (r^2), and the compressive strengths, σ_c, are shown, estimated by the regression models, for a critical side size cube, L_{crit} = 100 cm. The

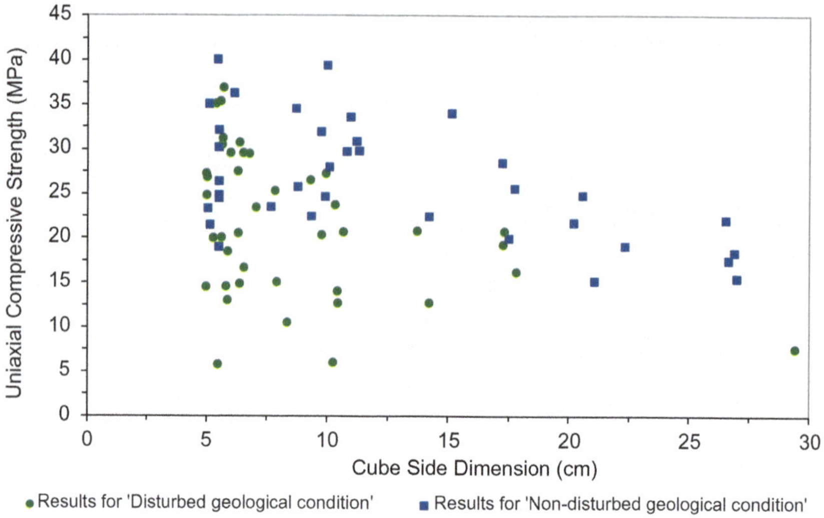

Fig. 23.11 Uniaxial compression tests carried out with cubic samples of different sizes of the Bonito coal seam

Table 23.6 Results of the regression analyses for the determination of the scale effect on the strength of the Bonito coal seam at the Fontanella Mine

	Model parameters $\sigma_C = k * L^{\alpha}$					
Geological Condition	df	k	α	p	r^2 (%)	σ_c at critical size $L_{crit = 100\ cm}$ (MPa)
Disturbed geological condition	40	43.92	−0.39	0.0160	13.9	7.3
Non-disturbed geological condition	36	44.07	−0.23	0.0009	27.2	15.5
Both conditions combined	77	34.19	−0.19	0.0230	6.6	14.1

σ_c uniaxial compressive strength of a cubic specimen at critical size, L_{crit} 100 cm, L cube side dimension,
df degrees of freedom, *a*, *b* coefficients of the regression model

coefficient of determination (r^2) is a measure of the model fit to the data set, that is, it characterizes the fraction of the total variability explained by the proposed model. The obtained values are extremely low. The p value is a measure of significance. When this value is less than 0.05, there is a statistically significant relationship between the two analyzed variables, at the 95% confidence level. In both cases, however, the values of σ_c (7.3 and 15.5 MPa) obtained by the models are unacceptable from the practical perspective, considering the mining experience in this carbonaceous layer and the in situ strength values estimated from back-analyses of pillar failure cases in the Fontanella mine itself, between 4.2 and 5.4 MPa (Gonzatti 2007).

Table 23.7 Summary of the results obtained with the application of the power function with $\alpha = -0.5$, for critical sizes of 100 and 150 cm

Geological condition	Model parameters $\sigma_C = k * L^{\alpha}$ df	k	α	σ_c at critical size $L_{crit = 100\ cm}$ (MPa)	σ_c at critical size $L_{crit = 150\ cm}$ (MPa)
Disturbed geological condition	62	58.19	−0.5	5.8	4.8
Non-disturbed geological condition	36	79.29	−0.5	7.9	6.5
Both conditions combined	77	67.05	−0.5	6.7	5.5

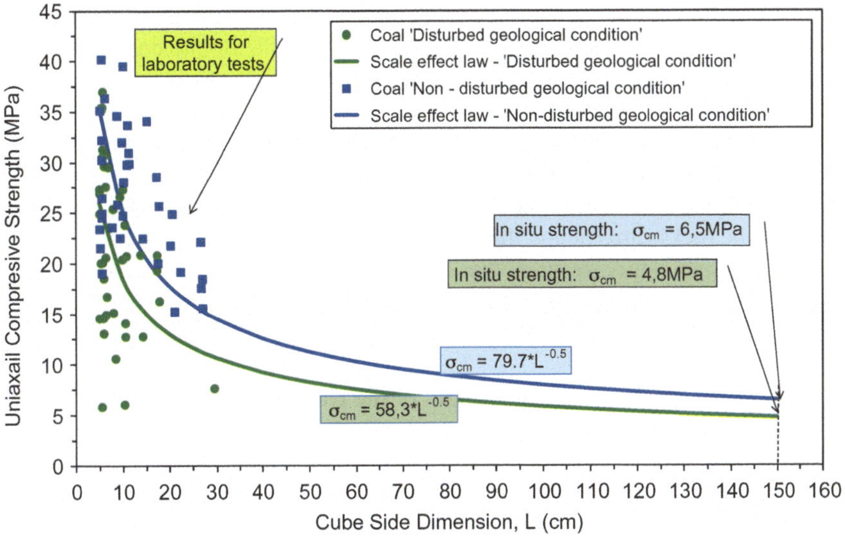

Fig. 23.12 Uniaxial compressive strength results in cubic test specimens of different sizes and exponential model to predict the scale effect for the Bonito coal layer

Although the α values, between −0.2 and − 0.4, are compatible with those obtained in several of the most important international studies, the unreal results and limitations in the size of the test specimens led to the adjustment of the power function with the coefficient $\alpha = -0.5$. In addition, a critical size of 150 cm was included. The obtained results are summarized in Table 23.7.

The analysis of the estimated results for σ_c in the critical size, $L_{crit} = 100$ cm of edge, shows values that are too high, even in comparison with the typical strengths of other more known carbonaceous layers of the South Catarinense basin. Therefore, and considering that the Bonito seam has a high thickness and a structural pattern with at least two families of very well characterized discontinuities, which is very unlike the other coal layers of this coal basin, it is advisable to adopt a critical size of 150 cm, to define their in situ strength(s). Based on the above, the in situ strength of the Bonito seam, σ_{cm}, based on the laboratory tests on the Fontanella mine samples, is between 4.8 and 6.5 MPa, as shown in Fig. 23.12.

23.5 Geotechnical Characterization of the Host Rocks of the Coal Seams

23.5.1 Sampling of Host Rocks

The selection of the sites sampled within the South Catarinense Coalfield considered preliminary geology and distribution studies of the different host rock lithotypes of the coal seams in the study area. In addition, the distribution of sampling sites includes the most important and representative mines in the mine, as seen in Fig. 23.4.

Host rock surveys of the Barro Branco and Irapuá coal seams were conducted in galleries located in the mining fronts. The host rocks of the Bonito coal seams were sampled from boreholes drilled from the surface and all samples are from the Fontanella Mine.

The samples were collected using a double tube (NX) core barrel, with recovery above 98%. Samples were collected at least 5 m above and below the coal seam for laboratory tests, as seen in Photo 23.7. Table 23.8 shows the quantitative sampling results of the rocks surrounding the different coal layers. Before the selection of the samples for laboratory testing, the different lithologies were macroscopically described and the Rock Quality Designation (RQD) was determined.

23.5.1.1 Lithographic Characteristics of the Host Rocks of the Barro Branco Coal Seam

In the roof of the Barro Branco seam, there could be only sandstones, only siltstones or a succession of sandstone and siltstone layers. The siltstone that makes up the immediate roof varies from dark gray or black, massive and consistent siltstone to a fissile and more micaceous rock (silty shale). At times it appears laminated, with lenses of fine sandstone. The sandstones, in turn, are fine-grained, sometimes coarse, quartzous, slightly micaceous, and coherent. They show clear or gradational contacts with the base and top siltstones. The immediate floor of the Barro Branco coal seam is formed, for the most part, by fine- to coarse-grained quartz sandstone. Near the floor of the Barro Branco seam, locally, there may be a layer of black, compact siltstone, easily alterable in the presence of water. Figure 23.13 shows typical geological profiles of the host rocks of the immediate roof and floor in the first 5 m above and below the coal layer.

23.5.1.2 Lithographic Characteristics of the Host Rocks of the Bonito Coal Seam

The immediate roof of the Bonito coal seam is mostly composed of laminated, fine- to medium-grained quartz sandstones. Similar to what is observed in the host rocks of the Barro Branco coal seam, these sandstones sometimes show lenticular to wavy

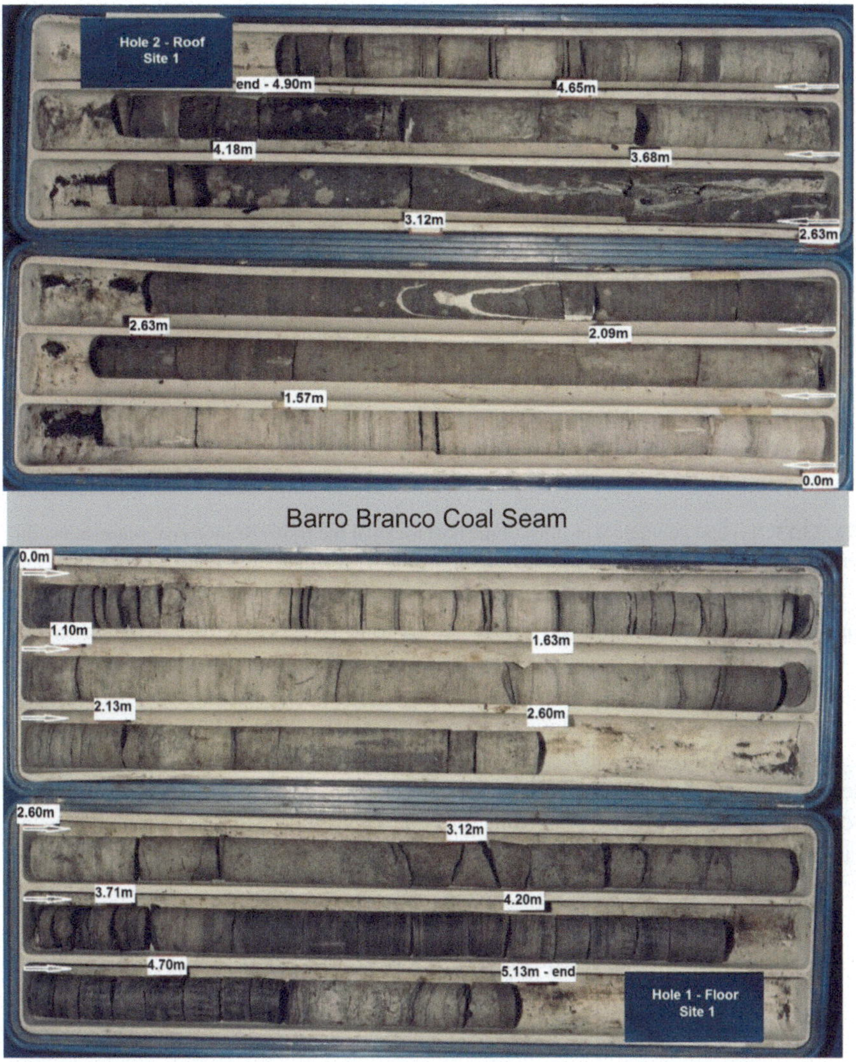

Photo 23.7 Drill core, NX diameter, used in the characterization tests of the host rocks of the Barro Branco coal seam—Site 1 in Fig. 23.4

Table 23.8 Quantitative boreholes drilled to sample the host rocks of the South Catarinense Coalfield

Hosts rocks of coal seams	Total boreholes	Total vertical boring used for sampling (m)
Barro Branco from the 4 sites in the south Catarinense coalfield	23	~148
Irapuá at Sangão mine-A (site 2)	2	~20
Bonito at Fontanella mine (site 4)	20	~150

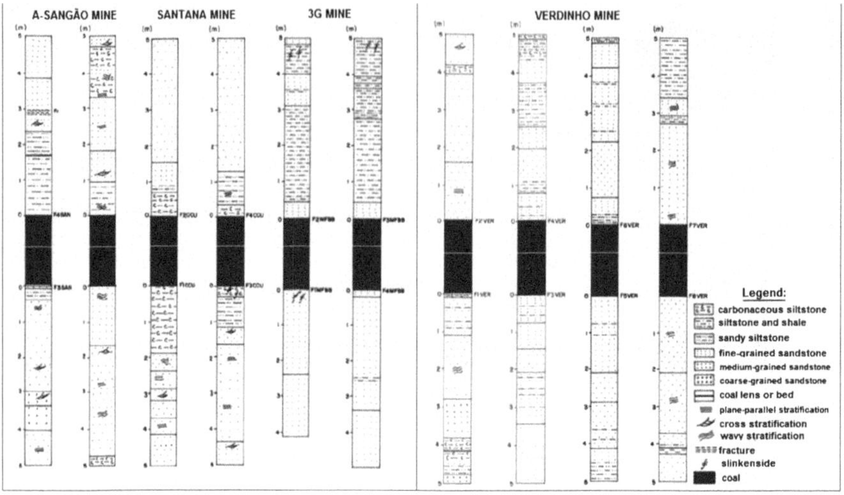

Fig. 23.13 Typical geological profiles of the host rocks of the Barro Branco coal seam, at the different mines sampled in the South Catarinense Coalfield (modified from Zorzi 1990)

lamination, with interbedding lenses of siltstone material with sandstone layers. Other lithologies present in the immediate ceiling, although less frequent, are the massive sandstones, fine- to coarse-grained, of essentially quartz composition.

The floor of the Bonito layer consists of interbedding siltstones and fine- to coarse-grained massive sandstones. In the immediate floor, with approximate thicknesses between 2 and 4 m, are massive or laminated siltstones, though they also occur in areas in which the fine- to coarse-grained sandstones are present. Figure 23.14 shows typical profiles of the host rocks of the Bonito coal seam.

23.5.2 Laboratory Tests on the Host Rocks of the Barro Branco Coal Seam

23.5.2.1 Petrographic Analyses

The macroscopic petrographic analyses, as well as the determinations of RQD of the different lithologies, were carried out before the mechanical laboratory tests. These lithologies can be macroscopically divided into five geomechanical classes, characterized by the features seen in Table 23.9. They appear both on the roof and on the floor of the coal layer, depending on the location within the coal deposit, as seen in Fig. 23.13.

23.5.2.2 Physical and Mechanical Testing

The preparation and finishing processes of the cylindrical test specimens, as well as the execution of the assays, followed the procedures suggested by the ISRM (1978a, 1978b, 1979b, 1983), specific for each type of test. Prior to the mechanical

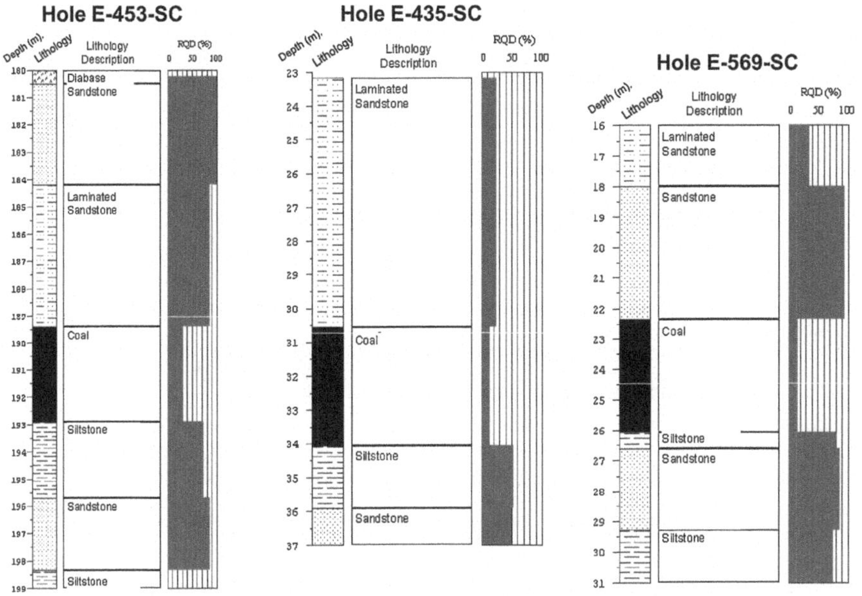

Fig. 23.14 Typical geological profiles of the host rocks of the Bonito coal seam, at the Fontanella Mine in Treviso/SC, in the South Catarinense Coalfield (modified from Zorzi et al. 2007)

tests, the bulk density and P-Wave velocity were determined. The latter was determined using a "PUNDIT" (portable ultrasonic non-destructive digital indicating tester) with transducers of 54 KHz, 1 MHz, and 2 MHz. In total, 702 assays were carried out, as seen in Table 23.10.

A Hoek cell was used in the triaxial tests. The assay procedure was type I—individual test or type II—multiple failure state test. The confining pressures used ranged from 10 to 30 MPa for all lithologies. The Brazilian tensile tests were conducted in a Losenhausen machine with a load capacity of 100 kN. The tensile strength was determined parallel to the bedding planes of the rock.

The uniaxial and triaxial compression tests were conducted on an MTS 816 loading system with a load capacity of 2 MN. In both testing modes, the application of the major stress was perpendicular to the rock bedding planes. The strain measurements on the test specimens were performed with electric strain gages and a Kyowa electronic system. Young's modulus was calculated using the average method, while Poisson's ratio was based on the secant method. All the tests were conducted according to the ISRM methods, with the samples at natural humidity, between 0.5% and 3%.

The geomechanical classification of the different lithological types studied was based on the macroscopic/microscopic petrographic analysis and statistical analysis of the P-wave velocity, tensile strength, and uniaxial compressive strength results. Finally, the results of all the types of mechanical tests separated by class were used for the verification of the average failure envelope. The procedures used to define the failure envelope of each lithology follow the Hoek–Brown failure

Table 23.9 Typical macroscopic petrographic features of the sampled geomechanical classes

Geomechanical classes	Macroscopic description
Sandstone *HS* high strength	Orthoquartzite sandstone, fine-grained, light gray color with a carbonate cement. Areas with light spots are the result of strong carbonate cementation. It is compact. The rock is very coherent, with good to excellent RQD (> 75%) in the massive areas and poor RQD (<25%) in interspersed zones. 
Sandstone *MS* medium strength	Quartz-rich arkose, fine to medium-grained, light gray color, interbedded with thin lenses. Quartz cement. Frequent contact fractures. Excellent (> 75%) to very poor (<25%) RQD. 
Sandstone *LS* low strength	Coarse-grained sandstone, light gray color with dark spots. Interspersed organic matter lenses cause frequent horizontal bedding fractures. The cement is carbonate, although the rock has a brittle aspect. Poor RQD (25–50%). 
Siltstone *HS* high strength	Dark gray siltstone, with thin lenses of interbedded sandstone. Very frequent bedding plane fractures. Poor RQD (<50%). 
Siltstone *LS* low strength	Carbonaceous siltstone interbedded with fine-grained sandstone forming small lenses. Plane-parallel stratification, extremely fractured on bedding planes. Very poor RQD (<25%). 

Table 23.10 Tests performed with materials from different geomechanical classes

Geomechanical classes	Test type						
	Density	P-Wave velocity	Indirect tensile strength	Uniaxial compressive strength	Young's modulus	Poisson's ratio	Triaxial compression
Sandstone HS	344	344	169	85	31	30	90
Sandstone MS	91	91	48	22	5	5	21
Sandstone LS	73	73	38	17	4	4	18
Siltstone HS	134	134	75	30	7	7	29
Siltstone LS	60	60	35	14	2	3	11
Total of tests	702	702	365	168	48	49	169

criterion version by Hoek et al. (2002). The calculations of the representative envelope parameters of the intact rock were performed with the RocLab software, version 1.032.

In the South Catarinense Coalfield, the materials belonging to the Sandstone HS and Siltstone HS geomechanical classes represent approximately 50% and 20%,

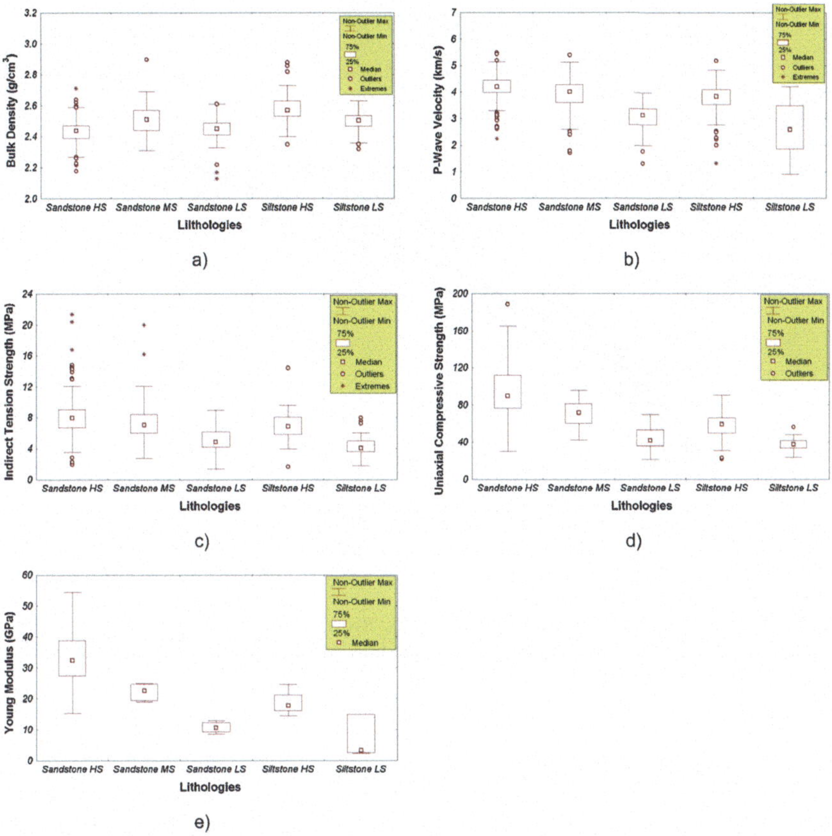

Fig. 23.15 Physical features—(**a**) bulk density and (**b**) the P-wave velocity—and mechanical properties—(**c**) tensile strength, (**d**) uniaxial compressive strength, (**e**) average Young's modulus (secant 50%)—for the main geomechanical classes of the host rocks of the Barro Branco coal seam in the South Catarinense Coalfield (Software Statgraphics Plus n.d.)

respectively, of the sampled lithologies. Each of the other three classes represents about 10%. The results of the physical and mechanical properties, after the statistical treatment of the data and definition of the five geomechanical classes, are shown in box-and-whisker diagrams in Fig. 23.15a–e. Figure 23.16 shows the failure envelopes of all 12 samples tested and classified as Sandstone HS, along with the average envelope for this lithology. An analogous analysis procedure was carried out to define the envelopes of the other lithologies. Table 23.11 shows the strength parameters, while Fig. 23.17 graphically represents the envelopes.

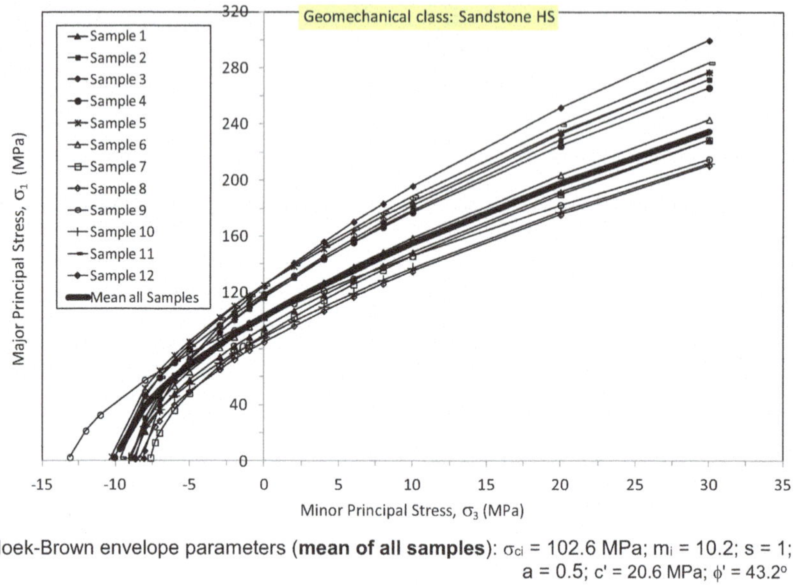

Fig. 23.16 Failure envelope according to the Hoek–Brown criterion of all intact rock samples, classified geomechanically as Sandstone HS, of the host rocks of the Barro Branco coal seam in the South Catarinense Coalfield

23.5.2.3 Analysis of Results

Two different lithologies were observed in the host rocks of the Barro Branco coal seam: sandstones and siltstones. The association of physical features and unconfined mechanical properties suggests the division of the sandstones into three classes: Sandstone HS, fine-grained, the most resistant one; Sandstone MS, medium- to fine-grained; Sandstone LS, medium- to coarse-grained, and the least resistant one. The siltstones, in turn, are divided into two classes: Siltstone HS and Siltstone LS. Macroscopically, these siltstones are distinguished only by the degree of fissility. From the shear strength perspective, the failure envelopes also suggest five geomechanical classes with different strength characteristics (see the parameters in Table 23.11), although the lithologies Sandstone LS and the Siltstone HS are similar in the compression zone (Fig. 23.17).

23.5.3 Laboratory Tests on the Host Rocks of the Bonito Coal Seam

23.5.3.1 Petrographic Analyses

The macroscopic petrographic analyses, as well as the determinations of Rock Quality Designation—RQD of the different lithologies, were carried out before the mechanical tests. This information, associated with physical and mechanical

Table 23.11 Mean parameters of the five geomechanical classes of the host rocks of the Barro Branco coal seam

Geomechanical classes	Typical properties of the intact rock from laboratory tests						Failure criterion parameters for intact rock			
							Hoek–Brown ($s = 1$, $a = 0.5$)		Mohr–Coulomb	
	ρ (g/cm^3)	Vp (km/s)	ITS (MPa)	UCS (MPa)	E_m (GPa)	$\nu_{sec\ 50\%}$	σ_{ci} (MPa)	m_i	c' (MPa)	ϕ' (o)
Sandstone HS	2.43 *(0.07)*[a]	4.2 *(0.45)*	8.2 *(2.67)*	93.0 *(26.47)*	34.0 *(9.61)*	0.31 *(0.12)*	102.6	10.2	20.6	43.2
Sandstone MS	2.51 *(0.09)*	3.9 *(0.65)*	7.6 *(2.97)*	71.0 *(14.56)*	22.1 *(2.84)*	0.37 *(0.12)*	75.9	7.0	16.8	38.9
Sandstone LS	2.44 *(0.08)*	3.1 *(0.51)*	5.1 *(1.74)*	42.3 *(13.71)*	10.8 *(1.90)*	0.34 *(0.11)*	56.1	7.9	12.0	40.3
Siltstone HS	2.58 *(0.08)*	3.8 *(0.56)*	6.8 *(1.79)*	57.3 *(16.64)*	18.7 *(3.37)*	0.24 *(0.05)*	57.9	6.2	13.5	37.5
Siltstone LS	2.50 *(0.06)*	2.6 (0.95)	4.3 (1.46)	38.3 (7.90)	2.9 (0.57)	0.33 (0.10)	32.7	4.1	8.4	32.5

ρ natural bulk density, *Vp* P-Wave velocity, *ITS* indirect tensile strength, *UCS* uniaxial compressive strength, E_m average Young's modulus, $\nu_{sec\ 50\%}$ Poison's ratio 50% secant method, σ_{ci} uniaxial compressive strength, m_i, s, a constant of the Hoek–Brown criterion, c' cohesive strength, ϕ' angle of friction of the Mohr–Coulomb criterion

[a]Values in parentheses = standard deviation

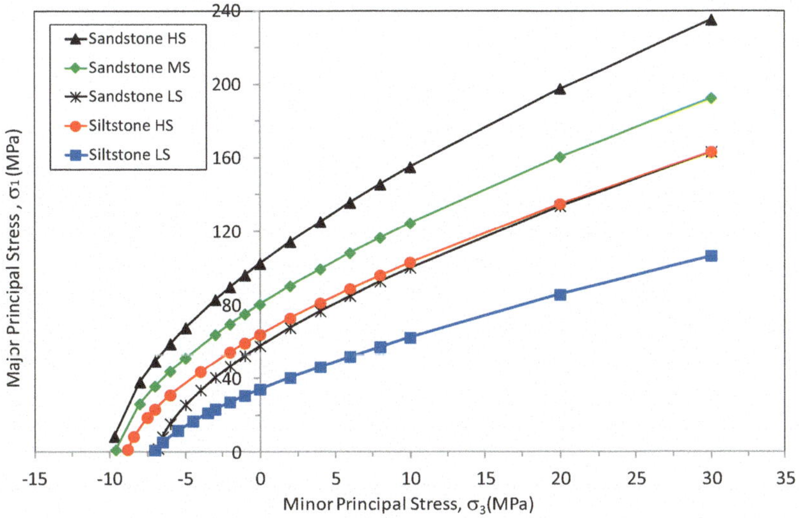

Fig. 23.17 Average failure envelopes according to the Hoek–Brown failure criterion, representing five different geomechanical classes of the host rocks of the Barro Branco coal seam

Table 23.12 Typical macroscopic petrographic features of the sampled geomechanical classes

Geomechanical classes	Macroscopic description
Sandstone *HS* high strength to *MS* medium strength	Massive high-strength (HS), fine-grained sandstone with silica and/or carbonate cementation, and good to excellent RQD (>75%). Massive medium-strength (MS), fine to medium-grained sandstone, with thin lenses of siltstone, with silica and/or carbonate cementation, and good to excellent RQD (>75%). 
Laminated sandstone *HS* high strength to *MS* medium strength	Laminated sandstone to laminated siltstone of high (HS) to medium (MS) strength: Layers of fine-grained, compact sandstone, alternating with layers of siltstone. Sometimes the sandstone layers predominate, sometimes the siltstone layers (zebra siltstone) predominate, with poor to good RQD (>50% to <75%). 
Siltstone *MS* medium strength	Massive siltstone to sandy fine-grained siltstone, of medium strength (MS), and poor RQD (>25% to <50%). 

properties, indicated five geomechanical classes, characterized by the petrographic features seen in Table 23.12. The sandstones and the laminated sandstone appear both on the roof and the floor of the coal layer, depending on the location within the coal deposit. On the other hand, the siltstones are usually found on the floor, as seen in the profiles in Fig. 23.14.

23.5.3.2 Physical and Mechanical Testing

All the physical and mechanical tests were conducted according to ISRM methods, with the samples at natural humidity between 0.1% and 2.8%. The lowest moisture contents were observed in the sandstones, followed by the laminated sandstones and the siltstones, which presented the highest levels. The Brazilian tests were conducted so that the tensile strength was determined parallel to the bedding planes of the rock. The uniaxial and triaxial compression tests were conducted with the application of the major stress perpendicularly to the rock bedding planes. In triaxial tests, the confining pressures ranged from 10 to 30 MPa for all lithologies and the procedure was Type I—individual test. In total, 270 trials were performed, as shown in Table 23.13. The bulk density and P-wave velocity values can be considered as the approximate expression of the proportion of each lithological type in the sampled area (approximately 36 km^2).

Table 23.13 Tests performed with materials from different geomechanical classes

Geomechanical classes	Test type					
	Bulk density	P-wave velocity	Indirect tensile strength	Uniaxial compressive strength	Young's modulus	Triaxial compression
Sandstone HS	45	45	18	27	4	9
Sandstone MS	53	53	30	23	3	–
Laminated sandstone HS	33	33	9	24	3	14
Laminated sandstone MS	75	75	38	37	11	–
Siltstone MS	64	64	10	53	20	8
Total of tests	270	270	105	164	41	31

The physical and mechanical property results, after the statistical data processing and the definition of five geomechanical classes (two massive sandstones—Sandstone HS and LS, two sandstones mixed with siltstones—Laminated Sandstone HS and MS, and one massive siltstone—Siltstone MS), are graphically represented in box–whisker diagrams in Fig. 23.18a through e. The mechanical test results (Brazilian, uniaxial compression, and triaxial compression tests), separated by class, were used to establish the average failure envelope. The procedures used to define the failure envelope of each lithology were analogous to those used for the host rocks of the Barro Branco coal seam described above.

Table 23.14 shows the main parameters that represent the average intact rock failure envelope of the different classes while the Fig. 23.19 shows graphically the average envelopes of the samples classified as Sandstone HS, Laminated Sandstone HS, and Siltstone MS.

23.5.3.3 Analysis of Results

The macroscopic petrographic analysis of the rock materials found in the Bonito seam of the South Catarinense Coalfield clearly shows three distinct lithologies: massive sandstones, laminated sandstones, and siltstones.

The combination of the petrographic, physical (Fig. 23.18b) and mechanical (Fig. 23.18c–e) characteristics suggest the division of the massive sandstones into two classes: Sandstone HS, fine-grained with predominant silica cementation and the strongest one, and Sandstone MS, also fine-grained but with more carbonate cement, and the weakest. In addition, the Laminated Sandstones can be divided into two classes: Laminated Siltstone HS, in which sandstone layers are predominant, interbedded with siltstone, and the strongest one, and Laminated Siltstone MS, the weakest one, with predominating siltstone layers interbedded with thin layers of sandstone. The Siltstone MS class is characterized by being massive and compact.

From the perspective of shear strength, only the geomechanical classes massive sandstones (HS), laminated sandstones (HS), and the massive siltstone were studied.

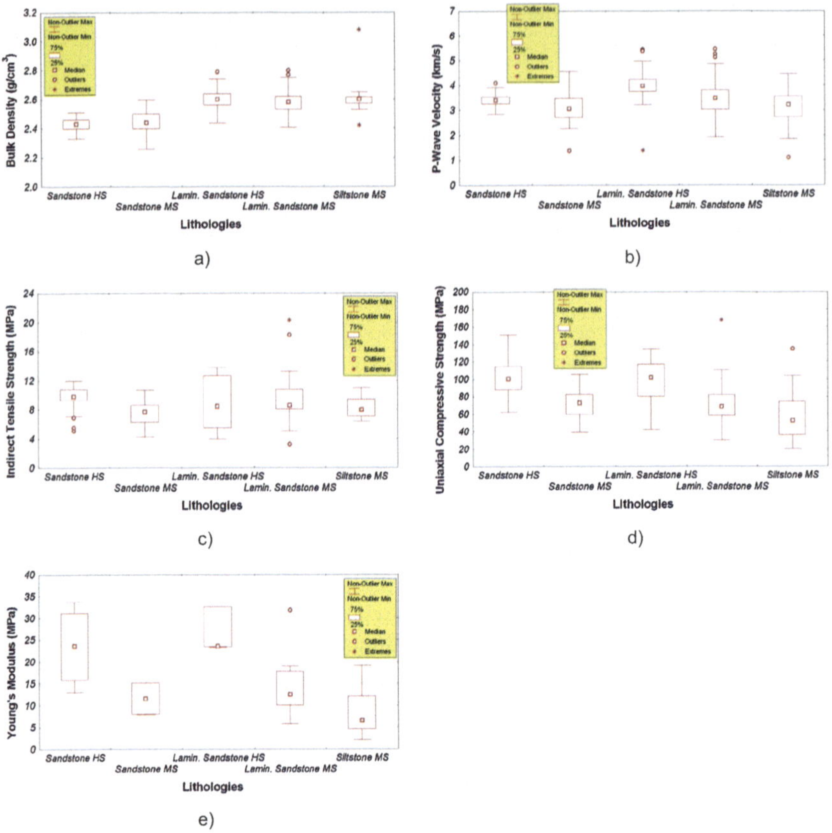

Fig. 23.18 Physical features—(**a**) bulk density and (**b**) P-wave velocity—and mechanical properties—(**c**) tensile strength, (**d**) uniaxial compressive strength and (**e**) average Young's modulus—for the main geomechanical classes of the host rocks of the Bonito coal seam in the South Catarinense Coalfield (Software Statgraphics Plus n.d.)

The failure envelope of these three different geomechanical classes differs, however, only in the tension and uniaxial compression zone. The failure envelopes suggest two geomechanical classes with different characteristics: the massive sandstones with the same characteristics of sandstones combined with the siltstone layers, and the siltstone lithology.

23.6 Behavior of Hosts Rocks in Coal Mines

23.6.1 Roof

The lithologies geomechanically classified as Sandstone HS and MS and Siltstone HS (host rocks of the Barro Branco coal seam—Table 23.11), and the lithologies Sandstones HS and MS and Laminated Sandstones HS and MS (host rocks of the

Table 23.14 Mean parameters of the five geomechanical classes of the host rocks of the Bonito coal seam

Geomechanical classes	Typical properties of the intact rock from laboratory tests								
						Failure criterion parameters for intact rock			
						Hoek–Brown ($s = 1$, $a = 0.5$)		Mohr–Coulomb	
	ρ (g/cm^3)	Vp (km/s)	ITS (MPa)	UCS (MPa)	E_m (GPa)	σ_{ci} (MPa)	m_i	c' (MPa)	ϕ' (o)
Sandstone HS	2.43 *(0.05)*[a]	3.42 *(0.26)*	9.4 *(1.98)*	102.1 *(21.17)*	23.5 *(9.43)*	89.7	6.5	20.3	37.9
Sandstone MS	2.45 *(0.07)*	3.14 *(0.59)*	7.4 *(1.61)*	71.8 *(17.65)*	11.6 *(3.62)*	–	–	–	–
Laminated sandstone HS	2.60 *(0.08)*	3.97 *(0.70)*	9.0 *(4.11)*	97.6 *(26.47)*	26.6 *(5.4)*	89.3	7.4	19.4	39.6
Laminated sandstone MS	2.58 *(0.08)*	3.48 *(0.71)*	9.5 *(3.15)*	72.2 *(23.67)*	14.1 *(7.1)*	–	–	–	–
Siltstone MS	2.59 *(0.12)*	3.10 *(0.68)*	8.18 *(1.45)*	57.05 *(24.62)*	8.30 *(5.10)*	57.8	5.1	14.0	35.1

ρ natural bulk density, *Vp* P-Wave velocity, *ITS* indirect tensile strength, *UCS* uniaxial compressive strength, E_m average Young's modulus, σ_{ci} uniaxial compressive strength, m_i, s, a constant of the Hoek–Brown criterion, c' cohesive strength, ϕ' angle of friction of the Mohr–Coulomb criterion
[a]Values in parentheses = standard deviation

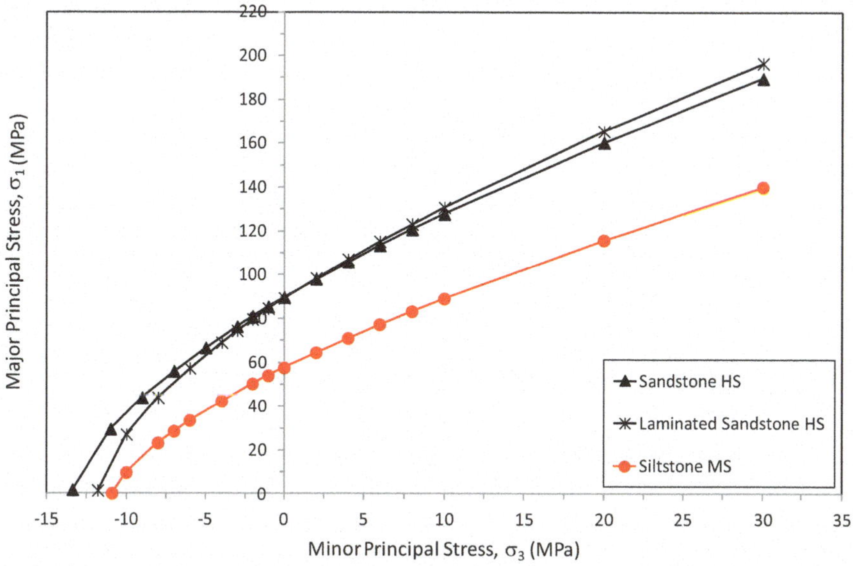

Fig. 23.19 Average failure envelopes according to the Hoek–Brown criterion representing three different geomechanical classes of the host rocks of the Bonito coal seam

Bonito coal seam—Table 23.14) comprise the best-quality rock massifs as to the roof stability of the galleries. When in the immediate roof, these rocks are almost self-sustaining (sandstones) or require a light support system (siltstones). When they are above the immediate roof, they comprise the anchor points of the tie rod system for the suspension of the bottom layers.

The greatest problems of roof stability in the coal mines in the South Catarinense Deposit are associated with the presence of the Sandstone LS or the Siltstone LS, present in the host rocks of the Barro Branco coal seam (see Table 23.11). These lower mechanical quality materials usually require interventions with high load-bearing support systems, that is, longer bolts and tighter meshes (higher bolt density per area). In this type of massif, tie rod systems are not efficient when the support mechanism is based solely on the beam effect.

23.6.2 Floor

Almost all the geomechanical classes of the host rocks of the Barro Branco coal seam (see Table 23.11) may be present on the floor, although Sandstone HS, Sandstone MS, and Siltstone HS predominate. Similar to the roof, the Sandstone LS and especially the Siltstone LS are the classes with the greatest amount of floor alteration and collapse problems. One of the problems is associated with the low load-bearing capacity to support the pillar loads. In extreme cases, pillar punching and the collapse of the gallery floor can occur. Another important aspect is associated with the alteration of the siltstone in the presence of water, and disaggregation, due to high traffic of heavy machinery. This type of problem occurs predominantly in areas where the coal layer is inclined. Under these conditions, the tensile stresses transferred by the machine tires are higher, leading to peeling of the rock.

This same type of problem has been observed on the floor of the Bonito coal seam, where the main concerns are associated with the Siltstone MS class (see Table 23.14). Although this siltstone has medium strength, it is characterized by being an alterable rock when in the presence of water and it breaks due to fatigue caused by the heavy equipment traffic.

23.6.3 Recommendations for the Dimensioning of Coal Pillars

According to the geological models found for the roof and floor host rocks, proposed by Zorzi et al. (1991), as shown in Fig. 23.20, the geomechanical classes Sandstones HS and MS and Siltstone HS, present in the host rocks of the Barro Branco coal seam, as well as all of the host lithologies of the Bonito layer, can be included in the geological model of competent rocks. These rocks do not require load capacity analysis when dimensioning the coal pillars. However, the

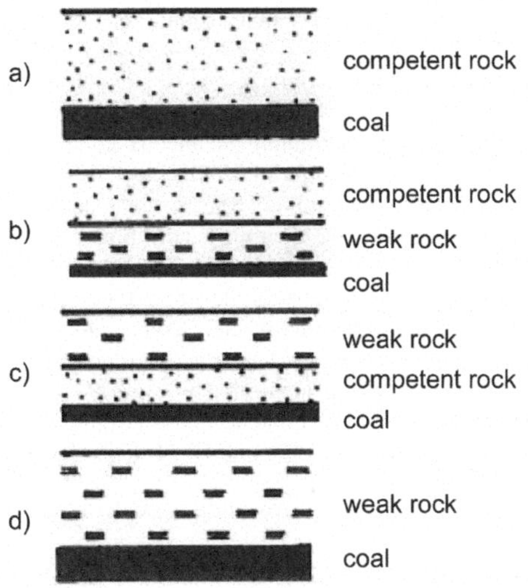

Fig. 23.20 **(a–d)** Basic geological models proposed for host rocks bearing capacity analysis in the Brazilian Coal Basin. The picture represents models for the immediate roof (about 5 m). The same model is valid for the floor condition (modified from Zorzi et al. 1991)

geomechanical classes of the host rocks of the Barro Branco coal seam classified as Sandstone LS and Siltstone LS have mechanical characteristics very close to those included in the model of weak rocks. For these two geomechanical classes, it is desirable to analyze the load capacity after the dimensioning of the pillars, considering the strength of the coal layer.

References

Costa JFCL (2000) Assessment of coal reserves—Fontanella Project—Carbonífera Metropolitana S.A. Department of Mining Engineering, Federal University of Rio Grande do Sul. Porto Alegre-RS, 79p. (in Portuguese)

Dobereiner L (1987) Geotechnical soft sandstone. Thesis summary 08. ABGE—Brazilian Association of Engineering Geology, Sao Paulo-SP, Brazil (in Portuguese)

Duarte JMG, Moraes RB, Oliveira CA, Cardoso RM (2010) Some specific geotechnical aspects of the UHE Mauá project. In.: Geotechnical Engineering Practice Symposium of the Southern Region, VII. GEOSUL 2010. Foz do Iguaçu PR, Brazil, 9p. (in Portuguese)

Gaddy FL (1956) A study of the ultimate strength of coal as related to the absolute size of the cubical samples tested. Virginia Polytechnic Institute. Blacksburg, Virginia, USA, 57p.

Gonzatti C (2007) Proposal for the estimation of the in situ strength of coal seams with the use of geophysics. School of Engineering of São Carlos, University of São Paulo (Doctoral thesis), 274p. (in Portuguese)

Hoek E, Carranza-Torres C, Corkum B, (2002) Hoek-Hrown failure criterion - 2002 edition. In: Proc. of the NARMS-TAC Conference, v. 1, pp 267-273

IPT–Technological Research Institute of Sao Paulo State (1987) Additional laboratory tests for geotechnical studies to support the optimization of bolting project of Esperança Mine. Report N° 25264. Sao Paulo-SP, Brazil, 41p. (in Portuguese)

ISRM (1978a) Suggested methods for determining sound velocity. Int J Rock Mech Min Sci Geomech Abstr 15(2):53–58
ISRM (1978b) Suggested methods for determining tensile strength of rock material. Int J Rock Mech Min Sci Geomech Abstr 15(3):99–103
ISRM (1979a) Suggested methods for determining the uniaxial compressive strength and deformability of rock materials. Int J Rock Mech Min Sci Geomech Abstr 16(2):135–140
ISRM (1979b) Suggested methods for determining water content, porosity, density, absorption and related properties and swelling and slake-durability index properties (part 1). Int J Rock Mech Min Sci Geomech Abstr 16(2):143–146
ISRM (1983) Suggested methods for determining the strength of rock materials in triaxial compression: revised version. Int J Rock Mech Min Sci Geomech Abstr 20(6):283–290
ISRM (1999) Suggested methods for the complete stress-strain curve for intact rock in uniaxial compression. Int J Rock Mech Min Sci Geomech Abstr 36:279–289
Midea NF, Yoshikawa NK, Marchi AJ (1985) Pillar design—A example of the role of rock mechanics. In: Brazilian Mining Congress, 1st. Brasília-DF, Brazil. IBRAM, Belo Horizonte-MG, Brazil, pp 198–214. (in Portuguese)
Software Statgraphics Plus (n.d.) Version 4.1—Professional System. S/N 4565002707. User manual (in Portuguese)
Steart FA (1954) Strength and stability of pillars in coal mines. J Metallurg Min Soc South Africa, 309–325
Wilson AH (1983) The stability of underground workings in the soft rocks of the coal measures. Int J Min Eng 1(2):107–108. London, Chapman and Hall
Zorzi L coord. (1990) PILAR Projet: Pillar design in coal mining. CIENTEC—Foundation for Science and Technology of the Rio Grande do Sul State. Porto Alegre-RS, Brazil, 2v. (in Portuguese)
Zorzi L coord. (1996) MULTICAMADAS Projet: Underground mining of multiple coal seams—Geomechanical study. CIENTEC—Foundation for Science and Technology of the Rio Grande do Sul State. Porto Alegre-RS, Brazil, 2v. (in Portuguese)
Zorzi L, Agostini IM, Benedett JV (1989) Geomechanical characteristics of sedimentary formations in the Santa Terezinha Coal Basin. In: Coal Brazilian Congress, 2nd. Porto Alegre-RS, Brazil, vol. 1, pp 13–43 (in Portuguese)
Zorzi L, Agostini IM, Gonzatti C (1991) Methodology for pillar sizing applied to southern Brazil coal seam conditions. CIENTEC—Foundation for Science and Technology of the Rio Grande do Sul State, Porto Alegre-RS, Brazil (Technical Bulletin nº 23) (in Portuguese)
Zorzi L, Agostni IM, Gonzatti C (2006) BONITO Project—Fontanella mine—Technical Report Nº 1285/65526. CIENTEC—Foundation for Science and Technology of the Rio Grande do Sul State, Porto Alegre-RS, Brazil, 37p. (in Portuguese)
Zorzi L, Agostni IM, Gonzatti C (2007) BONITO Project—Fontanella mine—Technical Report Nº 1285/66826. CIENTEC—Foundation for Science and Technology of the Rio Grande do Sul State, Porto Alegre-RS, Brazil, 22p. + 5 anexos (in Portuguese)

Chapter 24
Soft Rocks in Dam Foundation and Dam Sites

Ricardo Abrahão

24.1 General

For any structure with a purpose of storing water, the foundation has three important functions: (a) provide enough strength to assure the stability of the structure, (b) provide stiffness so that the resulting deformations are adequate to the structure behavior, and (c) provide sufficient tightness to limit the water flow to reasonable quantities and allow the seepage control. The design must consider the evaluation of these three functions to avoid hazard risk and failure. This concept is universal but applies to soft rocks as well. However, in these cases, the evaluation requires more attention since the values of the geomechanical parameters may be poor and closer to the limits of the design criteria.

Technical papers, as well as symposiums and congresses, always show few examples of real concrete dam foundation failures. Some other examples show that the failures have mostly been a consequence of hydrology problems, appurtenant earth dams failures or operation problems.

Up to now one could not count more than ten cases really related to concrete dam foundations. Their references show that the information collection started on nineteenth century with most cases occurring in the beginning of the twentieth century. It is worth mentioning the following ones:

1. Bouzey, France, 1895[1];
2. Austin, USA, 1911, USA, 1916[2];
3. Saint Francis, USA, 1928 (Rogers 2006);

[1] https://www.arct.cam.ac.uk/Downloads/chs/vol10/article4.pdf

[2] https://en.wikipedia.org/wiki/Austin_Dam_failure

R. Abrahão (✉)
São Paulo, Brazil
e-mail: ricardo@rageociencias.com.br

M. Kanji et al. (eds.), *Soft Rock Mechanics and Engineering*,
https://doi.org/10.1007/978-3-030-29477-9_24

4. Malpasset, France, 1959[3];
5. Shi Kang, Taiwan, 1999[4], and
6. Camará, Brazil (Abrahão and Degaspare 2016).

Accidents and dam foundation failures allow for learning lessons, teaching to avoid similar problems and situations. However, it is possible that the required care may have led to a situation in which actual structures are oversized with higher costs related to its shape and excess of concrete.

The idealization of failure modes, strength, and deformability problems has been almost all developed based on intuitions or mathematical models based solely on monitoring correlation, as discussed in this chapter, with the presentation of possible failure modes for concrete dams.

In these considerations, soft rock encompasses intact rock and rock mass with low strength, weathered rock, discontinuities with low shear strength and unconsolidated sediments.

Emphasis is given for concrete dams since earth or rock fill dams do not apply important loads on the foundation.

24.2 Soft Rock as Dam Foundation

Soft rocks unveil a lot of characteristics and properties that must be considered with whenever they are considered as a dam foundation.

Among the important properties it is worth mentioning the specific gravity (dry, natural, saturated), porosity, absorption, and water content. Usual mechanical properties are unconfined compressive strength (UCS), triaxial compression, direct shear on intact rock or on discontinuities, tensile strength, and deformability modulus and wave propagation velocities. Considering the influence of the water, some behaviors have also to be evaluated, as permeability, expansion volume and force, slaking, weathering, and so on. Finally, for certain types of soft rocks as slate, shales, siltstones, the influence of mechanical parameters anisotropy has also to be evaluated.

On the other side, it is necessary to formulate a geomechanical model which represent the behavior of the structural unit considering dam and foundation, required to analyze the dam stability and behavior. Therefore, those parameters must be translated to the model to allow it to run with appropriate parameters. Insufficient knowledge of the foundation characteristics could lead to a conservative design, impairing the economy of the project.

For design purposes, the failure modes that are currently used for certifying the dam stability will be considered as an unknown surface for continuum rock masses or as a predefined surface of geological discontinuities, including concrete–rock contact.

[3] http://www.ecolo.org/documents/documents_in_french/malpasset/malpasset.htm

[4] http://shake.iis.u-tokyo.ac.jp/home-new/projects/2003)/manuscripts/Ch02c.pdf

Usually, for practical reasons, the evaluation of the rock masses has been established considering the use of the following parameters and methods:

- Shear strength of the contact concrete x rock (in situ tests or empirical);
- Shear strength of discontinuities below the foundation (in situ tests or empirical);
- Weathering degree (ISRM, RMR, Q, …);
- Longitudinal velocity (celerity) of the terrain, obtained from site geophysical testing;
- Rock mass deformability modulus (in situ tests or empirical);
- Rock mass classifications: RMR, (Bieniawski 1973), Q (Barton 1988), GSI (Hoek 2007);
- UCS, unconfined compression strength as classification index.

24.3 Rock Mass Strength

With the purpose of classification, the strengths (UCS) of intact samples of soft rock are considered in the range of approximately 1–25 MPa. This range of values is used in empirical correlations to ascribe mechanical values to the rock mass.

However, considering dam design, all characteristics and parameters of soft rocks must be correlated to parameters that match the necessary ones to be used in dimensioning methods. The basic dimensioning methods that have been employed in dam design are the limit equilibrium method, the mathematical model and their combination. Therefore, the needed parameters are: friction angle, cohesion, deformability moduli, Poisson's' coefficient and discontinuities stiffnesses. Time dependent parameters are rarely employed.

The use of most known classification systems as RMR, GSI, Q and DMR and their combination has led to modelling in a safe side of the structural stability of dams.

Since a dam foundation is a huge area, actual load capacity formulation does not apply due to complex loads distribution. Therefore, considering the rock mass as a continuum, one must deal with the general strength of the rock mass which is nowadays derived from UCS associated to other characteristics such as weathering and fracturing.

Current concrete dam projects apply from 0.3 to 7 MPa of compressive stresses on the foundation. Recent experiences show concrete dam being built on soft rocks with UCS of the order of 10 MPa and the maximum compressive applied load of 0.5–1 MPa. As a first approach for recommendation, the relation UCS to maximum compression load should be higher than 10.

If the rupture surface is not anticipated for continuum rock masses, present modelling can only be assessed by means of mathematical models. Routinely, evaluation of these cases is made considering the contact concrete x rock as the primary failure mode. For gravity dams, whenever this surface presents at least some cohesion

(0.01 MPa) and 38° of friction angle, the corresponding structure meets most of the design criteria requirements.

24.4 Discontinuity Strength

As rock masses are jointed and fractured, it is more likely to model failure through lower strength surface than through continuum rock mass. These cases encompass both soft and hard rocks with weathered exfoliation joints, general jointing, bedding, faults, and so on, which are also considered soft rocks. Every planar surface or assemblage of surfaces occurring in shallow depths below foundation level without cohesion and friction angles below 40° might be worth analyzing, since for current design criteria, it may not match the assigned safety factors.

Almost all cases of occurrence of discontinuities below the foundation which encompass the whole size of the overlying structures require some intervention to bring the structure and its foundation to the design criteria requirements. Excavation to remove part or the whole discontinuity or inclusion of rigid bodies is the most used solution.

For example, in Fig. 24.1, the yellow horizontal line represents an open contact between basalt flows, weathered, without cohesion, with 28° of friction angle and high normal and tangential stiffnesses. According to the design and the construction, part of the discontinuity was removed and additional concrete was placed under the region of the draft tube.

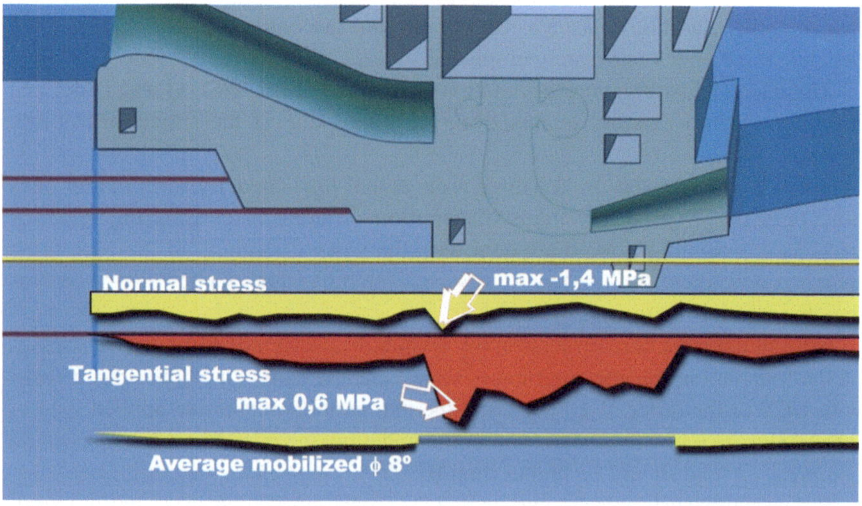

Fig. 24.1 Três Irmãos Power Plant, Brazil. Section through the Intake and Power House. Stress distribution along the discontinuity (shown as the upper yellow line). Tangential stresses are enhanced for better visualization

It is easy to see that the rigid inclusion takes great part of the shear stress and that the remaining area of the discontinuity bear a small part of it. In this case the relation arctan(τ/σ) for each element of the FEM discloses a curve (shown in yellow at the bottom of the figure) in which its clear that the discontinuity is being used beyond its capability, with only 8° of friction angle being mobilized.

To perform the limit equilibrium method in this case, the stress distribution must be evaluated by means of numerical methods and then, recalculated. In this case the equilibrium equation becomes:

$$1 \le \frac{\int \sigma_{\text{disc}} \cdot \tan\left(\dfrac{\theta_{\text{disc}}}{\gamma_\theta}\right) + \int \sigma_{\text{conc}} \cdot \tan\left(\dfrac{\theta_{\text{conc}}}{\gamma_\theta}\right) + \dfrac{c_{\text{disc}}}{\gamma_c} + \dfrac{c_{\text{conc}}}{\gamma_c}}{\int \tau} \tag{24.1}$$

where

$\int \sigma_{\text{disc}}$ = Normal force on discontinuity surface;
$\int \sigma_{\text{conc}}$= Normal force on concrete surface;
$\int \tau$ = Driving force;
θ_{disc} = Discontinuity surface friction angle;
θ_{conc} = Concrete surface friction angle;
γ_θ = friction reduction coefficient;
γ_c = cohesion reduction coefficient.

Further discussion on this subject can be found elsewhere (Souza Lima and Abrahão 1982; Souza Lima et al. 1982).

This example stresses the necessity of prospecting and investigating the soft rock mass and its geomechanical parameters as discussed above. It is recommended to test these parameters by means of field investigation. Lab and correlation of parameters reduce the accuracy of the analysis, inducing the design to be more conservative.

24.4.1 Discontinuity Stiffness

Shear stiffness K_s, which is the discontinuity stiffness in shear, usually taken as the average slope up to the stress-displacement peak, and normal stiffness K_n defined as the normal effective stress increment required for discontinuity closure are important parameters to define the behavior of a joint or fracture, whenever a mathematical model is required to anticipate stress distribution and displacements in dam foundation, mainly when contrasting stiffnesses along the discontinuity are present.

The use of discontinuities stiffnesses for modeling purposes is only recommended if the adopted parameters have been obtained by means of testing or consistent derivation based on state-of-the-art-correlations. Scale effects have been reported by Kanji 1990 and Barton 2007, which can modify the results. Therefore, whenever K_s or K_n should be used, their parameters must be carefully and accurately acquired.

For example, in the case showed in Fig. 24.1, if the K_s were smaller, more tangential load should be taken by the concrete intercept that, therefore, would be larger.

The range of discontinuities which really play important role on dam foundation are those closed but with weathered walls to those with soft filling up to 20 mm thick. Thicker fills must be considered individually as a separate material. The shear stiffness is mostly related to the characteristics of discontinuities gouge than the rock type itself.

For these cases, in what concerns dam foundations, some trends were observed and basically, as a first approach, the correlation of shear stiffness K_s to normal stress can be estimated from Fig. 24.2.

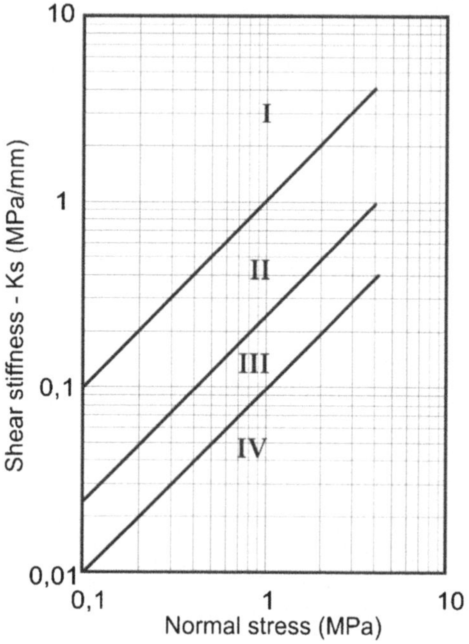

Fig. 24.2 Shear stiffness of normal and shear loaded rock joints and weathered, or clay filled discontinuities as a function of effective normal stress in the near-surface range from 0.1 to 4 MPa. (*I*) Rough unfilled joints, or with a thin clay film, $\delta_{peak} < 1$ mm. (*II*) Joints without filling, or with a thin clay film, δ_{peak} 1.0–2.5 mm. (*III*) Filled joints (thickness up to 12 mm) with δ_{peak} 2.5–5.0 mm. (*IV*) Smooth joints with fillings >20 mm (Infanti Jr and Kanji 1990)

Correlations of normal stiffness K_n are a little more complex to derive and are basically taken from field or lab tests. However, when modelling dam foundation, K_n does not play an important role, since the resulting displacements are small and do not compromise the model equilibrium.

24.4.2 *Deformability*

Some recent concrete dam structures have been founded on deformable rock masses, as low as 0.5 GPa, depending on the applied loads. The resulting deformation for various cases of loading must be dealt with, considering the stiffness contrast between the rock mass (E_m) and the concrete (E_c) and, primarily, the contrast inside the rock mass itself that should be avoided whenever possible. Prof. Rocha's (1976) recommendation, still valid for conceptual definitions states the following rule (Table 24.1):

Table 24.1 Deformability contrast between the rock mass and concrete (Rocha 1976)

Ec/Em	Influence on dam	Problems
<4	Negligible	None
4–8	Low Importance	Minor
8–16	Important	Some
>16	Very important	Moderate to big

Usually, a fast and comprehensive way to check rock mass deformability for dam foundations have been the evaluation of P-waves seismic velocity, measured by refraction geophysical method.

The following correlation is one of the possible available to evaluate the foundation deformability (Table 24.2).

Table 24.2 Correlation between seismic velocity and rock mass deformability. Modified from Barton (2007)

Seismic velocity (km/s)	E (GPa) (min)	E (GPa) (mean)
	100	100
6.0	53	68
	30	46
5.0	17	32
	9	22
4.0	5	15
	3	10
3.0	2	7
	1	5
2.0	0.5	3
	0.3	2
1.0	0.2	1.5
	0.1	1.0

To deal with deformability contrasts inside the rock mass a possible solution is to excavate to a reasonable depth until the contrast is eliminated. Enlarging concrete base area to dilute the loads is also possible.

However, whenever these solutions are impractical, the contraction joints for gravity dams can also be repositioned so as every block of concrete is founded in a rock mass with different deformability. For example, during the construction of a concrete gravity dam, a greater alluvial thickness than expected had occurred, impossible to excavate inside the cofferdammed area, had to be modified to be compatible with their foundations. The section shown in Fig. 24.3 illustrates the final section showing the position of the contraction joints to accommodate concrete blocks on contrasting deformability foundations.

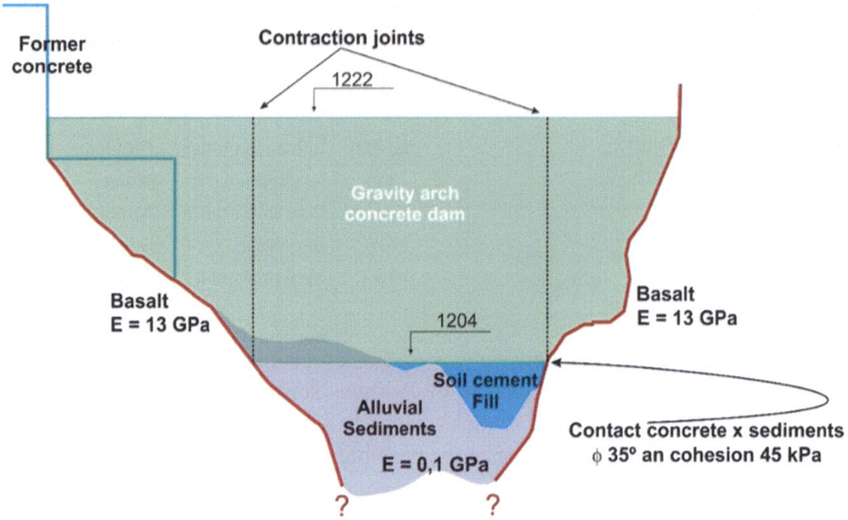

Fig. 24.3 Developed section of a concrete dam with contrasting deformability foundation

The former straight axis gravity dam had to be rearranged as a gravity arch to be accommodated into the remaining cofferdammed area as shown in Fig. 24.4.

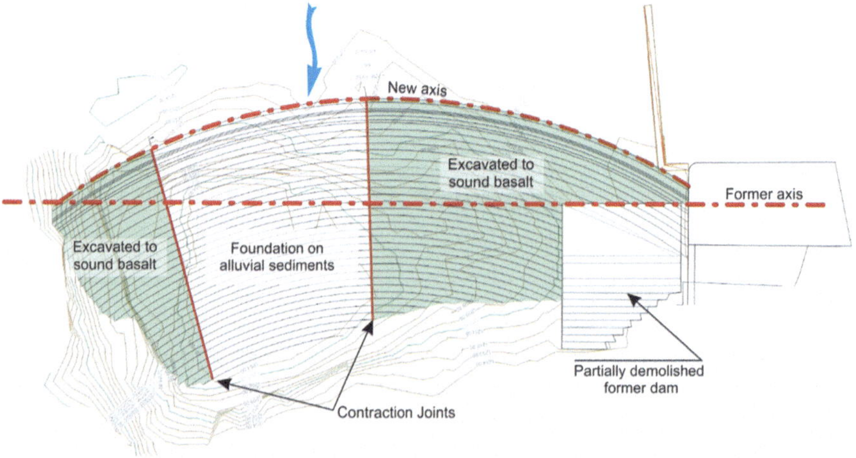

Fig. 24.4 Plan view of the new lay out of the dam, showing former and new axis development

The parameters indicated in Fig. 24.3 were adopted in the mathematical model whose results were also used in the limit equilibrium method. All loading cases, including seismic acceleration, resulted in Safety Factors within the requirements of the design criteria.

During construction, a three-rod extensometer was installed in the center of the arch, recording the foundation movements under construction loads. The measured deformation showed deformation moduli between 0.5 and 0.3 GPa, with some dispersion of values.

The dam is part of Morro Azul Power Plant in Colombia, operating without problems. The final situation to this date is shown in Fig. 24.5.

Fig. 24.5 Drone view of the completed dam under operation

24.5 Seepage

The seepage in the dam foundation has been a concern since long time ago and is fully described in hundreds of papers with actual cases and with mathematical developments. Seepage, based on theory has been evaluated by hand, by electrical analogies and now, by means of sophisticated mathematical models.

For soft rocks the same consideration applies and the parameters to be evaluated are the permeability of the rock mass for porous media or through pervious discontinuities whose persistence, opening, gouge materials and, finally the permeability have to be dealt with.

However, many cases still retain some old design criteria for evaluating seepage forces and consequent hydrostatic thrust and uplift loads.

As shown in Fig. 24.6, whenever a soft rock layer below the surface controls the sliding stability of the structure, it is possible to reduce the resulting forces, if adequate drainage and seepage analysis are considered.

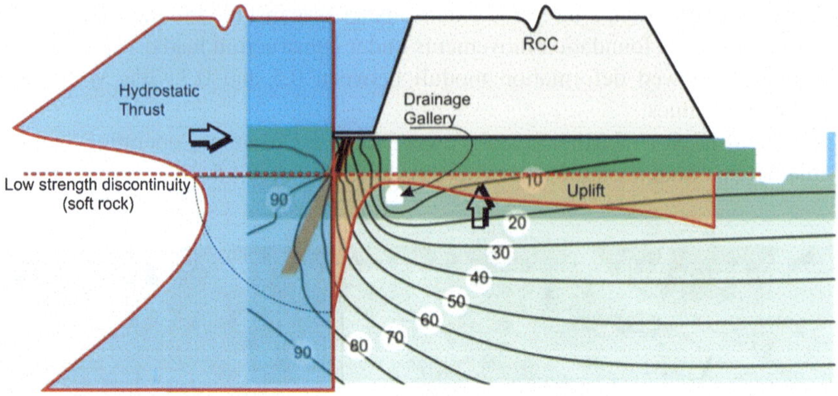

Fig. 24.6 Hydrostatic thrust and uplift in a controlling surface below the foundation of a concrete dam. Actual case for Serra do Facão Power Plant (Brazil, 212 MW)

The figure shows an actual case with a low strength chlorite filled discontinuity below the foundation and a drainage system including a drainage gallery excavated on chlorite schist.

The seepage mathematical model shows the shape of piezometric curves related to the percentage of head loss in the foundation. A cross section through the tension zone, immediately upstream of the concrete structure indicates that the hydrostatic horizontal thrust has a sudden reduction below the topographic surface. This effect has also been observed through monitoring of tens of concrete dams of various heights.

The same consideration can be made for the discontinuity present in the foundation. In the upstream limit the hydrostatic pressure is the same of the of the thrust and, consequently, smaller than the total upstream head. The integration of the curve delivers a smaller uplift force, compared to some criteria still being used. In the same manner, monitoring has been confirming this consideration, including that of the example.

As another example, in the case of Itaipu Power Plant (Brazil, 14.000 MW), the actual uplift reduction compared to some current criteria was almost 50% (Fig. 24.7).

This approach under consideration may lead to important reduction on concrete and excavation volumes.

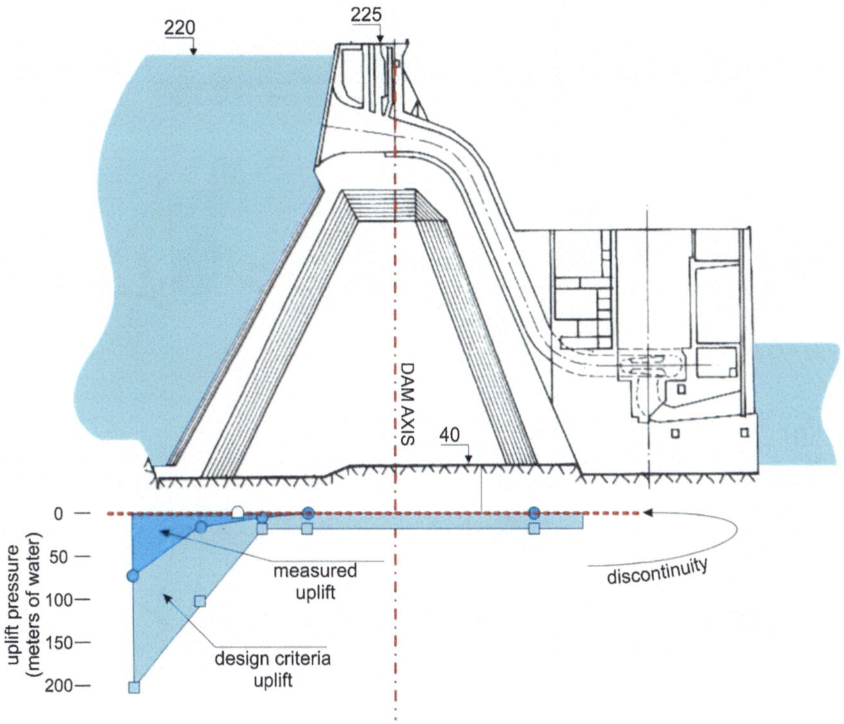

Fig. 24.7 Itaipu main dam uplift monitoring for blocks F19/20

24.6 Failure Modes

The dimensioning of the dam structure is based on a likely failure mode whose kinematic is assumed to be the worst case. All mathematical evaluation is developed to represent the failure mode introducing load variation and a hazard risk related to strength parameter reduction index. On the other hand, mathematical models can indicate the zones of plastification and whether this situation converges to stability or not. The engineering decision is based on the evaluation of all these possibilities.

Examples of hypothetical rupture modes of dam foundations are illustrated in Fig. 24.8, which can help analyzing the stability of the da and its foundation.

Modes (a) and (b) are current failure modes for embankment dams encompassing foundation on weathered soft rocks or soils or with conditioning rock discontinuities.

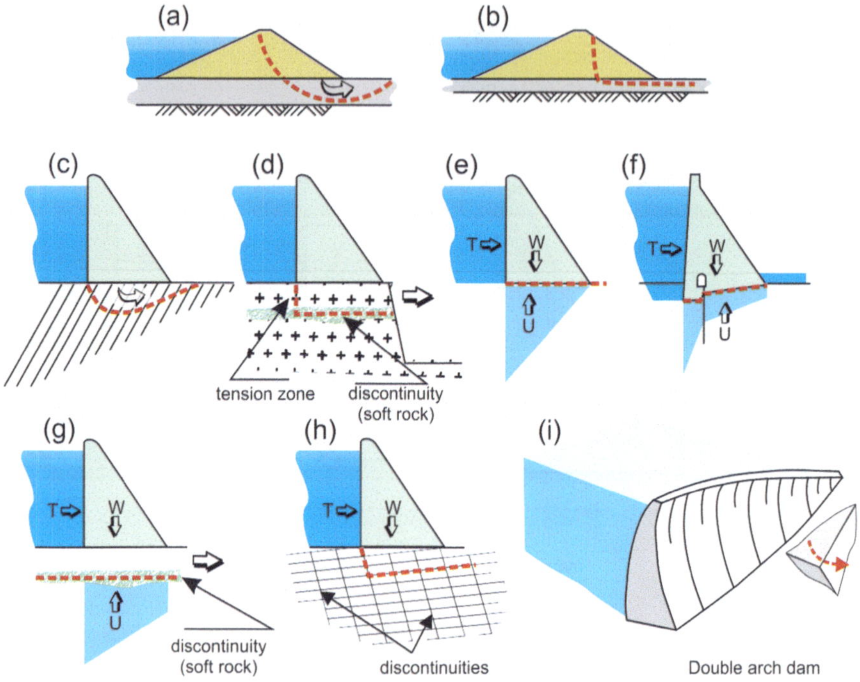

Fig. 24.8 Likely modes of rupture for modelling purposes (Modified from Kanji 1990)

Mode (c) is an unlikely case of failure for continuum rock masses in which the available load capacity is smaller than the needed, and it restricted to very low strength rock foundations.

Modes (d), (g), (h) and (i) are those related to the presence of discontinuities in an adverse position immediately below the foundation level. Modes (d) and (f) relates to a poor concrete—rock adherence, which is seldom the case if a proper construction is made. Mode (i) is a three-dimension modelling specially derived from Malpasset failure.

The above modes of potential foundation failure are the most common ones. Notwithstanding, since geologic conditions varies from site to site, new situations may arise, deserving special attention. New unwanted failures might bring more knowledge to this subject.

24.7 Case Histories

24.7.1 St. Francis Dam

"Was a concrete gravity arch dam, designed to create a reservoir as a storage point of the Los Angeles Aqueduct. It was located 40 miles (64 km) northwest of Los Angeles, California, near the present city of Santa Clarita (Fig. 24.9).

Fig. 24.9 St. Francis Dam was a curved concrete gravity arch, 209 feet high (64 m). The complete structure on the left side and the remnants on the right side (Rogers 2006)

The dam built between 1924 and 1926 under the supervision of William Mulholland, chief engineer and general manager of the Los Angeles Department of Water and Power, and then called the Bureau of Water Works and Supply. Three minutes before midnight on March 12, 1928, the dam catastrophically failed, and the resulting flood killed more than 450 people. The collapse of the St. Francis Dam is one of the worst American civil engineering failures of the twentieth century and remains the second-greatest loss of life in California's history, after the 1906 San Francisco Earthquake and fire. The disaster marked the end Mulholland's career.

Modern geologists know the type of rock found in the San Francisquito Canyon is unsuitable for supporting a dam and a reservoir. The dam was built squarely over the San Francisquito earthquake fault, although this fault has since been inactive.

J. David Rogers, a professor of geological engineering at Missouri University of Science and Technology, has published a comprehensive account of the dam's failure. The dam's failure can be attributed to three major factors: the instability of the paleo mega landslide on which the dam was built, the failure to compensate for the additional height added to the dam's design, and the design and construction being overseen by only one person." (Wikipedia)" (Fig. 24.10).

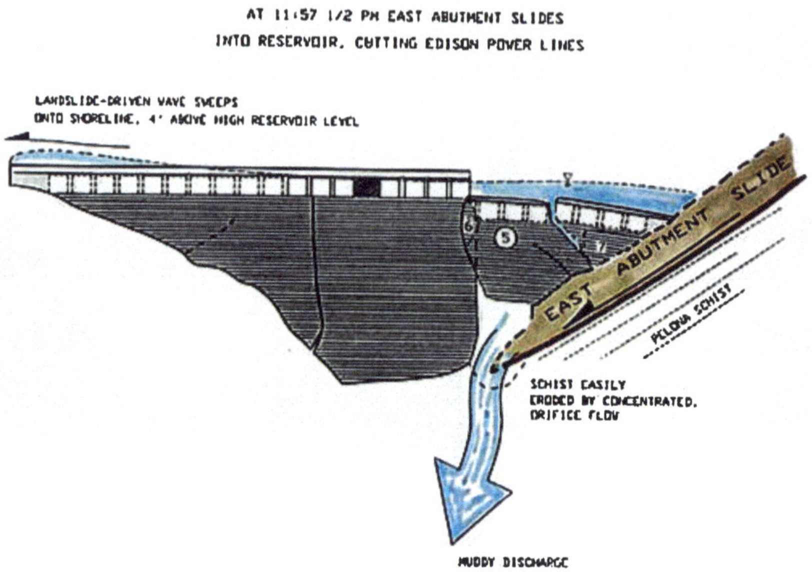

Fig. 24.10 St. Francis dam failure interpretation. East abutment is basically debris of a paleo slide (Rogers 2006)

24.7.2 Malpasset Dam[5]

"Fréjus sous les eaux du barrage: en pleine soirée, alors qu'il faisait déjà nuit, une vague de 40 mètres de haut déferla dans toute la vallée en aval de Malpasset, jusqu'à la ville de Fréjus.

Au début de l'hiver 1959, les pluies torrentielles vinrent remplir pour la première fois le nouveau barrage de Malpasset, en amont de Fréjus, dans le sud de la France. Lorsque celui-ci céda soudainement, le 2 décembre 1959 à 21h13, près de 50 millions de mètres cubes d'eau déferlèrent, ravageant campagnes et villages jusqu'à la mer. C'est la plus grande catastrophe de ce genre qui ait jamais touché la France" (Fig. 24.11)

[5] http://www.ecolo.org/documents/documents_in_french/malpasset/malpasset.htm

Fig. 24.11 Malpasset before and after the failure, Duffaut 2013

The construction of Malpasset dam on River Reyran, south of France, started in 1952 ending in 1954. The dam failed during the night of December 2, 1959, 5 years after the reservoir filling. The reservoir was approaching the final elevation when heavy rains pushed up the level very quickly forcing the opening of the bottom outlet discharging 40 m^3/s. Even so, after a few hours the dam failed flooding Frejus 7 km downstream. The wave killed 423 people instantly.

Malpasset was a double arch, 60-m high dam. Its crest was 223 m long and very thin at that time. In that period, it was considered the thinnest dam in whole Europe.

Formerly the causes were assigned to geology and lack of drainage as the primary cause of failure. The combination of low strength schistosity and a major fault formed a dihedral shape in the right abutment which led to failure caused by the growing hydraulic thrust (Figs. 24.12 and 24.13).

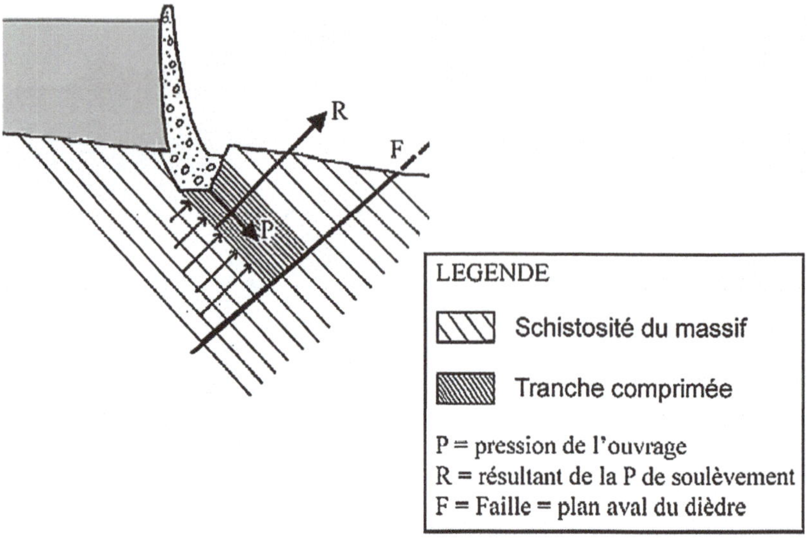

Fig. 24.12 Old simplified illustration of Malpasset right abutment cross section showing geological background for the failure. http://www.aria.developpement-durable.gouv.fr/wp-content/files_mf/FD_29490_malpasset_1959_ang.pdf

Fig. 24.13 Actual situation of the right abutment after failure. Edited from Fig. 24.5, Duffaut 2013

However Duffaut (2013), describes a more complex combination of events which eventually led to the disaster.

"The geology set the first traps; the mechanical behavior of the rock aggravated the dangerous forces; the practice of drainage only was of rule under thick dams; two fortuitous circumstances, the construction of a bridge and a flash flood, conjugated; all of those traps were in a way preparatory cause. Money inflation, lack of any state control, blindness in front of alarms, and absence of any qualified staff completed the scenery. One should stress the technical isolation of André Coyne as well, instead of the high level of implication of engineers of both EDF and the state in hydro-dams inside France."

24.7.3 Camará Dam

Camará dam is a Brazilian 50 m high roller compacted concrete structure founded on granite and migmatites with expressive weathered exfoliation joints, which develop parallel to the topography (Abrahão and Degaspare 2016). In 2004 the left abutment failed due to the presence of a very weathered exfoliation joint, which caused the dam to collapse, emptying the reservoir (Fig. 24.14).

Fig. 24.14 Left abutment of Camará Dam. Just after failure on the left side, and few days later, a complete concrete collapse

After all legal procedures, the decision was to rebuild the dam and the procedures started in 2011. After additional investigation, including TV inspection of boreholes, a new geomechanical model came up with a set of joints, parallel to the surface and some faults with a high degree of weathering producing a soft material resulting from rock mass relics. The general idealization is illustrated in Fig. 24.15.

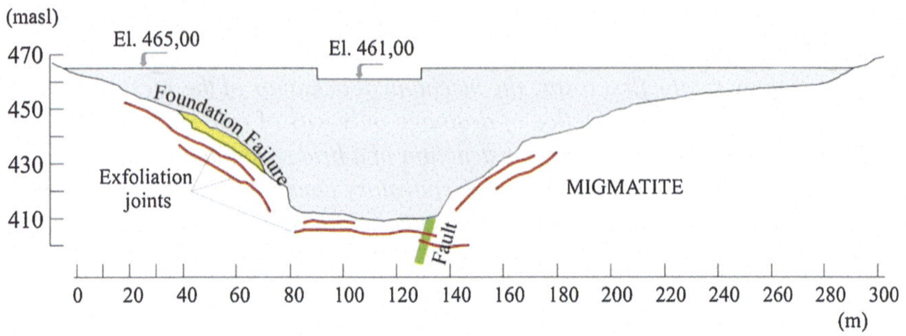

Fig. 24.15 Vertical cross section through dam axis showing the model adopted

To a certain depth the presence of those features compromised the sliding stability of the dam, since the available friction angle for the features was taken as 35°, with no cohesion. These parameters were insufficient to fulfil the design criteria requirements. Therefore, to bring the dam back to the design criteria some element should be added to assure the stability.

It was then decided to cut the remaining concrete block and the foundation to build a shear key at the toe of the dam. The part of the dam that had collapsed where completed rebuilt, with shear keys and plain vibrated concrete.

The model that represents the final solution for the highest concrete block is illustrated in Fig. 24.16.

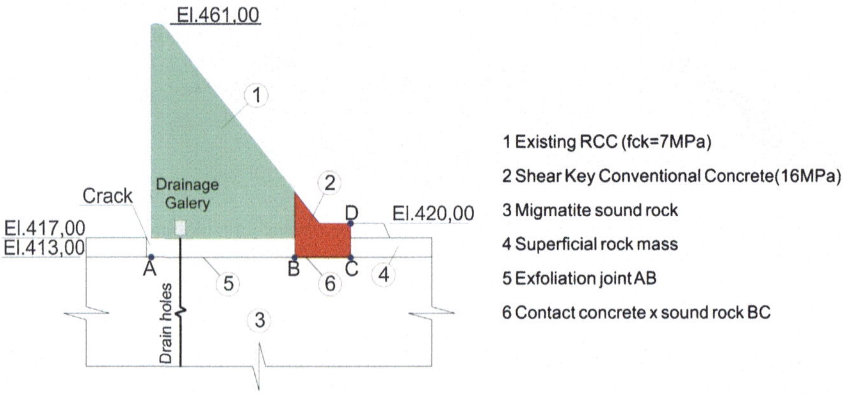

Fig. 24.16 Two-dimensional section showing the model used in calculating the stability of the dam

After a complete set of additional investigations, the geomechanical model based on this section considered the following parameters:

- Foundation migmatite sound rock mass: deformability modulus = 20 GPa; Poisson = 0.25;
- Disturbed surface rock mass: deformability modulus = 15 GPa; Poisson = 0.25;
- Exfoliation joints: average friction angle 35°, null cohesion; shear stiffness $K_s = 0.07$ GPa/m; normal stiffness $K_n = 2.87$ GPa/m;
- Contact concrete × sound rock: friction angle 45°; cohesion = 200 kN/m^2;
- Hundreds of water loss tests registered a rock mass almost impervious with its permeability varying from 5.10^{-5} cm/s to practically impervious. Exceptionally, in a few spots where the weathered joints were sampled, some total loss was recorded.

Finally, the dam was completed in middle 2017 and actual situation is shown on Fig. 24.17.

Fig. 24.17 Camará Dam after complete refurbishment

24.8 Conclusions

The design and construction of dams founded on soft rocks and rock mass with weak discontinuities can be considered in the same way of those built on competent rock mass. The difference is the care one should have in the evaluation of the strength and deformability parameters which always take the model much closer of the limits established in the design criteria.

Poor evaluation of the geological and geotechnical environment will always and inexorably lead to unwanted cost and construction time increase, or to the failure of the structure.

References

Abrahão RA, Degaspare JC (2016) Rebuilding Camará water supply dam, II International Specialized Conference of Soft Rocks, ISRM. Cartagena, Colombia

Barton N (1988) Rock mass classification and tunnel reinforcement selection using the Q-system. In: Kirkaldie L (ed) Rock classification systems foe engineering purposes, ASTM STP 984. American Society for Testing Materials, Philadelphia, pp 59–88

Barton N (2007) Rock quality, seismic velocity, attenuation and anisotropy. Taylor and Francis Group, London, ISBN 0-41539441-4

Bieniawski (1973) Engineering classification of jointed rock masses. Trans S Afr Inst Civ Engrs 15:335–344

Duffaut P (2013) The traps behind the failure of Malpasset arch dam, France, in 1959. J Rock Mech Geotech Eng 5:335–341

Hoek E (2007) Practical Rock Engineering. Course notes: https://www.rocscience.com/learning/hoek-s-corner/books

Infanti N Jr, Kanji MA (1990) Estimating the shear stiffness of rock joints. In: International Symposium on Rock Joints, Loen. International Symposium on Rock Joints, vol 1. N. Barton and O. Stephansson, Oslo, pp 799–804

Kanji MA (1990) Dam foundation on soft and fractured rocks: Problems and solutions. In: 3° Congresso Sulamericano de Mec. Rochas, Caracas, Soc. Venezolana Geotecnia, vol 1, pp 235–252 (in Spanish)

Rocha M (1976) Some problems related to Rock Mechanics of low strength materials. Geotecnia. J Portug Geotech Soc 18:3–27. (in Portuguese)

Rogers JD (2006) Lessons learned from the St. Francis dam failure. Geo—Strata Magazine of the Geo Institute

Souza Lima VM, Abrahão RA (1982) Two practical examples of numerical approaches for solving discontinuity problems in dam design. Int J Numer Anal Methods Geomech 6:461–481

Souza Lima VM, Abrahão RA, Pinheiro R, Degaspare JC (1982) Rock foundations with marked discontinuities. Criteria and assumptions for stability analyses. 14th. ICOLD congress, Rio de Janeiro, Brazil

Chapter 25
Soft Rock as a Dam Construction Material

Manoel de S. Freitas Jr

25.1 Introduction

The soft rock material becomes a very attractive application, when the soft rock material is available at dam site in large volumes, very often from the required excavations (power house, approach channel, spillway, tunnels). Geologically, soft rocks are conceptualized as a low-resistance rock, high porosity, low cementation and very sensitive to changes in natural weather—weathering dependence (Kanji 2014a, b). According to USSD (United States Society on Dams), the economic design and construction on dam rockfill materials depends on utilizing onsite materials.

A non-cohesive rockfill material with high grain size fines percentage (>40% below # 1″), and developing low pore pressures even those classified as "random" has been used on the upstream (ECRD) or at center and downstream zones (CFRD) for practical and construction purposes.

Several examples of soft rockfill dams obtained from blasting or required excavations areas from natural soft sedimentary rocks, such as mudstones, limestones, and metamorphic rocks, schists, phyllites, besides conglomerates, are presented. The current rockfill construction practice has shown that soft or weak rocks use for future high dams should fulfill and to seek quality practices and costs optimization seeking to construction schedule. Test fills performed at dam offset, prior to the construction works, are important required practices for testing equipment and material controls. Monitoring design programs are quite important for dam safety control and future very high (>300 m height) dam construction.

M. d. S. Freitas Jr (✉)
Hydrogeo Engenharia, São Paulo, SP, Brazil

M. Kanji et al. (eds.), *Soft Rock Mechanics and Engineering*,
https://doi.org/10.1007/978-3-030-29477-9_25

25.2 Soft and Weak Materials in Rockfill Construction

Terzaghi et al. (1996) presented a civil engineering definition for fill materials: "*'Soil'—a natural aggregate of mineral grains that can be separated by such gentle means as agitation in water. 'Rock on the other hand, is a natural aggregate of minerals connected by strong and permanent cohesive forces.'*" Boundary between soils and rock is necessary arbitrary. The traditional criteria to select a "proper rockfill," recommend a nonuniform grain size material (Coefficient of Uniformity, D_{60}/D_{10}, CU > 10) and having a percentage of fines that does not develop pore pressures during the construction (Penman and Charles 1976; Penman et al. 1982), that is. draining.

The rockfill material should be a dense and interlocking structure and being therefore important its specification for compaction, vibratory roller type, wetting facilities, and layer thickness control. Well-graded materials (CU > 10) have higher deformability modules than "uniform" rockfill ones. The rockfill deformability is directly proportional to the stresses state, that is, the dam height and inversely proportional to the compactness, a function of the grain size (CU) (Veiga Pinto 1979; Veiga Pinto et al. 1988). Rockfill grain size issues related on the "amount of fines" has often concerned several discussions among experts.

An advantage of a high presence of fines (non-cohesive) is to form a uniform surface and to seek reduction of the truck tires wearing during the transit facilities, that means improve trucks traffic ability. The rockfill premises of non-development of interstitial pressures and support for good handle proceedings were stated by Cooke and Sherard (1987): "*When a rockfill contains a fines content exceeding some limits, the final evaluation of suitability can be made based on trafficability of the rockfill surface when the material is thoroughly wetted. A stable construction surface under the travel of heavy trucks demonstrates that the wheel loads are being carried by a rockfill skeleton.*"

In soft (weak) rockfills with high coefficient of nonuniformity (CU) grading size, fines excess benefits filling voids and contribute to reduce large particle stressed and breakage contacts during spreading and vibratory roller passes, as consequence. Table 25.1 presents the geological and geotechnical engineering practice and strengths of several rock types and the weathering process (ICOLD 2008).

Table 25.1 Strength classification

Class	Description	Uniaxial compressive strength (MN/m^2)
A	Very high strength	Over 200
B	High strength	100–200
C[a]	Medium strength	50–100
D[a]	Low strength	25–50
E[a]	Very low strength	Less < than 25

[a]Rocks with these strengths would probably form "weak" rockfills

25.3 Cases of Successfully High Rockfill (Soft Rock) Dams

In the sequence, several cases of soft rock application in high dam construction are presented.

25.3.1 Emborcação ECRD Dam

Emborcação ECRD (earth core) hydropower, 1190 MW, 158 m height (CEMIG), which construction was concluded in Oct. 1981 (reservoir impounding in March, 31, 1982), is located in Southeast Brazil. The rockfill zoning was built with gneissic rock, schist slightly decomposed, obtained from local quarry located at right bank 1 km from dam site. The main rockfill was in general, founded on sound rock. Internal impervious blanket and abutment areas were founded on residual soils. Granite-gneissic rock foundation is the main foundation constitution, from 0 to 30 m thickness (Divino 2010; Divino et al. 2013a, b). Figure 25.1 shows Emborcação dam main cross section.

Post-construction and Operation period: During the first filling reservoir period until de completion full in April 1982, cracks were recorded at dam crest. These cracks were progressed by 1989. Remedial treatment proceedings were implemented at crest area showing an apparent settlements and cracks phenomenon stabilization. Dam movements were recorded from abutment to dam center part. Longitudinal cracks were monitored along the crest between stations 2+00 e 6+00 and 14+00 to 15+00 (abutment area). Cracks phenomenon causes were explained due to different modulus among dam zoning materials. In 1996, after 28 m reservoir

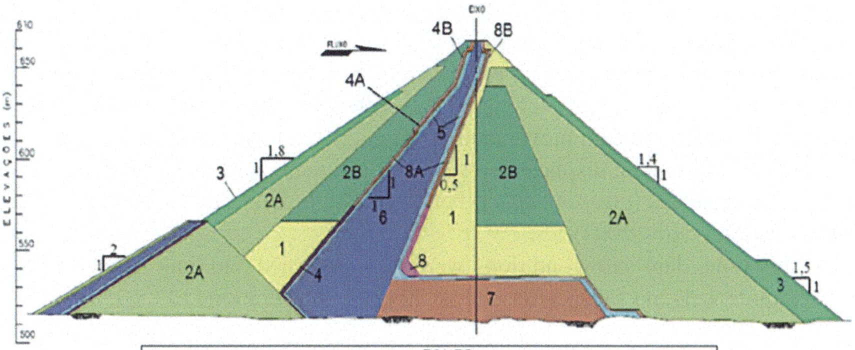

ZONES	
1 Compacted rockfill 0,60 m layers	5 Sand filter
2A Comapcted rockfill 1,20 m layers	6 Impervious core
2B Compacted rockfill 0,90 m layers	7 Random impervious material
3 Large rock blocks	8 Downstream transition
4 Upstream transition	8A Fine transition
4A Fine transition	8B Coarse transition
4B Coarse transition	

Fig. 25.1 Emborcação dam, main cross section

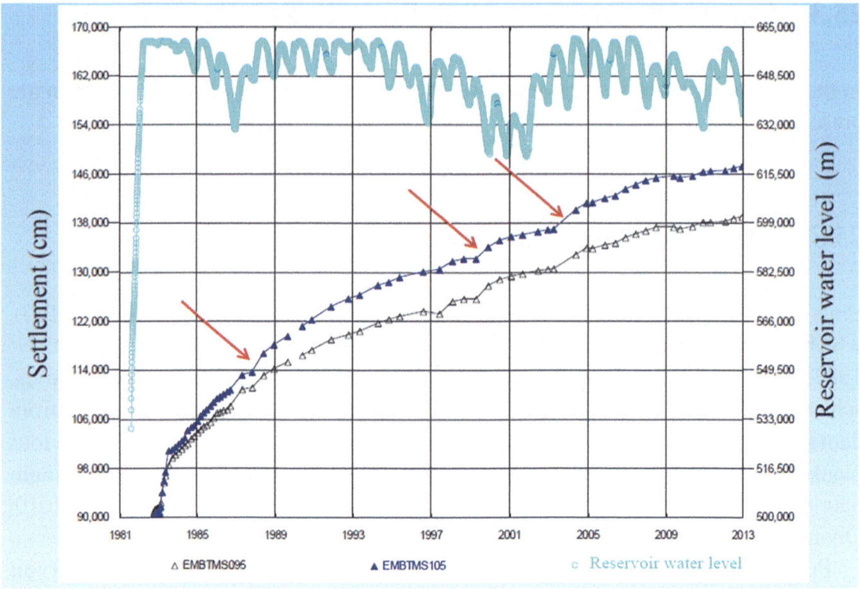

Fig. 25.2 Settlements increments in St. 9+00 and 10+00 due reservoir W.L. variation (Divino et al. 2013a, b)

water level drawdown, cracks at crest areas reappeared. In 1999 with the W.L. at el. 622 (Minimum Operation at El. 615), cracks phenomenon increased significantly. Reservoir drawdown variation oscillated from El. 640 (Jan.) to 622 m Dec.1999). During 2000 and 2001 reservoir W.L. was kept 661.0 m. at el. 622 m, e new remedial cracks treatment were implemented at crest area. From 2004 on the reservoir operated up to W.L. el. 661.0 m, that means, 1.0 m above the Normal W.L. (el. 660 m). In addition, longitudinal cracks were recorded at upstream crest part, from station 9 + 60 and 10 + 20, and 12 + 40 to 12 + 76 m. Figure 25.2 shows settlements at upstream dam crest in highest height (station 9+00 to 10+00) and the influence of reservoir oscillation in dam movements (Divino 2010; Divino et al. 2013a, b).

In 2007, longitudinal displacements at crest upstream zone were monitored and recorded movements (*creeping effect*) between stations 8+00m to 13+00m, according to Fig. 25.3 (Nov. 2007).

At this time, dam safe conditions were not concerned and dam behavior progressed quite well and safety after almost 30 years of operation.

25.3.2 *Irapé ECRD (Earth Core) Dam*

Irapé ECRD (earth clay and gravel core) rockfill dam, 208 m high, 360 MW (CEMIG), was completed in 2008, is located at Jequitinhonha River, MG, in a “canyon” valley, Brazil. The rock mass site is a graphitic quartz-mica-schist with

Fig. 25.3 Emborcação dam crest movements behavior after 28 years of operation

disseminated sulfide minerals. The rock is usually fine grained, dark grey colored, with a metamorphic milonitic. The dam site geology shows schist mica rock with competent foundation. The geological characteristics of the rock includes occurrence of sulfides, mainly pyrite and pyrrhotite in relatively high percentages.

Irapé dam core was built with sandy-clay with a mixture of gravel and clay in the deep channels. A layer of more plastic, self-healing soil was placed along the clay/rock contact on the valley walls to help the redistribution of stresses (Divino 2010).

Natural sand was used in the fine filters, and crushed rock in the coarse filters and transitions. Non-sulfide or sulfide-poor rocks were used in the transitions, only (Marques Filho et al. 2009).

In view a tight construction schedule, excavated materials including large amounts of weathered sulfide-rich rocks from the required excavations were implemented, with the exception of materials zones 7 and 9 (Fig. 25.4), using non-sulfide or sulfide-poor rocks. Test fills were carrying out and investigating the rockfill deposits.

Main Zone 6 slightly weathered rock (coefficient of uniformity, CU ~ 15), Dmax. 80 cm, a sulfide-rich material, placed in 0.80 m thick layers at the upstream zone (below El. 470.00), remained permanently underwater (not able to affect the primary rock sulfides).

Zone 5 L (random, Fig. 25.4) placed at upstream inner zone and close to the clay core–saprolite with decomposed rock fragments, well graded, CU around 140–160 (density of 27 kN/m^3). The amount of sulfides was limited to 0.1% . Materials 5 (CU ~ 20) and 5A (Dmax. 40 cm) are, respectively, slightly to medium weathered rockfill and medium to highly weathered rockfill. The 5A material was placed downstream from the filters, between the 5 material and the outer shell of material 6.

Irapé performance post construction: Core (processed mixture of natural and sandy gravel) settlements have recorded reasonable uniformity. The maximum

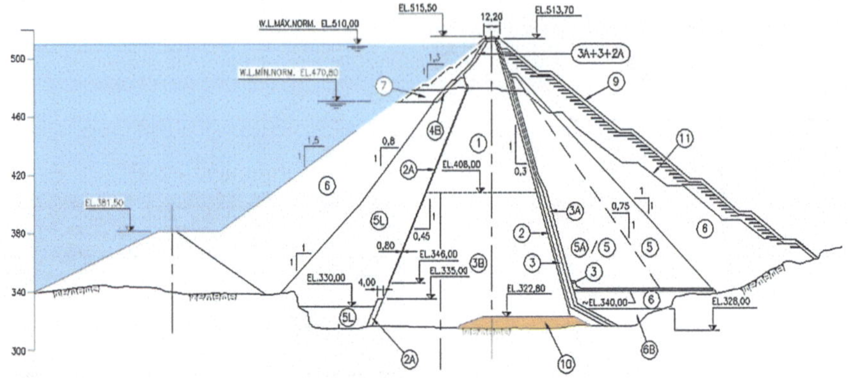

Zones	Description	Layer thicness (cm)	Zones	Description	Layer thicness (cm)
1	Sandy clayey earths/ Clayed Sandy earths	20-25	5L	Random. Highly weathered rock and saprolite	40
2	Filter of natural sand	40	6	Rockfill with slight to sound rock	80
2A	Fine Transition material	40	6A	Dumped rockfill	----
3A	Medium transition	40	6B	Rockfill of sound and slight weathered rock	80
4	Clayey gravel	40	7	Rockfill protection	120
4A	Coarse transition	25 – 30	9	Rockfill of revetment	120
5	Coarse dumped transition	40	10	RCC Base	
5A	Rockfill with slight and medium decomposed rock	- ---	11	Construction joint	

Fig. 25.4 Irapé dam main cross section and zoning materials

accumulated settlement was observed at dam axis, el. 442.60 (Crest at El. 516.7) with 2.1 m, 1% of the max. dam height. Rockfill dam have progressed safety and quite well and being monitored during operation, with a maximum accumulated settlement at rockfill zone was 1.50 m. The Irapé rockfill dam it was a good example of how a deep investigation, design and construction methods and instrumental support have achieved an economically and safe alternative.

25.3.3 Tianshengqiao 1 (TSQ.1) CFRD

Tianshengqiao 1 (TSQ.1) CFRD, is located on the Nanpan River, Southeast of P.R. of China, concluded in 2000 (1200 MW, 178 m max. height), where more than 4,000,000 m^3 of mudstone soft rock (30% of the material grain size < # 1 1/5″) was applied at III C downstream inner zone, in view its susceptible to weathering effect (Fig. 25.5). The mudstone material came from the spillway (approach channel) required excavations with a very significant optimization in cost and time. Figure 25.5 shows Tianshengqiao 1 dam main cross section and zoning materials.

During construction stage a significant volume (2,800,000 m^3) of the downstream III C (mudstone) was launched between end-August to early December, 1998, when this dam area rose by about 110 m, in order to fulfill the construction schedule (Figs. 25.6 and 25.7).

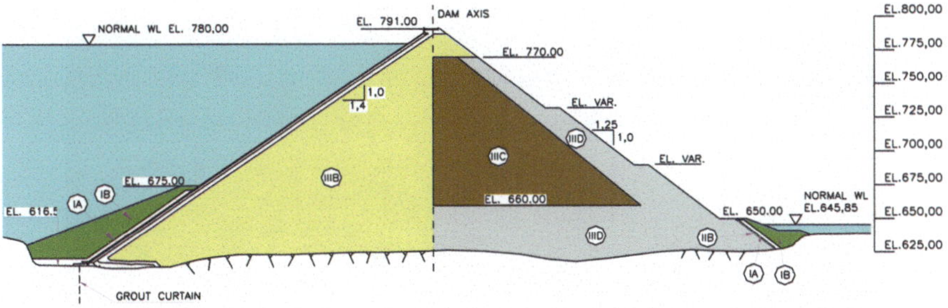

ZONE	DESCRIPTION	ZONE	DESCRIPTION
IA	CLAY. DUMPED IN THE UPSTREAM SIDE COMPACTED IN 30 cm THICK IN THE DOWNSTREAM SIDE	IIIB	MAIN ROCKFILL, SOUND TO SLIGHTLY WEATHERED LIMESTONE ø MAX. = 80cm. WETTING : 20% IN VOLUME
IB	RANDOM ROCKFILL. DUMPED	IIIC	ROCKFILL, MUDSTONE, ø MAX. = 80cm LAYER THICKNESS = 80cm
IIA	SEMI PERVIOUS ZONE. CRUSHER-RUN SOUND LIMESTONE ø MAX. = 80mm. LAYER THICKNESS = 40cm	IIID	ROCKFILL, LIMESTONE, ø MAX. = 1,00m LAYER THICKNESS = 1,00m
IIIA	TRANSITION. FINE ROCKFILL. SOUND TO SLIGHTLY WEATHERED LIMESTONE ø MAX. = 30cm. LAYER THICKNESS = 40cm	IIB	TRANSITION. FINE ROCKFILL, COMPACTED IN 30 cm THICK

Fig. 25.5 Tianshengqiao 1 CFRD, main cross section and zoning materials

Fig. 25.6 TSQ.1 downstream view in August 1998. Detail of slope of zone III C of "mudstone"

Fig. 25.7 TSQ.1 downstream view—zone III C progress in November 1998

A family of cracks were recorded at the upstream upper part "cushion zone (IIA)" before third face slab stage concrete construction. After cracks remedial treatments (grouting by gravity) the designer decided to reinforced the slab designing in this third stage with a double armor to confront future deformations ("bending") of the concrete face slab. Figures 25.8 and 25.9 show the excavation works progress of mudstone at the entrance of the spillway approach channel areas.

After blasting services the spillway excavation and rockfill dam progressed successfully, and even during the rainy season there was no significant decrease in dam volumes production. Table 25.2 shows mudstone geological characteristics tests results.

Workability: The behavioral mudstone characteristics during the launch, spreading and compaction activities showed no significant differences in the site fronts among limestone III B (upstream) and III C (downstream) mudstone zones. In restricted areas with excess of fines (<25 mm) the trucks had transit difficulties during heavy rainy days. However, after rainy period ending, the activities were restarted easily in few (2 or 3) hours later. The specification advised a rockfill wetting control limited to 10% of the mudstone rock fill volume, to 20% for the Zone IIIB (limestone). It was not possible to detect a separation between results between sound limestone (zones III B and III D) and the weathered from and the mudstone (Zone IIIC). The "finer threshold" shows a percentage of fines <25 mm (1″) ranging maximum 40% to 15% (minimum). For mudstone, an average value of in-situ density was found of 2.23 t/m^3 (22.3 kN/m^3). Figure 25.10 presents the grain sizes ranges of zones IIIB, IIIC, and IIID rockfills and comparison with specified particle size range.

Fig. 25.8 TSQ.1 "mudstone" excavation borrow areas

Fig. 25.9 TSQ.1 spillway approach channel entrance—mudstone excavation

End of Construction dam settlements behavior: The deformability modulus at the end of the construction estimated from settlement cells ranged between 41 and 48 MPa (upstream and center zones), and recorded values at downstream zone up to 30 MPa (Wu et al. 2000; ICOLD 2008). For the mudstone (Zone IIIC) an average value of in situ density was of 22.3 kN/m^3, according to in situ control records. The mudstone showed a good workability for use in the rockfill.

Table 25.2 Mudstone and sandstone geological characteristics tests results

Rock type	Specific gravity	Bulk density (g/cm³)	Porosity (%)	Unconfined dry compressive strength (kPa)	Saturated unconfined compressive strength (kPa)	Softening coefficient	Absorption rate (%)
Sandstone	2.74	2.68	2.00	83.3	80.4[a]	0.96	0.60
Mudstone	2.75	2.64	4.00	30.7	19.6	0.64	1.54

[a]Average value of 16 specimens only

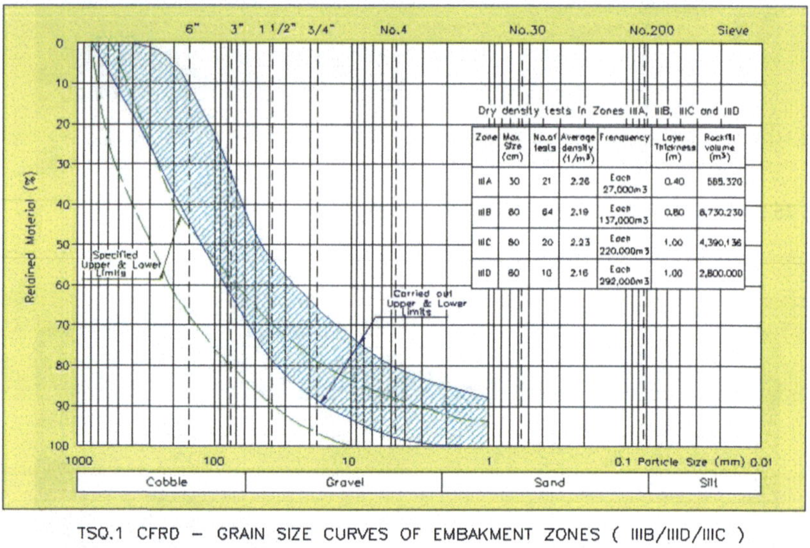

Fig. 25.10 TSQ.1 CFRD, grain size curves and dry tests results: IIIB zone (limestone), IIID and III C (mudstone) (Wu et al. 2000)

The maximum settlement in December 2000, at end of reservoir filling, was 3.40 m (1.9% max. Dam height rate) measured in at the Elev. 725, downstream at the "central" portion of Zone IIIC.

At the dam crest, settlements were recorded of 1.20 m and deformation towards downstream of the upper end of the slab of about 1.38 m. No slab cracks were recorded at the end of first reservoir filling (2000). Cracks and concrete spalling phenomenon at face slab upper part were recorded in 2003 and 2004, according to international literature. Leakage records ranged from 150 l/s (2000) to around 70 l/s in 2005, after remedial and mitigation measures showing a decreasing tendency.

TSQ.1 rockfill dam operation since slab remedial treatment works (2003–2004) have carried on quite safe, according to international practices.

Currently, more tight rockfill specifications and quality control have been applied from 2000s on, as international practice.

25.3.4 *Paute Mazar CFRD*

Paute Mazar Hydroelectric Project (HIDROPAUTE and CELEC S.A.), 166 m CFRD, 170 MW, is located in Ecuador, 100 km from Cuenca city, 340 m long crest, topped in Elev. 2164 m. According to design, upstream slopes 1.4H:1 V and downstream slopes 1.25H:1 V, rockfill volume of 5,350,000 m^3 at a remarkable narrow valley, in a seismic region. Mazar dam construction started in April 2005, and reservoir impounding and commissioning completed on first half 2010. At dam local, geologic investigations present metamorphic rocks corresponding to quartzitic schists, with intercalations of chloritic and sericitic schists. The main source of quartzitic schists material comes from quarries and from required spillway excavation areas (Costa et al. 2011; Freitas Jr and Marengo 2017). Figure 25.11 shows Mazar dam cross section.

In May 2009 the rockfill dam topped Elv. 2159 m. During the first stage of the dam construction, a 5 m wide chimney drain with select sound rocks (Fig. 25.11) was designed at the intersection of Zones 3B and 3C of the dam, up to the elevation 2140, due to that part of the material coming from spillway excavations and placed at downstream areas presented a weathered rock fines excess. In addition, the chimney drain works as a safety protection in view seismic hazard activity and in case of damage slab impervious barrier. Rockfill construction has adopted modern concept by a thinner layers thickness tide control (max. 50 cm at Zone 3B and 80 cm at Zone 3C), and high water fill (>300 l/m^3) supply at upstream zones, mainly, and 6 passes of a 13.6 ton vibratory compactor. In addition, to extend the section of the upstream Zone 3B to at least two-thirds of the cross section, concept.

Mazar postconstruction records: Dam performance by April/May 2011, 2 years after reservoir impounding is presented briefly as follows: Maximum rockfill

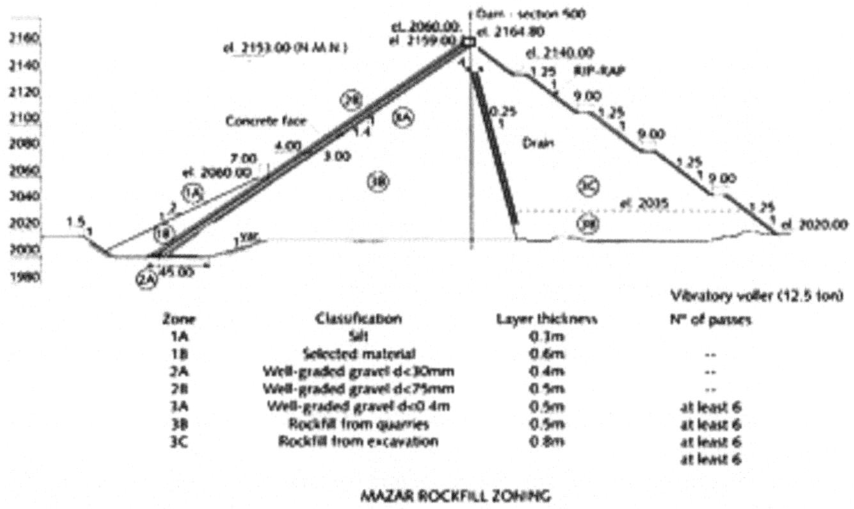

Fig. 25.11 Mazar CFRD main cross section

settlement by April 2011 was **1, 54 m** and was recorded at downstream part, Elv. 2090 m, that means 1 (one) % of the total dam height. Above Elv. 2125.0 a residual movement (creeping) was still being recorded. At lower levels there is a settlement tendency stabilization. At dam crest, parapet wall (upstream) a maximum settlement of 19 cm was recorded at center part and a horizontal downstream displacement of 9 cm close to left bank.

Slabs upper part movements (perpendicular to face slab surface) around 25 cm were measured at end of construction. After impounding, an additional 15 cm movement at Elv. 2158 m was recorded in April 2011, at center part. Leakage readings around 480 l/s were recorded during operation phase beginning at maximum operation water level, 2153.0 (May, 2011). During dry season period in March 2011 reservoir water level achieved at Elv. 2127 m, the leakage decreased to around 282 l/s.

25.3.5 Nuozhadu ECRD (Earth Core) Dam

Nuozhadu ECRD, 5850 MW, 261.5 m max. Height, is located on Lancang River at Yunnan Province, China, was another interesting experience of soft rock using in large volumes (4,780,000 m^3) at upstream shell zone (14.0% of the total dam rockfill volume, 34 × 10^6 m^3). Rockfill materials were obtained from required excavations with huge benefit costs for the project. Upstream slope 1.9 (H):1.0 (V) and 1.8 (H):1.0(V) downstream slope were implemented according to design, in an earthquake seismic area (Yuan et al. 2017). The soft (rockfill) materials came from excavated local dam sites, mainly composed of various rock types, such as argillaceous siltstone, siltstone, fine sandstone, glutenite, breccia, silty mudstone. Figure 25.12 shows Nuozhadu dam main cross section.

The core soil materials for Nuozhadu Rockfill Dam contains a lot of clay particles and has low strength, gravels are admixed for modification. Dam rockfill material was a mixture of soft, medium hard and hard rock, of which the soft accounts no more than 20% of total. It has been demonstrated that such kind of a mixture can be safely applied at upstream zones below the reservoir W. L. Figure 25.13 shows the soft rock grain size materials applied at rockfill zones.

Anti-seismic measures for earth-rockfill dam: According to the authors, conventional anti-seismic measures such as geogrids, concrete slab beams and prestressed anchorage reinforced concrete frame beams, cannot fully meet the anti-seismic requirements of the Nuozhadu Project, only.

Therefore, anti-seismic measures suitable for Scale 9 fortification were developed and applied for the high earth-rockfill dam in the Nuozhadu dam. Nevertheless, a combination of stainless steel bars and stainless steel flat grid, block revetment on the upstream and downstream surface of dam crest, as shown in Fig. 25.14, as seismic design protection.

Nuozhadu dam post construction behavior: Nuozhadu dam is in operation since August 2012, and has been performed in safety conditions with a maximum 0.826 m settlement at dam crest (total settlement in relation to dam height of 0.3%) in December 2016, and total seepage of the measuring weirs 11.6 l/s (Yuan et al. 2017).

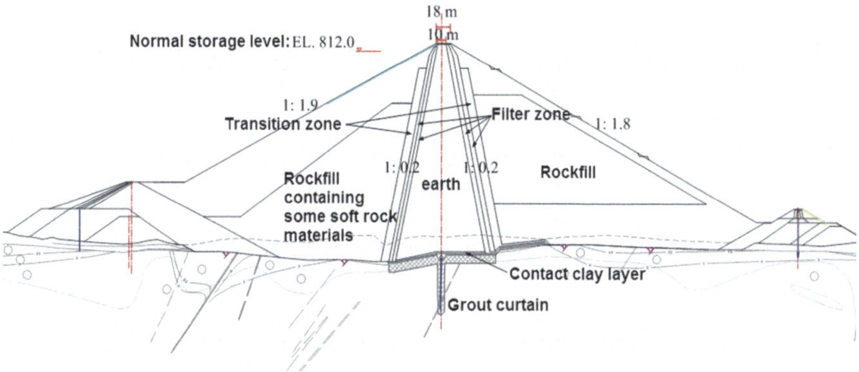

Fig. 25.12 Maximum cross section of Nuozhadu Earth-Rockfill Dam

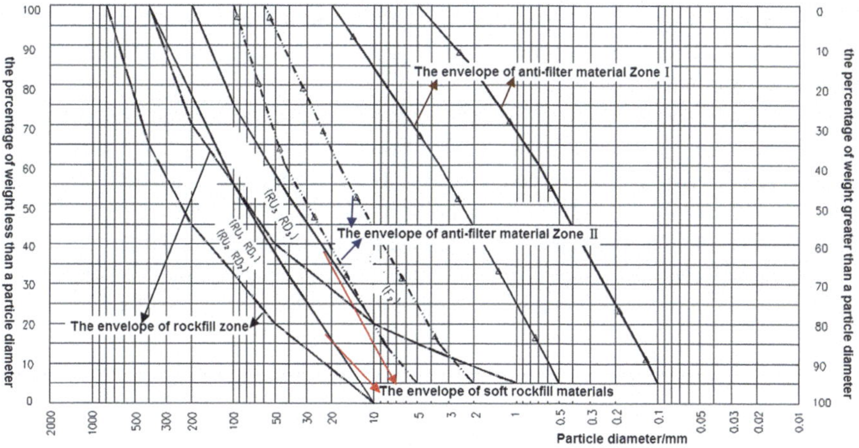

Fig. 25.13 Nuozhadu dam—Soft rock material grain gradation

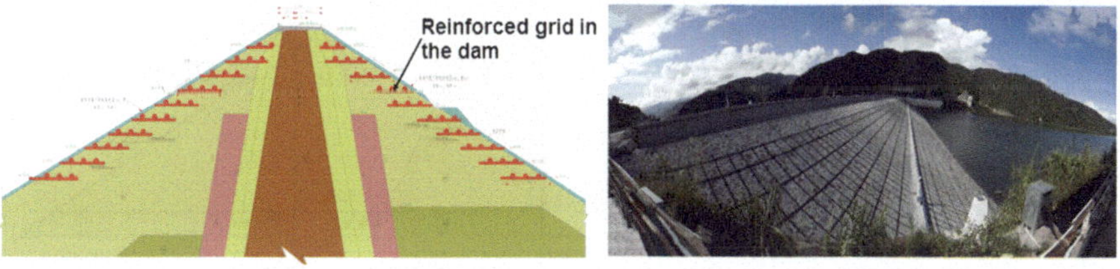

Fig. 25.14 New -seismic measures of Nuozhadu Earth-Rockfill Dam (Yuan et al. 2017)

25.3.6 Porce III CFRD

Porce III CFRD, a 154 m high dam, 688 MW, is located at Northeast of the city of Medellin, Colombia, which construction activities completed in 2010. Figure 25.15 shows the main cross section and dam zoning materials. The main dam source materials came from the spillway excavations, where different types of schists were found: quartzitic, sericitic, and graphitic, with several weathering levels: IIA, IIB, and III (Marulanda et al. 2007). Graphitic and feldspathic shales lead to soft rockfills (Marulanda et al. 2007).

Technical achievements: Quartz feldspathic shale, slightly weathered were applied for main rockfill. Materials zones 3B, 3C y 3D consisted of blocks, irregularly shaped, with maximum size of 60 cm. Coarse material varying from 85% to 90%. Material Passing # 4 sieve varying 10–15%.

Filters and transitions materials were crushed gneiss produced during the excavation of the pressure and intake tunnels. Materials zones 2A, 2B, and 3A used coarse particles ¾″, 3″, and 10″, respectively. Zone 1A consisted of non-plastic lime, which is protected with randomly material coming the excavations.

Figures 25.16 and 25.17 show materials IIB, IIA mixtures for zones 3D and 3C purposes (Marulanda 2016).

A geological survey of all surface and underground excavations was conducted for classification and characterization of all different types and degree of weathering schists. In addition, laboratory oedometric tests were conducted, and test fills from selected various types of schist mixes, layers thickness and compactor passes, were carried out at site. Complemented by plate load tests to obtain deformability modulus.

Soft rockfill required a strong quality control during construction, load plate tests in rockfills were developed and correlating with void relation (e), oedometers tests

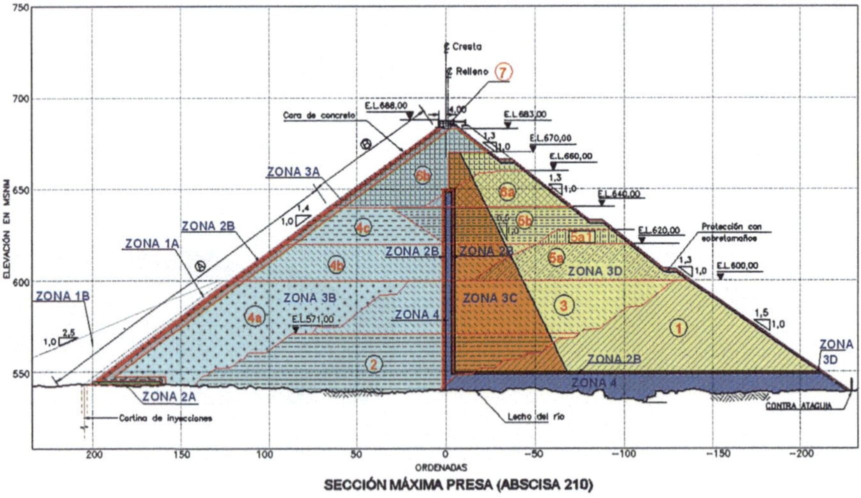

Fig. 25.15 Porce III CFRD (Marulanda 2016)

Fig. 25.16 Types of schists

Fig. 25.17 Zones 3D & 3C. Dam fill at el. 555.7 70% IIB + 30% IIA

and instrumentation in the body of the dam. According to authors (Marulanda et al. 2007), wetting facilities ranged 300 l/m^3 were specified for zones 3B, 3C, and 3D compaction works. Zones 3B, and 3C rockfill max. particles 60 cm were recommended and thickness layers of 0.60 cm (after spreading) during construction.

At 3D zone, 1.0 m thickness was recommended. At the 3B Zone (upstream) material mixtures: 80% III y 20% IIB were accepted. At 3C zone, mixtures 100% IIB and 3D zone 30% IIA y 70% IIB were recommended.

A rockfill central chimney drain denominated zone 4, composed of coarse material between 4″ and 12″, connected to a horizontal downstream rockfill dam, zone 4 (Fig. 25.15), was built as a safety element for high dams at seismic activity areas, due to a high earthquake can damaged the grouted plinth impermeability barrier or induce face slab ruptures and introduce seepage throughout upstream Zone 3B. This vertical drain concept was latter introduced in the following CFRDs: Mazar (Ecuador), Chaglla (Peru), and Reventazón (Costa Rica), all of them located in high seismic areas.

Porce III Post construction behavior: After 6 year working, the performance is outstanding, because the settlements in the crest are about 44 cm, the seepage is near 60 l/s in the downstream spillway, which confirms that CFRD built with soft rockfill coming from graphitic shales was an economical and safe solution, according to authors.

25.3.7 Jiangpinche CFRD

Jiangpinche CFRD dam, 450 MW, 219 m high, is located on Lishui River, Hubei Province, China, at a V-shaped narrow valley. The main construction material is a moraine conglomerate. Dam rockfill slopes are 1(V): 1.4(H): US/DS slopes. Figure 25.18 shows Jiangpinche main cross section.

Moraine conglomerate quarry was located on left bank at a distance ranged 4.2–5.3 km from dam site. The estimated available moraine volume was 16.9 × 10^6 m^3.

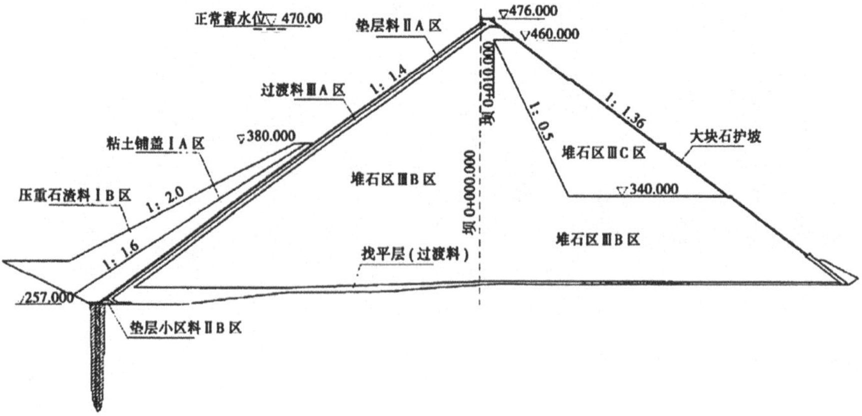

Fig. 25.18 Jiangpinche CFRD main cross section (Wang and OU 2013)

Table 25.3 Moraine conglomerate strength property tests (average values) (Wang et al. 2013)

Rock type	Weathering degree	Dry density (kN/m^3)	Grain density (kN/m^3)	Water absorption ratio (%)	Porosity %	Saturated compressive strength	Softening coefficient
Moraine conglomerate	Weakly weathered	2.70	2.72	0.17	0.44	71.1	0.65
	Slightly weathering	2.72	2.72	0.09	0.24	87.0	0.66

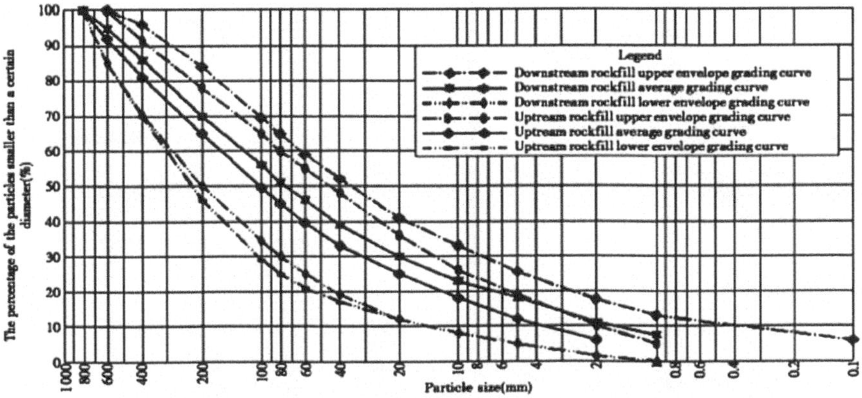

Fig. 25.19 Grading curves of upstream and downstream rockfill zones (Wang et al. 2013)

Table 25.3 presents Moraine conglomerate physical and mechanical property tests. Table 25.3 shows moraine strength property tests results.

Figure 25.19 shows grading curves of rockfill materials. (Coefficient of Uniformity) was estimated in 60, in average.

Dam rockfill zones features: (1) main (III B) rockfill moraine conglomerate applied shown a good gradation, with a maximum particle of 80 cm, and fine contents between 5 mm and 0.075 mm, ranged from 20% to 5%, respectively. Dry density in situ tests 22.0 kN/m^3 and 18.8% porosity. (2) Secondary rockfill (IIIC), maximum particle 80 cm, dry density tests ranged 21.6–22.0 kN/m^3, corresponding 19.7% porosity. (3) Vibratory roller 32 ton was handled in the compaction works and an amount 15%/m^3 for watering facilities.

25.3.8 Reventazón CFRD

Reventazón CFRD dam (350 MW, 130 m height) construction was completed in October, 2015, located on the Reventazón River, Costa Rica. The first reservoir filling stage began on November 27th, 2015, reaching the maximum level of operation on October 20th, 2016. Figure 25.20 shows dam main cross section and zoning materials. Dam slopes design, adopted upstream face slope 1.5H: 1.0 V and 1.6H: 1.0 V at downstream face. Zoning design consists of good quality alluvial gravel and conglomerate materials from river bed areas for upstream 3B, and low quality with high fines percentage materials at center and transition T Zone (Roy et al. 2017).

At 3C zone, alluvial gravel and from spillway excavations areas mixtures were used. Figures 25.21a, b show alluvial gravel and conglomerate materials at dam site.

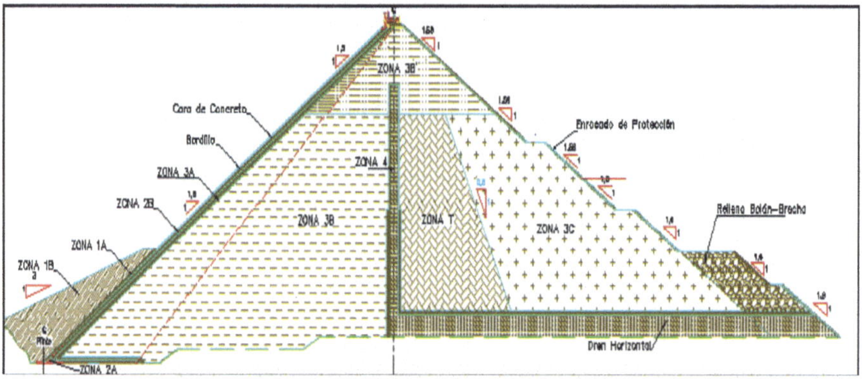

Zona	Description	Dmax. (m)	Thickness layer (m)
1A	Silty sand (no cohesive)	0.0025	0.40
1B	Random material	0.30	0.60
2A	Filter	0.019- 0.038	0.20
2B	Semi permeable filter	0.10	0.30
3A	Allluvial gravel	0.25	0.30
3B	Allluvial gravel	0.55	0.60
3B'	Alluvial and conglomerate gravel mixture 5:1 to 10:1	0.55	0.60
T	Alluvial and conglomerate breccia mixture 2: 1	0.55	0.60
3C	Conglomerate and Alluvial gravel and breccia mixture 3: 1	0.55	0.60
4	Processed Alluvial gravel	0.30	0.80 / 0.60
Riprap	Lavas	> 0.70	-

Fig. 25.20 Reventazón CFRD main cross section and zoning materials

(a) (b)

Fig. 25.21 Reventazón CFRD, alluvial gravel material, and conglomerate

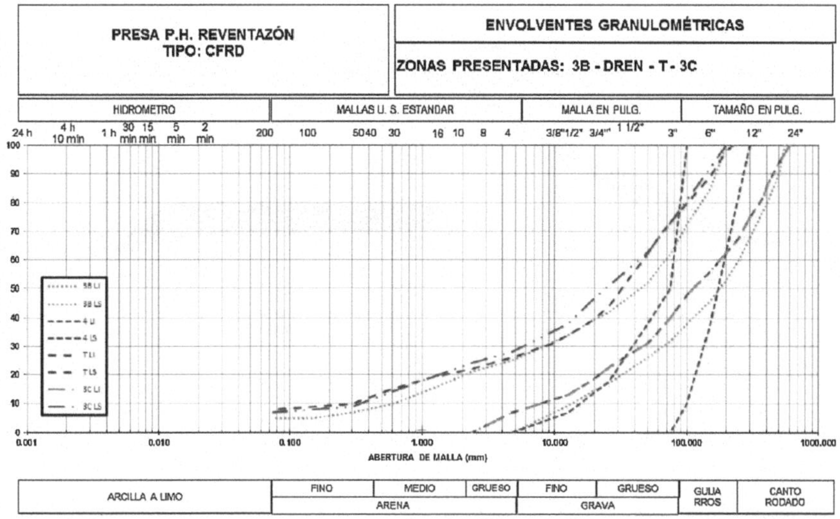

Fig. 25.22 Reventazón grain size grading 3B, T, 3C materials

Zone 3B (upstream) materials were compacted with 60 cm thickness layers and four passes of vibratory roller. At T (center) and 3C (downstream) zones, 60 cm thickness layers, six roller passes were required. Minimum specified density at zones 3B, 3C, and T were 22.5 kN/m^3, 19.5 kN/m^3 and 21.5 kN/m^3, respectively. Figure 25.22 shows rockfill 3B, T, and 3C materials grain size curves.

Reventazón dam post filling behavior: The maximum crest vertical settlement recorded was 5.36 cm (0.04% max. Height ratio), but rockfill movements creeping process is still progressing, too. Face slab vertical displacement recorded (inclinometer) was around 7.0 cm, at slab upper part.

The movements at dam crest were monitored by 18 benchmarks installed along the both parapet walls, upstream and downstream. The maximum vertical settlements was 54 mm. The leakage monitored at downstream dam part was 250 l/s.

Reventazón dam behavior after the first filling is according to good international literature records and practices.

25.3.9 *Nam Ngum 2 CFRD*

Nam Ngum 2 CFRD, 615 MW, a 182 m height dam is located in Laos PDR, which construction was completed in 2010. Face slab was built in two stages and performed from December 2008 to July 2009. Reservoir impounding started in March 2010 and commercial operation plant in December 2010. Nam Ngum 2 site is a narrow valley bedded to massive sandstone (fresh, hard and slight fractured),

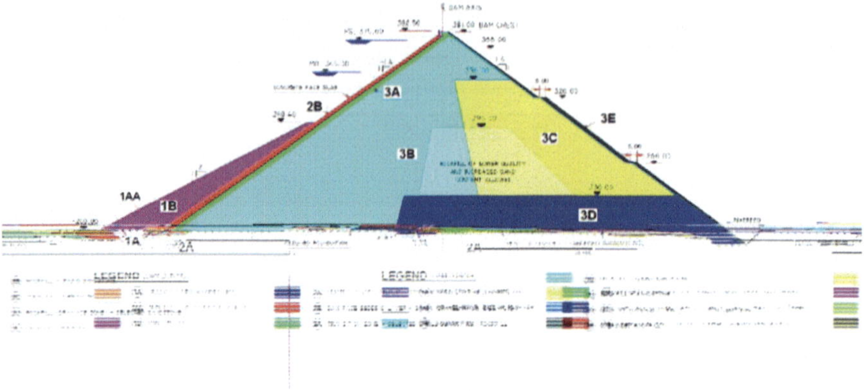

Fig. 25.23 Nam Ngum 2 CFRD Main cross section and rockfill zoning

(a) (b)

Fig. 25.24 Soft sandstones materials at dam center zone (Moll and Straubhaarr 2011)

interbedded with siltstone, fresh to weathered, soft (Moll and Straubhaarr 2011). Figure 25.23 presents Nam Ngum dam rockfill zoning design.

Dam rockfill design zones were adjusted to allow use a moderate weathered sandstone placed whereby the lower quality soft materials at central zone, according to the authors. Upstream and downstream slopes inclined 1.4(H):1(V) and 1.5 (H):1(V) downstream slope between two berms, according to design. Total rockfill volume, 9.7×10^6 m^3.

Rockfill layers were placed in 0.80 m (zone 3B) lifts and 1.2 m (zone 3C) thickness, and compacted with 6–8 passes of a 16ton vibratory roller and wetting facilities around 100–150 l/m^3. Rockfill deformation modulus ranged 50–70 MPa were recorded. In 2011, a leakage record of 260 l/s (at 97% of Head WL) was registered. Figures 25.24a, b show soft sandstones materials at dam area.

A monitoring program has registered a maximum settlements of 2.2 m at end of construction (1.2% of max. Dam height rate). The maximum slab deformation was 0.36 m (at 97% o Head W.L.). Downstream part movements recorded from settlement cells and geodetic survey showed a settlement rate 20–40 mm/month in 2011.

25.4 Soft and Weak Rockfill Dams Construction Features

Current rockfill dams specification, since 2000s, recommend 12–25 ton vibratory rollers and rockfill thickness layers $\leq$1.0 m, and wetting facilities >150 l/m^3 during compaction activities. Some examples of 1.20 m thickness layers at downstream zone are recorded too, for concrete face rockfill dams international literature. Due to weathering rocks soft materials (with fines excess) placed at center (T) and 3C areas, the highest dam rockfill settlements have been recorded at 3C downstream zones in concrete face rockfill dams, mainly.

An important rockfill soft or weak construction feature is the possibility of large releases daily production above 40,000 m^3 (>1,000,000 m^3 monthly), regardless of the weather (rainy season), limited only by the access conditions of and transport distances. In case of alluvial or conglomerate materials, where a maximum grain size very often ranged from 0.10 to 0.80 m, wetting facilities are not necessary in these cases. So alluvial and conglomerate materials requires less thickness layers, ranged 60–80 cm.

The recommendation for future high rockfill soft rocks dams must concerning to follow a tight specification compaction control, layers thickness layers maximum at 100 cm, and minimum six passes of a vibratory roller of at least 12 tons.

Wetting services during spreading and compaction activities should be restricted in rockfill soft rocks or random materials, in comparison to sound or less weathered rock.

The rockfill dam zoning design should accommodate overburden soft (weak) rock ("dirty rockfill") material focus on to minimize stockpiling, quarry exploitation and costs.

25.5 Deformability and Creeping Effect

Relevant factors have affected rockfill compressibility and dam behavior such as grain size distribution, grain shape, mineralogy or interlocking, and the state of confining stresses. Rockfill collapse phenomenon in ECRDs (earth core) has been recorded basically due to rock breakage contact and fracturing effect after wetting by the reservoir filling. In rockfill soft rock this breakage and fracturing effect by wetting is reduced.

In Emborcação ECRD, reservoir W.L. fluctuations have been influenced upstream dam settlements and cracks at dam crest. In long-term monitoring program, drying and wetting process and rockfill weathering acceleration can induce creeping effect. After 28 years operation, cracks at dam crest areas are continuously been monitored and progressing with reservoir level oscillation in Emborcação dam.

In Aguamilpa CFRD dam (187 m), built in Mexico in 1994, the upstream shell reached a deformation modulus of 310 MPa in the zone of the dam built with gravels (3B material), 130 MPa in transition material, but the 3C material downstream, had a greater compressibility of the rockfill (48 MPa) in comparison with the used upstream rockfill, generating a crack in the upper part of the concrete slab, Figs. 25.25 and 25.26 (Freitas Jr and Marengo 2017).

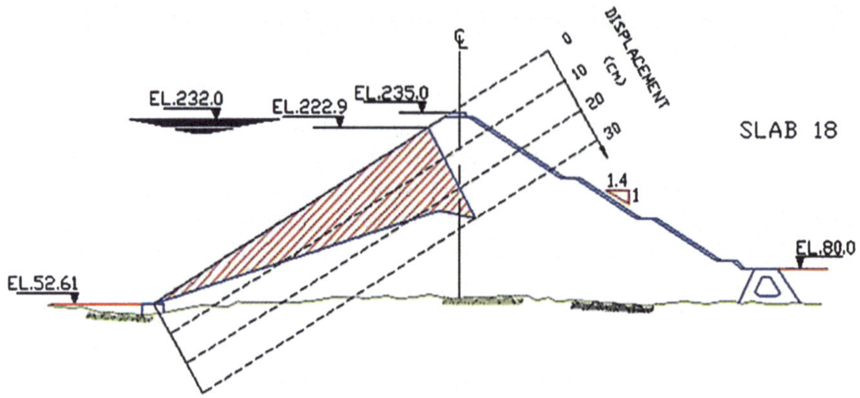

Fig. 25.25 Settlement of Aguamilpa dam

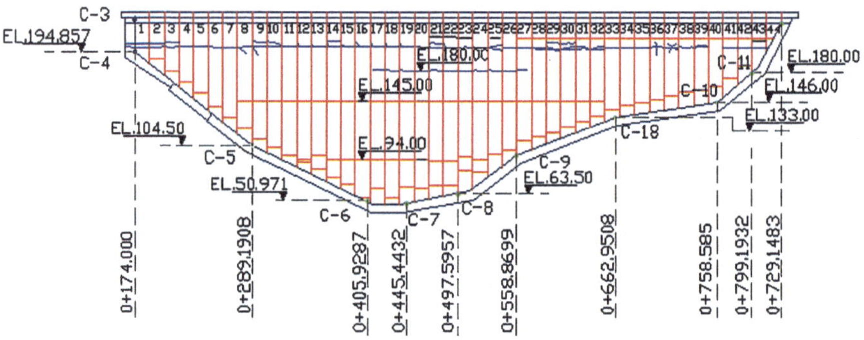

Fig. 25.26 Aguamilpa CFRD—Cracks in the Face slab (Freitas Jr and Marengo 2017)

25.6 Final Remarks and Conclusions

Soft Rockfill main points: (1) as usual and compressible materials which fit in the lower deformability range of other construction materials; (2) breakage and crushing behavior follow particular rules even at low average stresses; (3) soft rockfills are subject to creep phenomenon or slow deformation under a constant state of stresses.

In ECRD alternative the first flooding would cause significant upstream settlements from the core, and crest upstream deflection during early stages of the reservoir filling. At Infernillo Dam (Mexico), this phenomenon was pronounced because the rockfill was placed dry and the reservoir rose very rapidly (Marsal 1973).

Soft "weak" rock, is a suitable material and workability for dam zoning, as the international practice has been demonstrated. Local quarries exploitation or main structures required excavations (spillway, powerhouse, tunnels) very often result in considerable weathered and soft rocks and can supply rockfill dam purposes. Rockfill dam zoning design should accommodate these overburden soft (weak) rock ("dirty rockfill") materials focus on to minimize stockpiling and costs.

It is recommended to restrict the layer thickness 80 cm or to a maximum of 100 cm and six passed of a vibrating smooth roller of at least 10 ton and up to 25 ton. Some examples of 120 cm thickness at downstream shells have been applied, even in CFRD, according to international records (not recommended).

Although there was a restriction on the wetting of soft or "weak "rocks (10% of the volume), in general the material showed good workability immediately after the rainy seasons.

Rockfill trafficability concept as stated by Cooke and Sherard (1987) related rockfill fines content, must consider trafficability of the rockfill surface when the material is thoroughly wetted.

Test fills before main rockfill zones construction are mandatory for soft "weak" rocks proceedings control, such as vibratory roller passes, thickness layers, wetting facilities, and so on.

Brazilian rockfill CFRD and ERCD dams have been recorded "creeping effect," settlements along several decades due to slow wetting at downstream zones, during the rainy season, every year.

The owner's operation team must monitoring crest and downstream areas during all operation period, in addition instrumentation readings and dam back-analysis. Long-term dam inspections and behavior should be supported by periodic independent engineer inspectors.

References

Cooke B, Sherard J (1987) ASCE—Proceedings, Concrete Face, Rockfill dams—Design, Construction, Performance. Detroit, USA

Costa M, Vanegas SV, Estrella E, Freitas MS Jr (2011) MAZAR CFRD (ECUADOR) construction and performance after two years of operation, second international symposium on rockfill dams. Chincold-CBDB, Rio de Janeiro, RJ, Brazil

Divino PL (2010) Comportamento de Enrocamentos em Barragens—Estudo de Caso da Barragem de Emborcação, Dissertação de Mestrado, UFOP, outubro 2010

Divino PL et al (2013a) Emborcação dam stress-behavior in 28 years of operation, 3rd symposium on Rockfill dams. Kunming, China

Divino P, Saulo R, Gutemberg S (2013b) Emborcação dam stress-strain behavior in 28 years of operation—3rd International Symposium on Rockfill Dams, Kunming, China

Freitas MS Jr, Marengo H (2017) IV International Symposium on Rockfill Dams, General Report Theme 1, Belo Horizonte, MG, Brazil

ICOLD-Weak and Shales in Dams (2008) Committee on Material for Fill Dams

Kanji MA (2014a) Critical issues in soft rocks. J Rock Mechan Geotechn Eng. 6(3)

Kanji MA (2014b) Engineering works affected by soft rocks, Braz. Symp. Rock Mech., Rock mechanics for natural resources and infrastructure ISRM. Specialized Conference, Engineering Works Affected by Soft

Marques Filho PL, Wunder E, Calcina AM, Garcez Duarte JM, Côsso Lima AL (2009) IRAPÉ Power Plant—Design and Construction measures required by Sulfide rich rockmass. 20th ICOLD, Brasília, Brazil

Marsal R (1973) Mechanical properties of Rockfill, Embankment-Dam Engineering Casagrande Volume

Marulanda C (2016) INCA, Mexico, October. UNAM—USSD

Marulanda A et al (2007) III Symposium on CFRD-dams Honoring Barry Cooke, 2007, Florianopolis
Moll S, Straubhaarr R (2011) Nam Ngum 2 CFRD—Behavior during construction and first impounding, 6th International Conference on dam Engineering, Lisbon, Portugal
Penman ADM, Charles JA (1976) The quality and suitability of rockfill used in dam construction—12th International Congress on Large Dams, Mexico City, Mexico
Penman ADM, Charles JA, Humphreys JD (1982) Sandstone Rockfill in two dams, 14th International Congress on Large Dams, Rio de Janeiro, Brazil
Roy R, Darío C, Eduardo A (2017) Comportamento da Barragem de Reventazón durante seu Primeiro Enchimento, 4th International Symposium on Rockfill Dams, May 15—18th Belo Horizonte, Brazil
Terzaghi K, Peck RB, Mesri G (1996) Soil mechanics in engineering practice. Wiley, Hoboken, NJ
Veiga Pinto A (1979) Características de Resistência e Deformabilidade dos Materiais de Enrocamento (Portuguese), Geotecnia No. 27, Lisbon, Portugal
Veiga Pinto A., de Melo FG, Matias Ramos C (1988) Design criteria of upstream concrete facing rockfill dams, 16th International Congresso on Large Dams, San Francisco, USA
Wang G, OU H (2013) Design of High CFRD for Jiangpinche Hydropower Station, 3rd International Symposium on Rockfill dams. Kunming, China
Wang X, Liu W, Yin Y, Li Z (2013) Study on characteristics of moraine conglomerate in building Jiangpinche CFRD, 3rd International Symposium on Rockfill dams. Kunming, China
Wu GY, Freitas MS Jr et al (2000) Planning & Construction of Tianshengqiao 1, CFRD, 2000, ICOLD, 20th Congress, Beijing
Yuan X, AI Y, Chi F, Chen H (2017) Research and practice of key technologies for Nuozhadu high earth-rockfill dam. 4th International Symposium on Rockfill Dams, May 15—18th Belo Horizonte, Brazil

Correction to: Characterization of Soft Rocks in Brazilian Coal Beds

Clovis Gonzatti, João Alberto Fiorentini, Luiz Zorzi, and Ivone Maria Agostini

Correction to: Chapter 23 in M. Kanji et al. (eds.), *Soft Rock Mechanics and Engineering*, https://doi.org/10.1007/978-3-030-29477-9_23

Owing to the production team's oversight, the figures in the chapter were misplaced. The correction has been updated in the chapter. The correct order is now as follows:

IDENTIFICATION	IMAGE PRESENTED IN	AT PAGE
Photo 23.1	Fig. 23.14	686
Photo 23-2	Fig. 23.15	689
Photo 23.3	Fig. 23.16	689
Photo 23.4	Fig. 23.17	690
Photo 23.5	Fig. 23.18	692
Photo 23.6	Fig. 23.19	694
Photo 23.7	Fig. 23.20	696
Fig. 23.5	Photo 23.1	671
Fig. 23.6	Photo 23.2	671
Fig. 23.7	Photo 23.3	672
Fig. 23.8	Photo 23.4	672

The updated online version of this chapter can be found at
https://doi.org/10.1007/978-3-030-29477-9_23

M. Kanji et al. (eds.), *Soft Rock Mechanics and Engineering*,
https://doi.org/10.1007/978-3-030-29477-9_26

Fig. 23.9	Fig. 23.5	674
Fig. 23.10	Fig. 23.6	674
Fig. 23.11	Fig. 23.7	675
Fig. 23.12	Fig. 23.8	677
Fig. 23.13	Photo 23.5	677
Fig. 23.14	Fig. 23.9	678
Fig. 23.15	Fig. 23.10	680
Fig 23.16	Photo 23.6	681
Fig. 23.17	Fig. 23.11	682
Fig. 23.18	Fig. 23.12	684
Fig. 23.19	Photo 23.7	685
Fig. 23.20	Fig. 23.13	686

Index

M. Kanji et al. (eds.), *Soft Rock Mechanics and Engineering*,
https://doi.org/10.1007/978-3-030-29477-9

D

N

V

W

X

Y

The manufacturer's authorised representative in the EU is Springer Nature Customer Service Centre GmbH, Europaplatz 3, 69115 Heidelberg, Germany. If you have any concerns regarding our products, please contact ProductSafety@springernature.com

Printed and bound by CPI Group (UK) Ltd, Croydon, CR0 4YY
30/09/2026
02225657-0019